Mycotoxins and Their Metabolites in Humans and Animals

Martin Weidenbörner

Mycotoxins and Their Metabolites in Humans and Animals

Springer

Martin Weidenbörner
mweidenboerner@web.de

ISBN 978-1-4419-7432-7 e-ISBN 978-1-4419-7433-4
DOI 10.1007/978-1-4419-7433-4
Springer New York Dordrecht Heidelberg London

Library of Congress Control Number: 2011920683

Printed on acid-free paper

Springer is part of Springer Science+Business Media (www.springer.com)

Helene

Preface

Mycotoxins are secondary toxic mold products which are widespread in foods and feeds. The already published books *Mycotoxins in Feedstuffs* and *Mycotoxins in Foodstuffs* provide a good overview of mycotoxins. It is estimated that 4.5 billion of the world's population are exposed to mycotoxins, which can be found in temperate as well as in continental climates. However, especially in low-income countries (e.g., parts of Africa, Southeast Asia, Central and South America) people are chronically exposed to high levels of mycotoxins. In these countries, staple foods like groundnuts and other nuts, maize, as well as other cereals, are especially affected. For example, in West Africa aflatoxin contamination of humans starts in utero and continues throughout life. Besides the hepatitis B virus (HBV), exposure to high levels of dietary aflatoxins poses a major risk for developing human hepatocellular carcinoma (HCC) in these countries. However, even low levels of aflatoxin ingestion causes a suppression of the immune system and increases susceptibility to diseases in several animal species.

Besides their acute toxicity, mycotoxins have other harmful effects. They are, for example, cytotoxic, genotoxic, hepatotoxic, nephrotoxic, mutagenic, neurotoxic, and teratogenic. Human toxicoses due to mycotoxins have been reported, for example, in China, India, Japan, Kenya, Korea, and Russia. If optimal conditions of temperature, humidity, and a suitable substrate prevail, mycotoxins are produced on agricultural commodities in the field, in storage and/or during processing. Because mycotoxins are known to have these detrimental effects, many countries have set legal limits for these toxic fungal metabolites in order to limit their intake.

Contamination especially by aflatoxins, fumonisins, ochratoxin, deoxynivalenol, and zearalenone of a wide range of food products from around the world is of major concern. These food products are mainly of plant origin. Foodstuffs of animal origin, except milk and derived products, show a lower contamination rate. Furthermore, their mycotoxin concentration is usually low. Therefore, food items of animal origin generally pose a minor danger to consumers. However, the milk and breast milk mycotoxin AFM_1, which is also found in milk-derived products, can concentrate on foods. As a result, the contamination of babies via breast milk (mainly AFM_1) in different parts of the world should not be underestimated. The capacity of babies for biotransformation of carcinogens is generally slower than that in adults. By comparison, foodstuffs of plant origin play a major role in the mycotoxin contamination of human beings. This mycotoxin contamination is well documented. It is also proved by several publications, which show the presence of mycotoxins in human organs, tissues, and fluids.

Besides the above-mentioned mycotoxins, numerous other toxic fungal metabolites exist, which all pose either a minor or major danger. They are of great concern from a food perspective regarding human exposure.

This book summarizes the results of publications dealing with the natural and artificial contamination of humans and animals by mycotoxins, as well as mycotoxin experiments with animals. The major part of the book lists animal studies that investigate deposits and elimination of these toxic fungal metabolites. Furthermore, the results of articles documenting mycotoxin contamination of pets are also presented. In addition, information about detoxification products and the duration of a mycotoxin in and its clearance time from an animal are given. Moreover, the book gives advice on whether antimycotoxic substances are effective in reducing mycotoxin contamination in animals and humans.

This book provides physicians with a fast and comprehensive overview of the countries in which mycotoxin contamination of humans predominantly happens, as well as the concentration at which specific mycotoxins are found in human organs, tissues, and fluids. Veterinarians are informed about what mycotoxins, at what concentrations, can be found naturally in animals. More detailed information is presented if the index number referring to the corresponding publication at the end of the book is used.

This book may be suitable for physicians (global), pathologists (global), epidemiologists, veterinarians, nutritionists, livestock breeders, pet keepers, farmers, the food and feed industry, institutes (e.g., consumer production), ministries (global), libraries, hospitals, healthware stations, UNO, mycologists, mycotoxicologists, microbiologists, biologists, and students of corresponding fields.

For practical use, the different mycotoxins in humans, animals, organs, tissues, or fluids are listed showing natural or artificial mycotoxin contamination. Therefore, each mycotoxin can be looked up for natural or artificial presence at the end of the book.

The book exclusively comprises articles treating concentrations of mycotoxins in humans or animals. Publications or data which express mycotoxins in % values, radioactivity or in other ways are not considered. Articles dealing with *in vitro* data are also not presented. All articles presented are available as publications of German Scientific Libraries as well as the U.S. National Library of Medicine–National Institutes of Health. The most cited publications have been included. Articles cited in this book have been selected by preference, where a declaration of a mycotoxin concentration or any advice of it is given in the title. Nevertheless, some articles containing no concentration declaration in the title, but only in the running text, are also cited.

Each declaration of the mycotoxin contamination of humans or animals comprises five main categories, e.g.:

incidence: 3/7 - three positives for aflatoxin contamination in relation to seven investigated sample

sample constitution: origin of the test people and/or composition of the sample
contamination: natural or artificial (which concentration of a mycotoxin has been applied in an experiment)
concentration: residue values of the mycotoxin(s)
country: origin of the publication, in some cases, also origin of the test people.

If a sample shows a "natural contamination", information on the sample constitution is given briefly. In most cases, where a sample shows a "natural contamination", details were not available in the corresponding article so further comments are omitted. This may not be true for human beings. In the case of an "artificial contamination", a more precise definition of the sample constitution is presented.

Usually, the highest mycotoxin value or the highest and the lowest value of mycotoxin contamination in an experiment is given. The presented concentrations occur in the way they are presented in the published papers. If a variant of a trial is not listed, no mycotoxin contamination is recorded. However, in some cases, a variant may be stated although mycotoxin concentration is not detected. In general, HPLC values have been used for concentration declaration.

If concentration of milk mycotoxins is given, this milk more or less comes directly from cows (natural contamination). You will find additional information about natural mycotoxin contamination of milk, for example processed milk (pasteurized, UHT-milk, etc.) in *Mycotoxins in Foodstuffs*. In addition, data on the natural mycotoxin contamination of "cow milk", "human breast milk", "pig kidney", "pig serum", etc., can be found in the book *Mycotoxins in Foodstuffs*. For a comprehensive overview, these values as well as new data have also been published here.

Bonn, Germany

Martin Weidenbörner

Acknowledgments

I thank my family, especially my wife, for their patience and for giving me valuable advice and support in writing this book. I appreciate valuable advice of Dr. Ursula Monnerjahn-Karbach. I also wish to thank Mrs. Renate Frohn, Mrs. Ursula Kleinheyer-Thomas, Mrs. Verena Borman as well as the whole staff from the Bran Library for Medicine, Science, and Agricultural Science, Bonn University for very good support in searching and finding the corresponding publications. I also appreciate my father's very careful review of the total manuscript.

Abbreviations

AC	β-apo-8′-carotenal
ACs	activated carbons
af	affected
AF(s)	aflatoxin(s)
AFB-AA	aflatoxin B_1-albumin adducts
AFB_1	aflatoxin B_1
AFB_1 *endo*-epoxide	aflatoxin B_1 8,9-*endo*-epoxide
AFB_1 *exo*-epoxide	aflatoxin B_1 8,9-*exo*-epoxide
AFB_1-FAPy	8,9-dihydro-8-(2,6-diamino-4-oxo-3,4-dihydropyrimid-5-yl-formamido)-9-hydroxyaflatoxin B_1
AFB_1-FAPyr	2,3-dihydro-2-(N^5-formyl-2′,5′,6′-triamino-4′-oxo-N^5-pyrimidyl- 3-hydroxyaflatoxin B_1
AFB-N^7-FAPyr (minor)	8,9-dihydro-8-(2-amino-6-formamido-4-oxo-3, 4-dihydropyrimid- 5-yl amino)-9-hydroxyaflatoxin B_1
AFB-N^7-FAPyr (major)	8,9-dihydro-8-(2,6-diamino-4-oxo-3,4-dihydropyrimid-5-yl formamido)-9-hydroxyaflatoxin B_1
AFB-N^7-Gua	2,3-dihydro-2-(N^7-guanyl)-3-hydroxyaflatoxin B_1
AFB_1-N^7-Gua^1	2,3-dihydro-2-(N^7-guanyl)-9-hydroxyaflatoxin B_1
AFB_1-N^7-Gua^2	2,3-dihydro-2-(N^7-guanyl)-3-hydroxyaflatoxin B_1
AFB_1-N^7-Gua^3	8,9-dihydro-8-(N^7-guanyl)-9-hydroxyaflatoxin B_1
AFB_1-SG	aflatoxin B_1-glutathione conjugate
AFB-GuaI	2,3-dihydro-2-(7′-guanyl)-3-hydroxyaflatoxin B_1
AFB-NAC	AFB_1-mercapturic acid
exo-AFB_1-NAC	exo-AFB_1-mercapturic acids
AFL	aflatoxicol
AFL-g	aflatoxicol-glucuronide
$AFLM_1$	aflatoxicol M_1
$AFLM_1$-g	aflatoxicol M_1-glucuronide
AFM_1	aflatoxin M_1
AF-N^7-Gua	aflatoxin-N^7-guanine

AFP_1	aflatoxin P_1
AFQ_1	aflatoxin Q_1
AMB	amphotericin B
avg	average
b wt	bodyweight
B-I/B-II	barley cultures of *Penicillium viridicatum*
BC	β-carotene
BEN	Balkan endemic nephropathy
BHA	2(3)-*tert*-butyl-4-hydroxyanisole
BHT	butylated hydroxytoluene
bmi	body mass index
BNF/βNF	β-naphthoflavone
BSO	D,L-buthionine-*S*-sulfoximine
L-BSO	L-butionine-sulfoximine
ca	case(s)
CAC1	activated charcoal
CAC2	activated charcoal
CHL	chlorophyllin
CIN	chronic interstitial nephropathy
CIT	citrinin
CMD	choline/methionine-deficient diet
CMS	complete basal diet
conc	concentration
const	constitution
CP	calcium propionate
CPA	cyclopiazonic acid
CPFA	cyclopropenoid fatty acid(s)
CPL	clinoptilolite
CPR	chromatogram poorly resolved
CX	canthaxanthin
DAS	diacetoxyscirpenol
DEDON	deepoxydeoxynivalenol
DEM	diethyl maleate
DHBV	duck hepatitis B virus
DHEA	dehydroepiandrosterone
DIOL	2,3-dihydro-2,3-dihydroxyaflatoxin B_1
DMSO	dimethyl sulfoxide
DNA	desoxy nucleic acid
DOM/DOM-1	deepoxydeoxynivalenol = 3α,7α, 15-trihydroxytrichothec-9, 12-diene-8-one
DON	deoxynivalenol (vomitoxin)

3-aDON	3-acetyldeoxynivalenol
DYP	dried yeast product
EFDV	encephalopathy and fatty degeneration of the viscera
ELISA	enzyme-linked immunosorbent assay
EN	endemic nephropathy
eq	equivalent(s)
EQ	ethoxyquin
FA	fusaric acid
FB_1	fumonisin B_1
FB_2	fumonisin B_2
FB_3	fumonisin B_3
FPC	fish protein concentrate
FX	fusarenon-X
Gluc	glucuronide conjugate
GSH	reduced glutathione
GTP	green tea polyphenol
GUA/Gua	guanine
HbsAg	hepatitis B virus surface antigen
HBV	hepatitis B virus
HCC	hepatocellular carcinoma
HPLC-f	high-performance liquid chromatography with fluorescence detection
hr	hour(s)
HSCAS	hydrated sodium calcium aluminosilicate
hum	human(s)
I3C	indole-3-carbinol
ia	intra-aortal
IA	invasive aspergillosis
IDMS	isotope dilution mass spectrometry
ig	intragastric
in	intranasal
ip	intraperitoneal
it	intratracheal
iv	intravenous
ivs	intravascular
KIN	karyomegalic interstitial nephritis
LOD	limit of detection
LOQ	limit of quantification
Lys-AFB_1/AFB_1-lys	lysine-AFB_1/AFB_1-lysine

min	minute(s)
MOS	mannanoligosaccharide
MWF	micronized wheat fibers
3-MC	3-methylcholanthrene
na	not analyzed
NAC	mercapturic acid
nd	not detected
ndr	not determined
nec	no exact comment
neg	negative
NIV	nivalenol
NMB	nonmoldy barley
NMB+T	nonmoldy barley+toxin
no	number
NPC	nonparenchymal cells
NR	not reported
o	oral
OTA	ochratoxin A
OP-OTA	lactone opened ochratoxin A
OTA-OH	4-hydroxyochratoxin A
OTα	ochratoxin α
PA	penicillic acid
PB	phenobarbital/phenobarbitone
PC	parenchymal cells
PCB	polyclorinated biphenyls
peo	test people
PG	propylene glycol
PHC	primary hepatocellular carcinoma
PNA	penitrem A
pos	positive
pr	present(ed)
RBC	red blood cells
resp	respectively
rRNA	ribosomal ribonucleic acid
sa	sample(s)
sc	subcutaneous
SG	glutathione
t	topical
tr	traces
TRICHO	trichothecene

UTT	urinary tract tumors
VER	verrucarol
WHV	woodchuck hepatitis virus
wt	weight
YCW	yeast cell walls
ZEA	zearalenone
ZEA-Gluc	zearalenone-glucuronide
α-ZEAOL	α-zearalenol
α-ZEAOL-Gluc	α-zearalenol-glucuronide
β-ZEAOL	β-zearalenol
β-ZEAOL-Gluc	β-zearalenol-glucuronide
±	higher/lower values are reported

Notation

kg = Kilogram
mg = Milligram = 10^{-3} g; 1 mg/kg = 1:10^6 = ppm = parts per million
μg = Microgram = 10^{-6} g; 1 μg/kg = 1:10^9 = ppb = parts per billion

l = Liter
ml = Milliliter = 10^{-3} l; 1 ml/l = 1:10^3
μl = Microtliter = 10^{-3} ml; 1 μl/l = 1:10^6 = ppm = parts per million

Contents

Human

Human Natural Contamination

Human amniotic fluid may contain the following mycotoxins and/or their metabolites:

Ochratoxin A
incidence: 4/22, sa. const.: people of Germany, contamination: natural, conc. range: <0.06–0.13 ng/ml, Ø conc.: 0.11 ng/ml, country: Germany[341]

Human bile may contain the following mycotoxins and/or their metabolites:

Aflatoxin B_1
incidence: 3/4*, sa. const.: people of Thailand (children), contamination: natural, conc. range: tr**–8 µg/kg, country: USA/Thailand[10], *EFDV ca., **partly unconfirmed
incidence: 2/6*, sa. const.: people of Thailand (children), contamination: natural, conc. range: tr**, country: USA/Thailand[10], *dying from causes other than EFDV, **a blue fluorescent spot similar to that of AFB_1

Human blood may contain the following mycotoxins and/or their metabolites:

Aflatoxicol
incidence: 15/625*, sa. const.: people of Nigeria (babies), contamination: natural, conc. range: 22–2,383 ng/l, country: UK/Nigeria[205], *cord blood

incidence: 3/461*, sa. const.: people of Ghana, Nigeria, and Kenya (females), contamination: natural, conc. range: 177–280 ng/l, Ø conc.: 223.66 ng/l, country: UK[241], *cord blood

incidence: 1/40*, sa. const.: people of Nigeria (neonates), contamination: natural, conc.: 1,131 pg/ml**, country: Nigeria/UK[262], *non-jaundiced, **peripheral blood

incidence: 34/64*, sa. const.: people of Sierra Leone, contamination: natural, conc. range: 0.007–2.2 ng/ml, Ø conc.: 0.1 ng/ml, country: Sierra Leone/UK[390], *cord blood
incidence: 2/8*, sa. const.: people of Sierra Leone, contamination: natural, conc. range: 0.02–1.7 ng/ml, Ø conc.: 0.86 ng/ml, country: Sierra Leone/UK[390], *maternal blood
incidence: 4/6*, sa. const.: people of Sierra Leone (male infants), contamination: natural, conc. range: 0.02–0.04 ng/ml, country: Sierra Leone/UK[390], *low birthweight
incidence: 4/5*, sa. const.: people of Sierra Leone (female infants), contamination: natural, conc. range: 0.02–2.2 ng/ml, country: Sierra Leone/UK[390], *low birthweight
incidence: 10/21*, sa. const.: people of Sierra Leone (male infants), contamination: natural, conc. range: 0.007–0.5 ng/ml, country: Sierra Leone/UK[390], *normal birthweight
incidence: 8/20*, sa. const.: people of Sierra Leone (female infants), contamination: natural, conc. range: 0.008–1.6 ng/ml, country: Sierra Leone/UK[390], *normal birthweight

Aflatoxin B_1
incidence: 6/74*, sa. const.: people of the USA (patients), contamination: natural, conc. range: 2.0–12.0 ng/ml, country: USA[19], *6 Reye's syndrome ca. thereof 4 AFB_1-pos.

incidence: 2/5*, sa. const.: people of the USA (children), contamination: natural, conc. range: 11.93–31.3 ng/ml, Ø conc.: 21.615 ng/ml, country: USA[65], *Reye's syndrome ca.

incidence: 5/56*, sa. const.: people of Taiwan (females), contamination: natural, conc. range: 1.4–2.7 µmol/mol DNA**, Ø conc.: 1.94 µmol/mol DNA**, country: Taiwan, Republic of China[156], *cord blood, **AFB_1-DNA adducts

M. Weidenbörner, *Mycotoxins and Their Metabolites in Humans and Animals*,
DOI 10.1007/978-1-4419-7433-4_1,

incidence: 16/625*, sa. const.: people of Nigeria (babies), contamination: natural, conc. range: 168–69,973 ng/l, country: UK/Nigeria[205], *cord blood

incidence: 13/125*, sa. const.: people of Kenya (females), contamination: natural, conc. range: 89–11,574 ng/l, country: UK[241], *maternal blood
incidence: 6/77*, sa. const.: people of Nigeria (females), contamination: natural, conc. range: 553–10,390 ng/l, Ø conc.: 3,707.66 ng/l, country: UK[241], *maternal blood
incidence: 20/461*, sa. const.: people of Ghana, Nigeria, and Kenya (females), contamination: natural, conc. range: 185–43,822 ng/l, country: UK[241], *cord blood

incidence: 2/40*, sa. const.: people of Nigeria (neonates), contamination: natural, conc. range: 590–1,006 pg/ml, Ø conc.: 798 pg/ml, country: Nigeria/UK[262], *non-jaundiced, peripheral blood
incidence: 3/37*, sa. const.: people of Nigeria (neonates), contamination: natural, conc. range: 130–3,130 pg/ml, Ø conc.: 2,070.33 pg/ml, country: Nigeria/UK[262], *jaundiced, peripheral blood
incidence: 3/40*, sa. const.: people of Nigeria (neonates), contamination: natural, conc. range: 474–2,216 pg/ml, Ø conc.: 1,342 pg/ml, country: Nigeria/UK[262], *non-jaundiced, cord blood
incidence: 6/37*, sa. const.: people of Nigeria (neonates), contamination: natural, conc. range: 214–238,177 pg/ml, Ø conc.: 82,481.16 pg/ml, country: Nigeria/UK[262], *jaundiced, cord blood

incidence: 64/64, sa. const.: people of Ghana (34 males and 30 females), contamination: natural, conc. range: 0.3325–2.2703 pmol/mg albumin*, Ø conc.: 0.9972 pmol/mg albumin, country: USA/Ghana[321], *AFB_1-albumin adducts

incidence: 11/64*, sa. const.: people of Sierra Leone, contamination: natural, conc. range: 0.4–9.0 ng/ml, Ø conc.: 1.0 ng/ml, country: Sierra Leone/UK[390], *cord blood

incidence: 2/8*, sa. const.: people of Sierra Leone, contamination: natural, conc. range: 0.2–0.3 ng/ml, Ø conc.: 0.25 ng/ml, country: Sierra Leone/UK[390], *maternal blood
incidence: 6/6*, sa. const.: people of Sierra Leone (male infants), no contamination with AFB_1, conc.: nd, country: Sierra Leone/UK[390], *low birthweight
incidence: 5/5*, sa. const.: people of Sierra Leone (female infants), no contamination with AFB_1, conc.: nd, country: Sierra Leone/UK[390], *low birthweight
incidence: 4/21*, sa. const.: people of Sierra Leone (male infants), contamination: natural, conc. range: 0.5–2.4 ng/ml, country: Sierra Leone/UK[390], *normal birthweight
incidence: 3/20*, sa. const.: people of Sierra Leone (female infants), contamination: natural, conc. range: 0.8–9.0 ng/ml, country: Sierra Leone/UK[390], *normal birthweight

incidence: 64/423*, sa. const.: people of Singapore (342 males (53 af) and 81 females (11 af)), contamination: natural, conc. range: 3.0–17 pg/ml, Ø conc.: 5.4 pg/ml, country: Singapore[444], *normal subjects
incidence: 2/302*, sa. const.: people of Singapore (253 males (2 af.) and 49 females), contamination: natural, conc. range: 7.5–7.9 pg/ml, Ø conc.: 7.7 pg/ml, country: Singapore[444], *hepatitis B carriers
incidence: 1/58*, sa. const.: people of Singapore (49 males (1 af) and 9 females), contamination: natural, conc.: 7.4 pg/ml, country: Singapore[444], *PHC patients

incidence: 140/140, sa. const.: people of Ghana (males and females), age: 19–86 years, contamination: natural, conc. range: 0.12–3.00 pmol/mg albumin*, Ø conc.: 0.89 pmol/mg albumin, country: USA/Ghana[457], *AFB_1-albumin adducts

incidence: ?/150*, sa. const.: people of Nigeria (neonates), contamination: natural, conc. range: 32.3–35.6 ng/ml, country: Nigeria[458], *jaundiced, cord blood incidence: 27/201*, sa. const.: people of the UAE (females), contamination: natural, conc. range: 228–15,225 pg/ml, country: UAE[467], *umbilical cord blood

Aflatoxin B_2

incidence: 4/625*, sa. const.: people of Nigeria (babies), contamination: natural, conc. range: 15–144 ng/l, country: UK/Nigeria[205], *cord blood

incidence: 2/77*, sa. const.: people of Nigeria (females), contamination: natural, conc. range: 28–33 ng/l, Ø conc.: 30.5 ng/l, country: UK[241], *maternal blood

incidence: 19/461*, sa. const.: people of Ghana, Nigeria, and Kenya (females), contamination: natural, conc. range: 10–925 ng/l, country: UK[241], *cord blood

incidence: 1/37*, sa. const.: people of Nigeria (neonates), contamination: natural, conc.: 70 pg/ml, country: Nigeria/UK[262], *jaundiced, cord blood

incidence: 7/64*, sa. const.: people of Sierra Leone, contamination: natural, conc. range: 0.02–1.2 ng/ml, Ø conc.: 0.3 ng/ml, country: Sierra Leone/UK[390], *cord blood

incidence: 2/8*, sa. const.: people of Sierra Leone, contamination: natural, conc. range: 0.2–0.3 ng/ml, Ø conc.: 0.25 ng/ml, country: Sierra Leone/UK[390], *maternal blood

incidence: 6/6*, sa. const.: people of Sierra Leone (male infants), no contamination with AFB_2, conc.: nd, country: Sierra Leone/UK[390], *low birthweight

incidence: 5/5*, sa. const.: people of Sierra Leone (female infants), no contamination with AFB_2, conc.: nd, country: Sierra Leone/UK[390], *low birthweight

incidence: 3/21*, sa. const.: people of Sierra Leone (male infants), contamination: natural, conc. range: 0.07–0.1 ng/ml, country: Sierra Leone/UK[390], *normal birthweight

incidence: 1/20*, sa. const.: people of Sierra Leone (female infants), contamination: natural, conc.: 0.7 ng/ml, country: Sierra Leone/UK[390], *normal birthweight

Aflatoxin G_1

incidence: 1/14, sa. const.: people of The Sudan (females), contamination: natural, conc.: 787 pg/ml, country: UK/The Sudan[134]

incidence: 6/625*, sa. const.: people of Nigeria (babies), contamination: natural, conc. range: 97–16,543 ng/l, country: UK/Nigeria[205], *cord blood

incidence: 4/461*, sa. const.: people of Ghana, Nigeria, and Kenya (females), contamination: natural, conc. range: 611–2,086 ng/l, country: UK[241] *cord blood

incidence: 4/40*, sa. const.: people of Nigeria (neonates), contamination: natural, conc. range: 1,149–3,151 pg/ml, Ø conc.: 1,981.5 pg/ml, country: Nigeria/UK[262], *non-jaundiced, peripheral blood

incidence: 3/37*, sa. const.: people of Nigeria (neonates), contamination: natural, conc. range: 850–11,728 pg/ml, Ø conc.: 7,185 pg/ml, country: Nigeria/UK[262], *jaundiced, peripheral blood

incidence: 2/40*, sa. const.: people of Nigeria (neonates), contamination: natural, conc. range: 1,348–1,985 pg/ml, Ø conc.: 1,666.5 pg/ml, country: Nigeria/UK[262], *non-jaundiced, cord blood

incidence: 1/37*, sa. const.: people of Nigeria (neonates), contamination: natural, conc.: 2,053 pg/ml, country: Nigeria/UK[262], *jaundiced, cord blood

incidence: 12/64*, sa. const.: people of Sierra Leone, contamination: natural, conc. range: 0.004–8.8 ng/ml, Ø conc.: 1.6 ng/ml, country: Sierra Leone/UK[390], *cord blood

incidence: 1/8*, sa. const.: people of Sierra Leone, contamination: natural, conc.: 8.8 ng/ml, country: Sierra Leone/UK[390], *maternal blood

incidence: 1/6*, sa. const.: people of Sierra Leone (male infants), contamination: natural, conc.: 2.2 ng/ml, country: Sierra Leone/UK[390], *low birthweight

incidence: 3/5*, sa. const.: people of Sierra Leone (female infants), contamination:

natural, conc. range: 2.8–5.9 ng/ml, country: Sierra Leone/UK[390], *low birthweight
incidence: 1/21*, sa. const.: people of Sierra Leone (male infants), contamination: natural, conc.: 0.004 ng/ml, country: Sierra Leone/UK[390], *normal birthweight
incidence: 2/20*, sa. const.: people of Sierra Leone (female infants), contamination: natural, conc. range: 0.2–3.2 ng/ml, Ø conc.: 1.7 ng/ml, country: Sierra Leone/UK[390], *normal birthweight

incidence: 1/58*, sa. const.: people of Singapore (49 males (1 af) and 9 females), contamination: natural, conc.: 17 pg/ml, country: Singapore[444], *PHC patients

incidence: 1/15, sa. const.: people of Kenya (9 male patients (1 af*) and 6 female), age: 0.6–52 years?, contamination: natural, conc.: 13,230 pg/ml, country: Kenya/UK[466], *additionally stomach cancer

Aflatoxin G_2
incidence: 1/14, sa. const.: people of The Sudan (females), contamination: natural, conc.: 8 pg/ml, country: UK/The Sudan[134]

incidence: 13/625*, sa. const.: people of Nigeria (babies), contamination: natural, conc. range: 15–275 ng/l, country: UK/Nigeria[205], *cord blood

incidence: 1/461, sa. const.: people of Ghana, Nigeria, and Kenya (females), contamination: natural, conc.: 37 ng/l, country: UK[241], *cord blood

incidence: 1/40*, sa. const.: people of Nigeria (neonates), contamination: natural, conc.: 438 pg/ml, country: Nigeria/UK[262], *non-jaundiced, cord blood
incidence: 2/37*, sa. const.: people of Nigeria (neonates), contamination: natural, conc. range: 13 pg/ml, Ø conc.: 13 pg/ml, country: Nigeria/UK[262], *jaundiced, cord blood

incidence: 26/64*, sa. const.: people of Sierra Leone, contamination: natural, conc. range: 0.002–3.0 ng/ml, Ø conc.: 0.07 ng/ml, country: Sierra Leone/UK[390], *cord blood
incidence: 5/8*, sa. const.: people of Sierra Leone, contamination: natural, conc. range: 0.004–0.1 ng/ml, Ø conc.: 0.0408 ng/ml, country: Sierra Leone/UK[390], *maternal blood
incidence: 4/6*, sa. const.: people of Sierra Leone (male infants), contamination: natural, conc. range: 0.02–3.0 ng/ml, country: Sierra Leone/UK[390], *low birthweight
incidence: 2/5*, sa. const.: people of Sierra Leone (female infants), contamination: natural, conc. range: 0.03–0.4 ng/ml, Ø conc.: 0.215 ng/ml, country: Sierra Leone/UK[390], *low birthweight
incidence: 6/21*, sa. const.: people of Sierra Leone (male infants), contamination: natural, conc. range: 0.006–0.3 ng/ml, country: Sierra Leone/UK[390], *normal birthweight
incidence: 8/20*, sa. const.: people of Sierra Leone (female infants), contamination: natural, conc. range: 0.004–0.4 ng/ml, country: Sierra Leone/UK[390], *normal birthweight

Aflatoxin M_1
incidence: 1/14, sa. const.: people of The Sudan (females), contamination: natural, conc.: 5 pg/ml, country: UK/The Sudan[134]

incidence: 25/625*, sa. const.: people of Nigeria (babies), contamination: natural, conc. range: 32–11,354 ng/l, country: UK/Nigeria[205], *cord blood

incidence: 3/77*, sa. const.: people of Nigeria (females), contamination: natural, conc. range: 38–483 ng/l, Ø conc.: 262 ng/l, country: UK[241], *maternal blood
incidence: 63/461*, sa. const.: people of Ghana, Nigeria, and Kenya (females), contamination: natural, conc. range: 25–8,942 ng/l, country: UK[241], *cord blood

incidence: 4/37*, sa. const.: people of Nigeria (neonates), contamination: natural, conc. range: 24–8,464 pg/ml, Ø conc.: 3,301.75 pg/ml, country: Nigeria/UK[262], *jaundiced, peripheral blood

incidence: 1/40*, sa. const.: people of Nigeria (neonates), contamination: natural, conc.: 40 pg/ml, country: Nigeria/UK[262], *non-jaundiced, cord blood
incidence: 2/37*, sa. const.: people of Nigeria (neonates), contamination: natural, conc. range: 57–713 pg/ml, Ø conc.: 385 pg/ml, country: Nigeria/UK[262], *jaundiced, cord blood

incidence: 5/20, sa. const.: people of Egypt (females), contamination: natural, conc. range: 0.1–2.1 ng/ml, Ø conc.: 1.18 ng/ml, country: Egypt[352]

incidence: 36/64*, sa. const.: people of Sierra Leone, contamination: natural, conc. range: 0.007–5.1 ng/ml, Ø conc.: 0.4 ng/ml, country: Sierra Leone/UK[390], *cord blood
incidence: 3/8*, sa. const.: people of Sierra Leone, contamination: natural, conc. range: 0.09–0.8 ng/ml, Ø conc.: 0.397 ng/ml, country: Sierra Leone/UK[390], *maternal blood
incidence: 3/6*, sa. const.: people of Sierra Leone (male infants), contamination: natural, conc. range: 0.4–0.8 ng/ml, country: Sierra Leone/UK[390], *low birthweight
incidence: 3/5*, sa. const.: people of Sierra Leone (female infants), contamination: natural, conc. range: 0.2–5.1 ng/ml, country: Sierra Leone/UK[390], *low birthweight
incidence: 10/21*, sa. const.: people of Sierra Leone (male infants), contamination: natural, conc. range: 0.03–1.7 ng/ml, country: Sierra Leone/UK[390], *normal birthweight
incidence: 10/20*, sa. const.: people of Sierra Leone (female infants), contamination: natural, conc. range: 0.01–2.9 ng/ml, country: Sierra Leone/UK[390], *normal birthweight

incidence: 111/166*, sa. const.: people of the UAE, contamination: natural, conc. range: 0.05–10.44 ng/ml, country: UAE[428], *cord blood
incidence: 113/166*, sa. const.: people of the UAE (females), contamination: natural, conc. range: 0.03–8.49 ng/ml, country: UAE[428], *maternal blood
incidence: 1/1*, sa. const.: person of the UAE (premature baby boy), contamination: natural, conc.: 3.99 ng/ml, country: UAE[428], *cord blood

incidence: 6/15, sa. const.: people of Kenya (9 male (4 af*) and 6 female patients (2 af*)), age: 0.6–52 years?, contamination: natural, conc. range: 30–757 pg/ml, Ø conc.: 414.8 pg/ml, country: Kenya/UK[466], *additionally cirrhosis, hepatitis, or marasmic kwashiorkor

incidence: 107/201*, sa. const.: people of the UAE (females), contamination: natural, conc. range: 110–4,060 pg/ml, country: UAE[467], *umbilical cord blood

Aflatoxin M_2

incidence: 21/625*, sa. const.: people of Nigeria (babies), contamination: natural, conc. range: 14–3,644 ng/l, country: UK/Nigeria[205], *cord blood

incidence: 4/77*, sa. const.: people of Nigeria (females), contamination: natural, conc. range: 48–3,480 ng/l, Ø conc.: 948.25 ng/l, country: UK[241], *maternal blood
incidence: 47/461*, sa. const.: people of Ghana, Nigeria, and Kenya (females), contamination: natural, conc. range: 15–732 ng/l, country: UK[241], *cord blood

incidence: 5/40*, sa. const.: people of Nigeria (neonates), contamination: natural, conc. range: 51–664 pg/ml, Ø conc.: 313.2 pg/ml, country: Nigeria/UK[262], *non-jaundiced, peripheral blood
incidence: 3/37*, sa. const.: people of Nigeria (neonates), contamination: natural, conc. range: 233–3,557 pg/ml, Ø conc.: 1,666.66 pg/ml, country: Nigeria/UK[262], *jaundiced, peripheral blood
incidence: 3/40*, sa. const.: people of Nigeria (neonates), contamination: natural, conc. range: 32–649 pg/ml, Ø conc.: 261.67 pg/ml, country: Nigeria/UK[262], *non-jaundiced, cord blood

incidence: 3/37*, sa. const.: people of Nigeria (neonates), contamination: natural, conc. range: 374–974 pg/ml, Ø conc.: 687.67 pg/ml, country: Nigeria/UK[262], *jaundiced, cord blood

incidence: 19/64*, sa. const.: people of Sierra Leone, contamination: natural, conc. range: 0.02–5.4 ng/ml, Ø conc.: 0.7 ng/ml, country: Sierra Leone/UK[390], *cord blood
incidence: 1/8*, sa. const.: people of Sierra Leone, contamination: natural, conc.: 2.5 ng/ml, country: Sierra Leone/UK[390], *maternal blood
incidence: 5/6*, sa. const.: people of Sierra Leone (male infants), contamination: natural, conc. range: 0.1–5.4 ng/ml, country: Sierra Leone/UK[390], *low birthweight
incidence: 3/5*, sa. const.: people of Sierra Leone (female infants), contamination: natural, conc. range: 0.3–0.7 ng/ml, country: Sierra Leone/UK[390], *low birthweight
incidence: 1/21*, sa. const.: people of Sierra Leone (male infants), contamination: natural, conc.: 0.07 ng/ml, country: Sierra Leone/UK[390], *normal birthweight
incidence: 4/20*, sa. const.: people of Sierra Leone (female infants), contamination: natural, conc. range: 0.08–0.3 ng/ml, country: Sierra Leone/UK[390], *normal birthweight

incidence: 1/15, sa. const.: people of Kenya (9 male (1 af*) and 6 female), age: 0.6–52 years?, contamination: natural, conc.: 99 pg/ml, country: Kenya/UK[466], *additionally cirrhosis

incidence: 31/201*, sa. const.: people of the UAE (females), contamination: natural, conc. range: 210–3,700 pg/ml, country: UAE[467], *umbilical cord blood

Aflatoxin $M_1 + M_2$
incidence: ?/125*, sa. const.: people of Kenya (females), contamination: natural, conc. range: 12–1,689 pg/l, country: UK[241], *maternal blood

Aflatoxin
incidence: 3/10*, sa. const.: people of Kenya (children), contamination: natural, conc. range: 50–1,680 pg/ml, country: Kenya/UK[95], *control
incidence: 4/11*, sa. const.: people of Kenya (children), contamination: natural, conc. range: 99–9,571 pg/ml, country: Kenya/UK[95], *marasmus ca.
incidence: 2/4*, sa. const.: people of Kenya (children), contamination: natural, conc. range: 41–917 pg/ml, country: Kenya/UK[95], *marasmic kwashiorkor ca.
incidence: 9/14*, sa. const.: people of Kenya (children), contamination: natural, conc. range: 16–66,588 pg/ml, country: Kenya/UK[95], *kwashiorkor ca.

incidence: 475/479, sa. const.: people of Benin and Togo (children), contamination: natural, conc. range: 5–1,064 pg/mg albumin*, country: UK/Benin[140], *AF-albumin adducts

incidence: 2/35*, sa. const.: people of Thailand (females), contamination: natural, conc. range: pos, country: UK/USA[381], *maternal blood
incidence: 17/35*, sa. const.: people of Thailand (neonates), contamination: natural, conc. range: ≤4.2 ng/ml, country: UK/USA[381], *cord blood
incidence: 28/28, sa. const.: people of Nepal, contamination: natural, conc. range: 0.06–10 ng/ml, country: UK/USA[381]

incidence: 119/119*, sa. const.: people of The Gambia (females), contamination: natural, conc. range: 4.8–260.8 pg/mg, Ø conc.: 40.4 pg/mg, country: UK/The Gambia[420], *maternal blood
incidence: 48/99*, sa. const.: people of The Gambia (neonates), contamination: natural, conc. range: 5.0–89.6 pg/mg, Ø conc.: 10.1 pg/mg, country: UK/The Gambia[420], *cord blood
incidence: 13/118*, sa. const.: people of The Gambia (infants), contamination: natural, conc. range: 5.0–30.2 pg/mg, Ø conc.: 8.7 pg/mg, country: UK/The

Gambia[420], *infant blood after 16 weeks of birth

incidence: ?/?*, sa. const.: people of the Republic of Guinea (males and females), Ø age: 33.7 years, contamination: natural, Ø conc.: 5.5 pg/mg** ***, country: UK/Republic of Guinea[541], *control, **level at harvest (for detailed information please see the article), ***AF-albumin adducts
incidence: ?/?*, sa. const.: people of the Republic of Guinea (males and females), Ø age: 28.6 years, contamination: natural, Ø conc.: 7.2 pg/mg** ***, country: UK/Republic of Guinea[541], *intervention group, **level at harvest (for detailed information please see the article), ***AF-albumin adducts
incidence: ?/?*, sa. const.: people of the Republic of Guinea (males and females), Ø age: 33.7 years, contamination: natural, Ø conc.: 18.7 pg/mg** ***, country: UK/Republic of Guinea[541], *control, **level 5 months later at postharvest (for detailed information please see the article), ***AF-albumin adducts
incidence: ?/?*, sa. const.: people of the Republic of Guinea (males and females), Ø age: 28.6 years, contamination: natural, Ø conc.: 11.7 pg/mg** ***, country: UK/Republic of Guinea[541], *intervention group, **level 5 months later at postharvest (for detailed information please see the article), ***AF-albumin adducts
incidence: ?/?*, sa. const.: people of the Republic of Guinea (males and females), Ø age: 33.7 years, contamination: natural, Ø conc.: 18.7 pg/mg** ***, country: UK/Republic of Guinea[541], *control, **level at the end of the study (for detailed information please see the article), ***AF-albumin adducts
incidence: ?/?*, sa. const.: people of the Republic of Guinea (males and females), Ø age: 28.6 years, contamination: natural, Ø conc.: 8.0 pg/mg** ***, country: UK/Republic of Guinea[541], *intervention group, **level at the end of the study (for detailed information please see the article), ***AF-albumin adducts

incidence: 119/124, sa. const.: people of the Republic of Guinea (children), age: 2–5 years, contamination: natural, conc. range: 8.8–11.0 pg/mg albumin, Ø conc.: 9.9 pg/mg albumin*, country: UK/Republic of Guinea/USA[554], *AF-albumin adducts

Ochratoxin A
incidence: 14/20*, sa. const.: people of Tunisia (males and females), contamination: natural, conc. range: 0–7.5 ng/ml*, country: Tunisia/France[184], *healthy control
incidence: 52/60, sa. const.: people of Tunisia (males and females), contamination: natural, conc. range: 0–140.5 ng/ml*, country: Tunisia/France[184], *nephropathy patients
incidence: 13/20*, sa. const.: people of Tunisia (males and females), contamination: natural, conc. range: 0–3.2 ng/ml*, country: Tunisia/France[184], *healthy control
incidence: 36/40, sa. const.: people of Tunisia (males and females), contamination: natural, conc. range: 1.68–171.25 ng/ml*, country: Tunisia/France[184], *nephropathy patients

incidence: 73/140*, sa. const.: people of Tunisia (males and females), contamination: natural, conc. range: 0.1–8.8 ng/ml*, country: Tunisia/France[185], *control (general population)
incidence: 210/581*, sa. const.: people of Tunisia (males and females), contamination: natural, conc. range: 1.2–100 ng/ml* (12 values from 125 to 1,136 ng/ml), country: Tunisia/France[185], *with chronic renal failure

incidence: 52/100, sa. const.: people of Hungary (in part patients), contamination: natural, conc. range: 0.2–1 ng/ml (34 sa), 1–5 ng/ml (16 sa), 5–10 ng/ml (1 sa), 12.9 ng/ml (1 sa), country: Hungary[192]

incidence: 18/22*, sa. const.: people of France, contamination: natural, conc. range: 0.3–1,001 ng/ml, country: France[261], *patients (CIN ca.)
incidence: 38/71*, sa. const.: people of France, contamination: natural, conc. range: 0.28–6.72 ng/ml, country: France[261], *patients with renal diseases other than CIN

incidence: 39/39, sa. const.: people of Sweden (females), contamination: natural, conc. range: 90–940 ng/l, Ø conc.: 167 ng/l, country: Sweden[294]
incidence: 9/216, sa. const.: people of Poland, contamination: natural, conc. range: ≤4.8 ng/cm^3, country: Poland[303]

incidence: 2/2, sa. const.: people of France (male and female), contamination: natural, conc. range: 20.5–1,001 ng/ml, country: France[311]

incidence: 14/21*, sa. const.: people of Tunisia, contamination: natural, conc. range: 0.1–2.3 ng/ml, country: Tunisia/France[329], *healthy persons
incidence: 33/33*, sa. const.: people of Tunisia, contamination: natural, conc. range: 0.7–1,136 ng/ml, Ø conc.: 80.59 ng/ml, country: Tunisia/France[329], *nephropathy patients

incidence: 77/79*, sa. const.: people of Germany, contamination: natural, conc. range: 0.06–0.90 ng/ml, country: Germany[341], *umbilical cord blood

incidence: 2/13, sa. const.: people of Egypt (females), contamination: natural, conc. range: 3.22–4.12 ng/ml, Ø conc.: 3.67 ng/ml, country: Egypt[352]

incidence: 78/144, sa const.: people of Denmark, contamination: natural, conc. range: ≤13.2 µg/l, country: Denmark[386]
incidence: 16/64*, sa. const.: people of Sierra Leone, contamination: natural, conc. range: 0.2–3.5 ng/ml, Ø conc.: 0.9 ng/ml, country: Sierra Leone/UK[390], *cord blood
incidence: 1/8*, sa. const.: people of Sierra Leone, contamination: natural, conc.: 0.2 ng/ml, country: Sierra Leone/UK[390], *maternal blood
incidence: 6/6*, sa. const.: people of Sierra Leone (male infants), contamination: natural, conc.: nd, country: Sierra Leone/UK[390], *low birthweight
incidence: 2/5*, sa. const.: people of Sierra Leone (female infants), contamination: natural, conc. range: 0.4–0.5 ng/ml, Ø conc.: 0.45 ng/ml, country: Sierra Leone/UK[390], *low birthweight
incidence: 3/21*, sa. const.: people of Sierra Leone (male infants), contamination: natural, conc. range: 0.4–0.6 ng/ml, country: Sierra Leone/UK[390], *normal birthweight
incidence: 7/20*, sa. const.: people of Sierra Leone (female infants), contamination: natural, conc. range: 0.5–2.6 ng/ml, country: Sierra Leone/UK[390], *normal birthweight

incidence: 5/200, sa. const.: people of Sweden, contamination: natural, conc. range: ≤0.88 ng/ml, country: EU[396]
incidence: 130/160, sa. const.: people of Italy, contamination: natural, conc. range: <0.9 ng/ml (128 sa), ≤2.83 ng/ml (2 sa), country: EU[396]
incidence: 267/309, sa. const.: people of Germany, contamination: natural, conc. range: <0.9 ng/ml (254 sa), 1–1.9 ng/ml (9 sa), 2–4.9 ng/ml (3 sa), 7.9 ng/ml (1 sa), country: EU[396]

incidence: 22/63*, sa. const.: people of the Ivory Coast (males and females), age of affected persons: 19–50 years, contamination: natural, conc. range: 0.00992–5.81 µg/l, Ø conc.: 0.83 µg/l country: France/Ivory Coast/Tunisia[438], *apparently healthy volunteers
incidence: 8/39*, sa. const.: people of the Ivory Coast (27 males and 12 females), age of affected persons: 25–52 years, contamination: natural, conc. range: 0.167–2.42 µg/l, Ø conc.: 1.05 µg/l country: France/Ivory Coast/Tunisia[438], *nephropathy patients

incidence: 210/210, sa. const.: people of Norway (141 males and 69 females), contamination: natural, conc. range:

21–5,534 ng/l, Ø conc.: 397 ng/l, country: Norway[440]

incidence: 30/30*, sa. const.: people of Poland, contamination: natural, conc. range: 0.14–3.41 ng/ml, Ø conc.: 1.14 ng/ml, country: Poland[446], *maternal blood serum

incidence: 28/30*, sa. const.: people of Poland, contamination: natural, conc. range: 0.56–5.42 ng/ml, Ø conc.: 1.96 ng/ml, country: Poland[446], *fetal blood serum

incidence: 22/44*, sa. const.: people of Chile (healthy donors of Colbún, 16 males (6 af) and 28 females (16 af)), contamination: natural, conc. range: 0.07–2.75 ppb, country: Chile[504], *for detailed information please see the article

incidence: 40/44*, sa. const.: people of Chile (healthy donors of San Vicente Tagua-Tagua, 19 males (16 af) and 25 females (24 af)), contamination: natural, conc. range: 0.22–2.12 ppb, country Chile[504], *for detailed information please see the article

incidence: 202/202, sa. const.: people of Norway (98 males and 104 females), Ø age: 38 years (women), 41 years (men), Ø wt.: 84 kg (men), 64 kg (women), contamination: natural, Ø conc.: 0.18 ng/ml, country: Sweden/Norway[506]

incidence: 191/191, sa. const.: people of Sweden (133 males and 58 females), Ø age: 43 years (women), 44 years (men), Ø wt.: 84 kg (men), 68 kg (women), contamination: natural, conc. range: 0.03–1.16 ng/ml, Ø conc.: 0.21 ng/ml, country: Sweden/Norway[506]

incidence: ?/30*, sa. const.: people of Pakistan, contamination: natural, conc. range: 0.036–1.239 ng/ml, country: Pakistan/Germany[544], *non-diseased control

incidence: ?/87*, sa. const.: people of Pakistan, contamination: natural, conc. range: 0.032–3.409 ng/ml, country: Pakistan/Germany[544], *bladder cancer patients

incidence: 194/13,797*, sa. const.: people of Yugoslavia, contamination: natural, conc. range: 5–100 ng/ml, country: Yugoslavia/Sweden[568], *blood sa. from endemic villages for Balkan nephropathy

incidence: 47/3,378*, sa. const.: people of Yugoslavia, contamination: natural, conc. range: 5–50 ng/ml, country: Yugoslavia/Sweden[568], *blood sa. from nonendemic villages for Balkan nephropathy

incidence: 82/576*, sa. const.: people of Bulgaria, contamination: natural, conc. range: 1–2 ng/g serum (51 sa), >2–35 ng/g serum (31 sa), country: Bulgaria/France[573], *blood sa. from people living in areas with and without endemic villages nephropathy in Bulgaria, for detailed information please see the article

Zearalenone

incidence: 8/74*, sa. const.: people of Poland (females), contamination: natural, conc. range: tr–137 ng/ml, country: Poland[432], *showed neoplastic lesions in reproductive system

α-Zearalenol

incidence: 2/74*, sa. const.: people of Poland (females), contamination: natural, conc. range: 5 ng/ml, country: Poland[432], *showed neoplastic lesions in reproductive system

see also Human plasma and Human serum

Human brain may contain the following mycotoxins and/or their metabolites:

Aflatoxicol

incidence: 7/18*, sa. const.: people of Nigeria (children: 4 males and 3 females af), contamination: natural, conc. range: 22–1,785 pg/g, Ø conc.: 539.14 pg/g, country: Nigeria/UK[285], *dying of kwashiorkor

incidence: 6/19*, sa. const.: people of Nigeria (children: 4 males and 2 females af), contamination: natural, conc. range: 27–831 pg/g, Ø conc.: 367.83 pg/g, country: Nigeria/UK[285], *dying from miscellaneous diseases

Aflatoxin B_1
incidence: 13/18*, sa. const.: people of Thailand (children), contamination: natural, conc. range: tr**, country: USA/Thailand[10], *EFDV ca., **sa. showed a blue fluorescent spot similar to that of AFB_1
incidence: 7/13*, sa. const.: people of Thailand (children and 2 adolescents), contamination: natural, conc. range: tr**, country: USA/Thailand[10], *dying from causes other than EFDV **, sa. showed a blue fluorescent spot similar to that of AFB_1
incidence: 4/18*, sa. const.: people of Nigeria (children: 3 males and 1 female af.), contamination: natural, conc. range: 1,233–3,913 pg/g, Ø conc.: 2,699.75 pg/g, country: Nigeria/UK[285], *dying of kwashiorkor
incidence: 1/19*, sa. const.: people of Nigeria (children: 1 female af), contamination: natural, conc.: 12,423 pg/g, country: Nigeria/UK[285], *dying from miscellaneous diseases

Aflatoxin B_2
incidence: 1/18*, sa. const.: people of Thailand (children), contamination: natural, conc.: tr**, country: USA/Thailand[10], *EFDV ca., **1 of the sa. showed a blue fluorescent spot similar to that of AFB_2
incidence: 1/13*, sa. const.: people of Thailand (children and 2 adolescents), contamination: natural, conc.: tr**, country: USA/Thailand[10], * dying from causes other than EFDV, **1 of the sa. showed a blue fluorescent spot similar to that of AFB_2
incidence: 1/18*, sa. const.: people of Nigeria (children: 1 female af), contamination: natural, conc.: 21 pg/g, country: Nigeria/UK[285], *dying of kwashiorkor
incidence: 2/19*, sa. const.: people of Nigeria (children: 1 male and 1 female af), contamination: natural, conc. range: 4–113 pg/g, Ø conc.: 58.5 pg/g, country: Nigeria/UK[285], *dying from miscellaneous diseases

Aflatoxin G_1
incidence: 4/18*, sa. const.: people of Nigeria (children: 1 male and 3 females af), contamination: natural, conc. range: 395–107,700 pg/g, Ø conc.: 33,704.25 pg/g, country: Nigeria/UK[285], *dying of kwashiorkor
incidence: 5/19*, sa. const.: people of Nigeria (children: 3 males and 2 females af), contamination: natural, conc. range: 2,267–71,742 pg/g, Ø conc.: 25,755.4 pg/g, country: Nigeria/UK[285], *dying from miscellaneous diseases

Aflatoxin G_2
incidence: 2/18*, sa. const.: people of Nigeria (children: 2 males af), contamination: natural, conc. range: 193–212 pg/g, Ø conc.: 202.5 pg/g, country: Nigeria/UK[285], *dying of kwashiorkor
incidence: 2/19*, sa. const.: people of Nigeria (children: 2 males af), contamination: natural, conc. range: 8–13 pg/g, Ø conc.: 10.5 pg/g, country: Nigeria/UK[285], *dying from miscellaneous diseases

Aflatoxin M_1
incidence: 2/17, sa. const.: people of Malaysia (1 adult and 16 children: 12 males and 5 females), age: adult 49 years, children 2.5–11 years, contamination: natural, conc. range: 1,229–13,314 pg/g tissue, Ø conc.: 7,271.5 pg/g tissue, country: Singapore/UK[259]
incidence: 1/18*, sa. const.: people of Nigeria (children: 1 female af), contamination: natural, conc.: 3,943 pg/g, country: Nigeria/UK[285], *dying of kwashiorkor
incidence: 1/19*, sa. const.: people of Nigeria (children: 1 male af), contamination: natural, conc.: 5,092 pg/g, country: Nigeria/UK[285], *dying from miscellaneous diseases

Aflatoxin M_2
incidence: 3/17, sa. const.: people of Malaysia (1 adult and 16 children: 12 males and 5 females), age: adult 49 years, children 2.5–11 years, contamination: natural, conc. range: 348–5,244 pg/g tissue, Ø conc.: 2,337 pg/g tissue, country: Singapore/UK[259]

incidence: 3/18*, sa. const.: people of Nigeria (children: 2 males and 1 female af), contamination: natural, conc. range: 1,007–1,854 pg/g, Ø conc.: 1,503.33 pg/g, country: Nigeria/UK[285], *dying of kwashiorkor

incidence: 5/19*, sa. const.: people of Nigeria (children: 4 males and 1 female af), contamination: natural, conc. range: 717–5,290 pg/g, Ø conc.: 2,040.4 pg/g, country: Nigeria/UK[285], * dying from miscellaneous diseases

Human breast may contain the following mycotoxins and/or their metabolites:

Aflatoxin B_1
incidence: 3/5*, sa. const.: people of the UK (1 male and 4 females (3 af)), contamination: natural, conc. range: 0.43–3.36 AFB_1-DNA adducts/10^6 nucleotides, Ø conc.: 1.406 AFB_1-DNA adducts/10^6 nucleotides, country: UK/Mexico[58], *non-tumor ca.

Human breast milk may contain the following mycotoxins and/or their metabolites:

Aflatoxicol
incidence: 6/800, sa. const.: human breast milk from women of The Sudan, Kenya, and Ghana, contamination: natural, conc. range: 14–270 ng/l, country: UK[241]

incidence: 41/113, sa. const.: human breast milk from women of Sierra Leone, contamination: natural, conc. range: 0.005–50.9 ng/ml, country: Sierra Leone/UK[354]

Aflatoxin B_1
incidence: 41/800, human breast milk from women of The Sudan, Kenya, and Ghana contamination: natural, conc. range: 150–55,792 ng/l, country: UK[241]

incidence: 20/113, sa. const.: human breast milk from women of Sierra Leone contamination: natural, conc. range: 0.05–372 ng/ml, country: Sierra Leone/UK[354]

incidence: 1/231, sa. const.: human breast milk from women of Italy, contamination: natural, conc.: 11.4 ng/l, country: Italy[356]

incidence: 75/75, sa. const.: human breast milk from women of Turkey, contamination: natural, conc. range: 94–129 ng/l (17 sa), 130–149 ng/l (15 sa), 150–199 ng/l (25 sa), 200–300 ng/l (10 sa), >300–4,123.80 ng/l (8 sa), country: Turkey[588]

Aflatoxin B_2
incidence: 10/800, sa. const.: human breast milk from women of The Sudan, Kenya, and Ghana, contamination: natural, conc. range: 49–623 ng/l, country: UK[241]

Aflatoxin G_1
incidence: 4/800, sa. const.: human breast milk from women of The Sudan, Kenya, and Ghana, contamination: natural, conc. range: 1,890–5,180 ng/l, country: UK[241]

incidence: 3/5, sa. const.: human breast milk from women of The Gambia, contamination: natural, conc. range: 18–114 pg/ml, Ø conc.: 67.33 pg/ml, country: USA/France/UK[273]

incidence: 22/113, sa. const.: human breast milk from women of Sierra Leone, contamination: natural, conc. range: 0.005–139 ng/ml, country: Sierra Leone/UK[354]

Aflatoxin G_2
incidence: 3/800, sa. const.: human breast milk from women of The Sudan, Kenya, and Ghana contamination: natural, conc. range: 10–87 ng/l, country: UK[241]

incidence: 25/113, sa. const.: human breast milk from women of Sierra Leone, contamination: natural, conc. range: 0.003–366 ng/ml, country: Sierra Leone/UK[354]

Aflatoxin M_1

incidence: 129/140, sa. const.: human breast milk from women of UAE and other countries, contamination: natural, conc. range: ≤3,400 pg/ml, country: UAE[1]

incidence: 11/73, sa. const.: human breast milk from women of Australia, contamination: natural, conc. range: 0.028–1.031 ng/ml, country: Australia/UK[46]

incidence: 5/11, sa. const.: human breast milk from women of Thailand, contamination: natural, conc. range: 0.039–1.736 ng/ml, country: Australia/UK[46]

incidence: 8/61, sa. const.: human breast milk from women of Turkey, contamination: natural, conc. range: 5.10–6.90 ng/l, Ø conc.: 5.68 ng/l country: Turkey[99]

incidence: 13/99, sa. const.: human breast milk from women of The Sudan, contamination: natural, conc. range: 5–64 pg/ml, Ø conc.: 19 pg/ml, country: UK/The Sudan[134]

incidence: 4/82, sa. const.: human breast milk from women of Italy, contamination: natural, conc. range: 7 ng/l (1 sa), >10–50 ng/l (2 sa), 140 ng/l (1 sa), Ø conc.: 55.35 ng/l, country: Italy[233]

incidence: 121/800, sa. const.: human breast milk from women of The Sudan, Kenya, and Ghana, contamination: natural, conc. range: 5–1,379 ng/l, country: UK[241]

incidence: 138/388, sa. const.: human breast milk from women of Egypt, contamination: natural, conc. range: 5.6–5,131 pg/ml, country: Finland/UK/Egypt[244]

incidence: 10/64, sa. const.: human breast milk from women of the UAE, contamination: natural, conc. range: 0.3–1.3 ng/ml, Ø conc.: 0.77 ng/ml, country: UAE/UK[247]

incidence: 15/15, sa. const.: human breast milk from women of the UAE, contamination: natural, conc. range: 7–23 pg/ml, country: UAE/UK[247]

incidence: 443/445, sa. const.: human breast milk from women of Arabic, European, and Asiatic countries, contamination: natural, conc. range: 0.002–3 ng/ml, country: UAE/UK[248]

incidence: 5/5, sa. const.: human breast milk from women of The Gambia, contamination: natural, conc. range: ≤1.4 pg/ml, country: USA/France/UK[273]

incidence: 157/160, sa. const.: human breast milk from women of Iran, contamination: natural, conc. range: 0.3–26.7 ng/kg, country: Iran[332]

incidence: 2/10, sa. const.: human breast milk from women of Egypt, contamination: natural, conc. range: 0.5–5 ppb, Ø conc.: 2.75 ppb, country: Egypt[351]

incidence: 66/120, sa. const.: human breast milk from women of Egypt, contamination: natural, conc. range: 0.2–2.09 ng/ml, Ø conc.: 0.35 ng/ml, country: Egypt[352]

incidence: 35/113, sa. const.: human breast milk from women of Sierra Leone, contamination: natural, conc. range: 0.2–99 ng/ml, country: Sierra Leone/UK[354]

incidence: 1/22, sa. const.: human breast milk from women of Brazil, contamination: natural, conc.: 0.02 ng/ml, country: Brazil[355]

incidence: 1/231, sa. const.: human breast milk from women of Italy, contamination: natural, conc.: 194 ng/l, country: Italy[356]

incidence: 245/443, sa. const.: human breast milk from women of Egypt, contamination: natural, conc. range:

4.2–889 pg/ml, country: Finland/UK/Egypt[533]

incidence: 6/54, sa. const.: human breast milk from women of Zimbabwe, contamination: natural, Ø conc.: 3.6 pg/ml (mean value), country: Zimbabwe[569]

incidence: 75/75, sa. const.: human breast milk from women of Turkey, contamination: natural, conc. range: 60–79 ng/l (13 sa), 80–99 ng/l (24 sa), 100–299.99 ng/l (38 sa), country: Turkey[588]

Aflatoxin M_2
incidence: 11/99, sa. const.: human breast milk from women of The Sudan, contamination: natural, conc. range: 3–20 pg/ml, Ø conc.: 12.2 pg/ml, country: UK/The Sudan[134]

incidence: 103/800, sa. const.: human breast milk from women of The Sudan, Kenya, and Ghana, contamination: natural, conc. range: 3–6,368 ng/l, country: UK[241]

incidence: 70/113, sa. const.: human breast milk from women of Sierra Leone, contamination: natural, conc. range: 0.07–77.5 ng/ml, country: Sierra Leone/UK[354]

Aflatoxin M_1 & M_2
incidence: 13/99, sa. const.: human breast milk from women of The Sudan, contamination: natural, conc. range: 3–84 pg/ml, Ø conc.: 34.7 pg/ml, country: UK/The Sudan[134]

Aflatoxin
incidence: 8/18, sa. const.: human breast milk from women of The Sudan and Ghana, contamination: natural, conc. range: 1–45 pg/ml, country: France/Zimbabwe[271]
incidence: 6/54, sa. const.: human breast milk from women of Zimbabwe, contamination: natural, conc. range: 14.1–50.5 pg/ml, Ø conc.: 33.65 pg/ml, country: France/Zimbabwe[271]

Ochratoxin A
incidence: 9/50, sa. const.: human breast milk from women of Italy, contamination: natural, conc. range: 1.7–6.6 ng/ml, Ø conc.: 4.19 ng/ml, country: Italy[97]

incidence: 38/92, sa. const.: human breast milk from women of Hungary, contamination: natural, conc. range: 0.22–1 ng/ml (13 sa), 1–2 ng/ml (12 sa), 2–3 ng/ml (8 sa), 3–5 ng/ml (3 sa), 5 to ≤7.63 ng/ml (2 sa), country: Hungary[192]

incidence: 61/82, sa. const.: human breast milk from women of Italy, contamination: natural, conc. range: 5–10 ng/l (28 sa), >10–50 ng/l (27 sa), >50–405 ng/l (6 sa), Ø conc.: 30.43 ng/l, country: Italy[233]

incidence: 74/85, sa. const.: human breast milk from women of Italy, contamination: natural, conc. range: 0.02 ng/ml (26 sa), 0.1 ng/ml (34 sa), 0.5 ng/ml (7 sa), >1 ng/ml (7 sa), country: Italy[268]

incidence: 2/100, sa. const.: human breast milk from women of Australia, contamination: natural, conc. range: 3–3.6 ng/ml, Ø conc.: 3.3 ng/ml, country: Australia[290]

incidence: 4/36, sa. const.: human breast milk from women of Germany, contamination: natural, conc. range: 0.017–0.030 µg/kg, Ø conc.: 0.0238 µg/kg, country: Germany[291]

incidence: 23/40, sa. const.: human breast milk from women of Sweden, contamination: natural, conc. range: 10–40 ng/l, country: Sweden[294]

incidence: 4/36, sa. const.: human breast milk from women of Germany, contamination: natural, conc. range: 0.017–0.030 ng/ml, country: Germany[301]

incidence: 17/80, sa. const.: human breast milk from women of Norway, contamination: natural, conc. range: 10–182 ng/l, Ø conc.: 30 ng/l, country: Norway[346]

incidence: 38/115, sa. const.: human breast milk from women of Norway, contamination: natural, conc. range: 10–130 ng/l, Ø conc.: 37.56 ng/l, country: Norway[347]

incidence: 4/40, sa. const.: human breast milk from of from Switzerland, contamination: natural, conc. range: 5–14 pg/g, country: Switzerland[349]

incidence: 3/10, sa. const.: human breast milk from women of Egypt, contamination: natural, conc. range: 3–15 ppb, Ø conc.: 8.87 ppb, country: Egypt[351]

incidence: 43/120, sa. const.: human breast milk from women of Egypt, contamination: natural, conc. range: 5.07–45.01 ng/ml, Ø conc.: 21.06 ng/ml, country: Egypt[352]

incidence: 40/113, sa. const.: human breast milk from women of Sierra Leone, contamination: natural, conc. range: 0.2–337 ng/ml, country: Sierra Leone/UK[354]

incidence: 2/28, sa. const.: human breast milk from women of Brazil, contamination: natural, conc. range: 0.01–0.02 ng/ml, Ø conc.: 0.015 ng/ml, country: Brazil[355]

incidence: 198/231, sa. const.: human breast milk from women of Italy, contamination: natural, conc. range: ≤57 ng/l, country: Italy[356]

incidence: 5/13, sa. const.: human breast milk from women of Poland, contamination: natural, conc. range: 0.0053–0.017 ng/ml, Ø conc.: 0.01026 ng/ml country: Poland[446]

incidence: 36/50, sa. const.: human breast milk from women of Egypt, contamination: natural, Ø conc.: 1.89 ng/ml*, country: Egypt[449], *for all sa.

incidence: 23/76, sa. const.: human breast milk from women of Slovakia, contamination: natural, conc. range: LOQ (14 sa), 2.30–60.30 ng/l (9 sa), Ø conc.: 19.79 ng/ml, country: Slovakia[450]

incidence: 96/142, sa. const.: human breast milk from women of Italy, contamination: natural, conc. range: LOD/LOQ–0.9 μg/l (93 sa), 1.0–2.35 μg/l (3 sa), country: EU[508]

incidence: 75/75, sa. const.: human breast milk from women of Turkey, contamination: natural, conc. range: 600–1,499 ng/l (28 sa), 1,500–2,499 ng/l (31 sa), 2,500–2,999 ng/l (3 sa), 3,000–3,499 ng/l (3 sa), 3,500–13,111.30 ng/l (10 sa), country: Turkey[636]

Human cervix may contain the following mycotoxins and/or their metabolites:

AFLATOXIN B_1
incidence: 3/5*, sa. const.: people of the UK (females), contamination: natural, conc. range: 0.48–4.9 AFB_1-DNA adducts/10^6 nucleotides, Ø conc.: 2.69 AFB_1-DNA adducts/10^6 nucleotides, country: UK/Mexico[58], *non-tumor ca.

Human colon may contain the following mycotoxins and/or their metabolites:

AFLATOXIN B_1
incidence: 17/32* **, sa. const.: people of the UK (10 males (6 af), 12 females (6 af), and 10 of unknown sex (5 af), contamination: natural, conc. range: 0.04–56.96 AFB_1-DNA adducts/10^6 nucleotides, country: UK/Mexico[58], *sa. of colon and sigmoid, right, transverse, and left colon, **partly tumor ca. (for detailed information please see the article)

Human Colostrum see Human breast milk

Human endometrium may contain the following mycotoxins and/or their metabolites:

ZEARALENONE
incidence: 25/49, sa. const.: people of Poland (females), contamination: natural, conc. range: 47.8 ng/ml (mean value, 3 sa*), 167 ng/ml (mean value, 22 sa**), country: Poland[412], *endometrial hyperplasia, **endometrial adenocarcinoma

Human feces may contain the following mycotoxins and/or their metabolites:

Aflatoxin B_1
incidence: 2/40, sa. const.: people of Ghana (29 males and 11 females), age: 30–73 years, contamination: natural, Ø conc.: 47.9 pg/g (wet weight), country: Ghana[461]

Aflatoxin G_1
incidence: 34/40, sa. const.: people of Ghana (29 males and 11 females), age: 30–73 years, contamination: natural, Ø conc.: 129.6 pg/g (wet weight), country: Ghana[461]

Aflatoxin M_1
incidence: 3/40, sa. const.: people of Ghana (29 males and 11 females), age: 30–73 years, contamination: natural, Ø conc.: 58.7 pg/g (wet weight), country: Ghana[461]

Aflatoxin Q_1
incidence: 5/40, sa. const.: people of Ghana (29 males and 11 females), age: 30–73 years, contamination: natural, Ø conc.: 101.3 pg/g (wet weight), country: Ghana[461]

Aflatoxin
incidence: 11/20, sa. const.: people of Egypt (12 males and 8 females (11 af) of Egypt, age: 20–40 years, contamination: natural, conc. range: 1.8–6 µg/kg, country: Australia/Finland[217]

Fumonisin B_1
incidence: 7/20*, sa. const.: people of South Africa (school-children), age: 6–12 years, contamination: natural, conc. range: 6.0–19.56 mg/g, country: South Africa[463], *rural sa.
incidence: 2/23*, sa. const.: people of South Africa (adults), age: 12–60 years, contamination: natural, conc. range: 3.5–16.2 mg/g, Ø conc.: 9.85 mg/g, country: South Africa[463], *urban sa.

Human funiculum may contain the following mycotoxins and/or their metabolites:

Ochratoxin A
incidence: 9/40*, sa. const.: people of Italy (females), contamination: natural, conc. range: LOD/LOQ–0.9 µg/l (6 sa), 1.0–1.9 µg/l (2 sa), 9.4 µg/l (1 sa), country: EU[508], *pregnant women with (12 thereof 3 OTA-pos) and without (28 thereof 6 OTA-pos) pathologies

Human hair may contain the following mycotoxins and/or their metabolites:

Fumonisin B_1
incidence: 5/5, sa. const.: people of South Africa (composite bulk hair), contamination: natural, conc. range: tr–93.5 µg/kg, country: South Africa[436]

Fumonisin B_2
incidence: 4/5, sa. const.: people of South Africa (composite bulk hair), contamination: natural, conc. range: tr–23.5 µg/kg, country: South Africa[436]

Fumonisin B_3
incidence: 2/5, sa. const.: people of South Africa (composite bulk hair), contamination: natural, conc. range: tr, country: South Africa[436]

Human heart may contain the following mycotoxins and/or their metabolites:

Aflatoxin B_1
incidence: 3/17, sa. const.: people of Malaysia (1 adult and 16 children: 12 males and 5 females), age: adult 49 years, children 2.5–11 years, contamination: natural, conc. range: 830–2,355 pg/g tissue, Ø conc.: 1,454.66 pg/g tissue, country: Singapore/UK[259]

Aflatoxin B_2
incidence: 1/17, sa. const.: people of Malaysia (1 adult and 16 children: 12 males and 5 females), age: adult 49 years, children 2.5–11 years, contamination: natural, conc.: 19 pg/g tissue, country: Singapore/UK[259]

Human intestine may contain the following mycotoxins and/or their metabolites:

Aflatoxin B_1
incidence: 4/5*, sa. const.: people of Thailand (children), contamination: natural, conc. range: tr–81 µg/kg**, country: USA/ Thailand[10], *EFDV ca., **contents of intestine
incidence: 3/9*, sa. const.: people of Thailand (children), contamination: natural, conc. range: tr** ***, country: USA/Thailand[10], *dying from causes other than EFDV, **contents of intestine, ***sa. showed a blue fluorescent spot similar to that of AFB_1

Aflatoxin B_2
incidence: 1/5*, sa. const.: people of Thailand (children), contamination: natural, conc.: 10 µg/kg**, country: USA/Thailand[10], *EFDV ca., **in intestine contents

Human kidney may contain the following mycotoxins and/or their metabolites:

Aflatoxin B_1
incidence: 11/14*, sa. const.: people of Thailand (children), contamination: natural, conc. range: tr **, country: USA/Thailand[10], *EFDV ca., **sa. showed a blue fluorescent spot similar to that of AFB_1
incidence: 6/11*, sa. const.: people of Thailand (children), contamination: natural, conc. range: tr**, country: USA/Thailand[10], *dying from causes other than EFDV, ** showed a blue fluorescent spot similar to that of AFB_1

incidence: 2/17, sa. const.: people of Malaysia (1 adult and 16 children: 12 males and 5 females), age: adult 49 years, children 2.5–11 years, contamination: natural, conc. range: 200–336 pg/g tissue, Ø conc.: 268 pg/g tissue, country: Singapore/UK[259]

Aflatoxin B_2
incidence: 1/11*, sa. const.: people of Thailand (children), contamination: natural, conc.: tr**, country: USA/Thailand[10], *dying from causes other than EFDV, **a blue fluorescent spot similar to that of AFB_2

Aflatoxin M_1
incidence: 3/17, sa. const.: people of Malaysia (1 adult and 16 children: 12 males and 5 females), age: adult 49 years, children 2.5–11 years, contamination: natural, conc. range: 877–18,521 pg/g tissue, Ø conc.: 6,911.33 pg/g tissue, country: Singapore/UK[259]

Aflatoxin M_2
incidence: 1/17, sa. const.: people of Malaysia (1 adult and 16 children: 12 males and 5 females), age: adult 49 years, children 2.5–11 years, contamination: natural, conc.: 445 pg/g tissue, country: Singapore/UK[259]

Ochratoxin A
incidence: 3/46, sa. const.: people of Germany, contamination: natural, conc. range: 0.10–0.30 µg/kg, Ø conc.: 0.20 µg/kg, country: Germany[291]

Human liver may contain the following mycotoxins and/or their metabolites:

Aflatoxicol
incidence: 5/19*, sa. const.: people of Liberia, Nigeria, and South Africa (children), contamination: natural, conc.: 188 pg/g (1 kwashiorkor ca), 108–8,500 pg/g (4 marasmic kwashiorkor ca) country: UK[11], *10 kwashiorkor ca., 6 marasmic kwashiorkor ca., and 3 marasmus ca.

incidence: 2/16*, sa. const.: people of The Sudan (children), contamination: natural, conc. range: 1,004–4,370 pg/g, Ø conc.: 2,687 pg/g, country: The Sudan/UK[32], *kwashiorkor ca.

incidence: 1/13*, sa. const.: people of The Sudan (children), contamination: natural, conc.: 68,936 pg/g, country: The Sudan/UK[32], *miscellaneous liver disease ca.

incidence: 2/22*, sa. const.: people of Ghana (children), contamination: natural, conc. range: 12–99 pg/g, Ø conc.: 55.5 pg/g, country: Ghana/UK[61], *kwashiorkor ca.

incidence: 2/15, sa. const.: people of Kenya (patients: 9 males (2 af*) and 6 females), age: 0.6–52 years?, contamination: natural, conc. range: 280–2,157 pg/g, Ø conc.: 1,218.5 pg/g, country: Kenya/UK[466], *additionally cirrhosis or marasmic kwashiorkor

Aflatoxin B_1

incidence: 1/1, sa. const.: person of the USA (male), contamination: natural, conc.: 520 ng/g wet liver, country: USA[4]

incidence: 1/20*, sa. const.: people of the USA (adults and children: males and females (1 af)), contamination: natural, conc.: 22.5 µg/kg, country: USA[8], *8 Reye-syndrome ca. thereof 1 AFB_1-pos., 2 ca. of acute encephalopathy and 10 ca. without liver disease

incidence: 17/19*, sa. const.: people of Thailand (children), contamination: natural, conc. range: tr–93 µg/kg, country: USA/Thailand[10], *EFDV ca.

incidence: 8/13*, sa. const.: people of Thailand (children), contamination: natural, conc. range: tr, country: USA/Thailand[10], *dying from causes other than EFDV

incidence: 9/19*, sa. const.: people of Liberia, Nigeria, and South Africa (children), contamination: natural, conc. range: 391–8,350 pg/g (9 kwashiorkor ca), country: UK[11], *10 kwashiorkor ca., 6 marasmic kwashiorkor ca., and 3 marasmus ca.

incidence: 16/37*, sa. const.: people of the USA (patients), contamination: natural, conc. range: <1–62 ng/g, country: USA[19], *15 Reye's-syndrome ca. thereof 11 AFB_1-pos.

incidence: 5/15, sa. const.: people of Czechoslovakia (3 males and 2 females af*), contamination: natural, conc. range: 0.36–5.2 µg/kg, Ø conc.: 3.312 µg/kg, country: France/Czechoslovakia[28], *liver cancer ca.

incidence: 2/16*, sa. const.: people of The Sudan (children), contamination: natural, conc. range: 32,174–33,206 pg/g, Ø conc.: 32,690 pg/g, country: The Sudan/UK[32], *kwashiorkor ca

incidence: 2/2, sa. const.: people of New Zealand (children), contamination: natural, conc. range: 5–50 µg/kg wet weight (estimates), country: New Zealand[35]

incidence: 2/8*, sa. const.: people of Taiwan (1 male and 1 female af), contamination: natural, conc. range: 1.2–1.7 µmol/mol DNA**, Ø conc.: 1.45 µmol/mol DNA**, country: USA/Taiwan, Republic of China[49], *histology = adjacent-normal tissue, **AFB_1-FAPy adducts

incidence: 7*/9, sa. const.: people of Taiwan (6 males and 1 female af), contamination: natural, conc. range: 1.2–3.5 µmol/mol DNA**, Ø conc.: 2.23 µmol/mol DNA**, country: USA/Taiwan, Republic of China[49], *histology = tumor tissue/focal nodule, **AFB_1-FAPy adducts

incidence: 10/23*, sa. const.: people of the UK, Africa, and Southeast Asia (4 males (1 af), 11 females (6 af), and 8 of unknown sex (3 af)), contamination: natural, conc. range: 0.23–19.8 AFB_1-DNA adducts/10^6 nucleotides, country: UK/Mexico[58], *non-tumor ca. and unknown

incidence: 20/22*, sa. const.: people of Ghana (children), contamination: natural, conc. range: 62–4,409 pg/g, Ø conc.: 1,009.5 pg/g, country: Ghana/UK[61], *kwashiorkor ca.

incidence: 6/7*, sa. const.: people of the USA (children), contamination: natural, conc. range: 2.23–17.33 ng/g, Ø conc.: 9.18 ng/ml, country: USA[65], *Reye's-syndrome ca.

incidence: 6/100, sa. const.: people of France (males (4 af) and females (2 af)), contamination: natural, conc. range: tr–20 μg/kg, country: France/Senegal[122]

incidence: 7/8, sa. const.: people of Czechoslovakia (6 males and 1 female child af), contamination: natural, conc. range: 0.63–3.51 pmol AFB_1/mg DNA, Ø conc.: 1.72 pmol AFB_1/mg DNA, country: UK/Czechoslovakia[153]

incidence: 6/17, sa. const.: people of Malaysia (1 adult and 16 children: 12 males and 5 females), age: adult 49 years, children 2.5–11 years, contamination: natural, conc. range: 532–3,176 pg/g tissue, Ø conc.: 1,309.83 pg/g tissue, country: Singapore/UK[259]

incidence: 2/15, sa. const.: people of Kenya (patients: 9 males and 6 females (2 af*)), age: 0.6–52 years?, contamination: natural, conc. range: 2,232–92,978 pg/g, Ø conc.: 47,605 pg/g, country: Kenya/UK[466], *additionally HCC

incidence: 3?/23*, sa. const.: people of Nigeria, contamination: natural, conc. range: 3–15 μg/kg, Ø conc.: 7 μg/kg, country: Nigeria[521], *5 HCC ca. thereof AFB_1-3 to 4 pos.

incidence: 2/2, sa. const.: people of Kenya (children), contamination: natural, conc. range: 39–89 ppb, Ø conc.: 64 ppb, country: Kenya[545]

incidence: 6/100, sa. const.: people of France (55 males and 45 females), age: 24–84 years, contamination: natural, conc. range: tr–20 μg/kg, country: France[570]

incidence: 35/50*, sa. const.: people of Taiwan (42 males and 8 females), age: 41–84 years, contamination: natural, conc.: pr**, country: Taiwan, Republic of China/USA[603], *HCC patients, **AFB_1-DNA adducts

Aflatoxin B_2

incidence: 2/19*, sa. const.: people of Thailand (children), contamination: natural, conc. range: 6–11 μg/kg, Ø conc.: 8.5 μg/kg, country: USA/Thailand[10], *EFDV ca.

incidence: 1/16*, sa. const.: people of The Sudan (children), contamination: natural, conc.: 1,786 pg/g, country: The Sudan/UK[32], *kwashiorkor ca.

incidence: 3/17, sa. const.: people of Malaysia (1 adult and 16 children: 12 males and 5 females), age: adult 49 years, children 2.5–11 years, contamination: natural, conc. range: 43–121 pg/g tissue, Ø conc.: 69 pg/g tissue, country: Singapore/UK[259]

incidence: 1/15, sa. const.: people of Kenya (patients: 9 males and 6 females (1 af*)), age: 0.6–52 years?, contamination: natural, conc.: 13 pg/g, country: Kenya/UK[466], *additionally HCC

Aflatoxin G_1

incidence: 1/13*, sa. const.: people of The Sudan (children), contamination: natural, conc.: 154,817 pg/g, country: The Sudan/UK[32], *miscellaneous liver disease ca.

incidence: 1/17, sa. const.: people of Malaysia (1 adult and 16 children: 12 males and 5 females), age: adult 49 years, children 2.5–11 years, contamination: natural, conc.: 9,116 pg/g tissue, country: Singapore/UK[259]

incidence: 2/154*, sa. const.: people of Singapore (121 males and 33 females), contamination: natural, conc. range: 22–27 pg/g, Ø conc.: 24.5 pg/g, country: Singapore[444], *normal subjects

incidence: 2/15, sa. const.: people of Kenya (patients: 9 males (1 af*) and 6 females (1 af*)), age: 0.6–52 years?, contamination: natural, conc. range: 128–3,186 pg/g, Ø conc.: 1,657 pg/g, country: Kenya/UK[466], *additionally HCC or marasmic kwashiorkor

Aflatoxin G_2

incidence: 2/13*, sa. const.: people of The Sudan (children), contamination: natural, conc.: 158–274 pg/g, Ø conc.: 216 pg/g, country: The Sudan/UK[32], *miscellaneous liver disease ca incidence: 1/15, sa. const.: people of Kenya (patients: 9 males (1 af*) and 6 females), age: 0.6–52 years?, contamination: natural, conc.: 13 pg/g, country: Kenya/UK[466], *additionally peritonitis

AFLATOXIN M_1
incidence: 1/19*, sa. const.: people of Liberia, Nigeria, and South Africa (children), contamination: natural, conc.: 15 pg/g (1 marasmic kwashiorkor ca), country: UK[11], *10 kwashiorkor ca., 6 marasmic kwashiorkor ca., and 3 marasmus ca.

incidence: 5/17, sa. const.: people of Malaysia (1 adult and 16 children: 12 males and 5 females), age: adult 49 years, children 2.5–11 years, contamination: natural, conc. range: 284–14,537 pg/g tissue, Ø conc.: 4,900.4 pg/g tissue, country: Singapore/UK[259]

incidence: 8/154*, sa. const.: people of Singapore (121 males and 33 females, contamination: natural, conc. range: 22–142 pg/g, Ø conc.: 54.7 pg/g, country: Singapore[444], *normal subjects

incidence: 3/15, sa. const.: people of Kenya (patients: 9 males (3 af*) and 6 females), age: 0.6–52 years?, contamination: natural, conc. range: 423–15,909 pg/g, Ø conc.: 5,916.66 pg/g, country: Kenya/UK[466], *additionally cirrhosis, stomach cancer or HCC

AFLATOXIN M_2
incidence: 2/13*, sa. const.: people of The Sudan (children), contamination: natural, conc.: 1,474–3,158 pg/g, Ø conc.: 2,316 pg/g, country: The Sudan/UK[32], *miscellaneous liver disease ca.

AFLATOXIN
incidence: 1/13*, sa. const.: people of the USA (children), contamination: natural, conc.: 0.04 ppb?, country: USA[62], *thereof 12 Reye's-syndrome ca.

Human lung may contain the following mycotoxins and/or their metabolites:

AFLATOXICOL
incidence: 4/20*, sa. const.: people of Nigeria (children), contamination: natural, conc. range: 20–280 pg/g, Ø conc.: 111 pg/g, country: Nigeria/UK[224], *kwashiorkor ca.
incidence: 3/20*, sa. const.: people of Nigeria (children), contamination: natural, conc. range: 7–85 pg/g, Ø conc.: 52 pg/g, country: Nigeria/UK[224], *miscellaneous disease ca.

incidence: 1/17, sa. const.: people of Malaysia (1 adult and 16 children: 12 males and 5 females), age: adult 49 years, children 2.5–11 years, contamination: natural, conc.: 27 pg/g tissue, country: Singapore/UK[259]

AFLATOXIN B_1
incidence: 3/6, sa. const.: people of Czechoslovakia (males and females af), contamination: natural, conc. range: 10–54 ng/g, Ø conc.: 28 ng/g, country: Czechoslovakia[186]

incidence: 5/17, sa. const.: people of Malaysia (1 adult and 16 children: 12 males and 5 females), age: adult 49 years, children 2.5–11 years, contamination: natural, conc. range: 38–3,465 pg/g tissue, Ø conc.: 1,103 pg/g tissue, country: Singapore/UK[259]
incidence: 1/1, sa. const.: person of Japan (male), age: 41 years, contamination: natural, conc.: 0.635 µg/g dry weight, country: Japan[459]

AFLATOXIN B_2
incidence: 1/17, sa. const.: people of Malaysia (1 adult and 16 children: 12 males and 5 females), age: adult 49 years, children 2.5–11 years, contamination: natural, conc.: 48 pg/g tissue, country: Singapore/UK[259]

incidence: 1/1, sa. const.: person of Japan (male), age: 41 years, contamination: natural, conc.: 0.0273 µg/g dry weight, country: Japan[459]

AFLATOXIN G_1
incidence: 3/20*, sa. const.: people of Nigeria (children), contamination:

natural, conc. range: 3,414–52,099 pg/g, Ø conc.: 31,647.66 pg/g, country Nigeria/UK[224], *kwashiorkor ca.
incidence: 6/20*, sa. const.: people of Nigeria (children), contamination: natural, conc. range: 132–84,807 pg/g, Ø conc.: 20,400 pg/g, country: Nigeria/UK[224], *miscellaneous disease ca.

Aflatoxin G_2
incidence: 3/20*, sa. const.: people of Nigeria (children), contamination: natural, conc. range: 94–550 pg/g, Ø conc.: 334.66 pg/g, country: Nigeria/UK[224], *kwashiorkor ca.
incidence: 3/20*, sa. const.: people of Nigeria (children), contamination: natural, conc. range: 8–1,837 pg/g, Ø conc.: 636 pg/g, country: Nigeria/UK[224], *miscellaneous disease ca.

Aflatoxin M_1
incidence: 1/17, sa. const.: people of Malaysia (1 adult and 16 children: 12 males and 5 females), age: adult 49 years, children 2.5–11 years, contamination: natural, conc.: 1,289 pg/g tissue, country: Singapore/UK[259]

incidence: 1/1, sa. const.: person of Japan (male), age: 41 years, contamination: natural, conc.: 0.0525 µg/g dry weight, country: Japan[459]

Aflatoxin M_2
incidence: 11/20*, sa. const.: people of Nigeria (children), contamination: natural, conc. range: 210–2,723 pg/g, Ø conc.: 894.27 pg/g, country: Nigeria/UK[224], *kwashiorkor ca.
incidence: 10/20*, sa. const.: people of Nigeria (children), contamination: natural, conc. range: 100–3,058 pg/g, Ø conc.: 1,230.9 pg/g, country: Nigeria/UK[224], *miscellaneous disease ca.

incidence: 1/17, sa. const.: people of Malaysia (1 adult and 16 children: 12 males and 5 females), age: adult 49 years, children 2.5–11 years, contamination: natural, conc.: 1,595 pg/g tissue, country: Singapore/UK[259]

Aflatoxins
incidence: 1/17, sa. const.: people of Malaysia (1 adult and 16 children: 12 males and 5 females), age: adult 49 years, children 2.5–11 years, contamination: natural, conc.: 27 pg/g tissue*, country: Singapore/UK[259], *AFB_1, AFB_2, AFG_1, AFM_1, AFM_2, and AFL

Human Maternal Milk see Human breast milk

Human Milk see Human breast milk

Human pancreas may contain the following mycotoxins and/or their metabolites:

Aflatoxin B_1
incidence: 5/12*, sa. const.: people of the UK (3 males (2 af), 8 females (3 af), and 1 of unknown sex), contamination: natural, conc. range: 0.21–0.47 AFB_1-DNA adducts/10^6 nucleotides, country: UK/Mexico[58], *non-tumor ca.

Human placenta may contain the following mycotoxins and/or their metabolites:

Aflatoxin B_1
incidence: 69/120, sa. const.: people of Taiwan (females), contamination: natural, conc. range: 0.6–6.3 µmol/mol DNA*, Ø conc.: 2.34 µmol/mol DNA*, country: Taiwan, Republic of China[156], *AFB_1-DNA adducts

Ochratoxin A
incidence: 11/40*, sa. const.: people of Italy (females), contamination: natural, conc. range: LOD/LOQ–0.9 µg/l (2 sa), 1.0–1.9 µg/l (1 sa), 2.0–5.0 µg/l (3 sa), ≤10.57 µg/l (5 sa), country: EU[508], * pregnant women with (12 thereof 4 OTA-pos) and without (28 thereof 7 OTA-pos) pathologies

Human plasma may contain the following mycotoxins and/or their metabolites:

Aflatoxin B_1

incidence: 5/62*, sa. const.: people of Turkey (33 male and 29 female humans: healthy), contamination: natural, conc.: 101.2 pg/ml (mean value), country: Turkey[582], *control

incidence: 50/203*, sa. const.: people of Turkey (119 male and 84 female patients), contamination: natural, conc.: 36.1 pg/ml (mean value), country: Turkey[582], *93 chronic hepatitis ca., 64 cirrhosis ca., and 46 HCC ca.

Aflatoxin B_2

incidence: 3/62*, sa. const.: people of Turkey (33 male and 29 female humans: healthy), contamination: natural, conc.: 18.8 pg/ml (mean value), country: Turkey[582], *control

incidence: 35/203*, sa. const.: people of Turkey (119 male and 84 female patients), contamination: natural, conc.: 28.4 pg/ml (mean value), country: Turkey[582], *93 chronic hepatitis ca., 64 cirrhosis ca., and 46 HCC ca.

Aflatoxin G_1

incidence: 6/62*, sa. const.: people of Turkey (33 male and 29 female humans: healthy), contamination: natural, conc.: 32.5 pg/ml (mean value), country: Turkey[582], *control

incidence: 46/203*, sa. const.: people of Turkey (119 male and 84 female patients), contamination: natural, conc.: 92.0 pg/ml (mean value), country: Turkey[582], *93 chronic hepatitis ca., 64 cirrhosis ca., and 46 HCC ca.

Aflatoxin G_2

incidence: 2/62*, sa. const.: people of Turkey (33 male and 29 female humans: healthy), contamination: natural, conc.: 10.4 pg/ml (mean value), country: Turkey[582], *control

incidence: 37/203*, sa. const.: people of Turkey (119 male and 84 female patients), contamination: natural, conc.: 18.2 pg/ml (mean value), country: Turkey[582], *93 chronic hepatitis ca., 64 cirrhosis ca., and 46 HCC ca.

Ochratoxin A

incidence: 148/249, sa. const.: people of Croatia (males and females), contamination: natural, conc. range: 0.2–1.0 ng/ml (135 sa), >1.0–15.9 ng/ml (13 sa), country: Croatia[187]

incidence: 320/734, sa. const.: people of Croatia (males and females), contamination: natural, conc. range: 0.2–1.0 ng/ml (288 sa), >1.0 ng/ml (32 sa), country: Croatia[189]

incidence: 8/8, sa. const.: people of Germany/Switzerland? (4 males and 4 females, af.), age: 26–57 years, contamination: natural, conc. range: 0.2–0.88 ng/ml, country: Germany/ Switzerland[197]

incidence: 185/309*, sa. const.: people of Morocco (213 males and 96 females), age: 18–60 years, contamination: natural, conc. range: 0.08–6.59 ng/ml, Ø conc.: 0.29 ng/ml, country: Morocco/France[219], *healthy volunteers

incidence: 40/75*, sa. const.: people of Spain (44 males and 31 females), age: 27–80 years, contamination: natural, conc. range: ≤4.0 ng/ml, country: Spain/ France[221], *healthy donors

incidence: 56/72*, sa. const.: people of Spain (40 males and 32 females), age: 27–80 years, contamination: natural, conc. range: ≤11.70 ng/ml, country: Spain/France[221], *nephropathy patients

incidence: 88/198, sa. const.: people of Croatia (healthy inhabitants of Zagreb), contamination: natural, conc. range: >0.2–1.3 ng/ml, country: Croatia[226]

incidence: 156/184, sa. const.: people of Japan (130 males (114 af) and 54 females (42 af)), contamination: natural, conc. range: 4–278 pg/ml, Ø conc.: 68 pg/ml, country: Japan[235] incidence: 50/50*, sa. const.: people of UK, contamination: natural, conc. range: 0.4–3.11 ng/ml, country: UK[236], *32 volunteers (normal diet?), 11 vegetarians, 7 consumed ethnic diet

incidence: 38/297, sa. const.: people of Sweden, contamination: natural, conc. range: 0.3–0.8 ng/ml (33 sa), >0.8 to ≤6.7 ng/ml (5 sa), country: Sweden[293]

incidence: 144/144, sa. const.: people of Canada (72 males and 72 females), age for 134 donors: 19–68 years, contamination: natural, conc. range: 0.29–2.37 ng/ml, Ø conc.: 0.88 ng/ml, country: Canada[345]

incidence: 83?/250, sa. const.: people of Lebanon (164 males (51 af) and 86 females (31 af)), age: 16 to ≥60 years, contamination: natural, conc. range: 0.1–0.87 ng/ml, Ø conc.: 0.17 ng/ml, country: Lebanon/France[429]

incidence: 91/142, sa. const.: people from Argentina (male inhabitants of Mar del Plata), Ø age: 37 years, Ø weight: 80 kg, contamination: natural, conc. range: LOD to <LOQ (1 sa), LOQ–0.2 ng/ml (39 sa), >0.2–1.0 ng/ml (39 sa), >1.0–10.0 ng/ml (12 sa), country: Argentina/Sweden[502]
incidence: 36/57, sa. const.: people from Argentina (female inhabitants of Mar del Plata), Ø age: 40 years, Ø weight: 70 kg (females), contamination: natural, conc. range: LOQ–0.2 ng/ml (17 sa), >0.2–1.0 ng/ml (13 sa), >1.0–10.0 ng/ml (5 sa), 47.6 ng/ml (1 sa), country: Argentina/Sweden[502]
incidence: 122/193, sa. const.: people from Argentina (male inhabitants of General Rodriguez), Ø age: 35 years, Ø weight: 80 kg, contamination: natural, conc. range: LOQ–0.2 ng/ml (16 sa), >0.2–1.0 ng/ml (61 sa), >1.0–10.0 ng/ml (41 sa), >10.0–74.8 ng/ml (4 sa), country: Argentina/Sweden[502]
incidence: 25/42, sa. const.: people from Argentina (female inhabitants of General Rodriguez), Ø age: 36 years, Ø weight: 68 kg, contamination: natural, conc. range: LOQ–0.2 ng/ml (9 sa), >0.2–1.0 ng/ml (10 sa), >1.0–10.0 ng/ml (6 sa), country: Argentina/Sweden[502]

incidence: 202/202, sa. const.: people of The Netherlands (males and females), Ø age: 40 years, contamination: natural, conc. range: LOD/LOQ to ≤0.78 µg/l, country: EU[508]

incidence: 168/168, sa. const.: people of Spain (88 males and 80 females), contamination: natural, conc. range: LOD/LOQ–0.9 µg/l (90 sa), 1.0–1.9 µg/l (50 sa), 2.0–5.0 µg/l (27 sa), 5.58 µg/l (1 sa), country: EU[508]

incidence: 191/191, sa. const.: people of Sweden (133 males and 58 females), Ø age: 44 years, contamination: natural, conc. range: LOD/LOQ–0.9 µg/l (189 sa), 1.0–1.23 µg/l (2 sa), country: EU[508]
incidence: 7/7*, sa. const.: people of the UK (males and females), age: 18–55 years, contamination: natural, conc. range: LOD/LOQ–2.15? µg/l, country: EU[508], *eating ethnic diet
incidence: 32/32*, sa. const.: people of the UK (males and females), age: 18–55 years, contamination: natural, conc. range: LOD/LOQ–3.11? µg/l, country: EU[508], *eating normal diet
incidence: 11/11*, sa. const.: people of the UK (males and females), age: 18–55 years, contamination: natural, conc. range: LOD/LOQ–2.46? µg/l, country: EU[508], *vegetarians

incidence: 168/168, sa. const.: people of Spain (88 males and 80 females), age: 18–63 years, contamination: natural, conc. range: 0.120–5.580 ng/ml, Ø conc.: 1.192 ng/ml, country: Spain[542]

incidence: 130/132, sa. const.: people of Spain (male inhabitants of the province of Lleida), age: 18–45 years and more, contamination: natural, conc. range: 0.12–8.03 ng/ml, country: Spain[584]
incidence: 145/147, sa. const.: people of Spain (females inhabitants of the

province of Lleida), age: 18–45 years and more, contamination: natural, conc. range: 0.11–8.68 ng/ml, country: Spain[584]

see also Human blood and Human serum

Human rectum may contain the following mycotoxins and/or their metabolites:

Aflatoxin B_1
incidence: 5/9*, sa. const.: people of the UK (4 males (tumor ca) and 5 females (5 af)), contamination: natural, conc. range: 0.26–10.26 AFB_1-DNA adducts/10^6 nucleotides, country: UK/Mexico[58]

Human renal tissue may contain the following mycotoxins and/or their metabolites:

Ochratoxin A
incidence: 3/14*, sa. const.: people of Egypt (males and females), age: 38–70 years, contamination: natural, conc. range: 0.51–1.28 ng/g, country: Egypt/France[227], *with urothelial tumors

Human semen may contain the following mycotoxins and/or their metabolites:

Aflatoxin G_1
incidence: 20/50, sa. const.: people of Nigeria (males: infertile), age: adult, contamination: natural, conc. range: 0.50–2.80 μg/ml, country: Nigeria[443]
incidence: 4/50, sa. const.: people of Nigeria (males: fertile), age: adult, contamination: natural, conc. range: 0.30–0.50 μg/ml, country: Nigeria[443]

Aflatoxin M_1
incidence: 20/50, sa. const.: people of Nigeria (males: infertile), age: adult, contamination: natural, conc. range: 1.0–3.20 μg/ml, country: Nigeria[443]
incidence: 4/50, sa. const.: people of Nigeria (males: fertile), age: adult, contamination: natural, conc. range: 0.43–0.48 μg/ml, country: Nigeria[443]

Aflatoxin M_2
incidence: 20/50, sa. const.: people of Nigeria (males: infertile), age: adult, contamination: natural, conc. range: 0.90–3.60 μg/ml, country: Nigeria[443]
incidence: 4/50, sa. const.: people of Nigeria (males: fertile), age: adult, contamination: natural, conc. range: 0.45–0.62 μg/ml, country: Nigeria[443]

Human serum may contain the following mycotoxins and/or their metabolites:

Aflatoxicol
incidence: 1/16*, sa. const.: people of The Sudan (children), contamination: natural, conc.: 930 pg/ml, country: The Sudan/UK[32], * kwashiorkor ca.

incidence: 44/44*, sa. const.: people of The Sudan (male and female children), no contamination with AFL, conc.: nd, country: UK/The Sudan[36], *control
incidence: 1/57*, sa. const.: people of The Sudan (male and female children), contamination: natural, conc.: pr, country: UK/The Sudan[36], *marasmus ca.
incidence: 4/32*, sa. const.: people of The Sudan (male and female children), contamination: natural, conc.: pr, country: UK/The Sudan[36], *marasmic kwashiorkor ca.
incidence: 6/44*, sa. const.: people of The Sudan (male and female children), contamination: natural, conc.: pr, country: UK/The Sudan[36], *kwashiorkor ca.

incidence: 67/67*, sa. const.: people of Nigeria (60 infants and 7 of their mothers), no contamination with AFL, conc.: nd, country: Nigeria/UK[195], *control (non-jaundiced?!)
incidence: 14/340*, sa. const.: people of Nigeria (270 infants and 70 of their mothers), contamination: natural, conc. range: 26–750 ng/l, country: Nigeria/UK[195], *jaundiced

incidence: 15/30*, sa. const.: people of Egypt (19 male and 11 female infants), age: 7–20 months, contamination: natural, conc. range: 2–20 ng/100 ml, Ø conc.: 9.27 ng/100ml, country: Egypt[447], *kwashiorkor ca.
incidence: 3/30*, sa. const.: people of Egypt (16 male and 14 female infants), age: 6–13 months, contamination: natural, conc. range: 2–14 ng/100 ml, Ø conc.: 6.33 ng/100 ml, country: Egypt[447], *marasmus ca.

incidence: 10/36*, sa. const.: people of Sierra Leone (children: under 5), contamination: natural, conc. range: 0.03–0.9 ng/ml, country: Sierra Leone[540], *14 controls, 3 kwashiorkors, 9 underweights, 2 marasmics, and 8 "unspecified" (for detailed information please see the article)

Aflatoxin B_1
incidence: 356/357*, sa. const.: people of The Gambia (188 male and 169 female inhabitants of periurban areas), Ø age: 24 years, contamination: natural, conc. range: 14.9–33.4 pg AFB_1-lysine eq/mg albumin (mean values), country: UK/France/The Gambia/USA[13], *thereof 181 HBV ca. but no predominant contamination, for detailed information please see the article
incidence: 356/357*, sa. const.: people of The Gambia (188 male and 169 female inhabitants of rural areas), Ø age: 24 years, contamination: natural, conc. range: 28.5–42.8 pg AFB_1-lysine eq/mg albumin (mean values), country: UK/France/The Gambia/USA[13], *thereof 181 HBV ca. but no predominant contamination, for detailed information please see the article

incidence: 159/163, sa. const.: people of The Gambia (male and female children), contamination: natural, conc. range: 5–25 pg AFB_1-lysine eq/mg albumin (48 sa), 26–50 pg AFB_1-lysine eq/mg albumin (45 sa), 51–75 pg AFB_1-lysine eq/mg albumin (24 sa), 76–100 pg AFB_1-lysine eq/mg albumin (11 sa), 101–200 pg AFB_1-lysine eq/mg albumin (20 sa), ≤350 pg AFB_1-lysine eq/mg albumin (7 sa), country: France/The Gambia/People's Republic of China[15]
incidence: 29/29, sa. const.: people of Senegal (male and female children), contamination: natural, conc. range: 5–25 pg AFB_1-lysine eq/mg albumin (20 sa), 26–50 pg AFB_1-lysine eq/mg albumin (6 sa), 51–75 pg AFB_1-lysine eq/mg albumin (2 sa), 76–100 pg AFB_1-lysine eq/mg albumin (1 sa), country: France/The Gambia/People's Republic of China[15]
incidence: 15/30, sa. const.: people of Kenya (male and female children), contamination: natural, conc. range: 5–25 pg AFB_1-lysine eq/mg albumin (8 sa), 26–50 pg AFB_1-lysine eq/mg albumin (3 sa), 76–100 pg AFB_1-lysine eq/mg albumin (1 sa), 101–200 pg AFB_1-lysine eq/mg albumin (1 sa), ≤350 pg AFB_1-lysine eq/mg albumin (2 sa), country: France/The Gambia/People's Republic of China[15]
incidence: 4/26, sa. const.: people of Uganda (male and female children), contamination: natural, conc. range: 5–25 pg AFB_1-lysine eq/mg albumin (3 sa), 26–50 pg AFB_1-lysine eq/mg albumin (1 sa), country: France/The Gambia/People's Republic of China[15]
incidence: 20/20, sa. const.: people of The Gambia (males and females), contamination: natural, conc. range: 5–25 pg AFB_1-lysine eq/mg albumin (13 sa), 26–50 pg AFB_1-lysine eq/mg albumin (2 sa), 51–75 pg AFB_1-lysine eq/mg albumin (1 sa), 76–100 pg AFB_1-lysine eq/mg albumin (1 sa), 101–200 pg AFB_1-lysine eq/mg albumin (3 sa), country: France/The Gambia/People's Republic of China[15]
incidence: 28/61, sa. const.: people of Kenya (males and females), contamination: natural, conc. range: 5–25 pg AFB_1-lysine eq/mg albumin (18 sa), 26–50 pg AFB_1-lysine eq/mg albumin (2 sa), 51–75 pg AFB_1-lysine eq/mg albumin (1 sa), 76–100 pg AFB_1-lysine

eq/mg albumin (4 sa), 101–200 pg AFB_1-lysine eq/mg albumin (2 sa), >200 pg AFB_1-lysine eq/mg albumin (1 sa), country: France/The Gambia/People's Republic of China[15]

incidence: 11/84, sa. const.: people of Thailand (males and females), contamination: natural, conc. range: 5–25 pg AFB_1-lysine eq/mg albumin (10 sa), 26–50 pg AFB_1-lysine eq/mg albumin (1 sa), country: France/The Gambia/People's Republic of China[15]

incidence: 66/70, sa. const.: people of Guinea (males), age: 9–80 years, contamination: natural, conc. range: 4–50 AFB_1-lysine eq/mg albumin (51 sa), >50–385 pg AFB_1-lysine eq/mg albumin (15 sa), country: Guinea/France[17]

incidence: 1/1, sa. const.: person of the USA (female), age: 31 years, contamination: natural, conc.: 3.39 ng/ml, country: USA[20]

incidence: 5/20*, sa. const.: people of Japan (males: healthy), age: 20–63 years, contamination: natural, conc. range: 20–56 pg/ml, country: Japan[21], *sa. taken after fasting

incidence: 29/80*, sa. const.: people of Japan (males: healthy), age: 20–63 years, contamination: natural, conc. range: 20–1,169 pg/ml, country: Japan[21], *sa. taken after lunch

incidence: 1,187/1,188, sa. const.: people of the People's Republic of China, contamination: natural, conc. range: 0.14–4.39 pmol AFB_1/mg albumin, country: USA/People's Republic of China[22]

incidence: 115/117, sa. const.: people of The Gambia (children), age: 3–4 years, contamination: natural, conc. range: 2.2–250.4 pg AFB_1-lysine eq/mg albumin, country: France/The Gambia/Italy/UK[23]

incidence: 33/60, sa. const.: people of the People's Republic of China, contamination: natural, conc. range: ≤890 pmol AFB_1/g albumin, Ø conc.: 221 pmol AFB_1/g albumin, country: USA/People's Republic of China[26]

incidence: 1/8*, sa. const.: people of Thailand (6 males and 2 females (1 af)), age: 17–73 years, contamination: natural, conc.: 7.4 pg AFB_1-lysine eq/mg albumin, country: France/USA/Thailand[29], *HCC ca.

incidence: 3/16*, sa. const.: people of The Sudan (children), contamination: natural, conc. range: 353–588 pg/ml, Ø conc.: 447 pg/ml, country: The Sudan/UK[32], *kwashiorkor ca.

incidence: 1/1*, sa. const.: person of The Sudan (child), contamination: natural, conc.: 666 pg/ml, country: The Sudan/UK[32], *micronodular cirrhosis with portal hypertension

incidence: 16/16*, sa. const.: people of Mexico (patients: males and females), contamination: natural, conc. range: 0.54–4.65 pmol AF/mg albumin, Ø conc.: 2.70 pmol aflatoxin/mg albumin, country: USA/Finland/Singapore/Mexico[55], *HCC ca.

incidence: 59/78, sa. const.: people of Nigeria (males: healthy), age: 18–47 years, contamination: natural, conc. range: 20–3,100 pg/ml, Ø conc.: 665 pg/ml, country: UK/Nigeria[68]

incidence: ?/20*, sa. const.: people of Nigeria (male farmers), contamination: natural, conc. range: 0.025–0.57 μg/ml, country: Nigeria[69], *control

incidence: ?/15*, sa. const.: people of Nigeria (males and females), contamination: natural, conc. range: 0.005–0.130 μg/ml, country: Nigeria[69], * different disease ca.

incidence: 1/15, sa. const.: people of The Sudan (females), contamination: natural, conc.: 111 pg/ml, country: UK/The Sudan[134]

incidence: 4/67*, sa. const.: people of Nigeria (60 infants and 7 of their mothers), contamination: natural, conc. range: 1,956–20,371 ng/l**, country: Nigeria/UK[195], *control (non-jaundiced?!), **only in infants?

incidence: 21/340*, sa. const.: people of Nigeria (270 infants and 70 of their mothers), contamination: natural, conc. range: 256–58,239 ng/l, country: Nigeria/UK[195], *jaundiced

incidence: 2/7*, sa. const.: people of India, contamination: natural, conc. range: tr, country: India[237], *jaundiced

incidence: 3/60*, sa. const.: people of Nigeria (neonates), contamination: natural, conc. range: 182–2,094 pg/ml, Ø conc.: 958.33 pg/ml, country: Nigeria/UK[262], *non-jaundiced
incidence: 7/64*, sa. const.: people of Nigeria (neonates), contamination: natural, conc. range: 242–10,239 pg/ml, Ø conc.: 2,580.57 pg/ml, country: Nigeria/UK[262], *jaundiced

incidence: 2/78*, sa. const.: people of Nigeria, contamination: natural, conc. range: 2,676–6,532 pg/ml, Ø conc.: 4,604 pg/ml, country: Nigeria/UK[269], *urban population
incidence: 97/97*, sa. const.: people of Nigeria, no contamination with AFB_1, conc.: nd, country: Nigeria/UK[269], *rural population

incidence: 1/20, sa. const.: people of Argentina (patients: 13 males (1 af) and 7 females), age: 42–64 years, contamination: natural, conc.: 0.47 ng/cm³, country: Argentina[434]

incidence: 24/30*, sa. const.: people of Egypt (19 male and 11 female infants), age: 7–20 months, contamination: natural, conc. range: 4–69 ng/100 ml, Ø conc.: 32.38 ng/100 ml, country: Egypt[447], *kwashiorkor ca.
incidence: 13/30*, sa. const.: people of Egypt (16 male and 14 female infants), age: 6–13 months, contamination: natural, conc. range: 10–18 ng/100 ml, Ø conc.: 13.62 ng/100 ml, country: Egypt[447], *marasmus ca.

incidence: ?/40*, sa. const.: people of the People's Republic of China, age: 20–55 years, contamination: natural, conc. range: 0.16–1.70 pmol/mg albumin** ***, country: USA/People's Republic of China[451], *control (receiving placebos for 3 months), **AFB-AA, ***for detailed information please see the article
incidence: ?/40*, sa. const.: people of the People's Republic of China, age: 20–55 years, contamination: natural, conc. range: 0.26–1.63 pmol/mg albumin** ***, country: USA/People's Republic of China[451], *receiving GTP 500 mg for 3 months, **AFB-AA, ***for detailed information please see the article
incidence: ?/40*, sa. const.: people of the People's Republic of China, age: 20–55 years, contamination: natural, conc. range: 0.35–1.55 pmol/mg albumin** ***, country: USA/People's Republic of China[451], *receiving GTP 1,000 mg for 3 months, **AFB-AA, ***for detailed information please see the article

incidence: 2/40, sa. const.: people of Ghana (29 males and 11 females), age: 30–73 years, contamination: natural, Ø conc.: 12.5 ng/ml, country: Ghana[461]

incidence: ?/264*, sa. const.: people of Taiwan (males: healthy), age: 30.3–64.8 years, contamination: natural, conc. range: 5.0–355.8 pmol AFB_1/mg albumin** ***, country: USA/Taiwan, Republic of China[485], *132 HbsAg carriers, 132 non-carriers, **AFB_1-albumin adducts, ***measured at time point 1 (for detailed information please see the article)
incidence: ?/264*, sa. const.: people of Taiwan (males: healthy), age: 30.3–64.8 years, contamination: natural, conc. range: 5.0–205.2 pmol AFB_1/mg albumin** ***, country: USA/Taiwan, Republic of China[485], *132 HbsAg carriers, 132 non-carriers, **AFB_1-albumin adducts, ***measured at time point 2 (for detailed information please see the article)

incidence: ?/100*, sa. const.: people of Taiwan (100 males: school students),

age: 13–15 years, contamination: natural, conc. range: 2–138 fmol/mg* **, country: USA/Taiwan, Republic of China[486], *50 positive and 50 negative for HbsAg, **AFB_1-albumin adducts, ***for detailed information please see the article
incidence: ?/100*, sa. const.: people of Taiwan (100 females: school students), age: 13–15 years, contamination: natural, conc. range: 2–174 fmol/mg** ***, country: USA/Taiwan, Republic of China[486], *50 positive and 50 negative for HbsAg, **AFB_1-albumin adducts, ***for detailed information please see the article

incidence: 6/36*, sa. const.: people of Sierra Leone (children: under 5), contamination: natural, conc. range: 0.2–2.5 ng/ml, country: Sierra Leone[540], *14 controls, 3 kwashiorkors, 9 underweights, 2 marasmics, and 8 "unspecified" (for detailed information please see the article)

incidence: 2/45, sa. const.: people of Denmark (male workers in animal-feed production), age: 25–62 years, contamination: natural, conc. range: 50–54 pg AFB_1/mg albumin* **, Ø conc.: 52 pg AFB_1/mg albumin*, country: Denmark[559], *first blood sample, **AFB_1-adducts
incidence: 7/45, sa. const.: people of Denmark (male workers in animal-feed production), age: 25–62 years, contamination: natural, conc. range: 44–100 pg AFB_1/mg albumin* **, Ø conc.: 64.86 pg AFB_1/mg albumin*, country: Denmark[559], *second blood sample, **AFB_1-adducts

incidence: 20?/20*, sa. const.: people of Egypt (17 males and 3 females), Ø age: 53.17 years, contamination: natural, Ø conc.: 7.33 ng/ml (mean value), country: Egypt[592], *control, for detailed information please see the article
incidence: 80?/80*, sa. const.: people of Egypt (66 males and 14 females), Ø age: 52.88 years, contamination: natural, Ø conc.: 32.47 ng/ml (mean value), country: Egypt[592], *HCC patients, for detailed information please see the article

incidence: 42/42, sa. const.: people of the People's Republic of China (30 males and 12 females), age: 25–64 years, contamination: natural, conc. range: ≈≤344 ng AFB_1/g albumin, country: USA/People's Republic of China[607]

incidence: 98/104, sa. const.: people of UK (47 males and 57 females volunteers), age: 18–65 years, contamination: natural, conc. range: <5 pg/mg* (6 sa), 5–15 pg/mg* (18 sa), 16–25 pg/mg* (27 sa), 26–35 pg/mg* (25 sa), 36–45 pg/mg* (17 sa), 46–55 pg/mg* (5 sa), 56–65 pg/mg* (5 sa), 66–95 pg/mg* (1 sa), country: UK[627], *AFB_1-lysine (eq) pg/mg albumin

Aflatoxin B_2
incidence: 2/16*, sa. const.: people of The Sudan (children), contamination: natural, conc. range: 9–12 pg/ml, Ø conc.: 10.5 pg/ml, country: The Sudan/UK[32], *kwashiorkor ca.

incidence: ?/20*, sa. const.: people of Nigeria (male farmers), contamination: natural, conc. range: 0.010–0.390 µg/ml, country: Nigeria[69], *control
incidence: ?/15*, sa. const.: people of Nigeria (males and females), contamination: natural, conc. range: 0.008–0.240 µg/ml, country: Nigeria[69], *different disease ca.

incidence: 1*/12, sa. const.: people of The Sudan (children), contamination: natural, conc.: 3 pg/ml, country: UK/The Sudan[134], *marasmus/kwashiorkor ca.

incidence: 67/67*, sa. const.: people of Nigeria (60 infants and 7 of their mothers), no contamination with AFB_2, conc.: nd, country: Nigeria/UK[195], *control (non-jaundiced?!)
incidence: 4/340*, sa. const.: people of Nigeria (270 infants and 70 of their mothers), contamination: natural, conc. range: 17–5? ng/l, country: Nigeria/UK[195], *jaundiced

incidence: 1/60*, sa. const.: people of Nigeria (neonates), contamination: natural, conc.: 40 pg/ml, country: Nigeria/UK[262], *non-jaundiced
incidence: 1/64*, sa. const.: people of Nigeria (neonates), contamination: natural, conc.: 20 pg/ml, country: Nigeria/UK[262], *jaundiced

incidence: 1/78*, sa. const.: people of Nigeria, contamination: natural, conc.: 36 pg/ml, country: Nigeria/UK[269], *urban population

incidence: 97/97*, sa. const.: people of Nigeria, no contamination with AFB_2, conc.: nd, country: Nigeria/UK[269], *rural population

incidence: 7/30*, sa. const.: people of Egypt (19 male and 11 female infants), age: 7–20 months, contamination: natural, conc. range: 4–16 ng/100 ml, Ø conc.: 12.00 ng/100 ml, country: Egypt[447], *kwashiorkor ca.
incidence: 1/30*, sa. const.: people of Egypt (16 male and 14 female infants), age: 6–13 months, contamination: natural, conc.: 5 ng/100 ml, country: Egypt[447], *marasmus ca.

incidence: 4/36*, sa. const.: people of Sierra Leone (children: under 5), contamination: natural, conc. range: 0.04–4.0 ng/ml, country: Sierra Leone[540], *14 controls, 3 kwashiorkors, 9 underweights, 2 marasmics, and 8 "unspecified" (for detailed information please see the article)

Aflatoxin B_{2a}
incidence: 12/30*, sa. const.: people of Egypt (19 male and 11 female infants), age: 7–20 months, contamination: natural, conc. range: 4–35 ng/100 ml, Ø conc.: 15.58 ng/100 ml, country: Egypt[447], *kwashiorkor ca.
incidence: 4/30*, sa. const.: people of Egypt (16 male and 14 female infants), age: 6–13 months, contamination: natural, conc. range: 3–9 ng/100 ml, Ø conc.: 6.00 ng/100 ml, country: Egypt[447], *marasmus ca.

Aflatoxin B
incidence: 2/45, sa. const.: people of Denmark (males), age: 25–62 years, contamination: natural, conc. range: 50–54 pg AFB/mg albumin***, Ø conc.: 52 pg AFB/mg albumin***, country: Denmark[435], *AFB-albumin adducts, **first blood sa.
incidence: 7/45, sa. const.: people of Denmark (males), age: 25–62 years, contamination: natural, conc. range: 44–100 pg AFB/mg albumin* **, Ø conc.: 64.86 pg AFB/mg albumin* **, country: Denmark[435], *AFB-albumin adducts, **second blood sa.

incidence: 17/17, sa. const.: people of the People's Republic of China, contamination: natural, conc. range: 0.009–0.329 pmol AFB-lysine adduct/mg albumin, Ø conc.: 0.198 pmol AFB-lysine adduct/mg albumin, country: USA[517]
incidence: 20/20, sa. const.: people of The Gambia, contamination: natural, conc. range: 0.084–0.228 pmol AFB-lysine adduct/mg albumin, Ø conc.: 0.142 pmol AFB-lysine adduct/mg albumin, country: USA[517]
incidence: 20/20, sa. const.: people of the People's Republic of China, contamination: natural, conc. range: 0.065–0.142 pmol AFB-lysine adduct/mg albumin, Ø conc.: 0.098 pmol AFB-lysine adduct/mg albumin, country: USA[517]
incidence: 20/20, sa. const.: people of the People's Republic of China, contamination: natural, conc. range: 0.083–0.147 pmol AFB-lysine adduct/mg albumin, Ø conc.: 0.108 pmol AFB-lysine adduct/mg albumin, country: USA[517]

Aflatoxin G_1
incidence: 1/16*, sa. const.: people of The Sudan (children), contamination: natural, conc.: 975 pg/ml, country: The Sudan/UK[32], *kwashiorkor ca.

incidence: ?/20*, sa. const.: people of Nigeria (male farmers), contamination: natural, conc. range: 0.024–0.59 μg/ml, country: Nigeria[69], *control incidence: ?/15*, sa. const.: people of Nigeria (males and females), contamination: natural, conc. range: 0.005–0.180 µg/ml, country: Nigeria[69], *different disease ca.

incidence: 3/67*, sa. const.: people of Nigeria (60 infants and 7 of their mothers), contamination: natural, conc. range: 1,112–4,370 ng/l**, country: Nigeria/UK[195], *control (non-jaundiced?!), **only in infants?
incidence: 12/340*, sa. const.: people of Nigeria (270 infants and 70 of their mothers), contamination: natural, conc. range: 460–165,067 ng/l, country: Nigeria/UK[195], *jaundiced

incidence: 2/60*, sa. const.: people of Nigeria (neonates), contamination: natural, conc. range: 1,877–6,389 pg/ml, Ø conc.: 4,133 pg/ml, country: Nigeria/UK[262], *non-jaundiced
incidence: 2/64*, sa. const.: people of Nigeria (neonates), contamination: natural, conc. range: 293–1,074 pg/ml, Ø conc.: 683.5 pg/ml, country: Nigeria/UK[262], *jaundiced

incidence: 1/78*, sa. const.: people of Nigeria, contamination: natural, conc.: 8,828 pg/ml, country: Nigeria/UK[269], *urban population
incidence: 2/97*, sa. const.: people of Nigeria, contamination: natural, conc. range: 2,683–6,436 pg/ml, Ø conc.: 4,459.5 pg/ml, country: Nigeria/UK[269], *rural population

incidence: 19/30*, sa. const.: people of Egypt (19 male and 11 female infants), age: 7–20 months, contamination: natural, conc. range: 2–38 ng/100 ml, Ø conc.: 21.50 ng/100 ml, country: Egypt[447], *kwashiorkor ca.
incidence: 8/30*, sa. const.: people of Egypt (16 male and 14 female infants), age: 6–13 months, contamination: natural, conc. range: 1–12 ng/100 ml, Ø conc.: 7.75 ng/100 ml, country: Egypt[447], *marasmus ca.

incidence: 38/40, sa. const.: people of Ghana (29 males and 11 females), age: 30–73 years, contamination: natural, Ø conc.: 9.2 ng/ml, country: Ghana[461]

incidence: 1/150, sa. const.: people of Italy (33 females donors: healthy), 117 men, age: 40–75 years, contamination: natural, conc.: 0.40 ng/ml, country: Italy[465]

incidence: 5/36*, sa. const.: people of Sierra Leone (children: under 5), contamination: natural, conc. range: 0.5–1.6 ng/ml, country: Sierra Leone[540], *14 controls, 3 kwashiorkors, 9 underweights, 2 marasmics, and 8 "unspecified" (for detailed information please see the article)

Aflatoxin G_2

incidence: 1/16*, sa. const.: people of The Sudan (children), contamination: natural, conc.: 4 pg/ml, country: The Sudan/UK[32], *kwashiorkor ca.

incidence: ?/20*, sa. const.: people of Nigeria (male farmers), contamination: natural, conc. range: 0.012-0.192 µg/ml, country: Nigeria[69], *control
incidence: ?/15*, sa. const.: people of Nigeria (males and females), contamination: natural, conc. range: 0.002–0.080 µg/ml, country: Nigeria[69], *different disease ca.

incidence: 1/15, sa. const.: people of The Sudan (females), contamination: natural, conc.: 2 pg/ml, country: UK/The Sudan[134]
incidence: 2*/12, sa. const.: people of The Sudan (children), contamination: natural, conc.: 3–5 pg/ml, Ø conc.: 4 pg/ml, country: UK/The Sudan[134], *thereof 1 marasmus/kwashiorkor ca.

incidence: 1/67*, sa. const.: people of Nigeria (60 infants and 7 of their mothers), contamination: natural, conc.: 70 ng/l**, country: Nigeria/UK[195], *control (non-jaundiced?!), **only in infants?
incidence: 16/340*, sa. const.: people of Nigeria (270 infants and 70 of their mothers), contamination: natural, conc. range: 21–990 ng/l, country: Nigeria/UK[195], *jaundiced

incidence: 2/60*, sa. const.: people of Nigeria (neonates), contamination: natural, conc. range: 14–173 pg/ml, Ø conc.: 93.5 pg/ml, country: Nigeria/UK[262], *non-jaundiced

incidence: 78/78*, sa. const.: people of Nigeria, no contamination with AFG_2, conc.: nd, country: Nigeria/UK[269], *urban population
incidence: 1/97*, sa. const.: people of Nigeria, contamination: natural, conc.: 20 pg/ml, country: Nigeria/UK[269], *rural population

incidence: 6/36*, sa. const.: people of Sierra Leone (children: under 5), contamination: natural, conc. range: 0.01–1.2 ng/ml, country: Sierra Leone[540], *14 controls, 3 kwashiorkors, 9 underweights, 2 marasmics, and 8 "unspecified" (for detailed information please see the article)

Aflatoxin G_{2a}
incidence: 9/30*, sa. const.: people of Egypt (19 male and 11 female infants), age: 7–20 months, contamination: natural, conc. range: 2–17 ng/100 ml, Ø conc.: 8.22 ng/100ml, country: Egypt[447], *kwashiorkor ca.
incidence: 3/30*, sa. const.: people of Egypt (16 male and 14 female infants), age: 6–13 months, contamination: natural, conc. range: 2–28 ng/100 ml, Ø conc.: 11.33 ng/100ml, country: Egypt[447], *marasmus ca.

Aflatoxin M_1
incidence: 1/15, sa. const.: people of The Sudan (females), contamination: natural, conc.: 34 pg/ml, country: UK/The Sudan[134]
incidence: 1/12, sa. const.: people of The Sudan (children), contamination: natural, conc.: 12 pg/ml, country: UK/The Sudan[134]

incidence: 3/67*, sa. const.: people of Nigeria (60 infants and 7 of their mothers), contamination: natural, conc. range: 80–11,547 ng/l**, country: Nigeria/UK[195], *control (non-jaundiced?!), **only in infants?

incidence: 8/340*, sa. const.: people of Nigeria (270 infants and 70 of their mothers), contamination: natural, conc. range: 48–32,381 ng/l, country: Nigeria/UK[195], *jaundiced

incidence: 1/60*, sa. const.: people of Nigeria (neonates), contamination: natural, conc.: 417 pg/ml, country: Nigeria/UK[262], *non-jaundiced
incidence: 4/64*, sa. const.: people of Nigeria (neonates), contamination: natural, conc. range: 36–877 pg/ml, Ø conc.: 290.25 pg/ml, country: Nigeria/UK[262], *jaundiced

incidence: 1/78*, sa. const.: people of Nigeria, contamination: natural, conc.: 1,272 pg/ml, country: Nigeria/UK[269], *urban population
incidence: 4/97*, sa. const.: people of Nigeria, contamination: natural, conc. range: 20–4,984 pg/ml, Ø conc.: 1,324 pg/ml, country: Nigeria/UK[269], *rural population

incidence: 12/15*, sa. const.: people of Nepal (10 males (7 af) and 5 females (5 af)), contamination: natural, Ø conc.: 8.7 pg/ml, country: Japan/Nepal[360], *control
incidence: 12/23*, sa. const.: people of Nepal (8 males (4 af) and 15 females (8 af) patients, contamination: natural, conc. range: ≤15.4 pg/ml, Ø conc.: 8.9 pg/ml, country: Japan/Nepal[360], *patients

incidence: 1/1, sa. const.: person of the UAE (female), age: 32 years, contamination: natural, conc.: 1.82 ng/ml, country: UAE[428]

incidence: 4/30*, sa. const.: people of Egypt (19 male and 11 female infants), age: 7–20 months, contamination: natural, conc. range: 2–12 ng/100 ml, Ø conc.: 8.25 ng/100 ml, country: Egypt[447], *kwashiorkor ca.
incidence: 2/30*, sa. const.: people of Egypt (16 male and 14 female infants), age: 6–13 months, contamination: natural, conc. range: 10–15 ng/100 ml, Ø conc.: 12.50 ng/100 ml, country: Egypt[447], *marasmus ca.

incidence: 6/12*, sa. const.: people of Egypt (6 males and 6 females), contamination: natural, Ø conc.: 0.66 ng/ml, country: Egypt[455], *control patients
incidence: 27/46*, sa. const.: people of Egypt (30 males and 16 females), Ø age: 56 years, contamination: natural, Ø conc.: 5.61 ng/ml, country: Egypt[455], *HCC patients
incidence: 11/12*, sa. const.: people of Egypt (7 males and 5 females), Ø age: 48 years, contamination: natural, Ø conc.: 19.23 ng/ml, country: Egypt[455], *cirrhotic patients

incidence: 7/36*, sa. const.: people of Sierra Leone (children: under 5), contamination: natural, conc. range: 0.05–1.7 ng/ml, country: Sierra Leone[540], *14 controls, 3 kwashiorkors, 9 underweights, 2 marasmics, and 8 "unspecified" (for detailed information please see the article)

Aflatoxin M_2

incidence: 1/15, sa. const.: people of The Sudan (females), contamination: natural, conc.: 5 pg/ml, country: UK/The Sudan[134]

incidence: 3/67*, sa. const.: people of Nigeria (60 infants and 7 of their mothers), contamination: natural, conc. range: 3,262–4,350 ng/l**, country: Nigeria/UK[195], *control (non-jaundiced?!) group, **only in infants?
incidence: 11/340*, sa. const.: people of Nigeria (270 infants and 70 of their mothers), contamination: natural, conc. range: 407–9,280 ng/l, country: Nigeria/UK[195], *jaundiced

incidence: 6/60*, sa. const.: people of Nigeria (neonates), contamination: natural, conc. range: 89–7,962 pg/ml, Ø conc.: 1,515.33 pg/ml, country: Nigeria/UK[262], *non-jaundiced
incidence: 6/64*, sa. const.: people of Nigeria (neonates), contamination: natural, conc. range: 213–4,134 pg/ml, Ø conc.: 1,246.67 pg/ml, country: Nigeria/UK[262], *jaundiced

incidence: 1/78*, sa. const.: people of Nigeria, contamination: natural, conc.: 24,076 pg/ml, country: Nigeria/UK[269], *urban population
incidence: 1/97*, sa. const.: people of Nigeria, contamination: natural, conc.: 24 pg/ml, country: Nigeria/UK[269], *rural population

incidence: 12/36*, sa. const.: people of Sierra Leone (children: under 5), contamination: natural, conc. range: 0.04–4.0 ng/ml, country: Sierra Leone[540], *14 controls, 3 kwashiorkors, 9 underweights, 2 marasmics, and 8 "unspecified" (for detailed information please see the article)

Aflatoxin P

incidence: 30/30*, sa. const.: people of Egypt (19 male and 11 female infants), age: 7–20 months, no contamination with AFP, conc.: nd, country: Egypt[447], *kwashiorkor ca.
incidence: 1/30*, sa. const.: people of Egypt (16 male and 14 female infants), age: 6–13 months, contamination: natural, conc. range: 7 ng/100 ml, country: Egypt[447], *marasmus ca.

Aflatoxin Q_1

incidence: 3/40, sa. const.: people of Ghana (29 males and 11 females), age: 30–73 years, contamination: natural, Ø conc.: 7.5 ng/ml, country: Ghana[461]

Aflatoxin

incidence: 444/444, sa. const.: people of The Gambia (children), age: 3–4 years, contamination: natural, conc. range: 2.2–459 pg AF-lysine eq/mg albumin*, country: UK/The Gambia[5], *AF-albumin adducts

incidence: 323/391, sa. const.: people of The Gambia (male and female children), age: 3–8 years, contamination: natural, conc. range: 5–719.6 pg AF-lysine eq/mg albumin*, country: The Gambia/France/UK[33], *AF-albumin adducts incidence: 434/466, sa. const.: people of The Gambia (251 male and 221 female children), age: 6–9 years, contamination: natural, conc. range: 5–456 pg/mg*, Ø conc.: 22.3 pg/mg* (geometric mean), country: UK/The Gambia[37], *AF-albumin adducts

incidence: 5/108*, sa. const.: people of the USA, contamination: natural, conc. range: 2.43–4.68 ng/ml, Ø conc.: 3.012 ng/ml, country: USA[51], *thereof 17 Reye's-syndrome ca. but no predominant contamination

incidence: 27?/27, sa. const.: people of the UK, contamination: natural, conc. range: ≤64 pmol/l, country: UK[54]

incidence: 4/227, sa. const.: people of Czechoslovakia (patients), contamination: natural, conc. range: ≤74 ng/l, Ø conc.: 59.8 ng/l, country: Czechoslovakia[353]

incidence: 56/56, sa const.: people of the People's Republic of China (males and females), contamination: natural, conc. range: 0.81–2.41 pmol/mg albumin* **, Ø conc.: 1.24 pmol/mg albumin* **, country: People's Republic of China/USA[388], *AF-albumin adducts, **at the beginning of the study

incidence: 27?/27, sa const.: people of the People's Republic of China (males and females), contamination: natural, conc. range: 0.92–1.67 pmol/mg albumin* **, Ø conc.: 1.21 pmol/mg albumin* **, country: People's Republic of China/USA[388], *AF-albumin adducts, **at the end of the study

incidence: ?/102*, sa. const.: people of Kenya, contamination: natural, conc. range: 0.002–17.7 ng AF-lysine/mg albumin** ***, country: USA/UK[427], *thereof 19 patients with acute hepatic failures (Kenya aflatoxicosis outbreak), **analyzed by IDMS, ***AF-albumin adducts

incidence: ?/102*, sa. const.: people of Kenya, contamination: natural, conc. range: 0–13.6 ng/mg** ***, country: USA/UK[427], *thereof 19 patients with acute hepatic failures (Kenya aflatoxicosis outbreak), **analyzed by HPLC-f, ***AF-albumin adducts

incidence: ?/102*, sa. const.: people of Kenya, contamination: natural, conc. range: 0.018–67.0 ng/mg albumin** ***, country: USA/UK[427], *thereof 19 patients with acute hepatic failures (Kenya aflatoxicosis outbreak), **analyzed by ELISA, ***AF-albumin adducts

incidence: 2/28, sa. const.: people of the USA, contamination: natural, conc. range: 10.1–34.3 pg/mg* ** ***, country: USA/UK[427], *additionally four subjects with levels near or at the LOD, **analyzed by ELISA, ***AF-albumin adducts

incidence: 38/114, sa. const.: people of The Philippines (children), age: 0.08–12 years, weight for height: 6.6–23.1 kg/m, contamination: natural, conc. range: 20–5,600 pg/ml, Ø conc.: 462 pg/ml, country: UK/The Philippines[460]

incidence: ?/200, sa. const.: people of Benin (102 male and 98 female children), age: 16–37 months, contamination: natural, conc. range: 9.2–148.1 pg/mg*, country: UK/Benin[488], *AF-albumin adducts

incidence: 24/24*, sa. const.: people of Egypt (20 males and 4 females), age: 37–73 years, contamination: natural, conc. range: 3.5–25.8 pg/mg**, Ø conc.: 9.0 pg/mg** (geometric mean), country: UK/USA/Egypt[503], *control (without HCC), **AF-albumin adducts

incidence: 7/22*, sa. const.: people of Egypt (18 males and 4 females), age: 36–74 years, contamination: natural, conc. range: 0–32.8 pg/mg**, Ø conc.: 2.6 pg/mg** (geometric mean), country: UK/USA/Egypt[503], *HCC ca., **AF-albumin adducts

incidence: >402/423, sa. const.: people of the Republic of Guinea, contamination: natural, conc. range: 8.2–24.6 pg AFB_1-lysine eq/mg albumin*, country: Republic of Guinea/France/UK[564], *AF-albumin adducts (for detailed information please see the article)

incidence: 4/7, sa. const.: people of Kenya?, contamination: natural, conc. range: 175–670 pg AF/mg albumin*, Ø conc.: 390.25 pg AF/mg albumin*, country: France[629], *measured by ELISA

incidence: 4/7, sa. const.: people of Kenya?, contamination: natural, conc. range: 5.7–17.5 pg AF-lysine/mg albumin*, Ø conc.: 9.98 pg AF-lysine/mg albumin*, country: France[629], *measured by HPLC-f

incidence: 17/19, sa. const.: people of The Gambia, contamination: natural, conc. range: ~6.5–~190 pg AF-lysine/mg albumin*, country: France[629], *measured by hydrolysis-ELISA

incidence: 5/38, sa. const.: people of Thailand, contamination: natural, conc. range: ~4.7–~49 pg AF-lysine/mg albumin*, country: France[629], *measured by hydrolysis-ELISA

incidence: 14/14, sa. const.: people of France, no contamination with AF, conc.: nd*, country: France[629], *measured by hydrolysis-ELISA

Aflatoxins

incidence: 7/44*, sa. const.: people of The Sudan (male and female children), contamination: natural, conc.: 77 pg/ml (geometric means), country: UK/The Sudan[36], *control

incidence: 11/57*, sa. const.: people of The Sudan (male and female children), contamination: natural, conc.: 211 pg/ml (geometric means), country: UK/The Sudan[36], *marasmus ca.

incidence: 7/32*, sa. const.: people of The Sudan (male and female children), contamination: natural, conc.: 412 pg/ml (geometric means), country: UK/The Sudan[36], *marasmic kwashiorkor ca.

incidence: 16/44*, sa. const.: people of The Sudan (male and female children), contamination: natural, conc.: 706 pg/ml (geometric means), country: UK/The Sudan[36], *kwashiorkor ca.

incidence: 2/35*, sa. const.: people of Thailand (females), contamination: natural, conc. range: ≤1.22 nmol/ml, Ø conc.: 0.62 nmol/ml, country: UK[56], *maternal sera, (AF = AFB_1, AFG_1, AFQ_1)

incidence: 17/35*, sa. const.: people of Thailand (fetuses), contamination: natural, conc. range: 0.064–13.6 nmol/ml, Ø conc.: 3.1 nmol/ml, country: UK[56], *cord sera, (AF = AFB_1, AFG_1, AFQ_1)

incidence: 22/103*, sa. const.: people of The Sudan (children), contamination: natural, conc. range: 1–9 pg/ml** (4 sa), 10–99 pg/ml** (6 sa), 100–999 pg/ml** (10 sa), ≥1,000 pg/ml** (2 sa), country: UK/The Sudan[67], *control, **includes AFB_1, AFM_1

incidence: 52/138*, sa. const.: people of The Sudan (children), contamination: natural, conc. range: 1–9 pg/ml** (7 sa), 10–99 pg/ml** (16 sa), 100–999 pg/ml** (16 sa), ≥1,000 pg/ml** (13 sa), country: UK/The Sudan[67], *kwashiorkor, **includes AFB_1, AFM_1, AFL

incidence: 28/98*, sa. const.: people of The Sudan (children), contamination: natural, conc. range: 1–9 pg/ml** (7 sa), 10–99 pg/ml** (10 sa), 100–999 pg/ml** (5 sa), ≥1,000 pg/ml** (6 sa), country: UK/The Sudan[67], *marasmic kwashiorkor, **includes AFB_1, AFM_1, AFL

incidence: 31/118*, sa. const.: people of The Sudan (children), contamination: natural, conc. range: 1–9 pg/ml** (6 sa), 10–99 pg/ml** (7 sa), 100–999 pg/ml** (14 sa), ≥1,000 pg/ml** (4 sa), country: UK/The Sudan[67], *marasmus, **includes AFB_1, AFM_1, AFL

incidence: 28/28, sa. const.: people of Nepal (16 males and 12 females (all af)), contamination: natural, conc. range: 0.06–10 ng/ml*, country: UK/USA[381], *(AF = AFB_1, AFG_1, AFQ_1)

incidence: 7/100*, sa. const.: people of Italy (78 males and 22 females), age: 40–75 years, contamination: natural, conc. range: 0.40–1.20 ng/ml**, country: Italy[465], *lung cancer patients, **includes AFB_1, AFB_2

Gliotoxin

incidence: 2/11*, sa. const.: people of the USA (patients), contamination: natural, conc. range: 65–154 ng/ml, Ø conc.: 109.5 ng/ml, country: USA[78], *non-IA-patients
incidence: 4/5*, sa. const.: people of the USA (patients), contamination: natural, conc. range: 166–785 ng/ml, Ø conc.: 381.5 ng/ml, country: USA[78], *IA-patients

Ochratoxin A

incidence: 17/2,566*, sa. const.: people of Croatia (males and females), contamination: natural, conc. range: 2–10 ng/ml, Ø conc.: 2.8 ng/ml, country: Croatia[188], *control
incidence: 116/4,343*, sa. const.: people of Croatia (males and females), contamination: natural, conc. range: 2–50 ng/ml, country: Croatia[188], *EN ca.
incidence: 10/10* (8**), sa. const.: households of Croatia, contamination: natural, conc. range: 5****–50*** ng/ml, country: Croatia[188], *sa. sera of different households, **with OA in food samples, ***first and **** second blood sa., for detailed information please see the article

incidence: 5/65*, sa. const.: people of Italy (healthy), contamination: natural, conc. range: 0.12–2.0 ng/ml, Ø conc.: 0.53 ng/ml, country: Italy/Sweden[196], *control (no kidney disorders ca)
incidence: ?/40*, sa. const.: people of Italy (patients), contamination: natural, conc. range: 0.05–1.4 ng/ml, country: Italy/Sweden[196], *kidney disorders ca.
incidence: 9/28*, sa. const.: people of Italy (patients), contamination: natural, conc. range: 0.18–14 ng/ml, Ø conc.: 1.4 ng/ml, country: Italy/Sweden[196], *dialysis ca.

incidence: 62/62*, sa. const.: people of Tunisia (males (32 af) and females (30 af)), age: 21–80 years, contamination: natural, conc. range: 0.12–8.06 ng/ml, country: France/Tunisia[202], *control (no nephropathic trouble ca.)
incidence: 23/26*, sa. const.: people of Tunisia (12 males (10 af) and 14 females (13 af)), age: 20–71 years, contamination: natural, conc. range: 0.11–5.80 ng/ml, country: France/Tunisia[202], *nephropathic disease ca.
incidence: 15/21*, sa. const.: people of Tunisia (19 males (17 af) and 2 females (2 af)), age: 32–84 years, contamination: natural, conc. range: 0.14–0.74 ng/ml, country: France/Tunisia[202], *different nephropathic disease ca.

incidence: 57/408, sa. const.: people of Yugoslavia, contamination: natural, conc. range: 1–2 ng/ml (36 sa), 3–5 ng/ml (11 sa), 6–10 ng/ml (4 sa), 11–100 ng/ml (5 sa), 1,800 ng/ml (1 sa), country: Yugoslavia/Sweden/USA/Croatia[206]

incidence: 290/355, sa. const.: people of Hungary, contamination: natural, conc. range: 0.2–1.0 ng/ml (266 sa), >1.0 to ≤10.0 ng/ml (24 sa), country: Hungary[207]

incidence: 134/138, sa. const.: people of Italy (male and female adults: healthy), age: 35–65 years, Ø weight: 69.2 kg, contamination: natural, conc. range: 0.12–0.20 ng/ml (4 sa), 0.20–0.39 ng/ml (42 sa), 0.40–0.59 ng/ml (38 sa), 0.60–0.79 ng/ml (25 sa), 0.80–1.0 ng/ml (11 sa), >1.0–2.84 ng/ml (13 sa), 57.2 ng/ml (1 sa), country: Italy[208]

incidence: 11/60, sa. const.: people of France, contamination: natural, Ø conc.: 1.17 ng/ml, country: France/Tunisia[215]
incidence: 6/30*, sa. const.: people of Tunisia, contamination: natural, Ø conc.: 0.53 ng/ml, country: France/Tunisia[215], *control (no chronic nephropathic disease ca)
incidence: 10/30*, sa. const.: people of Tunisia, contamination: natural, Ø conc.: 0.28 ng/ml, country: France/Tunisia[215], *chronic nephropathic disease ca.

incidence: ?/40*, sa. const.: people of Turkey (17 males and 23 females), Ø age: 41 years, contamination: natural, conc. range: 0.19–1.43 ng/ml, country: Turkey[220], *control (no urinary disorder ca.)
incidence: ?/93*, sa. const.: people of Turkey (63 males and 30 females), Ø age:

42.5–56.8 years, contamination: natural, conc. range: 0.3–5.5 ng/ml, country: Turkey[220], *urinary disorder ca.

incidence: 80/644, sa. const.: people of the Czech Republic (305 males (39 af) and 339 females (41 af)), contamination: natural, conc. range: >1–12 µg/l, country: Czech Republic[222]

incidence: 1,138/1,222, sa. const.: people of the Czech Republic, Ø age: 32 years, contamination: natural, conc. range: 0.1–0.2 µg/l (798 sa), >0.2–1 µg/l (332 sa), >1–2 µg/l (7 sa), 13.7 µg/l (1 sa), country: Czech Republic/France[225]

incidence: 21/71*, sa. const.: people of Egypt (males and females), age: 5–70 years, contamination: natural, conc. range: 0.32–10.15 ng/ml, country: Egypt/France[227], *renal disease ca., for detailed information please see the article
incidence: 2/15*, sa. const.: people of Egypt (males and females), age: 22–50 years, contamination: natural, conc. range: 0.3–0.91 ng/ml, Ø conc.: 0.61 ng/ml, country: Egypt/France[227], *potential kidney donors, for detailed information please see the article

incidence: 213/277*, sa. const.: people of Hungary (volunteers: healthy), contamination: natural, conc. range: 0.1–0.499 ng/ml (160 sa), 0.5–1.40 ng/ml (53 sa), country: Hungary[234], *control
incidence: 103/150*, sa. const.: people of Hungary (patients: registered), contamination: natural, conc. range: 0.1–0.499 ng/ml (83 sa), 0.5–2.26 ng/ml (20 sa), country: Hungary[234], *kidney, liver, and tumor disease ca.

incidence: 87/100, sa. const.: people of Italy (females), contamination: natural, conc. range: 0.02 ng/ml (6 sa), 0.1 ng/ml (38 sa), 0.5 ng/ml (31 sa), >1 ng/ml (12 sa), country: Italy[268]

incidence: 173/306, sa const.: people of Germany, contamination: natural, conc. range: 0.1–14.4 µg/kg, Ø conc.: 0.6 µg/kg, country: Germany[291]

incidence: 728/785, sa const.: people of Germany, contamination: natural, conc. range: 0.06–2.03 ng/ml, Ø conc.: 0.267 ng/ml, country: Germany[302]

incidence: 15/639, sa const.: people of Yugoslavia, contamination: natural, conc. range: 1–57 ng/g, Ø conc.: 7.27 ng/g, country: Sweden/Yugoslavia[323]

incidence: 909/927, sa const.: people of Germany, contamination: natural, conc. range: 0.061–0.10 ng/ml, (41 sa), 0.11–0.30 ng/ml (607 sa), 0.31–0.50 ng/ml (205 sa), 0.51–0.70 ng/ml (31 sa), 0.71–0.90 ng/ml (9 sa), 0.91–≤2.03 ng/ml (16 sa), country: Germany[341]
incidence: 60/61, sa. const.: people of Germany (women: non-pregnant), contamination: natural, conc. range: <0.06–1.63 ng/ml, Ø conc.: 0.43 ng/ml, country: Germany[341]
incidence: 25/26, sa. const.: people of Germany (women: pregnant first trimenon), contamination: natural, conc. range: <0.06–0.88 ng/ml, Ø conc.: 0.30 ng/ml, country: Germany[341]
incidence: 7/9, sa. const.: people of Germany (mothers at birth), contamination: natural, conc. range: 0.06–0.42 ng/ml, country: Germany[341]
incidence: 26/27, sa. const.: people of Germany (puerperae 5th/6th day), contamination: natural, conc. range: <0.06–0.42 ng/ml, Ø conc.: 0.10 ng/ml, country: Germany[341]:
incidence: 77/79*, sa. const.: people of Germany, contamination: natural, conc. range: 0.06–0.90 ng/ml, country: Germany[341], *umbilical cord blood

incidence: 368/368, sa. const.: people of Switzerland (205 males and 163 females), age: 20–40 years, contamination: natural, conc. range: 0.06–6.02 ng/g, Ø conc.: 0.4 ng/g, country: Switzerland[349]

incidence: 35/143, sa. const.: people of Czechoslovakia (patients), contamination: natural, conc. range: 100–500 ng/l (19 sa), 500–1,000 ng/l (15 sa), 1,260 ng/l (1 sa), country: Czechoslovakia[353]

incidence: 30/30*, sa. const.: people of Poland, contamination: natural, conc. range: 0.14–3.41 ng/ml, Ø conc.: 1.14 ng/ml, country: Poland[446], *maternal blood serum
incidence: 28/30*, sa. const.: people of Poland, contamination: natural, conc. range: 0.56–5.42 ng/ml, Ø conc.: 1.96 ng/ml, country: Poland[446], *fetal blood serum

incidence: 36/50, sa. const.: people of Egypt (females), contamination: natural, Ø conc.: 4.28 ng/ml (all sa), country: Egypt[449]
incidence: 50?/50?, sa. const.: people of Egypt (infants), contamination: natural, Ø conc.: 1.26 ng/ml, country: Egypt[449]

incidence: 50/50, sa. const.: people of Portugal (27 male and 23 female nephropathic patients of Coimbra), Ø age: 66 years, Ø weight: 66 kg, contamination: natural, conc. range: 0.12–1.52 µg/l, country: Portugal[468], for detailed information please see the article
incidence: 45/45, sa. const.: people of Portugal (26 male and 19 female nephropathic patients of Aveiro), Ø age: 56 years, Ø weight: 65 kg, contamination: natural, conc. range: 0.15–1.03 µg/l, country: Portugal[468], for detailed information please see the article

incidence: ?/23*, sa. const.: people of Italy (males), age: 26–49 years, contamination: natural, conc. range: 0.03–0.95 ng/ml, country: Italy[470], *control
incidence: 6/6*, sa. const.: people of Italy (males), age: 26–49 years, contamination: natural, conc. range: 0.94–3.28 ng/ml, Ø conc.: 2.29 ng/ml, country: Italy[470], *people breathing OTA contaminated air

incidence: 29/29, sa. const.: people of Portugal (13 males and 16 females of Coimbra), age: 21–57 years, Ø weight: 69 kg, contamination: natural, conc. range: 0.19–0.96 µg/l, Ø conc.: 0.42 µg/l, country: Portugal[507]
incidence: 31/31*, sa. const.: people of Portugal (14 males and 17 females of Verride), age: 26–92 years, Ø weight: 73 kg, contamination: natural, conc. range: 0.25–2.49 µg/l, Ø conc.: 0.78 µg/l, country: Portugal[507]
incidence: 44/44*, sa. const.: people of Portugal (10 males and 34 females of Ereira), age: 19–88 years, Ø weight: 69 kg, contamination: natural, conc. range: 0.14–1.91 µg/l, Ø conc.: 0.44 µg/l, country: Portugal[507]

incidence: 1,596/1,732, sa. const.: people of Germany (male and female adults and children), contamination: natural, conc. range: LOD/LOQ–0.9 µg/l (1,578 sa), 1.0–1.9 µg/l (17 sa), 2.03 µg/l (1 sa), country: EU[508]
incidence: 231/273, sa. const.: people of Italy (males and females), contamination: natural, conc. range: LOD/LOQ–0.9 µg/l (190 sa), 1.0–1.9 µg/l (27 sa), 2.0–3.6 µg/l (14 sa), country: EU[508]

incidence: 240/594, sa. const.: people of the Czech Republic (496 males and 98 females), age: 18–58 years, contamination: natural, conc. range: 0.05–37 µg/l, country: Czech Republic[523]

incidence: 5/5, sa. const.: people of Bulgaria (volunteers of Gorno Peshtene, healthy), age: 20–30 years, contamination: natural, conc. range: 0.44–1.46 µg/l*, Ø conc.: 0.67 µg/l*, country: Bulgaria/France[538], *measured for 4 weeks (for detailed information please see the article)
incidence: 11/11, sa. const.: people of Bulgaria (volunteers of Beli Izvor, healthy), age: 20–30 years, contamination: natural, conc. range: 0.26–8.36 µg/l*, Ø conc.: 2.01 µg/l*, country: Bulgaria/France[538], *measured for 4 weeks (for detailed information please see the article)

incidence: 10/36*, sa. const.: people of Sierra Leone (children: under 5), contamination: natural, conc. range: 1.5–18.2 ng/ml, country: Sierra Leone[540], *14 controls, 3 kwashiorkors, 9 underweights, 2 marasmics, and 8 "unspecified" (for detailed information please see the article)

Ochratoxin A Methyl Ester
incidence: 4/639, sa const.: people of Yugoslavia, contamination: natural, conc. range: 5–42 ng/g, Ø conc.: 16.25 ng/g, country: Sweden/Yugoslavia[323]

Ochratoxin α
incidence: 20/639, sa const.: people of Yugoslavia, contamination: natural, conc. range: 2–44 ng/g, Ø conc.: 6.45 ng/g, country: Sweden/Yugoslavia[323]

Ochratoxin α Methyl Ester
incidence: 13/639, sa const.: people of Yugoslavia, contamination: natural, conc. range: 4–43 ng/g, Ø conc.: 9.38 ng/g, country: Sweden/Yugoslavia[323]

Ochratoxin B
incidence: 7/36*, sa. const.: people of Sierra Leone (children: under 5), contamination: natural, conc. range: 0.05–8.2 ng/ml, country: Sierra Leone[540], *14 controls, 3 kwashiorkors, 9 underweights, 2 marasmics, and 8 "unspecified" (for detailed information please see the article)

Trichothecenes
incidence: 7?/23, sa. const.: people of the USA (occupants of uncontaminated buildings), contamination: natural, conc. range: ≤0.11 ng/ml, country: USA/Canada[387]
incidence: 19?/40, sa. const.: people of the USA (occupants of uncontaminated buildings), contamination: natural, conc. range: ≤83.6 ng/ml, country: USA/Canada[387]

Zearalenone
incidence: 5/36, sa. const.: people of Hungary (patients), contamination: natural, conc. range: 18.9–103.5 µg/l, Ø conc.: 66.08 µg/l, country: Hungary[203] see also Human blood and Human plasma

Human serum/plasma may contain the following mycotoxins and/or their metabolites:

Ochratoxin A
incidence: 200/200, sa. const.: people of Sweden (males and females), contamination: natural, conc. range: LOD/LOQ–0.88 µg/l, country: EU[508]

Human spleen may contain the following mycotoxins and/or their metabolites:

Aflatoxin B_1
incidence: 1/17, sa. const.: people of Malaysia (1 adult and 16 children: 12 males and 5 females), age: adult 49 years, children 2.5–11 years, contamination: natural, conc.: 3,448 pg/g tissue, country: Singapore/UK[259]

Aflatoxin B_2
incidence: 1/17, sa. const.: people of Malaysia (1 adult and 16 children: 12 males and 5 females), age: adult 49 years, children 2.5–11 years, contamination: natural, conc.: 631 pg/g tissue, country: Singapore/UK[259]

Aflatoxin M_1
incidence: 3/17, sa. const.: people of Malaysia (1 adult and 16 children: 12 males and 5 females), age: adult 49 years, children 2.5–11 years, contamination: natural, conc. range: 20–4,746 pg/g tissue, Ø conc.: 2,348.66 pg/g tissue, country: Singapore/UK[259]

Aflatoxin M_2
incidence: 1/17, sa. const.: people of Malaysia (1 adult and 16 children: 12 males and 5 females), age: adult 49 years, children 2.5–11 years, contamination: natural, conc.: 1,479 pg/g tissue, country: Singapore/UK[259]

Human stomach may contain the following mycotoxins and/or their metabolites:

Aflatoxin B_1
incidence: 14/20*, sa. const.: people of Thailand (children), contamination:

natural, conc. range: tr–127 µg/kg**, country: USA/Thailand[10], *EFDV ca., **in stomach contents
incidence: 4/8*, sa. const.: people of Thailand (children), contamination: natural, conc. range: tr ** ***, country: USA/Thailand[10], *dying from causes other than EFDV, **in stomach contents, ***sa. showed a blue fluorescent spot similar to that of AFB_1

Aflatoxin B_2
incidence: 3/20*, sa. const.: people of Thailand (children), contamination: natural, conc. range: tr–19 µg/kg*, country: USA/Thailand[10], *EFDV ca., **in stomach contents

Human stool may contain the following mycotoxins and/or their metabolites:

Aflatoxicol
incidence: 1/5*, sa. const.: people of Kenya (children), contamination: natural, conc.: 1,711 ng after 9 days**, country: Kenya/UK[91], *kwashiorkor ca., **total excretion after 9 days, for detailed information please see the article
incidence: 15*/32**, sa. const.: people of Sierra Leone (children), contamination: natural, Ø conc.: 0.63 ng/ml, country: Sierra Leone[540], *5 controls, 1 underweight, 3 marasmics, and 6 kwashiorkors AFL-contaminated, **11 controls, 4 underweights, 5 marasmics, and 12 kwashiorkors

Aflatoxin B_1
incidence: 7/18*, sa. const.: people of Thailand (children), contamination: natural, conc. range: tr–123 µg/kg, country: USA/Thailand[10], *EFDV ca.
incidence: 3/5*, sa. const.: people of Thailand (children), contamination: natural, conc. range: tr**, country: USA/Thailand[10], *dying from causes other than EFDV, **sa. showed a blue fluorescent spot similar to that of AFB_1
incidence: 2/5*, sa. const.: people of Kenya (children), contamination: natural, conc. range: 2,436–3,147 ng after 9 days**, country: Kenya/UK[91], *kwashiorkor ca. **total excretion after 9 days, for detailed information please see the article
incidence: 1/7*, sa. const.: people of Kenya (children), contamination: natural, conc.: 1,584 ng after 9 days**, country: Kenya/UK[91], *marasmic kwashiorkor ca., **total excretion after 9 days, for detailed information please see the article

incidence: 5/11*, sa. const.: people of Sierra Leone (children), contamination: natural, Ø conc.: 29 ng/ml, country: Sierra Leone[540], *control
incidence: 1/4*, sa. const.: people of Sierra Leone (children), contamination: natural, conc.: 21 ng/ml, country: Sierra Leone[540], *underweight ca.
incidence: 3/5*, sa. const.: people of Sierra Leone (children), contamination: natural, Ø conc.: 26 ng/ml, country: Sierra Leone[540], *marasmus ca.
incidence: 7/12*, sa. const.: people of Sierra Leone (children), contamination: natural, Ø conc.: 19 ng/ml, country: Sierra Leone[540], *kwashiorkor ca.

Aflatoxin B_2
incidence: 4/18*, sa. const.: people of Thailand (children), contamination: natural, conc. range: 4–19 µg/kg, Ø conc.: 12.8 µg/kg, country: USA/Thailand[10], *EFDV ca.

incidence: 1/11*, sa. const.: people of Sierra Leone (children), contamination: natural, conc.: 1.0 ng/ml, country: Sierra Leone[540], *control
incidence: 1/4*, sa. const.: people of Sierra Leone (children), contamination: natural, conc.: 1.5 ng/ml, country: Sierra Leone[540], *underweight ca.
incidence: 5/5*, sa. const.: Sierra Leone (children), no contamination with AFB_2, conc.: nd, country: Sierra Leone[540], *marasmus ca.
incidence: 12/12*, sa. const.: Sierra Leone (children), no contamination with AFB_2,

conc.: nd, country: Sierra Leone[540], *kwashiorkor ca.

Aflatoxin G_1
incidence: 6/11*, sa. const.: people of Sierra Leone (children), contamination: natural, Ø conc.: 11 ng/ml, country: Sierra Leone[540], *control
incidence: 4/4*, sa. const.: people of Sierra Leone (children), no contamination with AFG_1, conc.: nd, country: Sierra Leone[540], *underweight ca.
incidence: ?/5*, sa. const.: people of Sierra Leone (children), Ø conc.: 4 ng/ml, country: Sierra Leone[540], *marasmus ca.
incidence: ?/12*, sa. const.: people of Sierra Leone (children), Ø conc.: 15 ng/ml, country: Sierra Leone[540], *kwashiorkor ca.

Aflatoxin G_2
incidence: 1/11*, sa. const.: people of Sierra Leone (children), contamination: natural, conc.: 0.4 ng/ml, country: Sierra Leone[540], *control
incidence: 4/4*, sa. const.: people of Sierra Leone (children), no contamination with AFG_2, conc.: nd, country: Sierra Leone[540], *underweight ca.
incidence: ?/5*, sa. const.: people of Sierra Leone (children), conc.: 2 ng/ml, country: Sierra Leone[540], marasmus ca.
incidence: 0.3?/12*, sa. const.: Sierra Leone (children), conc.: ?, country Sierra Leone[540], *kwashiorkor ca.

Aflatoxin M_1
incidence: 2/5*, sa. const.: people of Kenya (children), contamination: natural, conc. range: 20–86 ng after 9 days**, country: Kenya/UK[91], *kwashiorkor ca., **total excretion after 9 days, for detailed information please see the article
incidence: 2/7*, sa. const.: people of Kenya (children), contamination: natural, conc. range: 190–505 ng after 9 days**, country: Kenya/UK[91], *marasmic kwashiorkor ca., **total excretion after 9 days, for detailed information please see the article
incidence: 3/11*, sa. const.: people of Sierra Leone (children), contamination: natural, Ø conc.: 1.6 ng/ml, country: Sierra Leone[540], *control
incidence: 4/4*, sa. const.: people of Sierra Leone (children), no contamination with AFM_1, conc.: nd, country Sierra Leone[540], *underweight ca.
incidence: 1/5*, sa. const.: people of Sierra Leone (children), contamination: natural, conc.: 0.6 ng/ml, country: Sierra Leone[540], *marasmus ca.
incidence: 7/12*, sa. const.: people of Sierra Leone (children), contamination: natural, Ø conc.: 2.7 ng/ml, country: Sierra Leone[540], *kwashiorkor ca.

Aflatoxin M_2
incidence: 4/5*, sa. const.: people of Kenya (children), contamination: natural, conc. range: 515–52,522 ng after 9 days**, country: Kenya/UK[91], *kwashiorkor ca., **total excretion after 9 days, for detailed information please see the article
incidence: 2/7*, sa. const.: people of Kenya (children), contamination: natural, conc. range: 173–14,936 ng after 9 days**, country: Kenya/UK[91], *marasmic kwashiorkor ca., **total excretion after 9 days, for detailed information please see the article
incidence: 6/11*, sa. const.: people of Sierra Leone (children), contamination: natural, Ø conc.: 3.7 ng/ml, country: Sierra Leone[540], *control
incidence: 1/4*, sa. const.: people of Sierra Leone (children), contamination: natural, conc.: 0.1 ng/ml, country: Sierra Leone[540], *underweight ca.
incidence: 1/5*, sa. const.: people of Sierra Leone (children), contamination: natural, conc.: 0.6 ng/ml, country: Sierra Leone[540], *marasmus ca.
incidence: 5/12*, sa. const.: people of Sierra Leone (children), contamination: natural, Ø conc.: 1.1 ng/ml, country: Sierra Leone[540], *kwashiorkor ca.

Ochratoxin A
incidence: 1/11*, sa. const.: people of Sierra Leone (children), contamination: natural, conc.: 1.6 ng/ml, country: Sierra Leone[540], *control
incidence: 1?/4*, sa. const.: people of Sierra Leone (children), contamination: natural, conc.: 2 ng/ml, country: Sierra Leone[540], *underweight ca.
incidence: 1?/5*, sa. const.: people of Sierra Leone (children), contamination: natural, conc.: 8.3 ng/ml, country: Sierra Leone[540], *marasmus ca.
incidence: 5/12*, sa. const.: people of Sierra Leone (children), contamination: natural, Ø conc.: 2.5 ng/ml, country: Sierra Leone[540], *kwashiorkor ca.

4-Hydroxyochratoxin A
incidence: 1/11*, sa. const.: people of Sierra Leone (children), contamination: natural, conc.: 1.0 ng/ml, country: Sierra Leone[540], *control
incidence: 1/4*, sa. const.: people of Sierra Leone (children), contamination: natural, conc.: 1.2 ng/ml, country: Sierra Leone[540], *underweight ca.
incidence: 2/5*, sa. const.: people of Sierra Leone (children), contamination: natural, Ø conc.: 0.1 ng/ml, country: Sierra Leone[540], *marasmus ca.
incidence: 7/12*, sa. const.: people of Sierra Leone (children), contamination: natural, Ø conc.: 0.4 ng/ml, country: Sierra Leone[540], *kwashiorkor ca.

Ochratoxin B
incidence: 2/11*, sa. const.: people of Sierra Leone (children), contamination: natural, Ø conc.: 8.4 ng/ml, country: Sierra Leone[540], *control
incidence: 1/4*, sa. const.: people of Sierra Leone (children), contamination: natural, conc.: 31 ng/ml, country: Sierra Leone[540], *underweight ca.
incidence: 4/5*, sa. const.: people of Sierra Leone (children), contamination: natural, Ø conc.: 27 ng/ml, country: Sierra Leone[540], *marasmus ca.
incidence: 7/12*, sa. const.: people of Sierra Leone (children), contamination: natural, Ø conc.: 29 ng/ml, country: Sierra Leone[540], *kwashiorkor ca.

Human urine may contain the following mycotoxins and/or their metabolites:

Aflatoxicol
incidence: 99/99*, sa. const.: people of the UK (male volunteers), age: 17–40 years, no contamination with AFL, conc.: nd, country: UK[117], *control
incidence: 1/133, sa. const.: people of The Netherlands and UK (heroin abusers), contamination: natural, conc.: 0.32 nmol/l, country: UK[117], pos. sa. from people of The Netherlands and the UK
incidence: 1/12, sa. const.: people of The Sudan (children), contamination: natural, conc.: 36 pg/ml, country: UK/The Sudan[134]
incidence: 68/110, sa. const.: people of Sierra Leone (females), age: 5–14 years, contamination: natural, conc. range: 0.05–8.9 ng/ml*, country: Sierra Leone[401], *in dry season
incidence: 35/93, sa. const.: people of Sierra Leone (females), age: 5–14 years, contamination: natural, conc. range: 0.1–9.0 ng/ml*, country: Sierra Leone[401], *in rainy season
incidence: 83/134, sa. const.: people of Sierra Leone (males), age: 5–14 years, contamination: natural, conc. range: 0.04–14.2 ng/ml*, country: Sierra Leone[401], *in dry season
incidence: 52/97, sa. const.: people of Sierra Leone (males), contamination: natural, conc. range: 0.02–7.2 ng/ml*, country: Sierra Leone[401], *in rainy season
incidence: 14/30, sa. const.: people of Sierra Leone (male children), contamination: natural, conc. range: 0.02–14.0 ng/ml, country: Sierra Leone[540]
incidence: 15/24, sa. const.: people of Sierra Leone (female children),

contamination: natural, conc. range: 0.02–1.4 ng/ml, country: Sierra Leone[540]

Aflatoxin B_1

incidence: 71/317*, sa. const.: people of the People's Republic of China (males), age: mostly 45–64 years, contamination: natural, conc. range: ?, country: People's Republic of China/USA[12], *thereof 50 HCC ca. (contaminated: 56 control and 15 HCC ca., for detailed information please see the article)
incidence: 49/317*, sa. const.: people of the People's Republic of China (males), age: mostly 45–64 years, contamination: natural, conc. range: 0.3–1.81 ng/ml**, country: People's Republic of China/USA[12], *thereof 50 HCC ca. (contaminated: 31 control and 18 HCC ca., for detailed information please see the article), **AFB_1-N^7-Gua adducts

incidence: 34/85*, sa. const.: people of Taiwan (males), age: 33–66 years, contamination: natural, Ø conc.: 0.52 ng/ml**, country: Taiwan, Republic of China[14], *thereof 42 asymptotic HbsAg carriers, **AFB_1-N^7-Gua[1]

incidence: 6/81, sa. const.: people of Kenya (3 males and 3 females af), children, age: 5–75 years, contamination: natural, conc. range: 0.3–3 pmol AFB-GuaI in 25 ml, country: USA/Kenya[18]

incidence: 4/10*, sa. const.: people of the USA (patients), contamination: natural, conc. range: 2.7–8.9 ng/ml*, country: USA[19], *thereof 5 Reye's-syndrome ca. thereof 4 AFB_1-pos.

incidence: 2/10*, sa. const.: people of The Sudan (children), contamination: natural, conc. range: 86–806 pg/ml, Ø conc.: 446 pg/ml, country: The Sudan/UK[32], *1 marasmic kwashiorkor and 9 marasmus ca. thereof 2 AFB_1-pos.
incidence: 1/3*, sa. const.: people of The Sudan (children), contamination: natural, conc.: 667 pg/ml, country: The Sudan/UK[32], *miscellaneous liver disease ca.

incidence: 29/29, sa. const.: people of the People's Republic of China (males and females), contamination: natural, conc. range: 0.01–0.03 ng/ml, country: People's Republic of China/USA[41]

incidence: 41/72*, sa. const.: people of the People's Republic of China, contamination: natural, conc. range: ≤156.6 pg/mg creatinine**, median level: 7.1 pg/mg creatinine**, country: People's Republic of China/USA[48], *receiving a placebo for 4 weeks, **AFB_1-NAC
incidence: 45/57*, sa. const.: people of the People's Republic of China, contamination: natural, conc. range: ≤245.5 pg/mg creatinine**, median level: 18.6 pg/mg creatinine**, country: People's Republic of China/USA[48], *receiving 125 mg oltipraz daily for 4 weeks, **AFB-NAC
incidence: 39/60*, sa. const.: people of the People's Republic of China, contamination: natural, conc. range: ≤189.4 pg/mg creatinine**, median level: 8.3 pg/mg creatinine**, country: People's Republic of China/USA[48], *receiving 500 mg oltipraz weekly for 4 weeks, **AFB-NAC

incidence: 15/108*, sa. const.: people of the USA, contamination: natural, conc. range: 5–61 ppt, Ø conc.: 13.93 ppt, country: USA[51], *thereof 17 Reye's-syndrome ca. but no predominant contamination

incidence: 2/5*, sa. const.: people of Kenya (children), contamination: natural, conc. range: 726–1,044 ng after 9 days**, country: Kenya/UK[91], *kwashiorkor ca., **total 24 h urine production calculated, for detailed information please see the article
incidence: 3/7*, sa. const.: people of Kenya (children), contamination: natural, conc. range: 41–1,316 ng after 9 days**, country: Kenya/UK[91], *marasmic kwashiorkor ca., **total excretion after 9 days, for detailed information please see the article

incidence: 99/99*, sa. const.: people of the UK (male volunteers), age: 17–40 years,

no contamination with AFB_1, conc.: nd, country: UK[117], *control
incidence: 9/133, sa. const.: people of The Netherlands and UK (heroin abusers), contamination: natural, conc. range: 0.73–25.80 nmol/l*, country: UK[117], *pos. sa. from people of The Netherlands and the UK

incidence: 20/96*, sa. const.: ?, contamination: natural, conc. range: 5.3–52 pg/g, country: USA/Costa Rica/Japan[121], *several Reye's-syndrome ca., but no statement concerning higher contamination, indicative

incidence: 25/1,228, sa. const.: people of Zimbabwe, contamination: natural, Ø conc.: 1 ng/ml*, country: Zimbabwe[152], *national average

incidence: 7/29, sa. const.: people of The Philippines, contamination: natural, conc. range: ≤4.25 ng/ml*, country: France[190], *AFB_1-eq.

incidence: 24/27, sa const.: people of the People's Republic of China (males and females), contamination: natural, conc. range: 6.6–494.9 ng/24-h*, Ø conc.: 103.6 ng/24-h*, country: People's Republic of China/USA[388], *AFB-NAC
incidence: 11/27, sa const.: people of the People's Republic of China (males and females), contamination: natural, conc. range: 64.9–1,789.8 ng/24-h*, Ø conc.: 407.3 ng/24-h*, country: People's Republic of China/USA[388], *AFB-N^7-Gua

incidence: 47/134, sa. const.: people of Sierra Leone (males), age: 5–14 years, contamination: natural, conc. range: 0.6–188 ng/ml*, country: Sierra Leone[401], *in dry season
incidence: 32/97, sa. const.: people of Sierra Leone (males), age: 5–14 years, contamination: natural, conc. range: 1.2–115 ng/ml*, country: Sierra Leone[401], *in rainy season
incidence: 53/110, sa. const.: people of Sierra Leone (females), age: 5–14 years, contamination: natural, conc. range: 0.04–319 ng/ml*, country: Sierra Leone[401], *in dry season
incidence: 38/93, sa. const.: people of Sierra Leone (females), age: 5–14 years, contamination: natural, conc. range: 0.08–127 ng/ml*, country: Sierra Leone[401], *in rainy season

incidence: ?/18, sa. const.: people of India (9 male and 9 female volunteers, healthy), age: 20–40 years, weight: 45–50 kg, contamination: natural, conc. range: 9.30–13.43 ng/mg creatinine* **, country: India[437], *AFB_1-N^7-Gua adducts, **only 5 maize eating rurals (females) affected

incidence: 24/30*, sa. const.: people of Egypt (19 male and 11 female children), age: 7–20 months, contamination: natural, conc. range: 1–15 ng/100 ml, Ø conc.: 8.29 ng/100 ml, country: Egypt[447], *kwashiorkor ca.
incidence: 13/30*, sa. const.: people of Egypt (16 male and 14 female children), age: 6–13 months, contamination: natural, conc. range: 5–9 ng/100 ml, Ø conc.: 6.92 ng/100 ml, country: Egypt[447], *marasmus ca.

incidence: 16/20, sa. const.: people of the People's Republic of China, contamination: natural, conc. range: 0.9–7.2 pg/20 ml*, Ø conc.: 2.93 pg/20 ml*, country: USA[448], *AFB_1-N^7-Gua

incidence: ?/40*, sa. const.: people of the People's Republic of China, age: 20–55 years, contamination: natural, conc. range: 0.09–57.92 pg/mg creatinine** ***, country: USA/People's Republic of China[451], *control (receiving placebos for 3 months), **AFB-NAC, ***for detailed information please see the article
incidence: ?/40*, sa. const.: people of the People's Republic of China, age: 20–55 years, contamination: natural, conc. range: 0.38–501.48 pg/mg creatinine** ***, country: USA/People's Republic of China[451], *receiving GTP 500 mg for 3 months, **AFB-NAC, ***for detailed information please see the article

incidence: ?/40*, sa. const.: people of the People's Republic of China, age: 20–55 years, contamination: natural, conc. range: 0.30–560.30 pg/mg creatinine** ***, country: USA/People's Republic of China[451], *receiving GTP 1,000 mg for 3 months, **AFB-NAC, ***for detailed information please see the article

incidence: 2/40, sa. const.: people of Ghana (29 males and 11 females), age: 30–73 years, contamination: natural, Ø conc.: 7.5 ng/ml, country: Ghana[461]

incidence: 1/50, sa. const.: people of Egypt (children), age: 1–2.5 years, contamination: natural, conc.: 189 pg/ml, country: Finland/UK/Egypt/Republic of Guinea[481]

incidence: 8/50, sa. const.: people of the Republic of Guinea (children), age: 2–4 years, contamination: natural, conc. range: 179–18,000 pg/ml, Ø conc.: 2,682 pg/ml?, country: Finland/UK/Egypt/Republic of Guinea[481]

incidence: 6/50, sa. const.: people of Singapore (male volunteers), age: 20–66 years, contamination: natural, conc. range: 185–2,300 pg/ml*, country: Singapore/France[496], *AFB_1-eq.

incidence: 8/30, sa. const.: people of Sierra Leone (male children), contamination: natural, conc. range: 0.7–53.0 ng/ml, country: Sierra Leone[540]

incidence: 9/24, sa. const.: people of Sierra Leone (female children), contamination: natural, conc. range: 0.6–54.1 ng/ml, country: Sierra Leone[540]

incidence: 34/86, sa. const.: people of Taiwan, age: 33–66 years, contamination: natural, conc. range: 0.10–6.06 ng/mg* **, country: Taiwan, Republic of China[571], *AFB_1-N^7-Gua[1], **19 of 43 HbsAg carriers and 15 of 43 HbsAg non-carriers pos. for AFB_1

incidence: 5/161, sa. const.: people of Nigeria, contamination: natural, Ø conc.: 2.87 ng/100 ml (mean value), country: USA/Nigeria[596]

Aflatoxin B_2

incidence: 1/15*, sa.*[1] const.: people of The Sudan (children), contamination: natural, conc.: 56 pg/ml, country: The Sudan/UK[32], *kwashiorkor ca

incidence: 1/10*, sa. const.: people of The Sudan (children), contamination: natural, conc.: 46 pg/ml, country: The Sudan/UK[32], *1 marasmic kwashiorkor and 9 marasmus ca. thereof 1 AFB_2-pos.

incidence: 1/7*, sa. const.: people of Kenya (children), contamination: natural, conc.: 1 ng after 9 days**, country: Kenya/UK[91], *marasmic kwashiorkor ca., **total excretion after 9 days, for detailed information please see the article

incidence: 2/99*, sa. const.: people of the UK (male volunteers), age: 17–40 years, contamination: natural, conc. range: 0.13–0.24 nmol/l, Ø conc.: 0.185 nmol/l, country: UK[117], *control

incidence: 8*/133, sa. const.: people of The Netherlands and UK (heroin abusers), contamination: natural, conc. range: 0.09–1.53 nmol/l, country: UK[117], *pos. sa. from people of The Netherlands and the UK

incidence: 25/1,228, sa. const.: people of Zimbabwe, contamination: natural, Ø conc.: 1 ng/ml*, country: Zimbabwe[152], *national average

incidence: 40/134, sa. const.: people of Sierra Leone (males), age: 5–14 years, contamination: natural, conc. range: 0.01–15.5 ng/ml*, country: Sierra Leone[401], *in dry season

incidence: 9/97, sa. const.: people of Sierra Leone (males), age: 5–14 years, contamination: natural, conc. range: 0.2–48 ng/ml*, country: Sierra Leone[401], *in rainy season

incidence: 18/110, sa. const.: people of Sierra Leone (females), age: 5–14 years, contamination: natural, conc. range: 0.2–152 ng/ml*, country: Sierra Leone[401], *in dry season incidence: 19/93, sa. const.: people of Sierra Leone (females), age: 5–14 years, contamination: natural, conc. range: 0.1–12 ng/ml*, country: Sierra Leone[401], *in rainy season

incidence: 6/30*, sa. const.: people of Egypt (19 male and 11 female infants), age: 7–20 months, contamination: natural, conc. range: 2–4 ng/100 ml, Ø conc.: 2.67 ng/100 ml, country: Egypt[447], *kwashiorkor ca.

incidence: 1/30*, sa. const.: people of Egypt (16 male and 14 female infants), age: 6–13 months, contamination: natural, conc.: 2 ng/100 ml, country: Egypt[447], *marasmus ca.

incidence: 5/50, sa. const.: people of Egypt (children), age: 1–2.5 years, contamination: natural, conc. range: 0.8–2.2 pg/ml, Ø conc.: 1.4 pg/ml, country: Finland/UK/Egypt/Republic of Guinea[481]

incidence: 29/50, sa. const.: people of the Republic of Guinea (children), age: 2–4 years, contamination: natural, conc. range: 0.6–43 pg/ml, Ø conc.: 5.7 pg/ml, country: Finland/UK/Egypt/Republic of Guinea[481]

incidence: 2/30, sa. const.: people of Sierra Leone (male children), contamination: natural, conc. range: 4.4–6.3 ng/ml, Ø conc.: 5.35 ng/ml, country: Sierra Leone[540]

incidence: 1/24, sa. const.: people of Sierra Leone (female children), contamination: natural, conc.: 1.8 ng/ml, country: Sierra Leone[540]

incidence: 37/241, sa. const.: people of Thailand (males and females), age: 30–40 years, contamination: natural, conc. range: 50–4,776 ng AFB_1 eq/ml, country: Thailand/France[562]

Aflatoxin B_{2a}

incidence: 12/30*, sa. const.: people of Egypt (19 male and 11 female infants), age: 7–20 months, contamination: natural, conc. range: 1–9 ng/100 ml, Ø conc.: 3.58 ng/100 ml, country: Egypt[447], *kwashiorkor ca.

incidence: 3/30*, sa. const.: people of Egypt (16 male and 14 female infants), age: 6–13 months, contamination: natural, conc. range: 2–4 ng/100 ml, Ø conc.: 3.00 ng/100 ml, country: Egypt[447], *marasmus ca.

incidence: 72/161, sa. const.: people of Nigeria, contamination: natural, Ø conc.: 0.60 ng/100 ml (mean value), country: USA/Nigeria[596]

Aflatoxin B

incidence: 7/74*, sa. const.: people of India (children: healthy), age: 1–5 years, contamination: natural, conc. range: 0.02–0.05 μg/day, country: India[3], *control

incidence: 15/255*, sa. const.: people of India (children), age: 1–5 years, contamination: natural, conc. range: 0.02–0.05 μg/day, country: India[3], *Indian childhood cirrhosis ca.

Aflatoxin G_1

incidence: 1/5*, sa. const.: people of Kenya (children), contamination: natural, conc.: 98 ng after 9 days**, country: Kenya/UK[91], *kwashiorkor ca., **total excretion after 9 days, for detailed information please see the article

incidence: 282/1,228, sa. const.: people of Zimbabwe, contamination: natural, Ø conc.: 9 ng/ml*, country: Zimbabwe[152], *national average

incidence: 51/134, sa. const.: people of Sierra Leone (males), age: 5–14 years, contamination: natural, conc. range: 2.9–169 ng/ml*, country: Sierra Leone[401], *in dry season

incidence: 27/97, sa. const.: people of Sierra Leone (males), contamination: natural, conc. range: 0.8–57.4 ng/ml*, country: Sierra Leone[401], *in rainy season

incidence: 42/110, sa. const.: people of Sierra Leone (females), age: 5–14 years, contamination: natural, conc. range:

0.4–138 ng/ml*, country: Sierra Leone[401], *in dry season
incidence: 18/93, sa. const.: people of Sierra Leone (females), age: 5–14 years, contamination: natural, conc. range: 1.0–150 ng/ml*, country: Sierra Leone[401], *in rainy season

incidence: 18/30*, sa. const.: people of Egypt (19 male and 11 female infants), age: 7–20 months, contamination: natural, conc. range: 1–11 ng/100 ml, Ø conc.: 4.78 ng/100 ml, country: Egypt[447], *kwashiorkor ca.
incidence: 7/30*, sa. const.: people of Egypt (16 male and 14 female infants), age: 6–13 months, contamination: natural, conc. range: 2–8 ng/100 ml, Ø conc.: 3.57 ng/100 ml, country: Egypt[447], *marasmus ca.

incidence: 17/40, sa. const.: people of Ghana (29 males and 11 females), age: 30–73 years, contamination: natural, Ø conc.: 6.75 ng/ml, country: Ghana[461]

incidence: 2/50, sa. const.: people of Egypt (children), age: 1–2.5 years, contamination: natural, conc. range: 72.1–81.1 pg/ml, Ø conc.: 76.6 pg/ml, country: Finland/UK/Egypt/Republic of Guinea[481]
incidence: 1/50, sa. const.: people of the Republic of Guinea (children), age: 2–4 years, contamination: natural, conc.: 709 pg/ml, country: Finland/UK/Egypt/Republic of Guinea[481]

incidence: 2/30, sa. const.: people of Sierra Leone (male children), contamination: natural, conc. range: 23.0–39.4 ng/ml, Ø conc.: 31.2 ng/ml, country: Sierra Leone[540]
incidence: 3/24, sa. const.: people of Sierra Leone (female children), contamination: natural, conc. range: 0.01–17.0 ng/ml, country: Sierra Leone[540]

incidence: 16/161, sa. const.: people of Nigeria, contamination: natural, Ø conc.: 4.82 ng/100 ml (mean value), country: USA/Nigeria[596]

Aflatoxin G_2
incidence: 2/15*, sa. const.: people of The Sudan (children), contamination: natural, conc. range: 24–80 pg/g, Ø conc.: 52 pg/ml, country: The Sudan/UK[32], *kwashiorkor ca.

incidence: 1/5*, sa. const.: people of Kenya (children), contamination: natural, conc.: 1 ng after 9 days**, country: Kenya/UK[91], *kwashiorkor ca., **total excretion after 9 days, for detailed information please see the article

incidence: 184/1,228, sa. const.: people of Zimbabwe, contamination: natural, Ø conc.: 24 ng/ml*, country: Zimbabwe[152], *national average

incidence: 3/134, sa. const.: people of Sierra Leone (males), age: 5–14 years, contamination: natural, conc. range: 0.1–1.5 ng/ml*, country: Sierra Leone[401], *in dry season
incidence: 2/97, sa. const.: people of Sierra Leone (males), age: 5–14 years, contamination: natural, conc. range: 0.2–0.7 ng/ml*, country: Sierra Leone[401], *in rainy season
incidence: 110/110, sa. const.: people of Sierra Leone (females), age: 5–14 years, no contamination with AFG_2, conc.: nd*, country: Sierra Leone[401], *in dry season
incidence: 3/93, sa. const.: people of Sierra Leone (females), age: 5–14 years, contamination: natural, conc. range: 1.1–2.0 ng/ml*, country: Sierra Leone[401], *in rainy season

incidence: 12/50, sa. const.: people of Egypt (children), age: 1–2.5 years, contamination: natural, conc. range: 0.9–8.0 pg/ml, Ø conc.: 2.2 pg/ml, country: Finland/UK/Egypt/Guinea[481]
incidence: 18/50, sa. const.: people of the Republic of Guinea (children), age: 2–4 years, contamination: natural, conc. range: 1.4–199 pg/ml, Ø conc.: 19.0 pg/ml, country: Finland/UK/Egypt/Republic of Guinea[481]

Aflatoxin G_{2a}
incidence: 5/30*, sa. const.: people of Egypt (19 male and 11 female infants), age: 7–20 months, contamination: natural, conc. range: 1–3 ng/100 ml, Ø conc.: 1.60 ng/100 ml, country: Egypt[447], *kwashiorkor ca.
incidence: 2/30*, sa. const.: people of Egypt (16 male and 14 female infants), age: 6–13 months, contamination: natural, conc. range: 1–16 ng/100 ml, Ø conc.: 8.50 ng/100 ml, country: Egypt[447], *marasmus ca.

Aflatoxin L
incidence: 15/161, sa. const.: people of Nigeria, contamination: natural, Ø conc.: 0.38 ng/100 ml (mean value), country: USA/Nigeria[596]

Aflatoxin M_1
incidence: 67/317*, sa. const.: people of the People's Republic of China (males), age: mostly 45–64 years, contamination: natural, conc. range: 0.17–5.2 ng/ml, country: People's Republic of China/USA[12], *thereof 50 HCC ca. (contaminated: 49 control and 18 HCC ca., for detailed information please see the article)

incidence: 2/15*, sa. const.: people of The Sudan (children), contamination: natural, conc. range: 28–484 pg/ml, Ø conc.: 256 pg/ml, country: The Sudan/UK[32], *kwashiorkor ca.
incidence: 2/10*, sa. const.: people of The Sudan (children), contamination: natural, conc. range: 487–1,075 pg/ml, Ø conc.: 781 pg/ml, country: The Sudan/UK[32], *1 marasmic kwashiorkor and 9 marasmus ca. thereof 2 AFM_1-pos.

incidence: ?/252, sa. const.: people of the People's Republic of China (males and females), contamination: natural, conc. range: ≤3.2 ng/ml and 0.4–4.8 µg/day, country: People's Republic of China/USA[41]

incidence: 58/72*, sa. const.: people of the People's Republic of China, contamination: natural, conc. range: ≤144.8 pg/mg of creatinine, median level: 9.3 pg/mg of creatinine, country: People's Republic of China/USA[48], *receiving a placebo for 4 weeks
incidence: 47/57*, sa. const.: people of the People's Republic of China, contamination: natural, conc. range: ≤70.3 pg/mg of creatinine, median level: 7.1 pg/mg of creatinine, country: People's Republic of China/USA[48], *receiving 125 mg oltipraz daily for 4 weeks
incidence: 49/60*, sa. const.: people of the People's Republic of China, contamination: natural, conc. range: ≤25.3 pg/mg of creatinine, median level: 4.6 pg/mg of creatinine, country: People's Republic of China/USA[48], *receiving 500 mg oltipraz weekly for 4 weeks

incidence: 4/108*, sa. const.: people of the USA, contamination: natural, conc. range: 50–170 ppt, Ø conc.: 97.5 ppt, country: USA[51], *17 Reye's syndrome ca.

incidence: 88/138, sa. const.: people of the People's Republic of China (males), age: 35–64 years, contamination: natural, conc. range: ≤108 ng/12 h, Ø conc.: 3.2 ng/12 h, country: Taiwan, Republic of China/People's Republic of China/USA[80]
incidence: 21/32, sa. const.: people of Taiwan (males), age: 35–64 years, contamination: natural, conc. range: ≤17 ng/12 h, Ø conc.: 2.7 ng/12 h, country: Taiwan, Republic of China/People's Republic of China/USA[80]

incidence: 2/5*, sa. const.: people of Kenya (children), contamination: natural, conc. range: 339–580 ng/day**, country: Kenya/UK[91], *kwashiorkor ca., **total 24 h urine production calculated, for detailed information please see the article
incidence: 1/7*, sa. const.: people of Kenya (children), contamination: natural, conc.: 7,775 ng after 9 days**, country: Kenya/UK[91], *marasmic kwashiorkor ca., **total excretion after 9 days, for detailed information please see the article

incidence: 99/99*, sa. const.: people of the UK (male volunteers), age: 17–40 years, no contamination with AFM_2, conc.: nd, country: UK[117], *control
incidence: 12*/133, sa. const.: people of The Netherlands and UK (heroin abusers), contamination: natural, conc. range: 0.12–29.09 nmol/l, country: UK[117], *pos. sa. from people of The Netherlands and the UK

incidence: 1/12, sa. const.: people of The Sudan (children), contamination: natural, conc.: 313 pg/ml, country: UK/The Sudan[134]

incidence: 1,007/1,228, sa. const.: people of Zimbabwe, contamination: natural, Ø conc.: 4.2 ng/ml*, country: Zimbabwe[152], *national average

incidence: 78/145, sa. const.: people of the People's Republic of China (males), Ø age: 39.2 years, contamination: natural, conc. range: >3.6–243 ng/l, country: People's Republic of China/USA[223]

incidence: 1/20, sa. const.: people of Egypt (females), contamination: natural, conc.: 3.13 ng/ml, country: Egypt[352]

incidence: 24/27, sa const.: people of the People's Republic of China (males and females), contamination: natural, conc. range: 0.9–1,258.0 ng/24-h, Ø conc.: 192.3 ng/24-h, country: People's Republic of China/USA[388]

incidence: 56/134, sa. const.: people of Sierra Leone (males), age: 5–14 years, contamination: natural, conc. range: 0.5–374 ng/ml*, country: Sierra Leone[401], *in dry season
incidence: 42/97, sa. const.: people of Sierra Leone (males), age: 5–14 years, contamination: natural, conc. range: 0.1–35 ng/ml*, country: Sierra Leone[401], *in rainy season
incidence: 48/110, sa. const.: people of Sierra Leone (females), age: 5–14 years, contamination: natural, conc. range: 2.3–34 ng/ml*, country: Sierra Leone[401], *in dry season
incidence: 55/93, sa. const.: people of Sierra Leone (females), age: 5–14 years, contamination: natural, conc. range: 0.3–124 ng/ml*, country: Sierra Leone[401], *in rainy season

incidence: 4/30*, sa. const.: people of Egypt (19 male and 11 female infants), age: 7–20 months, contamination: natural, conc. range: 1–3 ng/100 ml, Ø conc.: 2.25 ng/100 ml, country: Egypt[447], *kwashiorkor ca.
incidence: 2/30*, sa. const.: people of Egypt (16 male and 14 female infants), age: 6–13 months, contamination: natural, conc. range: 4–7 ng/100 ml, Ø conc.: 5.50 ng/100 ml, country: Egypt[447], *marasmus ca.

incidence: ?/40*, sa. const.: people of the People's Republic of China, age: 20–55 years, contamination: natural, conc. range: 0.24–1,276.25 pg/mg creatinine**, country: People's Republic of China[451], *control (receiving placebos for 3 months), **for detailed information please see the article
incidence: ?/40*, sa. const.: people of the People's Republic of China, age: 20–55 years, contamination: natural, conc. range: 0.18–746.10 pg/mg creatinine**, country: People's Republic of China[451], *receiving GTP 500 mg for 3 months, **for detailed information please see the article
incidence: ?/40*, sa. const.: people of the People's Republic of China, age: 20–55 years, contamination: natural, conc. range: 0.12–338.85 pg/mg creatinine**, country: People's Republic of China[451], *receiving GTP 1,000 mg for 3 months, **for detailed information please see the article

incidence: 6/12*, sa. const.: people of Egypt (6 males and 6 females), contamination: natural, Ø conc.: 0.98 ng/ml, country: Egypt[455], *control patients
incidence: 19/46*, sa. const.: 16 women, 30 men, Ø age: 56 years, contamination: natural, Ø conc.: 3.82 ng/ml, country: Egypt[455], *HCC patients

incidence: 11/12*, sa. const.: people of Egypt (7 males and 5 females), Ø age: 48 years, contamination: natural, Ø conc.: 43.22 ng/ml, country: Egypt[455], *cirrhotic patients

incidence: 83/91, sa. const.: people of Ghana (males and females), age: 19–86 years?, contamination: natural, conc. range: ≤11,562.36 pg/mg creatinine*, country: USA/Ghana[457]

incidence: 47/49*, sa. const.: people of the People's Republic of China, contamination: natural, conc. range: 0.01–2.09 ppb, country: People's Republic of China[462], *people from high liver cancer incidence area
incidence: 48/50*, sa. const.: people of the People's Republic of China, contamination: natural, conc. range: 0.01–0.37 ppb, country: People's Republic of China[462], *people from low liver cancer incidence area
incidence: 94/96*, sa. const.: people of the People's Republic of China (children), contamination: natural, conc. range: ≤2.09 ppb, country: People's Republic of China[462], *people from high liver cancer incidence area
incidence: 87/96*, sa. const.: people of the People's Republic of China (children), contamination: natural, conc. range: ≤0.61 ppb, country: People's Republic of China[462], *people from low liver cancer incidence area

incidence: 1/7, sa. const.: people of Kenya (patients: 4 males and 3 females (1 af*)), age: 10–25 years?, contamination: natural, conc.: 139 pg/ml, country: Kenya/UK[466], *additionally cirrhosis

incidence: 4/50, sa. const.: people of Egypt (children), age: 1–2.5 years, contamination: natural, conc. range: 5.0–6.2 pg/ml, Ø conc.: 5.5 pg/ml, country: Finland/UK/Egypt/Republic of Guinea[481]
incidence: 32/50, sa. const.: people of the Republic of Guinea (children), age: 2–4 years, contamination: natural, conc. range: 8.0–801 pg/ml, Ø conc.: 97.0 pg/ml, country: Finland/UK/Egypt/Guinea[481]

incidence: 12/30, sa. const.: people of Sierra Leone (male children), contamination: natural, conc. range: 0.5–44.3 ng/ml, country: Sierra Leone[540]
incidence: 5/24, sa. const.: people of Sierra Leone (female children), contamination: natural, conc. range: 1.3–16.4 ng/ml, country: Sierra Leone[540]

incidence: 14/161, sa. const.: people of Nigeria, contamination: natural, Ø conc.: 0.69 ng/100 ml (mean value), country: USA/Nigeria[596]

incidence: 42/42, sa. const.: people of the People's Republic of China (30 males and 12 females), age: 25–64 years, contamination: natural, conc. range: ≈≤3.25 µg/3 days, country: USA/People's Republic of China[607]

Aflatoxin M_2

incidence: 1/15*, sa. const.: people of The Sudan (children), contamination: natural, conc.: 156 pg/ml, country: The Sudan/UK[32], *kwashiorkor ca.

incidence: 1/5*, sa. const.: people of Kenya (children), contamination: natural, conc.: 108 ng after 9 days**, country: Kenya/UK[91], *kwashiorkor ca., **total excretion after 9 days, for detailed information please see the article
incidence: 2/7*, sa. const.: people of Kenya (children), contamination: natural, conc. range: 239–4,710 ng after 9 days**, country: Kenya/UK[91], *marasmic kwashiorkor ca., **total 24 h urine production calculated (cumulative extraction after 9 days), for detailed information please see the article

incidence: 99/99*, sa. const.: people of the UK (male volunteers), age: 17–40 years, no contamination with AFM_2, conc.: nd, country: UK[117], *control
incidence: 6*/133, sa. const.: people of The Netherlands and UK (heroin abusers), contamination: natural, conc. range: 0.4–1.88 nmol/l, country: UK[117], *pos. sa. from people of The Netherlands and the UK

incidence: 71/134, sa. const.: people of Sierra Leone (males), age: 5–14 years, contamination: natural, conc. range: 4.5–130 ng/ml*, country: Sierra Leone[401], *in dry season
incidence: 62/97, sa. const.: people of Sierra Leone (males), age: 5–14 years, contamination: natural, conc. range: 1.3–41.3 ng/ml*, country: Sierra Leone[401], *in rainy season
incidence: 48/110, sa. const.: people of Sierra Leone (females), age: 5–14 years, contamination: natural, conc. range: 4.5–94 ng/ml*, country: Sierra Leone[401], *in dry season
incidence: 66/93, sa. const.: people of Sierra Leone (females), age: 5–14 years, contamination: natural, conc. range: 5.1–86 ng/ml*, country: Sierra Leone[401], *in rainy season

incidence: 2/7, sa. const.: people of Kenya (patients: 4 males (2 af*) and 3 females), age: 10–25 years?, contamination: natural, conc. range: 36–241 pg/ml, Ø conc.: 138.5 pg/ml, country: Kenya/UK[466], *additionally cirrhosis or HCC

incidence: 19/30, sa. const.: people of Sierra Leone (male children), contamination: natural, conc. range: 1.2–26.0 ng/ml, country: Sierra Leone[540]
incidence: 16/24, sa. const.: people of Sierra Leone (female children), contamination: natural, conc. range: 0.5–32.0 ng/ml, country: Sierra Leone[540]

Aflatoxin P_1
incidence: 53/317*, sa. const.: people of the People's Republic of China (males), age: mostly 45–64 years, contamination: natural, conc. range: 0.59–16.0 ng/ml, country: People's Republic of China/USA[12], *thereof 50 HCC ca. (contaminated: 39 control and 14 HCC ca., for detailed information please see the article)

incidence: 8/27, sa. const.: people of the People's Republic of China (males and females), contamination: natural, conc. range: 80.4–3,569.7 ng/24-h, Ø conc.: 664.9 ng/24-h, country: People's Republic of China/USA[388]

Aflatoxin P
incidence: 30/30*, people of Egypt (19 male and 11 female infants), age: 7–20 months, no contamination with AFP, conc.: nd, country: Egypt[447], *kwashiorkor ca.
incidence: 1/30*, sa. const.: people of Egypt (16 male and 14 female infants), age: 6–13 months, contamination: natural, conc.: 2 ng/100 ml, country: Egypt[447], *marasmus ca.

Aflatoxin Q_1
incidence: 7/27, sa. const.: people of the People's Republic of China (males and females), contamination: natural, conc. range: 77.3–137.5 ng/24-h, Ø conc.: 92.2 ng/24-h, country: People's Republic of China/USA[388]

Aflatoxin
incidence: ?/people from 48 counties?, sa. const.: people of the People's Republic of China (males), contamination: natural, conc. range: 0–611 ng/kg/4 h, country: USA/People's Republic of China[7]

incidence: 42?/42, sa. const.: people of the People's Republic of China (30 males and 12 females), age: 25–64 years, 3.3–6.6 µg/day* (mean value), country: USA/People's Republic of China[25], *total AF-metabolites
incidence: ?/12, sa. const.: people of the People's Republic of China (females), Ø age: 44.8 years, contamination: natural, conc.: 13.09 µg*/collection period** (mean value), country: USA/People's Republic of China[25], *total AF-metabolites, **3 days
incidence: ?/30, sa. const.: people of the People's Republic of China (males), Ø age: 46.6 years, contamination: natural, conc.: 12.98 µg*/collection period** (mean value), country: USA/People's Republic of China[25], *total AF metabolites, **3 days

incidence: ?/20*, sa. const.: people of The Gambia (10 males and 10 females), age: 15–56 years, contamination: natural, conc. range: 48.2–7,099.2 ng/day**, country: USA/France/UK[50], *thereof 10 HBV ca. but no predominant contamination, **AF-N^7-Gua

incidence: 6/8*, sa. const.: people of Kenya (children), contamination: natural, conc. range: 3–533 pg/ml, country: Kenya/UK[95], *control
incidence: 5/11*, sa. const.: people of Kenya (children), contamination: natural, conc. range: 6–986 pg/ml, country: Kenya/UK[95], *marasmus ca.
incidence: 3/5*, sa. const.: people of Kenya (children), contamination: natural, conc. range: 60–4,425 pg/ml, country: Kenya/UK[95], *marasmic kwashiorkor ca.
incidence: 5/12*, sa. const.: people of Kenya (children), contamination: natural, conc. range: 40–1,370 pg/ml, country: Kenya/UK[95], *kwashiorkor ca.

incidence: ?/112*, sa. const.: people of Taiwan (males), contamination: natural, conc. range: 27.0–107.7 pg/ml (mean values), country: USA/Taiwan, Republic of China[104], *partly HBV and/or HCC
incidence: ?/134*, sa. const.: people of Taiwan (females), contamination: natural, conc. range: 20.3–61.9 pg/ml (mean values), country: USA/Taiwan, Republic of China[104], *partly HBV and/or HCC

incidence: 64/65, sa. const.: people of The Philippines (children), age: 0.08–12 years, weight for height: 6.6–23.1 kg/m, contamination: natural, conc. range: 0.1–4.77 ng/ml*, country: UK/The Philippines[460], *AF-metabolites

incidence: 53/87*, sa. const.: people of the People's Republic of China (males and females), age: 25–65 years, contamination: natural, conc. range: <LOD–4.10 pg adduct/mg creatinine**, country: USA/People's Republic of China[495], *ingesting placebos for 4 months (for detailed information please see the article), **AF-N^7-Gua
incidence: 52/82*, sa. const.: people of the People's Republic of China (males and females), age: 25–65 years, contamination: natural, conc. range: <LOD–0.98 pg adduct/mg creatinine**, country: USA/People's Republic of China[495], *ingesting 100 mg chlorophyllin three times daily for 4 months (for detailed information please see the article), **AF-N^7-Gua

incidence: 153/2,553, sa. const.: people of Zimbabwe (outpatients, school children, farm laborers, mine workers, and middle class donors), contamination: natural, Ø conc.: 5.1 ng/ml (mean value), country: Zimbabwe[569]

Aflatoxins

incidence: 21/106*, sa. const.: people of The Sudan (male and female children), contamination: natural, conc.: 191 pg/ml (geometric mean), country: UK/The Sudan[36], *control
incidence: 18/70*, sa. const.: people of The Sudan (male and female children), contamination: natural, conc.: 508 pg/ml (geometric mean), country: UK/The Sudan[36], *marasmus ca.
incidence: 8/32*, sa. const.: people of The Sudan (male and female children), contamination: natural, conc.: 742 pg/ml (geometric mean), country: UK/The Sudan[36], *marasmic kwashiorkor ca.
incidence: 14/42*, sa. const.: people of The Sudan (male and female children), contamination: natural, conc.: 143 pg/ml (geometric mean), country: UK/The Sudan[36], *kwashiorkor ca.

incidence: 20?/20, sa. const.: people of the People's Republic of China, contamination: natural, conc. range: 0.1–10 ng/ml*, country: USA/People's Republic of China[59], *mainly AFM_1, AFB_1-N^7-Gua[2], AFP_1, and AFB_1

incidence: 44/155*, sa. const.: people of The Sudan (children), contamination: natural, conc. range: 1–9 pg/ml** (1 sa), 10–99 pg/ml** (14 sa), 100–999 pg/ml** (16 sa), ≥1,000 pg/ml** (13 sa), country: UK/The Sudan[67], *control, **includes AFB_1, AFM_1, AFL

incidence: 31/114*, sa. const.: people of The Sudan (children), contamination: natural, conc. range: 1–9 pg/ml** (1 sa), 10–99 pg/ml** (10 sa), 100–999 pg/ml** (11 sa), ≥1,000 pg/ml** (9 sa), country: UK/The Sudan[67], *kwashiorkor ca., **includes AFB_1, AFM_1, AFL
incidence: 30/77*, sa. const.: people of The Sudan (children), contamination: natural, conc. range: 10–99 pg/ml** (9 sa), 100–999 pg/ml** (8 sa), ≥1,000 pg/ml** (13 sa), country: UK/The Sudan[67], *marasmic kwashiorkor ca., **includes AFB_1, AFM_1, AFL
incidence: 31/119*, sa. const.: people of The Sudan (children), contamination: natural, conc. range: 10–99 pg/ml** (7 sa), 100–999 pg/ml** (13 sa), ≥1,000 pg/ml** (11 sa), country: UK/The Sudan[67], *marasmus ca., **includes AFB_1, AFM_1, AFL

Deoxynivalenol
incidence: 15/15, sa. const.: people of the People's Republic of China (females), age: 19–75 years, contamination: natural, conc. range: 4–94 ng/ml, Ø conc.: 30.33 ng/ml, country: UK/Canada/People's Republic of China/USA[415]
incidence: 296/300, sa. const.: people of the UK, contamination: natural, conc. range: ≤65.97 μg/day, country: UK[417]
incidence: 25/25*, sa. const.: people of the UK (9 males and 16 females), age: 21–59 years, height: 1.52–1.96 m, wt.: 51–93 kg, BMI: 19.2–32.0, contamination: natural, conc. range: 4.9–10.5 ng/mg creatinine, country: UK[442], *wheat-based food intake (normal diet), for detailed information please see the article
incidence: 9/25*, sa. const.: people of the UK (9 males and 16 females), age: 21–59 years, height: 1.52–1.96 m, wt.: 51–93 kg, BMI: 19.2–32.0, contamination: natural, conc. range: 0.4–0.9 ng/mg creatinine, country: UK[442], *wheat reduction intervention, for detailed information please see the article

Fumonisin B_1
incidence: ?/75, sa. const.: people of Mexico (females), contamination: natural, conc. range: 18.8–248.0 pg/ml, country: UK/Mexico/Germany[510], for detailed information please see article

Ochratoxin A
incidence: 1/25, sa. const.: people of Egypt (male and female humans: healthy), age: 21–49 years, contamination: natural, conc.: 0.31 ng/ml, country: Egypt/France[227], for detailed information please see the article
incidence: 16/67*, sa. const.: people of Egypt (males and females), age: 5–70 years, contamination: natural, conc. range: 0.43–8.19 ng/ml, country: Egypt/France[227], *renal disease ca., for detailed information please see the article
incidence: 2/14*, sa. const.: people of Egypt (males and females), age: 22–50 years, contamination: natural, conc. range: 0.22–3.42 ng/ml, Ø conc.: 1.82 ng/ml, country: Egypt/France[227], *potential kidney donors, for detailed information please see the article
incidence: 46/50*, sa. const.: people of the UK, contamination: natural, conc. range: <0.01–0.058 ng/ml, country: UK[236], *32 volunteers (normal diet?), 11 vegetarians, 7 consumed ethnic diet
incidence: 2/2, sa. const.: people of France (male and female), contamination: natural, conc. range: 367–1,801 ng/ml, country: France[311]
incidence: 29/134, sa. const.: people of Sierra Leone (males), age: 5–14 years, contamination: natural, conc. range: 0.07–59 ng/ml*, country: Sierra Leone[401], *in dry season
incidence: 26/97, sa. const.: people of Sierra Leone (males), age: 5–14 years, contamination: natural, conc. range: 0.6–72.2 ng/ml*, country: Sierra Leone[401], *in rainy season
incidence: 34/110, sa. const.: people of Sierra Leone (females), age: 5–14 years, contamination: natural, conc. range:

0.08–148 ng/ml*, country: Sierra Leone[401], *in dry season
incidence: 21/93, sa. const.: people of Sierra Leone (females), age: 5–14 years, contamination: natural, conc. range: 0.7–4.9 ng/ml*, country: Sierra Leone[401], *in rainy season

incidence: 13/60, sa. const.: people of Portugal (inhabitants of Coimbra), age: 15–67 years, contamination: natural, conc. range: 0.011–0.208 ng/ml*, country: Portugal/Spain[426], *morning sa. (for detailed information please see the article)
incidence: 14/60, sa. const.: people of Portugal (inhabitants of Coimbra), age: 15–67 years, contamination: natural, conc. range: 0.008–0.011 ng/ml*, country: Portugal/Spain[426], *afternoon sa. (for detailed information please see the article)
incidence: 25/62, sa. const.: people of Spain (inhabitants of Valencia), age: 18–53 years, contamination: natural, conc. range: 0.007–0.124 ng/ml*, country: Portugal/Spain[426], *morning sa. (for detailed information please see the article)
incidence: 26/62, sa. const.: people of Spain (inhabitants of Valencia), age: 18–53 years, contamination: natural, conc. range: 0.008–0.089 ng/ml*, country: Portugal/Spain[426], *afternoon sa. (for detailed information please see the article)

incidence: 122/155, sa. const.: people of Portugal, contamination: natural, conc. range: 0.008–0.069 ng/ml, Ø conc.: 0.022 ng/ml, country: Portugal[456]

incidence: 50/152*, sa. const.: people of Bulgaria, contamination: natural, conc. range: 5–604 ng/l, country: France/India/Bulgaria[464], *control families and BEN/UTT affected people (for detailed information please see the article)

incidence: 54/88, sa. const.: people of Hungary (healthy humans: 46 males (29 af) and 42 females (25 af)), age: 8–80 years, contamination: natural, conc. range: 0.006–0.065 ng/ml, Ø conc.: 0.013 ng/ml, country: Hungary[472] (for detailed information please see the article)

incidence: 42/60, sa. const.: people of Portugal (34 males and 26 females), age: 19–82 years, contamination: natural, conc. range: 0.021–0.105 ng/ml, country: Portugal/Czech Republic[505]

incidence: 8/24*, sa. const.: people of the UK (males and females), age: 18–55 years, contamination: natural, conc. range: LOD/LOQ–0.058 µg/l, country: EU[508], *eating normal diet
incidence: 4/7*, sa. const.: people of the UK (males and females), age: 18–55 years, contamination: natural, conc. range: LOD/LOQ–0.023 µg/l, country: EU[508], *eating ethnic diet
incidence: 2/9*, sa. const.: people of the UK (males and females), age: 18–55 years, contamination: natural, conc. range: LOD/LOQ–0.054 µg/l, country: EU[508], *vegetarians

incidence: 22/38*, sa. const.: people of Italy, contamination: natural, conc. range: 0.012–0.046 ng/ml, country: Italy[513], *healthy individuals
incidence: 3/3*, sa. const.: people of Italy, contamination: natural, conc. range: ≤1.40 ng/ml, country: Italy[513], *KIN patients

incidence: 5/5, sa. const.: people of Bulgaria (volunteer inhabitants of Gorno Peshtene, healthy), age: 20–30 years, contamination: natural, conc. range: 16–98 ng/l*, Ø conc.: 50.8 ng/l*, country: Bulgaria/France[538], *measured for 4 weeks (for detailed information please see the article)
incidence: 11/11, sa. const.: people of Bulgaria (volunteer inhabitants of Beli Izvor, healthy), age: 20–30 years, contamination: natural, conc. range: 36–860 ng/l*, Ø conc.: 168.64 ng/l*, country: Bulgaria/France[538], *measured for 4 weeks (for detailed information please see the article)

incidence: 10/30, sa. const.: people of Sierra Leone (male children),

contamination: natural, conc. range: 0.3–26.6 ng/ml, country: Sierra Leone[540]
incidence: 3/24, sa. const.: people of Sierra Leone (female children), contamination: natural, conc. range: 0.7–16.0 ng/ml, country: Sierra Leone[540]

incidence: 16/22, sa. const.: people of Portugal (male inhabitants of Lisbon), contamination: natural, conc. range: nd-0.071 ng/ml, country: Portugal[580]
incidence: 15/21, sa. const.: people of Portugal (female inhabitants of Lisbon), contamination: natural, conc. range: nd-0.055 ng/ml, country: Portugal[580]

4-Hydroxyochratoxin A
incidence: 50/134, sa. const.: people of Sierra Leone (males), age: 5–14 years, contamination: natural, conc. range: 0.1–29 ng/ml*, country: Sierra Leone[401], *in dry season
incidence: 28/97, sa. const.: people of Sierra Leone (males), age: 5–14 years, contamination: natural, conc. range: 0.2–37 ng/ml*, country: Sierra Leone[401], *in rainy season
incidence: 41/110, sa. const.: people of Sierra Leone (females), age: 5–14 years, contamination: natural, conc. range: 0.1–1.47 ng/ml*, country: Sierra Leone[401], *in dry season
incidence: 48/93, sa. const.: people of Sierra Leone (females), age: 5–14 years, contamination: natural, conc. range: 0.2–33 ng/ml*, country: Sierra Leone[401], *in rainy season

incidence: 12/30, sa. const.: people of Sierra Leone (male children), contamination: natural, conc. range: 0.04–21.0 ng/ml, country: Sierra Leone[540]
incidence: 12/24, sa. const.: people of Sierra Leone (female children), contamination: natural, conc. range: 0.1–18.0 ng/ml, country: Sierra Leone[540]

Ochratoxin B
incidence: 64/134, sa. const.: people of Sierra Leone (males), age: 5–14 years, contamination: natural, conc. range: 0.4–218 ng/ml*, country: Sierra Leone[401], *in dry season
incidence: 31/97, sa. const.: people of Sierra Leone (males), age: 5–14 years, contamination: natural, conc. range: 0.05–45 ng/ml*, country: Sierra Leone[401], *in rainy season
incidence: 51/110 sa. const.: people of Sierra Leone (females), age: 5–14 years, contamination: natural, conc. range: 0.6–124 ng/ml*, country: Sierra Leone[401], *in dry season
incidence: 41/93, sa. const.: people of Sierra Leone (females), age: 5–14 years, contamination: natural, conc. range: 0.06–81 ng/ml*, country: Sierra Leone[401], *in rainy season

incidence: 7/30, sa. const.: people of Sierra Leone (male children), contamination: natural, conc. range: 0.4–37.1 ng/ml, country: Sierra Leone[540]
incidence: 4/24, sa. const.: people of Sierra Leone (female children), contamination: natural, conc. range: 2.0–33.3 ng/ml, country: Sierra Leone[540]

Human Artificial Contamination

Human blood may contain the following mycotoxins and/or their metabolites:

Diacetoxyscirpenol
incidence: 1/44*, sa. const.: people of Kampuchea (Cambodia)/Laos, contamination: artificial**, conc.: 5 ppb, country: USA[288], *human victims of chemical warfare, **strong indication

incidence: 4/7, sa. const.: people of Iran (soldiers), contamination: artificial*, conc. range: 0.38–0.70 ppm, Ø conc.: 0. 54 ppm, country: Belgium[498], *very strong indication

HT-2 Toxin
incidence: 2/2*, sa. const.: people of Kampuchea (Cambodia), contamination: artificial, conc. range: 10–22 ppb, Ø conc.: 16 ppb, country: USA[194], *Yellow Rain victims

incidence: 2/4*, sa. const.: people of Kampuchea (Cambodia)/Laos, contamination: artificial**, conc. range: tr, country: USA[288], *victims exposed to biological warfare chemicals, **strong indication

incidence: 11/44*, sa. const.: people of Kampuchea (Cambodia)/Laos, contamination: artificial**, conc. range: 2–296.1 ppb, Ø conc.: 50.36 ppb, country: USA[288], *human victims of chemical warfare, **strong indication

incidence: 2/2*, sa. const.: people of Laos (1 male and 1 female), contamination: artificial**, conc. range: 19.2–21.1 ppb, Ø conc.: 20.15 ppb, country: USA[497], *victims of chemical attack, **strong indication

Nivalenol
incidence: 3/7, sa. const.: people of Iran (soldiers), contamination: artificial*, conc. range: 0.11–0.15 ppm, Ø conc.: 0.136 ppm, country: Belgium[498], *very strong indication

T-2 Toxin
incidence: 2/2*, sa. const.: people of Kampuchea (Cambodia), contamination: artificial, conc. range: 11–18 ppb, Ø conc.: 14.5 ppb, country: USA[194], *Yellow Rain victims

incidence: 17/44*, sa. const.: people of Kampuchea (Cambodia)/Laos, contamination: artificial**, conc. range: 1.4–110.4 ppb, Ø conc.: 34.04 ppb, country: USA[288], *human victims of chemical warfare, **very strong indication

incidence: 2/2*, sa. const.: people of Laos (1 male and 1 female), contamination: artificial*, conc. range: 14.5–66.95 ppb, Ø conc.: 40.725 ppb, country: USA[497], *victims of chemical attack, **strong indication

incidence: 1/7, sa. const.: people of Iran (soldiers), contamination: artificial*, conc.: 0.41 ppm, country: Belgium[498], *very strong indication

Verrucarol
incidence: 3/7, sa. const.: people of Iran (soldiers), contamination: artificial*, conc. range: 0.07–0.33 ppm, Ø conc.: 0.23 ppm, country: Belgium[498], *very strong indication

Human esophagus may contain the following mycotoxins and/or their metabolites:

HT-2 Toxin
incidence: 1/1*, sa. const.: person of Kampuchea (Cambodia)/Laos, contamination: artificial**, conc.: 4.02 ppm, country: USA[288], *victim of chemical attack, **strong indication

T-2 Toxin
incidence: 1/1*, sa. const.: person of Kampuchea (Cambodia)/Laos, contamination: artificial**, conc.: 25.1 ppb, country: USA[288], *victim of chemical attack, **strong indication

Human feces may contain the following mycotoxins and/or their metabolites:

Verrucarol
incidence: 1/1, sa. const.: person of Iran (soldier), contamination: artificial*, conc.: 0.30 ppm, country: Belgium[498], *very strong indication

Human heart may contain the following mycotoxins and/or their metabolites:

HT-2 Toxin
incidence: 1/1*, sa. const.: person of Kampuchea (Cambodia), contamination: artificial**, conc.: 1.2 ppm, country: USA[194], *victim of chemical attack, **strong indication

Human intestine may contain the following mycotoxins and/or their metabolites:

HT-2 Toxin
incidence: 1/1* **, sa. const.: person of Kampuchea (Cambodia), contamination: artificial***, conc.: 9.6 ppb, country: USA[194], *large intestine, **victim of chemical attack, ***strong indication

T-2 Toxin
incidence: 1/1* **, sa. const.: person of Kampuchea (Cambodia), contamination: artificial***, conc.: 88.0 ppb, country: USA[194], *large intestine, **victim of chemical attack, ***strong indication

Human kidney may contain the following mycotoxins and/or their metabolites:

Diacetoxyscirpenol
incidence: 1/1*, sa. const.: person of Kampuchea (Cambodia), contamination: artificial**, conc.: 2.55 ppm, country: USA[194], *victim of chemical attack, **strong indication

T-2 Toxin
incidence: 1/1*, sa. const.: person of Kampuchea (Cambodia), contamination: artificial**, conc.: 6.8 ppb, country: USA[194], *victim of chemical attack, **strong indication

Human lung may contain the following mycotoxins and/or their metabolites:

T-2 Toxin
incidence: 1/1*, sa. const.: sa. const.: person of Kampuchea (Cambodia), contamination: artificial**, conc.: 8.5 ppb, country: USA[194], *victim of chemical attack, **strong indication

Human stomach may contain the following mycotoxins and/or their metabolites:

HT-2 Toxin
incidence: 1/1*, sa. const.: person of Kampuchea (Cambodia), contamination: artificial**, conc.: 4.02 ppm, country: USA[194], *victim of chemical attack, **strong indication

T-2 Toxin
incidence: 1/1*, sa. const.: person of Kampuchea (Cambodia), contamination: artificial**, conc.: 25.1 ppb, country: USA[194], *victim of chemical attack, **strong indication

Human urine may contain the following mycotoxins and/or their metabolites:

Diacetoxyscirpenol
incidence: 1/6, sa. const.: people of Iran (soldiers), contamination: artificial*, conc.: 0.22 ppm, country: Belgium[498], *very strong indication

HT-2 Toxin
incidence: 1/2*, sa. const.: people of Kampuchea (Cambodia), contamination: artificial, conc.: 18 ppb, country: USA[194], *Yellow Rain victims

incidence: 3/3*, sa. const.: people of Kampuchea (Cambodia)/Laos, contamination: artificial**, conc. range: 1.3–7.4 ppb, Ø conc.: 3.5 ppb, country: USA[288], *victims of chemical warfare, **strong indication

incidence: 1/1*, sa. const.: person of Kampuchea (Cambodia) (male), contamination: artificial, conc.: 18 ppb, country: USA[497], *victim of chemical attack

incidence: 2/6, sa. const.: people of Iran (soldiers), contamination: artificial*, conc. range: 0.07–0.13 ppm, Ø conc.: 0.10 ppm, country: Belgium[498], *very strong indication

Nivalenol
incidence: 2/6, sa. const.: people of Iran (soldiers), contamination: artificial*, conc. range: 0.09–0.16 ppm, Ø conc.: 0.125 ppm, country: Belgium[498], *very strong indication

T-2 Toxin
incidence: 1/2*, sa. const.: people of Kampuchea (Cambodia), contamination: artificial, conc.: tr, country: USA[194], *Yellow Rain victims

incidence: 3/3*, sa. const.: people of Kampuchea (Cambodia)/Laos, contamination: artificial**, conc. range: 4.0–22.0 ppb, Ø conc.: 10.33 ppb, country: USA[288], *victims of chemical warfare, **strong indication

incidence: 1/1*, sa. const.: person of Kampuchea (Cambodia) (male), contamination: artificial, conc.: tr, country: USA[497], *victim of chemical attack

incidence: 1/6, sa. const.: people of Iran (soldiers), contamination: artificial*, conc.: 0.18 ppm, country: Belgium[498], *very strong indication

Verrucarol
incidence: 3/6, sa. const.: people of Iran (soldiers), contamination: artificial*, conc. range: 0.07–0.14 ppm, Ø conc.: 0.103 ppm, country: Belgium[498], *very strong indication

Barrow see Pig

Beef

Beef Natural Contamination see also Steer

Beef liver may contain the following mycotoxins and/or their metabolites:

Aflatoxin B_1
incidence: ?/?, sa. const.: livers from beefs of the USA?, contamination: probably natural, conc.: 0.72 ng/g, country: USA[75]

Aflatoxin G_1
incidence: ?/?, sa. const.: livers from beefs of the USA?, contamination: probably natural, conc.: 0.21 ng/g, country: USA[75]

Aflatoxin M_1
incidence: ?/?, sa. const.: livers from beefs of the USA?, contamination: probably natural, conc.: 0.25 ng/g, country: USA[75]

Boar see Pig

Bovine see Cattle, Cow or Steer

Bovine Milk see Cow milk

Bovine Udder see Cattle udder

Broiler see Chicken

Broiler Chicken see Chicken

Buffalo

Buffalo Natural Contamination

Buffalo milk, raw may contain the following mycotoxins and/or their metabolites:

Aflatoxin M_1
incidence: 59/66, sa. const.: milk from buffaloes of India, contamination: natural, conc. range: 0.074*–0.076** µg/l (mean values), country: India[43], *bulk, **individual

incidence: 60/207, sa. const.: milk from buffaloes of Italy, contamination: natural, conc. range: 4–676 ng/kg, country: Italy[161]

incidence: 108/116*, sa. const.: milk from buffaloes of India, contamination: natural, conc. range: 0.6–15 ng/ml (59 sa), 16–30 ng/ml (37 sa), 31–45 ng/ml (11 sa), 48 ng/ml (1 sa), country: India/UK[257], *periurban area
incidence: 2/100*, sa. const.: milk from buffaloes of India, contamination: natural, conc. range: 0.6–15 ng/ml (2 sa), country: India/UK[257], *rural area

incidence: 65/120*, sa. const.: milk from buffaloes of Pakistan, contamination: natural, conc. range: ?, country: Pakistan[277], *urban area
incidence: 51/120*, sa. const.: milk from buffaloes of Pakistan, contamination: natural, conc. range: ?, country: Pakistan[277], *semi-urban area
incidence: 37/120*, sa. const.: milk from buffaloes of Pakistan, contamination: natural, conc. range: ?, country: Pakistan[277], *rural area

Buffalo Artificial Contamination

Buffalo milk, raw may contain the following mycotoxins and/or their metabolites:

AFLATOXIN M_1
incidence: 27/50, sa. const.: milk from Murrah she-buffaloes, contamination: artificial (dose: up to 3 mg AFB_1, o., daily; for detailed information please see the article), conc. range: tr–4.8 μg/l, country: India[272]

Calf

Calf Natural Contamination see also Cattle

Calf liver may contain the following mycotoxins and/or their metabolites:

AFLATOXIN B_1
incidence: 1/3, sa. const.: Hereford calves of Australia, contamination: natural, conc.: 0.5 μg/kg, country: Australia[242]

Calf Artificial Contamination

Calf feces may contain the following mycotoxins and/or their metabolites:

OCHRATOXIN A
incidence: 2/2, sa. const.: Holstein-Friesian male calves, preruminant, age: 19–20 days, wt.: 44 kg, contamination: artificial (dose: **0.25 mg crystalline OTA**/kg b. wt., i.v., once; for detailed information please see the article), conc. range: 1.62* **-4.90** mg, country: Canada[179], *incomplete feces collection, **cumulative excretion
incidence: 2/2, sa. const.: Holstein-Friesian male calves, preruminant, age: 16–21 days, wt.: 60 kg, contamination: artificial (dose: **0.5 mg OTA**/kg b. wt., o., once; for detailed information please see the article), conc. range: 2.35–2.99 mg*, Ø conc.: 2.67 mg*, country: Canada[179], *cumulative excretion
incidence: 4/4, sa. const.: Holstein-Friesian male calves, ruminant, age: 46–69 days, wt.: 68–100 kg, contamination: artificial (dose: **2 mg OTA**/kg b. wt., o., once; for detailed information please see the article), conc. range: ≤3.6 mg*, country: Canada[179], *cumulative excretion

OCHRATOXIN α
incidence: 2/2, sa. const.: Holstein-Friesian male calves, preruminant, age: 19–20 days, wt.: 44 kg, contamination: artificial (dose: **0.25 mg crystalline OTA**/kg b. wt., i.v., once; for detailed information please see the article), conc.: nd*, country: Canada[179], *cumulative excretion
incidence: 2/2, sa. const.: Holstein-Friesian male calves, preruminant, age: 16–21 days, wt.: 60 kg, contamination: artificial (dose: **0.5 mg OTA**/kg b. wt., o., once; for detailed information please see the article), conc.: nd*, country: Canada[179], *cumulative excretion
incidence: 4/4, sa. const.: Holstein-Friesian male calves, ruminant, age: 46–69 days, wt.: 68–100 kg, contamination: artificial (dose: **2 mg OTA**/kg b. wt., o., once; for detailed information please see the article), conc. range: ≤30.0 mg*, country: Canada[179], *cumulative excretion

Calf kidney may contain the following mycotoxins and/or their metabolites:

Aflatoxin M_1
incidence: 6/6*, sa. const.: Friesian male calves, age: ≈12 weeks, wt.: ≈80 kg, contamination: no OTA and/or AFB_1 (for detailed information please see the article), conc.: nd, country: UK/USA[210], *control
incidence: 6/6, sa. const.: Friesian male calves, age: ≈12 weeks, wt.: ≈80 kg, contamination: artificial (dose: **155 mg OTA** (total amount ingested), o., for up to 87 days; for detailed information please see the article), conc.: nd*, country: UK/USA[210], *after 87 days of experimental period
incidence: 4/6, sa. const.: Friesian male calves, age: ≈12 weeks, wt.: ≈80 kg, contamination: artificial (dose: **143 mg OTA + 4.04 mg AFB_1** (total amount ingested), o., for up to 87 days; for detailed information please see the article), conc. range: tr–0.03 μg/kg*, country: UK/USA[210], *after 87 days of experimental period
incidence: 5/6, sa. const.: Friesian male calves, age: ≈12 weeks, wt.: ≈80 kg, contamination: artificial (dose: **3.87 mg AFB_1** (total amount ingested), o., for up to 87 days; for detailed information please see the article), conc. range: tr–0.03 μg/kg*, country: UK/USA[210], *after 87 days of experimental period

Ochratoxin A
incidence: 6/6*, sa. const.: Friesian male calves, age: ≈12 weeks, wt: ≈80 kg, contamination: no OTA and/or AFB_1 (for detailed information please see the article), conc.: nd, country: UK/USA[210], *control
incidence: 3/6, sa. const.: Friesian male calves, age: ≈12 weeks, wt.: ≈80 kg, contamination: artificial (dose: **155 mg OTA** (total amount ingested), o., for up to 87 days; for detailed information please see the article), conc. range: tr–5 μg/kg*, country: UK/USA[210], *after 87 days of experimental period
incidence: 2/6, sa. const.: Friesian male calves, age: ≈12 weeks, wt.: ≈80 kg, contamination: artificial (dose: **143 mg OTA + 4.04 mg AFB_1** (total amount ingested), o., for up to 87 days; for detailed information please see the article), conc. range: tr–5 μg/kg*, country: UK/USA[210], *after 87 days of experimental period
incidence: 6/6, sa. const.: Friesian male calves, age: ≈12 weeks, wt.: ≈80 kg, contamination: artificial (dose: **3.87 mg AFB_1** (total amount ingested), o., for up to 87 days; for detailed information please see the article), conc.: nd*, country: UK/USA[210], *after 87 days of experimental period

Ochratoxin α
incidence: 6/6*, sa. const.: Friesian male calves, age: ≈12 weeks, wt.: ≈80 kg, contamination: no OTA and/or AFB_1 (for detailed information please see the article), conc.: nd, country: UK/USA[210], *control
incidence: 6/6, sa. const.: Friesian male calves, age: ≈12 weeks, wt.: ≈80 kg, contamination: artificial (dose: **155 mg OTA** (total amount ingested), o., for up to 87 days; for detailed information please see the article), conc. range: tr–10 μg/kg*, country: UK/USA[210], *after 87 days of experimental period
incidence: 6/6, sa. const.: Friesian male calves, age: ≈12 weeks, wt.: ≈80 kg, contamination: artificial (dose: **143 mg OTA + 4.04 mg AFB_1** (total amount ingested), o., for up to 87 days; for detailed information please see the article), conc. range: tr–10 μg/kg*, country: UK/USA[210], *after 87 days of experimental period
incidence: 6/6, sa. const.: Friesian male calves, age: ≈12 weeks, wt.: ≈80 kg, contamination: artificial (dose: **3.87 mg AFB_1** (total amount ingested), o., for up to 87 days; for detailed information please see the article), conc.: nd*, country: UK/USA[210], *after 87 days of experimental period

Calf liver may contain the following mycotoxins and/or their metabolites:

Aflatoxin B_1
incidence: 6/6*, sa. const.: Friesian male calves, age: ≈12 weeks, wt.: ≈80 kg, contamination: no OTA and/or AFB_1 (for detailed information please see the article), conc.: nd, country: UK/USA[210], *control
incidence: 6/6, sa. const.: Friesian male calves, age: ≈12 weeks, wt.: ≈80 kg, contamination: artificial (dose: **155 mg OTA** (total amount ingested), o., for up to 87 days; for detailed information please see the article), conc.: nd*, country: UK/USA[210], *after 87 days of experimental period
incidence: 6/6, sa. const.: Friesian male calves, age: ≈12 weeks, wt.: ≈80 kg, contamination: artificial (dose: **143 mg OTA + 4.04 mg AFB_1** (total amount ingested), o., for up to 87 days; for detailed information please see the article), conc.: nd*, country: UK/USA[210], *after 87 days of experimental period
incidence: 1/6, sa. const.: Friesian male calves, age: ≈12 weeks, wt.: ≈80 kg, contamination: artificial (dose: **3.87 mg AFB_1** (total amount ingested), o., for up to 87 days; for detailed information please see the article), conc.: tr*, country: UK/USA[210], *after 87 days of experimental period

Aflatoxin M_1
incidence: 6/6*, sa. const.: Friesian male calves, age: ≈12 weeks, wt.: ≈80 kg, contamination: no OTA and/or AFB_1 (for detailed information please see the article), conc.: nd, country: UK/USA[210], *control
incidence: 6/6, sa. const.: Friesian male calves, age: ≈12 weeks, wt.: ≈80 kg, contamination: artificial (dose: **155 mg OTA** (total amount ingested), o., for up to 87 days; for detailed information please see the article), conc.: nd*, country: UK/USA[210], *after 87 days of experimental period
incidence: 6/6, sa. const.: Friesian male calves, age: ≈12 weeks, wt.: ≈80 kg, contamination: artificial (dose: **143 mg OTA + 4.04 mg AFB_1** (total amount ingested), o., for up to 87 days; for detailed information please see the article), conc.: nd*, country: UK/USA[210], *after 87 days of experimental period
incidence: 1/6, sa. const.: Friesian male calves, age: ≈12 weeks, wt.: ≈80 kg, contamination: artificial (dose: **3.87 mg AFB_1** (total amount ingested), o., for up to 87 days; for detailed information please see the article), conc.: tr*, country: UK/USA[210], *after 87 days of experimental period

Calf muscle may contain the following mycotoxins and/or their metabolites:

Ochratoxin A
incidence: 6/6*, sa. const.: Friesian male calves, age: ≈12 weeks, wt.: ≈80 kg, contamination: no OTA and/or AFB_1 (for detailed information please see the article), conc.: nd, country: UK/USA[210], *control
incidence: 6/6, sa. const.: Friesian male calves, age: ≈12 weeks, wt.: ≈80 kg, contamination: artificial (dose: **155 mg OTA** (total amount ingested), o., for up to 87 days; for detailed information please see the article), conc.: nd*, country: UK/USA[210], *after 87 days of experimental period
incidence: 1/6, sa. const.: Friesian male calves, age: ≈12 weeks, wt.: ≈80 kg, contamination: artificial (dose: **143 mg OTA + 4.04 mg AFB_1** (total amount ingested), o., for up to 87 days; for detailed information please see the article), conc.: tr*, country: UK/USA[210], *after 87 days of experimental period
incidence: 6/6, sa. const.: Friesian male calves, age: ≈12 weeks, wt.: ≈80 kg, contamination: artificial (dose: **3.87 mg AFB_1** (total amount ingested), o., for up to 87 days; for detailed information please see the article), conc.: nd*, country: UK/USA[210], *after 87 days of experimental period

Ochratoxin α
incidence: 6/6*, sa. const.: Friesian male calves, age: ≈12 weeks, wt.: ≈80 kg, contamination: no OTA and/or AFB_1 (for detailed information please see the article), conc.: nd, country: UK/USA[210], *control
incidence: 6/6, sa. const.: Friesian male calves, age: ≈12 weeks, wt.: ≈80 kg, contamination: artificial (dose: **155 mg OTA** (total amount ingested), o., for up to 87 days; for detailed information please see the article), conc.: nd*, country: UK/USA[210], *after 87 days of experimental period
incidence: 1/6, sa. const.: Friesian male calves, age: ≈12 weeks, wt.: ≈80 kg, contamination: artificial (dose: **143 mg OTA + 4.04 mg AFB_1** (total amount ingested), o., for up to 87 days; for detailed information please see the article), conc.: tr*, country: UK/USA[210], *after 87 days of experimental period
incidence: 6/6, sa. const.: Friesian male calves, age: ≈12 weeks, wt.: ≈80 kg, contamination: artificial (dose: **3.87 mg AFB_1** (total amount ingested), o., for up to 87 days; for detailed information please see the article), conc.: nd*, country: UK/USA[210], *after 87 days of experimental period

Calf plasma may contain the following mycotoxins and/or their metabolites:

Ochratoxin α
incidence: 6/6*, sa. const.: Friesian male calves, age: ≈12 weeks, wt.: ≈80 kg, contamination: no OTA and/or AFB_1 (for detailed information please see the article), conc.: nd, country: UK/USA[210], *control
incidence: 5/6, sa. const.: Friesian male calves, age: ≈12 weeks, wt.: ≈80 kg, contamination: artificial (dose: **155 mg OTA** (total amount ingested), o., for up to 87 days; for detailed information please see the article), conc.: tr*, country: UK/USA[210], *after 87 days of experimental period
incidence: 5/6, sa. const.: Friesian male calves, age: ≈12 weeks, wt.: ≈80 kg, contamination: artificial (dose: **143 mg OTA + 4.04 mg AFB_1** (total amount ingested), o., for up to 87 days; for detailed information please see the article), conc.: tr*, country: UK/USA[210], *after 87 days of experimental period
incidence: 6/6, sa. const.: Friesian male calves, age: ≈12 weeks, wt.: ≈80 kg, contamination: artificial (dose: **3.87 mg AFB_1** (total amount ingested), o., for up to 87 days; for detailed information please see the article), conc.: nd*, country: UK/USA[210], *after 87 days of experimental period

Calf serum may contain the following mycotoxins and/or their metabolites:

Ochratoxin A
incidence: 1/1, sa. const.: Holstein-Friesian male calve, preruminant, age: 19 or 20 days, wt.: 44 kg, contamination: artificial (dose: **0.25 mg OTA/kg** b. wt., i.v., once; for detailed information please see the article), conc. range: ≤3.0 µg/ml*, country: Canada[179], *after 0 h (also at other hour intervals up to 120 h measured, lowest value conc.: <0.1 µg/ml after 120 h)
incidence: 2?/2, sa. const.: Holstein-Friesian male calves, preruminant, age: 16 and 21 days, wt.: 60 kg, contamination: artificial (dose: **0.5 mg OTA/kg** b. wt., o., once; for detailed information please see the article), conc. range: ≤0.5 µg/ml*, country: Canada[179], *after ≈37 h (also at other hour intervals up to 120 h measured, lowest value conc.: ≈0.1 µg/ml after ≈83 h)
incidence: 4?/4, sa. const.: Holstein-Friesian male calves, ruminant, age: 46–69 days, wt.: 68–100 kg, contamination: artificial (dose: **2.0 mg OTA/kg** b. wt., o., once; for detailed information please see the article), conc. range: ≤2.2 µg/ml*, country: Canada[179], *after 2–4 h (also at other hour intervals

up to ≈96 h measured, lowest value conc.: ≈0.05 μg/ml after ≈96 h)

OCHRATOXIN α
incidence: 4?/4, sa. const.: Holstein-Friesian male calves, ruminant, age: 46–69 days, wt.: 68–100 kg, contamination: artificial (dose: 2.0 mg OTA/kg b. wt., o., once; for detailed information please see the article), conc. range: ≈≤0.32 μg/ml*, country: Canada[179], *after ≈35 h (also at other hour intervals up to 120 h measured, lowest value conc.: ≈0.07 μg/ml after 120 h)

Calf urine may contain the following mycotoxins and/or their metabolites:

OCHRATOXIN A
incidence: 2/2, sa. const.: Holstein-Friesian male calves, preruminant, age: 19–20 days, wt.: 44 kg, contamination: artificial (dose: **0.25 mg crystalline OTA**/kg b. wt., i.v., once; for detailed information please see the article) conc. range: 0.41* **–2.75** mg, country: Canada[179], *incomplete urine collection, **cumulative excretion
incidence: 2/2, sa. const.: Holstein-Friesian male calves, preruminant, age: 16–21 days, wt.: 60 kg, contamination: artificial (dose: **0.5 mg OTA**/kg b. wt., o., once; for detailed information please see the article), conc. range: 0.97–0.99 mg*, Ø conc.: 0.98 mg*, country: Canada[179], *cumulative excretion
incidence: 4/4, sa. const.: Holstein-Friesian male calves, ruminant, age: 46–69 days, wt.: 68–100 kg, contamination: artificial (dose: **2 mg OTA**/kg b.w., o., once; for detailed information please see the article), conc. range: ≤0.9 mg*, country: Canada[179], *cumulative excretion

OCHRATOXIN α
incidence: 2/2, sa. const.: Holstein-Friesian male calves, preruminant, age: 19–20 days, wt.: 44 kg, contamination: artificial (dose: **0.25 mg crystalline OTA**/kg b. wt., i.v., once; for detailed information please see the article), conc.: nd*, country: Canada[179], *cumulative excretion
incidence: 2/2, sa. const.: Holstein-Friesian male calves, preruminant, age: 16–21 days, wt.: 60 kg, contamination: artificial (dose: **0.5 mg OTA**/kg b. wt., o., once; for detailed information please see the article), conc. range: 24.05–26.68 mg*, Ø conc.: 25.365 mg*, country: Canada[179], *cumulative excretion
incidence: 4/4, sa. const.: Holstein-Friesian male calves, ruminant, age: 46–69 days, wt.: 68–100 kg, contamination: artificial (dose: **2 mg OTA**/kg b. wt., o., once; for detailed information please see the article), conc. range: ≤110.9 mg*, country: Canada[179], *cumulative excretion

Camel

Camel Natural Contamination

Camel fetus may contain the following mycotoxins and/or their metabolites:

GLIOTOXIN
incidence: 1/1*, sa. const.: fetus from camel of the UAE, contamination: natural, conc.: tr, country: Germany/UAE[149], *allantois from fetus

Camel intestine may contain the following mycotoxins and/or their metabolites:

GLIOTOXIN
incidence: 1/2, sa. const.: intestines from camels of the UAE, contamination: natural, conc.: tr*, country: Germany/UAE[149], *colon
incidence: 1/2, sa. const.: intestines from camels of the UAE, contamination: natural, conc.: tr*, country: Germany/UAE[149], *duodenum

Camel milk may contain the following mycotoxins and/or their metabolites:

Aflatoxin M_1
incidence: 6/20, sa. const.: milk from camels of the UAE, contamination: natural, conc. range: 0.25–0.8 ng/ml, Ø conc.: 0.46 ng/ml, country: UAE/UK[247]

Camel rumen may contain the following mycotoxins and/or their metabolites:

Gliotoxin
incidence: 2/2, sa. const.: rumens from camels of the UAE, contamination: natural, conc. range: tr, country: Germany/UAE[149]

Carp see Fish, carp

Cat

Cat Natural Contamination

Cat kidney may contain the following mycotoxins and/or their metabolites:

Ochratoxin A
incidence: 16/26, sa. const.: kidneys from cats of Austria, age: 7 weeks–13 years (for detailed information please see the article), contamination: natural, conc. range: 0.1–1.63 µg/kg, Ø conc.: 0.85 µg/kg, country: Austria[340]

Cat Artifical Contamination

Cat blood may contain the following mycotoxins and/or their metabolites:

HT-2 Toxin
incidence: 1/1, sa. const.: cat, contamination: artificial (dose: 200 µg T-2 toxin/kg on the 1st and 1.5 mg T-2 toxin/kg on the 2nd day; for detailed information please see the article), conc.: 0.46 ppb*, country: USA[497], *after 6 h of the 2nd dose

3′-Hydroxy HT-2 Toxin
incidence: 1/1, sa. const.: cat, contamination: artificial (dose: 200 µg T-2 toxin/kg on the 1st and 1.5 mg T-2 toxin/kg on the 2nd day; for detailed information please see the article), conc.: 0.82 ppb*, country: USA[497], *after 6 h of the 2nd dose

T-2 Tetraol
incidence: 1/1, sa. const.: cat, contamination: artificial (dose: 200 µg T-2 toxin/kg on the 1st and 1.5 mg T-2 toxin/kg on the 2nd day; for detailed information please see the article), conc.: 44.50 ppb*, country: USA[497], *after 6 h of the 2nd dose

Cat heart may contain the following mycotoxins and/or their metabolites:

HT-2 Toxin
incidence: 1/1, sa. const.: cat, contamination: artificial (dose: 200 µg T-2 toxin/kg on the 1st and 1.5 mg T-2 toxin/kg on the 2nd day; for detailed information please see the article), conc.: 199.90 ppb*, country: USA[497], *after 6 h of the 2nd dose

3′-Hydroxy HT-2 Toxin
incidence: 1/1, sa. const.: cat, contamination: artificial (dose: 200 µg T-2 toxin/kg on the 1st and 1.5 mg T-2 toxin/kg on the 2nd day; for detailed information please see the article), conc.: 58.75 ppb*, country: USA[497], *after 6 h of the 2nd dose

T-2 Toxin
incidence: 1/1, sa. const.: cat, contamination: artificial (dose: 200 µg T-2 toxin/kg on the 1st and 1.5 mg T-2 toxin/kg on the 2nd day; for detailed information please see the article), conc.: 0.99 ppb*, country: USA[497], *after 6 h of the 2nd dose

3′-Hydroxy T-2 Toxin
incidence: 1/1, sa. const.: cat, contamination: artificial (dose: 200 µg T-2 toxin/kg on the 1st and 1.5 mg T-2 toxin/kg on the 2nd day; for detailed information please see the article), conc.:

13.25 ppb*, country: USA[497], *after 6 h of the 2nd dose

T-2 Tetraol
incidence: 1/1, sa. const.: cat, contamination: artificial (dose: 200 μg T-2 toxin/kg on the 1st and 1.5 mg T-2 toxin/kg on the 2nd day; for detailed information please see the article), conc.: 10.14 ppb*, country: USA[497], *after 6 h of the 2nd dose

Cat kidney may contain the following mycotoxins and/or their metabolites:

HT-2 Toxin
incidence: 1/1, sa. const.: cat, contamination: artificial (dose: 200 μg T-2 toxin/kg on the 1st and 1.5 mg T-2 toxin/kg on the 2nd day; for detailed information please see the article), conc.: 138.58 ppb*, country: USA[497], *after 6 h of the 2nd dose

3′-Hydroxy HT-2 Toxin
incidence: 1/1, sa. const.: cat, contamination: artificial (dose: 200 μg T-2 toxin/kg on the 1st and 1.5 mg T-2 toxin/kg on the 2nd day; for detailed information please see the article), conc.: 55.88 ppb*, country: USA[497], *after 6 h of the 2nd dose

T-2 Toxin
incidence: 1/1, sa. const.: cat, contamination: artificial (dose: 200 μg T-2 toxin/kg on the 1st and 1.5 mg T-2 toxin/kg on the 2nd day; for detailed information please see the article), conc.: 34.07 ppb*, country: USA[497], *after 6 h of the 2nd dose

3′-Hydroxy T-2 Toxin
incidence: 1/1, sa. const.: cat, contamination: artificial (dose: 200 μg T-2 toxin/kg on the 1st and 1.5 mg T-2 toxin/kg on the 2nd day; for detailed information please see the article), conc.: 34.07 ppb*, country: USA[497], *after 6 h of the 2nd dose

T-2 Tetraol
incidence: 1/1, sa. const.: cat, contamination: artificial (dose: 200 μg T-2 toxin/kg on the 1st and 1.5 mg T-2 toxin/kg on the 2nd day; for detailed information please see the article), conc.: 57.57 ppb*, country: USA[497], *after 6 h of the 2nd dose

Cat liver may contain the following mycotoxins and/or their metabolites:

HT-2 Toxin
incidence: 1/1, sa. const.: cat, contamination: artificial (dose: 200 μg T-2 toxin/kg on the 1st and 1.5 mg T-2 toxin/kg on the 2nd day; for detailed information please see the article), conc.: 2.92 ppb*, country: USA[497], *after 6 h of the 2nd dose

3′-Hydroxy HT-2 Toxin
incidence: 1/1, sa. const.: cat, contamination: artificial (dose: 200 μg T-2 toxin/kg on the 1st and 1.5 mg T-2 toxin/kg on the 2nd day; for detailed information please see the article), conc.: 1.00 ppb*, country: USA[497], *after 6 h of the 2nd dose

T-2 Tetraol
incidence: 1/1, sa. const.: cat, contamination: artificial (dose: 200 μg T-2 toxin/kg on the 1st and 1.5 mg T-2 toxin/kg on the 2nd day; for detailed information please see the article), conc.: 9.56 ppb*, country: USA[497], *after 6 h of the 2nd dose

Cat lung may contain the following mycotoxins and/or their metabolites:

HT-2 Toxin
incidence: 1/1, sa. const.: cat, contamination: artificial (dose: 200 μg T-2 toxin/kg on the 1st and 1.5 mg T-2 toxin/kg on the 2nd day; for detailed information please see the article), conc.: 25.56 ppb*, country: USA[497], *after 6 h of the 2nd dose

3′-Hydroxy HT-2 Toxin
incidence: 1/1, sa. const.: cat, contamination: artificial (dose: 200 μg T-2 toxin/kg on the 1st and 1.5 mg T-2 toxin/kg on the 2nd day; for detailed information please see the article), conc.: 5.28 ppb*, country: USA[497], *after 6 h of the 2nd dose

T-2 Toxin
incidence: 1/1, sa. const.: cat, contamination: artificial (dose: 200 μg T-2 toxin/kg on the 1st and 1.5 mg T-2 toxin/kg on the 2nd day; for detailed information please see the article), conc.: 15.30 ppb*, country: USA[497], *after 6 h of the 2nd dose

3′-Hydroxy T-2 Toxin
incidence: 1/1, sa. const.: cat, contamination: artificial (dose: 200 μg T-2 toxin/kg on the 1st and 1.5 mg T-2 toxin/kg on the 2nd day; for detailed information please see the article), conc.: 5.42 ppb*, country: USA[497], *after 6 h of the 2nd dose

T-2 Tetraol
incidence: 1/1, sa. const.: cat, contamination: artificial (dose: 200 μg T-2 toxin/kg on the 1st and 1.5 mg T-2 toxin/kg on the 2nd day; for detailed information please see the article), conc.: 15.30 ppb*, country: USA[497], *after 6 h of the 2nd dose

Cat urine may contain the following mycotoxins and/or their metabolites:

HT-2 Toxin
incidence: 1/1, sa. const.: cat, contamination: artificial (dose: 200 μg T-2 toxin/kg on the 1st and 1.5 mg T-2 toxin/kg on the 2nd day; for detailed information please see the article), conc.: 0.21 ppb, country: USA[497], *after 6 h of the 2nd dose

3′-Hydroxy HT-2 Toxin
incidence: 1/1, sa. const.: cat, contamination: artificial (dose: 200 μg T-2 toxin/kg on the 1st and 1.5 mg T-2 toxin/kg on the 2nd day; for detailed information please see the article), conc.: 0.34 ppb, country: USA[497], *after 6 h of the 2nd dose

T-2 Tetraol
incidence: 1/1, sa. const.: cat, contamination: artificial (dose: 200 μg T-2 toxin/kg on the 1st and 1.5 mg T-2 toxin/kg on the 2nd day; for detailed information please see the article), conc.: 160.88 ppb*, country: USA[497], *after 6 h of the 2nd dose

Cattle

Cattle Natural Contamination

Cattle bile may contain the following mycotoxins and/or their metabolites:

Zeranol
incidence: 8/70, sa. const.: bile from castrated male cattles of Northern Ireland, contamination: natural, conc. range: ≥1.0 to ≤3.0 ng/ml*, Ø conc.: 1.65 ng/ml, country: Northern Ireland, UK[529], *most probable of *Fusarium* origin

α-Zearalenol
incidence: 8?/70, sa. const.: bile from castrated male cattles of Northern Ireland, contamination: natural, conc. range: ≤20.5 ng/ml*, country: Northern Ireland, UK[529], *most probable of *Fusarium* origin

β-Zearalenol
incidence: 8?/70, sa. const.: bile from castrated male cattles of Northern Ireland, contamination: natural, conc. range: ≤23.4 ng/ml*, country: Northern Ireland, UK[529], *most probable of *Fusarium* origin

Cattle udder may contain the following mycotoxins and/or their metabolites:

Gliotoxin
incidence: 1/1, sa. const.: udder from a Holstein-Friesian female of Germany, age: 8 years, contamination: natural, conc.: 9.2 mg/kg udder, country: Germany[126]

Cattle urine may contain the following mycotoxins and/or their metabolites:

ZEARALANOLS
incidence: 282/415*, sa. const.: urine from cattles of New Zealand, contamination: natural, conc. range: ≤12.3 ng/ml**, country: New Zealand[230], *export animals, **most probable of *Fusarium* origin

ZEARALENOLS
incidence: 282/415*, sa. const.: urine from cattles of New Zealand, contamination: natural, conc. range: ≤163 ng/ml, country: New Zealand[230], *export animals

Cattle Artificial Contamination

Cattle feces may contain the following mycotoxins and/or their metabolites:

FUMONISIN B_1
incidence: ?/5, sa. const.: cattles, contamination: artificial (dose: **200 µg FB_1/g**, o., once; for detailed information please see the article), conc.: 1 µg/g (mean value), country: USA[276]
incidence: ?/5, sa. const.: cattles, contamination: artificial (dose: **400 µg FB_1/g**, o., once; for detailed information please see the article), conc.: 6 µg/g (mean value), country: USA[276]

HYDROLIZED FUMONISIN B_1
incidence: ?/5, sa. const.: cattles, contamination: artificial (dose: **200 µg FB_1/g**, o., once; for detailed information please see the article), conc.: 14 µg/g (mean value), country: USA[276]
incidence: ?/5, sa. const.: cattles, contamination: artificial (dose: **400 µg FB_1/g**, o., once; for detailed information please see the article), conc.: 14 µg/g (mean value), country: USA[276]

Cattle liver may contain the following mycotoxins and/or their metabolites:

ZEARALENONE
incidence: 5/5*, sa. const.: Simmendal breed heifers, age: 1.5–2.0 years, Ø wt.: 392 kg, contamination: (dose: **158 µg ZEA/animal** besides other *Fusarium* mycotoxins, o., daily for 84 days; for detailed information please see the article), conc.: nr**, country: Austria[599], *control, **after 84 days of ZEA-administration
incidence: 5/5, sa. const.: Simmendal breed heifers, age: 1.5–2.0 years, Ø wt.: 392 kg, contamination: (dose: two **25 mg zeranol pellets** implanted and additionally **158 µg ZEA**/animal besides other *Fusarium* mycotoxins, o., daily for 84 days; for detailed information please see the article), conc.: nr*, country: Austria[599], *after 84 days of ZEA-administration
incidence: 5?/5, sa. const.: Simmendal breed heifers, age: 1.5–2.0 years, Ø wt.: 392 kg, contamination: artificial (dose: **2.74 mg ZEA**/animal besides other *Fusarium* mycotoxins, o., daily for 84 days; for detailed information please see the article), conc. range: tr–1.2 µg/kg*, country: Austria[599], *after 84 days of ZEA-administration

α-ZEARALENONE
incidence: 5/5*, sa. const.: Simmendal breed heifers, age: 1.5–2.0 years, Ø wt.: 392 kg, contamination: (dose: **158 µg ZEA**/animal besides other *Fusarium* mycotoxins, o., daily for 84 days; for detailed information please see the article), conc.: nr**, country: Austria[599], *control, **after 84 days of ZEA-administration
incidence: 5/5, sa. const.: Simmendal breed heifers, age: 1.5–2.0 years, Ø wt.: 392 kg, contamination: (dose: two **25 mg zeranol pellets** implanted and additionally **158 µg ZEA**/animal besides other *Fusarium* mycotoxins, o., daily for 84 days; for detailed information please see the article), conc.: nr*, country: Austria[599], *after 84 days of ZEA-administration

incidence: 5?/5, sa. const.: Simmendal breed heifers, age: 1.5–2.0 years, Ø wt.: 392 kg, contamination: artificial (dose: **2.74 mg ZEA**/animal besides other *Fusarium* mycotoxins, o., daily for 84 days; for detailed information please see the article), conc. range: tr–1.2 µg/kg*, country: Austria[599], *after 84 days of ZEA-administration

β-Zearalenone
incidence: 5?/5*, sa. const.: Simmendal breed heifers, age: 1.5–2.0 years, Ø wt.: 392 kg, contamination: (dose: **158 µg ZEA**/animal besides other *Fusarium* mycotoxins, o., daily for 84 days; for detailed information please see the article), conc.: tr**, country: Austria[599], *control, **after 84 days of ZEA-administration
incidence: 5/5, sa. const.: Simmendal breed heifers, age: 1.5–2.0 years, Ø wt.: 392 kg, contamination: (dose: two **25 mg zeranol** pellets implanted and additionally **158 µg ZEA**/animal besides other *Fusarium* mycotoxins, o., daily for 84 days; for detailed information please see the article), conc.: nr*, country: Austria[599], *after 84 days of ZEA-administration
incidence: 5?/5, sa. const.: Simmendal breed heifers, age: 1.5–2.0 years, Ø wt.: 392 kg, contamination: artificial (dose: **2.74 mg ZEA**/animal besides other *Fusarium* mycotoxins, o., daily for 84 days; for detailed information please see the article), conc. range: 5–11.5 µg/kg*, country: Austria[599], *after 84 days of ZEA-administration

Cattle omasum may contain the following mycotoxins and/or their metabolites:

T-2 Toxin
incidence: 1/1, sa. const.: female calve, wt.: 201–268 kg, contamination: no T-2 toxin (for detailed information please see the article), conc.: nr*, country: USA[403], *in omasum contents
incidence: 1/6, sa. const.: female calves, wt.: 201–268 kg, contamination: artificial (dose: **2.4 mg T-2 toxin**/kg, o., once (the remaining calves received different applications of T-2 toxin in different conc.); for detailed information please see the article), conc.: 40 ppb* **, country: USA[403], *in omasum contents, **after 24 h

Cattle plasma may contain the following mycotoxins and/or their metabolites:

T-2 Toxin
incidence: 2?/2, sa. const.: female calves, wt.: 201–268 kg, contamination: artificial (dose: **0.6 mg T-2 toxin**/kg, i.v., once), conc.: ≈≤1,300 ppb* (mean value), country: USA[403], *after 0 min (also at other min intervals up to 25 min measured, lowest conc.: ≈80 ppb after 25 min)
incidence: 2?/2, sa. const.: female calves, wt.: 201–268 kg, contamination: artificial (dose: **1.2 mg T-2 toxin**/kg, i.v., once), conc.: ≈≤4,000 ppb* (mean value), country: USA[403], *after 0 min (also at other min intervals up to 60 min measured, lowest conc.: ≈55 ppb after 60 min)

Cattle rumen may contain the following mycotoxins and/or their metabolites:

T-2 Toxin
incidence: 1/1, sa. const.: female calve, wt.: 201–268 kg, contamination: no T-2 toxin (for detailed information please see the article), conc.: nr*, country: USA[403], *in rumen contentsincidence: 1/6, sa. const.: female calves, wt.: 201–268 kg, contamination: artificial (dose: 2.4 mg T-2 toxin/kg, o., once (the remaining calves received different applications of **T-2 toxin** in **different conc.**); for detailed information please see the article), conc.: 62 ppb* **, country: USA[403], *in rumen contents, **after 24 h

Cattle urine may contain the following mycotoxins and/or their metabolites:

Fumonisin B_1
incidence: ?/5, sa. const.: cattles, contamination: artificial (dose: **200 µg FB_1/g**, o., once; for detailed information please see the article), conc. range: 0.1–0.5 µg/g (mean values), country: USA[276]
incidence: ?/5, sa. const.: cattles, contamination: artificial (dose: **400 µg FB_1/g**, o., once; for detailed information please see the article), conc. range: 0.1–0.7 µg/g (mean values), country: USA[276]

Diacetoxyscirpenol
incidence: 3/3*, sa. const.: heifers, wt.: 74.5–139.1 kg, contamination: no DAS (for detailed information please see the article), conc.: nd, country: USA[512], *control
incidence: 1/4, sa. const.: heifers, wt.: 74.5–139.1 kg, contamination: artificial (dose: **0.5 mg DAS/kg** b. wt., injection, once; for detailed information please see the article), conc.: 0.14 ng/ml*, country: USA[512], *collected after 15 min

Monoacetoxyscirpenol
incidence: 3/3*, sa. const.: heifers, wt.: 74.5–139.1 kg, contamination: no DAS (for detailed information please see the article), conc.: nd, country: USA[512], *control
incidence: 1/4, sa. const.: heifers, wt.: 74.5–139.1 kg, contamination: artificial (dose: **0.5 mg DAS/kg** b. wt., injection, once; for detailed information please see the article), conc.: 2.10 ng/ml, country: USA[512], *collected after 15 min

Scirpentriol
incidence: 3/3*, sa. const.: heifers, wt.: 74.5–139.1 kg, contamination: no DAS (for detailed information please see the article), conc.: nd, country: USA[512], *control
incidence: 1/4, sa. const.: heifers, wt.: 74.5–139.1 kg, contamination: artificial (dose: **0.5 mg DAS/kg** b. wt., injection, once; for detailed information please see the article), conc.: 1.17 ng/ml, country: USA[512], *collected after 15 min

Taleranol
incidence: 5?/5*, sa. const.: Simmendal breed heifers, age: 1.5–2.0 years, Ø wt.: 392 kg, contamination: (dose: **158 µg ZEA**/animal besides other *Fusarium* mycotoxins, o., daily for 84 days; for detailed information please see the article), conc.: nr**, country: Austria[599], *control, **after 84 days of ZEA-administration
incidence: 5?/5, sa. const.: Simmendal breed heifers, age: 1.5–2.0 years, Ø wt.: 392 kg, contamination: (dose: two **25 mg zeranol** pellets implanted and additionally **158 µg ZEA**/animal besides other *Fusarium* mycotoxins, o., daily for 84 days; for detailed information please see the article), conc. range: 2–5 µg/l*, country: Austria[599], *after 84 days of ZEA-administration
incidence: 5?/5, sa. const.: Simmendal breed heifers, age: 1.5–2.0 years, Ø wt.: 392 kg, contamination: artificial (dose: **2.74 mg ZEA**/animal besides other *Fusarium* mycotoxins, o., daily for 84 days; for detailed information please see the article), conc. range: 2–3 µg/l*, country: Austria[599], *after 84 days of ZEA-administration

Zearalanone
incidence: 5/5*, sa. const.: Simmendal breed heifers, age: 1.5–2.0 years, Ø wt.: 392 kg, contamination: (dose: **158 µg ZEA**/animal besides other *Fusarium* mycotoxins, o., daily for 84 days; for detailed information please see the article), conc.: nr**, country: Austria[599], *control, **after 84 days of ZEA-administration
incidence: 5?/5, sa. const.: Simmendal breed heifers, age: 1.5–2.0 years, Ø wt.: 392 kg, contamination: (dose: two **25 mg zeranol** pellets implanted and additionally **158 µg ZEA**/animal besides other *Fusarium* mycotoxins, o., daily for

84 days; for detailed information please see the article), conc.: tr*, country: Austria[599], *after 84 days of ZEA-administration

incidence: 5?/5, sa. const.: Simmendal breed heifers, age: 1.5–2.0 years, Ø wt.: 392 kg, contamination: artificial (dose: **2.74 mg ZEA**/animal besides other *Fusarium* mycotoxins, o., daily for 84 days; for detailed information please see the article), conc.: tr*, country: Austria[599], *after 84 days of ZEA-administration

Zearalenone

incidence: 5?/5*, sa. const.: Simmendal breed heifers, age: 1.5–2.0 years, Ø wt.: 392 kg, contamination: (dose: **158 µg ZEA**/animal besides other *Fusarium* mycotoxins, o., daily for 84 days; for detailed information please see the article), conc.: tr**, country: Austria[599], *control, **after 84 days of ZEA-administration

incidence: 5?/5, sa. const.: Simmendal breed heifers, age: 1.5–2.0 years, Ø wt.: 392 kg, contamination: (dose: two **25 mg zeranol** pellets implanted and additionally **158 µg ZEA**/animal besides other *Fusarium* mycotoxins, o., daily for 84 days; for detailed information please see the article), conc.: tr*, country: Austria[599], *after 84 days of ZEA-administration

incidence: 5?/5, sa. const.: Simmendal breed heifers, age: 1.5–2.0 years, Ø wt.: 392 kg, contamination: artificial (dose: **2.74 mg ZEA**/animal besides other *Fusarium* mycotoxins, o., daily for 84 days; for detailed information please see the article), conc. range: 5–8 µg/l*, country: Austria[599], *after 84 days of ZEA-administration

α-Zearalenol

incidence: 5/5*, sa. const.: Simmendal breed heifers, age: 1.5–2.0 years, Ø wt.: 392 kg, contamination: (dose: **158 µg ZEA**/animal besides other *Fusarium* mycotoxins, o., daily for 84 days; for detailed information please see the article), conc.: nr**, country: Austria[599], *control, **after 84 days of ZEA-administration

incidence: 5?/5, sa. const.: Simmendal breed heifers, age: 1.5–2.0 years, Ø wt.: 392 kg, contamination: (dose: two **25 mg zeranol** pellets implanted and additionally **158 µg ZEA**/animal besides other *Fusarium* mycotoxins, o., daily for 84 days; for detailed information please see the article), conc.: tr*, country: Austria[599], *after 84 days of ZEA-administration

incidence: 5?/5, sa. const.: Simmendal breed heifers, age: 1.5–2.0 years, Ø wt.: 392 kg, contamination: artificial (dose: **2.74 mg ZEA**/animal besides other *Fusarium* mycotoxins, o., daily for 84 days; for detailed information please see the article), conc. range: 3–5 µg/l*, country: Austria[599], *after 84 days of ZEA-administration

β-Zearalenol

incidence: 5?/5*, sa. const.: Simmendal breed heifers, age: 1.5–2.0 years, Ø wt.: 392 kg, contamination: (dose: **158 µg ZEA**/animal besides other *Fusarium* mycotoxins, o., daily for 84 days; for detailed information please see the article), conc.: tr**, country: Austria[599], *control, **after 84 days of ZEA-administration

incidence: 5/5, sa. const.: Simmendal breed heifers, age: 1.5–2.0 years, Ø wt.: 392 kg, contamination: (dose: two **25 mg zeranol** pellets implanted and additionally **158 µg ZEA**/animal besides other *Fusarium* mycotoxins, o., daily for 84 days; for detailed information please see the article), conc.: nr*, country: Austria[599], *after 84 days of ZEA-administration

incidence: 5?/5, sa. const.: Simmendal breed heifers, age: 1.5–2.0 years, Ø wt.: 392 kg, contamination: artificial (dose: **2.74 mg ZEA**/animal besides other *Fusarium* mycotoxins, o., daily for 84 days; for detailed information please see the article), conc. range: 20–65 µg/l*, country: Austria[599], *after 84 days of ZEA-administration

Zeranol
incidence: 5/5*, sa. const.: Simmendal breed heifers, age: 1.5–2.0 years, Ø wt.: 392 kg, contamination: (dose: **158 µg ZEA**/animal besides other *Fusarium* mycotoxins, o., daily for 84 days; for detailed information please see the article), conc.: nr**, country: Austria[599], *control, **after 84 days of ZEA-administration
incidence: 5?/5, sa. const.: Simmendal breed heifers, age: 1.5–2.0 years, Ø wt.: 392 kg, contamination: (dose: two **25 mg zeranol** pellets implanted and additionally **158 µg ZEA**/animal besides other *Fusarium* mycotoxins, o., daily for 84 days; for detailed information please see the article), conc. range: 2–5 µg/l*, country: Austria[599], *after 84 days of ZEA-administration
incidence: 5?/5, sa. const.: Simmendal breed heifers, age: 1.5–2.0 years, Ø wt.: 392 kg, contamination: artificial (dose: **2.74 mg ZEA**/animal besides other *Fusarium* mycotoxins, o., daily for 84 days; for detailed information please see the article), conc. range: 2–3 µg/l*, country: Austria[599], *after 84 days of ZEA-administration

Channel Catfish see Fish, channel catfish

Chicken

Chicken Natural Contamination

see also Hen

Chicken liver may contain the following mycotoxins and/or their metabolites:

Aflatoxin B_1
incidence: 122/225, sa. const.: livers from chickens of Thailand, contamination: natural, conc. range: 0.003–35.45 ppb, Ø conc.: 2.473 ppb, country: Thailand[577]

Chicken muscle may contain the following mycotoxins and/or their metabolites:

Aflatoxin B_1
incidence: 97/225, sa. const.: muscles from chickens of Thailand, contamination: natural, conc. range: 0.024–24.34 ppb, Ø conc.: 0.744 ppb, country: Thailand[577]

Chicken Artificial Contamination

Chicken bile may contain the following mycotoxins and/or their metabolites:

Zearalenone
incidence: 11/11, sa. const.: male broilers, age: 25–27 days, contamination: artificial (dose: 5.8 µg ZEA/kg b. wt., o., once; for detailed information please see the article), conc. range: ≈≤160 ng/g* (mean value), country: Germany[543], *≈4 h elapsed after bolus (also at other hour intervals up to 48 h measured, lowest conc.: ≈20 ng/g after 48 h)

α-Zearalenol
incidence: 11/11, sa. const.: male broilers, age: 25–27 days, contamination: artificial (dose: 5.8 µg ZEA/kg b. wt., o., once; for detailed information please see the article), conc. range: ≈≤100 ng/g* (mean value), country: Germany[543], *≈7 h elapsed after bolus (also at other hour intervals up to 48 h measured, lowest conc.: ≈10 ng/g after 48 h)

Chicken blood may contain the following mycotoxins and/or their metabolites:

Aflatoxin B_1
incidence: 2/4, sa. const.: chickens, age: 3 weeks, contamination: artificial (dose: 1 mg AFB_1 (labeled and unlabeled), o., once; for detailed information please

see the article), conc. range: ≈0.45–0.5 ppm*, country: USA[617], ***after 12 h** (also measured after 48 and 72 h)
incidence: 4/4, sa. const.: chickens, age: 3 weeks, contamination: artificial (dose: 1 mg AFB_1 (labeled and unlabeled), o., once; for detailed information please see the article), conc.: nd*, country: USA[617], ***after 24 h** (also measured after 48 and 72 h)

Zearalenone
incidence: 6/6*, sa. const.: broiler chickens (Ross male X Arbor Acre female), age: 4 weeks, contamination: no ZEA (for detailed information please see the article), conc.: nr, country: USA[106], *control
incidence: 4?/4, sa. const.: broiler chickens (Ross male X Arbor Acre female), age: 4 weeks, contamination: artificial (dose: **5.0 mg ZEA** (labeled)/kg b. wt., intubated into the crop, once; for detailed information please see the article), conc. range: ≤143.6 ppb* ** *** (mean value), country: USA[106], *eq. conc., **after 8 h (also measured after 0.5, 4, 12, 24 and 48 h, lowest conc.: 70.8 ppb after 48 h), *** ZEA and/or metabolites

Chicken Breast see Chicken muscle, breast

Chicken Breast Muscle see Chicken muscle, breast

Chicken chest may contain the following mycotoxins and/or their metabolites:

Ochratoxin A
incidence: 4?/4, sa. const.: New Hampshire-Leghorn cross chicks, age: 36 days, contamination: artificial (dose: feeding 1 ppm OTA for 5 weeks and then given a single dose of 50 µg ^{3}H-OTA per chick by intubation (OTA labeled and unlabeled); for detailed information please see the article), conc. range: ≤0.17 ppb* (mean value), country: USA[316], *after 8 h (also measured after 24 and 48 h, lowest conc.: 0.10 ppb after 48 h)

Chicken crop may contain the following mycotoxins and/or their metabolites:

Aflatoxin B_1
incidence: 20/20*, sa. const.: White Leghorn chicks, age: 14 days, contamination: no AFB_1 (for detailed information please see the article), conc.: nd, country: USA[53], *control
incidence: 20?/20, sa. const.: White Leghorn chicks, age: 14 days, contamination: artificial (dose: **120 ng AFB_1/g** ration for 28 days; for detailed information please see the article), conc.: 0.033 ng/g* (mean value), country: USA[53], *after 28 days of AFB_1-administration
incidence: 20?/20, sa. const.: White Leghorn chicks, age: 14 days, contamination: artificial (dose: **525 ng AFB_1/g** ration **+ 25% soil** for 28 days; for detailed information please see the article), conc.: 0.019 ng/g* (mean value), country: USA[53], *after 28 days of AFB_1-administration
incidence: 20?/20, sa. const.: White Leghorn chicks, age: 14 days, contamination: artificial (dose: **630 ng AFB_1/g** ration **+ 10% soil** for 28 days; for detailed information please see the article), conc.: 0.21 ng/g* (mean value), country: USA[53], *after 28 days of AFB_1-administration
incidence: 20?/20, sa. const.: White Leghorn chicks, age: 14 days, contamination: artificial (dose: **700 ng AFB_1/g** ration for 28 days; for detailed information please see the article), conc.: 0.74 ng/g* (mean value), country: USA[53], *after 28 days of AFB_1-administration

Aflatoxin B_2

incidence: 20/20*, sa. const.: White Leghorn chicks, age: 14 days, contamination: no AFB_1 (for detailed information please see the article), conc.: nd, country: USA[53], *control

incidence: 20/20, sa. const.: White Leghorn chicks, age: 14 days, contamination: artificial (dose: **120 ng AFB_1/g** ration for 28 days; for detailed information please see the article), conc.: nd*, country: USA[53], *after 28 days of AFB_1-administration

incidence: 20/20, sa. const.: White Leghorn chicks, age: 14 days, contamination: artificial (dose: **525 ng AFB_1/g** ration **+ 25% soil** for 28 days; for detailed information please see the article), conc.: nd*, country: USA[53], *after 28 days of AFB_1-administration

incidence: 20/20, sa. const.: White Leghorn chicks, age: 14 days, contamination: artificial (dose: **630 ng AFB_1/g** ration **+ 10% soil** for 28 days; for detailed information please see the article), conc.: nd*, country: USA[53], *after 28 days of AFB_1-administration

incidence: 20?/20, sa. const.: White Leghorn chicks, age: 14 days, contamination: artificial (dose: **700 ng AFB_1/g** ration for 28 days; for detailed information please see the article), conc.: 0.029 ng/g* (mean value), country: USA[53], *after 28 days of AFB_1-administration

Aflatoxin M_1

incidence: 20/20*, sa. const.: White Leghorn chicks, age: 14 days, contamination: no AFB_1 (for detailed information please see the article), conc.: nd, country: USA[53], *control

incidence: 20/20, sa. const.: White Leghorn chicks, age: 14 days, contamination: artificial (dose: **120 ng AFB_1/g** ration for 28 days; for detailed information please see the article), conc.: nd*, country: USA[53], *after 28 days of AFB_1-administration

incidence: 20/20, sa. const.: White Leghorn chicks, age 14 days, contamination: artificial (dose: **525 ng AFB_1/g** ration **+ 25% soil** for 28 days; for detailed information please see the article), conc.: nd*, country: USA[53], *after 28 days of AFB_1-administration

incidence: 20/20, sa. const.: White Leghorn chicks, age: 14 days, contamination: artificial (dose: **630 ng AFB_1/g** ration **+ 10% soil** for 28 days; for detailed information please see the article), conc.: nd*, country: USA[53], *after 28 days of AFB_1-administration

incidence: 20?/20, sa. const.: White Leghorn chicks, age: 14 days, contamination: artificial (dose: **700 ng AFB_1/g** ration for 28 days; for detailed information please see the article), conc.: 0.57 ng/g* (mean value), country: USA[53], *after 28 days of AFB_1-administration

Chicken Egg see Hen egg

Chicken excreta may contain the following mycotoxins and/or their metabolites:

Aflatoxin B_1

incidence: 10/10*, sa. const.: male commercial broilers, age: 3 weeks, contamination: no AF; for detailed information please see the article), conc.: nr, country: Australia[400], *control

incidence: ?/10, sa. const.: male commercial broilers, age: 3 weeks, contamination: artificial (dose: **1.5 mg total AF/kg** diet, o., for 5 weeks; for detailed information please see the article), conc.: 0.53 mg/kg* dry matter (mean value), country: Australia[400], *after 5 weeks of AF-administration

incidence: 2/4, sa. const.: chickens, age: 3 weeks, contamination: artificial (dose: 1 mg AFB_1 (labeled and unlabeled), o., once; for detailed information please see the article), conc. range: ≈1.25–2.75 ppm*, country: USA[617], ***after 12 h** (also measured after 48 and 72 h)

incidence: 2/4, sa. const.: chickens, age: 3 weeks, contamination: artificial (dose: 1 mg AFB_1 (labeled and unlabeled), o., once; for detailed information please see the article), conc. range: ≈0.5–8.35 ppm*, country: USA[617], ***after 24 h** (also measured after 48 and 72 h)

Aflatoxin B_2
incidence: 10/10*, sa. const.: male commercial broilers, age: 3 weeks, contamination: no AF; for detailed information please see the article), conc.: nr, country: Australia[400], *control
incidence: ?/10, sa. const.: male commercial broilers, age: 3 weeks, contamination: artificial (dose: **1.5 mg total AF**/kg diet, o., for 5 weeks; for detailed information please see the article), conc.: tr*, country: Australia[400], *after 5 weeks of AF-administration

Aflatoxin G_1
incidence: 10/10*, sa. const.: male commercial broilers, age: 3 weeks, contamination: no AF; for detailed information please see the article), conc.: nr, country: Australia[400], *control
incidence: ?/10, sa. const.: male commercial broilers, age: 3 weeks, contamination: artificial (dose: **1.5 mg total AF**/kg diet, o., for 5 weeks; for detailed information please see the article), conc.: 0.21 mg/kg* dry matter (mean value), country: Australia[400], *after 5 weeks of AF-administration

Aflatoxin G_2
incidence: 10/10*, sa. const.: male commercial broilers, age: 3 weeks, contamination: no AF; for detailed information please see the article), conc.: nr, country: Australia[400], *control
incidence: ?/10, sa. const.: male commercial broilers, age: 3 weeks, contamination: artificial (dose: **1.5 mg total AF**/kg diet, o., for 5 weeks; for detailed information please see the article), conc.: tr*, country: Australia[400], *after 5 weeks of AF-administration

HT-2 Toxin
incidence: 12?/12, sa. const.: chickens, age: 5 weeks, contamination: artificial (dose: 3.5 mg T-2 toxin/kg b. wt., i.p., once), conc.: 3,606 ppb* (mean value), country: USA[145], *after 18 h

3-Acetoxy-3´-Hydroxy HT-2 Toxin
incidence: 12?/12, sa. const.: chickens, age: 5 weeks, contamination: artificial (dose: 3.5 mg T-2 toxin/kg b. wt., i.p.., once), conc.: <1 ppb* (mean value), country: USA[145], *after 18 h

3′-Hydroxy HT-2 Toxin
incidence: 12?/12, sa. const.: chickens, age: 5 weeks, contamination: artificial (dose: 3.5 mg T-2 toxin/kg b. wt.., i.p., once), conc.: 8,224 ppb* (mean value), country: USA[145], *after 18 h

Ochratoxin A
incidence: 10/10*, sa. const.: male commercial broilers, age: 3 weeks, contamination: no OTA; for detailed information please see the article), conc.: nr, country: Australia[400], *control
incidence: ?/10, sa. const.: male commercial broilers, age: 3 weeks, contamination: artificial (dose: **1.0 mg OTA**/kg diet, o., for 5 weeks), conc.: 0.48 mg/kg* dry matter (mean value), country: Australia[400], *after 5 weeks of AF-administration

T-2 Toxin
incidence: 12?/12, sa. const.: chickens, age: 5 weeks, contamination: artificial (dose: 3.5 mg T-2 toxin/kg b. wt., i.p., once), conc.: 353 ppb* (mean value), country: USA[145], *after 18 h

3′-Hydroxy T-2 Toxin
incidence: 12?/12, sa. const.: chickens, age: 5 weeks, contamination: artificial (dose: 3.5 mg T-2 toxin/kg b. wt., i.p., once), conc.: 3,284 ppb* (mean value), country: USA[145], *after 18 h

T-2 Tetraol
incidence: 12?/12, sa. const.: chickens, age: 5 weeks, contamination: artificial

(dose: 3.5 mg T-2 toxin/kg b. wt., i.p., once), conc.: 682 ppb* (mean value), country: USA[145], *after 18 h

4-Acetoxy T-2 Tetraol
incidence: 12?/12, sa. const.: chickens, age: 5 weeks, contamination: artificial (dose: 3.5 mg T-2 toxin/kg b. wt., i.p., once), conc.: 1,041 ppb* (mean value), country: USA[145], *after 18 h

8-Acetoxy T-2 Tetraol
incidence: 12?/12, sa. const.: chickens, age: 5 weeks, contamination: artificial (dose: 3.5 mg T-2 toxin/kg b. wt., i.p., once), conc.: 603 ppb* (mean value), country: USA[145], *after 18 h

15-Acetoxy T-2 Tetraol
incidence: 12?/12, sa. const.: chickens, age: 5 weeks, contamination: artificial (dose: 3.5 mg T-2 toxin/kg b. wt., i.p., once), conc.: 2,065 ppb* (mean value), country: USA[145], *after 18 h

T-2 Triol
incidence: 12?/12, sa. const.: chickens, age: 5 weeks, contamination: artificial (dose: 3.5 mg T-2 toxin/kg b. wt., i.p., once), conc.: 40 ppb* (mean value), country: USA[145], *after 18 h

Zearalenone
incidence: 6/6, sa. const.: broiler chickens (Ross male X Arbor Acre female), age: 4 weeks, contamination: no ZEA (for detailed information please see the article), conc.: nr, country: USA[106]
incidence: 8/11, sa. const.: broiler chickens (Ross male X Arbor Acre female), age: 4 weeks, contamination: artificial (dose: **5 mg ZEA** (labeled)/kg b. wt., intubated into the crop, once; for detailed information please see the article), conc. range: ≤68.7 ppm*, country: USA[106], *after 8 h (also measured after 0.5, 4, 12, 24 and 48 h, lowest conc.: nd after 0.5, 4 and 24 h)

incidence: ?/16*, sa. const.: male broilers, age: 25 days, contamination: artificial (dose: 6.5 mg ZEA/kg b. wt., o., once; for detailed information please see the article), conc. range: ≈≤160 ng/g** *** (mean value), country: Germany[543], *dried excreta, ****without Mycofix®-Plus,** ***≈3.5 h elapsed after bolus (also at other hour intervals up to 48 h measured, lowest conc.: ≈10 ng/g after 48 h)
incidence: ?/16*, sa. const.: male broilers, age: 25 days, contamination: artificial (dose: 6.5 mg ZEA/kg b. wt., o., once; for detailed information please see the article), conc. range: ≈≤153 ng/g** *** (mean value), country: Germany[543], *dried excreta, ****with Mycofix®-Plus,** ***≈3.5 h elapsed after bolus (also at other hour intervals up to 48 h measured, lowest conc.: ≈10 ng/g after 48 h)

α-Zearalenol
incidence: 6/6, sa. const.: broiler chickens (Ross male X Arbor Acre female), age: 4 weeks, contamination: no ZEA (for detailed information please see the article), conc.: nr, country: USA[106]
incidence: 11/11, sa. const.: broiler chickens (Ross male X Arbor Acre female), age: 4 weeks, contamination: artificial (dose: **5 mg ZEA** (labeled)/kg b. wt., intubated into the crop, once; for detailed information please see the article), conc. range: ≤57.7 ppm*, country: USA[106], *after 8 h (also measured after 0.5, 4, 12, 24 and 48 h, lowest conc.: 0.2 ppm after 0.5 h)

incidence: ?/16*, sa. const.: male broilers, age: 25 days, contamination: artificial (dose: 6.5 mg ZEA/kg b. wt., o., once; for detailed information please see the article), conc. range: ≈≤50 ng/g** *** (mean value), country: Germany[543], *dried excreta, ****without Mycofix®-Plus,** ***≈3.5 h elapsed after bolus (also at other hour intervals up to 48 h measured, lowest conc.: ≈5 ng/g after 48 h)
incidence: ?/16*, sa. const.: male broilers, age: 25 days, contamination: artificial (dose: 6.5 mg ZEA/kg b. wt., o., once; for detailed information please see the article), conc. range: ≈≤47 ng/g** *** (mean value), country: Germany[543], *dried

excreta, **with Mycofix®-Plus, ***≈3.5 h elapsed after bolus (also at other hour intervals up to 48 h measured, lowest conc.: ≈5 ng/g after 48 h)

β-Zearalenol
incidence: 6/6, sa. const.: broiler chickens (Ross male X Arbor Acre female), age: 4 weeks, contamination: no ZEA (for detailed information please see the article), conc.: nr, country: USA[106]
incidence: 11/16, sa. const.: broiler chickens (Ross male X Arbor Acre female), age: 4 weeks, contamination: artificial (dose: **5 mg ZEA** (labeled)/kg b. wt., intubated into the crop, once; for detailed information please see the article), conc. range: ≤53.3 ppm*, country: USA[106], *after 8 h (also measured after 0.5, 4, 12, 24 and 48 h, lowest conc.: 0.3 ppm after 0.5 h)

Chicken fat may contain the following mycotoxins and/or their metabolites:

Aflatoxin M_1
incidence: ?/?*, sa. const.: Hubbard strain male broilers, age: 14 days, contamination: no AFB_1 (for detailed information please see the article), conc.: nr, country: Italy[232], *control
incidence: 3?/3, sa. const.: Hubbard strain male broilers, age: 14 days, contamination: artificial (dose: **50 μg AFB_1**/kg feed, o., for up to 64 days; for detailed information please see the article), conc.: 0.70 μg/kg* (mean value), country: Italy[232], *after 64 days (also measured after 36 days but conc.: nd)

Zearalenone
incidence: 6/6*, sa. const.: broiler chickens (Ross male X Arbor Acre female), age: 4 weeks, contamination: no ZEA (for detailed information please see the article), conc.: nr, country: USA[106], *control
incidence: 4?/4, sa. const.: broiler chickens (Ross male X Arbor Acre female), age: 4 weeks, contamination: artificial (dose: 5.0 mg ZEA (labeled)/kg b. wt., intubated into the crop, once; for detailed information please see the article), conc. range: ≤57.6 ppb* ** *** (mean value), country: USA[106], *eq. conc., **after 0.5 h (also measured after 4, 8, 12, 24 and 48 h, lowest conc.: 20.3 ppb after 12 h), *** ZEA and/or metabolites

Chicken Feces see Chicken excreta

Chicken female repro tract may contain the following mycotoxins and/or their metabolites:

Ochratoxin A
incidence: 4?/4, sa. const.: New Hampshire-Leghorn cross chicks, age: 36 days, contamination: artificial (dose: feeding 1 ppm OTA for 5 weeks and then given a single dose of 50 μg ^{3}H-OTA per chick by intubation (OTA labeled and unlabeled); for detailed information please see the article), conc. range: ≤6.81 ppb* (mean value), country: USA[316], *after 8 h (also measured after 24 and 48 h, lowest conc.: nd after 48 h)

Chicken gastrointestinal tract may contain the following mycotoxins and/or their metabolites:

Aflatoxin B_1
incidence: 4/4, sa. const.: chickens, age: 3 weeks, contamination: artificial (dose: 1 mg AFB_1 (labeled and unlabeled), o., once; for detailed information please see the article), conc. range: ≈0.3–1.3 ppm*, country: USA[617], ***after 12 h** (also measured after 48 and 72 h)
incidence: 4/4, sa. const.: chickens, age: 3 weeks, contamination: artificial (dose: 1 mg AFB_1 (labeled and unlabeled), o., once; for detailed information please see the article), conc. range: ≈0.35–4.85 ppm*, country: USA[617], ***after 24 h** (also measured after 48 and 72 h)

Chicken gizzard may contain the following mycotoxins and/or their metabolites:

Aflatoxin B_1

incidence: 6/6*, sa. const.: Hubbard White Mountain broilers, age: 1 week, contamination: no AFB_1 and AFB_2, conc.: nd, country: USA[357], *control
incidence: 6/6, sa. const.: Hubbard White Mountain broilers, age: 1 week, contamination: artificial (dose: **2,057 µg AFB_1/kg** feed and **1,323 µg AFB_2/kg** feed, o., for 5 weeks), conc. range: ≤0.50 µg/kg*, country: USA[357], *3 h after withdrawal of the contaminated feed (also measured after 1, 2 and 4 days, lowest conc.: nd after 4 days)

incidence: 16?/16?*, sa. const.: Hybro broiler chickens, age: 23 days, contamination: no AF (for detailed information please see the article), conc.: nd, country: Spain[359], *control
incidence: 16?/16, sa. const.: Hybro broiler chickens, age: 23 days, contamination: artificial (dose: **2.5 mg AF/kg** feed, o., for 32 days; for detailed information please see the article), conc. range: ≤8.05 µg/kg* (mean value), country: Spain[359] (measured at 4th, 8th*, 16th and 32nd day, lowest conc.: 0.58 µg/kg after 32 days, intoxication period)

incidence: 16?/16, sa. const.: Hybro broiler chickens, age: 23 days, contamination: artificial (dose: **5.0 mg AF/kg** feed, o., for 32 days; for detailed information please see the article), conc. range: ≤17.8 µg/kg* (mean value), country: Spain[359] (measured at 4th, 8th*, 16th and 32nd day, lowest conc.: 1.84 µg/kg after 32 days, intoxication period)

incidence: 4/4, sa. const.: chickens, age: 3 weeks, contamination: artificial (dose: 1 mg AFB_1 (labeled and unlabeled), o., once; for detailed information please see the article), conc. range: ≈0.8–1.85 ppm*, country: USA[617], ***after 12 h** (also measured after 48 and 72 h)
incidence: 4/4, sa. const.: chickens, age: 3 weeks, contamination: artificial (dose: 1 mg AFB_1 (labeled and unlabeled), o., once; for detailed information please see the article), conc. range: ≈3.3–7.75 ppm* **, country: USA[617], ***after 12 h** (also measured after 48 and 72 h), **in gizzard contents
incidence: 4/4, sa. const.: chickens, age: 3 weeks, contamination: artificial (dose: 1 mg AFB_1 (labeled and unlabeled), o., once; for detailed information please see the article), conc. range: ≈0.5–1.55 ppm*, country: USA[617], ***after 24 h** (also measured after 48 and 72 h)
incidence: 4/4, sa. const.: chickens, age: 3 weeks, contamination: artificial (dose: 1 mg AFB_1 (labeled and unlabeled), o., once; for detailed information please see the article), conc. range: ≈1–7.7 ppm* **, country: USA[617], ***after 24 h** (also measured after 48 and 72 h), **in gizzard contents

Aflatoxin B_2

incidence: 6/6*, sa. const.: Hubbard White Mountain broilers, age: 1 week, contamination: no AFB_1 + AFB_2, conc.: nd, country: USA[357], *control
incidence: 6/6, sa. const.: Hubbard White Mountain broilers, age: 1 week, contamination: artificial (dose: **2,057 µg AFB_1/kg** feed and **1,323 µg AFB_2/kg** feed, o., for 5 weeks), conc. range: ≤0.33 µg/kg*, country: USA[357], *3 h after withdrawal of the contaminated feed (also measured after 1, 2 and 4 days, lowest conc.: nd after 4 days)

incidence: 16?/16?*, sa. const.: Hybro broiler chickens, age: 23 days, contamination: no AF (for detailed information please see the article), conc.: nd, country: Spain[359], *control
incidence: 16?/16, sa. const.: Hybro broiler chickens, age: 23 days, contamination: artificial (dose: **2.5 mg AF/kg** feed, o., for

32 days; for detailed information please see the article), conc. range: ≤0.2 μg/kg* (mean value), country: Spain[359] (measured at 4th, 8th, 16th* and 32nd day, lowest conc.: 0.01 μg/kg after 32 days, intoxication period)
incidence: 16?/16, sa. const.: Hybro broiler chickens, age: 23 days, contamination: artificial (dose: **5.0 mg AF/kg** feed, o., for 32 days; for detailed information please see the article), conc. range: ≤0.45 μg/kg* (mean value), country: Spain[359] (measured at 4th, 8th*, 16th and 32nd day, lowest conc.: 0.04 μg/kg after 8 days, intoxication period)

Aflatoxin G_1
incidence: 16?/16?*, sa. const.: Hybro broiler chickens, age: 23 days, contamination: no AF (for detailed information please see the article), conc.: nd, country: Spain[359], *control
incidence: 16?/16, sa. const.: Hybro broiler chickens, age: 23 days, contamination: artificial (dose: **2.5 mg AF/kg** feed, o., for 32 days; for detailed information please see the article), conc. range: ≤1.84 μg/kg* (mean value), country: Spain[359] (measured at 4th, 8th*, 16th and 32nd day, lowest conc.: 0.25 μg/kg after 32 days, intoxication period)
incidence: 16?/16, sa. const.: Hybro broiler chickens, age: 23 days, contamination: artificial (dose: **5.0 mg AF/kg** feed, o., for 32 days; for detailed information please see the article), conc. range: ≤4.3 μg/kg* (mean value), country: Spain[359] (measured at 4th, 8th*, 16th and 32nd day, lowest conc.: 0.32 μg/kg after 32 days, intoxication period)

Aflatoxin M_1
incidence: 6/6*, sa. const.: Hubbard White Mountain broilers, age: 1 week, contamination: no AFB_1 + AFB_2, conc.: nd, country: USA[357], *control
incidence: 4/6, sa. const.: Hubbard White Mountain broilers, age: 1 week, contamination: artificial (dose: **2,057 μg AFB_1/kg** feed + **1,323 μg AFB_2/kg** feed, o., for 5 weeks), conc. range: ≤0.03 μg/kg*, country: USA[357], *3 h after withdrawal of the contaminated feed (also measured after 1, 2 and 4 days, lowest conc.: nd after 4 days)

Aflatoxin M_2
incidence: 6/6*, sa. const.: Hubbard White Mountain broilers, age: 1 week, contamination: no AFB_1 + AFB_2, conc.: nd, country: USA[357], *control
incidence: 5/6, sa. const.: Hubbard White Mountain broilers, age: 1 week, contamination: artificial (dose: **2,057 μg AFB_1/kg** feed + **1,323 μg AFB_2/kg** feed, o., for 5 weeks), conc. range: ≤0.31 μg/kg*, country: USA[357], *3 h after withdrawal of the contaminated feed (also measured after 1, 2 and 4 days, lowest conc.: nd after 4 days)

Ochratoxin A
incidence: 4?/4, sa. const.: New Hampshire-Leghorn cross chicks, age: 36 days, contamination: artificial (dose: feeding 1 ppm OTA for 5 weeks and then given a single dose of 50 μg ^{3}H-OTA per chick by intubation (OTA labeled and unlabeled); for detailed information please see the article), conc. range: ≤2.95 ppb* (mean value), country: USA[316], *after 8 h (also measured after 24 and 48 h, lowest conc.: 0.01 ppb after 48 h)

Penicillic Acid
incidence: 3/3*, sa. const.: Hyline W-36 male laying strain chickens, age: 7 days, contamination: no PA, conc.: nr, country: USA[524], *control
incidence: 3/3, sa. const.: Hyline W-36 male laying strain chickens, age: 7 days, contamination: artificial (dose: **50 mg PA/kg** b. wt., o. by intubation, once), conc.: nd* (mean value), country: USA[524], *after 4 h
incidence: 3?/3, sa. const.: Hyline W-36 male laying strain chickens, age: 7 days, contamination: artificial (dose: **100 mg PA/kg** b. wt., o. by intubation, once), conc.: 2.8 ng/g* (mean value), country: USA[524], *after 4 h

incidence: 3?/3, sa. const.: Hyline W-36 male laying strain chickens, age: 7 days, contamination: artificial (dose: **200 mg PA**/kg b. wt., o. by intubation, once), conc.: 3.6 ng/g* (mean value), country: USA[524], *after 4 h
incidence: 3?/3, sa. const.: Hyline W-36 male laying strain chickens, age: 7 days, contamination: artificial (dose: **400 mg PA**/kg b. wt., o. by intubation, once), conc.: 19 ng/g* (mean value), country: USA[524], *after 4 h

Zearalenone
incidence: 6/6*, sa. const.: broiler chickens (Ross male X Arbor Acre female), age: 4 weeks, contamination: no ZEA (for detailed information please see the article), conc.: nr, country: USA[106], *control
incidence: 4?/4*, sa. const.: broiler chickens (Ross male X Arbor Acre female), age: 4 weeks, contamination: artificial (dose: **5.0 mg ZEA** (labeled)/kg b. wt., intubated into the crop, once; for detailed information please see the article), conc. range: ≤1,228.8 ppb** *** **** (mean value), country: USA[106], *gizzard without contents or mucosa, **eq. conc., ***after 4 h (also measured after 0.5, 8, 12, 24 and 48 h, lowest conc.: 95.6 ppb after 48 h), ****ZEA and/or metabolites

Chicken heart may contain the following mycotoxins and/or their metabolites:

Aflatoxin B_1
incidence: 6/6*, sa. const.: Hubbard White Mountain broilers, age: 1 week, contamination: no AFB_1 + AFB_2, conc.: nd, country: USA[357], *control
incidence: ?/6, sa. const.: Hubbard White Mountain broilers, age: 1 week, contamination: artificial (dose: **2,057 µg AFB_1**/kg feed + **1,323 µg AFB_2**/kg feed, o., for 5 weeks), conc.: 0.08 µg/kg*, country: USA[357], *3 h after withdrawal of the contaminated feed (also measured after 1, 2 and 4 days, lowest conc.: nd after 1, 2 and 4 days)

incidence: 1/4, sa. const.: chickens, age: 3 weeks, contamination: artificial (dose: 1 mg AFB_1 (labeled and unlabeled), o., once; for detailed information please see the article), conc.: ≈0.45 ppm*, country: USA[617], ***after 12 h** (also measured after 48 and 72 h)
incidence: 4/4, sa. const.: chickens, age: 3 weeks, contamination: artificial (dose: 1 mg AFB_1 (labeled and unlabeled), o., once; for detailed information please see the article), conc.: nd*, country: USA[617], ***after 24 h** (also measured after 48 and 72 h)

Aflatoxin B_2
incidence: 6/6*, sa. const.: Hubbard White Mountain broilers, age: 1 week, contamination: no AFB_1 + AFB_2, conc.: nd, country: USA[357], *control
incidence: ?/6, sa. const.: Hubbard White Mountain broilers, age: 1 week, contamination: artificial (dose: **2,057 µg AFB_1**/kg feed + **1,323 µg AFB_2**/kg feed, o., for 5 weeks), conc.: 0.06 µg/kg*, country: USA[357], *3 h after withdrawal of the contaminated feed (also measured after 1, 2 and 4 days, lowest conc.: nd after 1, 2 and 4 days)

Aflatoxin M_1
incidence: 6/6*, sa. const.: Hubbard White Mountain broilers, age: 1 week, contamination: no AFB_1 + AFB_2, conc.: nd, country: USA[357], *control
incidence: 6/6, sa. const.: Hubbard White Mountain broilers, age: 1 week, contamination: artificial (dose: **2,057 µg AFB_1**/kg feed + **1,323 µg AFB_2**/kg feed, o., for 5 weeks), conc.: nd, country: USA[357]

Aflatoxin M_2
incidence: 6/6*, sa. const.: Hubbard White Mountain broilers, age: 1 week, contamination: no AFB_1 + AFB_2, conc.: nd, country: USA[357], *control
incidence: ?/6, sa. const.: Hubbard White Mountain broilers, age: 1 week, contamination: artificial (dose: **2,057 µg**

AFB_1/kg feed + **1,323 µg AFB_2/kg** feed, o., for 5 weeks), conc.: 0.19 µg/kg*, country: USA[357], *3 h after withdrawal of the contaminated feed (also measured after 1, 2 and 4 days, lowest conc.: nd after 1, 2 and 4 days)

Aflatoxin
incidence: 25/25*, sa. const.: Hubbard chickens, age: 1 day, contamination: no AFB_1, (for detailed information please see the article), conc.: nd, country: Egypt[372], *control
incidence: ?/25, sa. const.: Hubbard chickens, age: 1 day, contamination: artificial (dose: **100 ppb AFB_1**, o., for 6 weeks; for detailed information please see the article), conc. range: ≤3.90* ng/g* ** ***, country: Egypt[372], *AF-residues, **after 5 weeks of AFB_1-administration (also measured after 3 and 4 weeks, lowest conc.: nr after 3 and 4 weeks), ***in heart muscle
incidence: ?/25, sa. const.: Hubbard chickens, age: 1 day, contamination: artificial (dose: **250 ppb AFB_1**, o., for 6 weeks; for detailed information please see the article), conc. range: ≤23.64* ng/g* ** ***, country: Egypt[372], *AF-residues, **after 5 weeks of AFB_1-administration (also measured after 3 and 4 weeks, lowest conc.: 15.26 ng/g after 3 weeks), ***in heart muscle
incidence: ?/25, sa. const.: Hubbard chickens, age: 1 day, contamination: artificial (dose: **500 ppb AFB_1**, o., for 6 weeks; for detailed information please see the article), conc. range: ≤36.36* ng/g* ** ***, country: Egypt[372], *AF-residues, **after 3 weeks of AFB_1-administration (also measured after 4 and 5 weeks, lowest conc.: 34.33 ng/g after 4 weeks), ***in heart muscle
incidence: ?/25, sa. const.: Hubbard chickens, age: 1 day, contamination: artificial (dose: **750 ppb AFB_1**, o., for 6 weeks; for detailed information please see the article), conc. range: ≤58.82* ng/g* ** ***, country: Egypt[372], *AF-residues, **after 5 weeks of AFB_1-administration (also measured after 3 and 4 weeks, lowest conc.: 39.22 ng/g after 3 weeks), ***in heart muscle

Ochratoxin A
incidence: 4?/4, sa. const.: New Hampshire-Leghorn cross chicks, age: 36 days, contamination: artificial (dose: feeding 1 ppm OTA for 5 weeks and then given a single dose of 50 µg ^{3}H-OTA per chick by intubation (OTA labeled and unlabeled); for detailed information please see the article), conc. range: ≤0.45 ppb* (mean value), country: USA[316], *after 8 h (also measured after 24 and 48 h, lowest conc.: nd after 48 h)

Penicillic Acid
incidence: 3/3*, sa. const.: Hyline W-36 male laying strain chickens, age: 7 days, contamination: no PA, conc.: nr, country: USA[524], *control
incidence: 3/3, sa. const.: Hyline W-36 male laying strain chickens, age: 7 days, contamination: artificial (dose: **50 mg PA**/kg b. wt., o. by intubation, once), conc.: nd* (mean value), country: USA[524], *after 4 h
incidence: 3/3, sa. const.: Hyline W-36 male laying strain chickens, age: 7 days, contamination: artificial (dose: **100 mg PA**/kg b. wt., o. by intubation, once), conc.: nd* (mean value), country: USA[524], *after 4 h
incidence: 3/3, sa. const.: Hyline W-36 male laying strain chickens, age: 7 days, contamination: artificial (dose: **200 mg PA**/kg b. wt., o. by intubation, once), conc.: nd* (mean value), country: USA[524], *after 4 h
incidence: 3?/3, sa. const.: Hyline W-36 male laying strain chickens, age: 7 days, contamination: artificial (dose: **400 mg PA**/kg b. wt., o. by intubation, once), conc.: 590 ng/g* (mean value), country: USA[524], *after 4 h

Zearalenone
incidence: 6/6*, sa. const.: broiler chickens (Ross male X Arbor Acre female), age: 4 weeks, contamination: no ZEA

(for detailed information please see the article), conc.: nr, country: USA[106], *control
incidence: 4?/4, sa. const.: broiler chickens (Ross male X Arbor Acre female), age: 4 weeks, contamination: artificial (dose: **5.0 mg ZEA** (labeled)/kg b. wt., intubated into the crop, once; for detailed information please see the article), conc. range: ≤140 ppb* ** *** (mean value), country: USA[106], *eq. conc., **after 0.5 h (also measured after 4, 8, 12, 24 and 48 h, lowest conc.: 64.0 ppb after 48 h), ***ZEA and/or metabolites

Chicken intestine may contain the following mycotoxins and/or their metabolites:

OCHRATOXIN A
incidence: 4?/4, sa. const.: New Hampshire-Leghorn cross chicks, age: 36 days, contamination: artificial (dose: feeding 1 ppm OTA for 5 weeks and then given a single dose of 50 µg ^{3}H-OTA per chick by intubation (OTA labeled and unlabeled); for detailed information please see the article), conc. range: ≤7.05 ppb* (mean value), country: USA[316], *after 8 h (also measured after 24 and 48 h, lowest conc.: 0.33 ppb after 48 h)

Chicken kidney may contain the following mycotoxins and/or their metabolites:

AFLATOXICOL
incidence: ?/?*, sa. const.: Hubbard strain male broilers, age: 14 days, contamination: no AFB_1 (for detailed information please see the article), conc.: nr**, country: Italy[232], *control
incidence: 6?/6, sa. const.: Hubbard strain male broilers, age: 14 days, contamination: artificial (dose: **50 µg AFB_1**/kg feed, o., for up to 64 days; for detailed information please see the article), conc. range: 0.05*–0.60** µg/kg (mean values), country: Italy[232], after 64* and 36** days

incidence: ?/?*, sa. const.: Hubbard strain male broiler chickens, age: 14 days, contamination: no AFB_1 + OTA (for detailed information please see the article), conc.: nr, country: Italy[515], *control
incidence: 3?/3, sa. const.: Hubbard strain male broiler chickens, age: 14 days, contamination: artificial (dose: **50 µg AFB_1**/kg feed + 50 µg OTA/kg feed, o., for up to 64 days; for detailed information please see the article), conc. range: 0.50*–0.60** µg/kg (mean values), country: Italy[515], after 36* and 64** days

AFLATOXIN B_1
incidence: ?/?*, sa. const.: Hubbard strain male broilers, age: 14 days, contamination: no AFB_1 (for detailed information please see the article), conc.: nr, country: Italy[232], *control
incidence: 6?/6, sa. const.: Hubbard strain male broilers, age: 14 days, contamination: artificial (dose: **50 µg AFB_1**/kg feed, o., for up to 64 days; for detailed information please see the article), conc. range: 0.05*–0.06** µg/kg (mean values), country: Italy[232], after 64** and 36* days

incidence: 6/6*, sa. const.: Hubbard White Mountain broilers, age: 1 week, contamination: no AFB_1 + AFB_2, conc.: nd, country: USA[357], *control
incidence: ?/6, sa. const.: Hubbard White Mountain broilers, age: 1 week, contamination: artificial (dose: **2,057 µg AFB_1**/kg feed + **1,323 µg AFB_2**/kg feed, o., for 5 weeks), conc.: 0.05 µg/kg*, country: USA[357], *3 h after withdrawal of the contaminated feed (also measured after 1, 2 and 4 days, lowest conc.: nd after 1, 2 and 4 days)

incidence: 16?/16?*, sa. const.: Hybro broiler chickens, age: 23 days, contamination: no AF (for detailed information please see the article), conc.: nd, country: Spain[359], *control
incidence: 12?/16, sa. const.: Hybro broiler chickens, age: 23 days,

contamination: artificial (dose: **2.5 mg AF**/kg feed, o., for 32 days; for detailed information please see the article), conc. range: ≤0.37 μg/kg*, country: Spain[359] (measured at 4th, 8th*, 16th and 32nd day, lowest conc.: nd after 32 days, intoxication period)
incidence: 16?/16, sa. const.: Hybro broiler chickens, age: 23 days, contamination: artificial (dose: **5.0 mg AF**/kg feed, o., for 32 days; for detailed information please see the article), conc. range: ≤0.46 μg/kg*, country: Spain[359] (measured at 4th, 8th, 16th* and 32nd day, lowest conc.: 0.09 μg/kg after 4 days, intoxication period)

incidence: 10/10*, sa. const.: male commercial broilers, age: 3 weeks, contamination: no AF (for detailed information please see the article), conc.: nr, country: Australia[400], *control
incidence: ?/10, sa. const.: male commercial broilers, age: 3 weeks, contamination: artificial (dose: **1.5 mg total AF**/kg diet, o., for 5 weeks; for detailed information please see the article), conc. range: <0.2–0.6 μg/kg* (mean values), country: Australia[400], *after 5 weeks of AF-administration

incidence: ?/?*, sa. const.: Hubbard strain male broiler chickens, age: 14 days, contamination: no AFB_1 + OTA (for detailed information please see the article), conc.: nr, country: Italy[515], *control
incidence: 3?/3, sa. const.: Hubbard strain male broiler chickens, age: 14 days, contamination: artificial (dose: **50 μg** $\mathbf{AFB_1}$/kg feed + **50 μg OTA**/kg feed, o., for up to 64 days; for detailed information please see the article), conc. range: 0.40*–0.60** μg/kg (mean values), country: Italy[515], after 36* and 64** days

incidence: 30/30*, sa. const.: broiler chicks, age: 1 day, contamination: no AFB_1 (for detailed information please see the article), conc.: nd, country: India[595], *control
incidence: ?/30, sa. const.: broiler chicks, age: 1 day, contamination: artificial (dose: **20, 40, 60, 80 or 100* ppb** $\mathbf{AFB_1}$, o., for 15, 30 or 45 days; for detailed information please see the article), conc.: ≤3.21 ng/g**, country: India[595], **after 45 days (also measured after 15 and 30 days, lowest conc.: 3.1 ng/g after 30 days)

incidence: 2/4, sa. const.: chickens, age: 3 weeks, contamination: artificial (dose: 1 mg AFB_1 (labeled and unlabeled), o., once; for detailed information please see the article), conc. range: ≈0.15–0.25 ppm*, country: USA[617], ***after 12 h** (also measured after 48 and 72 h)
incidence: 1/4, sa. const.: chickens, age: 3 weeks, contamination: artificial (dose: 1 mg AFB_1 (labeled and unlabeled), o., once; for detailed information please see the article), conc.: ≈0.15 ppm*, country: USA[617], ***after 24 h** (also measured after 48 and 72 h)

Aflatoxin B_2
incidence: 6/6*, sa. const.: Hubbard White Mountain broilers, age: 1 week, contamination: no AFB_1 + AFB_2, conc.: nd, country: USA[357], *control
incidence: ?/6, sa. const.: Hubbard White Mountain broilers, age: 1 week, contamination: artificial (dose: **2,057 μg** $\mathbf{AFB_1}$/kg feed + **1,323 μg** $\mathbf{AFB_2}$/kg feed, o., for 5 weeks), conc. range: ≤0.05 μg/kg*, country: USA[357], *3 h after withdrawal of the contaminated feed (also measured after 1*, 2 and 4 days, lowest conc.: nd after 2 and 4 days)

Aflatoxin B_{2a}
incidence: 16?/16?*, sa. const.: Hybro broiler chickens, age: 23 days, contamination: no AF (for detailed information please see the article), conc.: nd, country: Spain[359], *control
incidence: 4?/16, sa. const.: Hybro broiler chickens, age: 23 days, contamination: artificial (dose: **2.5 mg AF**/kg feed, o., for 32 days; for detailed information

please see the article), conc. range: ≤0.15 μg/kg* (mean value), country: Spain[359] (measured at 4th, 8th, 16th* and 32nd day, lowest conc.: nd after 4, 8 and 32 days, intoxication period)
incidence: 8?/16, sa. const.: Hybro broiler chickens, age: 23 days, contamination: artificial (dose: **5.0 mg AF**/kg feed, o., for 32 days; for detailed information please see the article), conc. range: ≤0.17 μg/kg (mean value), country: Spain[359] (measured at 4th, 8th, 16th and 32nd day, lowest conc.: nd after 4 and 32 days, intoxication period)

incidence: 10/10*, sa. const.: male commercial broilers, age: 3 weeks, contamination: no AF; for detailed information please see the article), conc.: nr, country: Australia[400], *control
incidence: ?/10, sa. const.: male commercial broilers, age: 3 weeks, contamination: artificial (dose: **1.5 mg total AF**/kg diet, o., for 5 weeks; for detailed information please see the article), conc. range: 0.4–0.5 μg/kg* (mean values), country: Australia[400], *after 5 weeks of AF-administration

Aflatoxin M_1
incidence: 6/6*, sa. const.: Hubbard White Mountain broilers, age: 1 week, contamination: no AFB_1 + AFB_2, conc.: nd, country: USA[357], *control
incidence: ?/6, sa. const.: Hubbard White Mountain broilers, age: 1 week, contamination: artificial (dose: **2,057 μg AFB_1**/kg feed + **1,323 μg AFB_2**/kg feed, o., for 5 weeks), conc.: 0.10 μg/kg*, country: USA[357], *3 h after withdrawal of the contaminated feed (also measured after 1, 2 and 4 days, lowest conc.: nd after 1, 2 and 4 days)

incidence: 16?/16?*, sa. const.: Hybro broiler chickens, age: 23 days, contamination: no AF (for detailed information please see the article), conc.: nd, country: Spain[359], *control
incidence: 16?/16, sa. const.: Hybro broiler chickens, age: 23 days, contamination: artificial (dose: **2.5 mg AF**/kg feed, o., for 32 days; for detailed information please see the article), conc. range: ≤0.18 μg/kg* (mean value), country: Spain[359] (measured at 4th, 8th, 16th* and 32nd day, lowest conc.: nd after 4 days, intoxication period)
incidence: 16?/16, sa. const.: Hybro broiler chickens, age: 23 days, contamination: artificial (dose: **5.0 mg AF**/kg feed, o., for 32 days; for detailed information please see the article), conc. range: ≤0.13 μg/kg* (mean value), country: Spain[359] (measured at 4th, 8th, 16th* and 32nd day, lowest conc.: nd after 4 and 32 days, intoxication period)

incidence: ?/?*, sa. const.: Hubbard strain male broiler chickens, age: 14 days, contamination: no AFB_1 + OTA (for detailed information please see the article), conc.: nr, country: Italy[515], *control
incidence: 3?/3, sa. const.: Hubbard strain male broiler chickens, age: 14 days, contamination: artificial (dose: **50 μg AFB_1**/kg feed + **50 μg OTA**/kg feed, o., for up to 64 days; for detailed information please see the article), conc. range: <0.01 μg/kg* (mean value), country: Italy[515], *after 36 and 64 days

Aflatoxin M_2
incidence: 6/6*, sa. const.: Hubbard White Mountain broilers, age: 1 week, contamination: no AFB_1 + AFB_2, conc.: nd, country: USA[357], *control
incidence: ?/6, sa. const.: Hubbard White Mountain broilers, age: 1 week, contamination: artificial (dose: **2,057 μg AFB_1**/kg feed + **1,323 μg AFB_2**/kg feed, o., for 5 weeks), conc.: ≤2.07 μg/kg*, country: USA[357], *3 h after withdrawal of the contaminated feed (also measured after 1, 2 and 4 days, lowest conc.: nd after 2 and 4 days)

Ochratoxin A
incidence: 4?/4, sa. const.: New Hampshire-Leghorn cross chicks, age: 36 days, contamination: artificial

(dose: feeding 1 ppm OTA for 5 weeks and then given a single dose of 50 μg ^{3}H-OTA per chick by intubation (OTA labeled and unlabeled); for detailed information please see the article), conc. range: ≤11.77 ppb* (mean value), country: USA[316], *after 8 h (also measured after 24 and 48 h, lowest conc.: 0.58 ppb after 48 h)

incidence: 30/30*, sa. const.: male and female Hubbard chicks, age: 1 day, contamination: no OTA (for detailed information please see the article), conc.: nd, country: Canada[376], *control
incidence: 6?/6, sa. const.: male and female Hubbard chicks, age: 1 day, contamination: artificial (dose: **2.0 ppm OTA**, o., for 8 weeks; for detailed information please see the article), conc.: 41 ppb* ** (mean value), country: Canada[376], *in broilers, ****0 h** after withdrawal from treated feed
incidence: 6?/6, sa. const.: male and female Hubbard chicks, age: 1 day, contamination: artificial (dose: **2.0 ppm OTA**, o., for 8 weeks; for detailed information please see the article), conc.: 16 ppb* ** (mean value), country: Canada[376], *in broilers, ****24 h** after withdrawal from treated feed
incidence: 6/6, sa. const.: male and female Hubbard chicks, age: 1 day, contamination: artificial (dose: **2.0 ppm OTA**, o., for 8 weeks; for detailed information please see the article), conc.: nd* **, country: Canada[376], *in broilers, ****48 h** after withdrawal from treated feed

incidence: 10/10*, sa. const.: male commercial broilers, age: 3 weeks, contamination: no OTA; for detailed information please see the article), conc.: nr, country: Australia[400], *control
incidence: ?/10, sa. const.: male commercial broilers, age: 3 weeks, contamination: artificial (dose: **1.0 mg OTA**/kg diet, o., for 5 weeks), conc. range: 3.0–10.0 μg/kg* (mean values), country: Australia[400], *after 5 weeks of OTA-administration

incidence: 4/4*, sa. const.: wingbanded Ross broiler chicks, age: 28 days, Ø wt.: 835.19 g, contamination: no OTA (for detailed information please see the article), conc.: nd, country: The Netherlands/Hungary[473], *control
incidence: 4?/4, sa. const.: wingbanded Ross broiler chicks, age: 28 days, Ø wt.: 835.19 g, contamination: artificial (dose: a total of **0.5 mg OTA/week**, o., for 4 weeks; for detailed information please see the article), conc. range: ≤1.25 ng/g* (mean value), country: The Netherlands/Hungary[473], *after 7 days of OTA-administration (also measured after 14, 21 and 28 days, lowest conc.: nd after 28 days)

incidence: 3/3*, sa. const.: male Hubbard strain chicks, age: 35 days, contamination: no OTA alone/or OTA/PA (for detailed information please see the article), conc.: nd, country: Italy[491], *control
incidence: 3?/3, sa. const.: male Hubbard strain chicks, age: 35 days, contamination: artificial (dose: **100 μg OTA/kg** feed, o., for 31 days; for detailed information please see the article), conc.: 0.6 μg/kg* (mean value), country: Italy[491], *after 31 days
incidence: 3?/3, sa. const.: male Hubbard strain chicks, age: 35 days, contamination: artificial (dose: **100 μg OTA + 1 mg PA/kg** feed, o., for 31 days; for detailed information please see the article), conc.: 4.1 μg/kg* (mean value), country: Italy[491], *after 31 days

incidence: ?/?*, sa. const.: Hubbard strain male broiler chickens, age: 14 days, contamination: no AFB_1 + OTA (for detailed information please see the article), conc.: nr, country: Italy[515], *control
incidence: 3?/3, sa. const.: Hubbard strain male broiler chickens, age: 14 days, contamination: artificial (dose: **50 μg AFB_1/kg** feed + **50 μg OTA/kg** feed, o., for up to 64 days; for detailed information please see the article), conc. range: 0.5*–2.1** μg/kg (mean values), country: Italy[515], after 36* and 64** days

incidence: ?/?*, sa. const.: Hubbard strain male broiler chickens, age: 14 days, contamination: no AFB_1 + OTA (for detailed information please see the article), conc.: nr, country: Italy[515], *control
incidence: 3?/3, sa. const.: Hubbard strain male broiler chickens, age: 14 days, contamination: artificial (dose: **50 μg AFB_1/kg** feed + **50 μg OTA/kg** feed, o., for 36 days; for detailed information please see the article), conc. range: <0.5*–0.8** μg/kg (mean values), country: Italy[515], 14** or 28* days after withdrawal from treatment

incidence: 27/27*, sa. const.: White Leghorn chickens, age: 1 day, contamination: no OTA (for detailed information please see the article), conc.: nd, country: Denmark[566], *control
incidence: 27?/27, sa. const.: White Leghorn chickens, age: 1 day, contamination: artificial (dose: **0.3 mg OTA/kg** feed, for 341 days), conc.: ≈13.5 μg/kg* (mean value), country: Denmark[566], *after 341 days
incidence: 27?/27, sa. const.: White Leghorn chickens, age: 1 day, contamination: artificial (dose: **1 mg OTA/** kg feed, for 341 days), conc.: ≈18 μg/kg* (mean value), country: Denmark[566], *after 341 days
incidence: 5?/5, sa. const.: White Leghorn chickens, age: 1 day, contamination: artificial (dose: **1 mg OTA/kg** feed, for the **last 14 of 314 days;** for detailed information please see the article), conc.: ≈32.5 μg/kg* (mean value), country: Denmark[566], *after 341 days

incidence: ?/?*, sa. const.: newly hatched chicks of Rhode Island Red strain, contamination: no OTA (for detailed information please see the article), conc.: nd, country: Poland[590], *control
incidence: ?/?, sa. const.: newly hatched chicks of Rhode Island Red strain, contamination: artificial (dose: cockerels and hens received **2.1 ppm OTA** in the diet, o., for 5 weeks; for detailed information please see the article), conc.: 60.6 ppb*, country: Poland[590], *after ? weeks
incidence: ?/?, sa. const.: newly hatched chicks of Rhode Island Red strain, contamination: artificial (dose: cockerels and hens received **4.1 ppm OTA** in the diet, o., for 5 weeks; for detailed information please see the article), conc.: 189.0 ppb*, country: Poland[590], *after ? weeks

Penicillic Acid
incidence: 3/3*, sa. const.: Hyline W-36 male laying strain chickens, age: 7 days, contamination: no PA, conc.: nr, country: USA[524], *control
incidence: 3?/3, sa. const.: Hyline W-36 male laying strain chickens, age: 7 days, contamination: artificial (dose: **50 mg PA/kg** b. wt., o. by intubation, once), conc.: 1,480 ng/g* (mean value), country: USA[524], *after 4 h
incidence: 3?/3, sa. const.: Hyline W-36 male laying strain chickens, age: 7 days, contamination: artificial (dose: **100 mg PA/kg** b. wt., o. by intubation, once), conc.: 1,920 ng/g* (mean value), country: USA[524], *after 4 h
incidence: 3?/3, sa. const.: Hyline W-36 male laying strain chickens, age: 7 days, contamination: artificial (dose: **200 mg PA/kg** b. wt., o. by intubation, once), conc.: 2,160 ng/g* (mean value), country: USA[524], *after 4 h
incidence: 3?/3, sa. const.: Hyline W-36 male laying strain chickens, age: 7 days, contamination: artificial (dose: **400 mg PA/kg** b. wt., o. by intubation, once), conc.: 2,910 ng/g* (mean value), country: USA[524], *after 4 h

Chicken Leg see Chicken muscle, leg

Chicken Leg Muscle see Chicken muscle, leg

Chicken liver may contain the following mycotoxins and/or their metabolites:

Aflatoxicol

incidence: ?/?*, sa. const.: Hubbard strain male broilers, age: 14 days, contamination: no AFB_1 (for detailed information please see the article), conc.: nr, country: Italy[232], *control

incidence: 6?/6, sa. const.: Hubbard strain male broilers, age: 14 days, contamination: artificial (dose: **50 µg AFB_1/kg** feed, o., for up to 64 days; for detailed information please see the article), conc. range: 0.60*–1.10** µg/kg (mean values), country: Italy[232], after 64* and 36** days

incidence: 16?/16?*, sa. const.: Hybro broiler chickens, age: 23 days, contamination: no AF (for detailed information please see the article), conc.: nd, country: Spain[359], *control

incidence: 8?/16, sa. const.: Hybro broiler chickens, age: 23 days, contamination: artificial (dose: **2.5 mg AF/kg** feed, o., for 32 days; for detailed information please see the article), conc. range: ≤0.61 µg/kg* (mean value), country: Spain[359] (measured at 4th, 8th, 16th* and 32nd day, lowest conc.: nd after 4 and 32 days, intoxication period)

incidence: 4?/16, sa. const.: Hybro broiler chickens, age: 23 days, contamination: artificial (dose: **5.0 mg AF/kg** feed, o., for 32 days; for detailed information please see the article), conc. range: ≤tr* (mean value), country: Spain[359] (measured at 4th, 8th, 16th* and 32nd day, lowest conc.: nd after 4, 8 and 32 days, intoxication period)

incidence: ?/?*, sa. const.: Hubbard strain male broiler chickens, age: 14 days, contamination: no AFB_1 + OTA (for detailed information please see the article), conc.: nr, country: Italy[515], *control

incidence: 3?/3, sa. const.: Hubbard strain male broiler chickens, age: 14 days, contamination: artificial (dose: **50 µg AFB_1/kg** feed + **50 µg OTA/kg** feed, o., for up to 64 days; for detailed information please see the article), conc. range: 3.40*–3.50** µg/kg (mean value), country: Italy[515], after 36* and 64** days

Aflatoxin B_1

incidence: 20/20*, sa. const.: White Leghorn chicks, age: 14 days, contamination: no AFB_1 (for detailed information please see the article), conc.: nd, country: USA[53], *control

incidence: 20?/20, sa. const.: White Leghorn chicks, age: 14 days, contamination: artificial (dose: **120 ng AFB_1/g** ration for 28 days; for detailed information please see the article), conc.: 0.005 ng/g* (mean value), country: USA[53], *after 28 days of AFB_1-administration

incidence: 20?/20, sa. const.: White Leghorn chicks, age: 14 days, contamination: artificial (dose: **525 ng AFB_1/g** ration + **25% soil** for 28 days; for detailed information please see the article), conc.: 0.018 ng/g* (mean value), country: USA[53], *after 28 days of AFB_1-administration

incidence: 20?/20, sa. const.: White Leghorn chicks, age: 14 days, contamination: artificial (dose: **630 ng AFB_1/g** ration + **10% soil** for 28 days; for detailed information please see the article), conc.: 0.007 ng/g* (mean value), country: USA[53], *after 28 days of AFB_1-administration

incidence: 20?/20, sa. const.: White Leghorn chicks, age: 14 days, contamination: artificial (dose: **700 ng AFB_1/g** ration for 28 days; for detailed information please see the article), conc.: 1.29 ng/g* (mean value), country: USA[53], *after 28 days of AFB_1-administration

incidence: 17/17*, sa. const.: broiler chickens, contamination: no AFB_1 (for detailed information please see the article), conc.: nd, country: Germany[90], *control incidence: ?/7, sa. const.: broiler chickens, contamination: artificial (dose: **2.0 ppm AFB_1**, o., for 56 days; for detailed information please see the article), conc. range: ≤22.10 ppb* **, country: Germany[90], *AFB_1-residues, **after 8 weeks of AFB_1-administration (up to 15.0 ppm AFB_1 applicated but lower residue values recorded)

incidence: ?/?*, sa. const.: Hubbard strain male broilers, age: 14 days, contamination: no AFB_1 (for detailed information please see the article), conc.: nr, country: Italy[232], *control
incidence: 6?/6, sa. const.: Hubbard strain male broilers, age: 14 days, contamination: artificial (dose: **50 µg AFB_1/kg** feed, o., for up to 64 days; for detailed information please see the article), conc. range: 0.01*–0.02** µg/kg (mean values), country: Italy[232], after 64* and 36** days

incidence: 6/6*, sa. const.: Hubbard White Mountain broilers, age: 1 week, contamination: no AFB_1 + AFB_2, conc.: nd, country: USA[357], *control
incidence: 6/6, sa. const.: Hubbard White Mountain broilers, age: 1 week, contamination: artificial (dose: **2,057 µg AFB_1/kg** feed + **1,323 µg AFB_2/kg** feed, o., for 5 weeks), conc. range: ≤0.29 µg/kg*, country: USA[357], *3 h after withdrawal of the contaminated feed (also measured after 1, 2 and 4 days, lowest conc.: nd after 4 days)

incidence: 16?/16?*, sa. const.: Hybro broiler chickens, age: 23 days, contamination: no AF (for detailed information please see the article), conc.: nd, country: Spain[359], *control
incidence: 16?/16, sa. const.: Hybro broiler chickens, age: 23 days, contamination: artificial (dose: **2.5 mg AF/kg** feed, o., for 32 days; for detailed information please see the article), conc. range: ≤0.52 µg/kg* (mean value), country: Spain[359] (measured at 4th, 8th, 16th* and 32nd day, lowest conc.: 0.09 µg/kg after 4 days, intoxication period)
incidence: 16?/16, sa. const.: Hybro broiler chickens, age: 23 days, contamination: artificial (dose: **5.0 mg AF/kg** feed, o., for 32 days; for detailed information please see the article), conc. range: ≤0.44 µg/kg* (mean value), country: Spain[359] (measured at 4th, 8th, 16th* and 32nd day, lowest conc.: 0.05 µg/kg after 32 days, intoxication period)

incidence: 3?/3, sa. const.: broiler chicks, contamination: artificial (dose: 3 µg AFB_1/g b. wt., o. gavage, daily for up to 20 days; for detailed information please see the article), conc. range: ≤25.2 ng/g* ** wet matter basis, country: Spain[373], *AFB_1-residues, **after 12 days and 11 doses

incidence: ?/?*, sa. const.: broiler chicks, contamination: no AF (for detailed information please see the article), conc.: nd, country: India[398], *control
incidence: ?/?, sa. const.: broiler chicks, contamination: artificial (dose: **0.5 ppm AF**, for ?; for detailed information please see the article), conc. range: ≤4.7 ppb*, country: India[398], *after 45 days

incidence: ?/10*, sa. const.: male commercial broilers, age: 3 weeks, contamination: no AF; for detailed information please see the article), conc.: nr, country: Australia[400], *control
incidence: ?/10, sa. const.: male commercial broilers, age: 3 weeks, contamination: artificial (dose: **1.5 mg total AF/kg** diet, o., for 5 weeks; for detailed information please see the article), conc. range: 0.4–1.4 µg/kg* (mean values), country: Australia[400], *after 5 weeks of AF-administration

incidence: ?/?*, sa. const.: Hubbard strain male broiler chickens, age: 14 days, contamination: no AFB_1 + OTA (for detailed information please see the article), conc.: nr, country: Italy[515], *control incidence: 3?/3, sa. const.: Hubbard strain male broiler chickens, age: 14 days, contamination: artificial (dose: **50 µg AFB_1 + 50 µg OTA/kg** feed, o., for up to 64 days; for detailed information please see the article), conc. range: 0.04*–0.15** µg/kg (mean values), country: Italy[515], after 64* and 36** days

incidence: 30/30*, sa. const.: hybrid male and female Arbor Acor broiler chickens, age: 1 day, contamination: no AFB_1 and/or CP (for detailed information please see the article), conc.: nd, country: Thailand[574], *control
incidence: 30?/30, sa. const.: hybrid male and female Arbor Acor broiler chickens, age: 1 day, contamination: artificial (dose: **different conc. of AFB_1** with/without CP fed for different periods; for detailed information please see the article), conc. range: ≤0.13 ppb* ** (mean value), country: Thailand[574], *AFB_1-residues, **after 6 weeks of AFB_1-administration

incidence: 30/30*, sa. const.: broiler chicks, age: 1 day, contamination: no AFB_1 (for detailed information please see the article), conc.: nd, country: India[595], *control
incidence: ?/30, sa. const.: broiler chicks, age: 1 day, contamination: artificial (dose: **20, 40, 60, 80* or 100 ppb AFB_1**, o., for 15, 30 or 45 days; for detailed information please see the article), conc.: ≤6.740 ng/g* **, country: India[595], **after 45 days (also measured after 15 and 30 days, lowest conc.: nd after 15 days)

incidence: 2/4, sa. const.: chickens, age: 3 weeks, contamination: artificial (dose: 1 mg AFB_1 (labeled and unlabeled), o., once; for detailed information please see the article), conc. range: ≈0.25–0.3 ppm*, country: USA[617], ***after 12 h** (also measured after 48 and 72 h)
incidence: 2/4, sa. const.: chickens, age: 3 weeks, contamination: artificial (dose: 1 mg AFB_1 (labeled and unlabeled), o., once; for detailed information please see the article), conc. range: ≈0.5–0.55 ppm*, country: USA[617], ***after 24 h** (also measured after 48 and 72 h)

Aflatoxin B_2
incidence: 20/20*, sa. const.: White Leghorn chicks, age: 14 days, contamination: no AFB_1 (for detailed information please see the article), conc.: nd, country: USA[53], *control
incidence: 20/20, sa. const.: White Leghorn chicks, age: 14 days, contamination: artificial (dose: **120 ng AFB_1/g** ration for 28 days; for detailed information please see the article), conc.: nd*, country: USA[53], *after 28 days of AFB_1-administration
incidence: 20/20, sa. const.: White Leghorn chicks, age: 14 days, contamination: artificial (dose: **525 ng AFB_1/g** ration **+ 25% soil** for 28 days; for detailed information please see the article), conc.: nd*, country: USA[53], *after 28 days of AFB_1-administration
incidence: 20/20, sa. const.: White Leghorn chicks, age: 14 days, contamination: artificial (dose: **630 ng AFB_1/g** ration **+ 10% soil** for 28 days; for detailed information please see the article), conc.: nd*, country: USA[53], *after 28 days of AFB_1-administration
incidence: 20?/20, sa. const.: White Leghorn chicks, age: 14 days, contamination: artificial (dose: **700 ng AFB_1/g** ration for 28 days; for detailed information please see the article), conc.: 0.028 ng/g* (mean value), country: USA[53], *after 28 days of AFB_1-administration

incidence: 6/6*, sa. const.: Hubbard White Mountain broilers, age: 1 week, contamination: no AFB_1 + AFB_2, conc.: nd, country: USA[357], *control
incidence: 6/6, sa. const.: Hubbard White Mountain broilers, age: 1 week, contamination: artificial (dose: **2,057 µg AFB_1/kg** feed + **1,323 µg AFB_2/kg** feed, o.,

for 5 weeks), conc. range: ≤0.13 μg/kg*, country: USA[357], *3 h after withdrawal of the contaminated feed (also measured after 1, 2 and 4 days, lowest conc.: nd after 4 days)

Aflatoxin B_{2a}
incidence: 16?/16?*, sa. const.: Hybro broiler chickens, age: 23 days, contamination: no AF (for detailed information please see the article), conc.: nd, country: Spain[359], *control
incidence:12?/16, sa. const.: Hybro broiler chickens, age: 23 days, contamination: artificial (dose: **2.5 mg AF**/kg feed, o., for 32 days; for detailed information please see the article), conc. range: ≤0.22 μg/kg* (mean value), country: Spain[359] (measured at 4th, 8th, 16th* and 32nd day, lowest conc.: nd after 32 days, intoxication period)
incidence:12?/16, sa. const.: Hybro broiler chickens, age: 23 days, contamination: artificial (dose: **5.0 mg AF**/kg feed, o., for 32 days; for detailed information please see the article), conc. range: ≤0.16 μg/kg* (mean value), country: Spain[359] (measured at 4th*, 8th, 16th* and 32nd day, lowest conc.: nd after 32 days, intoxication period)

incidence: 10/10*, sa. const.: male commercial broilers, age: 3 weeks, contamination: no AF; for detailed information please see the article), conc.: nr, country: Australia[400], *control
incidence: ?/10, sa. const.: male commercial broilers, age: 3 weeks, contamination: artificial (dose: **1.5 mg total AF**/kg diet, o., for 5 weeks; for detailed information please see the article), conc. range: 0.9–2.1 μg/kg* (mean values), country: Australia[400], *after 5 weeks of AF-administration

Aflatoxin G_1
incidence: 16?/16?*, sa. const.: Hybro broiler chickens, age: 23 days, contamination: no AF (for detailed information please see the article), conc.: nd, country: Spain[359], *control
incidence: 2?/16, sa. const.: Hybro broiler chickens, age: 23 days, contamination: artificial (dose: **2.5 mg AF**/kg feed, o., for 32 days; for detailed information please see the article), conc. range: 0.15*–0.16** μg/kg (mean value), country: Spain[359] (measured at 4th, 8th, 16th** and 32nd* day)
incidence: ?/16, sa. const.: Hybro broiler chickens, age: 23 days, contamination: artificial (dose: **5.0 mg AF**/kg feed, o., for 32 days; for detailed information please see the article), conc.: 0.1 μg/kg* (mean value), country: Spain[359] (measured at 4th, 8th, 16th* and 32nd day)

incidence: 10/10*, sa. const.: male commercial broilers, age: 3 weeks, contamination: no AF; for detailed information please see the article), conc.: nr, country: Australia[400], *control
incidence: ?/10, sa. const.: male commercial broilers, age: 3 weeks, contamination: artificial (dose: **1.5 mg total AF**/kg diet, o., for 5 weeks; for detailed information please see the article), conc. range: <0.2–0.6 μg/kg* (mean values), country: Australia[400], *after 5 weeks of AF-administration

Aflatoxin G_{2a}
incidence: 10/10*, sa. const.: male commercial broilers, age: 3 weeks, contamination: no AF; for detailed information please see the article), conc.: nr, country: Australia[400], *control
incidence: ?/10, sa. const.: male commercial broilers, age: 3 weeks, contamination: artificial (dose: **1.5 mg total AF**/kg diet, o., for 5 weeks; for detailed information please see the article), conc. range: <0.2–0.6 μg/kg* (mean values), country: Australia[400], *after 5 weeks of AF-administration

Aflatoxin M_1
incidence: 20/20*, sa. const.: White Leghorn chicks, age: 14 days, contamination: no AFB_1 (for detailed

information please see the article), conc.: nd, country: USA[53], *control
incidence: 20/20, sa. const.: White Leghorn chicks, age: 14 days, contamination: artificial (dose: **120 ng AFB_1/g** ration for 28 days; for detailed information please see the article), conc.: nd*, country: USA[53], *after 28 days of AFB_1-administration
incidence: 20/20, sa. const.: White Leghorn chicks, age: 14 days, contamination: artificial (dose: **525 ng AFB_1/g** ration **+ 25% soil** for 28 days; for detailed information please see the article), conc.: nd*, country: USA[53], *after 28 days of AFB_1-administration
incidence: 20/20, sa. const.: White Leghorn chicks, age: 14 days, contamination: artificial (dose: **630 ng AFB_1/g** ration **+ 10% soil** for 28 days; for detailed information please see the article), conc.: nd*, country: USA[53], *after 28 days of AFB_1-administration
incidence: 20?/20, sa. const.: White Leghorn chicks, age: 14 days, contamination: artificial (dose: **700 ng** AFB_1/g ration for 28 days; for detailed information please see the article), conc.: 0.095 ng/g* (mean value), country: USA[53], *after 28 days of AFB_1-administration

incidence: ?/?*, sa. const.: Hubbard strain male broilers, age: 14 days, contamination: no AFB_1 (for detailed information please see the article), conc.: nr, country: Italy[232], *control
incidence: 3?/3, sa. const.: Hubbard strain male broilers, age: 14 days, contamination: artificial (dose: **50 μg AFB_1/kg** feed, o., for up to 64 days; for detailed information please see the article), conc.: 0.36 μg/kg*, country: Italy[232], *after 64 days (also measured after 36 days but conc.: nd)

incidence: 6/6*, sa. const.: Hubbard White Mountain broilers, age: 1 week, contamination: no AFB_1 + AFB_2, conc.: nd, country: USA[357], *control
incidence: 5/6, sa. const.: Hubbard White Mountain broilers, age: 1 week, contamination: artificial (dose: **2,057 μg AFB_1/kg** feed + **1,323 μg AFB_2/kg** feed, o., for 5 weeks), conc. range: ≤0.14 μg/kg*, country: USA[357], *3 h after withdrawal of the contaminated feed (also measured after 1, 2 and 4 days, lowest conc.: nd after 4 days)

incidence: 16?/16?*, sa. const.: Hybro broiler chickens, age: 23 days, contamination: no AF (for detailed information please see the article), conc.: nd, country: Spain[359], *control
incidence: 12?/16?, sa. const.: Hybro broiler chickens, age: 23 days, contamination: artificial (dose: **2.5 mg AF/kg** feed, o., for 32 days; for detailed information please see the article), conc. range: ≤0.09 μg/kg* (mean value), country: Spain[359] (measured at 4th, 8th, 16th* and 32nd day, lowest conc.: nd after 4 and 32 days, intoxication period)
incidence: 16?/16?, sa. const.: Hybro broiler chickens, age: 23 days, contamination: artificial (dose: **5.0 mg AF/kg** feed, o., for 32 days; for detailed information please see the article), conc. range: ≤0.03 μg/kg* (mean value), country: Spain[359] (measured at 4th*, 8th*, 16th* and 32nd* day, lowest conc.: nd after 8 days, intoxication period)

incidence: ?/?*, sa. const.: Hubbard strain male broiler chickens, age: 14 days, contamination: no AFB_1 + OTA (for detailed information please see the article), conc.: nr, country: Italy[515], *control
incidence: 3?/3, sa. const.: Hubbard strain male broiler chickens, age: 14 days, contamination: artificial (dose: **50 μg AFB_1 + 50 μg OTA/kg** feed, o., for up to 64 days; for detailed information please see the article), conc. range: <0.01 μg/kg* (mean value), country: Italy[515], *after 36 and 64 days

incidence: 30/30*, sa. const.: hybrid male and female Arbor Acor broiler chickens, age: 1 day, contamination: no AFB_1 and/or

CP (for detailed information please see the article), conc.: nd, country: Thailand[574], *control
incidence: 30?/30, sa. const.: hybrid male and female Arbor Acor broiler chickens, age: 1 day, contamination: artificial (dose: **different conc. of AFB_1** with/without CP fed for different periods; for detailed information please see the article), conc. range: ≤0.32 ppb* ** (mean value), country: Thailand[574], *AFM_1-residues, **after 6 weeks of AFB_1-administration

Aflatoxin M_2
incidence: 6/6*, sa. const.: Hubbard White Mountain broilers, age: 1 week, contamination: no AFB_1 + AFB_2, conc.: nd, country: USA[357], *control
incidence: 6/6, sa. const.: Hubbard White Mountain broilers, age: 1 week, contamination: artificial (dose: **2,057 µg AFB_1/kg** feed and **1,323 µg AFB_2/kg** feed, o., for 5 weeks), conc. range: ≤6.53 µg/kg*, country: USA[357], *3 h after withdrawal of the contaminated feed (also measured after 1, 2 and 4 days, lowest conc.: nd after 4 days)

Aflatoxin
incidence: 25/25*, sa. const.: Hubbard chickens, age: 1 day, contamination: no AFB_1, (for detailed information please see the article), conc.: nd, country: Egypt[372], *control
incidence: ?/25, sa. const.: Hubbard chickens, age: 1 day, contamination: artificial (dose: **100 ppb AFB_1**, o., for 6 weeks; for detailed information please see the article), conc. range: ≤10.20 ng/g* **, country: Egypt[372], *AF-residues, **after 3 weeks of AFB_1-administration (also measured after 4 and 5 weeks, lowest conc.: 3.93 ng/g after 4 weeks)
incidence: ?/25, sa. const.: Hubbard chickens, age: 1 day, contamination: artificial (dose: **250 ppb AFB_1**, o., for 6 weeks; for detailed information please see the article), conc. range: ≤13.42 ng/g***, country: Egypt[372], *AF-residues, **after 3 weeks of AFB_1-administration (also measured after 4 and 5 weeks, lowest conc.: 10.0 ng/g after 5 weeks)
incidence: ?/25, sa. const.: Hubbard chickens, age: 1 day, contamination: artificial (dose: **500 ppb AFB_1**, o., for 6 weeks; for detailed information please see the article), conc. range: ≤18.44 ng/g***, country: Egypt[372], *AF-residues, **after 4 weeks of AFB_1-administration (also measured after 3 and 5 weeks, lowest conc.: 16.42 ng/g after 3 weeks)
incidence: ?/25, sa. const.: Hubbard chickens, age: 1 day, contamination: artificial (dose: **750 ppb AFB_1**, o., for 6 weeks; for detailed information please see the article), conc. range: ≤34.93 ng/g***, country: Egypt[372], *AF-residues, **after 3 weeks of AFB_1-administration (also measured after 4 and 5 weeks, lowest conc.: 21.28 ng/g after 5 weeks)

HT-2 Toxin
incidence: 12?/12, sa. const.: chickens, age: 5 weeks, contamination: artificial (dose: 3.5 mg T-2 toxin/kg b. wt., i.p., once), conc.: 233 ppb* (mean value), country: USA[145], *after 18 h

3′-Hydroxy HT-2 Toxin
incidence: 12?/12, sa. const.: chickens, age: 5 weeks, contamination: artificial (dose: 3.5 mg T-2 toxin/kg b. wt., i.p., once), conc.: 1,370 ppb* (mean value), country: USA[145], *after 18 h

Ochratoxin A
incidence: 4?/4, sa. const.: New Hampshire-Leghorn cross chicks, age: 36 days, contamination: artificial (dose: feeding 1 ppm OTA for 5 weeks and then given a single dose of 50 µg ^{3}H-OTA per chick by intubation (OTA labeled and unlabeled); for detailed information please see the article), conc. range: ≤4.07 ppb* (mean value), country: USA[316], *after 8 h (also measured after 24 and 48 h, lowest conc.: 0.14 ppb after 48 h)

incidence: 30/30*, sa. const.: male and female Hubbard chicks, age: 1 day,

contamination: no OTA (for detailed information please see the article), conc.: nd, country: Canada[376], *control
incidence: 6?/6, sa. const.: male and female Hubbard chicks, age: 1 day, contamination: artificial (dose: **2.0 ppm OTA**, o., for 8 weeks; for detailed information please see the article), conc.: 24 ppb* ** (mean value), country: Canada[376], *in broilers, ****0 h** after withdrawal from treated feed
incidence: 6/6, sa. const.: male and female Hubbard chicks, age: 1 day, contamination: artificial (dose: **2.0 ppm OTA**, o., for 8 weeks; for detailed information please see the article), conc.: nd* **, country: Canada[376], *in broilers, ****24 h** after withdrawal from treated feed
incidence: 6/6, sa. const.: male and female Hubbard chicks, age: 1 day, contamination: artificial (dose: **2.0 ppm OTA**, o., for 8 weeks; for detailed information please see the article), conc.: nd* **, country: Canada[376], *in broilers, ****48 h** after withdrawal from treated feed

incidence: 10/10*, sa. const.: male commercial broilers, age: 3 weeks, contamination: no OTA; for detailed information please see the article), conc.: nr, country: Australia[400], *control
incidence: ?/10, sa. const.: male commercial broilers, age: 3 weeks, contamination: artificial (dose: **1.0 mg OTA/kg** diet, o., for 5 weeks), conc. range: 1.5–2.5 µg/kg* (mean values), country: Australia[400], *after 5 weeks of OTA-administration

incidence: 4/4*, sa. const.: wingbanded Ross broiler chicks, age: 28 days, Ø wt.: 835.19 g, contamination: no OTA (for detailed information please see the article), conc.: nd, country: The Netherlands/Hungary[473], *control
incidence: 4?/4, sa. const.: wingbanded Ross broiler chicks, age: 28 days, Ø wt.: 835.19 g, contamination: artificial (dose: a total of **0.5 mg OTA**/week, o., for 4 weeks; for detailed information please see the article), conc. range: ≤1.45 ng/g* (mean value), country: The Netherlands/Hungary[473], *after 7 days of OTA-administration (also measured after 14, 21 and 28 days, lowest conc.: 0.2 ng/g after 28 days)

incidence: 3/3*, sa. const.: male Hubbard strain chicks, age: 35 days, contamination: no OTA alone or OTA/PA (for detailed information please see the article), conc.: nd, country: Italy[491], *control
incidence: 3?/3, sa. const.: male Hubbard strain chicks, age: 35 days, contamination: artificial (dose: **100 µg OTA/kg** feed, o., for 31 days; for detailed information please see the article), conc.: 2.1 µg/kg* (mean value), country: Italy[491], *after 31 days
incidence: 3?/3, sa. const.: male Hubbard strain chicks, age: 35 days, contamination: artificial (dose: **100 µg OTA + 1 mg PA**/kg feed, o., for 31 days; for detailed information please see the article), conc.: 5.0 µg/kg* (mean value), country: Italy[491], *after 31 days

incidence: ?/?*, sa. const.: Hubbard strain male broiler chickens, age: 14 days, contamination: no AFB_1 + OTA (for detailed information please see the article), conc.: nr, country: Italy[515], *control
incidence: 3?/3, sa. const.: Hubbard strain male broiler chickens, age: 14 days, contamination: artificial (dose: **50 µg AFB_1**/kg feed + **50 µg OTA/kg** feed, o., for up to 64 days; for detailed information please see the article), conc. range: 38.0*–40.0** µg/kg (mean value), country: Italy[515], after 64* and 36** days
incidence: ?/?*, sa. const.: Hubbard strain male broiler chickens, age: 14 days, contamination: no AFB_1 + OTA (for detailed information please see the article), conc.: nr, country: Italy[515], *control
incidence: 3?/3, sa. const.: Hubbard strain male broiler chickens, age: 14 days, contamination: artificial (dose: **50 µg**

AFB$_1$/kg feed + **50 µg OTA/kg** feed, o., for 36 days; for detailed information please see the article), conc. range: 1.2*–2.0** µg/kg (mean values), country: Italy[515], 14** or 28* days after withdrawal from treatment

incidence: ?/?*, sa. const.: male Euribird broiler chickens, age: 1 day, contamination: no OTA (for detailed information please see the article), conc.: nr, country: Poland[531], *control
incidence: ?/?, sa. const.: male Euribird broiler chickens, age: 1 day, contamination: artificial (dose: **0.5 mg OTA/kg** feed, o., for 8 weeks; for detailed information please see the article), conc.: nd*, country: Poland[531], *after 8 weeks of OTA-administration
incidence: ?/?, sa. const.: male Euribird broiler chickens, age: 1 day, contamination: artificial (dose: **1.0 mg OTA/kg** feed, o., for 8 weeks; for detailed information please see the article), conc.: 32.8 µg/kg* (mean value), country: Poland[531], *after 8 weeks of OTA-administration
incidence: ?/?, sa. const.: male Euribird broiler chickens, age: 1 day, contamination: artificial (dose: **1.5 mg OTA/kg** feed, o., for 8 weeks; for detailed information please see the article), conc.: 14.2 µg/kg* (mean value), country: Poland[531], *after 8 weeks of OTA-administration
incidence: ?/?, sa. const.: male Euribird broiler chickens, age: 1 day, contamination: artificial (dose: **2.0 mg OTA/kg** feed, o., for 8 weeks; for detailed information please see the article), conc.: 34.3 µg/kg* (mean value), country: Poland[531], *after 8 weeks of OTA-administration

incidence: ?/?*, sa. const.: female Euribird broiler chickens, age: 1 day, contamination: no OTA (for detailed information please see the article), conc.: nr, country: Poland[531], *control
incidence: ?/?, sa. const.: female Euribird broiler chickens, age: 1 day, contamination: artificial (dose: **0.5 mg OTA/kg** feed, o., for 8 weeks; for detailed information please see the article), conc.: nd*, country: Poland[531], *after 8 weeks of OTA-administration
incidence: ?/?, sa. const.: female Euribird broiler chickens, age: 1 day, contamination: artificial (dose: **1.0 mg OTA/kg** feed, o., for 8 weeks; for detailed information please see the article), conc.: 22.8 µg/kg* (mean value), country: Poland[531], *after 8 weeks of OTA-administration
incidence: ?/?, sa. const.: female Euribird broiler chickens, age: 1 day, contamination: artificial (dose: **1.5 mg OTA/kg** feed, o., for 8 weeks; for detailed information please see the article), conc.: 39.8 µg/kg* (mean value), country: Poland[531], *after 8 weeks of OTA-administration
incidence: ?/?, sa. const.: female Euribird broiler chickens, age: 1 day, contamination: artificial (dose: **2.0 mg OTA/kg** feed, o., for 8 weeks; for detailed information please see the article), conc.: 58.6 µg/kg* (mean value), country: Poland[531], *after 8 weeks of OTA-administration

incidence: 27/27*, sa. const.: White Leghorn chickens, age: 1 day, contamination: no OTA (for detailed information please see the article), conc.: nd, country: Denmark[566], *control
incidence: 27?/27, sa. const.: White Leghorn chickens, age: 1 day, contamination: artificial (dose: **0.3 mg OTA/kg** feed, for 341 days), conc.: ≈2 µg/kg* (mean value), country: Denmark[566], *after 341 days
incidence: 27?/27, sa. const.: White Leghorn chickens, age: 1 day, contamination: artificial (dose: **1 mg OTA/kg** feed, for 341 days), conc.: ≈5 µg/kg* (mean value), country: Denmark[566], *after 341 days
incidence: 5?/5, sa. const.: White Leghorn chickens, age: 1 day,

contamination: artificial (dose: **1 mg OTA/** kg feed, for the **last 14 of 314 days;** for detailed information please see the article), conc.: ≈8.5 μg/kg* (mean value), country: Denmark[566], *after 341 days

incidence: ?/?*, sa. const.: newly hatched chicks of Rhode Island Red strain, contamination: no OTA (for detailed information please see the article), conc.: nd, country: Poland[590], *control
incidence: ?/?, sa. const.: newly hatched chicks of Rhode Island Red strain, contamination: artificial (dose: cockerels and hens received **2.1 ppm OTA** in the diet, o., for 5 weeks; for detailed information please see the article), conc.: 26.0 ppb*, country: Poland[590], *after ? weeks
incidence: ?/?, sa. const.: newly hatched chicks of Rhode Island Red strain, contamination: artificial (dose: cockerels and hens received **4.1 ppm OTA** in the diet, o., for 5 weeks; for detailed information please see the article), conc.: 83.0 ppb*, country: Poland[590], *after ? weeks

incidence: ?/?*, sa. const.: broilers, contamination: no OTA (for detailed information please see the article), conc.: 0.03 μg/kg** (mean value), country: Germany[601], *control, **after 6 weeks
incidence: ?/?, sa. const.: broilers, contamination: artificial (dose: **1.5 mg OTA**/kg in the diet, o., for 3 weeks **afterwards OTA-free diet** for 3 weeks (for detailed information please see the article), conc.: 0.02 μg/kg* (mean value), country: Germany[601], *after 6 weeks
incidence: ?/?, sa. const.: broilers, contamination: artificial (dose: **OTA-free diet** for 3 weeks **afterwards 1.5 mg OTA/kg in the diet,** o., for 3 weeks (for detailed information please see the article), conc.: 16.16 μg/kg* (mean value), country: Germany[601], *after 6 weeks
incidence: ?/?, sa. const.: broilers, contamination: artificial (dose: **1.5 mg OTA/kg in the diet,** o., for **6 weeks** (for detailed information please see the article), conc.: 10.98 μg/kg* (mean value), country: Germany[601], *after 6 weeks

Penicillic Acid
incidence: 3/3*, sa. const.: Hyline W-36 male laying strain chickens, age: 7-days, contamination: no PA, conc.: nr, country: USA[524], *control
incidence: 3/3, sa. const.: Hyline W-36 male laying strain chickens, age: 7-days, contamination: artificial (dose: **50 mg PA**/kg b. wt., o. by intubation, once), conc.: nd*, country: USA[524], *after 4 h
incidence: 3/3, sa. const.: Hyline W-36 male laying strain chickens, age: 7-days, contamination: artificial (dose: **100 mg PA**/kg b. wt., o. by intubation, once), conc.: nd* (mean value), country: USA[524], *after 4 h
incidence: 3?/3, sa. const.: Hyline W-36 male laying strain chickens, age: 7 days, contamination: artificial (dose: **200 mg PA**/kg b. wt., o. by intubation, once), conc.: 5.1 ng/g* (mean value), country: USA[524], *after 4 h
incidence: 3?/3, sa. const.: Hyline W-36 male laying strain chickens, age: 7 days, contamination: artificial (dose: **400 mg PA**/kg b. wt., o. by intubation, once), conc.: 6.8 ng/g* (mean value), country: USA[524], *after 4 h

T-2 Toxin
incidence: 12?/12, sa. const.: chickens, age: 5 weeks, contamination: artificial (dose: 3.5 mg T-2 toxin/kg b. wt., i.p., once), conc.: 4 ppb* (mean value), country: USA[145], *after 18 h

T-2 Tetraol
incidence: 12?/12, sa. const.: chickens, age: 5 weeks, contamination: artificial (dose: 3.5 mg T-2 toxin/kg b. wt., i.p., once), conc.: 18 ppb* (mean value), country: USA[145], *after 18 h

4-Acetoxy T-2 Tetraol
incidence: 12?/12, sa. const.: chickens, age: 5 weeks, contamination: artificial (dose: 3.5 mg T-2 toxin/kg b. wt., i.p.,

once), conc.: 20 ppb* (mean value), country: USA[145], *after 18 h

15-Acetoxy T-2 Tetraol

incidence: 12?/12, sa. const.: chickens, age: 5 weeks, contamination: artificial (dose: 3.5 mg T-2 toxin/kg b. wt., i.p., once), conc.: 22 ppb* (mean value), country: USA[145], *after 18 h

T-2 Triol

incidence: 12?/12, sa. const.: chickens, age: 5 weeks, contamination: artificial (dose: 3.5 mg T-2 toxin/kg b. wt., i.p., once), conc.: 210 ppb* (mean value), country: USA[145], *after 18 h

Zearalenone

incidence: 6/6*, sa. const.: broiler chickens (Ross male X Arbor Acre female), age: 4 weeks, contamination: no ZEA (for detailed information please see the article), conc.: nr, country: USA[106], *control
incidence: 4?/4, sa. const.: broiler chickens (Ross male X Arbor Acre female), age: 4 weeks, contamination: artificial (dose: **5.0 mg ZEA** (labeled)/kg b. wt., intubated into the crop, once; for detailed information please see the article), conc. range: ≤2,095.0 ppb* ** *** (mean value), country: USA[106], *eq. conc., **after 0.5 h (also measured after 4, 8, 12, 24 and 48 h, lowest conc.: 161.2 ppb after 48 h), ***ZEA and/or metabolites
incidence: 5?/11, sa. const.: broiler chickens (Ross male X Arbor Acre female), age: 4 weeks, contamination: artificial (dose: **10 mg ZEA**/bird for 6 days and intubation of **20.92 mg ZEA**/ml at day 9 (ZEA labeled and unlabeled); for detailed information please see the article), conc. range: ≤681 ppb*, country: USA[106], *after 0.5 h (also measured after 4, 8 and 24 h, lowest conc.: 0.0 ppm after 24 h)

incidence: 5/5*, sa. const.: Leghorn broilers, age: 2 weeks, contamination: no ZEA (for detailed information please see the article), conc.: nr, country: Romania/France[635], *control
incidence: 5?/5, sa. const.: Leghorn broilers, age: 2 weeks, contamination: artificial (dose: **25 mg ZEA**/kg b. wt., i.p., for 3 days (for detailed information please see the article), conc.: 3.01 μg/g*, country: Romania/France[635], after 24 h of the last treatment
incidence: 5?/5, sa. const.: Leghorn broilers, age: 2 weeks, contamination: artificial (dose: **25 mg ZEA**/kg b. wt., i.p., once and before **80 mg PB**/kg b. wt., i.p., for 3 days (for detailed information please see the article), conc.: 3.69 μg/g*, country: Romania/France[635], after 24 h of the last treatment

α-Zearalenol

incidence: 6/6*, sa. const.: broiler chickens (Ross male X Arbor Acre female), age: 4 weeks, contamination: no ZEA (for detailed information please see the article), conc.: nr, country: USA[106], *control
incidence: 3?/11, sa. const.: broiler chickens (Ross male X Arbor Acre female), age: 4 weeks, contamination: artificial (dose: **10 mg ZEA**/bird for 6 days and intubation of **20.92 mg ZEA**/ml at day 9 (ZEA labeled and unlabeled); for detailed information please see the article), conc. range: ≤1,200 ppb*, country: USA[106], *after 0.5 h (also measured after 4, 8, 24 and 48 h, lowest conc.: 0.0 ppm after 48 h)

incidence: 5/5*, sa. const.: Leghorn broilers, age: 2 weeks, contamination: no ZEA (for detailed information please see the article), conc.: nr, country: Romania/France[635], *control
incidence: 5?/5, sa. const.: Leghorn broilers, age: 2 weeks, contamination: artificial (dose: **25 mg ZEA**/kg b. wt., i.p., for 3 days (for detailed information please see the article), conc.: 26.46 μg/g*, country: Romania/France[635], after 24 h of the last treatment

incidence: 5?/5, sa. const.: Leghorn broilers, age: 2 weeks, contamination: artificial (dose: **25 mg ZEA**/kg b. wt., i.p., once and before **80 mg PB**/kg b. wt., i.p., for 3 days (for detailed information please see the article), conc.: 5.48 μg/g*, country: Romania/France[635], after 24 h of the last treatment

β-Zearalenol
incidence: 6/6*, sa. const.: broiler chickens (Ross male X Arbor Acre female), age: 4 weeks, contamination: no ZEA (for detailed information please see the article), conc.: nr, country: USA[106], *control
incidence: 3?/11, sa. const.: broiler chickens (Ross male X Arbor Acre female), age: 4 weeks, contamination: artificial (dose: **10 mg ZEA**/bird for 6 days and intubation of **20.92 mg ZEA**/ml at day 9 (ZEA labeled and unlabeled); for detailed information please see the article), conc. range: ≤662 ppb*, country: USA[106], *after 0.5 h (also measured after 4, 8, 24 and 48 h, lowest conc.: 0.0 ppm after 48 h)

incidence: 5/5*, sa. const.: Leghorn broilers, age: 2 weeks, contamination: no ZEA (for detailed information please see the article), conc.: nr, country: Romania/France[635], *control
incidence: 5?/5, sa. const.: Leghorn broilers, age: 2 weeks, contamination: artificial (dose: **25 mg ZEA**/kg b. wt., i.p., for 3 days (for detailed information please see the article), conc.: 23.69 μg/g*, country: Romania/France[635], after 24 h of the last treatment
incidence: 5?/5, sa. const.: Leghorn broilers, age: 2 weeks, contamination: artificial (dose: **25 mg ZEA**/kg b. wt., i.p., once and before **80 mg PB**/kg b. wt., i.p., for 3 days (for detailed information please see the article), conc.: 9.57 μg/g*, country: Romania/France[635], after 24 h of the last treatment

Chicken lung may contain the following mycotoxins and/or their metabolites:

Aflatoxin B_1
incidence: 2/4, sa. const.: chickens, age: 3 weeks, contamination: artificial (dose: 1 mg AFB_1 (labeled and unlabeled), o., once; for detailed information please see the article), conc. range: ≈0.3 ppm*, country: USA[617], *after **12 h** (also measured after 48 and 72 h)
incidence: 1/4, sa. const.: chickens, age: 3 weeks, contamination: artificial (dose: 1 mg AFB_1 (labeled and unlabeled), o., once; for detailed information please see the article), conc.: ≈0.3 ppm*, country: USA[617], *after **24 h** (also measured after 48 and 72 h)

HT-2 Toxin
incidence: 12?/12, sa. const.: chickens, age: 5 weeks, contamination: artificial (dose: 3.5 mg T-2 toxin/kg b. wt., i.p., once), conc.: <1 ppb* (mean value), country: USA[145], *after 18 h

3′-Hydroxy HT-2 Toxin
incidence: 12?/12, sa. const.: chickens, age: 5 weeks, contamination: artificial (dose: 3.5 mg T-2 toxin/kg b. wt.., i.p.., once), conc.: 3 ppb* (mean value), country: USA[145], *after 18 h

Ochratoxin A
incidence: 4?/4, sa. const.: New Hampshire-Leghorn cross chicks, age: 36 days, contamination: artificial (dose: feeding 1 ppm OTA for 5 weeks and then given a single dose of 50 μg ^{3}H-OA per chick by intubation (OTA unlabeled and labeled); for detailed information please see the article), conc. range: ≤4.99 ppb* (mean value), country: USA[316], *after 48 h (also measured after 8 and 24 h, lowest conc.: 0.49 ppb after 24 h)

Chicken meat may contain the following mycotoxins and/or their metabolites:

Zearalanone
incidence: 5/5*, sa. const.: Leghorn broilers, age: 2 weeks, contamination: no

ZEA (for detailed information please see the article), conc.: nr, country: Romania/France[635], *control
incidence: 5?/5, sa. const.: Leghorn broilers, age: 2 weeks, contamination: artificial (dose: **25 mg ZEA**/kg b. wt., i.p., for 3 days (for detailed information please see the article), conc.: 1.72 ng/g* dry matter, country: Romania/France[635], after 24 h of the last treatment
incidence: 5?/5, sa. const.: Leghorn broilers, age: 2 weeks, contamination: artificial (dose: **25 mg ZEA**/kg b. wt., i.p., once and before **80 mg PB**/kg b. wt., i.p., for 3 days (for detailed information please see the article), conc.: 0.56 ng/g* dry matter, country: Romania/France[635], after 24 h of the last treatment

β-Zearalanol
incidence: 5/5*, sa. const.: Leghorn broilers, age: 2 weeks, contamination: no ZEA (for detailed information please see the article), conc.: nr, country: Romania/France[635], *control
incidence: 5?/5, sa. const.: Leghorn broilers, age: 2 weeks, contamination: artificial (dose: **25 mg ZEA**/kg b. wt., i.p., for 3 days (for detailed information please see the article), conc.: 5.61 ng/g* dry matter, country: Romania/France[635], after 24 h of the last treatment
incidence: 5?/5, sa. const.: Leghorn broilers, age: 2 weeks, contamination: artificial (dose: **25 mg ZEA**/kg b. wt., i.p., once and before **80 mg PB**/kg b. wt., i.p., for 3 days (for detailed information please see the article), conc.: 7.10 ng/g* dry matter, country: Romania/France[635], after 24 h of the last treatment

Zearalenone
incidence: 5/5*, sa. const.: Leghorn broilers, age: 2 weeks, contamination: no ZEA (for detailed information please see the article), conc.: nr, country: Romania/France[635], *control
incidence: 5?/5, sa. const.: Leghorn broilers, age: 2 weeks, contamination: artificial (dose: **25 mg ZEA**/kg b. wt., i.p., for 3 days (for detailed information please see the article), conc.: 35.72 ng/g* dry matter, country: Romania/France[635], after 24 h of the last treatment
incidence: 5?/5, sa. const.: Leghorn broilers, age: 2 weeks, contamination: artificial (dose: **25 mg ZEA**/kg b. wt., i.p., once and before **80 mg PB**/kg b. wt., i.p., for 3 days (for detailed information please see the article), conc.: 22.98 ng/g* dry matter, country: Romania/France[635], after 24 h of the last treatment

α-Zearalenol
incidence: 5/5*, sa. const.: Leghorn broilers, age: 2 weeks, contamination: no ZEA (for detailed information please see the article), conc.: nr, country: Romania/France[635], *control
incidence: 5?/5, sa. const.: Leghorn broilers, age: 2 weeks, contamination: artificial (dose: **25 mg ZEA**/kg b. wt., i.p., for 3 days (for detailed information please see the article), conc.: 13.42 ng/g* dry matter, country: Romania/France[635], after 24 h of the last treatment
incidence: 5?/5, sa. const.: Leghorn broilers, age: 2 weeks, contamination: artificial (dose: **25 mg ZEA**/kg b. wt., i.p., once and before **80 mg PB**/kg b. wt., i.p., for 3 days (for detailed information please see the article), conc.: 4.08 ng/g* dry matter, country: Romania/France[635], after 24 h of the last treatment

β-Zearalenol
incidence: 5/5*, sa. const.: Leghorn broilers, age: 2 weeks, contamination: no ZEA (for detailed information please see the article), conc.: nr, country: Romania/France[635], *control
incidence: 5?/5, sa. const.: Leghorn broilers, age: 2 weeks, contamination: artificial (dose: **25 mg ZEA**/kg b. wt., i.p., for 3 days (for detailed information please see the article), conc.: 11.47 ng/g* dry matter, country: Romania/France[635], after 24 h of the last treatment

incidence: 5?/5, sa. const.: Leghorn broilers, age: 2 weeks, contamination: artificial (dose: **25 mg ZEA**/kg b. wt., i.p., once and before **80 mg PB**/kg b. wt., i.p., for 3 days (for detailed information please see the article), conc.: 5.48 ng/g* dry matter, country: Romania/France[635], after 24 h of the last treatment

Chicken muscle may contain the following mycotoxins and/or their metabolites:

Aflatoxin B_1
incidence: 20/20*, sa. const.: White Leghorn chicks, age: 14 days, contamination: no AFB_1 (for detailed information please see the article), conc.: nd, country: USA[53], *control
incidence: 20/20, sa. const.: White Leghorn chicks, age: 14 days, contamination: artificial (dose: **120 ng AFB_1/g** ration for 28 days; for detailed information please see the article), conc.: nd*, country: USA[53], *after 28 days of AFB_1-administration
incidence: 20?/20, sa. const.: White Leghorn chicks, age: 14 days, contamination: artificial (dose: **525 ng AFB_1/g** ration **+ 25% soil** for 28 days; for detailed information please see the article), conc.: 0.004 ng/g* (mean value), country: USA[53], *after 28 days of AFB_1-administration
incidence: 20?/20, sa. const.: White Leghorn chicks, age: 14 days, contamination: artificial (dose: **630 ng AFB_1/g** ration **+ 10% soil** for 28 days; for detailed information please see the article), conc.: 0.003 ng/g* (mean value), country: USA[53], *after 28 days of AFB_1-administration
incidence: 20?/20, sa. const.: White Leghorn chicks, age: 14 days, contamination: artificial (dose: **700 ng AFB_1/g** ration for 28 days; for detailed information please see the article), conc.: 0.014 ng/g* (mean value), country: USA[53], *after 28 days of AFB_1-administration

incidence: 16?/16?*, sa. const.: Hybro broiler chickens, age: 23 days, contamination: no AF (for detailed information please see the article), conc.: nd, country: Spain[359], *control
incidence: 16?/16, sa. const.: Hybro broiler chickens, age: 23 days, contamination: artificial (dose: **2.5 mg AF**/kg feed, o., for 32 days; for detailed information please see the article), conc. range: ≤0.21* µg/kg (mean value), country: Spain[359] (measured at 4th*, 8th*, 16th and 32nd day, lowest conc.: 0.07 µg/kg after 16 days, intoxication period)
incidence: 16?/16, sa. const.: Hybro broiler chickens, age: 23 days, contamination: artificial (dose: **5.0 mg AF**/kg feed, o., for 32 days; for detailed information please see the article), conc. range: ≤0.09 µg/kg* (mean value), country: Spain[359] (measured at 4th, 8th, 16th* and 32nd day, lowest conc.: 0.01 µg/kg after 8 days, intoxication period)

incidence: ?/?*, sa. const.: broiler chicks, contamination: no AF (for detailed information please see the article), conc.: nd, country: India[398], *control
incidence: ?/?, sa. const.: broiler chicks, contamination: artificial (dose: **0.5 ppm AF**; for detailed information please see the article), conc. range: ≤2.0 ppb*, country: India[398], *after 45 days

incidence: 30/30*, sa. const.: hybrid male and female Arbor Acor broiler chickens, age: 1 day, contamination: no AFB_1 and/or CP (for detailed information please see the article), conc.: nd, country: Thailand[574], *control
incidence: 30?/30, sa. const.: hybrid male and female Arbor Acor broiler chickens, age: 1 day, contamination: artificial (dose: **different conc. of AFB_1** with/without CP fed for different periods; for detailed information please see the article), conc.: 0.02 ppb* ** (mean value), country: Thailand[574], *AFB_1-residues, **after 6 weeks of AFB_1-administration

incidence: 30/30*, sa. const.: broiler chicks, age: 1 day, contamination: no AFB_1 (for detailed information please see the article), conc.: nd, country: India[595], *control

incidence: ?/30, sa. const.: broiler chicks, age: 1 day, contamination: artificial (dose: **20, 40, 60, 80 or 100* ppb AFB_1**, o., for 15, 30 or 45 days (for detailed information please see the article), conc.: ≤5.129 ng/g* **, country: India[595], **after 45 days (also measured after 15 and 30 days, lowest conc.: nd after 15 days)

Aflatoxin B_2

incidence: 20/20*, sa. const.: White Leghorn chicks, age: 14 days, contamination: no AFB_1 (for detailed information please see the article), conc.: nd, country: USA[53], *control

incidence: 20/20, sa. const.: White Leghorn chicks, age: 14 days, contamination: artificial (dose: **120 ng AFB_1/g** ration for 28 days; for detailed information please see the article), conc.: nd*, country: USA[53], *after 28 days of AFB_1-administration

incidence: 20/20, sa. const.: White Leghorn chicks, age: 14 days, contamination: artificial (dose: **525 ng AFB_1/g** ration **+ 25% soil** for 28 days; for detailed information please see the article), conc.: nd*, country: USA[53], *after 28 days of AFB_1-administration

incidence: 20/20, sa. const.: White Leghorn chicks, age: 14 days, contamination: artificial (dose: **630 ng AFB_1/g** ration **+ 10% soil** for 28 days; for detailed information please see the article), conc.: nd*, country: USA[53], *after 28 days of AFB_1-administration

incidence: 20?/20, sa. const.: White Leghorn chicks, age: 14 days, contamination: artificial (dose: **700 ng AFB_1/g** ration for 28 days; for detailed information please see the article), conc.: 0.002 ng/g* (mean value), country: USA[53], *after 28 days of AFB_1-administration

Aflatoxin M_1

incidence: 20/20*, sa. const.: White Leghorn chicks, age: 14 days, contamination: no AFB_1 (for detailed information please see the article), conc.: nd, country: USA[53], *control

incidence: 20?/20, sa. const.: White Leghorn chicks, age: 14 days, contamination: artificial (dose: **120 ng AFB_1/g** ration for 28 days; for detailed information please see the article), conc.: 0.064 ng/g* (mean value), country: USA[53], *after 28 days of AFB_1-administration

incidence: 20?/20, sa. const.: White Leghorn chicks, age: 14 days, contamination: artificial (dose: **525 ng AFB_1/g** ration **+ 25% soil** for 28 days; for detailed information please see the article), conc.: 0.11 ng/g* (mean value), country: USA[53], *after 28 days of AFB_1-administration

incidence: 20/20, sa. const.: White Leghorn chicks, age: 14 days, contamination: artificial (dose: **630 ng AFB_1/g** ration **+ 10% soil** for 28 days; for detailed information please see the article), conc.: nd*, country: USA[53], *after 28 days of AFB_1-administration

incidence: 20/20, sa. const.: White Leghorn chicks, age: 14 days, contamination: artificial (dose: **700 ng AFB_1/g** ration for 28 days; for detailed information please see the article), conc.: nd*, country: USA[53], *after 28 days of AFB_1-administration

incidence: 30/30*, sa. const.: hybrid male and female Arbor Acor broiler chickens, age: 1 day, contamination: no AFB_1 and/or CP (for detailed information please see the article), conc.: nd, country: Thailand[574], *control

incidence: 30?/30, sa. const.: hybrid male and female Arbor Acor broiler chickens, age: 1 day, contamination: artificial (dose: **different conc. of AFB_1** with/without CP fed for different periods; for detailed information please see the article), conc. range: ≤0.08 ppb* ** (mean value), country: Thailand[574], *AFM_1-residues, **after 6 weeks of AFB_1-administration

Cyclopiazonic Acid

incidence: 10/10*, sa. const.: unsexed White Leghorn chickens, age: 200 days, contamination: no CPA (for detailed information please see the article), conc.: nr, country: USA[557], *control

incidence: 10?/10, sa. const.: unsexed White Leghorn chickens, age: 200 days, contamination: artificial (dose: **0.5 mg CPA/kg** b. wt., by crop gavage, once; for detailed information please see the article), conc.: 350 ppb* (mean value), country: USA[557], *after 3 h (also measured after 24, 48 and 96 h but lower residue values recorded, lowest conc.: ≈nd after 48 h)

incidence: 10?/10, sa. const.: unsexed White Leghorn chickens, age: 200 days, contamination: artificial (dose: **5.0 mg CPA/kg** b. wt., by crop gavage, once; for detailed information please see the article), conc.: 2,000 ppb* (mean value), country: USA[557], *after 3 h (also measured after 24, 48 and 96 h but lower residue values recorded, lowest conc.: ≈50 ppb after 96 h)

incidence: 10?/10, sa. const.: unsexed White Leghorn chickens, age: 200 days, contamination: artificial (dose: **10.0 mg CPA/kg** b. wt., by crop gavage, once; for detailed information please see the article), conc.: 2,500 ppb* (mean value), country: USA[557], *after 3 h (also measured after 24, 48 and 96 h but lower residue values recorded, lowest conc.: ≈200 ppb after 96 h)

Ochratoxin A

incidence: 27/27*, sa. const.: White Leghorn chickens, age: 1 day, contamination: no OTA (for detailed information please see the article), conc.: nd, country: Denmark[566], *control

incidence: 27?/27, sa. const.: White Leghorn chickens, age: 1 day, contamination: artificial (dose: **0.3 mg OTA/kg** feed, for 341 days), conc.: ≈2 μg/kg* (mean value), country: Denmark[566], *after 341 days

incidence: 27?/27, sa. const.: White Leghorn chickens, age: 1 day, contamination: artificial (dose: **1 mg OTA/kg** feed, for 341 days), conc.: ≈2 μg/kg* (mean value), country: Denmark[566], *after 341 days

incidence: 5?/5, sa. const.: White Leghorn chickens, age: 1 day, contamination: artificial (dose: **1 mg OTA/kg** feed, for the **last 14 of 314 days;** for detailed information please see the article), conc.: ≈6 μg/kg* (mean value), country: Denmark[566], *after 341 days

incidence: ?/?*, sa. const.: broilers, contamination: no OTA (for detailed information please see the article), conc.: nd, country: Germany[601], *control

incidence: ?/?, sa. const.: broilers, contamination: artificial (dose: **1.5 mg OTA/kg** in the diet, o., for 3 weeks **afterwards OTA-free diet** for 3 weeks (for detailed information please see the article), conc.: nd*, country: Germany[601], *after 6 weeks

incidence: ?/?, sa. const.: broilers, contamination: artificial (dose: **OTA-free diet** for 3 weeks **afterwards 1.5 mg OTA/kg in the diet**, o., for 3 weeks (for detailed information please see the article), conc.: 7.46 μg/kg* (mean value), country: Germany[601], *after 6 weeks

incidence: ?/?, sa. const.: broilers, contamination: artificial (dose: **1.5 mg OTA/kg in the diet**, o., for **6 weeks** (for detailed information please see the article), conc.: 3.02 μg/kg* (mean value), country: Germany[601], *after 6 weeks

Zearalenone

incidence: 6/6*, sa. const.: broiler chickens (Ross male X Arbor Acre female), age: 4 weeks, contamination: no ZEA (for detailed information please see the article), conc.: nr, country: USA[106], *control

incidence: 4?/4, sa. const.: broiler chickens (Ross male X Arbor Acre female), age: 4 weeks, contamination: artificial (dose: **5.0 mg ZEA** (labeled)/kg b.w., intubated into the crop, once; for detailed information please see the article), conc. range: ≤111.1 ppb* ** *** (mean value), country: USA[106], *eq. conc., **after 24 h (also measured after 0.5, 4, 8, 12 and 48 h, lowest conc.: 35.0 ppb after 0.5 h), ***ZEA and/or metabolites

Chicken muscle, breast may contain the following mycotoxins and/or their metabolites:

AFLATOXIN B_1
incidence: 17/17*, sa. const.: broiler chickens, contamination: no AFB_1 (for detailed information please see the article), conc.: nd, country: Germany[90], *control
incidence: ?/113, sa. const.: broiler chickens, contamination: artificial (dose: **15.0 ppm AFB_1**, o., for 56 days; for detailed information please see the article), conc. range: ≤12.20 ppb* **, country: Germany[90], *AFB_1-residues, **after 8 weeks of AFB_1-administration (lower doses also applicated and lower residue values recorded)

incidence: 6/6*, sa. const.: Hubbard White Mountain broilers, age: 1 week, contamination: no AFB_1 + AFB_2, conc.: nd, country: USA[357], *control
incidence: 5/6, sa. const.: Hubbard White Mountain broilers, age: 1 week, contamination: artificial (dose: **2,057 µg AFB_1/kg feed + 1,323 µg AFB_2/kg feed**, o., for 5 weeks), conc. range: ≤0.08 µg/kg*, country: USA[357], *3 h after withdrawal of the contaminated feed (also measured after 1, 2 and 4 days, lowest conc.: nd after 4 days)

incidence: 2/4, sa. const.: chickens, age: 3 weeks, contamination: artificial (dose: 1 mg AFB_1 (labeled and unlabeled), o., once; for detailed information please see the article), conc. range: ≈0.15–0.8 ppm*, country: USA[617], *after **12 h** (also measured after 48 and 72 h)
incidence: 1/4, sa. const.: chickens, age: 3 weeks, contamination: artificial (dose: 1 mg AFB_1 (labeled and unlabeled), o., once; for detailed information please see the article), conc.: ≈0.15 ppm*, country: USA[617], *after **24 h** (also measured after 48 and 72 h)

AFLATOXIN B_2
incidence: 6/6*, sa. const.: Hubbard White Mountain broilers, age: 1 week, contamination: no AFB_1 + AFB_2, conc.: nd, country: USA[357], *control
incidence: 6/6, sa. const.: Hubbard White Mountain broilers, age: 1 week, contamination: artificial (dose: **2,057 µg AFB_1/kg feed + 1,323 µg AFB_2/kg feed**, o., for 5 weeks), conc. range: ≤0.05 µg/kg*, country: USA[357], *3 h after withdrawal of the contaminated feed (also measured after 1, 2 and 4 days, lowest conc.: nd after 4 days)

AFLATOXIN M_1
incidence: 6/6*, sa. const.: Hubbard White Mountain broilers, age: 1 week, contamination: no AFB_1 + AFB_2, conc.: nd, country: USA[357], *control
incidence: 5/6, sa. const.: Hubbard White Mountain broilers, age: 1 week, contamination: artificial (dose: **2,057 µg AFB_1/kg feed + 1,323 µg AFB_2/kg feed**, o., for 5 weeks), conc. range: ≤0.05 µg/kg*, country: USA[357], *3 h after withdrawal of the contaminated feed (also measured after 1, 2 and 4 days, lowest conc.: nd after 4 days)

AFLATOXIN M_2
incidence: 6/6*, sa. const.: Hubbard White Mountain broilers, age: 1 week, contamination: no AFB_1 + AFB_2, conc.: nd, country: USA[357], *control
incidence: 2/6, sa. const.: Hubbard White Mountain broilers, age: 1 week, contamination: artificial (dose: **2,057 µg AFB_1/kg feed + 1,323 µg AFB_2/kg feed**, o., for 5 weeks), conc. range: ≤0.03 µg/kg*, Ø conc.: 0.025 µg/kg*, country: USA[357], *2 days after withdrawal of the contaminated feed (also measured after 0, 1 and 4 days, lowest conc.: nd after 4 days)

Aflatoxin
incidence: 25/25*, sa. const.: Hubbard chickens, age: 1 day, contamination: no AFB_1, (for detailed information please see the article), conc.: nd, country: Egypt[372], *control
incidence: ?/25, sa. const.: Hubbard chickens, age: 1 day, contamination: artificial (dose: **100 ppb AFB_1**, o., for 6 weeks; for detailed information please see the article), conc. range: ≤8.03 ng/g* **, country: Egypt[372], *AF-residues, **after 5 weeks of AFB_1-administration (also measured after 3 and 4 weeks, lowest conc.: 2.42 ng/g after 3 weeks)
incidence: ?/25, sa. const.: Hubbard chickens, age: 1 day, contamination: artificial (dose: **250 ppb AFB_1**, o., for 6 weeks; for detailed information please see the article), conc. range: ≤11.03 ng/g***, country: Egypt[372], *AF-residues, **after 5 weeks of AFB_1-administration (also measured after 3 and 4 weeks, lowest conc.: 7.66 ng/g after 3 weeks)
incidence: ?/25, sa. const.: Hubbard chickens, age: 1 day, contamination: artificial (dose: **500 ppb AFB_1**, o., for 6 weeks; for detailed information please see the article), conc. range: ≤18.71 ng/g***, country: Egypt[372], *AF-residues, **after 5 weeks of AFB_1-administration (also measured after 3 and 4 weeks, lowest conc.: 14.55 ng/g after 3 weeks)
incidence: ?/25, sa. const.: Hubbard chickens, age: 1 day, contamination: artificial (dose: **750 ppb AFB_1**, o., for 6 weeks; for detailed information please see the article), conc. range: ≤23.45 ng/g* **, country: Egypt[372], *AF-residues, **after 4 weeks of AFB_1-administration (also measured after 3 and 5 weeks, lowest conc.: 18.18 ng/g after 3 weeks)

Chicken muscle, leg may contain the following mycotoxins and/or their metabolites:

Aflatoxin B_1
incidence: 6/6*, sa. const.: Hubbard White Mountain broilers, age: 1 week, contamination: no AFB_1 + AFB_2, conc.: nd, country: USA[357], *control
incidence: 6/6, sa. const.: Hubbard White Mountain broilers, age: 1 week, contamination: artificial (dose: **2,057 µg AFB_1/kg feed + 1,323 µg AFB_2/kg feed**, o., for 5 weeks), conc. range: ≤0.16 µg/kg*, country: USA[357], *3 h after withdrawal of the contaminated feed (also measured after 1, 2 and 4 days, lowest conc.: nd after 4 days)

Aflatoxin B_2
incidence: 6/6*, sa. const.: Hubbard White Mountain broilers, age: 1 week, contamination: no AFB_1 + AFB_2, conc.: nd, country: USA[357], *control
incidence: 2/6, sa. const.: Hubbard White Mountain broilers, age: 1 week, contamination: artificial (dose: **2,057 µg AFB_1/kg feed + 1,323 µg AFB_2/kg feed**, o., for 5 weeks), conc. range: ≤0.06 µg/kg*, country: USA[357], *1 day after withdrawal of the contaminated feed (also measured after 0, 2 and 4 days, lowest conc.: nd after 4 days)

Aflatoxin M_1
incidence: 6/6*, sa. const.: Hubbard White Mountain broilers, age: 1 week, contamination: no AFB_1 + AFB_2, conc.: nd, country: USA[357], *control
incidence: 6/6, sa. const.: Hubbard White Mountain broilers, age: 1 week, contamination: artificial (dose: **2,057 µg AFB_1/kg feed + 1,323 µg AFB_2/kg feed**, o., for 5 weeks), conc. range: ≤0.06 µg/kg*, country: USA[357], *3 h after withdrawal of the contaminated feed (also measured after 1, 2 and 4 days, lowest value.: nd after 4 days)

Aflatoxin M_2
incidence: 6/6*, sa. const.: Hubbard White Mountain broilers, age: 1 week, contamination: no AFB_1 + AFB_2, conc.: nd, country: USA[357], *control
incidence: 3/6, sa. const.: Hubbard White Mountain broilers, age: 1 week, contamination: artificial (dose: **2,057 µg AFB_1/kg feed + 1,323 µg AFB_2/kg feed**, o., for 5 weeks), conc. range: ≤0.08 µg/kg*, country: USA[357], *1 day after withdrawal of the contaminated feed (also measured

after 0, 2 and 4 days, lowest conc.: nd after 4 days)

Ochratoxin A
incidence: 4?/4, sa. const.: New Hampshire-Leghorn cross chicks, age: 36 days, contamination: artificial (dose: feeding 1 ppm OTA for 5 weeks and then given a single dose of 50 μg ^{3}H-OTA per chick by intubation (OTA labeled and unlabeled); for detailed information please see the article), conc. range: ≤0.21 ppb* (mean value), country: USA[316], *after 8 h (also measured after 24 and 48 h, lowest conc.: 0.04 ppb after 48 h)

Chicken muscle, red may contain the following mycotoxins and/or their metabolites:

Ochratoxin A
incidence: ?/?*, sa. const.: male Euribird broiler chickens, age: 1 day, contamination: no OTA (for detailed information please see the article), conc.: nr, country: Poland[531], *control
incidence: ?/?, sa. const.: male Euribird broiler chickens, age: 1 day, contamination: artificial (dose: **0.5 mg OTA/kg** feed, o., for 8 weeks; for detailed information please see the article), conc.: nd*, country: Poland[531], *after 8 weeks of OTA-administration
incidence: ?/?, sa. const.: male Euribird broiler chickens, age: 1 day, contamination: artificial (dose: **1.0 mg OTA/kg** feed, o., for 8 weeks; for detailed information please see the article), conc.: 0.8 μg/kg* (mean value), country: Poland[531], *after 8 weeks of OTA-administration
incidence: ?/?, sa. const.: male Euribird broiler chickens, age: 1 day, contamination: artificial (dose: **1.5 mg OTA/kg** feed, o., for 8 weeks; for detailed information please see the article), conc.: nd*, country: Poland[531], *after 8 weeks of OTA-administration
incidence: ?/?, sa. const.: male Euribird broiler chickens, age: 1 day, contamination: artificial (dose: **2.0 mg OTA/kg** feed, o., for 8 weeks; for detailed information please see the article), conc.: 3.5 μg/kg* (mean value), country: Poland[531], *after 8 weeks of OTA-administration

incidence: ?/?*, sa. const.: female Euribird broiler chickens, age: 1 day, contamination: no OTA (for detailed information please see the article), conc.: nr, country: Poland[531], *control
incidence: ?/?, sa. const.: female Euribird broiler chickens, age: 1 day, contamination: artificial (dose: **0.5 mg OTA/kg** feed, o., for 8 weeks; for detailed information please see the article), conc.: nd*, country: Poland[531], *after 8 weeks of OTA-administration
incidence: ?/?, sa. const.: female Euribird broiler chickens, age: 1 day, contamination: artificial (dose: **1.0 mg OTA/kg** feed, o., for 8 weeks; for detailed information please see the article), conc.: 2.8 μg/kg* (mean value), country: Poland[531], *after 8 weeks of OTA-administration
incidence: ?/?, sa. const.: female Euribird broiler chickens, age: 1 day, contamination: artificial (dose: **1.5 mg OTA/kg** feed, o., for 8 weeks; for detailed information please see the article), conc.: 1.0 μg/kg* (mean value), country: Poland[531], *after 8 weeks of OTA-administration
incidence: ?/?, sa. const.: female Euribird broiler chickens, age: 1 day, contamination: artificial (dose: **2.0 mg OTA/kg** feed, o., for 8 weeks; for detailed information please see the article), conc.: 7.5 μg/kg* (mean value), country: Poland[531], *after 8 weeks of OTA-administration

Chicken muscle, thigh may contain the following mycotoxins and/or their metabolites:

Aflatoxicol
incidence: ?/?*, sa. const.: Hubbard strain male broiler chickens, age: 14 days, contamination: no AFB_1 + OTA

(for detailed information please see the article), conc.: nr, country: Italy[515], *control
incidence: 3?/3, sa. const.: Hubbard strain male broiler chickens, age: 14 days, contamination: artificial (dose: **50 µg AFB_1/kg feed + 50 µg OTA/kg feed**, o., for up to 64 days; for detailed information please see the article), conc. range: <0.04 µg/kg* (mean value), country: Italy[515], *after 36 and 64 days

Aflatoxin B_1
incidence: 17/17*, sa. const.: broiler chickens, contamination: no AFB_1 (for detailed information please see the article), conc.: nd, country: Germany[90], *control
incidence: ?/113, sa. const.: broiler chickens, contamination: artificial (dose: **15 ppm AFB_1**, o., for 56 days; for detailed information please see the article), conc. range: ≤24.20 ppb* **, country: Germany[90], *AFB_1-residues, **after 8 weeks of AFB_1-administration (lower doses also applicated and lower residue values recorded)

incidence: ?/?*, sa. const.: Hubbard strain male broilers, age: 14 days, contamination: no AFB_1 (for detailed information please see the article), conc.: nr, country: Italy[232], *control
incidence: 3?/3, sa. const.: Hubbard strain male broilers, age: 14 days, contamination: artificial (dose: **50 µg AFB_1/kg feed**, o., for up to 64 days; for detailed information please see the article), conc.: 0.01 µg/kg* (mean value), country: Italy[232], *after 64 days (also measured after 36 days but conc.: nd)
incidence: ?/?*, sa. const.: Hubbard strain male broiler chickens, age: 14 days, contamination: no AFB_1 + OTA (for detailed information please see the article), conc.: nr, country: Italy[515], *control
incidence: 3?/3, sa. const.: Hubbard strain male broiler chickens, age: 14 days, contamination: artificial (dose: **50 µg AFB_1/kg feed + 50 µg OTA/kg feed**, o., for up to 64 days; for detailed information please see the article), conc. range: 0.04*–0.05** µg/kg (mean values), country: Italy[515], after 36* and 64** days

Aflatoxin M_1
incidence: ?/?*, sa. const.: Hubbard strain male broilers, age: 14 days, contamination: no AFB_1 (for detailed information please see the article), conc.: nr, country: Italy[232], *control
incidence: 3?/3, sa. const.: Hubbard strain male broilers, age: 14 days, contamination: artificial (dose: **50 µg AFB_1/kg feed**, o., for up to 64 days; for detailed information please see the article), conc.: 0.02 µg/kg* (mean value), country: Italy[232], *after 36 and 64 days
incidence: ?/?*, sa. const.: Hubbard strain male broiler chickens, age: 14 days, contamination: no AFB_1 + OTA (for detailed information please see the article), conc.: nr, country: Italy[515], *control
incidence: 3?/3, sa. const.: Hubbard strain male broiler chickens, age: 14 days, contamination: artificial (dose: **50 µg AFB_1/kg feed + 50 µg OTA/kg feed**, o., for up to 64 days; for detailed information please see the article), conc. range: <0.01 µg/kg (mean value), country: Italy[515], *after 36 and 64 days

Ochratoxin A
incidence: 4/4*, sa. const.: wingbanded Ross broiler chicks, age: 28 days, Ø wt.: 835.19 g, contamination: no OTA (for detailed information please see the article), conc.: nd, country: The Netherlands/Hungary[473], *control
incidence: 4?/4, sa. const.: wingbanded Ross broiler chicks, age: 28 days, Ø wt.: 835.19 g, contamination: artificial (dose: a total of **0.5 mg OTA/week**, o., for 4 weeks; for detailed information please see the article), conc. range: ≈≤0.06 µg/kg* (mean value), country: The Netherlands/Hungary[473], *after 21 days of OTA-administration (also measured after 7 and 14 days, lowest value conc.: nd after 7 and 14 days)

incidence: 3/3*, sa. const.: male Hubbard strain chicks, age: 35 days, contamination: no OTA alone/or OTA/PA (for detailed information please see the article), conc.: nd, country: Italy[491], *control
incidence: 3?/3, sa. const.: male Hubbard strain chicks, age: 35 days, contamination: artificial (dose: **100 µg OTA**/kg feed, o., for 31 days; for detailed information please see the article), conc.: tr*, country: Italy[491], *after 31 days
incidence: 3?/3, sa. const.: male Hubbard strain chicks, age: 35 days, contamination: artificial (dose: **100 µg OTA + 1 mg PA**/kg feed, o., for 31 days; for detailed information please see the article), conc.: tr*, country: Italy[491], *after 31 days

incidence: ?/?*, sa. const.: Hubbard strain male broiler chickens, age: 14 days, contamination: no AFB_1 + OTA (for detailed information please see the article), conc.: nr, country: Italy[515], *control
incidence: 3?/3, sa. const.: Hubbard strain male broiler chickens, age: 14 days, contamination: artificial (dose: **50 µg** $\mathbf{AFB_1}$/kg feed + **50 µg OTA**/kg feed, o., for up to 64 days; for detailed information please see the article), conc. range: 1.9*–4.2** µg/kg (mean value), country: Italy[515], after 36* and 64** days
incidence: ?/?*, sa. const.: Hubbard strain male broiler chickens, age: 14 days, contamination: no AFB_1 + OTA (for detailed information please see the article), conc.: nr, country: Italy[515], *control
incidence: 3?/3, sa. const.: Hubbard strain male broiler chickens, age: 14 days, contamination: artificial (dose: **50 µg** $\mathbf{AFB_1}$/kg feed + **50 µg OTA**/kg feed, o., for 36 days; for detailed information please see the article), conc. range: <0.5*–0.8** µg/kg (mean values), country: Italy[515], 14** or 28* days after withdrawal from treatment

Chicken muscle, white may contain the following mycotoxins and/or their metabolites:

Ochratoxin A
incidence: ?/?*, sa. const.: male Euribird broiler chickens, age: 1 day, contamination: no OTA (for detailed information please see the article), conc.: nd, country: Poland[531], *control
incidence: ?/?, sa. const.: male Euribird broiler chickens, age: 1 day, contamination: artificial (dose: **0.5 mg OTA**/kg feed, o., for 8 weeks; for detailed information please see the article), conc.: nd*, country: Poland[531], *after 8 weeks of OTA-administration
incidence: ?/?, sa. const.: male Euribird broiler chickens, age: 1 day, contamination: artificial (dose: **1.0 mg OTA**/kg feed, o., for 8 weeks; for detailed information please see the article), conc.: nd*, country: Poland[531], *after 8 weeks of OTA-administration
incidence: ?/?, sa. const.: male Euribird broiler chickens, age: 1 day, contamination: artificial (dose: **1.5 mg OTA**/kg feed, o., for 8 weeks; for detailed information please see the article), conc.: 3.0 µg/kg* (mean value), country: Poland[531], *after 8 weeks of OTA-administration
incidence: ?/?, sa. const.: male Euribird broiler chickens, age: 1 day, contamination: artificial (dose: **2.0 mg OTA**/kg feed, o., for 8 weeks; for detailed information please see the article), conc.: 4.5 µg/kg* (mean value), country: Poland[531], *after 8 weeks of OTA-administration

incidence: ?/?*, sa. const.: female Euribird broiler chickens, age: 1 day, contamination: no OTA (for detailed information please see the article), conc.: nr, country: Poland[531], *control
incidence: ?/?, sa. const.: female Euribird broiler chickens, age: 1 day, contamination: artificial (dose: **0.5 mg OTA**/kg feed, o., for 8 weeks; for detailed information please see the article), conc.: nd*, country: Poland[531], *after 8 weeks of OTA-administration

incidence: ?/?, sa. const.: female Euribird broiler chickens, age: 1 day, contamination: artificial (dose: **1.0 mg OTA**/kg feed, o., for 8 weeks; for detailed information please see the article), conc.: nd*, country: Poland[531], *after 8 weeks of OTA-administration
incidence: ?/?, sa. const.: female Euribird broiler chickens, age: 1 day, contamination: artificial (dose: **1.5 mg OTA**/kg feed, o., for 8 weeks; for detailed information please see the article), conc.: 5.0 µg/kg* (mean value), country: Poland[531], *after 8 weeks of OTA-administration
incidence: ?/?, sa. const.: female Euribird broiler chickens, age: 1 day, contamination: artificial (dose: **2.0 mg OTA**/kg feed, o., for 8 weeks; for detailed information please see the article), conc.: 8.5 µg/kg* (mean value), country: Poland[531], *after 8 weeks of OTA-administration

Chicken plasma may contain the following mycotoxins and/or their metabolites:

FUSARENON-X
incidence: ?/6, sa. const.: broiler chickens, age: 4 weeks, Ø wt.: 1.35 kg, contamination: artificial (dose: 2.2 mg FX/kg b. wt., **i.v.**, once), conc. range: ≈≤270 ng/ml* (mean value), country: Thailand/Japan[474], *after ≈5 min (also at other min intervals up to 180 min measured, lowest conc.: ≈1.1 ng/ml after 180 min)
incidence: ?/6, sa. const.: broiler chickens, age: 4 weeks, Ø wt.: 1.35 kg, contamination: artificial (dose: 2.2 mg FX/kg b. wt., **o.**, once), conc. range: ≈≤5.5 ng/ml* (mean value), country: Thailand/Japan[474], *after ≈10 min (also at other min intervals up to 120 min measured, lowest conc.: ≈1.1 ng/ml after 120 min)

NIVALENOL
incidence: ?/6, sa. const.: broiler chickens, age: 4 weeks, Ø wt.: 1.35 kg, contamination: artificial (dose: 2.2 mg FX/kg b. wt., **i.v.**, once), conc. range: ≈≤419 ng/ml* (mean value), country: Thailand/Japan[474], *after ≈5 min (also at other min intervals up to 240 min measured, lowest conc.: ≈10 ng/ml after 240 min)
incidence: ?/6, sa. const.: broiler chickens, age: 4 weeks, Ø wt.: 1.35 kg, contamination: artificial (dose: 2.2 mg FX/kg b. wt., **o.**, once), conc. range: ≈≤240 ng/ml* (mean value), country: Thailand/Japan[474], *after ≈10 min (also at other min intervals up to 180 min measured, lowest conc.: nd? after 180 min)

OCHRATOXIN A
incidence: ?/6, sa. const.: Leghorn chickens, Ø wt.: 1.84 kg, contamination: artificial (dose: 2 mg OTA/kg, **o.** (by gavage), once; for detailed information please see the article), conc. range: ≤0.78 µg/ml* (mean value), country: France[172], *after 0.33 h (also at other hour intervals up to 4 h measured, lowest conc.: ≈0.1 µg/ml after 4 h)
incidence: ?/6, sa. const.: Leghorn chickens, Ø wt.: 1.84 kg, contamination: artificial (dose: 2 mg OTA/kg, **i.v.**, once; for detailed information please see the article), conc. range: ≈≤18 µg/ml* (mean value), country: France[172], *after 0.5 h? (also at other hour intervals up to 4 h measured, lowest conc.: ≈0.3 µg/ml after 4 h)

incidence: 4/4*, sa. const.: wingbanded Ross broiler chicks, age: 28 days, Ø wt.: 835.19 g, contamination: no OTA (for detailed information please see the article), conc.: nd, country: The Netherlands/Hungary[473], *control
incidence: 4?/4, sa. const.: wingbanded Ross broiler chicks, age: 28 days, Ø wt.: 835.19 g, contamination: artificial (dose: a total of **0.5 mg OTA**/week, o., for 4 weeks; for detailed information please see the article), conc. range: ≈≤1.6 ng/ml* (mean value), country: The Netherlands/Hungary[473], *after 14 days of OTA-administration (also measured after

7, 21 and 28 days, lowest conc.: 0.25 ng/g after 28 days)

incidence: ?/?*, sa. const.: newly hatched chicks of Rhode Island Red strain, contamination: no OTA (for detailed information please see the article), conc.: nd, country: Poland[590], *control
incidence: ?/?, sa. const.: newly hatched chicks of Rhode Island Red strain, contamination: artificial (dose: cockerels and hens received **2.1 ppm OTA** in the diet, o., for 5 weeks; for detailed information please see the article), conc.: 2.7 ppb*, country: Poland[590], *after ? weeks
incidence: ?/?, sa. const.: newly hatched chicks of Rhode Island Red strain, contamination: artificial (dose: cockerels and hens received **4.1 ppm OTA** in the diet, o., for 5 weeks; for detailed information please see the article), conc.: 5.5 ppb*, country: Poland[590], *after ? weeks

ZEARALENONE
incidence: 6/6*, sa. const.: broiler chickens (Ross male X Arbor Acre female), age: 4 weeks, contamination: no ZEA (for detailed information please see the article), conc.: nr, country: USA[106], *control
incidence: 4?/4, sa. const.: broiler chickens (Ross male X Arbor Acre female), age: 4 weeks, contamination: artificial (dose: **5.0 mg ZEA** (labeled)/kg b. wt., intubated into the crop, once; for detailed information please see the article), conc. range: ≤164.3 ppb* ** *** (mean value), country: USA[106], *eq. conc., **after 0.5 h (also measured after 4, 8, 12, 24 and 48 h, lowest conc.: 75.4 ppb after 48 h), ***ZEA and/or metabolites

incidence: 6/32, sa. const.: male broilers, age: 25–27 days, contamination: artificial (dose: 5.8 μg ZEA/kg b. wt., o. once; for detailed information please see the article), conc. range: ≈10 ng/g* (mean value), country: Germany[543], *probably up to 48 h measured

incidence: 5/5*, sa. const.: Leghorn broilers, age: 2 weeks, contamination: no ZEA (for detailed information please see the article), conc.: nr, country: Romania/France[635], *control
incidence: 5?/5, sa. const.: Leghorn broilers, age: 2 weeks, contamination: artificial (dose: **25 mg ZEA**/kg b. wt., i.p., for 3 days (for detailed information please see the article), conc.: 6.62 ng/ml*, country: Romania/France[635], after 24 h of the last treatment
incidence: 5?/5, sa. const.: Leghorn broilers, age: 2 weeks, contamination: artificial (dose: **25 mg ZEA**/kg b. wt., i.p., once and before **80 mg PB**/kg b. wt., i.p., for 3 days (for detailed information please see the article), conc.: 3.40 ng/ml*, country: Romania/France[635], after 24 h of the last treatment

α-ZEARALENOL
incidence: 6/32, sa. const.: male broilers, age: 25–27 days, contamination: artificial (dose: 5.8 μg ZEA/kg b. wt., o., once; for detailed information please see the article), conc. range: ≈5 ng/g* (mean value), country: Germany[543], *probably up to 48 h measured

Chicken serum may contain the following mycotoxins and/or their metabolites:

OCHRATOXIN A
incidence: ?/?*, sa. const.: broilers, contamination: no OTA (for detailed information please see the article), conc.: nd, country: Germany[601], *control
incidence: ?/?, sa. const.: broilers, contamination: artificial (dose: **1.5 mg OTA/kg** in the diet, o., for 3 weeks **afterwards OTA-free diet** for 3 weeks (for detailed information please see the article), conc.: nd*, country: Germany[601], *after 6 weeks
incidence: ?/?, sa. const.: broilers, contamination: artificial (dose: **OTA-free diet** for 3 weeks **afterwards 1.5 mg OTA/kg in the diet**, o., for 3 weeks (for detailed information please see the

article), conc.: 1.24 μg/l* (mean value), country: Germany[601], *after 6 weeks
incidence: ?/?, sa. const.: broilers, contamination: artificial (dose: **1.5 mg OTA/kg in the diet**, o., for **6 weeks** (for detailed information please see the article), conc.: 4.56 μg/l* (mean value), country: Germany[601], *after 6 weeks

Chicken skin may contain the following mycotoxins and/or their metabolites:

Aflatoxicol
incidence: ?/?*, sa. const.: Hubbard strain male broiler chickens, age: 14 days, contamination: no AFB_1 + OTA (for detailed information please see the article), conc.: nr, country: Italy[515], *control
incidence: 3?/3, sa. const.: Hubbard strain male broiler chickens, age: 14 days, contamination: artificial (dose: **50 μg AFB_1/kg feed** + **50 μg OTA/kg** feed, o., for up to 64 days; for detailed information please see the article), conc. range: <0.04 μg/kg* (mean value), country: Italy[515], *after 36 and 64 days

Aflatoxin B_1
incidence: ?/?*, sa. const.: Hubbard strain male broilers, age: 14 days, contamination: no AFB_1 (for detailed information please see the article), conc.: nr, country: Italy[232], *control
incidence: 3?/3, sa. const.: Hubbard strain male broilers, age: 14 days, contamination: artificial (dose: **50 μg AFB_1/kg** feed, o., for up to 64 days; for detailed information please see the article), conc.: 0.70 μg/kg* (mean value), country: Italy[232], *after 36 days (also measured after 64 days but conc.: nd)

incidence: ?/?*, sa. const.: Hubbard strain male broiler chickens, age: 14 days, contamination: no AFB_1 + OTA (for detailed information please see the article), conc.: nr, country: Italy[515], *control
incidence: 3?/3, sa. const.: Hubbard strain male broiler chickens, age: 14 days, contamination: artificial (dose: **50 μg AFB_1/kg** feed + **50 μg OTA/kg** feed, o., for up to 64 days; for detailed information please see the article), conc. range: 0.03*–0.06** μg/kg (mean values), country: Italy[515], after 64* and 36** days

Aflatoxin M_1
incidence: ?/?*, sa. const.: Hubbard strain male broiler chickens, age: 14 days, contamination: no AFB_1 + OTA (for detailed information please see the article), conc.: nr, country: Italy[515], *control
incidence: 3?/3, sa. const.: Hubbard strain male broiler chickens, age: 14 days, contamination: artificial (dose: **50 μg AFB_1/kg** feed + **50 μg OTA/kg** feed, o., for up to 64 days; for detailed information please see the article), conc. range: <0.01 μg/kg* (mean value), country: Italy[515], *after 36 and 64 days

Ochratoxin A
incidence: ?/?*, sa. const.: Hubbard strain male broiler chickens, age: 14 days, contamination: no AFB_1 + OTA (for detailed information please see the article), conc.: nr, country: Italy[515], *control
incidence: 3?/3, sa. const.: Hubbard strain male broiler chickens, age: 14 days, contamination: artificial (dose: **50 μg AFB_1/kg** feed + **50 μg OTA/kg** feed, o., for up to 64 days; for detailed information please see the article), conc. range: 1.6*–1.8** μg/kg (mean values), country: Italy[515], after 64* and 36** days
incidence: ?/?*, sa. const.: Hubbard strain male broiler chickens, age: 14 days, contamination: no AFB_1 + OTA (for detailed information please see the article), conc.: nr, country: Italy[515], *control
incidence: 3?/3, sa. const.: Hubbard strain male broiler chickens, age: 14 days, contamination: artificial (dose: **50 μg AFB_1/kg** feed + **50 μg OTA/kg** feed, o., for 36 days; for detailed information

please see the article), conc. range: <0.5*–0.8** μg/kg (mean values), country: Italy[515], 14** or 28* days after withdrawal from treatment

ZEARALENONE
incidence: 6/6*, sa. const.: broiler chickens (Ross male X Arbor Acre female), age: 4 weeks, contamination: no ZEA (for detailed information please see the article), conc.: nr, country: USA[106], *control

incidence: 4?/4, sa. const.: broiler chickens (Ross male X Arbor Acre female), age: 4 weeks, contamination: artificial (dose: **5.0 mg ZEA** (labeled)/kg b. wt., intubated into the crop, once; for detailed information please see the article), conc. range: ≤70.2 ppb* ** ***, country: USA[106], *eq. conc., **after 24 h (also measured after 0.5, 4, 8, 12 and 48 h, lowest conc.: 43.4 ppb after 48 h), ***ZEA and/or metabolites

Chicken stomach may contain the following mycotoxins and/or their metabolites:

OCHRATOXIN A
incidence: 4?/4, sa. const.: New Hampshire-Leghorn cross chicks, age: 36 days, contamination: artificial (dose: feeding 1 ppm OTA for 5 weeks and then given a single dose of 50 μg ^{3}H-OA per chick by intubation (OTA labeled and unlabeled); for detailed information please see the article), conc. range: ≤1.93 ppb* (mean value), country: USA[316], *after 8 h (also measured after 24 and 48 h, lowest conc.: 0.13 ppb after 48 h)

Chicken Thigh see Chicken muscle, thigh

Cock see Chicken

Cockerel see Hen

Coho salmon see Fish, coho salmon

Cow

Cow Natural Contamination

Cow liver may contain the following mycotoxins and/or their metabolites:

AFLATOXIN B_1
incidence: 1 (investigated)/68, sa. const.: livers from crossbred adult cows of the USA, contamination: natural, conc.: 5 ng/g, country: USA[76]

Cow milk, raw may contain the following mycotoxins and/or their metabolites:

AFLATOXIN B_1
incidence: 56/120, sa. const.: milk from cows of Iran, contamination: natural, conc. range: 0.25–22.00 ng/l, country: Iran[453]

AFLATOXIN M_1
incidence: 12/50, sa. const.: milk, probably from cows, of Egypt, contamination: natural, conc. range: 0.25–3.72 μg/l, Ø conc.: 1 μg/l, country: Egypt[2]

incidence: 61/66, sa. const.: milk from cows of India, contamination: natural, conc. range: 0.110*–0.143** μg/l (mean values), country: India[43], *bulk, **individual

incidence: 6/30* sa. const.: milk from cows of Poland, contamination: natural, conc. range: 3.6–10.6 ng/kg, country: Poland/Germany[44]

incidence: 303/6,246*, sa. const.: milk from cows of France, contamination: natural, conc. range: 0.05–0.5 μg/l (284 sa), >0.5 to ≤5 μg/l (19 sa), country: France[45], *and milk powder

incidence: ?/?, sa. const.: milk, probably from cows, of the USA?, contamination: probably natural, conc.: 0.07 ng/g, country: USA[75]

incidence: 40/40, sa. const.: milk from cows of the USA, contamination: natural, conc. range: 0.001–0.273 ppb, country: USA[96] incidence: 46/376, sa. const.: milk from cows of Czechoslovakia, contamination: natural, conc. range: 0.025–0.1 μg/l (44 sa), >0.1 μg/l (2 sa), country: Czechoslovakia[158]

incidence: 27/89, sa. const.: milk from cows of Czechoslovakia, contamination: natural, conc. range: 0.025–0.1 μg/l (21 sa), 0.1–0.5 μg/l (6 sa), country: Czechoslovakia[159]

incidence: 11/22, sa. const.: milk, probably from cows, of Poland, contamination: natural, conc. range: 0.010–0.250 μg/l, country: Poland[160]

incidence: 126/235, sa. const.: milk from cows of Italy, contamination: natural, conc. range: 4–1,262 ng/kg, country: Italy[161]

incidence: 24/27, sa. const.: milk from cows of Italy, contamination: natural, Ø conc.: 27.05 ppt, country: Italy[162]

incidence: 24/409, sa. const.: milk, probably from cows, of the UK, contamination: natural, conc. range: 0.02–0.05 μg/kg (10 sa), 0.05–0.1 μg/kg (6 sa), >0.1 μg/kg (8 sa), country: UK[163]

incidence: 3/71, sa. const.: milk, probably from cows, of the Republic of Cyprus, contamination: natural, conc. range: 0.03–0.04 μg/l, Ø conc.: 0.035 μg/l, country: Republic of Cyprus[164]

incidence: 12/61, sa. const.: milk, probably from cows, of Spain, contamination: natural, conc. range: 0.01–0.02 μg/kg (10 sa), 0.02–0.04 μg/kg (2 sa), country: Spain[165]

incidence: 4/99, sa. const.: milk from cows of Greece, contamination: natural, conc. range: 0.10–0.13 μg/kg, country: Greece[166]

incidence: 317/332, sa. const.: milk from cows of Italy, contamination: natural, conc. range: 1–10 ng/kg (168 sa), 11–50 ng/kg (121 sa), 51–100 ng/kg (25 sa), >100–406 ng/kg (3 sa), country: Italy[170]

incidence: ?/78*, sa. const.: milk from cows of Italy, contamination: natural, conc. range: <5–93 ng/l, country: Italy[178], *organic milk

incidence: ?/78*, sa. const.: milk from cows of Italy, contamination: natural, conc. range: <5–66 ng/l, country: Italy[178], *conventionally produced milk

incidence: 122/214, sa. const.: milk from cows of Italy, contamination: natural, conc. range: 3–10 ppt (18 sa), 11–50 ppt (92 sa), 51–101 ppt (12 sa), country: Italy[216]

incidence: 6/56, sa. const.: milk from cows of Argentina, contamination: natural, conc. range: 0.012–0.030 μg/l, Ø conc.: 0.016 μg/l country: Argentina[238]

incidence: 50/324, sa. const.: milk from cows of the UK, contamination: natural, conc. range: 0.01–0.04 μg/kg (39 sa), 0.05–0.10 μg/kg (9 sa), >0.10–0.18 μg/kg (2 sa), country: UK[239]

incidence: 25/31, sa. const.: milk, probably from cows of Portugal, contamination: natural, conc. range: 0.005–0.010 μg/l (17 sa), 0.011–0.020 μg/l (2 sa), 0.021–0.050 μg/l (6 sa), country: Portugal[240]

incidence: 1/10*, sa. const.: milk, probably from cows of Turkey, contamination: natural, conc.: 10.8 ng/l, country: Turkey[243], *5 raw and 5 pasteurized

incidence: 10/12, sa. const.: milk from cows of Austria, contamination: natural, conc. range: 0–2 ng/kg (1 sa), 2–5 ng/kg (3 sa), 5–10 ng/kg (2 sa), 10–50 ng/kg (3 sa), 51.8 ng/kg (1 sa), country: Austria[245]

incidence: 22/30*, sa. const.: milk from cows of Greece, contamination: natural, conc. range: 5–10 ng/l (7 sa), 11–20 ng/l (10 sa), 21–50 ng/l (4 sa), 55 ng/l (1 sa), country: Greece[246], * from December 1999 to May 2000

incidence: 18/28*, sa. const.: milk from cows of Greece, contamination: natural, conc. range: 5–10 ng/l (3 sa), 11–20 ng/l (10 sa), 21–50 ng/l (4 sa), 60 ng/l (1 sa), country: Greece[246], *from December 2000 to May 2001

incidence: 7/50, sa. const.: milk from cows of Brazil, contamination: natural, conc. range: 0.1–1.68 μg/l, country: Brazil[250] incidence: 66/67, sa. const.: milk from cows of Thailand, contamination: natural, conc. range: >0–0.05 ppb (9 sa), >0.05–0.125 (16 sa), >0.125–0.25 ppb (19 sa), >0.25–0.5 ppb (5 sa), >0.5 ppb (17 sa), country: Thailand[251]

incidence: 50/85, sa. const.: milk, probably from cows, of Egypt, contamination: natural, conc. range: nd-15 ppt, country: Egypt[252]

incidence: 10/42, sa. const.: milk, probably from cows, of Brazil, contamination: natural, conc. range: 0.29505–1.9749 μg/l, Ø conc.: 0.68485 μg/l, country: Brazil[253]

incidence: 84/105*, sa. const.: milk, probably from cows, of The Netherlands, contamination: natural, conc. range: 0.015–0.090 μg/l, country: The Netherlands[254], *raw and heat-treated milk sa.

incidence: 5/9*, sa. const.: milk from cows of Kuwait, contamination: natural, conc. range: 0.20–0.21 μg/l, Ø conc.: 0.206 μg/l, country: Kuwait[255], *fresh

incidence: 36*/67**, sa. const.: milk from cows of Iran, contamination: natural, conc. range: 50–500 μg/kg, country: Iran[256], *all contaminated sa. contained AFM_1 and additionally 5 with AFM_1 and AFM_2, 3 with AFM_2, 2 with AFB_1 and AFM_1, **milk sa. only from villages

incidence: 8/31, sa. const.: milk from cows of Italy, contamination: natural, conc. range: 5–24 ng/kg (7 sa), 91 ng/kg (1 sa), country: Italy[270]

incidence: 23/40*, sa. const.: milk from cows of Pakistan, contamination: natural, conc. range: ?, country: Pakistan[277], *urban area

incidence: 22/40*, sa. const.: milk from cows of Pakistan, contamination: natural, conc. range: ?, country: Pakistan[277], *semi-urban area

incidence: 18/40*, sa. const.: milk from cows of Pakistan, contamination: natural, conc. range: ?, country: Pakistan[277], *rural area

incidence: 72/72, sa. const.: milk from cows of Iran, contamination: natural, conc. range: 4.3–91.8 ng/l, Ø conc.: 24.21 ng/l, country: Iran[452]

incidence: 120/120, sa. const.: milk from cows of Iran, contamination: natural, conc. range: 4–352.3 ng/l, Ø conc.: 102.73 ng/l, country: Iran[453]

incidence: 13/22, sa. const.: milk from cows of Brazil, contamination: natural, conc. range: >0.01 μg/l (5 sa), 0.02–0.05 μg/l (6 sa), >0.05–<0.5 μg/l (2 sa), country: Brazil[454]

incidence: 9/264*, sa. const.: milk from cows of France, contamination: natural, conc. range: tr (6 sa), 8–26 ng/l (3 sa), country: France[478], *raw bulk milk

incidence: 98/98, sa. const.: milk from cows of Iran, contamination: natural, conc. range: ≤0.050 μg/l (61 sa), 0.05–0.10 μg/l (29 sa), 0.1–0.392 μg/l (8 sa), country: Iran[479]

incidence: 35/49, sa. const.: milk from cows of Lybia, contamination: natural, conc. range: 0.03–3.13 ng/ml, Ø conc.: 0.377 ng/ml, country: Scotland, UK[514]

incidence: 85/111, sa. const.: milk from cows of Iran, contamination: natural, conc. range: 0.015–0.28 μg/l, country: Iran[518]

incidence: 5/92, sa. const.: milk from cows of Spain, contamination: natural, conc. range: 14.0–24.9 ng/l*, Ø conc.: 20.5 ng/l*, country: Spain[532], *using ELISA

incidence: 168/168, sa. const.: milk, probably from cows, of Pakistan, contamination: natural, conc. range: 0.01–0.70 μg/l (mean values), country: Pakistan[539]

incidence: 60/94, sa. const.: milk from cows of Argentina, contamination: natural, conc. range: nd-0.07 μg/l, country: Argentina[589]

Aflatoxin M_2

incidence: 55/214, sa. const.: milk from cows of Italy, contamination: natural, conc. range: 3–10 ppt (44 sa), 11–17 ppt (11 sa), country: Italy[216]

Aflatoxin $M_1 + M_2$
incidence: 23/26, sa. const.: milk from cows of the USA, contamination: natural, conc. range: 0.21–1.43 ppb, Ø conc.: 0.59 ppb, country: USA[47]

Aflatoxin M
incidence: 17/20, sa. const.: milk from cows of Germany, contamination: natural, conc. range: <1 μg/l (13 sa), >1 μg/l (4 sa), country: Germany[89]

Fumonisin B_1
incidence: 1/155, sa. const.: milk from cows of the USA, contamination: natural, conc.: 1.29 ng/ml, country: USA[275]

Ochratoxin A
incidence: 3/264*, sa. const.: milk from cows of France, contamination: natural, conc. range: 5.0–6.6 ng/l, country: France[478], *raw bulk milk

Cow Artificial Contamination

Cow bile may contain the following mycotoxins and/or their metabolites:

Aflatoxicol
incidence: 1/1, sa. const.: Holstein cow (midlactation period), age: 5 years, wt.: 600 kg, contamination: artificial (dose: 0.5 mg AFB_1/kg b. wt., o., once), conc.: 0.36 ng/g* **, country: USA[92], *in cow that died, **after 60 h

Aflatoxin B_1
incidence: 1/1, sa. const.: Holstein cow (midlactation period), age: 5 years, wt.: 600 kg, contamination: artificial (dose: 0.5 mg AFB_1/kg b. wt., o., once), conc.: 1.6 ng/g* **, country: USA[92], *in cow that died, **after 60 h

incidence: 1/2, sa. const.: Holstein cows (midlactation), wt.: 408.2–544.3 kg, contamination: artificial (dose: 0.35 mg AFB_1/kg b. wt., o., for 3 days, (one cow was fed AF-rations for 3 days, the other cow, after 3 days of dosing, was fed AF-free rations for 7 additional days); for detailed information please see the article), conc.: 3.35 ng/ml*, country: USA[93], *after 4 days (thereof 3 days with AFB_1-administration)

incidence: 1/1, sa. const.: lactating dairy cows, contamination: artificial (dose: **10 ppb AFB_1** (labeled and unlabeled), o., twice daily over a 14-day period; for detailed information please see the article), conc.: na, country: USA[319]
incidence: 1/1, sa. const.: lactating dairy cows, contamination: artificial (dose: **50 ppb AFB_1** (labeled and unlabeled), o., twice daily over a 14-day period; for detailed information please see the article), conc.: na, country: USA[319]
incidence: 1/1, sa. const.: lactating dairy cows, contamination: artificial (dose: **250 ppb AFB_1** (labeled and unlabeled), o., twice daily over a 14-day period; for detailed information please see the article), conc.: nd* **, country: USA[319], *after 15 days (thereof 14 days with AFB_1-administration), **but significant radioactivity occurred
incidence: 1/1, sa. const.: lactating dairy cows, contamination: artificial (dose: **1,250 ppb AFB_1** (labeled and unlabeled), o., twice daily over a 14-day period; for detailed information please see the article), conc.: 0.26 μg/kg*, country: USA[319], *after 15 days (thereof 14 days with AFB_1-administration)

Aflatoxin M_1
incidence: 1/1, sa. const.: Holstein cow (midlactation period), age: 5 years, wt.: 600 kg, contamination: artificial (dose: 0.5 mg AFB_1/kg b. wt., o., once), conc.: 16 ng/g* **, country: USA[92], *in cow that died, **after 60 h

incidence: 1/2, sa. const.: Holstein cows (midlactation), wt.: 408.2–544.3 kg, contamination: artificial (dose: 0.35 mg AFB_1/kg b. wt., o., 3 days, (one cow was fed AF-rations for 3 days, the other cow, after 3 days of dosing, was fed AF-free rations for 7 additional days); for detailed information please see the article), conc.: 12.37 ng/ml*, country: USA[93], *after 4 days (thereof 3 days with AFB_1-administration)

Cow blood may contain the following mycotoxins and/or their metabolites:

Aflatoxin B_1
incidence: 2/2, sa. const.: Holstein cows (midlactation), wt.: 408.2–544.3 kg, contamination: artificial (dose: 0.35 mg AFB_1/kg b. wt., o., 3 days, (one cow was fed AF-rations for 3 days, the other cow, after 3 days of dosing, was fed AF-free rations for 7 additional days); for detailed information please see the article), conc.: ≤0.26* and ≤0.47** ng/ml, country: USA[93], these highest values measured at the 1st day in the case of cow 1* and cow 2** (always lower values measured up to 4 days (then slaughtered) in the case of cow 1 and always lower values measured up to 10 days (then slaughtered) in the case of cow 2), for overall information please see the article

Aflatoxin M_1
incidence: 2/2, sa. const.: Holstein cows (midlactation), wt.: 408.2–544.3 kg, contamination: artificial (dose: 0.35 mg AFB_1/kg b. wt., o., 3 days, (one cow was fed AF-rations for 3 days, the other cow, after 3 days of dosing, was fed AF-free rations for 7 additional days); for detailed information please see the article), conc. range: ≤0.79* and ≤1.49** ng/ml, country: USA[93], these highest values measured at the *4th day in the case of cow 1 and at the **1st day in the case of cow 2 (always lower values measured up to 4 days (then slaughtered) in the case of cow 1 and always lower values measured up to 10 days (then slaughtered) in the case of cow 2), for overall information please see the article

HT-2 Toxin
incidence: 1/1*, sa. const.: Holstein cow, wt.: 365 kg, contamination: artificial (dose: 2 × 200 mg T-2 toxin, by balling gun; for detailed information please see the article), conc.: pos**, country: USA[180], *whole blood, **6 h after administration of 2nd dosing

3′-Hydroxy-HT-2 Toxin
incidence: 1/1*, sa. const.: Holstein cow, wt.: 365 kg, contamination: artificial (dose: 2 × 200 mg T-2 toxin, by balling gun; for detailed information please see the article), conc.: pos**, country: USA[180], *whole blood, **6 h after administration of 2nd dosing

T-2 Toxin
incidence: 1/1*, sa. const.: Holstein cow, wt.: 365 kg, contamination: artificial (dose: 2 × 200 mg T-2 toxin, by balling gun; for detailed information please see the article), conc. range: ≤7.7 ppb**, country: USA[180], *whole blood, **6 h after administration of 1st dosing (also at other hour intervals up to 48 h measured, lowest conc.: 2.2 ppb after 24 h of 1st dosing)

T-2 Tetraol
incidence: 1/1*, sa. const.: Holstein cow, wt.: 365 kg, contamination: artificial (dose: 2 × 200 mg T-2 toxin, by balling gun; for detailed information please see the article), conc. range: ≤1.0 ppb**, country: USA[180], *whole blood, **6 h after administration of 2nd dosing (also at other hour intervals up to 48 h measured, lowest conc.: 0.2 ppb after 24 h of 1st dosing)

Deepoxy-T-2-Tetraol
incidence: 1/1*, sa. const.: Holstein cow, wt.: 365 kg, contamination: artificial (dose: 2 × 200 mg T-2 toxin, by balling gun; for detailed information please see the article), conc. range: ≤1.3 ppb**, country: USA[180], *whole blood, **24 h after administration of 1st dosing (also at other hour intervals up to 48 h measured, lowest conc.: 0.6 ppb after 6 h of 1st dosing)

Cow brain may contain the following mycotoxins and/or their metabolites:

Aflatoxin B_1
incidence: 1/2, sa. const.: Holstein cows (midlactation), wt.: 408.2–544.3 kg, contamination: artificial (dose: 0.35 mg AFB_1/kg b. wt., o., for 3 days (one cow was fed AF-rations for 3 days, the other cow,

after 3 days of dosing, was fed AF-free rations for 7 additional days); for detailed information please see the article), conc.: 0.70 ng/g*, country: USA[93], *after 4 days (thereof 3 days with AFB_1-administration)

Aflatoxin M_1
incidence: 1/2, sa. const.: Holstein cows (midlactation), wt.: 408.2–544.3 kg, contamination: artificial (dose: 0.35 mg AFB_1/kg b. wt., o., for 3 days (one cow was fed AF-rations for 3 days, the other cow, after 3 days of dosing, was fed AF-free rations for 7 additional days); for detailed information please see the article), conc.: 0.17 ng/g*, country: USA[93], *after 4 days (thereof 3 days with AFB_1-administration)

Cow fat may contain the following mycotoxins and/or their metabolites:

Aflatoxin B_1
incidence: 2/2, sa. const.: lactating Francaise Frisonne Pie Noire cows, age: 3–4 years, wt.: 500–550 kg, contamination: artificial (dose: peanut meal ((3.5 mg AFB_1/kg) spiked with labeled AFB_1) addition, o., for 10 days; for detailed information please see the article), conc. range: 2.6–3.2 ng AFB_1 eq/g* **, Ø conc.: 2.9 ng AFB_1 eq/g*, country: The Netherlands/UK/France[397], *after a 10-day period of consumption of **contaminated peanut meal**
incidence: 2/2, sa. const.: lactating Francaise Frisonne Pie Noire cows, age: 3–4 years, wt.: 500–550 kg, contamination: artificial (dose: peanut meal ((3.5 mg AFB_1/kg) spiked with labeled AFB_1; but decontaminated afterwards), o., for 10 days; for detailed information please see the article), conc.: nd*, country: The Netherlands/UK/France[397], *after a 10-day period of **consumption of decontaminated peanut meal**

Cow feces may contain the following mycotoxins and/or their metabolites:

Aflatoxin B_1
incidence: 1/1, sa. const.: lactating cow, wt.: 600 kg, contamination: artificial (dose: 300 mg AFs: 44% AFB_1, 2% AFB_2, 44% AFG_1?, o., once), conc.: 3,930 µg* (total amount found), country: UK[40], *collected over a period of 9 days

incidence: 2/2, sa. const.: Holstein cows (midlactation), wt.: 408.2–544.3 kg, contamination: artificial (dose: 0.35 mg AFB_1/kg b. wt., o., 3 days (one cow was fed AF-rations for 3 days, the other cow, after 3 days of dosing, was fed AF-free rations for 7 additional days); for detailed information please see the article), conc.: ≤2,719.60* and ≤3,597.40 ng/g**, country: USA[93], these highest values measured at the 3rd day in the case of cow 1** and cow 2* (always lower values measured up to 4 days (then slaughtered) in the case of cow 1 and always lower values measured up to 10 days (then slaughtered) in the case of cow 2), for overall information please see the article

Aflatoxin G_1
incidence: 1/1, sa. const.: lactating cow, wt.: 600 kg, contamination: artificial (dose: 300 mg AFs: 44% AFB_1, 2% AFB_2, 44% AFG_1?, o., once), conc.: 630 µg* (total amount found), country: UK[40], *collected over a period of 9 days

Aflatoxin M_1
incidence: 1/1, sa. const.: lactating cow, wt.: 600 kg, contamination: artificial (dose: 300 mg AFs: 44% AFB_1, 2% AFB_2, 44% AFG_1 ?, o., once), conc.: 2,960 µg* (total amount found), country: UK[40], *collected over a period of 9 days

incidence: 2/2, sa. const.: Holstein cows (midlactation), wt.: 408.2–544.3 kg, contamination: artificial (dose: 0.35 mg AFB_1/kg b. wt., o., 3 days (one cow was fed AF-rations for 3 days, the other cow, after 3 days of dosing, was fed AF-free rations for 7 additional days); for detailed information please see the article), conc.:

≤158.20* and ≤200.30** ng/g, country: USA[93], these highest values measured at the *3rd day in the case of cow 1 and at the **4th day in the case of cow 2 (always lower values measured up to 4 days (then slaughtered) in the case of cow 1 and always lower values measured up to 10 days (then slaughtered) in the case of cow 2), for overall information please see the article

Deoxynivalenol
incidence: 3/3, sa. const.: 2 Holstein dairy cows, 1 Ayrshire dairy cow, wt: ≈450 kg, contamination: artificial (dose: 66 mg DON/kg, o., for 5 days; for detailed information please see the article), conc. range: ≤660 ng/g* (37.4 mg in total, mean value), country: USA[278], *after 2 days of application (also at other day intervals up to 12 days measured, lowest conc.: nd at the beginning and the end of the experiment; for detailed information please see the article)

Deepoxydeoxynivalenol
incidence: 3/3, sa. const.: 2 Holstein dairy cows, 1 Ayrshire dairy cow, wt.: ≈450 kg, contamination: artificial (dose: 66 mg DON/kg, o., for 5 days; for detailed information please see the article), conc. range: ≤13,000 ng/g* ** (679.8 mg** in total, mean value), country: USA[278], *after 6 days of application (also at other day intervals up to 12 days measured, lowest conc.: nd at the beginning (1–4 days) of the experiment; for detailed information please see the article), **unconjugated DEDON

Cow gallbladder may contain the following mycotoxins and/or their metabolites:

Aflatoxin B_1
incidence: 1/2, sa. const.: Holstein cows (midlactation), wt.: 408.2–544.3 kg, contamination: artificial (dose: 0.35 mg AFB_1/kg b. wt., o., 3 days (one cow was fed AF-rations for 3 days, the other cow, after 3 days of dosing, was fed AF-free rations for 7 additional days); for detailed information please see the article), conc.: 1.98 ng/g*, country: USA[93], *after 4 days (thereof 3 days with AFB_1-administration)

Aflatoxin M_1
incidence: 1/2, sa. const.: Holstein cows (midlactation), wt.: 408.2–544.3 kg, contamination: artificial (dose: 0.35 mg AFB_1/kg b. wt., o., 3 days (one cow was fed AF-rations for 3 days, the other cow, after 3 days of dosing, was fed AF-free rations for 7 additional days); for detailed information please see the article), conc.: 8.52 ng/g*, country: USA[93], *after 4 days (thereof 3 days with AFB_1-administration)

Cow heart may contain the following mycotoxins and/or their metabolites:

Aflatoxin B_1
incidence: 1/2, sa. const.: Holstein cows (midlactation), wt.: 408.2–544.3 kg, contamination: artificial (dose: 0.35 mg AFB_1/kg b. wt., o., 3 days (one cow was fed AF-rations for 3 days, the other cow, after 3 days of dosing, was fed AF-free rations for 7 additional days); for detailed information please see the article), conc.: 0.43 ng/g* **, country: USA[93], *left ventricle, **after 4 days (thereof 3 days with AFB_1-administration)
incidence: 1/2, sa. const.: Holstein cows (midlactation), wt.: 408.2–544.3 kg, contamination: artificial (dose: 0.35 mg AFB_1/kg b. wt., o., 3 days (one cow was fed AF-rations for 3 days, the other cow, after 3 days of dosing, was fed AF-free rations for 7 additional days); for detailed information please see the article), conc.: 0.50 ng/g* **, country: USA[93], *right ventricle, **after 4 days (thereof 3 days with AFB_1-administration)

incidence: 6/6*, sa. const.: nonlactating Holstein dairy cows, contamination: no AF (for detailed information please see the article), conc.: nd, country: USA[150], control
incidence: 6/6, sa. const.: nonlactating Holstein dairy cows, contamination: artificial (dose: 148.54 mg AFB_1 in cottonseed (**but ammoniated afterwards**), o., over 14 days; for detailed information please see the article), conc.: nd*, country: USA[150], *after 14 days of AFB_1-administration
incidence: 4/6, sa. const.: nonlactating Holstein dairy cows, contamination: artificial (dose: 148.54 mg AFB_1 in cottonseed, o., over 14 days; for detailed information please see the article), conc.: nd*, country: USA[150], *after 14 days of AFB_1-administration

incidence: 2/2, sa. const.: lactating Francaise Frisonne Pie Noire cows, age: 3–4 years, wt.: 500–550 kg, contamination: artificial (dose: peanut meal ((3.5 mg AFB_1/kg) spiked with labeled AFB_1), o., for 10 days; for detailed information please see the article), conc. range: 4.5–5.1 ng AFB_1 eq/g*, Ø conc.: 4.8 ng AFB_1 eq/g*, country: The Netherlands/UK/France[397], *after a 10-day period of consumption of **contaminated peanut meal**
incidence: 2/2, sa. const.: lactating Francaise Frisonne Pie Noire cows, age: 3–4 years, wt.: 500–550 kg, contamination: artificial (dose: peanut meal ((3.5 mg AFB_1/kg) spiked with labeled AFB_1; but decontaminated afterwards), o., for 10 days; for detailed information please see the article), conc.: nd*, country: The Netherlands/UK/France[397], *after a 10-day period of consumption of **decontaminated peanut meal**

Aflatoxin B_2
incidence: 6/6*, sa. const.: nonlactating Holstein dairy cows, contamination: no AF (for detailed information please see the article), conc.: nd, country: USA[150], control
incidence: 6/6, sa. const.: nonlactating Holstein dairy cows, contamination: artificial (dose: 148.54 mg AFB_1 in cottonseed (**but ammoniated afterwards**), o., over 14 days; for detailed information please see the article), conc.: nd*, country: USA[150], *after 14 days of AFB_1-administration
incidence: 4/6, sa. const.: nonlactating Holstein dairy cows, contamination: artificial (dose: 148.54 mg AFB_1 in cottonseed, o., over 14 days; for detailed information please see the article), conc.: nd*, country: USA[150], *after 14 days of AFB_1-administration

Aflatoxin M_1
incidence: 1/2, sa. const.: Holstein cows (midlactation), wt.: 408.2–544.3 kg, contamination: artificial (dose: 0.35 mg AFB_1/kg b. wt., o., 3 days (one cow was fed AF-rations for 3 days, the other cow, after 3 days of dosing, was fed AF-free rations for 7 additional days), conc.: 1.73 ng/g*, country: USA[93], *left ventricle, **after 4 days (thereof 3 days with AFB_1-administration)
incidence: 1/2, sa. const.: Holstein cows (midlactation), wt.: 408.2–544.3 kg, contamination: artificial (dose: 0.35 mg AFB_1/kg b. wt., o., 3 days (one cow was fed AF-rations for 3 days, the other cow, after 3 days of dosing, was fed AF-free rations for 7 additional days); for detailed information please see the article), conc.: 0.78 ng/g*, country: USA[93], *right ventricle, **after 4 days (thereof 3 days with AFB_1-administration)

incidence: 6/6*, sa. const.: nonlactating Holstein dairy cows, contamination: no AF (for detailed information please see the article), conc.: nd, country: USA[150], control
incidence: 6/6, sa. const.: nonlactating Holstein dairy cows, contamination: artificial (dose: 148.54 mg AFB_1 in cottonseed (**but ammoniated afterwards**), o., over 14 days; for detailed information please see the article), conc.: nd*, country: USA[150], *after 14 days of AFB_1-administration

incidence: 4/6, sa. const.: nonlactating Holstein dairy cows, contamination: artificial (dose: 148.54 mg AFB_1 in cottonseed, o., over 14 days; for detailed information please see the article), conc. range: <0.05 µg/kg*, country: USA[150], *after 14 days of AFB_1-administration

T-2 Toxin
incidence: 1/1, sa. const.: Holstein cow, wt.: 365 kg, contamination: artificial (dose: 2 × 200 mg T-2 toxin, by balling gun; for detailed information please see the article), conc.: 0.8 ppb*, country: USA[180], *no statement when found

Cow intestine may contain the following mycotoxins and/or their metabolites:

Aflatoxin B_1
incidence: 2/2, sa. const.: Holstein cows (midlactation), wt.: 408.2–544.3 kg, contamination: artificial (dose: 0.35 mg AFB_1/kg b. wt., o., 3 days (one cow was fed AF-rations for 3 days, the other cow, after 3 days of dosing, was fed AF-free rations for 7 additional days); for detailed information please see the article), conc. range: 0.025*–1.34** ng/g***, country: USA[93], after 10* or 4** days, ***in intestine, small (caudal part)

Aflatoxin M_1
incidence: 1/2, sa. const.: Holstein cows (midlactation), wt.: 408.2–544.3 kg, contamination: artificial (dose: 0.35 mg AFB_1/kg b. wt., o., 3 days (one cow was fed AF-rations for 3 days, the other cow, after 3 days of dosing, was fed AF-free rations for 7 additional days); for detailed information please see the article), conc.: 2.34* ng/g**, country: USA[93], *after 4 days (thereof 3 days with AFB_1-administration), **in intestine, small (caudal part)

Cow kidney may contain the following mycotoxins and/or their metabolites:

Aflatoxicol
incidence: 1/1, sa. const.: Holstein cow (midlactation period), age: 5 years, wt.: 600 kg, contamination: artificial (dose: 0.5 mg AFB_1/kg b. wt., o., once), conc.: 2.6 ng/g* **, country: USA[92], *in cow that died, **after 60 h

Aflatoxin B_1
incidence: 1/1, sa. const.: Holstein cow (midlactation period), age: 5 years, wt.: 600 kg, contamination: artificial (dose: 0.5 mg AFB_1/kg b. wt., o., once), conc.: 3.3 ng/g* **, country: USA[92], *in cow that died, **after 60 h

incidence: 2/2, sa. const.: Holstein cows (midlactation), wt.: 408.2–544.3 kg, contamination: artificial (dose: 0.35 mg AFB_1/kg b. wt., o., 3 days (one cow was fed AF-rations for 3 days, the other cow, after 3 days of dosing, was fed AF-free rations for 7 additional days); for detailed information please see the article), conc. range: 0.02*–1.32** ng/g, country: USA[93], after 10* or 4** days

incidence: 6/6*, sa. const.: nonlactating Holstein dairy cows, contamination: no AF (for detailed information please see the article), conc.: nd, country: USA[150], *control
incidence: 6/6, sa. const.: nonlactating Holstein dairy cows, contamination: artificial (dose: 148.54 mg AFB_1 in cottonseed (**but ammoniated afterwards**), o., over 14 days; for detailed information please see the article), conc.: nd*, country: USA[150], *after 14 days of AFB_1-administration
incidence: 2/5, sa. const.: nonlactating Holstein dairy cows, contamination: artificial (dose: 148.54 mg AFB_1 in cottonseed, o., over 14 days; for detailed information please see the article), conc. range: tr – <0.005 µg/kg*, country: USA[150], *after 14 days of AFB_1-administration

incidence: 1/1, sa. const.: lactating dairy cow, contamination: artificial (dose:

10 ppb AFB_1 (labeled and unlabeled), o., twice daily over a 14-day period; for detailed information please see the article), conc.: na, country: USA[319]
incidence: 1/1, sa. const.: lactating dairy cow, contamination: artificial (dose: **50 ppb AFB_1** (labeled and unlabeled), o., twice daily over a 14-day period; for detailed information please see the article), conc.: na, country: USA[319]
incidence: 1/1, sa. const.: lactating dairy cow, contamination: artificial (dose: **250 ppb AFB_1** (labeled and unlabeled), o., twice daily over a 14-day period; for detailed information please see the article), conc.: nd* **, country: USA[319], *after 15 days (thereof 14 days with AFB_1-administration), **but significant radioactivity occurred
incidence: 1/1, sa. const.: lactating dairy cow, contamination: artificial (dose: **1,250 ppb AFB_1** (labeled and unlabeled), o., twice daily over a 14-day period; for detailed information please see the article), conc.: 0.22 µg/kg*, country: USA[319], *after 15 days (thereof 14 days with AFB_1-administration)

incidence: 2/2, sa. const.: lactating Francaise Frisonne Pie Noire cows, age: 3–4 years, wt.: 500–550 kg, contamination: artificial (dose: peanut meal (3.5 mg AFB_1/kg) spiked with labeled AFB_1), o., for 10 days; for detailed information please see the article), conc. range: 36.8–38.9 ng AFB_1 eq/g*, Ø conc.: 37.85 ng AFB_1 eq/g*, country: The Netherlands/UK/France[397], *after a 10-day period of consumption of **contaminated peanut meal**
incidence: 2/2, sa. const.: lactating Francaise Frisonne Pie Noire cows, age: 3–4 years, wt.t: 500–550 kg, contamination: artificial (dose: peanut meal (3.5 mg AFB_1/kg) spiked with labeled AFB_1; but decontaminated afterwards), o., for 10 days; for detailed information please see the article), conc. range: 2.7–3.0 ng AFB_1 eq/g*, Ø conc.: 2.85 ng AFB_1 eq/g*, country: The Netherlands/UK/France[397], *after a 10-day period of consumption of **decontaminated peanut meal**

Aflatoxin B_2
incidence: 6/6*, sa. const.: nonlactating Holstein dairy cows, contamination: no AF (for detailed information please see the article), conc.: nd, country: USA[150], *control
incidence: 6/6, sa. const.: nonlactating Holstein dairy cows, contamination: artificial (dose: 148.54 mg AFB_1 in cottonseed **(but ammoniated afterwards)**, o., over 14 days; for detailed information please see the article), conc.: nd*, country: USA[150], *after 14 days of AFB_1-administration
incidence: 4/6, sa. const.: nonlactating Holstein dairy cows, contamination: artificial (dose: 148.54 mg AFB_1 in cottonseed, o., over 14 days; for detailed information please see the article), conc.: nd*, country: USA[150], *after 14 days of AFB_1-administration

Aflatoxin M_1
incidence: 1/1, sa. const.: Holstein cow (midlactation period), age: 5 years, wt.: 600 kg, contamination: artificial (dose: 0.5 mg AFB_1/kg b. wt., o., once), conc.: 20 ng/g* **, country: USA[92], *in cow that died, **after 60 h

incidence: 2/2, sa. const.: Holstein cows (midlactation), wt.: 408.2–544.3 kg, contamination: artificial (dose: 0.35 mg AFB_1/kg b. wt., o., 3 days (one cow was fed AF-rations for 3 days, the other cow, after 3 days of dosing, was fed AF-free rations for 7 additional days); for detailed information please see the article), conc. range: 0.11*–56.64** ng/g, country: USA[93], after 10* or 4** days

incidence: 6/6*, sa. const.: nonlactating Holstein dairy cows, contamination: no AF (for detailed information please see the article), conc.: nd, country: USA[150], *control
incidence: 6/6, sa. const.: nonlactating Holstein dairy cows, contamination:

artificial (dose: total of 148.54 mg AFB_1 in cottonseed (**but ammoniated afterwards**), o., over 14 days; for detailed information please see the article), conc.: nd*, country: USA[150], after 14 days of AFB_1-administration
incidence: 5/5, sa. const.: nonlactating Holstein dairy cows, contamination: artificial (dose: total of 148.54 mg AFB_1 in cottonseed, o., over 14 days; for detailed information please see the article), conc. range: ≤0.3 µg/kg*, country: USA[150], *after 14 days of AFB_1-administration

incidence: 2/2, sa. const.: Jersey milking cows, age: adult, contamination: artificial (dose: **385–1,925 µg ZEA/kg** + (20 µg AFB_1/kg), o., for 7 weeks; for detailed information please see the article), conc. range: tr–0.7 µg/kg*, country: UK[267], *after 7 weeks of (AFB_1-) and ZEA-administration;
AFB_1 accidentally in the diet
incidence: 2/2, sa. const.: Jersey milking cows, age: adult, contamination: artificial (dose: **317–1,125 µg OTA/kg** + (20 µg AFB_1/kg), o., for 11 weeks; for detailed information please see the article), conc. range: 0.2 µg/kg*, country: UK[267], *after 11 weeks of (AFB_1-) and OTA-administration;
AFB_1 accidentally in the diet

incidence: 1/1, sa. const.: lactating dairy cows, contamination: artificial (dose: **10 ppb AFB_1** (labeled and unlabeled), o., twice daily over a 14-day period; for detailed information please see the article), conc.: na, country: USA[319]
incidence: 1/1, sa. const.: lactating dairy cows, contamination: artificial (dose: **50 ppb AFB_1** (labeled and unlabeled), o., twice daily over a 14-day period; for detailed information please see the article), conc.: na, country: USA[319]
incidence: 1/1, sa. const.: lactating dairy cows, contamination: artificial (dose: **250 ppb AFB_1** (labeled and unlabeled), o., twice daily over a 14-day period; for detailed information please see the article), conc.: nd* **, country: USA[319], *after15 days (thereof 14 days with AFB_1-administration), **but significant radioactivity occurred
incidence: 1/1, sa. const.: lactating dairy cow, contamination: artificial (dose: **1,250 ppb AFB_1** (labeled and unlabeled)), o., twice daily over a 14-day period; for detailed information please see the article), conc.: 0.72 µg/kg*, country: USA[319], *after 15 days (thereof 14 days with AFB_1-administration)

HT-2 Toxin
incidence: 1/1, sa. const.: Holstein cow, weight: 365 kg, contamination: artificial (dose: 2 × 200 mg T-2 toxin, by balling gun; for detailed information please see the article), conc.: 2.2 ppb*, country: USA[180], *no statement when found

Ochratoxin A
incidence: 1/2, sa. const.: Jersey milking cows, age: adult, contamination: artificial (dose: 317–1,125 µg OTA/kg + (20 µg AFB_1/kg), o., for 11 weeks; for detailed information please see the article), conc.: pr*, country: UK[267], *after 11 weeks of (AFB_1-) and OTA-administration; AFB_1 accidentally in the diet

Cow liver may contain the following mycotoxins and/or their metabolites:

Aflatoxicol
incidence: 1/1, sa. const.: Holstein cow (midlactation period), age: 5 years, wt.: 600 kg, contamination: artificial (dose: 0.5 mg AFB_1/kg b. wt., o., once), conc.: 0.88 ng/g* **, country: USA[92], *in cow that died, **after 60 h

Aflatoxin B_1
incidence: 1/1, sa. const.: Holstein cow (midlactation period), age: 5 years, wt.: 600 kg, contamination: artificial (dose: 0.5 mg AFB_1/kg b. wt., o., once), conc.: 5.1 ng/g* **, country: USA[92], *in cow that died, **after 60 h

incidence: 2/2, sa. const.: Holstein cows (midlactation), wt.: 408.2–544.3 kg, contamination: artificial (dose: 0.35 mg AFB_1/kg b. wt., o., 3 days (one cow was fed AF-rations for 3 days, the other cow, after 3 days of dosing, was fed AF-free rations for 7 additional days); for detailed information please see the article), conc. range: 0.03*–7.12** ng/g, country: USA[93], after 10* or 4** days

incidence: 6/6*, sa. const.: nonlactating Holstein dairy cows, contamination: no AF (for detailed information please see the article), conc.: nd, country: USA[150], *control
incidence: 6/6, sa. const.: nonlactating Holstein dairy cows, contamination: artificial (dose: total of 148.54 mg AFB_1 in cottonseed (**but ammoniated afterwards**), o., over 14 days; for detailed information please see the article), conc.: nd*, country: USA[150], *after 14 days of AFB_1-administration
incidence: 3/6, sa. const.: nonlactating Holstein dairy cows, contamination: artificial (dose: total of 148.54 mg AFB_1 in cottonseed, o., over 14 days; for detailed information please see the article), conc. range: 0.1–>0.1 µg/kg, country: USA[150], *after 14 days of AFB_1-administration

incidence: 1/1, sa. const.: lactating dairy cow, contamination: artificial (dose: **10 ppb AFB_1** (labeled and unlabeled), o., twice daily over a 14-day period; for detailed information please see the article), conc.: na, country: USA[319]
incidence: 1/1, sa. const.: lactating dairy cow, contamination: artificial (dose: **50 ppb AFB_1** (labeled and unlabeled), o., twice daily over a 14-day period; for detailed information please see the article), conc.: na, country: USA[319]
incidence: 1/1, sa. const.: lactating dairy cow, contamination: artificial (dose: **250 ppb AFB_1** (labeled and unlabeled), o., twice daily over a 14-day period; for detailed information please see the article), conc.: nd* **, country: USA[319], *after 15 days (thereof 14 days with AFB_1-administration), **but significant radioactivity occurred
incidence: 1/1, sa. const.: lactating dairy cow, contamination: artificial (dose: **1,250 ppb AFB_1** (labeled and unlabeled), o., twice daily over a 14-day period; for detailed information please see the article), conc.: 0.09 µg/kg*, country: USA[319], *after 15 days (thereof 14 days with AFB_1-administration)

incidence: 2/2, sa. const.: lactating Francaise Frisonne Pie Noire cows, age: 3–4 years, wt.: 500–550 kg, contamination: artificial (dose: peanut meal ((3.5 mg AFB_1/kg) spiked with labeled AFB_1), o., for 10 days; for detailed information please see the article), conc. range: 252.2–278.6 ng AFB_1 eq/g*, Ø conc.: 265.4 ng AFB_1 eq/g*, country: The Netherlands/UK/France[397], *after a 10-day period of consumption of **contaminated peanut meal**
incidence: 2/2, sa. const.: lactating Francaise Frisonne Pie Noire cows, age: 3–4 years, wt.: 500–550 kg, contamination: artificial (dose: peanut meal ((3.5 mg AFB_1/kg) spiked with labeled AFB_1; but decontaminated afterwards), o., for 10 days; for detailed information please see the article), conc. range: 0.2–11.6 ng AFB_1 eq/g*, Ø conc.: 5.9 ng AFB_1 eq/g*, country: The Netherlands/UK/France[397], *after a 10-day period of consumption of decontaminated peanut meal

Aflatoxin B_2
incidence: 6/6*, sa. const.: nonlactating Holstein dairy cows, contamination: no AF (for detailed information please see the article), conc.: nd, country: USA[150], *control
incidence: 6/6, sa. const.: nonlactating Holstein dairy cows, contamination: artificial (dose: 148.54 mg AFB_1 in cottonseed (**but ammoniated afterwards**), o., over 14 days; for detailed information please see the article), conc.: nd*, country: USA[150], *after 14 days of AFB_1-administration

incidence: 3/6, sa. const.: nonlactating Holstein dairy cows, contamination: artificial (dose: 148.54 mg AFB_1 in cottonseed, o., over 14 days; for detailed information please see the article), conc. range: 0.025–>0.025 µg/kg*, country: USA[150], *after 14 days of AFB_1-administration

Aflatoxin M_1

incidence: 1/1, sa. const.: Holstein cow (midlactation period), age: 5 years, wt.: 600 kg, contamination: artificial (dose: 0.5 mg AFB_1/kg b. wt., o., once), conc.: 4.3 ng/g* **, country: USA[92], *in cow that died, **after 60 h

incidence: 2/2, sa. const.: Holstein cows (midlactation), wt.: 408.2–544.3 kg, contamination: artificial (dose: 0.35 mg AFB_1/kg b. wt., o., 3 days (one cow was fed AF-rations for 3 days, the other cow, after 3 days of dosing, was fed AF-free rations for 7 additional days); for detailed information please see the article), conc. range: 0.023*–6.09** ng/g, country: USA[93], after 10* or 4** days

incidence: 6/6*, sa. const.: nonlactating Holstein dairy cows, contamination: no AF (for detailed information please see the article), conc.: nd, country: USA[150], *control

incidence: 6/6, sa. const.: nonlactating Holstein dairy cows, contamination: artificial (dose: total of 148.54 mg AFB_1 in cottonseed (**but ammoniated afterwards**), o., over 14 days; for detailed information please see the article), conc.: nd*, country: USA[150], *after 14 days of AFB_1-administration

incidence: 5/6, sa. const.: nonlactating Holstein dairy cows, contamination: artificial (dose: total of 148.54 mg AFB_1 in cottonseed, o., over 14 days; for detailed information please see the article), conc. range: tr–0.15 µg/kg*, country: USA[150], *after 14 days of AFB_1-administration

incidence: 1/1, sa. const.: lactating dairy cow, contamination: artificial (dose: **10 ppb AFB_1** (labeled and unlabeled), o., twice daily over a 14-day period; for detailed information please see the article), conc.: na, country: USA[319]

incidence: 1/1, sa. const.: lactating dairy cow, contamination: artificial (dose: **50 ppb AFB_1** (labeled and unlabeled), o., twice daily over a 14-day period; for detailed information please see the article), conc.: na, country: USA[319]

incidence: 1/1, sa. const.: lactating dairy cow, contamination: artificial (dose: **250 ppb AFB_1** (labeled and unlabeled), o., twice daily over a 14-day period; for detailed information please see the article), conc.: nd* **, country: USA[319], *after 15 days (thereof 14 days with AFB_1-administration), **but significant radioactivity occurred

incidence: 1/1, sa. const.: lactating dairy cow, contamination: artificial (dose: **1,250 ppb AFB_1** (labeled and unlabeled), o., twice daily over a 14-day period; for detailed information please see the article), conc.: 0.16 µg/kg*, country: USA[319], *after 15 days (thereof 14 days with AFB_1-administration)

Cow lung may contain the following mycotoxins and/or their metabolites:

Aflatoxin B_1

incidence: 1/2, sa. const.: Holstein cows (midlactation), wt.: 408.2–544.3 kg, contamination: artificial (dose: 0.35 mg AFB_1/kg b. wt., o., for 3 days (one cow was fed AF-rations for 3 days, the other cow, after 3 days of dosing, was fed AF-free rations for 7 additional days); for detailed information please see the article), conc.: 0.60 ng/g*, country: USA[93], *after 4 days (thereof 3 days with AFB_1-administration)

incidence: 2/2, sa. const.: lactating Francaise Frisonne Pie Noire cows, age:

3–4 years, wt.: 500–550 kg, contamination: artificial (dose: peanut meal (3.5 mg AFB_1/kg) spiked with labeled AFB_1), o., for 10 days; for detailed information please see the article), conc. range: 10.7–13.2 ng AFB_1 eq/g*, Ø conc.: 11.95 ng AFB_1 eq/g*, country: The Netherlands/UK/France[397], *after a 10-day period of consumption of **contaminated peanut meal**
incidence: 2/2, sa. const.: lactating Francaise Frisonne Pie Noire cows, age: 3–4 years, wt.: 500–550 kg, contamination: artificial (dose: peanut meal (3.5 mg AFB_1/kg) spiked with labeled AFB_1; but decontaminated afterwards), o., for 10 days; for detailed information please see the article), conc.: nd*, country: The Netherlands/UK/France[397], *after a 10-day period of consumption of **decontaminated peanut meal**

Aflatoxin M_1
incidence: 1/2, sa. const.: Holstein cows (midlactation), wt.: 408.2–544.3 kg, contamination: artificial (dose: 0.35 mg AFB_1/kg b. wt., o., for 3 days (one cow was fed AF-rations for 3 days, the other cow, after 3 days of dosing, was fed AF-free rations for 7 additional days); for detailed information please see the article), conc.: 0.90 ng/g*, country: USA[93], *after 4 days (thereof 3 days with AFB_1-administration)

HT-2 Toxin
incidence: 1/1, sa. const.: Holstein cow, wt.: 365 kg, contamination: artificial (dose: 2 × 200 mg T-2 toxin, by balling gun; for detailed information please see the article), conc.: 1.2 ppb*, country: USA[180], *no statement when found

Cow lymph may contain the following mycotoxins and/or their metabolites:

Aflatoxin B_1
incidence: 1/2, sa. const.: Holstein cows (midlactation), wt.: 408.2–544.3 kg, contamination: artificial (dose: 0.35 mg AFB_1/kg b. wt., o., for 3 days (one cow was fed AF-rations for 3 days, the other cow, after 3 days of dosing, was fed AF-free rations for 7 additional days); for detailed information please see the article), conc.: 0.30 ng/g* **, country: USA[93], *in supramammary lymph nodes, **after 4 days (thereof 3 days with AFB_1-administration)

Aflatoxin M_1
incidence: 1/2, sa. const.: Holstein cows (midlactation), wt.: 408.2–544.3 kg, contamination: artificial (dose: 0.35 mg AFB_1/kg b. wt., o., for 3 days (one cow was fed AF-rations for 3 days, the other cow, after 3 days of dosing, was fed AF-free rations for 7 additional days); for detailed information please see the article), conc.: 0.70 ng/g* **, country: USA[93], *in supramammary lymph nodes, **after 4 days (thereof 3 days with AFB_1-administration)

Cow mammary gland may contain the following mycotoxins and/or their metabolites:

Aflatoxin B_1
incidence: 1/2, sa. const.: Holstein cows (midlactation), wt.: 408.2–544.3 kg, contamination: artificial (dose: 0.35 mg AFB_1/kg b. wt., o., for 3 days (one cow was fed AF-rations for 3 days, the other cow, after 3 days of dosing, was fed AF-free rations for 7 additional days); for detailed information please see the article), conc.: 0.60 ng/g*, country: USA[93], *after 4 days (thereof 3 days with AFB_1-administration)

Aflatoxin M_1
incidence: 1/2, sa. const.: Holstein cows (midlactation), wt.: 408.2–544.3 kg, contamination: artificial (dose: 0.35 mg AFB_1/kg b. wt., o., for 3 days (one cow was fed AF-rations for 3 days, the other cow, after 3 days of dosing, was fed AF-free rations for 7 additional days); for detailed information please see the article), conc.:

24.5 ng/g*, country: USA[93], *after 4 days (thereof 3 days with AFB_1-administration)

incidence: 1/1, sa. const.: lactating dairy cows, contamination: artificial (dose: **10 ppb AFB_1** (labeled and unlabeled), o., twice daily over a 14-day period; for detailed information please see the article), conc.: na, country: USA[319]
incidence: 1/1, sa. const.: lactating dairy cows, contamination: artificial (dose: **50 ppb AFB_1** (labeled and unlabeled), o., twice daily over a 14-day period; for detailed information please see the article), conc.: na, country: USA[319]
incidence: 1/1, sa. const.: lactating dairy cows, contamination: artificial (dose: **250 ppb AFB_1** (labeled and unlabeled), o., twice daily over a 14-day period; for detailed information please see the article), conc.: nd* **, country: USA[319], *after 15 days (thereof 14 days with AFB_1-administration), **but significant radioactivity occurred
incidence: 1/1, sa. const.: lactating dairy cow, contamination: artificial (dose: **1,250 ppb AFB_1** (labeled and unlabeled), o., twice daily over a 14-day period; for detailed information please see the article), conc.: 0.27 µg/kg*, country: USA[319], *after 15 days (thereof 14 days with AFB_1-administration)

Cow milk, raw may contain the following mycotoxins and/or their metabolites:

Aflatoxicol
incidence: 12/12 milk sa., sa. const.: milk from Holstein cow (midlactation period), age: 5 years, wt.: 600 kg, contamination: artificial (dose: 0.5 mg AFB_1/kg b. wt., o., once), conc. range: ≈≤1.1 ng/g* **, country: USA[92], *in milk of cow that **died**, **after 60 h (also at other hour intervals up to 60 h measured, lowest conc.: ≈<0.05 ng/g after 24 h) (mycotoxin values very approximately)
incidence: 16/17 milk sa., sa. const.: milk from Holstein cow (midlactation period), age: 5 years, wt.: 600 kg, contamination: artificial (dose: 0.5 mg AFB_1/kg b. wt., o., once), conc. range: ≈≤0.2 ng/g* **, country: USA[92], *in milk of cow that **survived**, **after 60 h (also at other hour intervals up to 120 h measured, lowest conc.: ≈0.01 ng/g after 120 h) (mycotoxin values very approximately)

Aflatoxin B_1
incidence: 12/12 milk sa., sa. const.: milk from Holstein cow (midlactation period), age: 5 years, wt.: 600 kg, contamination: artificial (dose: 0.5 mg AFB_1/kg b. wt., o., once), conc. range: ≈≤10 ng/g* **, country: USA[92], *in milk of cow that died, **after 60 h (also at other hour intervals up to 60 h measured, lowest conc.: ≈1 ng/g after 24 h) (mycotoxin values very approximately)
incidence: 26/27 milk sa., sa. const.: milk from Holstein cow (midlactation period), age: 5 years, wt.: 600 kg, contamination: artificial (dose: 0.5 mg AFB_1/kg b. wt., o., once), conc. range: ≈≤8 ng/g* **, country: USA[92], *in milk of cow that **survived**, **after ≈70 h (also at other hour intervals up to 240 h measured, lowest conc.: ≈0.01 ng/g after 240 h) (mycotoxin values very approximately)

incidence: 16*/26*, sa. const.: milk from Holstein cows (midlactation), wt.: 408.2–544.3 kg, contamination: artificial (dose: 0.35 mg AFB_1/kg b. wt., o., 3 days (one cow was fed AF-rations for 3 days, the other cow, after 3 days of dosing, was fed AF-free rations for 7 additional days); for detailed information please see the article), conc. range: ≤1.00 and ≤1.18 ng/ml**, country: USA[93], *milk sa. from both cows, **these highest values measured at the 3rd day in the case of cow 1 and cow 2 (always lower values measured up to 4 days (then slaughtered) in the case of cow 1 and always lower values measured up to 10 days (then slaughtered) in the case of cow 2), for overall information please see the article

incidence: 1/1 animal, sa. const.: milk from cows at peak lactation, contamination: artificial (dose: 80 mg AFB_1, o., daily for 3 weeks; for detailed information please see the article), conc.: 290 ppb* **, country: USA[169], *in dried milk, **1 day after withdrawal (also measured 2, 3 and 4 days after withdrawal, lowest conc.: 175 ppb after 4 days)

incidence: 20/20 milk sa., sa. const.: milk from lactating Francaise Frisonne Pie Noire cows, age: 3–4 years, wt.: 500–550 kg, contamination: artificial (dose: peanut meal ((3.5 mg AFB_1/kg) spiked with labeled AFB_1), o., for 10 days; for detailed information please see the article), conc. range: 4.6–6.0 μg AFB_1 eq/l*, country: The Netherlands/UK/France[397], *fed **contaminated peanut meal**

incidence: 20/20 milk sa., sa. const.: milk from lactating Francaise Frisonne Pie Noire cows, age: 3–4 years, wt.: 500–550 kg, contamination: artificial (dose: peanut meal ((3.5 mg AFB_1/kg) spiked with labeled AFB_1; but decontaminated afterwards), o., for 10 days; for detailed information please see the article), conc. range: ≤0.8 μg AFB_1 eq/l*, country: The Netherlands/UK/France[397], *fed **decontaminated peanut meal**

Aflatoxin M_1
incidence: ?/?, sa. const.: milk from Friesian and Friesian X dairy cows, contamination: artificial (dose: between 155 and 244 μg AFB_1, o., daily; for detailed information please see the article), conc. range: 0.15–0.26 μg/l*, Ø conc.: 0.21 μg/l*, country: UK[39], *in bulk milk collected for 7 days
incidence: ?/?, sa. const.: milk from Friesian and Friesian X dairy cows, contamination: artificial (dose: between 155 and 244 μg AFB_1, o., daily; for detailed information please see the article), conc. range: <0.01–0.33 μg/l, country: UK[39]

incidence: ?/18? milk sa., sa. const.: milk from lactating cow, wt.: 600 kg, contamination: artificial (dose: 300 mg AFs: 44% AFB_1, 2% AFB_2, 44% AFG_1?, o., once), conc.: 492 μg* (total amount found), country: UK[40], *collected over a period of 9 days

incidence: ?/?, sa. const.: milk from late-lactation Holstein dairy cows, Ø wt.: 544 kg, contamination: artificial (dose: 100 μg AF/kg feed and 1.0% HSCAS in period 2 or 200 μg AF/kg feed and 0.5% HSCAS in period 2*, conc. range: 0.51–1.99 μg/l** (mean values), country: USA[77], *period = 7 days, 3 periods in all (1st and 3rd period only AF given) **values of all trials (for detailed information please see the article)

incidence: ?/?, sa. const.: milk from Holstein cows (midlactation), contamination: artificial (dose: 13 mg AFB_1 (pure/impure), via rumen orifice, daily for 7 days; for detailed information please see the article) conc. range: ≤10.58 ppb* ** (mean value), country: USA[83], *6 animals received pure and 3 animals impure AFB_1 (13 mg), **after 2–7 days of treatment
incidence: ?/?, sa. const.: milk from a Holstein cow (midlactation), contamination: artificial (dose: 25 mg AFB_1 (pure), via rumen orifice, daily for 7 days; for detailed information please see the article) conc.: 1.35 ppb* (mean value), country: USA[83], *after 2–7 days of treatment

incidence: ?/?, sa. const.: milk from Danish black and white breed, contamination: artificial (dose: **57 μg AFB_1** in feed/day, for 8 weeks?; for detailed information please see the article), conc. range: 27–74 ng/kg*, country: Denmark[85], *during 8 weeks a milk sa. from each cow analysed once a week
incidence: ?/?, sa. const.: milk from Danish black and white breed, contamination:

artificial (dose: **142 μg AFB_1** in feed/day, for 8 weeks?; for detailed information please see the article), conc. range: 38–128 ng/kg*, country: Denmark[85], *during 8 weeks a milk sa. from each cow analysed once a week
incidence: ?/?, sa. const.: milk from Danish black and white breed, contamination: artificial (dose: **226 μg AFB_1** in feed/day, for 8 weeks?; for detailed information please see the article), conc. range: 60–271 ng/kg*, country: Denmark[85], *during 8 weeks a milk sa. from each cow analysed once a week
incidence: ?/?, sa. const.: milk from Danish black and white breed, contamination: artificial (dose: **311 μg AFB_1** in feed/day, for 8 weeks?; for detailed information please see the article), conc. range: 96–138 ng/kg*, country: Denmark[85], *during 8 weeks a milk sa. from each cow analysed once a week

incidence: ?/8 animals, sa. const.: milk from Holstein-Friesian lactating cows, contamination: artificial (dose: 120 μg AFB_1/kg in ammoniated peanut meal, o., daily for 9 days (1,080 μg in total); for detailed information please see the article), conc. range: ≤0.070 μg/l* **, country: France[86], *total amount found: 52 μg AFM_1, **at day 8
incidence: ?/?, sa. const.: milk from Holstein-Friesian lactating cows, contamination: artificial (dose: ≈8–9 μg AFB_1/kg in untreated peanut meal, o., for 16 months; for detailed information please see the article), conc. range: ≤0.510 μg/l*, country: France[86], *total amount found: 1,247 μg AFM_1

incidence: 12/12 milk sa., sa. const.: milk from Holstein cow (midlactation period), age: 5 years, wt.: 600 kg, contamination: artificial (dose: 0.5 mg AFB_1/kg b. wt., o., once), conc. range: ≈≤200 ng/g* **, country: USA[92], *in milk of cow that died, **after ≈10 h (also at other hour intervals up to 60 h measured, lowest conc.: ≈<40 ng/g after ≈35 h) (mycotoxin values very approximately)

incidence: 27/27 milk sa., sa. const.: milk from Holstein cow (midlactation period), age: 5 years, wt.: 600 kg, contamination: artificial (dose: 0.5 mg AFB_1/kg b. wt., o., once), conc. range: ≈≤70 ng/g* **, country: USA[92], *in milk of cow that **survived**, **after ≈43 h (also at other hour intervals up to 240 h measured, lowest conc.: ≈<0.05 ng/g after 240 h) (mycotoxin values very approximately)

incidence: 24*/26*, sa. const.: milk from Holstein cows (midlactation), wt.: 408.2–544.3 kg, contamination: artificial (dose: 0.35 mg AFB_1/kg b. wt., o., for 3 days (one cow was fed AF-rations for 3 days, the other cow, after 3 days of dosing, was fed AF-free rations for 7 additional days); for detailed information please see the article), conc. range: ≤23.40** and ≤32.80*** ng/ml, country: USA[93], *milk sa. from both cows, these highest values measured at the **2nd day in the case of cow 2 and at the ***3rd day in the case of cow 1 (always lower values measured up to 4 days (then slaughtered) in the case of cow 1 and always lower values measured up to 10 days (then slaughtered) in the case of cow 2), for overall information please see the article

incidence: ?/?, sa. const.: milk from Holstein dairy cattles, contamination: artificial (dose: 650 μg AF/kg in **non-ammoniated cottonseeds** fed; for detailed information please see the article), conc. range: ≤1.8 μg/l (mean value), country: USA[107]
incidence: ?/?, sa. const.: milk from Holstein dairy cattles, contamination: artificial (dose: 15–80 μg AF/kg in **ammoniated cottonseeds** fed; for detailed information please see the article), conc. range: ≤0.18 μg/l (mean value), country: USA[107]

incidence: ?/?, sa. const.: milk from lactating cows, contamination: artificial (dose: **10 ppb AFB_1** (labeled and unlabeled) in concentrate, o., twice daily

for 14 days; for detailed information please see the article), conc.: nd*, country: USA[137], *at day 4 and 8 of feeding AFB_1
incidence: ?/?, sa. const.: milk from lactating cows, contamination: artificial (dose: **50 ppb AFB_1** (labeled and unlabeled) in concentrate, o., twice daily for 14 days; for detailed information please see the article), conc. range: 0.1 μg/l* (mean value), country: USA[137], *at day 4 and 8 of feeding AFB_1
incidence: ?/?, sa. const.: milk from lactating cows, contamination: artificial (dose: **250 ppb AFB_1** (labeled and unlabeled) in concentrate, o., twice daily for 14 days; for detailed information please see the article), conc. range: 0.23*and 0.26** μg/l (mean values), country: USA[137], at day 8* or 4** of feeding AFB_1
incidence: ?/?, sa. const.: milk from lactating cows, contamination: artificial (dose: **1,250 ppb AFB_1** (labeled and unlabeled) in concentrate, o., twice daily for 14 days; for detailed information please see the article), conc. range: 0.82* and 0.88** μg/l (mean values), country: USA[137], at day 4* or 8** of feeding AFB_1

incidence: ?/20, sa. const.: milk from a Holstein cow (mid-lactation), contamination: artificial (dose: 26 ppm AFB_1, into fistulated rumen, twice daily for 7 days), conc.: ≈10 ppb*, country: USA[147], *on day 3

incidence: 5/5 animals, sa. const.: milk from Holstein dairy cows, contamination: artificial (dose: 5.82 mg AFB_1/day in cottonseed (**but ammoniated afterwards**), o., for 9 days; for detailed information please see the article), conc.: nd, country: USA[150] (also at other day intervals over 9 days measured)
incidence: ?/4 animals, sa. const.: milk from Holstein dairy cows, contamination: artificial
(dose: 5.82 mg AFB_1/day in cottonseed, o., for 9 days; for detailed information please see the article), conc.: 5.15 μg/l (mean value), country: USA[150] (also at other day intervals over 9 days measured, lowest conc.: 2.5 μg/l after 2 days)

incidence: ?/?*, sa. const.: milk from lactating Holstein cows, contamination: artificial (dose: no AFB_1; for detailed information please see the article), conc.: ≈0.2 μg/l (mean value), country: USA[157], *control
incidence: ?/4 animals, sa. const.: milk from lactating Holstein cows, contamination: artificial (dose: **20 μg AFB_1/kg** complete feed, o., for 192 h; for detailed information please see the article), conc.: 0.33 μg/l* ** (mean value), country: USA[157], *after 120 h (also at other hour intervals up to 264 h measured), **cows with low milk production level
incidence: ?/4 animals, sa. const.: milk from lactating Holstein cows, contamination: artificial (dose: **20 μg AFB_1/kg** complete feed, o., for 192 h; for detailed information please see the article), conc.: 0.36 μg/l* ** (mean value), country: USA[157], *after 120 h (also at other hour intervals up to 264 h measured), **cows with high milk production level
incidence: ?/4 animals, sa. const.: milk from lactating Holstein cows, contamination: artificial (dose: **48 μg AFB_1/kg** complete feed, o., for 192 h; for detailed information please see the article), conc.: 0.51 μg/l* ** (mean value), country: USA[157], *after 132 h (also at other hour intervals up to 264 h measured), **cows with low milk production level
incidence: ?/4 animals, sa. const.: milk from lactating Holstein cows, contamination: artificial (dose: **48 μg AFB_1/kg** complete feed, o., for 192 h; for detailed information please see the article), conc.: 0.74 μg/l* ** (mean value), country: USA[157], *after 132 h (also at other hour intervals up to 264 h

measured), **cows with high milk production level
incidence: ?/4 animals, sa. const.: milk from lactating Holstein cows, contamination: artificial (dose: **104 μg AFB_1/kg** complete feed, o., for 192 h; for detailed information please see the article), conc.: 1.72 μg/l* ** (mean value), country: USA[157], *after 120 h (also at other hour intervals up to 264 h measured), **cows with low milk production level
incidence: ?/4 animals, sa. const.: milk from lactating Holstein cows, contamination: artificial (dose: **104 μg AFB_1/kg** complete feed, o., for 192 h; for detailed information please see the article), conc.: 1.50 μg/l* ** (mean value), country: USA[157], *after 120 h (also at other hour intervals up to 264 h measured), **cows with high milk production level
incidence: ?/3 animals, sa. const.: milk from lactating Holstein cows, contamination: artificial (dose: **44 μg AFB_1/kg** in contaminated cottonseed meal, o., for 192 h?; for detailed information please see the article), conc.: 0.59 μg/l* (mean value), country: USA[157], *cows with low milk production level
incidence: ?/4 animals, sa. const.: milk from lactating Holstein cows, contamination: artificial (dose: **44 μg AFB_1/kg** in contaminated cottonseed meal, o., for 192 h?; for detailed information please see the article), conc.: 0.62 μg/l* (mean value), country: USA[157], *cows with high milk production level
incidence: ?/4 animals, sa. const.: milk from lactating Holstein cows, contamination: artificial (dose: **49 μg AFB_1/kg** in contaminated corn, o., for 192 h?; for detailed information please see the article), conc.: 0.55 μg/l* (mean value), country: USA[157], *cows with low milk production level
incidence: ?/4 animals, sa. const.: milk from lactating Holstein cows, contamination: artificial (dose: **49 μg AFB_1/kg** in contaminated corn, o., for 192 h?; for detailed information please see the article), conc.: 0.51 μg/l* (mean value), country: USA[157], *cows with high milk production level

incidence: ?/?, sa. const.: milk from cows, contamination: artificial (dose: 5.5 ppm AFB_1 in groundnut meal (in various portions fed, **here 13.5% portion**, AFB_1-intake 5.0 mg/day) + hay + concentrate mix, o., for ? days; for detailed information please see the article), conc.: 3.0 μg/l* ** *** (mean value), country: UK[171], *in liquid milk, **within period of AFB_1-administration, ***lower AFB_1-intake values (mg/day) resulted in lower residue values
incidence: ?/?, sa. const.: milk from cows, contamination: artificial (dose: 18.8 ppm AFB_1 in groundnut meal (in various portions fed, **here 20% portion**, AFB_1-intake 24.5 mg/day) + hay + concentrate mix, o., for ? days; for detailed information please see the article), conc.: 13.3 μg/l* ** *** (mean value), country: UK[171], *in liquid milk, **within period of AFB_1-administration, ***lower AFB_1-intake values (mg/day) resulted in lower residue values
incidence: ?/?, sa. const.: milk from cows, contamination: artificial (dose: 5.5 ppm AFB_1 in groundnut meal (in various portions fed, **here 7.5% portion**, AFB_1-intake 0.90 mg/day) + grass + concentrate mix, o., for ? days; for detailed information please see the article), conc.: 0.33 μg/l* ** *** (mean value), country: UK[171], *in liquid milk, **within period of AFB_1-administration, ***lower AFB_1-intake values (mg/day) resulted in lower residue values

incidence: 12/12 animals, sa. const.: milk from cows in early lactation (2–4 weeks), contamination: artificial (dose: ≈39 μg AFB_1/day, o., for 14 days; for detailed information please see the article), conc. range: ≤0.07 μg/kg* (≤3.0 μg/day*), country: The Netherlands[211], *within

period of AFB_1-administration, early lactation
incidence: 8/8 animals, sa. const.: milk from cows in late lactation (34–36 weeks), contamination: artificial (dose: ≈34 µg AFB_1/day, o., for 14 days; for detailed information please see the article), conc. range: ≤0.04 µg/kg* (≤1.0 µg/day*), country: The Netherlands[211], *within period of AFB_1-administration, late lactation
incidence: 8/8 animals, sa. const.: milk from individual low milk-yielding cows, contamination: artificial (dose: **≈35 µg AFB_1**/day, o., for 14 days; for detailed information please see the article), conc. range: ≤0.08 µg/kg* (≤1.7 µg/day*), country: The Netherlands[211], *within period of AFB_1-administration
incidence: 8/8 animals, sa. const.: milk from individual low milk-yielding cows, contamination: artificial (dose: **≈56 µg AFB_1**/day, o., for 14 days; for detailed information please see the article), conc. range: ≤0.16 µg/kg* (≤2.4 µg/day*), country: The Netherlands[211], *within period of AFB_1-administration
incidence: 8/8 animals, sa. const.: milk from individual high milk-yielding cows, contamination: artificial (dose: **≈35 µg AFB_1**/day, o., for 14 days; for detailed information please see the article), conc. range: ≤0.04 µg/kg* (≤1.9 µg/day*), country: The Netherlands[211], *within period of AFB_1-administration
incidence: 8/8 animals, sa. const.: milk from individual high milk-yielding cows, contamination: artificial (dose: **≈56 µg AFB_1**/day, o., for 14 days; for detailed information please see the article), conc. range: ≤0.12 µg/kg* (≤3.8 µg/day*), country: The Netherlands[211], *within period of AFB_1-administration

incidence: ?/?, sa. const.: milk from multiparous late-lactation Friesian cows, age: 30–33 weeks, contamination: artificial (dose: **56.40 µg AFB_1**/day, o., for 7 days; for detailed information please see the article), conc.: 15.52 ng/l* (mean value), country: Italy[231], *on days 3 to 7 of the 1st week
incidence: ?/?, sa. const.: milk from multiparous late-lactation Friesian cows, age: 30–33 weeks, contamination: artificial (dose: **56.40 µg AFB_1/day + HSCAS (2%)**, o., for 7 days; for detailed information please see the article), conc.: 10.48 ng/l* (mean value), country: Italy[231], *on days 3 to 7 of the 2nd week
incidence: ?/?, sa. const.: milk from multiparous late-lactation Friesian cows, age: 30–33 weeks, contamination: artificial (dose: **67.15 µg AFB_1**/day, o., for 7 days; for detailed information please see the article), conc.: 17.25 ng/l* (mean value), country: Italy[231], *on days 3 to 7 of the 3rd week
incidence: ?/?, sa. const.: milk from multiparous late-lactation Friesian cows, age: 30–33 weeks, contamination: artificial (dose: **56.40 µg AFB_1**/day, o., for 7 days; for detailed information please see the article), conc.: 15.88 ng/l* (mean value), country: Italy[231], *on days 3 to 7 of the 1st week

incidence: ?/?, sa. const.: milk from multiparous late-lactation Friesian cows, age: 30–33 weeks, contamination: artificial (dose: **56.40 µg AFB_1/day + CAC1 (2%)**, o., for 7 days; for detailed information please see the article), conc.: 8.68 ng/l* (mean value), country: Italy[231], *on days 3 to 7 of the 2nd week
incidence: ?/?, sa. const.: milk from multiparous late-lactation Friesian cows, age: 30–33 weeks, contamination: artificial (dose: **67.15 µg AFB_1**/day, o., for 7 days; for detailed information please see the article), conc.: 17.83 ng/l* (mean value), country: Italy[231], *on days 3 to 7 of the 3rd week
incidence: ?/?, sa. const.: milk from multiparous late-lactation Friesian cows, age: 30–33 weeks, contamination: artificial (dose: **56.40 µg AFB_1**/day, o., for 7 days; for detailed information please see the article), conc.: 15.83 ng/l* (mean value), country: Italy[231], *on days 3 to 7 of the 1st week
incidence: ?/?, sa. const.: milk from multiparous late-lactation Friesian cows,

age: 30–33 weeks, contamination: artificial (dose: **56.40 µg AFB_1/day + CAC2 (2%)**, o., for 7 days; for detailed information please see the article), conc.: 12.35 ng/l* (mean value), country: Italy[231], *on days 3 to 7 of the 2nd week
incidence: ?/?, sa. const.: milk from multiparous late-lactation Friesian cows, age: 30–33 weeks, contamination: artificial (dose: **67.15 µg AFB_1/day**, o., for 7 days; for detailed information please see the article), conc.: 18.41 ng/l* (mean value), country: Italy[231], *on days 3 to 7 of the 3rd week

incidence: ?/?, sa. const.: milk from Jersey milking cows, age: adult, contamination: artificial (dose: **385–1,925 µg ZEA/kg** + (20 µg AFB_1/kg), o., for 7 weeks; for detailed information please see the article), conc. range: tr–0.6 µg/kg*, country: UK[267], *after 7 weeks, *after 7 weeks of (AFB_1-) and ZEA-administration; AFB_1 accidentally in the diet
incidence: ?/?, sa. const.: milk from Jersey milking cows, age: adult, contamination: artificial (dose: **317–1,125 µg OTA/kg** + (20 µg AFB_1/kg), o., for 11 weeks; for detailed information please see the article), conc. range: 0.06 µg/kg*, country: UK[267], *after 11 weeks, *after 11 weeks of (AFB_1-) and OTA-administration; AFB_1 accidentally in the diet

incidence: ?/?, sa. const.: milk from lactating cows, contamination: no AFB_1 (for detailed information please see the article), conc.: ?, country: Germany[279]
incidence: ?/2? animals, sa. const.: milk from lactating cows, contamination: artificial (dose: **0.2 mg AFB_1/day + 7.0 mg PCB/day**, o., for 10 and 30 days, resp.?; for detailed information please see the article), conc. range: ≤2.17 nmol*, country: Germany[279], *in **anamnesis phase**
incidence: ?/2? animals, sa. const.: milk from lactating cows, contamination: artificial (dose: **0.2 mg AFB_1/day + 7.0 mg PCB/day**, o., for 10 and 30 days, resp.?; for detailed information please see the article), conc. range: ≤12.64 nmol*, country: Germany[279], *in **dose phase** (20th–44th day)

incidence: ?/?*, sa. const.: milk from crossbred lactating cows, contamination: no AFB_1 (for detailed information please see the article), conc.: 2.53 ppt**, country: India[286], *control, **after 14 days of feeding
incidence: ?/?, sa. const.: milk from crossbred lactating cows, contamination: artificial (dose: **25 ppb AFB_1** in concentrate mixture "B", o., for 11 days; for detailed information please see the article), conc. range: ≤47.79 ppt* (total: 0.446 µg AFM_1/cow*), country: India[286], *after 7 days
incidence: ?/?, sa. const.: milk from crossbred lactating cows, contamination: artificial (dose: **50 ppb AFB_1** in concentrate mixture "C", o., for 6 days; for detailed information please see the article), conc. range: ≤68.14 ppt (total: 0.586 µg AFM_1/cow), country: India[286], *after 6 days
incidence: ?/?, sa. const.: milk from crossbred lactating cows, contamination: artificial (dose: **75 ppb AFB_1** in concentrate mixture "D", o., for 13 days; for detailed information please see the article), conc. range: ≤127.44 ppt* (total: 0.814 µg AFM_1/cow)*, country: India[286], *after 13 days
incidence: ?/?, sa. const.: milk from crossbred lactating cows, contamination: artificial (dose: **100 ppb AFB_1** in concentrate mixture "E", o., for 11 days; for detailed information please see the article), conc. range: ≤273.22 ppt* (total: 1,996 µg AFM_1/cow*), country: India[286], *after 11 days

incidence: 20/20 milk sa., sa. const.: milk from lactating Francaise Frisonne Pie Noir cows, age: 3–4 years, wt.: 500–550 kg, contamination: artificial (dose: peanut meal ((3.5 mg AFB_1/kg) spiked with labeled AFB_1), o., for 10 days; for detailed information please see the article), conc. range: 2.3–3.5 µg/l*, country: The

Netherlands/UK/France[397], *fed **contaminated peanut meal**
incidence: 20/20 milk sa., sa. const.: milk from lactating Francaise Frisonne Pie Noir cows, age: 3–4 years, wt.: 500–550 kg, contamination: artificial (dose: peanut meal ((3.5 mg AFB_1/kg) spiked with labeled AFB_1; but decontaminated afterwards), o., for 10 days; for detailed information please see the article), conc. range: 0.1 µg/l*, country: The Netherlands/UK/France[397], *fed **decontaminated peanut meal**

incidence: ?/8 animals, sa. const.: milk from crossbreeds of Dutch Friesian cows and Holstein Friesians (early-mid lactation), contamination: artificial (dose: **4 µg AFB_1/kg**, o., for 5 days; for detailed information please see the article), conc.: 0.01 µg/kg* (mean value) (0.3 µg/day*, mean value), country: The Netherlands[567], *after 6 and 7 days
incidence: ?/8 animals, sa. const.: milk from crossbreeds of Dutch Friesian cows and Holstein Friesians (early-mid lactation), contamination: artificial (dose: **12 µg AFB_1/kg**, o., for 5 days; for detailed information please see the article), conc.: 0.08 µg/kg* (mean value) (2.08 µg/day*, mean value), country: The Netherlands[567], *after 6 and 7 days
incidence: ?/8 animals, sa. const.: milk from crossbreeds of Dutch Friesian cows and Holstein Friesians (early-mid lactation), contamination: artificial (dose: **12 µg AFB_1/kg + 1% bentonite**, o., for 5 days; for detailed information please see the article), conc.: 0.05 µg/kg* (mean value) (1.36 µg/day*, mean value), country: The Netherlands[567], *after 6 and 7 days
incidence: ?/3 animals*, sa. const.: milk from from crossbreeds of Dutch Friesian cows and Holstein Friesians (early-mid lactation), contamination: artificial (dose: **2.8 µg AFB_1/kg** compound feed, o., for 14 days; for detailed information please see the article), conc.: 0.03 µg/kg** (mean value) (1.0 µg/day**, mean value), country: The Netherlands[567], *control, **sa. taken on day 12 and 14
incidence: ?/3 animals, sa. const.: milk from from crossbreeds of Dutch Friesian cows and Holstein Friesians (early-mid lactation), contamination: artificial (dose: **2.8 µg AFB_1/kg** compound feed **+ 1% HSCAS**, o., for 14 days; for detailed information please see the article), conc.: 0.03 µg/kg* (mean value) (1.0 µg/day*, mean value), country: The Netherlands[567], *sa. taken on day 12 and 14
incidence: ?/3 animals, sa. const.: milk from from crossbreeds of Dutch Friesian cows and Holstein Friesians (early-mid lactation), contamination: artificial (dose: **2.1 µg AFB_1/kg** compound feed **+ 1% bentonite**, o., for 14 days; for detailed information please see the article), conc.: 0.02 µg/kg* (mean value) (0.7 µg/day*, mean value), country: The Netherlands[567], *sa. taken on day 12 and 14

incidence: 1/1 animals, sa. const.: milk from a cow, age: 5 years, wt.: 200 kg, contamination: artificial (dose: up to 4.468 mg* AFB_1/kg ration, o., at day 9* (5 different dose applicated over 9 days); for detailed information please see the article), conc.: ≤0.088 µg/ml**, country: Egypt[575], **at the 9th day (also at other day intervals up to 15 days measured, lowest conc.: nd after 15 days)

Aflatoxin M

incidence: ?/?, sa. const.: milk from cows at peak lactation, contamination: artificial (dose: 80 mg AFB_1, o., daily for 3 weeks; for detailed information please see the article), conc.: 1,500 ppb* ** ***, country: USA[169], *in dried milk, **1 week before withdrawal (up to 7 days after withdrawal measured, lowest conc.: 160 ppb after 4 days), ***lower AFB_1-amounts given in combination with longer application periods resulted in lower residue values

Deepoxydeoxynivalenol

incidence: 20*/72*, sa. const.: 2 Holstein dairy cows, 1 Ayrshire dairy cow, wt.: ≈450 kg, contamination: artificial (dose: 66 mg DON/kg, o., for 5 days; for detailed information please see the article), conc. range: ≤26 ng/ml** (1.12 mg** in total, mean value), country: USA[278], *milk sa. from all 3 cows, **after 5 days of application (also at other day intervals up to 12 days measured, lowest conc.: nd at the beginning and the end of the experiment; for detailed information please see the article)

Fumonisin B_1

incidence: 3/3 animals, sa. const.: milk from black-pied Holstein Friesian cows, contamination: artificial (dose: 30 mg FB_1/cow (≈50 µg FB_1/kg b. wt.), i.v., once; for detailed information please see the article), conc. range: 0.27*–1.16** µg/kg, country: Germany[274], after 10* and **4 h of application (also at other hour intervals up to 58 h measured, lowest conc.: nd after 24 h for all 3 cows)

Zearalenone

incidence: 1/1 animal, sa. const.: milk from Holstein cross-breed cows, wt.: 480–580 kg, contamination: artificial (dose: **544.5 mg ZEA**/day, o., **for 21 days**), conc. range: ≤2.5 ng/ml*, country: Canada[79], *on day 2 (also at other day intervals up to 9 days measured, lowest conc.: ≈0.2 ng/ml after 5 days)

incidence: 1/1 animal, sa. const.: milk from Holstein cross-breed cows, wt.: 480–580 kg, contamination: artificial (dose: **1.8 g ZEA**/day, o., **for 1 day**), conc. range: ≤4.0 ng/ml*, country: Canada[79], *on day 1 (also at other day intervals up to 9 days measured, lowest conc.: nd after 3–9 days)

incidence: 1/1 animal, sa. const.: milk from Holstein cross-breed cows, wt.: 480–580 kg, contamination: artificial (dose: **6 g ZEA**/day, o., **for 1 day**), conc. range: ≤6.1 ng/ml*, country: Canada[79], *on day 2 (also at other day intervals up to 9 days measured, lowest conc.: nd after 5–9 days)

incidence: 3/11 milk sa., sa. const.: milk from cow, contamination: artificial (dose: 5 g crystalline ZEA, o., once), conc. range: tr*, country: Hungary/USA[123], *at 96, 108 and 120 h (also at other hour intervals up to 120 h measured, lowest conc.: nd up to 84 h)

α-Zearalenol

incidence: 1/1 animal, sa. const.: milk from Holstein cross-breed cows, wt.: 480–580 kg, contamination: artificial (dose: **544.5 mg ZEA**/day, o., **for 21 days**), conc. range: ≤3.0 ng/ml*, country: Canada[79], *on day 2 (also at other day intervals up to 9 days measured, lowest conc.: ≈0.45 ng/ml after 5 days)

incidence: 1/1 animal, sa. const.: milk from Holstein cross-breed cows, wt.: 480–580 kg, contamination: artificial (dose: **1.8 g ZEA**/day, o., **for 1 day**), conc. range: ≤1.5 ng/ml*, country: Canada[79], *on day 2 (also at other day intervals up to 9 days measured, lowest conc.: nd after 5–9 days)

incidence: 1/1 animal, sa. const.: milk from Holstein cross-breed cows, wt.: 480–580 kg, contamination: artificial (dose: **6 g ZEA**/day, o., **for 1 day**), conc. range: ≤4.0 ng/ml*, country: Canada[79], *on day 2 (also at other day intervals up to 9 days measured, lowest conc.: nd after 5 and 9 days)

β-Zearalenol

incidence: 1/1 animal, sa. const.: milk from Holstein cross-breed cows, wt.: 480–580 kg, contamination: artificial (dose: **544.5 mg ZEA**/day, o., **for 21 days**), conc.: nd-tr, country: Canada[79] (also at other day intervals up to 9 days measured)

incidence: 1/1 animal, sa. const.: milk from Holstein cross-breed cows, wt.: 480–580 kg, contamination: artificial (dose: **1.8 g ZEA**/day, o., **for 1 day**), conc. range: ≤4.1 ng/ml*, country: Canada[79], *on day 1 (also at other day intervals up to 9 days measured, lowest conc.: nd after 3–9 days)

incidence: 1/1 animal, sa. const.: milk from Holstein cross-breed cows, wt.:

480–580 kg, contamination: artificial (dose: **6 g ZEA/day**, o., **for 1 day**), conc. range: ≤6.6 ng/ml*, country: Canada[79], *on day 3 (also at other day intervals up to 9 days measured, lowest conc.: nd after 4–9 days)

incidence: 2/11 milk sa. const.: milk from cow, contamination: artificial (dose: 5 g crystalline ZEA, o., once), conc.: tr*, country: Hungary/USA[123], *at 108 and 120 h (also at other hour intervals up to 120 h measured, lowest conc.: nd up to 96 h)

Cow muscle may contain the following mycotoxins and/or their metabolites:

Aflatoxin B_1
incidence: 1/2, sa. const.: Holstein cows (midlactation), wt.: 408.2–544.3 kg, contamination: artificial (dose: 0.35 mg AFB_1/kg b. wt., o., for 3 days (one cow was fed AF-rations for 3 days, the other cow, after 3 days of dosing, was fed AF-free rations for 7 additional days); for detailed information please see the article), conc.: 0.56 ng/g* **, country: USA[93], *in ***biceps*** muscle, **after 4 days (thereof 3 days with AFB_1-administration)
incidence: 1/2, sa. const.: Holstein cows (midlactation), wt.: 408.2–544.3 kg, contamination: artificial (dose: 0.35 mg AFB_1/kg b. wt., o., for 3 days (one cow was fed AF-rations for 3 days, the other cow, after 3 days of dosing, was fed AF-free rations for 7 additional days); for detailed information please see the article), conc.: 0.31 ng/g* **, country: USA[93], *in ***iliopsoas*** muscle, **after 4 days (thereof 3 days with AFB_1-administration)
incidence: 1/2, sa. const.: Holstein cows (midlactation), wt.: 408.2–544.3 kg, contamination: artificial (dose: 0.35 mg AFB_1/kg b. wt., o., for 3 days (one cow was fed AF-rations for 3 days, the other cow, after 3 days of dosing, was fed AF-free rations for 7 additional days); for detailed information please see the article), conc.: 0.53 ng/g* **, country: USA[93], *in ***semitendinosus*** muscle, **after 4 days (thereof 3 days with AFB_1-administration)

incidence: 2/2, sa. const.: lactating Francaise Frisonne Pie Noire cows, age: 3–4 years, wt.: 500–550 kg, contamination: artificial (dose: peanut meal ((3.5 mg AFB_1/kg) spiked with labeled AFB_1), o., for 10 days; for detailed information please see the article), conc. range: 1.7–2.1 ng AFB_1 eq/g*, Ø conc.: 1.9 ng AFB_1 eq/g*, country: The Netherlands/UK/France[397], *after a 10-day period of consumption of **contaminated peanut meal**
incidence: 2/2, sa. const.: lactating Francaise Frisonne Pie Noire cows, age: 3–4 years, wt.: 500–550 kg, contamination: artificial (dose: peanut meal ((3.5 mg AFB_1/kg) spiked with labeled AFB_1; but decontaminated afterwards), o., for 10 days; for detailed information please see the article), conc.: nd*, country: The Netherlands/UK/France[397], *after a 10-day period of consumption of **decontaminated peanut meal**

Aflatoxin M_1
incidence: 1/2, sa. const.: Holstein cows (midlactation), wt.: 408.2–544.3 kg, contamination: artificial (dose: 0.35 mg AFB_1/kg b. wt., o., for 3 days (one cow was fed AF-rations for 3 days, the other cow, after 3 days of dosing, was fed AF-free rations for 7 additional days); for detailed information please see the article), conc.: 1.19 ng/g* **, country: USA[93], *in ***biceps*** muscle, **after 4 days (thereof 3 days with AFB_1-administration)
incidence: 1/2, sa. const.: Holstein cows (midlactation), wt.: 408.2–544.3 kg, contamination: artificial (dose: 0.35 mg AFB_1/kg b. wt., o., for 3 days (one cow was fed AF-rations for 3 days, the other cow, after 3 days of dosing, was fed AF-free rations for 7 additional days); for detailed information please see the article), conc.: 1.18 ng/g* **, country: USA[93], *in ***iliopsoas*** muscle, **after 4 days (thereof 3 days with AFB_1-administration)

incidence: 1/2, sa. const.: Holstein cows (midlactation), wt.: 408.2–544.3 kg, contamination: artificial (dose: 0.35 mg AFB_1/kg b. wt., o., for 3 days (one cow was fed AF-rations for 3 days, the other cow, after 3 days of dosing, was fed AF-free rations for 7 additional days); for detailed information please see the article), conc.: 1.03 ng/g* **, country: USA[93], *in ***semitendinosus*** muscle, **after 4 days (thereof 3 days with AFB_1-administration)

Cow plasma may contain the following mycotoxins and/or their metabolites:

Aflatoxicol
incidence: 1/1, sa. const.: Holstein cow (midlactation period), age: 5 years, wt.: 600 kg, contamination: artificial (dose: 0.5 mg AFB_1/kg b. wt., o., once), conc. range: ≈≤0.4 ng/g?* **, country: USA[92], *in cow that **died**, **after 60 h (also at other hour intervals up to 60 h measured, lowest conc.: ≈<0.02 ng/g after 24 h) (mycotoxin values very approximately)
incidence: 1/1, sa. const.: Holstein cow (midlactation period), age: 5 years, wt.: 600 kg, contamination: artificial (dose: 0.5 mg AFB_1/kg b. wt., o., once), conc. range: ≈≤0.2 ng/g* **, country: USA[92], *in cow that **survived**, **after ≈60 h (also at other hour intervals up to 120 h measured, lowest conc.: ≈0.01 ng/g after ≈35, ≈130, ≈145 and ≈155 h) (mycotoxin values very approximately)

Aflatoxin B_1
incidence: 1/1, sa. const.: Holstein cow (midlactation period), age: 5 years, wt.: 600 kg, contamination: artificial (dose: 0.5 mg AFB_1/kg b. wt., o., once), conc. range: ≈≤3 ng/g* **, country: USA[92], *in cow that **died**, **after ≈33 h (also at other hour intervals up to 60 h measured, lowest conc.: ≈<0.1 ng/g after ≈8 h) (mycotoxin values very approximately)
incidence: 1/1, sa. const.: Holstein cow (midlactation period), age: 5 years, wt.: 600 kg, contamination: artificial (dose: 0.5 mg AFB_1/kg b. wt., o., once), conc. range: ≈≤7 ng/g* **, country: USA[92], *in cow that **survived**, **after ≈60 h (also at other hour intervals up to 190 h measured, lowest conc.: ≈<0.05 ng/g after ≈190 h) (mycotoxin values very approximately)

Aflatoxin M_1
incidence: 1/1, sa. const.: Holstein cow (midlactation period), age: 5 years, wt.: 600 kg, contamination: artificial (dose: 0.5 mg AFB_1/kg b. wt., o., once), conc. range: ≈≤10 ng/g* **, country: USA[92], *in cow that **died**, **after ≈10 h (also at other hour intervals up to 60 h measured, lowest conc.: ≈<5 ng/g after 60 h) (mycotoxin values very approximately)
incidence: 1/1, sa. const.: Holstein cow (midlactation period), age: 5 years, wt.: 600 kg, contamination: artificial (dose: 0.5 mg AFB_1/kg b. wt., o., once), conc. range: ≈≤6 ng/g* **, country: USA[92], *in cow that **survived**, **after ≈60 h (also at other hour intervals up to 220 h measured, lowest conc.: ≈<0.05 ng/g after 220 h) (mycotoxin values very approximately)

Fumonisin B_1
incidence: 3/3, sa. const.: black-pied Hostein Friesian cows, contamination: artificial (dose: 30 mg FB_1/cow (≈50 µg FB_1/kg b. wt.), i.v., once; for detailed information please see the article), conc. range: 0.95*–5.15** µg/kg, country: Germany[274], after 48* and **8 h of application (also at other hour intervals up to 58 h measured, lowest conc.: nd after 58 h for all 3 cows)

incidence: 1/1, sa. const.: dairy cow, wt.: 452–630 kg, contamination: artificial (dose: 50 µg FB_1/kg b. wt.), i.v., once; for detailed information please see the article), conc. range: ≈≤210 ng/ml, country: Canada[619], *after ≈10 min (also at other min, hour up to 14 days measured, lowest conc.: nd after 120 min) incidence: 1/1, sa. const.: dairy cow, wt.: 452–630 kg, contamination: artificial (dose: 200 μg FB_1/kg b. wt.), i.v., once; for detailed information please see the article), conc. range: ≈≤650 ng/ml, country: Canada[619], *after ≈10 min (also at other min, hour up to 14 days measured, lowest conc.: nd after 120 min)

Zearalenone

incidence: 1/1, sa. const.: Holstein cross-breed cow, wt.: 480–580 kg, contamination: artificial (dose: **544.5 mg ZEA/day, o., for 21 days**), conc.: ≤3.0 ng/ml*, country: Canada[79] *on day 3 (also at other day intervals up to 5 days measured, lowest conc.: nd after 4 and 5 days)
incidence: 1/1, sa. const.: Holstein cross-breed cow, wt.: 480–580 kg, contamination: artificial (dose: **1.8 g ZEA/day, o., for 1 day**), conc.: ≤9.0 ng/ml*, country: Canada[79], *on day 1 (also at other day intervals up to 6 days measured, lowest conc.: nd after 5 and 6 days)
incidence: 1/1, sa. const.: Holstein cross-breed cow, wt.: 480–580 kg, contamination: artificial (dose: **6 g ZEA/day, o., for 1 day**), conc.: ≤13.0 ng/ml*, country: Canada[79], *on day 1 (also at other day intervals up to 7 days measured, lowest conc.: nd after 6 and 7 days)

Cow red blood cells may contain the following mycotoxins and/or their metabolites:

Aflatoxicol

incidence: 1/1, sa. const.: Holstein cow (midlactation period), age: 5 years, wt.: 600 kg, contamination: artificial (dose: 0.5 mg AFB_1/kg b. wt., o., once), conc. range: ≈≤0.1 ng/g* **, country: USA[92], *in cow that **died**, **after ≈25 h (also at other hour intervals up to 60 h measured, lowest conc.: ≈<0.05 ng/g after ≈35 h) (mycotoxin values very approximately)
incidence: 1/1, sa. const.: Holstein cow (midlactation period), age: 5 years, wt.: 600 kg, contamination: artificial (dose: 0.5 mg AFB_1/kg b. wt., o., once), conc. range: ≈≤0.07 ng/g* **, country: USA[92], *in cow that **survived**, **after ≈25 h (also at other hour intervals up to ≈144 h measured, lowest conc.: ≈0.01 ng/g after ≈145 h) (mycotoxin values very approximately)

Aflatoxin B_1

incidence: 1/1, sa. const.: Holstein cow (midlactation period), age: 5 years, wt.: 600 kg, contamination: artificial (dose: 0.5 mg AFB_1/kg b. wt., o., once), conc. range: ≈≤0.2 ng/g* **, country: USA[92], *in cow that **died**, **after ≈8 h (also at other hour intervals up to 60 h measured, lowest conc.: ≈<0.05 ng/g after ≈10 and ≈24 h) (mycotoxin values very approximately)
incidence: 1/1, sa. const.: Holstein cow (midlactation period), age: 5 years, wt.: 600 kg, contamination: artificial (dose: 0.5 mg AFB_1/kg b. wt., o., once), conc. range: ≈≤1.5 ng/g* **, country: USA[92], *in cow that **survived**, **after ≈50 h (also at other hour intervals up to 190 h measured, lowest conc.: ≈<0.05 ng/g after 190 h) (mycotoxin values very approximately)

Aflatoxin M_1

incidence: 1/1, sa. const.: Holstein cow (midlactation period), age: 5 years, wt.: 600 kg, contamination: artificial (dose: 0.5 mg AFB_1/kg b. wt., o., once), conc. range: ≈≤10 ng/g* **, country: USA[92], *in cow that **died**, **after ≈5 h (also at other hour intervals up to 60 h measured, lowest conc.: ≈<3 ng/g after 60 h) (mycotoxin values very approximately)
incidence: 1/1, sa. const.: Holstein cow (midlactation period), age: 5 years, wt.: 600 kg, contamination: artificial (dose: 0.5 mg AFB_1/kg b. wt., o., once), conc. range: ≈≤3 ng/g* **, country: USA[92], *in cow that **survived**, **after ≈10 h (also at other hour intervals up to 180 h measured, lowest conc.: ≈<0.05 ng/g after 180 h) (mycotoxin values very approximately)

Cow rumen may contain the following mycotoxins and/or their metabolites:

Aflatoxicol

incidence: 1/1, sa. const.: Holstein cow (midlactation period), age: 5 years, wt.: 600 kg, contamination: artificial (dose: 0.5 mg AFB_1/kg b. wt., o., once), conc.: 4.9 ng/g* ** ***, country: USA[92], *in cow that died, **in rumen contents, ***after 60 h

Aflatoxin B_1

incidence: 1/1, sa. const.: Holstein cow (midlactation period), age: 5 years, wt.: 600 kg, contamination: artificial (dose: 0.5 mg AFB_1/kg b. wt., o., once), conc.: 320 ng/g* ** ***, country: USA[92], *in cow that died, **in rumen contents, ***after 60 h

incidence: 2/2, sa. const.: Holstein cows (midlactation), wt.: 408.2–544.3 kg, contamination: artificial (dose: 0.35 mg AFB_1/kg b. wt., o., 3 days (one cow was fed AF-rations for 3 days, the other cow, after 3 days of dosing, was fed AF-free rations for 7 additional days); for detailed information please see the article), conc. range: ≤10.50* and ≤163.00** ng/g*** ****, country: USA[93], these highest values measured at the *4th day in the case of cow 1 and at the **1st day in the case of cow 2 (always nd measured up to 3 days (after 4 days slaughtered) in the case of cow 1 and always lower values measured up to 10 days (then slaughtered) in the case of cow 2), for overall information please see the article, ***in rumen content, ****wet

Aflatoxin M_1

incidence: 1/1, sa. const.: Holstein cow (midlactation period), age: 5 years, wt.: 600 kg, contamination: artificial (dose: 0.5 mg AFB_1/kg b. wt., o., once), conc.: 8.6 ng/g* ** ***, country: USA[92], *cow died, **in rumen contents, ***after 60 h

incidence: 2/2, sa. const.: Holstein cows (midlactation), wt.: 408.2–544.3 kg, contamination: artificial (dose: 0.35 mg AFB_1/kg b. wt., o., for 3 days (one cow was fed AF-rations for 3 days, the other cow, after 3 days of dosing, was fed AF-free rations for 7 additional days); for detailed information please see the article), conc. range: ≤47.70* and ≤848.90** ng/g*** ****, country: USA[93], these highest values measured at the *4th day in the case of cow 1 and at the **1st day in the case of cow 2 (always nd measured up to 3 days (after 4 days slaughtered) in the case of cow 1 and always lower values measured up to 10 days (then slaughtered) in the case of cow 2), for overall information please see the article, ***in rumen content, ****dry

Cow spleen may contain the following mycotoxins and/or their metabolites:

Aflatoxin B_1

incidence: 1/2, sa. const.: Holstein cows (midlactation), wt.: 408.2–544.3 kg, contamination: artificial (dose: 0.35 mg AFB_1/kg b. wt., o., for 3 days (one cow was fed AF-rations for 3 days, the other cow, after 3 days of dosing, was fed AF-free rations for 7 additional days); for detailed information please see the article), conc.: 0.75 ng/g*, country: USA[93], *after 4 days (thereof 3 days with AFB_1-administration)

incidence: 1/1, sa. const.: lactating dairy cow, contamination: artificial (dose: **10 ppb AFB_1** (labeled and unlabeled), o., twice daily over a 14-day period; for detailed information please see the article), conc.: na, country: USA[319]
incidence: 1/1, sa. const.: lactating dairy cow, contamination: artificial (dose: **50 ppb AFB_1** (labeled and unlabeled), o., twice daily over a 14-day period; for detailed information please see the article), conc.: na, country: USA[319]
incidence: 1/1, sa. const.: lactating dairy cow, contamination: artificial (dose: **250 ppb AFB_1** (labeled and unlabeled), o., twice daily over a 14-day

period; for detailed information please see the article), conc.: nd* **, country: USA[319], *after 15 days (thereof 14 days with AFB_1-administration), **but significant radioactivity occurred

incidence: 1/1, sa. const.: lactating dairy cow, contamination: artificial (dose: **1,250 ppb AFB_1** (labeled and unlabeled), o., twice daily over a 14-day period; for detailed information please see the article), conc.: 0.17 μg/kg**, country: USA[319], *after 15 days (thereof 14 days with AFB_1-administration)

incidence: 2/2, sa. const.: lactating Francaise Frisonne Pie Noire cows, age: 3–4 years, wt.: 500–550 kg, contamination: artificial (dose: peanut meal ((3.5 mg AFB_1/kg) spiked with labeled AFB_1), o., for 10 days; for detailed information please see the article), conc. range: 6.7–7.9 ng AFB_1 eq/g*, Ø conc.: 7.3 ng AFB_1 eq/g*, country: The Netherlands/UK/France[397], *after a 10-day period of consumption of **contaminated peanut meal**

incidence: 2/2, sa. const.: lactating Francaise Frisonne Pie Noire cows, age: 3–4 years, wt.: 500–550 kg, contamination: artificial (dose: peanut meal ((3.5 mg AFB_1/kg) spiked with labeled AFB_1; but decontaminated afterwards), o., for 10 days; for detailed information please see the article), conc.: nd*, country: The Netherlands/UK/France[397], *after a 10-day period of consumption of **decontaminated peanut meal**

Aflatoxin M_1

incidence: 1/2, sa. const.: Holstein cows (midlactation), wt.: 408.2–544.3 kg, contamination: artificial (dose: 0.35 mg AFB_1/kg b. wt., o., 3 days (one cow was fed AF-rations for 3 days, the other cow, after 3 days of dosing, was fed AF-free rations for 7 additional days); for detailed information please see the article), conc.: 1.4 ng/g*, country: USA[93], *after 4 days (thereof 3 days with AFB_1-doses)

Cow tongue may contain the following mycotoxins and/or their metabolites:

Aflatoxin B_1

incidence: 1/2, sa. const.: Holstein cows (midlactation), wt.: 408.2–544.3 kg, contamination: artificial (dose: 0.35 mg AFB_1/kg b. wt., o., for 3 days (one cow was fed AF-rations for 3 days, the other cow, after 3 days of dosing, was fed AF-free rations for 7 additional days); for detailed information please see the article), conc.: 0.87 ng/g*, country: USA[93], *after 4 days (thereof 3 days with AFB_1-doses)

Aflatoxin M_1

incidence: 1/2, sa. const.: Holstein cows (midlactation), wt.: 408.2–544.3 kg, contamination: artificial (dose: 0.35 mg AFB_1/kg b. wt., o., for 3 days (one cow was fed AF-rations for 3 days, the other cow, after 3 days of dosing, was fed AF-free rations for 7 additional days); for detailed information please see the article), conc.: 1.74 ng/g*, country: USA[93], *after 4 days (thereof 3 days with AFB_1-doses)

Cow urine may contain the following mycotoxins and/or their metabolites:

Aflatoxicol

incidence: 1/1, sa. const.: Holstein cow (midlactation period), age: 5 years, wt.: 600 kg, contamination: artificial (dose: 0.5 mg AFB_1/kg b. wt., o., once), conc.: 0.10 ng/g* **, country: USA[92], *cow died, **after 60 h

Aflatoxin B_1

incidence: 1/1, sa. const.: lactating cow, wt.: 600 kg, contamination: artificial (dose: 300 mg AFs: 44% AFB_1, 2% AFB_2, 44% AFG_1?, o., once), conc.: 760 μg* (total amount found), country: UK[40], *collected over a period of 9 days

incidence: 1/1, sa. const.: Holstein cow (midlactation period), age: 5 years, wt.: 600 kg, contamination: artificial (dose: 0.5 mg AFB_1/kg b. wt., o., once), conc.: 4.1 ng/g* **, country: USA[92], *cow died, **after 60 h

incidence: 2/2, sa. const.: Holstein cows (midlactation), wt.: 408.2–544.3 kg, contamination: artificial (dose: 0.35 mg AFB_1/kg b. wt., o., for 3 days (one cow was fed AF-rations for 3 days, the other cow, after 3 days of dosing, was fed AF-free rations for 7 additional days); for detailed information please see the article), conc.: ≤20.60* and ≤39.20** ng/ml, country: USA[93], these highest values measured at the *3rd day in the case of cow 1 and at the **1st day in the case of cow 2 (always lower values measured up to 4 days (then slaughtered) in the case of cow 1 and always lower values measured up to 10 days (then slaughtered) in the case of cow 2), for overall information please see the article

Aflatoxin G_1
incidence: 1/1, sa. const.: lactating cow, wt.: 600 kg, contamination: artificial (dose: 300 mg AFs: 44% AFB_1, 2% AFB_2, 44% AFG_1?, o., once), conc.: 1,220 µg* (total amount found), country: UK[40], *collected over a period of 9 days

Aflatoxin M_1
incidence: 1/1, sa. const.: lactating cow, wt.: 600 kg, contamination: artificial (dose: 300 mg AFs: 44% AFB_1, 2% AFB_2, 44% AFG_1?, o., once), conc.: 2,230 µg* (total amount found), country: UK[40], *collected over a period of 9 days

incidence: 1/1, sa. const.: Holstein cow (midlactation period), age: 5 years, wt.: 600 kg, contamination: artificial (dose: 0.5 mg AFB_1/kg b. wt., o., once), conc.: 37 ng/g* **, country: USA[92], *cow died, **after 60 h

incidence: 2/2, sa. const.: Holstein cows (midlactation), wt.: 408.2–544.3 kg, contamination: artificial (dose: 0.35 mg AFB_1/kg b. wt., o., for 3 days (one cow was fed AF-rations for 3 days, the other cow, after 3 days of dosing, was fed AF-free rations for 7 additional days); for detailed information please see the article), conc.: ≤588.0* and ≤600.80** ng/ml, country: USA[93], these highest values measured at the *2nd day in the case of cow 2 and at the **3rd day in the case of cow 1 (always lower values measured up to 4 days (then slaughtered) in the case of cow 1 and always lower values measured up to 10 days (then slaughtered) in the case of cow 2), for overall information please see the article

incidence: 2/2, sa. const.: Jersey milking cows, age: adult, contamination: artificial (dose: **385–1,925 µg ZEA**/kg + (20 µg AFB_1/kg), o., for 7 weeks; for detailed information please see the article), conc. range: 0.12–0.60 µg/kg*, Ø conc.: 0.36 µg/kg*, country: UK[267], *after 7 weeks of (AFB_1-) and ZEA-administration; AFB_1 accidentally in the diet
incidence: 2/2, sa. const.: Jersey milking cows, age: adult, contamination: artificial (dose: **317–1,125 µg OTA**/kg + (20 µg AFB_1/kg), o., for 11 weeks; for detailed information please see the article), conc. range: 0.09–0.22 µg/kg*, Ø conc.: 0.36 µg/kg*, country: UK[267], *after 11 weeks of (AFB_1-) and OTA-administration; AFB_1 accidentally in the diet

Deoxynivalenol
incidence: 3/3, sa. const.: 2 Holstein dairy cows, 1 Ayrshire dairy cow, wt.: ≈450 kg, contamination: artificial (dose: 66 mg DON/kg, o., for 5 days; for detailed information please see the article), conc. range: ≤390* ng/ml** (19.4 mg** in total, mean value), country: USA[278] *after 4 days of application (also at other day intervals up to 12 days measured, lowest conc.: nd at the beginning and the end of the experiment; for detailed information please see the article) **unconjugated DON

Deepoxydeoxynivalenol
incidence: 3/3, sa. const.: 2 Holstein dairy cows, 1 Ayrshire dairy cow, wt.: ≈450 kg, contamination: artificial (dose: 66 mg DON/kg, o., for 5 days; for detailed information please see the article), conc.

range: ≤18,000 ng/ml* (882.1 mg in total, mean value), country: USA[278], *after 4 days of application (also at other day intervals up to 12 days measured, lowest conc.: nd at the beginning (1–4 days) of the experiment; for detailed information please see the article)

HT-2 Toxin
incidence: 1/1, sa. const.: Holstein cow, wt.: 365 kg, contamination: artificial (dose: 2 × 200 mg T-2 toxin, by balling gun; for detailed information please see the article), conc. range: ≤19.5 ppb*, country: USA[180], *13 h after administration of 2nd dosing (also at other hour intervals up to 48 h measured, lowest conc.: 1.7 ppb after 3 h of 1st dosing)

3′-Hydroxy HT-2 Toxin
incidence: 1/1, sa. const.: Holstein cow, wt.: 365 kg, contamination: artificial (dose: 2 × 200 mg T-2 toxin, by balling gun; for detailed information please see the article), conc. range: ≤462.2 ppb*, country: USA[180], *13 h after administration of 2nd dosing (also at other hour intervals up to 48 h measured, lowest conc.: 5.3 ppb after 48 h of 2nd dosing)

Deepoxy-3′-Hydroxy HT-2 Toxin
incidence: 1/1, sa. const.: Holstein cow, wt.: 365 kg, contamination: artificial (dose: 2 × 200 mg T-2 toxin, by balling gun; for detailed information please see the article), conc. range: ≤461.9 ppb*, country: USA[180], *13 h after administration of 2nd dosing (also at other hour intervals up to 48 h measured, lowest conc.: 2.2 ppb after 48 h of 2nd dosing)

3′-Hydroxy T-2 Toxin
incidence: 1/1, sa. const.: Holstein cow, wt.: 365 kg, contamination: artificial (dose: 2 × 200 mg T-2 toxin, by balling gun; for detailed information please see the article), conc. range: ≤573.8 ppb*, country: USA[180], *6 h after administration of 2nd dosing (also at other hour intervals up to 48 h measured, lowest conc.: tr after 20 h of 1st dosing)

3′-Hydroxy-Iso-T-2 Toxin
incidence: 1/1, sa. const.: Holstein cow, wt.: 365 kg, contamination: artificial (dose: 2 × 200 mg T-2 toxin, by balling gun; for detailed information please see the article), conc. range: ≤7,221.7 ppb*, country: USA[180], *13 h after administration of 2nd dosing (also at other hour intervals up to 48 h measured, lowest conc.: nd after 3 h of 1st and 48 h of 2nd dosing)

T-2 Tetraol
incidence: 1/1, sa. const.: Holstein cow, wt.t: 365 kg, contamination: artificial (dose: 2 × 200 mg T-2 toxin, by balling gun; for detailed information please see the article), conc. range: ≤91.4 ppb*, country: USA[180], *3 h after administration of 1st dosing (also at other hour intervals up to 48 h measured, lowest conc.: tr after 1, 20, 25 h of 1st and 13, 24 and 48 h of 2nd dosing)

Deepoxy T-2 Tetraol
incidence: 1/1, sa. const.: Holstein cow, wt.: 365 kg, contamination: artificial (dose: 2 × 200 mg T-2 toxin, by balling gun; for detailed information please see the article), conc. range: ≤1,017.5 ppb*, country: USA[180], *13 h after administration of 2nd dosing (also at other hour intervals up to 48 h measured, lowest conc.: 9.1 ppb after 3 h of 1st dosing)

Deer

Deer Natural Contamination

Deer urine may contain the following mycotoxins and/or their metabolites:

Zearalanols
incidence: 14/41*, sa. const.: urine from deers of New Zealand, contamination: natural, conc. range: ≤0.94 ng/ml**, country: New Zealand[230], *export animals, **most probable of *Fusarium* origin

Zearalenols
incidence: 14/41*, sa. const.: urine from deers of New Zealand, contamination: natural, conc. range: ≤15 ng/ml, country: New Zealand[230], *export animals

Deer Artificial Contamination

Deer liver may contain the following mycotoxins and/or their metabolites:

Aflatoxin M_1
incidence: 7/7*, sa. const.: fawns (white-tailed deer), age: 16–22 weeks, contamination: no AF (for detailed information please see the article), conc.: nr, country: USA[634], *control
incidence: 5/6, sa. const.: fawns (white-tailed deer), age: 16–22 weeks, contamination: artificial (dose: **800 ppb AF**, o., for 8 weeks; for detailed information please see the article), conc. range: 1–2 ppb*, country: USA[634], *(1 liver contained >2 ppb)

Deer muscle may contain the following mycotoxins and/or their metabolites:

Aflatoxin M_1
incidence: 7/7*, sa. const.: fawns (white-tailed deer), age: 16–22 weeks, contamination: no AF (for detailed information please see the article), conc.: nr, country: USA[634], *control
incidence: 1/6, sa. const.: fawns (white-tailed deer), age: 16–22 weeks, contamination: artificial (dose: **800 ppb AF**, o., for 8 weeks; for detailed information please see the article), conc.: 0.5–1 ppb*, country: USA[634]

Dog

Dog Natural Contamination

Dog brain may contain the following mycotoxins and/or their metabolites:

Penitrem A
incidence: 1/1, sa. const.: brain from a male Welsh springer spaniel of Norway/Sweden?, age: 5 years, contamination: natural, conc.: pr, country: Norway/Sweden[586]

Roquefortine C
incidence: 1/1, sa. const.: brain from a male Welsh springer spaniel of Norway/Sweden?, age: 5 years, contamination: natural, conc.: pr, country: Norway/Sweden[586]

Dog kidney may contain the following mycotoxins and/or their metabolites:

Ochratoxin A
incidence: 3/3, sa. const.: kidneys from dogs of Austria?, contamination: natural, conc. range: 0.26–0.35 μg/kg, Ø conc.: 0.32 μg/kg, country: Austria[340]

Penitrem A
incidence: 1/1, sa. const.: kidney from a male Welsh springer spaniel of Norway/Sweden?, age: 5 years, contamination: natural, conc.: pr, country: Norway/Sweden[586]

Penitrem E
incidence: 1/1, sa. const.: kidney from a male Welsh springer spaniel of Norway/Sweden?, age: 5 years, contamination: natural, conc.: pr, country: Norway/Sweden[586]

Roquefortine C
incidence: 1/1, sa. const.: kidney from a male Welsh springer spaniel of Norway/Sweden?, age: 5 years, contamination: natural, conc.: pr, country: Norway/Sweden[586]

Dog liver may contain the following mycotoxins and/or their metabolites:

Aflatoxin M_1
incidence: 7/9, sa. const.: livers from 2 male and 4 female Basset hounds, 1 female Australian shepherd, 1 female spayed Airedale terrier and 1 male neutered Labrador Mix of the USA, age: 1.25–6 years, contamination: natural,

conc. range: 0.60–4.40 ppb, Ø conc.: 1.99 ppb, country: USA[471]

Penitrem A
incidence: 1/1, sa. const.: liver from a male Welsh springer spaniel of Norway/Sweden?, age: 5 years, contamination: natural, conc.: pr, country: Norway/Sweden[586]

Penitrem E
incidence: 1/1, sa. const.: liver from a male Welsh springer spaniel of Norway/Sweden?, age: 5 years, contamination: natural, conc.: pr, country: Norway/Sweden[586]

Roquefortine C
incidence: 1/1, sa. const.: liver from a male Welsh springer spaniel of Norway/Sweden?, age: 5 years, contamination: natural, conc.: pr, country: Norway/Sweden[586]

Dog stomach may contain the following mycotoxins and/or their metabolites:

Penitrem A
incidence: ?/2*, sa. const.: vomitus from a Schnauzer and a Miniatur Schnauzer of South Africa, age: 2–3 years, wt.: ≈20–25 kg, contamination: natural (for detailed information please see the article), conc.: 2.6 μg/g wet mass, country: South Africa/Norway[500], *vomitus

incidence: 1/1, sa. const.: vomitus from a male English setter, age: 10 years, contamination: natural, conc.: ca. 30,000 μg/kg*, country: Norway/Sweden[586], *in stomach contents (vomit)

Roquefortine C
incidence: ?/2*, sa. const.: vomitus from a Schnauzer and a Miniatur Schnauzer of South Africa, age: 2–3 years, wt.: ≈20–25 kg, contamination: natural (for detailed information please see the article), conc.: 34 μg/g wet mass, country: South Africa/Norway[500], *vomitus

Thomitrems
incidence: 1/1, sa. const.: vomitus from a male English setter of Norway/Sweden?, age: 10 years, contamination: natural, conc.: ca. 40,000 μg/kg*, country: Norway/Sweden[586], *in stomach contents (vomit)

Dog Vomit see Dog stomach

Dog Artificial Contamination

Dog plasma may contain the following mycotoxins and/or their metabolites:

HT-2 Toxin
incidence: 1/1, sa. const.: mongrel, wt.: 20 kg, contamination: artificial (dose: 0.4 mg T-2 Toxin/kg, i.v., once), conc.: ≈≤103 ng/ml*, country: Israel[487], *after ≈25 min (also at other min intervals up to ≈180 min measured, lowest conc.: ≈6 ng/ml after ≈180 min)

T-2 Toxin
incidence: 1/1, sa. const.: mongrel, wt.: 20 kg, contamination: artificial (dose: 0.4 mg T-2 Toxin/kg, i.v., once), conc.: ≈≤1,450 ng/ml*, country: Israel[487], *after 0 min (also measured after 5, 10 and 16 min, lowest conc.: nd after 16 min)

Verrucarol
incidence: ?/8, sa. const.: male and female monogrels, wt.: 20–25 kg, contamination: artificial (dose: 0.4 mg VER/kg, **i.v.**, once; for detailed information please see the article), conc. range: ≈≤590 ng/ml (mean value), country: Israel[408], *after <10 min (also at other min intervals up to ≈420 min measured, lowest conc.: ≈20 ng/ml after 420 min)
incidence: ?/8, sa. const.: male and female monogrels, wt.: 20–25 kg, contamination: artificial (dose: 0.8 mg VER/kg, **o.**, once; for detailed information please see the article), conc. range: ≈≤180 ng/ml* (mean value), country: Israel[408], *after ≈50 min (also at other min intervals up to

≈420 min measured, lowest conc.: ≈5 ng/ml after 420 min)

Duck

Duck Natural Contamination

Duck liver may contain the following mycotoxins and/or their metabolites:

Aflatoxin B_1
incidence: 1/8*, sa. const.: livers from domestic Chinese brown ducks, age: adult (at least 3 years old), contamination: natural, conc.: 3.71 ng AFB_1-FAPy/mg DNA** (mean value) or 6.38 pmol/mg DNA** (mean value), country: France/People's Republic of China[31], *7 were infected with HCC (1 additionally with AFB_1) **liver DNA

Duck Artificial Contamination

Duck bile may contain the following mycotoxins and/or their metabolites:

Zearalenone
incidence: ?/54*, sa. const.: Pekin ducks, age: 1 day, contamination: no DON + ZEA (for detailed information please see the article), conc.: 59 μg/kg** (mean value), country: Germany[469], *control, **after 49 days
incidence: ?/54, sa. const.: Pekin ducks, age: 1 day, contamination: artificial (dose: successively increased up to **6–7 mg DON/kg** and successively increased up to **0.05–0.06 mg ZEA/kg** contaminated wheat (proportion: 40%), for 49 days; for detailed information please see the article), conc. range: ≤296 μg/kg* (mean value), country: Germany[469], *after 49 days (0, 10, 20, 30, 40*, 50 or 60% *Fusarium*-contaminated wheat in the diet, always lower residue values recorded)

α-Zearalenol
incidence: ?/54*, sa. const.: Pekin ducks, age: 1 day, contamination: no DON + ZEA (for detailed information please see the article), conc.: 12 μg/kg** (mean value), country: Germany[469], *control, **after 49 days
incidence: ?/54, sa. const.: Pekin ducks, age: 1 day, contamination: artificial (dose: successively increased up to **6–7 mg DON/kg** and successively increased up to **0.05–0.06 mg ZEA/kg** contaminated wheat (proportion: 60%), for 49 days; for detailed information please see the article), conc. range: ≤57 μg/kg* (mean value), country: Germany[469], after 49 days (0, 10, 20, 30, 40, 50 or 60%* *Fusarium*-contaminated wheat in the diet, always equal or lower residue values recorded)

β-Zearalenol
incidence: ?/54*, sa. const.: Pekin ducks, age: 1 day, contamination: no DON + ZEA (for detailed information please see the article), conc.: nd, country: Germany[469], *control
incidence: ?/54, sa. const.: Pekin ducks, age: 1 day, contamination: artificial (dose: successively increased up to **6–7 mg DON/kg** and successively increased up to **0.05–0.06 mg ZEA/kg** contaminated wheat (proportion: 40%), for 49 days; for detailed information please see the article), conc. range: ≤25 μg/kg* (mean value), country: Germany[469], *after 49 days (0, 10, 20, 30, 40*, 50 or 60% *Fusarium*-contaminated wheat in the diet, always lower residue values recorded)

Duck liver may contain the following mycotoxins and/or their metabolites:

Aflatoxin
incidence: 4/4*, sa. const.: 3 male and 1 female Pekin ducklings, age: ≈1 year, contamination: artificial (dose: 0.02 mg AFB_1 (labeled)/kg b. wt., i.p., once), conc. range: 204–1,057 pg AF/mg DNA**, Ø conc.: 764 pg AF/mg DNA** ***, country: France[6], ***uninfected with DHBV,** **liver DNA, ***after 48 h
incidence: 4/4*, sa. const.: 3 male and 1 female Pekin ducklings, age: ≈1 year, contamination: artificial (dose: 0.02 mg AFB_1 (labeled)/kg b. wt., i.p., once), conc. range: 318–420 pg AF/mg DNA**, Ø conc.: 379 pg AF/mg DNA** ***, country:

France[6], ***infected with DHBV**, **liver DNA, ***after 48 h

incidence: 5?/5, sa. const.: Pekin ducks, contamination: artificial (dose: 2 µg AFB_1/kg b. wt., i.p., for 8 days and additionally 2 µg AFB_1 (labeled)/kg b. wt. 4 days later, i.p., once, **HBV pos.**; for detailed information please see the article), conc.: 34.94 pg AF/mg DNA* ** (mean value), country: France[30], *AF-liver DNA adducts, **after 24 h
incidence: 5?/5, sa. const.: Pekin ducks, contamination: artificial (dose: 2 µg AFB_1/kg b. wt., i.p., for 8 days and additionally 2 µg AFB_1 (labeled)/kg b. wt. 4 days later, i.p., once, **HBV neg.**; for detailed information please see the article), conc.: 32.14 pg AF/mg DNA* ** (mean value), country: France[30], *AF-liver DNA adducts, **after 24 h

incidence: 5?/5, sa. const.: Pekin ducks, contamination: artificial (dose: 2 µg AFB_1 (labeled)/kg b. wt., i.p., once, **HBV pos.**; for detailed information please see the article), conc.: 23.71 pg AF/mg DNA* ** (mean value), country: France[30], *AF-liver DNA adducts, **after 24 h
incidence: 5?/5, sa. const.: Pekin ducks, contamination: artificial (dose: 2 µg AFB_1 (labeled)/kg b. wt., i.p., once, **HBV neg.**; for detailed information please see the article), conc.: 26.52 pg AF/mg DNA* ** (mean value), country: France[30], *AF-liver DNA adducts, **after 24 h

Duck plasma may contain the following mycotoxins and/or their metabolites:

Aflatoxin
incidence: 4/4*, sa. const.: 3 male and 1 female Pekin ducklings, age: 1 year, contamination: artificial (dose: 0.02 mg AFB_1 (labeled)/kg b. wt., i.p., once), conc. range: 73–134 pg AF/mg plasma protein*, Ø conc.: 108.5 pg AF/mg plasma protein**, country: France[6], ***uninfected with DHBV**, **after 48 h

incidence: 4/4*, sa. const.: 3 male and 1 female Pekin ducklings, age: 1 year, contamination: artificial (dose: 0.02 mg AFB_1 (labeled)/kg b. wt., i.p., once), conc. range: 61–83 pg AF/mg plasma protein*, Ø conc.: 73 pg AF/mg plasma protein**, country: France[6], ***infected with DHBV**, **after 48 h

Fusarenon-X
incidence: ?/6, sa. const.: ducks, age: 4 weeks, Ø wt.: 1.16 kg, contamination: artificial (dose: 2.2 mg FX/kg b. wt., **i.v.**, once), conc. range: ≈≤230 ng/ml* (mean value), country: Thailand/Japan[474], *after ≈5 min (also at other min intervals up to 180 min measured, lowest conc.: ≈3.6 ng/ml after 180 min)
incidence: ?/6, sa. const.: ducks, age: 4 weeks, Ø wt.: 1.16 kg, contamination: artificial (dose: 2.2 mg FX/kg b. wt., **o.**, once), conc. range: ≈≤13 ng/ml* (mean value), country: Thailand/Japan[474], *≈10 min (also at other min intervals up to 120 min measured, lowest conc.: ≈1.7 ng/ml after 120 min)

Nivalenol
incidence: ?/6, sa. const.: ducks, age: 4 weeks, Ø wt.: 1.16 kg, contamination: artificial (dose: 2.2 mg FX/kg b. wt., **i.v.**, once), conc. range: ≈≤480 ng/ml (mean value), country: Thailand/Japan[474], *after ≈5 min (also at other min intervals up to 240 min measured, lowest conc.: ≈1 ng/ml after 240 min)
incidence: ?/6, sa. const.: ducks, age: 4 weeks, Ø wt.: 1.16 kg, contamination: artificial (dose: 2.2 mg FX/kg b. wt., **o.**, once), conc. range: ≈≤212 ng/ml* (mean value), country: Thailand/Japan[474], *≈10 min (also at other min intervals up to 180 min measured, lowest conc.: ≈2 ng/ml after 180 min)

Duck serum may contain the following mycotoxins and/or their metabolites:

Aflatoxin B_1
incidence: 5?/5, sa. const.: Pekin ducks, contamination: artificial (dose: 2 µg AFB_1/kg b. wt., i.p., for 8 days and additionally 2 µg

AFB_1 (labeled)/kg b. wt. 4 days later, i.p., once, **HBV pos.**; for detailed information please see the article), conc.: 15.1 pg AF/mg protein* ** (mean value), country: France[30], *AF-serum protein adducts, **after 24 h
incidence: 5?/5, sa. const.: Pekin ducks, contamination: artificial (dose: 2 μg AFB_1/kg b. wt., i.p., for 8 days and additionally 2 μg AFB_1 (labeled)/kg b. wt. 4 days later, i.p., once, **HBV neg.**; for detailed information please see the article), conc.: 11.9 pg AF/mg protein* ** (mean value), country: France[30], *AF-serum protein adducts, **after 24 h

incidence: 5?/5, sa. const.: Pekin ducks, contamination: artificial (dose: 2 μg AFB_1 (labeled)/kg b. wt., i.p., once, **HBV pos.**; for detailed information please see the article), conc.: 15.2 pg AF/mg protein* ** (mean value), country: France[30], *AF-serum protein adducts, **after 24 h
incidence: 5?/5, sa. const.: Pekin ducks, contamination: artificial (dose: 2 μg AFB_1 (labeled)/kg b. wt., i.p., once, **HBV neg.**; for detailed information please see the article), conc.: 14.0 pg AF/mg protein* ** (mean value), country: France[30], *AF-serum protein adducts, **after 24 h

Egg White see Hen egg

Egg Yolk see Hen egg

Ewe

Ewe Artificial Contamination

Ewe feces may contain the following mycotoxins and/or their metabolites:

Aflatoxin B_1
incidence: 1/1, sa. const.: lactating ewe, contamination: artificial (dose: 78 mg AFs: 36% AFB_1, 3% AFB_2, 52% AFG_1, 2% AFG_2?, via stomach tube, once), conc.: 123 μg*, country: UK[38], *total amount found

Aflatoxin G_1
incidence: 1/1, sa. const.: lactating ewe, contamination: artificial (dose: 78 mg AFs: 36% AFB_1, 3% AFB_2, 52% AFG_1, 2% AFG_2?, via stomach tube, once), conc.: 138 μg*, country: UK[38], *total amount found

Aflatoxin M_1
incidence: 1/1, sa. const.: lactating ewe, contamination: artificial (dose: 78 mg AFs: 36% AFB_1, 3% AFB_2, 52% AFG_1, 2% AFG_2?, via stomach tube, once), conc.: 768 μg*, country: UK[38], *total amount found

Ewe milk, raw may contain the following mycotoxins and/or their metabolites:

Aflatoxin B_1
incidence: 1/1, sa. const.: lactating ewe, contamination: artificial (dose: 78 mg AFs: 36% AFB_1, 3% AFB_2, 52% AFG_1, 2% AFG_2?, via stomach tube, once), conc.: 1.5 μg*, country: UK[38], *total amount found

Aflatoxin G_1
incidence: 1/1, sa. const.: lactating ewe, contamination: artificial (dose: 78 mg AFs: 36% AFB_1, 3% AFB_2, 52% AFG_1, 2% AFG_2?, via stomach tube, once), conc.: 1.4 μg*, country: UK[38], *total amount found

Aflatoxin M_1
incidence: 1/1, sa. const.: lactating ewe, contamination: artificial (dose: 78 mg AFs: 36% AFB_1, 3% AFB_2, 52% AFG_1, 2% AFG_2?, via stomach tube, once), conc.: 70.0 μg*, country: UK[38], *total amount found

incidence: 4/4, sa. const.: multiparous Sarda ewes (early lactation), wt.: ≈43.5 kg, contamination: artificial (dose: 2 mg AFB_1, o., once), conc. range: ≈≤0.39 μg/kg* (mean value), country: Italy[476], *after 6 h (also at other hour intervals up to 96 h measured, lowest conc.: nd after 96 h)
incidence: 4/4*, sa. const.: multiparous Sarda ewes (late lactation), wt.: ≈40.2 kg, contamination: no AFB_1 (for detailed information please see the article), conc.: nd, country: Italy[476], *control

incidence: 4/4, sa. const.: multiparous Sarda ewes (late lactation), wt.: ≈40.2 kg, contamination: artificial (dose: **32 µg AFB_1/day**, o., for 13 days; for detailed information please see the article), conc. range: ≈≤0.057 µg/kg* (mean value), country: Italy[476], *after 144 h of AFB_1-administration (up to 20 days measured, lowest conc.: nd at 72 h post withdrawal)
incidence: 4/4, sa. const.: multiparous Sarda ewes (late lactation), wt.: ≈40.2 kg, contamination: artificial (dose: **64 µg AFB_1/day**, o., for 13 days; for detailed information please see the article), conc. range: ≈≤0.226 µg/kg* (mean value), country: Italy[476], *after 144 h of AFB_1-administration (up to 20 days measured, lowest conc.: nd at 72 h post withdrawal)
incidence: 4/4, sa. const.: multiparous Sarda ewes (late lactation), wt.: ≈40.2 kg, contamination: artificial (dose: **128 µg AFB_1/day**, o., for 13 days; for detailed information please see the article), conc. range: ≈≤0.331 µg/kg* (mean value), country: Italy[476], *after 144 h of AFB_1-administration (up to 20 days measured, lowest conc.: nd at 72 h post withdrawal)

incidence: 5/5*, sa. const.: Sarda ewes (late lactation), wt.: ≈48 kg, contamination: no AFB_1 (for detailed information please see the article), conc.: nd, country: Italy[477], *control
incidence: 5?/5, sa. const.: Sarda ewes (late lactation), wt.: ≈48 kg, contamination: artificial (dose: **32 µg AFB_1/day**, o., for 7 days; for detailed information please see the article), conc. range: ≈≤200 ng/kg* (mean value), country: Italy[477], *after 7 days of first AFB_1-administration (also at other day intervals up to 11 days measured, lowest conc.: nd after 11 days)
incidence: 5?/5, sa. const.: Sarda ewes (late lactation), wt.: ≈48 kg, contamination: artificial (dose: **64 µg AFB_1/day**, o., for 7 days; for detailed information please see the article), conc. range: ≈≤380 ng/kg* (mean value), country: Italy[477], *after 4 days of first AFB_1-administration (also at other day intervals up to 11 days measured, lowest conc.: nd after 11 days)
incidence: 5?/5, sa. const.: Sarda ewes (late lactation), wt.: ≈48 kg, contamination: artificial (dose: **128 µg AFB_1/day**, o., for 7 days; for detailed information please see the article), conc. range: ≈≤640 ng/kg* (mean value), country: Italy[477], *after 4 days of first AFB_1-administration (also at other day intervals up to 11 days measured, lowest conc.: nd after 11 days)

incidence: 6/6*, sa. const.: Sarda ewes (late lactation), contamination: no AF contaminated wheat meal in diet (for detailed information please see the article), conc.: nd, country: Italy[583], *control
incidence: 6?/6, sa. const.: Sarda ewes (late lactation), contamination: artificial (dose: **1.13 µg AFB_1/kg feed = L-AF** (low portion of wheat meal naturally contaminated with AFs in the diet), o., for 14 days; for detailed information please see the article), conc. range: ≈≤43 ng/ml*, country: Italy[583], *after 8 days of AFs-administration (also at other day intervals up to 18 days measured, lowest conc.: nd after 18 days, no effect of DYP (12 g/day) observed given for 7 days starting on day 8)
incidence: 6?/6, sa. const.: Sarda ewes (late lactation), contamination: artificial (dose: **2.30 µg AFB_1/kg feed = M-AF** (medium portion of wheat meal naturally contaminated with AFs in the diet), o., for 14 days; for detailed information please see the article), conc. range: ≈≤52 ng/ml*, country: Italy[583], *after 8 days of AFs-administration (also at other day intervals up to 18 days measured, lowest conc.: nd after 18 days, no effect of DYP (12 g/day) observed given for 7 days starting on day 8)
incidence: 6?/6, sa. const.: Sarda ewes (late lactation), contamination: artificial (dose: **5.03 µg AFB_1/kg feed = H-AF** (high

portion of wheat meal naturally contaminated with AFs in the diet), o., for 14 days; for detailed information please see the article), conc. range: ≈≤81 ng/ml*, country: Italy[583], *after 10 days of AFs-administration (also at other day intervals up to 18 days measured, lowest conc.: nd after 18 days, no effect of DYP (12 g/day) observed given for 7 days starting on day 8)

Cyclopiazonic Acid
incidence: ?/3, sa. const.: crossbred lactating ewes, contamination: artificial (dose: 5 mg CPA/kg, o., daily for 2 days), conc. range: ≤568 ng/g*, country: Australia[367], *after ≈2 days (also at other day intervals up to 9 days measured, lowest conc.: nd after 9 days)

Ochratoxin A
incidence: 3?/3, sa. const.: dairy ewes, contamination: artificial (dose: 30 µg OTA + 1.8 µg OTB/kg b. wt., o., once), conc.: ≈≤158 µg/l* (mean value), country: France[555], *after 1 day (also at other day intervals up to 8 days measured, lowest conc.: nd after 6 days)

Ochratoxin α
incidence: 3?/3, sa. const.: dairy ewes, contamination: artificial (dose: 30 µg OTA + 1.8 µg OTB/kg b. wt., o., once), conc.: ≈≤1,200 µg/l* (mean value), country: France[555], *after 1 day (also at other day intervals up to 8 days measured, lowest conc.: nd after 6 days)

Ewe plasma may contain the following mycotoxins and/or their metabolites:

Ochratoxin A
incidence: 3?/3, sa. const.: dairy ewes, contamination: artificial (dose: 30 µg OTA + 1.8 µg OTB/kg b. wt., o., once), conc.: ≈≤17 µg/l* (mean value), country: France[555], *after ≈4 h (also at other hour intervals up to 168 h measured, lowest conc.: nd after 72 h)

Ochratoxin α
incidence: 3?/3, sa. const.: dairy ewes, contamination: artificial (dose: 30 µg OTA + 1.8 µg OTB/kg b. wt., o., once), conc.: ≈≤9 µg/l* (mean value), country: France[555], *after ≈4 h (also at other hour intervals up to 168 h measured, lowest conc.: nd after 72 h)

Ewe urine may contain the following mycotoxins and/or their metabolites:

Aflatoxin B_1
incidence: 1/1, sa. const.: lactating ewe, contamination: artificial (dose: 78 mg AFs: 36% AFB_1, 3% AFB_2, 52% AFG_1, 2% AFG_2?, via stomach tube, once), conc.: tr*, country: UK[38], *total amount found

Aflatoxin G_1
incidence: 1/1, sa. const.: lactating ewe, contamination: artificial (dose: 78 mg AFs: 36% AFB_1, 3% AFB_2, 52% AFG_1, 2% AFG_2?, via stomach tube, once), conc.: 2,411 µg*, country: UK[38], *total amount found

Aflatoxin M_1
incidence: 1/1, sa. const.: lactating ewe, contamination: artificial (dose: 78 mg AFs: 36% AFB_1, 3% AFB_2, 52% AFG_1, 2% AFG_2?, via stomach tube, once), conc.: 1,662 µg*, country: UK[38], *total amount found

Fish

Fish Artificial Contamination

Fish, carp

Fish, carp plasma may contain the following mycotoxins and/or their metabolites:

Ochratoxin A
incidence: 6?/6, sa. const.: carps, wt.: 1,000 g, contamination: artificial (dose: 50 ng OTA/g b. wt., o., once), conc.: 14 ng/ml (mean value), country: Sweden/Yugoslavia[193] (at other min intervals up to 270 min measured)

incidence: 6?/6, sa. const.: carps, wt: 1,000 g, contamination: artificial (dose: 50 ng OTA/g b. wt., **i.v.**, once), conc.: 200 ng/ml (mean value), country: Sweden/Yugoslavia[193] (at other min intervals up to 270 min measured)

Fish, Channel Catfish

Fish, channel catfish bile may contain the following mycotoxins and/or their metabolites:

Aflatoxin B_1
incidence: ?/?, sa. const.: channel catfishes, wt.: 0.3–0.5 kg, contamination: artificial (dose: 250 µg/kg AFB_1 (labeled), o., once; for detailed information please see the article), conc. range: ≤2,019 ppb* ** (mean value), country: USA[516], *after 24 h (also measured after 2, 4, 48 and 96 h, lowest conc.: 149 ppb after 2 h), **AFB_1 and its metabolites (total residues)

Fish, channel catfish fat may contain the following mycotoxins and/or their metabolites:

Aflatoxin B_1
incidence: ?/?, sa. const.: channel catfishes, wt.: 0.3–0.5 kg, contamination: artificial (dose: 250 µg/kg AFB_1 (labeled), o., once; for detailed information please see the article), conc. range: ≤58 ppb* ** (mean value), country: USA[516], *after 4 h (also measured after 2, 24, 48 and 96 h, lowest conc.: tr after 48 and 96 h), **AFB_1 eq.

Fish, channel catfish kidney may contain the following mycotoxins and/or their metabolites:

Aflatoxin B_1
incidence: ?/?, sa. const.: channel catfishes, wt.: 0.3–0.5 kg, contamination: artificial (dose: 250 µg/kg AFB_1 (labeled), o., once; for detailed information please see the article), conc. range: ≤220 ppb* ** *** (mean value), country: USA[516], *after 4 h (also measured after 2, 24, 48 and 96 h, lowest conc.: tr after 96 h), **AFB_1 eq., ***in **head kidney**
incidence: ?/?, sa. const.: channel catfishes, wt.: 0.3–0.5 kg, contamination: artificial (dose: 250 µg/kg AFB_1 (labeled), o., once; for detailed information please see the article), conc. range: ≤287 ppb* ** *** (mean value), country: USA[516], *after 4 h (also measured after 2, 24, 48 and 96 h, lowest conc.: 15 ppb after 96 h), **AFB_1 eq., ***in **trunk kidney**

Fish, channel catfish liver may contain the following mycotoxins and/or their metabolites:

Aflatoxin B_1
incidence: ?/?, sa. const.: channel catfishes, wt.: 0.3–0.5 kg, contamination: artificial (dose: 250 µg/kg AFB_1 (labeled), o., once; for detailed information please see the article), conc. range: ≤421 ppb* ** (mean value), country: USA[516], *after 4 h (also measured after 2, 24, 48 and 96 h, lowest conc.: 37 ppb after 48 h), **AFB_1 eq.

Fish, channel catfish muscle may contain the following mycotoxins and/or their metabolites:

Aflatoxin B_1
incidence: ?/?, sa. const.: channel catfishes, wt.: 0.3–0.5 kg, contamination: artificial (dose: 250 µg/kg AFB_1 (labeled), o., once; for detailed information please see the article), conc. range: ≤40 ppb* ** (mean value), country: USA[516], *after 4 h (also measured after 2, 24, 48 and 96 h, lowest conc.: tr after 24, 48 and 96 h), **AFB_1 eq.

Fish, channel catfish plasma may contain the following mycotoxins and/or their metabolites:

Aflatoxin B_1
incidence: ?/?, sa. const.: channel catfishes, wt.: 0.3–0.5 kg, contamination: artificial (dose: 250 µg/kg AFB_1 (labeled), o., once; for detailed information please see the article), conc. range: ≤596 ppb* **

(mean value), country: USA[516], *after 4 h (also measured after 2, 24, 48 and 96 h, lowest conc.: 6 ppb after 96 h), **AFB_1 eq.

Fish, channel catfish skin may contain the following mycotoxins and/or their metabolites:

AFLATOXIN B_1
incidence: ?/?, sa. const.: channel catfishes, wt.: 0.3–0.5 kg, contamination: artificial (dose: 250 µg/kg AFB_1 (labeled), o., once; for detailed information please see the article), conc. range: ≤82 ppb* ** (mean value), country: USA[516], *after 4 h (also measured after 2, 24, 48 and 96 h, lowest conc.: tr after 96 h), **AFB_1 eq.

Fish, channel catfish spleen may contain the following mycotoxins and/or their metabolites:

AFLATOXIN B_1
incidence: ?/?, sa. const.: channel catfishes, wt.: 0.3–0.5 kg, contamination: artificial (dose: 250 µg/kg AFB_1 (labeled), o., once; for detailed information please see the article), conc. range: ≤234 ppb* ** (mean value), country: USA[516], *after 4 h (also measured after 2, 24, 48 and 96 h, lowest conc.: tr after 48 and 96 h), **AFB_1 eq.

Fish, channel catfish urine may contain the following mycotoxins and/or their metabolites:

AFLATOXIN B_1
incidence: ?/?, sa. const.: channel catfishes, wt.: 0.3–0.5 kg, contamination: artificial (dose: 250 µg/kg AFB_1 (labeled), o., once; for detailed information please see the article), conc. range: ≤51 ppb* ** (mean value), country: USA[516], *after 4–6 h (also measured after 2, 8, 10, 12, 18 and 24 h, except the start value lowest conc.: 6 ppb after 24 h), **AFB_1 and its metabolites (total residues)

Fish, Coho Salmon

Fish, coho salmon bile may contain the following mycotoxins and/or their metabolites:

AFLATOXICOL
incidence: 2?/2*, sa. const.: coho salmons, age: 15 months, wt.: 44–184 g, contamination: artificial (dose: 50 µg AFB_1 (labeled)/kg fish, i.p., once; for detailed information please see the article), conc.: 1.49 µM AFL-g ** (mean value), country: USA[628], *control, **after 24 h, compare with **Fish, rainbow trout bile** AFL no.[628]

AFLATOXICOL M_1
incidence: 2?/2*, sa. const.: coho salmons, age: 15 months, wt.: 44–184 g, contamination: artificial (dose: 50 µg AFB_1 (labeled)/kg fish, i.p., once; for detailed information please see the article), conc.: 0.07 µM $AFLM_1$-g ** (mean value), country: USA[628], *control, **after 24 h, compare with **Fish, rainbow trout bile** $AFLM_1$ no.[628]

AFLATOXIN B_1
incidence: 2?/2*, sa. const.: coho salmons, age: 15 months, wt.: 44–184 g, contamination: artificial (dose: 50 µg AFB_1 (labeled)/kg fish, i.p., once; for detailed information please see the article), conc.: nd** ***, country: USA[628], *control, **AFB_1-SG, ***after 24 h, compare with **Fish, rainbow trout bile** AFB_1 no.[628]

Fish, coho salmon embryo may contain the following mycotoxins and/or their metabolites:

AFLATOXIN B_1
incidence: ?/12, sa. const.: coho salmon eggs/embryos, age: 21 days, contamination: artificial (dose: 0.5 µg AFB_1 (labeled)/ml, for 1 h; for detailed information please see the article), conc.: 0.07 pmol AFB_1/mg DNA (mean value), country: USA[115], *in embryos ≈48 h to hatching,

compare with **Fish, rainbow trout embryo** AFB_1 no.[115]

Fish, coho salmon liver may contain the following mycotoxins and/or their metabolites:

Aflatoxin B_1
incidence: 1/1, sa. const.: coho salmon, wt.: 30.6 g, contamination: artificial (dose: 80 ppb AFB_1 (labeled), o., for 3 weeks), conc.: 1.6 pmol AFB_1/mg DNA*, country: USA[115], *after 3 weeks,
compare with **Fish, rainbow trout liver** AFB_1, 1st statement of no.[115]

incidence: 3?/3, sa. const.: coho salmons, contamination: artificial (dose: 23.6 μg AFB_1 (labeled)/kg, i.p., once; for detailed information please see the article), conc.: 4.0 pmol AFB_1/mg DNA* (mean value), country: USA[115], *after 2 days, **AFB_1-N^7-Gua[3] adducts, compare with **Fish, rainbow trout liver** AFB_1, 2nd statement of no.[115]
incidence: 3?/3, sa. const.: coho salmons, contamination: artificial (dose: 23.6 μg AFB_1 (labeled)/kg, i.p., once; for detailed information please see the article), conc.: 1.2 pmol AFB_1/mg DNA* (mean value), country: USA[115], *after 21 days, **AFB_1-N^7-Gua[3] adducts,
compare with **Fish, rainbow trout liver** AFB_1, 3rd statement of no.[115]
incidence: 3?/3, sa. const.: coho salmons, contamination: artificial (dose: 23.6 μg AFB_1 (labeled)/kg, i.p., once; for detailed information please see the article), conc.: 0.5 pmol AFB_1/mg DNA* ** (mean value), country: USA[115], *after 2 days, **AFB_1-FAPyr adducts,
compare with **Fish, rainbow trout liver** AFB_1, 4th statement of no.[115]
incidence: 3?/3, sa. const.: coho salmons, contamination: artificial (dose: 23.6 μg AFB_1 (labeled)/kg, i.p., once; for detailed information please see the article), conc.: 0.8 pmol AFB_1/mg DNA* ** (mean value), country: USA[115], *after 21 days, **AFB_1-FAPyr adducts,
compare with **Fish, rainbow trout liver** AFB_1, 5th statement of no.[115]
incidence: 3?/3, sa. const.: coho salmons, contamination: artificial (dose: 23.6 μg AFB_1 (labeled)/kg, i.p., once; for detailed information please see the article), conc.: 0.2 pmol AFB_1/mg DNA* (mean value), country: USA[115], *after 2 days, **AFB_1-minor FAPyr isomer adducts,
compare with **Fish, rainbow trout liver** AFB_1, 6th statement of no.[115]
incidence: 3?/3, sa. const.: coho salmons, contamination: artificial (dose: 23.6 μg AFB_1 (labeled)/kg, i.p., once; for detailed information please see the article), conc.: 0.2 pmol AFB_1/mg DNA* (mean value), country: USA[115], *after 21 days, **AFB_1-minor FAPyr isomer adducts, compare with **Fish, rainbow trout liver** AFB_1, 7th statement of no.[115]

incidence: ?/?, sa. const.: coho salmons, age: 13 months, wt.: 61 g, contamination: artificial (dose: 50 μg AFB_1 (labeled)/kg b. wt., i.p., once; for detailed information please see the article), conc.: 10.6 pmol AFB_1 binding/mg DNA* (mean value), country: USA[177], *after 24 h,
compare with **Fish, rainbow trout liver** AFB_1, 1st statement of no.[177]

Fish, Medaka

Fish, medaka liver may contain the following mycotoxins and/or their metabolites:

Aflatoxin B_1
incidence: 6?/6, sa. const.: Japanese medaka fishes, age: adult, wt.: 4–5 g, contamination: artificial (dose: 70, 140, 275 or 550* μg AFB_1 (labeled)/kg b. wt., i.p., once; for detailed information please see the article), conc. range: ≈≤23 pmol AFB_1/μmol DNA* ** (mean value), country: USA[625], **after 24 h
incidence: 6?/6, sa. const.: Japanese medaka fishes, age: adult, wt.: 4–5 g, contamination:

artificial (dose: 550 μg AFB_1 (labeled)/kg b. wt., i.p., once; for detailed information please see the article), conc. range: ≈≤21.5 pmol AFB_1/μmol DNA* (mean value), country: USA[625], *after 1 day (also at other day intervals up to 21 days measured, lowest conc.: ≈3.5 pmol AFB_1/μmol DNA after 21 days)

Fish, Rainbow Trout

Fish, rainbow trout bile may contain the following mycotoxins and/or their metabolites:

Aflatoxicol

incidence: ?/?*, sa. const.: Mt. Shasta strain rainbow trouts, wt.: ≈400 g, contamination: artificial (dose: 25 μl containing 0.32 μmol AFB_1 (labeled)/ml, i.p., once; for detailed information please see the article), conc. range: ≈≤1.18 μM AFL-g in bile** (mean value), country: USA[212], *control, **after 24 h post-injection (also measured after 5, 10 and 16 h, lowest conc.: ≈0.48 μM AFL-g in bile after 5 h)

incidence: ?/?, sa. const.: Mt. Shasta strain rainbow trouts, wt.: ≈400 g, contamination: artificial (dose: 25 μl containing 0.32 μmol AFB_1 (labeled)/ml, i.p., once and **I3C-diet (0.2%)** for 12 weeks prior to AFB_1 injection; for detailed information please see the article), conc. range: ≈≤1.18 μM AFL-g in bile* (mean value), country: USA[212], *after 24 h post-injection (also measured after 5, 10 and 16 h, lowest conc.: ≈0.50 μM AFL-g in bile after 5 h)

incidence: ?/?*, sa. const.: Shasta strain rainbow trouts, age: 9 months, contamination: artificial (dose: 2.47 μg AFB_1 (labeled), i.p. injection, once; for detailed information please see the article), conc. range: ≈≤1.2 μM AFL-g in bile** (mean value), country: USA[213], *control, **after 24 h post-injection (also measured after 5, 10 and 16 h, lowest conc.: ≈0.45 μM AFL-g in bile after 5 h)

incidence: ?/?, sa. const.: Shasta strain rainbow trouts, age: 9 months, contamination: artificial (dose: **100 ppm PCB** for 2 month prior to 2.47 μg AFB_1 (labeled), i.p. injection, once; for detailed information please see the article), conc. range: ≈≤0.8 μM AFL-g in bile* (mean value), country: USA[213], *after 24 h post-injection (also measured after 5, 10 and 16 h, lowest conc.: ≈0.40 μM AFL-g in bile after 5 h)

incidence: ?/?*, sa. const.: Mt Shasta strain rainbow trouts, wt.: ≈50 g, contamination: artificial (dose: 25 μl containing 0.32 μmol AFB_1 (labeled)/ml, i.p., once; for detailed information please see the article), conc. range: ≈≤1.23 μM AFL-g in bile** (mean value), country: USA[260], *control, **after 24 h (also measured after 5, 10 and 16 h, lowest conc.: ≈0.45 μM AFL-g in bile after 5 h)

incidence: ?/?, sa. const.: Mt Shasta strain rainbow trouts, wt.: ≈50 g, contamination: artificial (dose: 25 μl containing 0.32 μmol AFB_1 (labeled)/ml, i.p., once and **BNF-diet (0.05%)** for 12 weeks prior to AFB_1-injection; for detailed information please see the article), conc. range: ≈≤0.5 μM AFL-g in bile* (mean value), country: USA[260], *after 10 h (also measured after 5, 16 and 24 h, lowest conc.: ≈0.33 μM AFL-g in bile after 5 h)

incidence: 2?/2*, sa. const.: Mt Shasta strain rainbow trouts, age: 16 months, wt.: 216–416 g, contamination: artificial (dose: 50 μg AFB_1 (labeled)/kg fish, i.p., once; for detailed information please see the article), conc.: 5.66 μM AFL-g** (mean value), country: USA[628], *control, **after 24 h,
compare with **Fish, coho salmon bile** AFL no.[628]

incidence: 2?/2, sa. const.: Mt Shasta strain rainbow trouts, age: 16 months, wt.: 216–416 g, contamination: artificial (dose: **βNF-diet (500 ppm)** fed for 3 weeks prior to 50 μg AFB_1 (labeled)/kg fish, i.p., once;

for detailed information please see the article), conc.: 0.92 μM AFL-g* (mean value), country: USA[628], *after 24 h
incidence: 2?/2, sa. const.: Mt Shasta strain rainbow trouts, age: 16 months, wt.: 216–416 g, contamination: artificial (dose: **I3C-diet (2,000 ppm)** fed for 3 weeks prior to 50 μg AFB_1 (labeled)/kg fish, i.p., once; for detailed information please see the article), conc.: 5.6 μM AFL-g* (mean value), country: USA[628], *after 24 h
incidence: 2?/2, sa. const.: Mt Shasta strain rainbow trouts, age: 16 months, wt.: 216–416 g, contamination: artificial (dose: **PCB-diet (100 ppm)** fed for 3 weeks prior to 50 μg AFB_1 (labeled)/kg fish, i.p., once; for detailed information please see the article), conc.: 2.64 μM AFL-g* (mean value), country: USA[628], *after 24 h

Aflatoxicol M_1
incidence: ?/?*, sa. const.: Mt. Shasta strain rainbow trouts, wt.: ≈400 g, contamination: artificial (dose: 25 μl containing 0.32 μmol AFB_1 (labeled)/ml, i.p., once; for detailed information please see the article), conc. range: ≈≤0.245 μM $AFLM_1$-g in bile** (mean value), country: USA[212], *control, **after 16 h post-injection (also measured after 5, 10 and 24 h, lowest conc.: ≈0.15 μM $AFLM_1$-g in bile after 5 h)
incidence: ?/?, sa. const.: Mt. Shasta strain rainbow trouts, wt.: ≈400 g, contamination: artificial (dose: 25 μl containing 0.32 μmol AFB_1 (labeled)/ml, i.p., once and **I3C-diet (0.2%)** for 12 weeks prior to AFB_1 injection; for detailed information please see the article), conc. range: ≈≤1.50 μM $AFLM_1$-g in bile* (mean value), country: USA[212], *after 16 h post-injection (also measured after 5, 10 and 24 h, lowest conc.: ≈0.60 μM $AFLM_1$-g in bile after 5 h)

incidence: ?/?*, sa. const.: Shasta strain rainbow trouts, age: 9 months, contamination: artificial (dose: 2.47 μg AFB_1 (labeled), i.p. injection, once; for detailed information please see the article), conc. range: ≈≤0.2 μM $AFLM_1$-g in bile** (mean value), country: USA[213], *control, **after 16 h post-injection (also measured after 5, 10 and 24 h, lowest conc.: ≈0.10 μM $AFLM_1$-g in bile after 5 h)
incidence: ?/?, sa. const.: Shasta strain rainbow trouts, age: 9 months, contamination: artificial (dose: **100 ppm PCB** for 2 month prior to 2.47 μg AFB_1 (labeled), i.p. injection, once; for detailed information please see the article), conc. range: ≈≤3.35 μM $AFLM_1$-g in bile* (mean value), country: USA[213], *after 16 h post-injection (also measured after 5, 10 and 24 h, lowest conc.: ≈2.1 μM $AFLM_1$-g in bile after 10 h)

incidence: ?/?*, sa. const.: Mt Shasta strain rainbow trouts, wt.: ≈50 g, contamination: artificial (dose: 25 μl containing 0.32 μmol AFB_1 (labeled)/ml, i.p., once; for detailed information please see the article), conc. range: ≈≤0.23 μM $AFLM_1$-g in bile** (mean value), country: USA[260], *control, **after 16 and 24 h (also measured after 5 and 10 h, lowest conc.: ≈0.1 μM $AFLM_1$-g in bile after 5 h)
incidence: ?/?, sa. const.: Mt Shasta strain rainbow trouts, wt.: ≈50 g, contamination: artificial (dose: 25 μl containing 0.32 μmol AFB_1 (labeled)/ml, i.p., once and **BNF-diet (0.05%)** for 12 weeks prior to AFB_1-injection; for detailed information please see the article), conc. range: ≈≤3.3 μM $AFLM_1$-g in bile* (mean value), country: USA[260], *after 10 h (also measured after 5, 16 and 24 h, lowest conc.: ≈1.26 μM $AFLM_1$-g in bile after 5 h)

incidence: 2?/2*, sa. const.: Mt Shasta strain rainbow trouts, age: 16 months, wt.: 216–416 g, contamination: artificial (dose: 50 μg AFB_1 (labeled)/kg fish, i.p., once; for detailed information please see the article), conc.: 1.47 μM $AFLM_1$-g ** (mean value), country: USA[628], *control, **after 24 h,
compare with **Fish, coho salmon bile** $AFLM_1$ no.[628]
incidence: 2?/2, sa. const.: Mt Shasta strain rainbow trouts, age: 16 months, wt.: 216–416 g, contamination: artificial (dose:

βNF-diet (500 ppm) for 3 weeks prior to 50 μg AFB_1 (labeled)/kg fish, i.p., once; for detailed information please see the article), conc.: 26.68 μM $AFLM_1$-g* (mean value), country: USA[628], *after 24 h
incidence: 2?/2, sa. const.: Mt Shasta strain rainbow trouts, age: 16 months, wt.: 216–416 g, contamination: artificial (dose: **I3C-diet (2,000 ppm)** for 3 weeks prior to 50 μg AFB_1 (labeled)/kg fish, i.p., once; for detailed information please see the article), conc.: 9.5 μM $AFLM_1$-g* (mean value), country: USA[628], *after 24 h
incidence: 2?/2, sa. const.: Mt Shasta strain rainbow trouts, age: 16 months, wt.: 216–416 g, contamination: artificial (dose: **PCB-diet (100 ppm)** for 3 weeks prior to 50 μg AFB_1 (labeled)/kg fish, i.p., once; for detailed information please see the article), conc.: 15.4 μM $AFLM_1$-g* (mean value), country: USA[628], *after 24 h

Aflatoxin B_1
incidence: 3?/3*, sa. const.: Mt. Shasta strain rainbow trouts, wt.: ≈400 g, contamination: artificial (dose: 25 μl containing 0.32 μmol AFB_1 (labeled)/ml, i.p., once; for detailed information please see the article), conc. range: ≈≤2.8 μM AFB_1 in bile** (mean value), country: USA[212], *control, **after 24 h post-injection (also measured after 5, 10 and 16 h, lowest conc.: ≈1.3 μM AFB_1 in bile after 5 h)
incidence: 3?/3, sa. const.: Mt. Shasta strain rainbow trouts, wt.: ≈400 g, contamination: artificial (dose: 25 μl containing 0.32 μmol AFB_1 (labeled)/ml, i.p., once and **I3C-diet (0.2%)** for 12 weeks prior to AFB_1 injection; for detailed information please see the article), conc. range: ≈≤3.7 μM AFB_1 in bile* (mean value), country: USA[212], *after 16 and 24 h post-injection (also measured after 5 and 10 h, lowest conc.: ≈1.6 μM AFB_1 in bile after 5 h)

incidence: 9?/9*, sa. const.: Shasta strain rainbow trouts, age: 9 months, contamination: artificial (dose: 2.47 μg AFB_1 (labeled), i.p. injection, once; for detailed information please see the article), conc. range: ≈≤2.8 μM AFB_1 in bile** (mean value), country: USA[213], *control, **after 24 h post-injection (also measured after 5, 10 and 16 h, lowest conc.: ≈1.3 μM AFB_1 in bile after 5 h)
incidence: 9?/9, sa. const.: Shasta strain rainbow trouts, age: 9 months, contamination: artificial (dose: **100 ppm PCB** for 2 month prior to 2.47 μg AFB_1 (labeled), i.p. injection, once; for detailed information please see the article), conc. range: ≈≤5.8 μM AFB_1 in bile* (mean value), country: USA[213], *after 16 h post-injection (also measured after 5, 10 and 24 h, lowest conc.: ≈4.5 μM AFB_1 in bile after 10 h)

incidence: ?/3*, sa. const.: Mt Shasta strain rainbow trouts, wt.: ≈50 g, contamination: artificial (dose: 25 μl containing 0.32 μmol AFB_1 (labeled)/ml, i.p., once; for detailed information please see the article), conc. range: ≈≤2.8 μM AFB_1 in bile** (mean value), country: USA[260], *control, **after 24 h (also measured after 5, 10 and 16 h, lowest conc.: ≈1.3 μM AFB_1 in bile after 5 h)
incidence: ?/3, sa. const.: Mt Shasta strain rainbow trouts, wt.: ≈50 g, contamination: artificial (dose: 25 μl containing 0.32 μmol AFB_1 (labeled)/ml, i.p., once and **BNF-diet (0.05%)** for 12 weeks prior to AFB_1-injection; for detailed information please see the article), conc. range: ≈≤5.6 μM AFB_1 in bile* (mean value), country: USA[260], *after 10 h (also measured after 5, 16 and 24 h, lowest conc.: ≈4.5 μM AFB_1 in bile after 5 h)

incidence: 2?/2*, sa. const.: Mt Shasta strain rainbow trouts, age: 16 months, wt.: 216–416 g, contamination: artificial (dose: 50 μg AFB_1 (labeled)/kg fish, i.p., once; for detailed information please see the article), conc.: 0.07 μM AFB_1-SG** (mean value), country: USA[628], *control, **after 24 h, compare with **Fish, coho salmon bile** AFB_1 no.[628]

incidence: 2?/2, sa. const.: Mt Shasta strain rainbow trouts, age: 16 months, wt.: 216–416 g, contamination: artificial (dose: **βNF-diet (500 ppm)** for 3 weeks prior to 50 μg AFB_1 (labeled)/kg fish, i.p., once; for detailed information please see the article), conc.: 0.02 μM AFB_1-SG* (mean value), country: USA[628], *after 24 h
incidence: 2?/2, sa. const.: Mt Shasta strain rainbow trouts, age: 16 months, wt.: 216–416 g, contamination: artificial (dose: **I3C-diet (2,000 ppm)** for 3 weeks prior to 50 μg AFB_1 (labeled)/kg fish, i.p., once; for detailed information please see the article), conc.: 0.02 μM AFB_1-SG* (mean value), country: USA[628], *after 24 h
incidence: 2?/2, sa. const.: Mt Shasta strain rainbow trouts, age: 16 months, wt.: 216–416 g, contamination: artificial (dose: **PCB-diet (100 ppm)** for 3 weeks prior to 50 μg AFB_1 (labeled)/kg fish, i.p., once; for detailed information please see the article), conc.: nd* **, country: USA[628], *AFB_1-SG, **after 24 h

Fish, rainbow trout blood may contain the following mycotoxins and/or their metabolites:

Aflatoxin B_1
incidence: 9?/9*, sa. const.: Mt. Shasta strain rainbow trouts, wt.: ≈400 g, contamination: artificial (dose: 25 μl containing 0.32 μmol AFB_1 (labeled)/ml, i.p., once; for detailed information please see the article), conc. range: ≈≤29 nM AFB_1 in blood** (mean value), country: USA[212], *control, **after 5 h post-injection (also measured after 10, 16 and 24 h, lowest conc.: ≈13.5 nM AFB_1 in blood after 24 h)
incidence: 9?/9, sa. const.: Mt. Shasta strain rainbow trouts, wt.: ≈400 g, contamination: artificial (dose: 25 μl containing 0.32 μmol AFB_1 (labeled)/ml, i.p., once and **I3C-diet (0.2%)** for 12 weeks prior to AFB_1 injection; for detailed information please see the article), conc. range: ≈≤26.5 nM AFB_1 in blood* (mean value), country: USA[212], *after 5 h post-injection (also measured after 10, 16 and 24 h, lowest conc.: ≈8 nM AFB_1 in blood after 24 h)

incidence: 9?/9*, sa. const.: Shasta strain rainbow trouts, age: 9 months, contamination: artificial (dose: 2.47 μg AFB_1 (labeled), i.p. injection, once; for detailed information please see the article), conc. range: ≈≤29 nM AFB_1 in blood** (mean value), country: USA[213], *control, **after 5 h post-injection (also measured after 10, 16 and 24 h, lowest conc.: ≈13.5 nM AFB_1 in blood after 24 h)
incidence: 9?/9, sa. const.: Shasta strain rainbow trouts, age: 9 months, contamination: artificial (dose: **100 ppm PCB** for 2 month prior to 2.47 μg AFB_1 (labeled), i.p. injection, once; for detailed information please see the article), conc. range: ≈≤34.5 nM AFB_1 in blood* (mean value), country: USA[213], *after 5 h post-injection (also measured after 10, 16 and 24 h, lowest conc.: ≈14.5 nM AFB_1 in blood after 24 h)

incidence: ?/9*, sa. const.: Mt Shasta strain rainbow trouts, wt.: ≈50 g, contamination: artificial (dose: 25 μl containing 0.32 μmol AFB_1 (labeled)/ml, i.p., once; for detailed information please see the article), conc. range: ≈≤29 nM AFB_1 in blood** (mean value), country: USA[260], *control, after 5 h** (also measured after 10, 16 and 24 h, lowest conc.: ≈13 nM AFB_1 in blood after 16 and 24 h)
incidence: ?/9, sa. const.: Mt Shasta strain rainbow trouts, wt.: ≈50 g, contamination: artificial (dose: 25 μl containing 0.32 μmol AFB_1 (labeled)/ml, i.p., once and **BNF-diet (0.05%)** for 12 weeks prior to AFB_1-injection; for detailed information please see the article), conc. range: ≈≤20 nM AFB_1 in blood* (mean value), country: USA[260], after 5 h* (also measured after 10, 16 and 24 h, lowest conc.: ≈5.5 nM AFB_1 in blood after 24 h)

Fish, rainbow trout carcass may contain the following mycotoxins and/or their metabolites:

Aflatoxin B_1
incidence: 3?/3*, sa. const.: Mt. Shasta strain rainbow trouts, wt.: ≈400 g, contamination: artificial (dose: 25 µl containing 0.32 µmol AFB_1 (labeled)/ml, i.p., once; for detailed information please see the article), conc. range: ≈≤27 pmol AFB_1/g carcass** (mean value), country: USA[212], *control, **after 5 h post-injection (also measured after 10, 16 and 24 h, lowest conc.: ≈16 pmol AFB_1/g carcass after 24 h)
incidence: 3?/3, sa. const.: Mt. Shasta strain rainbow trouts, wt.: ≈400 g, contamination: artificial (dose: 25 µl containing 0.32 µmol AFB_1 (labeled)/ml, i.p., once and **I3C-diet (0.2%)** for 12 weeks prior to AFB_1 injection; for detailed information please see the article), conc. range: ≈≤31 pmol AFB_1/g carcass* (mean value), country: USA[212], *after 5 h post-injection (also measured after 10, 16 and 24 h, lowest conc.: ≈14 pmol AFB_1/g carcass after 24 h)

incidence: 9?/9*, sa. const.: Shasta strain rainbow trouts, age: 9 months, contamination: artificial (dose: 2.47 µg AFB_1 (labeled), i.p. injection, once; for detailed information please see the article), conc. range: ≈≤28 pmol AFB_1/g carcass** (mean value), country: USA[213], *control, **after 5 h post-injection (also measured after 10, 16 and 24 h, lowest conc.: ≈17 pmol AFB_1/g carcass after 24 h)
incidence: 9?/9, sa. const.: Shasta strain rainbow trouts, age: 9 months, contamination: artificial (dose: **100 ppm PCB** for 2 month prior to 2.47 µg AFB_1 (labeled), i.p. injection, once; for detailed information please see the article), conc. range: ≈≤30 pmol AFB_1/g carcass* (mean value), country: USA[213], *after 5 h post-injection (also measured after 10, 16 and 24 h, lowest conc.: ≈13 pmol AFB_1/g carcass after 24 h)

incidence: ?/3*, sa. const.: Mt Shasta strain rainbow trouts, wt.: ≈50 g, contamination: artificial (dose: 25 µl containing 0.32 µmol AFB_1 (labeled)/ml, i.p., once; for detailed information please see the article), conc. range: ≈≤28 pmol AFB_1/g carcass** (mean value), country: USA[260], *control, after 5 h** (also measured after 10, 16 and 24 h, lowest conc.: ≈17 pmol AFB_1/g carcass after 24 h)
incidence: ?/3, sa. const.: Mt Shasta strain rainbow trouts, wt.: ≈50 g, contamination: artificial (dose: 25 µl containing 0.32 µmol AFB_1 (labeled)/ml, i.p., once and **BNF-diet (0.05%)** for 12 weeks prior to AFB_1-injection; for detailed information please see the article), conc. range: ≈≤23 pmol AFB_1/g carcass* (mean value), country: USA[260], after 5 h* (also measured after 10, 16 and 24 h, lowest conc.: ≈9.5 pmol AFB_1/g carcass after 24 h)

Fish, rainbow trout egg may contain the following mycotoxins and/or their metabolites:

Aflatoxicol
incidence: ?/72, sa. const.: Shasta strain rainbow trouts, contamination: artificial (dose: 0.01–0.05 µg AFB_1 + AFL (labeled)/ml solution, 1 h bath; for detailed information please see the article), conc.: 10.2 pmol AFL per part/nmol/ml bath solution* ** (mean value), country: USA[384], *in shell, ****immediately after exposure**
incidence: ?/36, sa. const.: Shasta strain rainbow trouts, contamination: artificial (dose: 0.01–0.05 µg AFB_1 + AFL (labeled)/ml solution, 1 h bath; for detailed information please see the article), conc.: 0.778 pmol AFL per part/nmol/ml bath solution* ** (mean value), country: USA[384], *in shell, ****after 24 h**

incidence:?/72, sa. const.: Shasta strain rainbow trouts, contamination: artificial (dose: 0.01–0.05 µg AFB_1 + AFL (labeled)/ml solution, 1 h bath; for detailed information please see the

article), conc.: 11.7 pmol AFL per part/nmol/ml bath solution* ** (mean value), country: USA[384], *in yolk, ****immediately after exposure**
incidence:?/36, sa. const.: Shasta strain rainbow trouts, contamination: artificial (dose: 0.01–0.05 µg AFB_1 + AFL (labeled)/ml solution, 1 h bath; for detailed information please see the article), conc.: 5.33 pmol AFL per part/nmol/ml bath solution* ** (mean value), country: USA[384], *in yolk, ****after 24 h**

Aflatoxin B_1
incidence: ?/?, sa. const.: Rainbow trout eggs, contamination: artificial (dose: 1.0 ppm AFB_1 (labeled), for 1 h; for detailed information please see the article), conc.: 2.7 ng/egg* (mean value), country: USA[229], *absorbed during 1 h exposure

incidence: ?/72, sa. const.: Shasta strain rainbow trouts, contamination: artificial (dose: 0.01–0.05 µg AFB_1 + AFL (labeled)/ml solution, 1 h bath; for detailed information please see the article), conc.: 22.3 pmol AFB_1 per part/nmol/ml bath solution* ** (mean value), country: USA[384], *in shell, ****immediately after exposure**
incidence: ?/36, sa. const.: Shasta strain rainbow trouts, contamination: artificial (dose: 0.01–0.05 µg AFB_1 + AFL (labeled)/ml solution, 1 h bath; for detailed information please see the article), conc.: 0.503 pmol AFB_1 per part/nmol/ml bath solution* ** (mean value), country: USA[384], *in shell, ****after 24 h**

incidence:?/72, sa. const.: Shasta strain rainbow trouts, contamination: artificial (dose: 0.01–0.05 µg AFB_1 + AFL (labeled)/ml solution, 1 h bath; for detailed information please see the article), conc.: 5.32 pmol AFB_1 per part/nmol/ml bath solution* ** (mean value), country: USA[384], *in yolk, ****immediately after exposure**
incidence:?/36, sa. const.: Shasta strain rainbow trouts, contamination: artificial (dose: 0.01–0.05 µg AFB_1 + AFL (labeled)/ml solution, 1 h bath; for detailed information please see the article), conc.: 1.84 pmol AFB_1 per part/nmol/ml bath solution* ** (mean value), country: USA[384], *in yolk, ****after 24 h**

Fish, rainbow trout embryo may contain the following mycotoxins and/or their metabolites:

Aflatoxicol
incidence: ?/72, sa. const.: Shasta strain rainbow trouts, contamination: artificial (dose: 0.01–0.05 µg AFB_1 + AFL (labeled)/ml solution, 1 h bath; for detailed information please see the article), conc.: 7.34 pmol AFL per part/nmol/ml bath solution* (mean value), country: USA[384], ***immediately after exposure**
incidence: ?/36, sa. const.: Shasta strain rainbow trouts, contamination: artificial (dose: 0.01–0.05 µg AFB_1 + AFL (labeled)/ml solution, 1 h bath; for detailed information please see the article), conc.: 3.01 pmol AFL per part/nmol/ml bath solution* (mean value), country: USA[384], ***after 24 h**

incidence: ?/72, sa. const.: Shasta strain rainbow trouts, contamination: artificial (dose: 0.01–0.05 µg AFB_1 + AFL (labeled)/ml solution, 1 h bath; for detailed information please see the article), conc.: 19.2 pmol AFL per part/nmol/ml bath solution* ** (mean value), country: USA[384], *in embryo + yolk, ****immediately after exposure**
incidence: ?/36, sa. const.: Shasta strain rainbow trouts, contamination: artificial (dose: 0.01–0.05 µg AFB_1 + AFL (labeled)/ml solution, 1 h bath; for detailed information please see the article), conc.: 8.47 pmol AFL per part/nmol/ml bath solution* ** (mean value), country: USA[384], *in embryo + yolk, ****after 24 h**

incidence: ?/?*, sa. const.: Shasta strain rainbow trouts, contamination: artificial (dose: 0.01–0.05 µg AFB_1 + AFL (labeled)/ml solution, 1 h bath; for detailed information please see the article), conc. range: ≈≤3.9 pmol AFL/mg DNA**, country: USA[384], *rainbow trout embryo DNA, **after 24 h, compare with **Fish, rainbow trout embryo** AFB_1, 5th statement of no.[384]

Aflatoxin B_1

incidence: ?/6, sa. const.: Shasta strain rainbow trout eggs/embryos, age: 21 days, contamination: artificial (dose: 0.5 µg AFB_1 (labeled)/ml, for 1 h; for detailed information please see the article), conc.: 1.38 pmol AFB_1/mg DNA* (mean value), country: USA[115], *in embryos ≈48 h to hatching, compare with **Fish, coho salmon embryo** AFB_1 no.[115]

incidence: ?/?*, sa. const.: fertilized rainbow trout 21-day-old eggs, contamination: artificial (dose: 0.5 ppm AFB_1 (labeled) exposure of 21-day-old eggs for 1 h; for detailed information please see the article), conc. range: ≤8.4 × 10^{-7} mmol AFB_1/mmol DNA nucleotides** (mean value), country: USA[229], *rainbow trout embryo DNA, **after 24 h

incidence: ?/72, sa. const.: Shasta strain rainbow trouts, contamination: artificial (dose: 0.01–0.05 µg AFB_1 + AFL (labeled)/ml solution, 1 h bath; for detailed information please see the article), conc.: 3.62 pmol AFB_1 per part/nmol/ml bath solution* (mean value), country: USA[384], ***immediately after exposure**
incidence: ?/36, sa. const.: Shasta strain rainbow trouts, contamination: artificial (dose: 0.01–0.05 µg AFB_1 + AFL (labeled)/ml solution, 1 h bath; for detailed information please see the article), conc.: 1.12 pmol AFB_1 per part/nmol/ml bath solution* (mean value), country: USA[384], ***after 24 h**

incidence: ?/72, sa. const.: Shasta strain rainbow trouts, contamination: artificial (dose: 0.01–0.05 µg AFB_1 + AFL (labeled)/ml solution, 1 h bath; for detailed information please see the article), conc.: 9.09 pmol AFL per part/nmol/ml bath solution* ** (mean value), country: USA[384], *in embryo + yolk, ****immediately after exposure**
incidence: ?/36, sa. const.: Shasta strain rainbow trouts, contamination: artificial (dose: 0.01–0.05 µg AFB_1 + AFL (labeled)/ml solution, 1 h bath; for detailed information please see the article), conc.: 3.05 pmol AFL per part/nmol/ml bath solution* ** (mean value), country: USA[384], *in embryo + yolk, ****after 24 h**

incidence: ?/?*, sa. const.: Shasta strain rainbow trouts, contamination: artificial (dose: 0.01–0.05 µg AFB_1 + AFL (labeled)/ml solution, 1 h bath; for detailed information please see the article), conc. range: ≈≤1.3 pmol AFB_1/mg DNA**, country: USA[384], *rainbow trout embryo DNA, ****after 24 h**, compare with **Fish, rainbow trout embryo** AFL 5th statement of no.[384]

Fish, Rainbow Trout Embryo and Yolk

see Fish, rainbow trout embryo

Fish, rainbow trout liver

may contain the following mycotoxins and/or their metabolites:

Aflatoxicol

incidence: ?/?*, sa. const.: Mt. Shasta strain rainbow trouts, wt.: ≈400 g, contamination: artificial (dose: 25 µl containing 0.32 µmol AFB_1 (labeled)/ml, i.p., once; for detailed information please see the article), conc. range: 13**–34*** pmol/g liver (mean values), country: USA[212], *control, after 24** and 5 h*** (also measured after 10 and 16 h but values between values of 5 and 24 h)
incidence: ?/?, sa. const.: Mt. Shasta strain rainbow trouts, wt.: ≈400 g,

contamination: artificial (dose: 25 µl containing 0.32 µmol AFB_1 (labeled)/ml, i.p., once and **I3C-diet (0.2%)** for 12 weeks prior to AFB_1-injection; for detailed information please see the article), conc. range: 10*–31** pmol/g liver (mean values), country: USA[212], after 24* and 5 h** (also measured after 10 and 16 h but values between values of 5 and 24 h)

incidence: ?/?*, sa. const.: Shasta strain rainbow trouts, age: 9 months, contamination: artificial (dose: 2.47 µg AFB_1 (labeled), i.p. injection, once; for detailed information please see the article), conc. range: 13**–34*** pmol/g liver (mean values), country: USA[213], *control, after 24** and 5 h*** (also measured after 10 and 16 h but values between values of 5 and 24 h)
incidence: ?/?, sa. const.: Shasta strain rainbow trouts, age: 9 months, contamination: artificial (dose: **100 ppm PCB**) for 2 months prior to 2.47 µg AFB_1 (labeled), i.p. injection, once; for detailed information please see the article), conc. range: 6*–32** pmol/g liver (mean values), country: USA[213], after 24* and 5 h** (also measured after 10 and 16 h but values between values of 5 and 24 h)

incidence: 6?/6, sa. const.: Shasta strain rainbow trouts, age: 3 months, contamination: artificial (dose: 164.5 pmol AF (labeled)/g diet, o., for 2 weeks; for detailed information please see the article), conc.: 3.57 pmol AFL/mg DNA* (mean value), country: USA[228], *after 15 days (thereof 14 days with AF-administration)

incidence: 9?/9*, sa. const.: Shasta strain rainbow trouts, wt.: ≈50 g, contamination: artificial (dose: 25 µl containing 0.32 µmol AFB_1 (labeled)/ml, i.p., once; for detailed information please see the article), conc. range: 13**–34*** pmol/g liver (mean values), country: USA[260], *control, after 24** and 5 h*** (also measured after 10 and 16 h but values between values of 5 and 24 h)
incidence: 9?/9, sa. const.: Shasta strain rainbow trouts, wt.: ≈50 g, contamination: artificial (dose: 25 µl containing 0.32 µmol AFB_1 (labeled)/ml, i.p., once and **BNF-diet (0.05%)** for 12 weeks prior to AFB_1-injection; for detailed information please see the article), conc. range: 1*–12** pmol/g liver (mean values), country: USA[260], after 24* and 5 h** (also measured after 10 and 16 h but values between values of 5 and 24 h)

incidence: ?/?, sa. const.: Shasta strain rainbow trouts, age: 2 months, Ø wt.: 1.5 g, contamination: no AFL; for detailed information please see the article), conc.: nr, country: USA[384]
incidence: ?/4 or 8, sa. const.: Shasta strain rainbow trouts, age: 2 months, Ø wt.: 1.5 g, contamination: artificial (dose: 12.4, 25, 50 and 75* ng AFL (labeled)/g dry wt. diet, for 2 weeks; for detailed information please see the article), conc. range: ≈≤6.0* ** pmol AFB_1/mg DNA, country: USA[384], **after 2 weeks

incidence: ?/3, sa. const.: Mt. Shasta strain rainbow trouts, age: 11 months, wt.: ≈66 g, contamination: artificial (dose: 0.00995 µg? AFL (labeled)/g b. wt., i.p., once; for detailed information please see the article), conc. range: ≤15.4 pmol/mg DNA* ** (mean value), country: USA[405], *after 2 days (also measured after 1, 4, and 7 days, lowest conc.: 14.3 pmol/mg DNA after 1 day), **[^{3}H]AFL-conc.

incidence: 3?/3*, sa. const.: rainbow trouts, wt.: 5–6 g b. wt., contamination: artificial (dose: 80 ppb AFB_1 (labeled) in the diet for 2 days; for detailed information please see the article), conc.: ≈4.05 pmol AFL/g liver** (mean value), country: USA[546], *control, **after AFB_1-administration
incidence: 3?/3, sa. const.: rainbow trouts, wt.: 5–6 g b. wt., contamination: artificial (dose: **pretreatment with CHL**

(4,000 ppm) for 2 weeks then 80 ppb AFB_1 (labeled) + **CHL (4,000 ppm)** in the diet for 2 days; for detailed information please see the article), conc.: ≈2.3 pmol AFL/g liver* (mean value), country: USA[546], *after AFB_1- and CHL-administration
incidence: 3?/3*, sa. const.: rainbow trouts, wt.: 5–6 g b. wt., contamination: artificial (dose: 80 ppb AFB_1 (labeled) in the diet for 2 days; for detailed information please see the article), conc.: ≈0.40 pmol AFL-g/g liver** (mean value), country: USA[546], *control, **after AFB_1-administration
incidence: 3?/3, sa. const.: rainbow trouts, wt.: 5–6 g b. wt., contamination: artificial (dose: **pretreatment with CHL (4,000 ppm)** for 2 weeks then 80 ppb AFB_1 (labeled) + **CHL (4,000 ppm)** in the diet for 2 days; for detailed information please see the article), conc.: ≈0.45 pmol AFL-g/g liver* (mean value), country: USA[546], *after AFB_1- and CHL-administration

Aflatoxicol M_1
incidence: 6?/6, sa. const.: Shasta strain rainbow trouts, age: 3 months, contamination: artificial (dose: 1,571 pmol AF (labeled)/g diet, o., for 2 weeks; for detailed information please see the article), conc.: 2.74 pmol $AFLM_1$/mg DNA* (mean value), country: USA[228], *after 15 days (thereof 14 days with AF-administration)

incidence: 3?/3*, sa. const.: rainbow trouts, weight: 5–6 g b. wt., contamination: artificial (dose: 80 ppb AFB_1 (labeled) in the diet for 2 days; for detailed information please see the article), conc.: ≈0.035 pmol $AFLM_1$/g liver** (mean value), country: USA[546], *control, **after AFB_1-administration
incidence: 3?/3, sa. const.: rainbow trouts, wt.: 5–6 g, contamination: artificial (dose: **pretreatment with CHL (4,000 ppm)** for 2 weeks then 80 ppb AFB_1 (labeled) + **CHL (4,000 ppm)** in the diet for 2 days; for detailed information please see the article), conc.: ≈0.027 pmol $AFLM_1$/g liver* (mean value), country: USA[546], *after AFB_1- and CHL-administration

incidence: 3?/3*, sa. const.: rainbow trouts, wt.: 5–6 g, contamination: artificial (80 ppb AFB_1 (labeled) in the diet for 2 days; for detailed information please see the article), conc.: ≈0.029 pmol $AFLM_1$-g/g liver** (mean value), country: USA[546], *control, **after AFB_1-administration
incidence: 3?/3, sa. const.: rainbow trouts, wt.: 5–6 g, contamination: artificial (dose: **pretreatment with CHL (4,000 ppm)** for 2 weeks then 80 ppb AFB_1 (labeled) + **CHL (4,000 ppm)** in the diet for 2 days; for detailed information please see the article), conc.: ≈0.017 pmol $AFLM_1$-g/g liver* (mean value), country: USA[546], *after AFB_1- and CHL-administration

Aflatoxin B_1
incidence: 1/1, sa. const.: Shasta strain rainbow trouts, wt.: 38.4 g, contamination: artificial (dose: 80 ppb AFB_1 (labeled), o., for 3 weeks), conc.: 29.0 pmol AFB_1/mg DNA*, country: USA[115], *after 3 weeks, compare with **Fish, coho salmon liver** AFB_1, 1st statement of no.[115]

incidence: 3?/3, sa. const.: Shasta strain rainbow trouts, contamination: artificial (dose: 28.1 µg AFB_1 (labeled)/kg, i.p., once; for detailed information please see the article), conc.: 74.0 pmol AFB_1/mg DNA* **(mean value), country: USA[115], *after 2 days, **AFB_1-N^7-Gua[3] adducts,
compare with **Fish, coho salmon liver** AFB_1, 2nd statement of no.[115]
incidence: ?/3 sa. const.: Shasta strain rainbow trouts, contamination: artificial (dose: 28.1 µg AFB_1 (labeled)/kg, i.p., once; for detailed information please see the article), conc.: 17.6 pmol AFB_1/mg DNA* **(mean value), country: USA[115], *after 21 days, **AFB_1-N^7-Gua[3] adducts, compare with **Fish, coho salmon liver** AFB_1, 3rd statement of no.[115]

incidence: 3?/3, sa. const.: Shasta strain rainbow trouts, contamination: artificial (dose: 28.1 μg AFB_1 (labeled)/kg, i.p., once; for detailed information please see the article), conc.: 7.7 pmol AFB_1/mg DNA* **(mean value), country: USA[115], *after 2 days, **AFB_1-FAPyr adducts, compare with **Fish, coho salmon liver** AFB_1, 4th statement of no.[115]
incidence: 3?/3, sa. const.: Shasta strain rainbow trouts, contamination: artificial (dose: 28.1 μg AFB_1 (labeled)/kg, i.p., once; for detailed information please see the article), conc.: 14.1 pmol AFB_1/mg DNA* **(mean values), country: USA[115], *after 21 days, **AFB_1-FAPyr adducts,
compare with **Fish, coho salmon liver** AFB_1, 5th statement of no.[115]
incidence: 3?/3, sa. const.: Shasta strain rainbow trouts, contamination: artificial (dose: 28.1 μg AFB_1 (labeled)/kg, i.p., once; for detailed information please see the article), conc.: 5.8 pmol AFB_1/mg DNA* **(mean value), country: USA[115], *after 2 days, **AFB_1-minor FAPyr isomer adducts,
compare with **Fish, coho salmon liver** AFB_1, 6th statement of no.[115]
incidence: 3?/3, sa. const.: Shasta strain rainbow trouts, contamination: artificial (dose: 28.1 μg AFB_1 (labeled)/kg, i.p., once; for detailed information please see the article), conc.: 2.6 pmol AFB_1/mg DNA* **(mean value), country: USA[115], *after 21 days, **AFB_1-minor FAPyr isomer adducts,
compare with **Fish, coho salmon liver** AFB_1, 7th statement of no.[115]

incidence: ?/?, sa. const.: Mt. Shasta strain rainbow trouts, age: 13 months, wt.: 74 g, contamination: artificial (dose: 50 μg AFB_1 (labeled)/kg b. wt., i.p., once; for detailed information please see the article), conc.: 243.0 pmol AFB_1/mg DNA* (mean value), country: USA[177], *after 24 h, compare with **Fish, coho salmon liver** AFB_1 no.[177]

incidence: 3?/3, sa. const.: Mt. Shasta strain rainbow trouts, age: 10 months, weight: ≈45 g, contamination: artificial (dose: 83 μg AFB_1 (labeled)/kg b. wt., once), conc. range: ≤263.0 pmol AFB_1 binding/mg DNA* (mean value), country: USA[177], *after 24 h (also measured after 4, 12 and 48 h but lower residue values recorded), for overall information please see the article

incidence: 3?/3, sa. const.: Mt. Shasta strain rainbow trouts, age: 11 months, wt.: ≈53 g, contamination: artificial (dose: **5 μg AFB_1** (labeled)/kg b. wt., once), conc.: 9.23 pmol AFB_1 binding/mg DNA* (mean value), country: USA[177], *after 24 h (for overall information please see the article)
incidence: 3?/3, sa. const.: Mt. Shasta strain rainbow trouts, age: 11 months, wt.: ≈53 g, contamination: artificial (dose: **25 μg AFB_1** (labeled)/kg b. wt., once), conc.: 55.6 pmol AFB_1 binding/mg DNA* (mean value), country: USA[177], *after 24 h (for overall information please see the article)
incidence: 3?/3, sa. const.: Mt. Shasta strain rainbow trouts, age: 11 months, wt.: ≈53 g, contamination: artificial (dose: **100 μg AFB_1** (labeled)/kg b. wt., once), conc.: 315 pmol AFB_1 binding/mg DNA* (mean value), country: USA[177], *after 24 h (for overall information please see the article)
incidence: 3?/3, sa. const.: Mt. Shasta strain rainbow trouts, age: 11 months, wt.: ≈53 g, contamination: artificial (dose: **300 μg AFB_1** (labeled)/kg b. wt., once), conc.: 1,240 pmol AFB_1 binding/mg DNA* (mean value), country: USA[177], *after 24 h (for overall information please see the article)

incidence: 3?/3, sa. const.: Mt. Shasta strain rainbow trouts, age: 11 months, wt.: 58–71 g, contamination: artificial (dose: **40% semipurified protein diets (casein)** for 5 months and afterwards 50 μg AFB_1 (labeled)/kg b. wt., i.p., once; for detailed information please see the article), conc.:

258 pmol AFB_1 binding/mg DNA* (mean value), country: USA[177], *after 24 h
incidence: 3?/3, sa. const.: Mt. Shasta strain rainbow trouts, age: 11 months, wt.: 58–71 g, contamination: artificial (dose: **50% semipurified protein diets (casein)** for 5 months and afterwards 50 µg AFB_1 (labeled)/kg b. wt., i.p., once; for detailed information please see the article), conc.: 253 pmol AFB_1 binding/mg DNA* (mean value), country: USA[177], *after 24 h
incidence: 3?/3, sa. const.: Mt. Shasta strain rainbow trouts, age: 11 months, wt.: 58–71 g, contamination: artificial (dose: **60% semipurified protein diets (casein)** for 5 months and afterwards 50 µg AFB_1 (labeled)/kg b. wt., i.p., once; for detailed information please see the article), conc.: 329 pmol AFB_1 binding/mg DNA* (mean value), country: USA[177], *after 24 h
incidence: 3?/3, sa. const.: Mt. Shasta strain rainbow trouts, age: 11 months, wt.: 58–71 g, contamination: artificial (dose: **70% semipurified protein diets (casein)** for 5 months and afterwards 50 µg AFB_1 (labeled)/kg b. wt., i.p., once; for detailed information please see the article), conc.: 299 pmol AFB_1 binding/mg DNA* (mean value), country: USA[177], *after 24 h
incidence: 3?/3, sa. const.: Mt. Shasta strain rainbow trouts, age: 11 months, wt.: 58–71 g, contamination: artificial (dose: **40% semipurified protein diets (FPC)** for 5 months and afterwards 50 µg AFB_1 (labeled)/kg b. wt., i.p., once; for detailed information please see the article), conc.: 228 pmol AFB_1 binding/mg DNA* ** (mean value), country: USA[177], *after 24 h
incidence: 3?/3, sa. const.: Mt. Shasta strain rainbow trouts, age: 11 months, wt.: 58–71 g, contamination: artificial (dose: **50% semipurified protein diets (FPC)** for 5 months and afterwards 50 µg AFB_1 (labeled)/kg b. wt., i.p., once; for detailed information please see the article), conc.: 190 pmol AFB_1 binding/mg DNA* ** (mean value), country: USA[177], *after 24 h
incidence: 3?/3, sa. const.: Mt. Shasta strain rainbow trouts, age: 11 months, wt.: 58–71 g, contamination: artificial (dose: **60% semipurified protein diets (FPC)** for 5 months and afterwards 50 µg AFB_1 (labeled)/kg b. wt., i.p., once; for detailed information please see the article), conc.: 309 pmol AFB_1 binding/mg DNA* ** (mean value), country: USA[177], *after 24 h
incidence: 3?/3, sa. const.: Mt. Shasta strain rainbow trouts, age: 11 months, wt.: 58–71 g, contamination: artificial (dose: **70% semipurified protein diets (FPC)** for 5 months and afterwards 50 µg AFB_1 (labeled)/kg b. wt., i.p., once; for detailed information please see the article), conc.: 189 pmol AFB_1 binding/mg DNA* ** (mean value), country: USA[177], *after 24 h

incidence: 3?/3*, sa. const.: Mt. Shasta strain rainbow trouts, age: 14 months, wt.: 95–145 g, contamination: artificial (dose: 50 µg AFB_1 (labeled)/kg b. wt., i.p., once, protein source **(40% casein diet)**; for detailed information please see the article), conc.: 2.72 nmol AFB_1/g liver ** (mean value), country: USA[177], *control (compare with 40% casein and 100 ppm CPFA-treated trouts), **after 24 h
incidence: ?/?, sa. const.: Mt. Shasta strain rainbow trouts, age: 14 months, wt.: 95–145 g, contamination: artificial (dose: 50 µg AFB_1 (labeled)/kg b. wt., i.p., once, protein source **(40% casein diet + 100 ppm CPFA)**; for detailed information please see the article), conc.: 2.47 nmol AFB_1/g liver* (mean value), country: USA[177], *after 24 h
incidence: ?/?*, sa. const.: Mt. Shasta strain rainbow trouts, age: 14 months, wt.: 95–145 g, contamination: artificial (dose: 50 µg AFB_1 (labeled)/kg b. wt., i.p., once, protein source **(40% casein diet)**; for detailed information please see the article), conc.: 275.0 pmol AFB_1 binding/mg DNA** (mean value), country: USA[177], *control (compare with 40% casein and 100 ppm CPFA-treated trouts), **after 24 h
incidence: ?/?, sa. const.: Mt. Shasta strain rainbow trouts, age: 14 months, wt.: 95–145 g, contamination: artificial

(dose: 50 μg AFB_1 (labeled)/kg b. wt., i.p., once, protein source **(40% casein diet + 100 ppm CPFA)**; for detailed information please see the article), conc.: 219.0 pmol AFB_1 binding/mg DNA* (mean value), country: USA[177], *after 24 h
incidence: ?/?*, sa. const.: Mt. Shasta strain rainbow trouts, age: 14 months, wt.: 95–145 g, contamination: artificial (dose: 50 μg AFB_1 (labeled)/kg b. wt., i.p., once, protein source **(70% casein diet)**; for detailed information please see the article), conc.: 2.57 nmol AFB_1/g liver** (mean value), country: USA[177], *control (compare with 70% casein and 100 ppm CPFA-treated trouts), **after 24 h
incidence: ?/?, sa. const.: Mt. Shasta strain rainbow trouts, age: 14 months, wt.: 95–145 g, contamination: artificial (dose: 50 μg AFB_1 (labeled)/kg b. wt., i.p., once, protein source **(70% casein diet + 100 ppm CPFA)**; for detailed information please see the article), conc.: 2.78 nmol AFB_1/g liver* ** (mean value), country: USA[177], *after 24 h
incidence: ?/?*, sa. const.: Mt. Shasta strain rainbow trouts, age: 14 months, wt.: 95–145 g, contamination: artificial (dose: 50 μg AFB_1 (labeled)/kg b. wt., i.p., once, protein source **(70% casein diet)**; for detailed information please see the article), conc.: 283.0 pmol AFB_1 binding/mg DNA** (mean value), country: USA[177], *control (compare with 70% casein and 100 ppm CPFA-treated trouts), **after 24 h
incidence: ?/?, sa. const.: Mt. Shasta strain rainbow trouts, age: 14 months, wt.: 95–145 g, contamination: artificial (dose: 50 μg AFB_1 (labeled)/kg b. wt., i.p., once, protein source **(70% casein diet + 100 ppm CPFA)**; for detailed information please see the article), conc.: 246.0 pmol AFB_1 binding/mg DNA* (mean value), country: USA[177], *after 24 h

incidence: ?/?*, sa. const.: Mt. Shasta strain rainbow trouts, age: 14 months, wt.: 95–145 g, contamination: artificial (dose: 50 μg AFB_1 (labeled)/kg b. wt., i.p., once, protein source **(40% FPC diet)**; for detailed information please see the article), conc.: 2.30 nmol AFB_1/g liver* ** (mean value), country: USA[177], *control (compare with 40% FPC and 100 ppm CPFA-treated trouts), **after 24 h
incidence: ?/?, sa. const.: Mt. Shasta strain rainbow trouts, age: 14 months, wt.: 95–145 g, contamination: artificial (dose: 50 μg AFB_1 (labeled)/kg b. wt., i.p., once, protein source **(40% FPC diet + 100 ppm CPFA)**; for detailed information please see the article), conc.: 2.60 nmol AFB_1/g liver* (mean value), country: USA[177], *after 24 h
incidence: ?/?*, sa. const.: Mt. Shasta strain rainbow trouts, age: 14 months, wt.: 95–145 g, contamination: artificial (dose: 50 μg AFB_1 (labeled)/kg b. wt., i.p., once, protein source **(40% FPC diet)**; for detailed information please see the article), conc.: 270.0 pmol AFB_1 binding/mg DNA* ** (mean value), country: USA[177], *control (compare with 40% FPC and 100 ppm CPFA-treated trouts), **after 24 h
incidence: ?/?, sa. const.: Mt. Shasta strain rainbow trouts, age: 14 months, wt.: 95–145 g, contamination: artificial (dose: 50 μg AFB_1 (labeled)/kg b. wt., i.p., once, protein source **(40% FPC diet + 100 ppm CPFA)**; for detailed information please see the article), conc.: 260.0 pmol AFB_1 binding/mg DNA* (mean value), country: USA[177], *after 24 h
incidence: ?/?*, sa. const.: Mt. Shasta strain rainbow trouts, age: 14 months, wt.: 95–145 g, contamination: artificial (dose: 50 μg AFB_1 (labeled)/kg b. wt., i.p., once, protein source **(70% FPC diet)**; for detailed information please see the article), conc.: 1.83 nmol AFB_1/g liver* ** (mean value), country: USA[177], *control (compare with 70% FPC and 100 ppm CPFA-treated trouts), **after 24 h
incidence: ?/?, sa. const.: Mt. Shasta strain rainbow trouts, age: 14 months, wt.: 95–145 g, contamination: artificial (dose: 50 μg AFB_1 (labeled)/kg b. wt., i.p., once, protein source **(70% FPC diet + 100 ppm**

CPFA); for detailed information please see the article), conc.: 1.99 nmol AFB_1/g liver* (mean value), country: USA[177], *after 24 h
incidence: ?/?*, sa. const.: Mt. Shasta strain rainbow trouts, age: 14 months, wt.: 95–145 g, contamination: artificial (dose: 50 µg AFB_1 (labeled)/kg b. wt., i.p., once, protein source **(70% FPC diet)**; for detailed information please see the article), conc.: 210.0 pmol AFB_1 binding/mg DNA* ** (mean value), country: USA[177], *control (compare with 70% FPC and 100 ppm CPFA-treated trouts), **after 24 h
incidence: ?/?, sa. const.: Mt. Shasta strain rainbow trouts, age: 14 months, wt.: 95–145 g, contamination: artificial (dose: 50 µg AFB_1 (labeled)/kg b. wt., i.p., once, protein source **(70% FPC diet + 100 ppm CPFA)**; for detailed information please see the article), conc.: 189.0 pmol AFB_1 binding/mg DNA* ** ***(mean value), country: USA[177], *after 24 h

incidence: ?/?*, sa. const.: Mt. Shasta strain rainbow trouts, age: 14 months, wt.: 80–125 g, contamination: artificial (dose: fasted 4 days then 50 µg AFB_1 (labeled)/kg b. wt., i.p., once; for detailed information please see the article), conc.: 2.09 nmol AFB_1/g liver** (mean value), country: USA[177], *control (compare with BNF-treated trouts), **after 24 h
incidence: ?/?, sa. const.: Mt. Shasta strain rainbow trouts, age: 14 months, wt.: 80–125 g, contamination: artificial (dose: **pretreatment** with **100 mg BNF**/kg b. wt., i.p., once afterwards 4 days fasting then 50 µg AFB_1 (labeled)/kg b. wt., i.p., once; for detailed information please see the article), conc.: 0.941 nmol AFB_1/g liver* (mean value), country: USA[177], *after 24 h

incidence: ?/?*, sa. const.: Mt. Shasta strain rainbow trouts, age: 14 months, wt.: 80–125 g, contamination: artificial (dose: fasted 4 days then 50 µg AFB_1 (labeled)/kg b. wt., i.p., once; for detailed information please see the article), conc.: 162.0 pmol AFB_1 binding/mg DNA** (mean value), country: USA[177], *control (compare with BNF-treated trouts), **after 24 h
incidence: ?/?, sa. const.: Mt. Shasta strain rainbow trouts, age: 14 months, wt.: 80–125 g, contamination: artificial (dose: **pretreatment** with **100 mg BNF**/kg b. wt., i.p., once afterwards 4 days fasting then 50 µg AFB_1 (labeled)/kg b. wt., i.p., once; for detailed information please see the article), conc.: 96.6 pmol AFB_1 binding/mg DNA* (mean value), country: USA[177], *after 24 h

incidence: ?/?*, sa. const.: Mt. Shasta strain rainbow trouts, wt.: ≈400 g, contamination: artificial (dose: 25 µl containing 0.32 µmol AFB_1 (labeled)/ml, i.p., once; for detailed information please see the article), conc. range: 46**–139*** pmol/g liver (mean values), country: USA[212], *control, after 24** and 5 h*** (also measured after 10 and 16 h but values between values of 5 and 24 h)
incidence: ?/?, sa. const.: Mt. Shasta strain rainbow trouts, wt.: ≈400 g, contamination: artificial (dose: 25 µl containing 0.32 µmol AFB_1 (labeled)/ml, i.p., once and **I3C-diet (0.2%)** for 12 weeks prior to AFB_1 injection; for detailed information please see the article), conc. range: 33*–104** pmol/g liver (mean values), country: USA[212], after 24* and 5 h** (also measured after 10 and 16 h but values between values of 5 and 24 h)

incidence: 3?/3*, sa. const.: Mt. Shasta strain rainbow trouts, wt.: ≈400 g, contamination: artificial (dose: 25 µl containing 0.32 µmol AFB_1 (labeled)/ml, i.p., once; for detailed information please see the article), conc. range: ≈≤9.5 nmol AFB_1/g liver** (mean value), country: USA[212], *control, **after 5 h post-injection (also measured after 10, 16 and 24 h, lowest conc.: ≈6.1 nmol AFB_1/g liver after 24 h)
incidence: 3?/3, sa. const.: Mt. Shasta strain rainbow trouts, wt.: ≈400 g, contamination: artificial (dose: 25 µl

containing 0.32 µmol AFB_1 (labeled)/ml, i.p., once and **I3C-diet (0.2%)** for 12 weeks prior to AFB_1 injection; for detailed information please see the article), conc. range: ≈≤6.3 nmol AFB_1/g liver* (mean value), country: USA[212], *after 5 h post-injection (also measured after 10, 16 and 24 h, lowest conc.: ≈3.8 nmol AFB_1/g liver after 24 h)

incidence: 3?/3*, sa. const.: Mt. Shasta strain rainbow trouts, wt.: ≈400 g, contamination: artificial (dose: 25 µl containing 0.32 µmol AFB_1 (labeled)/ml, i.p., once; for detailed information please see the article), conc. range: ≈≤29.5 nmol AFB_1/g DNA** *** (mean value), country: USA[212], *control, **after 5 h post-injection (also measured after 10, 16 and 24 h, lowest conc.: ≈24 nmol AFB_1/g DNA after 10 h), ***short-term binding of AFB_1
incidence: 3?/3, sa. const.: Mt. Shasta strain rainbow trouts, wt.: ≈400 g, contamination: artificial (dose: 25 µl containing 0.32 µmol AFB_1 (labeled)/ml, i.p., once and **I3C-diet (0.2%)** for 12 weeks prior to AFB_1 injection; for detailed information please see the article), conc. range: ≈≤14 nmol AFB_1/g DNA* ** (mean value), country: USA[212], *after 16 h post-injection (also measured after 5, 10, and 24 h, lowest conc.: ≈10.5 nmol AFB_1/g DNA after 5 h), **short-term binding of AFB_1

incidence: 5–8?/5–8*, sa. const.: Mt. Shasta strain rainbow trouts, wt.: ≈400 g, contamination: artificial (dose: 2 days fasted then 0.1 µmol AFB_1 (labeled)/kg b. wt., i.p., once; for detailed information please see the article), conc. range: ≈≤82 nmol AFB_1/g DNA** *** (mean value), country: USA[212], *control, **after 1 day post-injection (also measured after 2, 7 and 21 days, lowest conc.: ≈17 nmol AFB_1/g DNA after 21 days), ***long-term binding of AFB_1
incidence: 5–8?/5–8, sa. const.: Mt. Shasta strain rainbow trouts, wt.: ≈400 g, contamination: artificial (dose: 0.1 µmol AFB_1 (labeled)/kg b. wt., i.p., once and **I3C-diet (0.2%)** for 3 weeks and 2 days fasted prior to AFB_1 injection; for detailed information please see the article), conc. range: ≈≤20 nmol AFB_1/g DNA* ** (mean value), country: USA[212], *after 1 day post-injection (also measured after 2, 7 and 21 days, lowest conc.: ≈9.5 nmol AFB_1/g DNA after 21 days), **long-term binding of AFB_1

incidence: ?/?*, sa. const.: Shasta strain rainbow trouts, age: 9 months, contamination: artificial (dose: 2.47 µg AFB_1 (labeled), i.p. injection, once; for detailed information please see the article), conc. range: 46**–139*** pmol/**g liver** (mean values), country: USA[213], *control, after 24** and 5 h*** (also measured after 10 and 16 h but values between values of 5 and 24 h)
incidence: ?/?, sa. const.: Shasta strain rainbow trouts, age: 9 months, contamination: artificial (dose: **PCB (100 ppm)** for 2 month prior to 2.47 µg AFB_1 (labeled), i.p. injection, once; for detailed information please see the article), conc. range: 33*–119** pmol/**g liver** (mean values), country: USA[213], after 24* and 5 h** (also measured after 10 and 16 h but values between values of 5 and 24 h)

incidence: 3?/3*, sa. const.: Shasta strain rainbow trouts, age: 9 months, contamination: artificial (dose: 2.47 µg AFB_1 (labeled), i.p. injection, once; for detailed information please see the article), conc. range: 27.9**–30.0*** nmol AFB_1-derived adducts/**g DNA**** *** (mean values), country: USA[213], *control, after 24** and 5 h*** (also measured after 10 and 16 h but values below or between values of 5 and 24 h, lowest conc.: 23.9 nmol AFB_1-derived adducts/g DNA after 10 h), ***short-term AFB_1-DNA adducts
incidence: 3?/3, sa. const.: Shasta strain rainbow trouts, age: 9 months, contamination: artificial (dose: **PCB (100 ppm)** for 2 month prior to 2.47 µg AFB_1 (labeled), i.p. injection, once; for detailed information please see the article), conc. range: 35.3*–36.3** nmol

AFB_1-derived adducts/**g DNA***** (mean values), country: USA[213], after 5* and 24** h (also measured after 10 and 16 h but values below values of 5 and 24 h, lowest conc.: 30.8 nmol AFB_1-derived adducts/g DNA after 16 h), ***short-term AFB_1-DNA adducts

incidence: ?/?*, sa. const.: Shasta strain rainbow trouts, age: 9 months, contamination: artificial (dose: 2.0 μg AFB_1 (labeled), i.p. injection, once; for detailed information please see the article), conc. range: ≈≤36 nmol AFB_1**/g hepatic DNA**** *** (mean value), country: USA[213], *control, **after 1 day (also measured after 0.5, 2, 7 and 21 days, lowest conc.: ≈16 nmol AFB_1/g hepatic DNA after 21 days), ***long-term AFB_1-DNA adducts
incidence: ?/?, sa. const.: Shasta strain rainbow trouts, age: 9 months, contamination: artificial (dose: **PCB (100 ppm)** for 3 month prior to 2.0 μg AFB_1 (labeled), i.p. injection, once; for detailed information please see the article), conc. range: ≈≤25 nmol AFB_1**/g hepatic DNA*** ** (mean value), country: USA[213], *after 0.5 day (also measured after 1, 2, 7 and 21 days, lowest conc.: ≈9 nmol AFB_1/g hepatic DNA after 21 days), **long-term AFB_1-DNA adducts

incidence: 9?/9*, sa. const.: Shasta strain rainbow trouts, age: 9 months, contamination: artificial (dose: 2.47 μg AFB_1 (labeled), i.p. injection, once; for detailed information please see the article), conc. range: ≈≤9.5 nmol AFB_1**/g liver**** (mean value), country: USA[213], *control, **after 5 h post-injection (also measured after 10, 16 and 24 h, lowest conc.: ≈6.1 nmol AFB_1/g liver after 24 h)
incidence: 9?/9, sa. const.: Shasta strain rainbow trouts, age: 9 months, contamination: artificial (dose: **PCB (100 ppm)** for 2 month prior to 2.47 μg AFB_1 (labeled), i.p. injection, once; for detailed information please see the article), conc. range: ≈≤11.3 nmol AFB_1**/g liver*** (mean value), country: USA[213], *after 5 h post-injection (also measured after 10, 16 and 24 h, lowest conc.: ≈6.8 nmol AFB_1/g liver after 24 h)

incidence: 6?/6, sa. const.: Shasta strain rainbow trouts, age: 3 months, contamination: artificial (dose: 168.7 pmol AF (labeled)/g diet, o., for 2 weeks; for detailed information please see the article), conc.: 3.82 pmol AFB_1/mg DNA* (mean value), country: USA[228], *after 15 days (thereof 14 days with AF-administration)

incidence: 9?/9*, sa. const.: Mt Shasta strain rainbow trouts, wt.: 50 g, contamination: artificial (dose: 25 μl containing 0.32 μmol AFB_1 (labeled)/ml, i.p., once; for detailed information please see the article), conc. range: 46**–139*** pmol AFB_1/g liver (mean values), country: USA[260], *control, after 24** and 5 h*** (also measured after 10 and 16 h but values between values of 5 and 24 h)
incidence: 9?/9, sa. const.: Mt Shasta strain rainbow trouts, wt.: 50 g, contamination: artificial (dose: 25 μl containing 0.32 μmol AFB_1 (labeled)/ml, i.p., once and **BNF-diet (0.05%)** for 12 weeks prior to AFB_1-injection; for detailed information please see the article), conc. range: 11*–90** pmol AFB_1/g liver (mean values), country: USA[260], after 24* and 5 h** (also measured after 10 and 16 h but values between values of 5 and 24 h)

incidence: ?/3*, sa. const.: Mt Shasta strain rainbow trouts, wt.: ≈50 g, contamination: artificial (dose: 25 μl containing 0.32 μmol AFB_1 (labeled)/ml, i.p., once; for detailed information please see the article), conc. range: ≈≤9.5 nmol AFB_1/g liver** (mean value), country: USA[260], *control, after 5 h** (also measured after 10, 16 and 24 h, lowest conc.: ≈6.2 nmol AFB_1/g liver after 24 h)
incidence: ?/3, sa. const.: Mt Shasta strain rainbow trouts, wt.: ≈50 g, contamination: artificial (dose: 25 μl containing 0.32 μmol AFB_1 (labeled)/ml, i.p., once

and **BNF-diet (0.05%)** for 12 weeks prior to AFB_1-injection; for detailed information please see the article), conc. range: ≈≤8 nmol AFB_1/g liver* (mean value), country: USA[260], after 5 h* (also measured after 10, 16 and 24 h, lowest conc.: ≈3.7 nmol AFB_1/g liver after 24 h)

incidence: ?/3*, sa. const.: Mt Shasta strain rainbow trouts, contamination: artificial (dose: 64 nmol AFB_1 (labeled), i.p., once; for detailed information please see the article), conc. range: ≈≤60 nmol AFB_1/g hepatic DNA** (mean value), country: USA[260], *control, **after 48 h (also measured after 6, 24 and 96 h, lowest conc.: ≈43 nmol AFB_1/g hepatic DNA after 0.2 days)
incidence: ?/3, sa. const.: Mt Shasta strain rainbow trouts, contamination: artificial (dose: 64 nmol AFB_1 (labeled), i.p., once and **BNF-diet (0.5%)** for 3 weeks prior to AFB_1-injection, i.p., once; for detailed information please see the article), conc. range: ≈≤27 nmol AFB_1/g hepatic DNA* (mean value), country: USA[260], **after 24 h (also measured after 6, 48 and 96 h, lowest conc.: ≈20 nmol AFB_1/g hepatic DNA after 2 days)

incidence: 6?/6*, sa. const.: Mt Shasta strain rainbow trouts, contamination: artificial (dose: 180 nmol AFB_1 (labeled), i.p., once; for detailed information please see the article), conc. range: ≈≤145 nmol AFB_1/g hepatic DNA** (mean value), country: USA[260], *control, **after 1 day (also measured after 0.5, 2 and 7 days, lowest conc.: ≈75 nmol AFB_1/g hepatic DNA after 7 days)
incidence: 6?/6, sa. const.: Mt Shasta strain rainbow trouts, contamination: artificial (dose: 180 nmol AFB_1 (labeled), i.p., once and **BHA-diet (0.3%)** for 3 weeks prior to AFB_1-injection; for detailed information please see the article), conc. range: ≈≤195 nmol AFB_1/g hepatic DNA* (mean value), country: USA[260], *after 0.5 days (also measured after 1, 2 and 7 days, lowest conc.: ≈85 nmol AFB_1/g hepatic DNA after 7 days)

incidence: ?/3*, sa. const.: fingerling rainbow trouts, contamination: artificial (dose: **0 ppm I3C** in the diet for 6 weeks + 10, 20, 40 or 80** ppb AFB_1 (labeled and unlabeled) in the diet for the last 2 weeks; for detailed information please see the article), conc. range: ≈≤840** pg AFB_1/mg DNA*** (mean value), country: USA[313], *control, ***after 7 days of AFB_1-administration
incidence: ?/3, sa. const.: fingerling rainbow trouts, contamination: artificial (dose: **I3C (1,000 ppm)** in the diet for 6 weeks + 10, 20, 40 or 80* ppb AFB_1 (labeled and unlabeled) in the diet for the last 2 weeks; for detailed information please see the article), conc. range: ≈≤490* pg AFB_1/mg DNA** (mean value), country: USA[313], **after 7 days of AFB_1-administration
incidence: ?/3, sa. const.: fingerling rainbow trouts, contamination: artificial (dose: **I3C (2,000 ppm)** in the diet for 6 weeks + 20, 40, 80 or 160* ppb AFB_1 (labeled and unlabeled) in the diet for the last 2 weeks; for detailed information please see the article), conc. range: ≈≤310* pg AFB_1/mg DNA** (mean value), country: USA[313], **after 7 days of AFB_1-administration
incidence: ?/3, sa. const.: fingerling rainbow trouts, contamination: artificial (dose: **I3C (3,000 ppm)** in the diet for 6 weeks + 20, 40, 80 or 160* ppb AFB_1 (labeled and unlabeled) in the diet for the last 2 weeks; for detailed information please see the article), conc. range: ≈≤180* pg AFB_1/mg DNA** (mean value), country: USA[313], **after 7 days of AFB_1-administration
incidence: ?/3, sa. const.: fingerling rainbow trouts, contamination: artificial (dose: **I3C (4,000 ppm)** in the diet for 6 weeks + 40, 80, 160 or 320* ppb AFB_1 (labeled and unlabeled) in the diet for the last 2 weeks; for detailed information please see the article), conc. range: ≈≤215* pg AFB_1/mg DNA** (mean value), country: USA[313], **after 7 days of AFB_1-administration

incidence: ?/3*, sa. const.: fingerling rainbow trouts, contamination: artificial (dose: **0 ppm I3C**) in the diet for 6 weeks + 10, 20, 40 or 80** ppb AFB_1 (labeled and unlabeled) in the diet for the last 2 weeks; for detailed information please see the article), conc. range: ≈≤1,750** pg AFB_1/mg DNA*** (mean value), country: USA[313], *control, ***after 14 days of AFB_1-administration
incidence: ?/3, sa. const.: fingerling rainbow trouts, contamination: artificial (dose: **I3C (1,000 ppm)** in the diet for 6 weeks + 10, 20, 40 or 80* ppb AFB_1 (labeled and unlabeled) in the diet for the last 2 weeks; for detailed information please see the article), conc. range: ≈≤750* pg AFB_1/mg DNA** (mean value), country: USA[313], **after 14 days of AFB_1-administration
incidence: ?/3, sa. const.: fingerling rainbow trouts, contamination: artificial (dose: **I3C (2,000 ppm)** in the diet for 6 weeks + 20, 40, 80* or 160 ppb AFB_1 (labeled and unlabeled) in the diet for the last 2 weeks; for detailed information please see the article); for detailed information please see the article), conc. range: ≈≤300* pg AFB_1/mg DNA** (mean value), country: USA[313], **after 14 days of AFB_1-administration
incidence: ?/3, sa. const.: fingerling rainbow trouts, contamination: artificial (dose: **I3C (3,000 ppm)** in the diet for 6 weeks + 20, 40, 80 or 160* ppb AFB_1 (labeled and unlabeled) in the diet for the last 2 weeks; for detailed information please see the article); for detailed information please see the article), conc. range: ≈≤430* pg AFB_1/mg DNA** (mean value), country: USA[313], **after 14 days of AFB_1-administration
incidence: ?/3, sa. const.: fingerling rainbow trouts, contamination: artificial (dose: **I3C (4,000 ppm)** in the diet for 6 weeks + 40, 80, 160 or 320* ppb AFB_1 (labeled and unlabeled) in the diet for the last 2 weeks; for detailed information please see the article); for detailed information please see the article), conc. range: ≈≤450* pg AFB_1/mg DNA** (mean value), country: USA[313], **after 14 days of AFB_1-administration

incidence: ?/?, sa. const.: Shasta strain rainbow trouts, age: 2 months, Ø wt.: 1.5 g, contamination: no AFB_1; for detailed information please see the article), conc.: nr, country: USA[384]
incidence: ?/4 or 8, sa. const.: Shasta strain rainbow trouts, age: 2 months, Ø wt.: 1.5 g, contamination: artificial (dose: 12.4, 25, 50 and 75* ng AFB_1 (labeled)/g dry weight diet, for 2 weeks; for detailed information please see the article), conc. range: ≈≤5.8* pmol AFB_1/mg DNA**, country: USA[384], **after 2 weeks

incidence: ?/3, sa. const.: Mt. Shasta strain rainbow trouts, age: 11 months, wt.: ≈66 g, contamination: artificial (dose: 0.675 µg? AFB_1 (labeled)/g b. wt., i.p., once; for detailed information please see the article), conc. range: ≤40.4 pmol/mg DNA* ** (mean value), country: USA[405], *after 2 days (also measured after 1, 4, and 7 days, lowest conc.: 31.4 pmol/mg DNA after 1 day), **[^3H]AFB_1-conc.

incidence: ?/?*, sa. const.: fingerling rainbow trouts (9,490), wt.: 1.5–2.0 g, contamination: artificial (dose: 10, 20, 40, 80 or 160** ppb AFB_1 (labeled and unlabeled) and no CHL for 2 weeks; for detailed information please see the article), conc. range: ≈≤28 µmol AFB_1/mole DNA** *** (mean value), country: USA[536], *control, ***after 2 weeks
incidence: ?/?, sa. const.: fingerling rainbow trouts (9,490), wt.: 1.5–2.0 g, contamination: artificial (dose: 10, 20, 40 or 80* ppb AFB_1 (labeled and unlabeled) and **CHL (500 ppm)** added to the diet for 2 weeks; for detailed information please see the article), conc. range: ≈≤10.8 µmol AFB_1/mole DNA* ** (mean value), country: USA[536], **after 2 weeks
incidence: ?/?, sa. const.: fingerling rainbow trouts (9,490), wt.: 1.5–2.0 g, contamination: artificial (dose: 10, 20, 40,

80 or 160* ppb AFB_1 (labeled and unlabeled) and **CHL (2,000 ppm)** added to the diet for 2 weeks; for detailed information please see the article), conc. range: ≈≤19.5* μmol AFB_1/mole DNA** (mean value), country: USA[536], **after 2 weeks
incidence: ?/?, sa. const.: fingerling rainbow trouts (9,490), wt.: 1.5–2.0 g, contamination: artificial (dose: 10, 20, 40, 80 or 160* ppb AFB_1 (labeled and unlabeled) and **CHL (4,000 ppm)** added to the diet for 2 weeks; for detailed information please see the article), conc. range: ≈≤17.5 * μmol AFB_1/ mole DNA** (mean value), country: USA[536], **after 2 weeks

incidence: 3?/3*, sa. const.: rainbow trouts, wt.: 5–6 g, contamination: artificial (dose: 80 ppb AFB_1 (labeled) in the diet for 2 days; for detailed information please see the article), conc.: ≈2.25 pmol AFB_1/g liver** (mean value), country: USA[546], *control, **after AFB_1-administration
incidence: 3?/3, sa. const.: rainbow trouts, weight: 5–6 g, contamination: artificial (dose: **pretreatment CHL (4,000 ppm)** for 2 weeks then 80 ppb AFB_1 (labeled) + **CHL (4,000 ppm)** in the diet for another 2 days; for detailed information please see the article), conc.: ≈1.25 pmol AFB_1/g liver* (mean value), country: USA[546], *after AFB_1- and CHL-administration

incidence: 5/5, sa. const.: fingerling fishes (rainbow trout), wt.: ≈0.7 g, contamination: neither AFB_1- nor CHL-treatment (for detailed information please see the article), conc.: nd, country: USA[546]
incidence: 5/5, sa. const.: fingerling fishes (rainbow trout), wt.: ≈0.7 g b. wt., contamination: artificial (dose: CHL (500 ppm) for 30 min, no pre- and posttreatment; for detailed information please see the article), conc.: nd, country: USA[546]
incidence: 5?/5, sa. const.: fingerling fishes (rainbow trout), wt.: ≈0.7 g, contamination: artificial (dose: **pretreatment CHL (4,000 ppm)** for 7 days then bath treatment: 0.1 ppm AFB_1 (labeled in part) + **CHL (500 ppm)** for 30 min, **posttreatment: 4,000 ppm** CHL for another 2 days; for detailed information please see the article), conc.: 7.9 pg AFB_1/mg DNA* ** (mean value), country: USA[546], *AFB_1-DNA adducts, **within 1 h after bath treatment
incidence: 5?/5, sa. const.: fingerling fishes (rainbow trout), wt.: ≈0.7 g, contamination: artificial (dose: bath treatment: 0.1 ppm AFB_1 (labeled in part) + **CHL (500 ppm)** for 30 min, no pre- and posttreatment; for detailed information please see the article), conc.: 7.7 pg AFB_1/mg DNA* ** (mean value), country: USA[546], *AFB_1-DNA adducts, **within 1 h after bath treatment
incidence: 5?/5, sa. const.: fingerling fishes (rainbow trout), wt.: ≈0.7 g, contamination: artificial (dose: **pretreatment CHL (4,000 ppm)** for 7 days then bath treatment: 0.1 ppm AFB_1 (labeled in part) for 30 min, **posttreatment: CHL (4,000 ppm)** for another 2 days; for detailed information please see the article), conc.: 211 pg AFB_1/mg DNA* ** (mean value), country: USA[546], *AFB_1-DNA adducts, **within 1 h after bath treatment
incidence: 5?/5, sa. const.: fingerling fishes (rainbow trout), wt.: ≈0.7 g, contamination: artificial (dose: bath treatment: 0.1 ppm AFB_1 (labeled in part) for 30 min, no pre- and posttreatment; for detailed information please see the article), conc.: 194 pg AFB_1/mg DNA* ** (mean value), country: USA[546], *AFB_1-DNA adducts, **within 1 h after bath treatment

incidence: ?/5*, sa. const.: fingerling trouts, age: 3–5 months, wt.: 8–12 g, contamination: artificial (dose: no pretreatment with BNF (0.2–10 ppm) for 7 days, on 8th day 10 μg AFB_1 (labeled)/kg, i.p. injection, once; for detailed information please see the

article), conc.: ≈7.1 pmol AFB_1/mg DNA** (mean value), country: USA[558], *control, **after 24 h
incidence: ?/5, sa. const.: fingerling trouts, age: 3–5 months, wt.: 8–12 g, contamination: artificial (dose: **pretreatment** with 0.2, 0.5, 1.0, 2.0, **5.0*** or 10 ppm **BNF** for 7 days, on 8th day 10 μg AFB_1 (labeled)/kg, i.p. injection, once; for detailed information please see the article), conc.: ≈2.4 pmol AFB_1/mg DNA* ** (mean value), country: USA[558], **lowest value of all treatments after 24 h (also all other BNF-treatment values lower than the control)

incidence: ?/7*, sa. const.: fingerling trouts, age: 3–5 months, wt.: 3–4 g, contamination: artificial (dose: no pretreatment with BNF (10–200 ppm) for 7 days, on 8th day 10 μg AFB_1 (labeled)/kg, i.p. injection, once; for detailed information please see the article), conc.: ≈9.0 pmol AFB_1/mg DNA** (mean value), country: USA[558], *control, **after 24 h
incidence: ?/7, sa. const.: fingerling trouts, age: 3–5 months, wt.: 3–4 g, contamination: artificial (dose: **pretreatment** with 10, 20, 50, 100 or **200* ppm BNF** for 7 days, on 8th day 10 μg AFB_1 (labeled)/kg, i.p. injection, once; for detailed information please see the article), conc.: ≈3.7 pmol AFB_1/mg DNA* ** (mean value), country: USA[558], **lowest value of all treatments after 24 h (also all other BNF-treatment values lower than the control)

incidence: ?/7*, sa. const.: fingerling trouts, age: 3–5 months, wt.: 8–12 g, contamination: artificial (dose: no pretreatment with BNF (100–700 ppm) for 7 days, on 8th day 10 μg AFB_1 (labeled)/kg, i.p. injection, once; for detailed information please see the article), conc.: ≈19.8 pmol AFB_1/mg DNA** (mean value), country: USA[558], *control, **after 24 h
incidence: ?/7, sa. const.: fingerling trouts, age: 3–5 months, wt.: 8–12 g, contamination: artificial (dose: **pretreatment** with 100, 200, 300, 500 or **700* ppm BNF** for 7 days, on 8th day 10 μg AFB_1 (labeled)/kg, i.p. injection, once; for detailed information please see the article), conc.: ≈3.9 pmol AFB_1/mg DNA* ** (mean value), country: USA[558], **lowest value of all treatments after 24 h (also all other BNF-treatment values lower than the control)

incidence: ?/5*, sa. const.: fingerling trouts, wt.: 1–2 g, contamination: artificial (dose: 20 ppb AFB_1 (labeled) in the diet for 1, 3, 5 or 7 days and no CHL; for detailed information please see the article), conc. range: ≈≤685 pg AFB_1/mg DNA** (mean value), country: USA[560], *control (AFB_1 administered up to 7 days), **after 7 days of feeding AFB_1 (also measured after 1, 3 and 5 days)
incidence: ?/5, sa. const.: fingerling trouts, wt.: 1–2 g, contamination: artificial (dose: 20 ppb AFB_1 (labeled) in the diet for 1, 3, 5 or 7 days and **CHL (500 ppm)** co-administered to AFB_1; for detailed information please see the article), conc.: ≈≤475 pg AFB_1/mg DNA* (mean value), country: USA[560], *after 7 days of feeding AFB_1 and CHL (also measured after 1, 3 and 5 days)
incidence: ?/5, sa. const.: fingerling trouts, wt.: 1–2 g, contamination: artificial (dose: 20 ppb AFB_1 (labeled) in the diet for 1, 3, 5 or 7 days and **CHL (1,000 ppm)** co-administered to AFB_1; for detailed information please see the article), conc.: ≈≤340 pg AFB_1/mg DNA* (mean value), country: USA[560], *after 7 days of feeding AFB_1 and CHL (also measured after 1, 3 and 5 days)
incidence: ?/5, sa. const.: fingerling trouts, wt.: 1–2 g, contamination: artificial (dose: 20 ppb AFB_1 (labeled) in the diet for 1, 3, 5 or 7 days and **CHL (2,000 ppm)** co-administered to AFB_1; for detailed information please see the article), conc.: ≈≤200 pg AFB_1/mg DNA* (mean value), country: USA[560], *after 7 days of

feeding AFB_1 and CHL (also measured after 1, 3 and 5 days)

Aflatoxin M_1

incidence: ?/?*, sa. const.: Mt. Shasta strain rainbow trouts, wt.: ≈400 g, contamination: artificial (dose: 25 µl containing 0.32 µmol AFB_1 (labeled)/ml, i.p., once; for detailed information please see the article), conc. range: 0**–4*** pmol/g liver (mean value), country: USA[212], *control, after 24** and 5 h*** (also measured after 10 and 16 h but values between values of 5 and 24 h)
incidence: ?/?, sa. const.: Mt. Shasta strain rainbow trouts, wt.: ≈400 g, contamination: artificial (dose: 25 µl containing 0.32 µmol AFB_1 (labeled)/ml, i.p., once and **I3C (0.2%) diet** for 12 weeks prior to AFB_1 injection; for detailed information please see the article), conc. range: 1*–4** pmol/g liver (mean value), country: USA[212], after 24* and 5 h** (also measured after 10 and 16 h but values between values of 5 and 24 h)

incidence: ?/?*, sa. const.: Shasta strain rainbow trouts, age: 9 months, contamination: artificial (dose: 2.47 µg AFB_1 (labeled), i.p. injection, once; for detailed information please see the article), conc. range: 0**–4*** pmol/g liver (mean values), country: USA[213], *control, after 24** and 5 h*** (also measured after 10 and 16 h but values between values of 5 and 24 h)
incidence: ?/?, sa. const.: Shasta strain rainbow trouts, age: 9 months, contamination: artificial (dose: **100 ppm PCB** for 2 month prior to 2.47 µg AFB_1 (labeled), i.p. injection, once; for detailed information please see the article), conc. range: 2*–21** pmol/g liver (mean values), country: USA[213], after 24* and 5 h** (also measured after 10 and 16 h but values between values of 5 and 24 h)

incidence: 6?/6, sa. const.: Shasta strain rainbow trouts, age: 3 months, contamination: artificial (dose: 2,442 pmol AF (labeled)/g diet, o., for 2 weeks; for detailed information please see the article), conc.: 8.07 pmol AFM_1/mg DNA* (mean value), country: USA[228], *after 15 days (thereof 14 days with AF-administration)
incidence: 6?/6, sa. const.: Shasta strain rainbow trouts, age: 3 months, contamination: artificial (dose: 812.6 pmol AF (labeled)/g diet, o., for 2 weeks; for detailed information please see the article), conc.: 3.16 pmol AFM_1/mg DNA* (mean value), country: USA[228], *after 15 days (thereof 14 days with AF-administration)

incidence: 9?/9*, sa. const.: Mt Shasta strain rainbow trouts, wt.: ≈50 g, contamination: artificial (dose: 25 µl containing 0.32 µmol AFB_1 (labeled)/ml, i.p., once; for detailed information please see the article), conc. range: 0**–4*** pmol AFM_1/g liver (mean values), country: USA[260], *control, after 24** and 5 h*** (also measured after 10 and 16 h but values between values of 5 and 24 h)
incidence: 9?/9, sa. const.: Mt Shasta strain rainbow trouts, wt.: ≈50 g, contamination: artificial (dose: 25 µl containing 0.32 µmol AFB_1 (labeled)/ml, i.p., once and **BNF-diet (0.05%)** for 12 weeks prior to AFB_1-injection; for detailed information please see the article), conc. range: 1*–16** pmol AFM_1/g liver (mean values), country: USA[260], after 24 h* and 5 h** (also measured after 10 and 16 h but values between values of 5 and 24 h)

incidence: 3?/3*, sa. const.: rainbow trouts, wt.: 5–6 g b. wt., contamination: artificial (dose: 80 ppb AFB_1 (labeled) in the diet for another 2 days; for detailed information please see the article), conc. range: ≈0.034 pmol AFM_1/g liver **(mean value), country: USA[546], *control, *after AFB_1-administration
incidence: 3?/3, sa. const.: rainbow trouts, wt.: 5–6 g b. wt., contamination: artificial

(dose: **pretreatment CHL (4,000 ppm)** for 2 weeks then 80 ppb AFB_1 (labeled) + **CHL (4,000 ppm)** in the diet for another 2 days; for detailed information please see the article), conc. range: ≈0.017 pmol AFM_1/g liver* (mean value), country: USA[546], *after AFB_1- and CHL-administration

Fish, rainbow trout plasma may contain the following mycotoxins and/or their metabolites:

Aflatoxicol
incidence: ?/36*, sa. const.: Mt Shasta strain rainbow trouts, age: 9 months, contamination: artificial (dose: 25 µl containing 0.32 µmol AFB_1 (labeled)/ml, i.p. injection; for detailed information please see the article), conc. range: 0.2**–2.2*** nM in plasma (mean values), country: USA[212], *control, after 24** and 5 h*** (also measured after 10 and 16 h but values between values of 5 and 24 h)
incidence: ?/36, sa. const.: Mt Shasta strain rainbow trouts, age: 9 months, contamination: artificial (dose: **I3C-diet (0.2%)** for 12 weeks prior to 25 µl containing 0.32 µmol AFB_1 (labeled)/ml, i.p. injection; for detailed information please see the article), conc. range: 0.1*–1.6** nM in plasma (mean values), country: USA[212], after 24* and 5 h** (also measured after 10 and 16 h but values between values of 5 and 24 h)

incidence: ?/?*, sa. const.: Shasta strain rainbow trouts, contamination: artificial (dose: 2. 47 µg AFB_1 (labeled), i.p. injection, once; for detailed information please see the article), conc.: 2.2 nmol** in plasma (mean value), country: USA[213], *control, **5 h post-injection
incidence: ?/?, sa. const.: Shasta strain rainbow trouts, contamination: artificial (dose: **100 ppm PCB** for 2 months prior to 2.47 µg AFB_1 (labeled), i.p. injection, once; for detailed information please see the article), conc.: 1.8 nmol* in plasma (mean value), country: USA[213], *5 h post-injection

Aflatoxin B_1
incidence: ?/36*, sa. const.: Mt Shasta strain rainbow trouts, age: 9 months, contamination: artificial (dose: 25 µl containing 0.32 µmol AFB_1 (labeled)/ml, i.p. Injection, once; for detailed information please see the article), conc. range: 3.8**–19.7*** nM AFB_1 in plasma (mean values), country: USA[212], *control, after 24** and 5 h*** (also measured after 10 and 16 h but values between values of 5 and 24 h)
incidence: ?/36, sa. const.: Mt Shasta strain rainbow trouts, age: 9 months, contamination: artificial (dose: **I3C-diet (0.2%)** for 12 weeks prior to 25 µl containing 0.32 µmol AFB_1 (labeled)/ml, i.p. injection, once; for detailed information please see the article), conc. range: 2.7*–16.9** nM AFB_1 in plasma (mean values), country: USA[212], after 24* and 5 h** (also measured after 10 and 16 h but values between values of 5 and 24 h)

incidence: ?/?*, sa. const.: Shasta strain rainbow trouts, contamination: artificial (dose: 2.47 µg AFB_1 (labeled), i.p. injection, once; for detailed information please see the article), conc.: 19.7 nmol** in plasma (mean value), country: USA[213], *control, **5 h post-injection
incidence: ?/?, sa. const.: Shasta strain rainbow trouts, contamination: artificial (dose: **PCB (100 ppm)** for 2 month prior to 2.47 µg AFB_1 (labeled), i.p. injection, once; for detailed information please see the article), conc.: 11.0 nmol* in plasma (mean value), country: USA[213], *5 h post-injection

incidence: ?/?*, sa. const.: Mt Shasta strain rainbow trouts, wt.: ≈50 g, contamination: artificial (dose: 25 µl containing 0.32 µmol AFB_1 (labeled)/ml, i.p., once; for detailed information please see the article), conc. range: 3.8**–19.7*** nM

AFB_1 in plasma**** (mean values), country: USA[260], *control, after 24** and 5 h*** (also measured after 10 and 16 h but values between values of 5 and 24 h), ****values represent unmetabolized AFB_1 only
incidence: ?/?, sa. const.: Mt Shasta strain rainbow trouts, wt.: ≈50 g, contamination: artificial (dose: 25 µl containing 0.32 µmol AFB_1 (labeled)/ml, i.p., once and **BNF-diet (0.05%)** for 12 weeks prior to AFB_1-injection; for detailed information please see the article), conc. range: 0.6*–5.2** nM AFB_1 in plasma***(mean values), country: USA[260], after 24* and 5 h** (also measured after 10 and 16 h but values between values of 5 and 24 h), ***values represent unmetabolized AFB_1 only

Aflatoxin M_1
incidence: ?/36*, sa. const.: Mt Shasta strain rainbow trouts, age: 9 months, contamination: artificial (dose: 25 µl containing 0.32 µmol AFB_1 (labeled)/ml, i.p. injection, once; for detailed information please see the article), conc. range: 0.6**–1.0*** nM in plasma (mean values), country: USA[212], *control, after 5** and 24 h*** (also measured after 10 and 16 h but values between or higher values of 5 and 24 h, highest conc.: 1.1 nM in plasma after 16 h)
incidence: ?/36, sa. const.: Mt Shasta strain rainbow trouts, age: 9 months, contamination: artificial (dose: **I3C-diet (0.2%)** for 12 weeks prior to 25 µl containing 0.32 µmol AFB_1 (labeled)/ml, i.p. injection, once; for detailed information please see the article), conc. range: 0.2*–4.2** nM in plasma (mean values), country: USA[212], after 24* and 5 h** (also measured after 10 and 16 h but values between values of 5 and 24 h)

incidence: ?/?*, sa. const.: Shasta strain rainbow trouts, contamination: artificial (dose: 2.47 µg AFB_1 (labeled), i.p. injection, once; for detailed information please see the article), conc.: 0.6 nmol** in plasma (mean value), country: USA[213], *control, **5 h post-injection
incidence: ?/?, sa. const.: Shasta strain rainbow trouts, contamination: artificial (dose: **100 ppm PCB** for 2 month prior to 2.47 µg AFB_1 (labeled), i.p. injection, once; for detailed information please see the article), conc.: 6.1 nmol* in plasma (mean value), country: USA[213], *5 h post-injection

Fish, rainbow trout red blood cells may contain the following mycotoxins and/or their metabolites:

Aflatoxin B_1
incidence: ?/36*, sa. const.: Mt Shasta strain rainbow trouts, age: 9 months, contamination: artificial (dose: 25 µl containing 0.32 µmol AFB_1 (labeled)/ml, i.p. injection, once; for detailed information please see the article), conc. range: 0.705**–0.910*** nmol AFB_1/g RBC DNA (mean values), country: USA[212], *control, after 5 h** and 24 h*** (also measured after 10 and 16 h but values below values of 5 and 24 h, lowest conc.: 0.679 nmol AFB_1/g RBC DNA after 16 h)
incidence: ?/36, sa. const.: Mt Shasta strain rainbow trouts, age: 9 months, contamination: artificial (dose: **I3C-diet (0.2%)** for 12 weeks prior to 25 µl containing 0.32 µmol AFB_1 (labeled)/ml, i.p. injection, once; for detailed information please see the article), conc. range: 0.247*–0.481** nmol AFB_1/g RBC DNA** (mean values), country: USA[212], after 5* and 24 h** (also measured after 10 and 16 h but values between values of 5 and 24 h)

incidence: 1/1*, sa. const.: Shasta strain rainbow trouts, age: 9 months, contamination: artificial (dose: 2.47 µg AFB_1 (labeled), i.p. injection, once; for detailed information please see the article), conc. range: 0.708**–0.908*** nmol

AFB_1-derived adducts/g DNA**** (mean values), country: USA[213], *control, after 5** and 24*** h (also measured after 10 and 16 h but values below values of 5 and 24 h, lowest conc.: 0.682 nmol AFB_1-derived adducts/g DNA after 16 h), ****short-term AFB_1-DNA adducts
incidence: 1/1, sa. const.: Shasta strain rainbow trouts, age: 9 months, contamination: artificial (dose: **PCB (100 ppm)** for 2 month prior to 2.47 μg AFB_1 (labeled), i.p. injection, once; for detailed information please see the article), conc. range: 0.641*–0.895** nmol AFB_1-derived adducts/g DNA*** (mean values), country: USA[213], after 5* and 24** h (also measured after 10 and 16 h but values between and below values of 5 and 24 h, lowest conc.: 0.632 nmol AFB_1-derived adducts/g DNA after 16 h), ***short-term AFB_1-DNA adducts

incidence: ?/?*, sa. const.: Mt Shasta strain rainbow trouts, wt.: ≈50 g, contamination: artificial (dose: 25 μl containing 0.32 μmol AFB_1 (labeled)/ml, i.p., once; for detailed information please see the article), conc. range: 0.705**–0.910*** nM AFB_1/g RBC DNA**** (mean values), country: USA[260], *control, after 5** and 24 h*** (also measured after 10 and 16 h but values below values of 5 and 24 h), ****total AFB_1 and metabolites bound to DNA
incidence: ?/?, sa. const.: Mt Shasta strain rainbow trouts, wt.: ≈50 g, contamination: artificial (dose: 25 μl containing 0.32 μmol AFB_1 (labeled)/ml, i.p., once and **BNF-diet (0.05%)** for 12 weeks prior to AFB_1-injection; for detailed information please see the article), conc. range: 0.352*–0.515** nM AFB_1/g RBC DNA**** (mean values), country: USA[260], after 24* and 5 h** (also measured after 10 and 16 h but values between values of 5 and 24 h), ****total AFB_1 and metabolites bound to DNA

Fish, Rainbow Trout Shell see Fish, rainbow trout egg

Fish, Rainbow Trout Yolk see Fish, rainbow trout egg

Fish, Walleye Fish

Fish, walleye fish muscle may contain the following mycotoxins and/or their metabolites:

Aflatoxin B_1
incidence: 12/12*, sa. const.: walleyes, length: 290 mm, contamination: no AF, conc.: nd, country: USA[94], *control, for overall information please see the article
incidence: ?/6, sa. const.: walleyes, length: 290 mm, contamination: artificial (dose: **50 ppb AF**, o., for 30 days), conc.: 5 ppb* (mean value), country: USA[94], *after 30 days (for overall information please see the article)
incidence: ?/6, sa. const.: walleyes, length: 290 mm, contamination: artificial (dose: **100 ppb AF**, o., for 30 days), conc.: 10 ppb* (mean value), country: USA[94], *after 30 days (for overall information please see the article)

Aflatoxin G_1
incidence: 12/12*, sa. const.: walleyes, length: 290 mm, contamination: no AF, conc.: nd, country: USA[94], *control, for overall information please see the article
incidence: ?/6, sa. const.: walleyes, length: 290 mm, contamination: artificial (dose: **50 ppb AF**, o., for 30 days), conc.: 15 ppb* (mean value), country: USA[94], *after 30 days (for overall information please see the article)
incidence: ?/6, sa. const.: walleyes, length: 290 mm, contamination: artificial (dose: **100 ppb AF**, o., for 30 days), conc.: 20 ppb* (mean value), country: USA[94], *after 30 days (for overall information please see the article)

Aflatoxin G_2
incidence: 12/12*, sa. const.: walleyes, length: 290 mm, contamination: no AF, conc.: nd, country: USA[94], *control, for overall information please see the article

incidence: ?/6, sa. const.: walleyes, length: 290 mm, contamination: artificial (dose: **50 ppb AF**, o., for 30 days), conc.: 15 ppb* (mean value), country: USA[94], *after 30 days (for overall information please see the article)
incidence: ?/6, sa. const.: walleyes, length: 290 mm, contamination: artificial (dose: **100 ppb AF**, o., for 30 days), conc.: 20 ppb* (mean value), country: USA[94], *after 30 days (for overall information please see the article)

Fish, Zebrafish

Fish, zebrafish liver may contain the following mycotoxins and/or their metabolites:

AFLATOXIN B_1
incidence: 3/3*, sa. const.: **female** zebrafishes, age: adult, wt.: 0.5 g, contamination: no AFB_1 (for detailed information please see the article), conc.: nd, country: USA[626], *control
incidence: 3?/3, sa. const.: **female** zebrafishes, age: adult, wt.: 0.5 g, contamination: artificial (dose: 50, 100, 200 or 400* μg AFB_1 (labeled)/kg b. wt., i.p., once; for detailed information please see the article), conc. range: ≈≤235 pmol AFB_1/mg DNA* ** (mean value), country: USA[626], **after 24 h
incidence: 3?/3*, sa. const.: **male** zebrafishes, age: adult, wt.: 0.5 g, contamination: no AFB_1 (for detailed information please see the article), conc.: nd, country: USA[626], *control
incidence: 3?/3, sa. const.: **male** zebrafishes, age: adult, wt.: 0.5 g, contamination: artificial (dose: 50, 100, 200 or 400* μg AFB_1 (labeled)/kg b. wt., i.p., once; for detailed information please see the article), conc. range: ≈≤145 pmol AFB_1/mg DNA* ** (mean value), country: USA[626], **after 24 h
incidence: 3?/3, sa. const.: **female** zebrafishes, age: adult, wt.: 0.5 g, contamination: artificial (dose: 400 μg AFB_1 (labeled)/kg b. wt., i.p., once; for detailed information please see the article), conc. range: ≈≤400 pmol AFB_1/mg DNA* (mean value), country: USA[626], *after 7 days (also measured after 1, 2, 4, 14 and 21 days, lowest conc.: 240 pmol AFB_1/mg DNA after 4 days)
incidence: 3?/3, sa. const.: **male** zebrafishes, age: adult, wt.: 0.5 g, contamination: artificial (dose: 400 μg AFB_1 (labeled)/kg b. wt., i.p., once; for detailed information please see the article), conc. range: ≈≤230 pmol AFB_1/mg DNA* ** (mean value), country: USA[626], *after 7 days (also measured after 1, 2, 4, 14 and 21 days, lowest conc.: 75 pmol AFB_1/mg DNA after 1 day)

Fowlsee Turkey

Goat

Goat Natural Contamination

Goat milk, raw may contain the following mycotoxins and/or their metabolites:

AFLATOXIN M_1
incidence: 4/10*, sa. const.: milk from goats of Greece, contamination: natural, conc. range: 11–20 ng/l (2 sa), 21–50 ng/l (2 sa), country: Greece[246], *from December 1999 to May 2000
incidence: 8/12*, milk from goats of Greece, contamination: natural, conc. range: 5–10 ng/l (7 sa), 11–20 ng/l (1 sa), country: Greece[246], *from December 2000 to May 2001

Goat urine may contain the following mycotoxins and/or their metabolites:

ZEARALANOLS
incidence: 9/27*, sa. const.: urine from goats of New Zealand, contamination: natural, conc. range: ≤0.56 ng/ml**, country: New Zealand[230], *export animals, **most probable of *Fusarium* origin

Zearalenols
incidence: 9/27*, sa. const.: urine from goats of New Zealand, contamination: natural, conc. range: ≤19 ng/ml, country: New Zealand[230], *export animals

Goat Artificial Contamination

Goat lung may contain the following mycotoxins and/or their metabolites:

Macrocyclic Trichothecene
incidence: 6/6*, sa. const.: female and male Spanish-Boer cross-bred weanling goats, age: 10–16 weeks, contamination: no exposure to spore suspension of *Stachybotrys chartarum* (for detailed information please see the article), conc.: nr, country: USA[587], *control
incidence: 6?/6, sa. const.: female and male Spanish-Boer cross-bred weanling goats, age: 10–16 weeks, contamination: artificial (dose: six times exposure to dust and fluid of spore suspension of *Stachybotrys chartarum*, **SC1 group**; for detailed information please see the article), conc.: 34.9 ng/g* (mean value), country: USA[587], *after 72 h of last treatment
incidence: 6/6*, sa. const.: female and male Spanish-Boer cross-bred weanling goats, age: 10–16 weeks, contamination: no exposure to spore suspension of *Stachybotrys chartarum* (for detailed information please see the article), conc.: nr, country: USA[587], *control
incidence: 6?/6, sa. const.: female and male Spanish-Boer cross-bred weanling goats, age: 10–16 weeks, contamination: artificial (dose: single exposure to dust and fluid of spore suspension of *Stachybotrys chartarum*, **SC2 group**; for detailed information please see the article), conc.: 158.4 ng/g* (mean value), country: USA[587], *after 24 h of single treatment

Goat Lymph Nodes see Goat lymph

Goat lymph may contain the following mycotoxins and/or their metabolites:

Macrocyclic Trichothecene
incidence: 6/6*, sa. const.: female and male Spanish-Boer cross-bred weanling goats, age: 10–16 weeks, contamination: no exposure to spore suspension of *Stachybotrys chartarum* (for detailed information please see the article), conc.: nr, country: USA[587], *control
incidence: 6?/6, sa. const.: female and male Spanish-Boer cross-bred weanling goats, age: 10–16 weeks, contamination: artificial (dose: six times exposure to dust and fluid of spore suspension of *Stachybotrys chartarum*, **SC1 group**; for detailed information please see the article), conc.: 35.2 ng/g* ** (mean value), country: USA[587], *after 72 h of last treatment, **in lymph nodes
incidence: 6/6*, sa. const.: female and male Spanish-Boer cross-bred weanling goats, age: 10–16 weeks, contamination: no exposure to spore suspension of *Stachybotrys chartarum* (for detailed information please see the article), conc.: nr, country: USA[587], *control
incidence: 6?/6, sa. const.: female and male Spanish-Boer cross-bred weanling goats, age: 10–16 weeks, contamination: artificial (dose: single exposure to dust and fluid of spore suspension of *Stachybotrys chartarum*, **SC2 group**; for detailed information please see the article), conc.: 344.8 ng/g* ** (mean value), country: USA[587], *after 24 h of single treatment, **in lymph nodes

Goat milk, raw may contain the following mycotoxins and/or their metabolites:

Aflatoxin M_1
incidence: 8?/8*, sa. const.: lactating Alpine goats, contamination: neither AF nor HSCAS; for detailed information

please see the article), conc. range: ≤0.037 ppb** (mean value), country: USA[282], *control, **after 2 days (also measured after 4, 6 and 8 days, lowest conc.: 0.005 ppb after 8 days)
incidence: 8/8, sa. const.: lactating Alpine goats, contamination: no AF but **HSCAS (4%)** addition, o., for 8 days? (for detailed information please see the article), conc.: nd, country: USA[282] (measured after 2, 4, 6 and 8 days)
incidence: 8?/8, sa. const.: lactating Alpine goats, contamination: artificial (dose: **200 ppb AF** but no HSCAS addition, o., for 8 days? for detailed information please see the article), conc. range: ≤1.619 ppb* (mean value), country: USA[282], *after 6 days of AF-administration (also measured after 2, 4 and 8 days, lowest conc.: 1.046 ppb after 4 days)
incidence: 8?/8, sa. const.: lactating Alpine goats, contamination: artificial (dose: **200 ppb AF** and additionally **HSCAS (4%)** addition, o., for 8 days?; for detailed information please see the article), conc. range: ≤0.189 ppb* (mean value), country: USA[282], *after 2 days of AF-administration (also measured after 4, 6 and 8 days, lowest conc.: 0.163 ppb after 4 days)

incidence: 3?/3*, sa. const.: lactating Alpine goats, contamination: neither AF nor HSCAS (for detailed information please see the article), conc. range: ≤0.022 ppb** (mean value), country: USA[282], *control, **after 10 days of AF-administration (also measured after 2, 4, 6 and 12 days, lowest conc.: nd after 2 and 4 days)
incidence: 3/3, sa. const.: lactating Alpine goats, contamination: no AF but **HSCAS (1%)** addition, o., for 12 days? (for detailed information please see the article), conc.: nd, country: USA[282] (measured after 2, 4, 6, 10 and 12 days)
incidence: 3/3, sa. const.: lactating Alpine goats, contamination: no AF but **HSCAS (2%)** addition (for detailed information please see the article), conc.: nd, country: USA[282] (measured after 2, 4, 6, 10 and 12 days)
incidence: 3?/3, sa. const.: lactating Alpine goats, contamination: artificial (dose: **100 ppb AF** but no HSCAS addition, o., for 12 days?; for detailed information please see the article), conc. range: ≤0.649 ppb*, country: USA[282], *after 2 days of AF-administration (also measured after 4, 6, 10 and 12 days, lowest conc.: 0.481 ppb after 6 days)
incidence: 3?/3, sa. const.: lactating Alpine goats, contamination: artificial (dose: **100 ppb AF** and **additionally HSCAS (1%)** addition, o., for 12 days?; for detailed information please see the article), conc. range: ≤0.332 ppb, country: USA[282], *after 10 days of AF-administration (also measured after 2, 4, 6 and 12 days, lowest conc.: 0.198 ppb after 4 days)
incidence: 3?/3, sa. const.: lactating Alpine goats, contamination: artificial (dose: **100 ppb AF** and additionally **HSCAS (2%)** addition, o., for 12 days?; for detailed information please see the article), conc. range: ≤0.111 ppb, country: USA[282], *after 4 days of AF-administration (also measured after 2, 6, 10 and 12 days, lowest conc.: 0.098 ppb after 12 days)

incidence: 1?/1?, sa. const.: milk from a goat, age: 3 years, wt.: 20 kg, contamination: artificial (dose: up to 3.254 mg* AFB_1/kg ration, o., at day 1* (three different doses applicated over 8 days); for detailed information please see the article), conc.: ≤0.281 μg/ml*, country: Egypt[575], *at the first day (also at other day intervals up to 15 days measured, lowest conc.: nd after 15 days)

Goat serum may contain the following mycotoxins and/or their metabolites:

Macrocyclic trichothecene
incidence: ?/?*, sa. const.: female and male Spanish-Boer cross-bred weanling goats, age: 10–16 weeks, contamination: no exposure to spore suspension of *Stachybotrys chartarum* (for detailed

information please see the article), conc.: nr, country: USA[587], *control
incidence: 3/6, sa. const.: female and male Spanish-Boer cross-bred weanling goats, age: 10–16 weeks, contamination: artificial (dose: single exposure to dust and fluid of spore suspension of *Stachybotrys chartarum*, **SC1 group**; for detailed information please see the article), conc.: 1.69* ng/ml (mean value), country: USA[587], *after 24 h (also at other hour intervals up to 24 h measured)
incidence: ?/?*, sa. const.: female and male Spanish-Boer cross-bred weanling goats, age: 10–16 weeks, contamination: no exposure to spore suspension of *Stachybotrys chartarum* (for detailed information please see the article), conc.: nr, country: USA[587], *control
incidence: 6/6, sa. const.: female and male Spanish-Boer cross-bred weanling goats, age: 10–16 weeks, contamination: artificial (dose: single exposure to dust and fluid of spore suspension of *Stachybotrys chartarum*, **SC2 group**; for detailed information please see the article), conc. range: 2.02* to ≈≤8.3** ng/ml (mean value), country: USA[587], after 24* and 0.25 h** (also at other hour intervals up to 24 h measured)

Goat spleen may contain the following mycotoxins and/or their metabolites:

Macrocyclic trichothecene
incidence: 6/6*, sa. const.: female and male Spanish-Boer cross-bred weanling goats, age: 10–16 weeks, contamination: no exposure to spore suspension of *Stachybotrys chartarum* (for detailed information please see the article), conc.: nr, country: USA[587], *control
incidence: 6?/6, sa. const.: female and male Spanish-Boer cross-bred weanling goats, age: 10–16 weeks, contamination: artificial (dose: six times exposure to dust and fluid of spore suspension of *Stachybotrys chartarum*, **SC1 group**; for detailed information please see the article), conc.: 33.7 ng/g* (mean value), country: USA[587], *after 72 h of last treatment
incidence: 6/6*, sa. const.: female and male Spanish-Boer cross-bred weanling goats, age: 10–16 weeks, contamination: no exposure to spore suspension of *Stachybotrys chartarum* (for detailed information please see the article), conc.: nr, country: USA[587], *control
incidence: 6?/6, sa. const.: female and male Spanish-Boer cross-bred weanling goats, age: 10–16 weeks, contamination: artificial (dose: single exposure to dust and fluid of spore suspension of *Stachybotrys chartarum*, **SC2 group**; for detailed information please see the article), conc.: 147.0 ng/g* (mean value), country: USA[587], *after 24 h of single treatment

Guinea Pig

Guinea Pig Artificial Contamination

Guinea pig adrenal may contain the following mycotoxins and/or their metabolites:

Trichothecenes
incidence: 6?/6, sa. const.: male guinea pigs (Hartley Strain), wt.: 450–500 g, contamination: artificial (dose: 1.04 mg T-2 toxin (labeled)/kg, injected, once; for detailed information please see the article), conc. range: ≤1,276 pmol/mg wet weight* ** (mean value), country: USA[389], *TRICHO mole eq., **after 0.5 h (also at other hour intervals up to 672 h measured, lowest conc.: 23 pmol/mg wet weight after 672 h)

Guinea pig bile may contain the following mycotoxins and/or their metabolites:

Trichothecenes
incidence: 6?/6, sa. const.: male guinea pigs (Hartley Strain), wt.: 450–500 g, contamination: artificial (dose: 1.04 mg T-2

toxin (labeled)/kg, injected, once; for detailed information please see the article), conc. range: ≤535,760 pmol/ml* ** (mean value), country: USA[389], *TRICHO mole eq., **after 12 h (also at other hour intervals up to 672 h measured, lowest conc.: 10 pmol/mg wet weight after 672 h)

Guinea pig brain may contain the following mycotoxins and/or their metabolites:

TRICHOTHECENES
incidence: 6?/6, sa. const.: male guinea pigs (Hartley Strain), wt.: 450–500 g, contamination: artificial (dose: 1.04 mg T-2 toxin (labeled)/kg, injected, once; for detailed information please see the article), conc. range: ≤355 pmol/mg wet weight* ** (mean value), country: USA[389], *TRICHO mole eq., **after 0.5 h (also at other hour intervals up to 672 h measured, lowest conc.: 2 pmol/mg wet weight after 336 and 672 h)

Guinea pig fat may contain the following mycotoxins and/or their metabolites:

TRICHOTHECENES
incidence: 6?/6, sa. const.: male guinea pigs (Hartley Strain), wt.: 450–500 g, contamination: artificial (dose: 1.04 mg T-2 toxin (labeled)/kg, injected, once; for detailed information please see the article), conc. range: ≤1,171 pmol/mg wet weight* ** (mean value), country: USA[389], *TRICHO mole eq., **after 0.5 h (also at other hour intervals up to 672 h measured, lowest conc.: 3 pmol/mg wet weight after 672 h)

Guinea pig heart may contain the following mycotoxins and/or their metabolites:

TRICHOTHECENES
incidence: 6?/6, sa. const.: male guinea pigs (Hartley Strain), wt.: 450–500 g, contamination: artificial (dose: 1.04 mg T-2 toxin (labeled)/kg, injected, once; for detailed information please see the article), conc. range: ≤1,115 pmol/mg wet weight* ** (mean value), country: USA[389], *TRICHO mole eq., **after 0.5 h (also at other hour intervals up to 672 h measured, lowest conc.: 2 pmol/mg wet weight after 672 h)

Guinea pig kidney may contain the following mycotoxins and/or their metabolites:

TRICHOTHECENES
incidence: 6?/6, sa. const.: male guinea pigs (Hartley Strain), wt.: 450–500 g, contamination: artificial (dose: 1.04 mg T-2 toxin (labeled)/kg, injected, once; for detailed information please see the article), conc. range: ≤2,675 pmol/mg wet weight* ** (mean value), country: USA[389], *TRICHO mole eq., **after 0.5 h (also at other hour intervals up to 672 h measured, lowest conc.: 6 pmol/mg wet weight after 672 h)

Guinea pig liver may contain the following mycotoxins and/or their metabolites:

AFLATOXIN B_1
incidence: ?/?*, sa. const.: male Hartley guinea pigs, wt.: 235–450 g, contamination: no AFB_1, conc.: nr, country: France/Japan[24], *control
incidence: ?/?, sa. const.: male Hartley guinea pigs, wt.: 235–450 g, contamination: artificial (dose: **20 µg AFB_1/kg/day**, by gavage, daily for up to 14 days), conc.: 1.40 pmol AFB_1-FAPy/mg DNA* ** (mean value), country: France/Japan[24], *animals killed after 24 h of final treatment, **AFB_1-DNA adducts
incidence: ?/?, sa. const.: male Hartley guinea pigs, wt.: 235–450 g, contamination: artificial (dose: **80 µg AFB_1/kg/day**, by gavage, daily for up to 14 days), conc. range: ≤12.0 pmol AFB_1-FAPy/mg DNA* ** (mean values), country: France/Japan[24],

*animals killed after 24 h of final treatment (also measured after 1, 3 and 7 days, lowest conc.: ≈2.5 pmol AFB_1-FAPy/mg DNA after 1 day), **AFB_1-DNA adducts

TRICHOTHECENES
incidence: 6?/6, sa. const.: male guinea pigs (Hartley Strain), wt.: 450–500 g, contamination: artificial (dose: 1.04 mg T-2 toxin (labeled)/kg, injected, once; for detailed information please see the article), conc. range: ≤1,826 pmol/mg wet weight* ** (mean value), country: USA[389], *TRICHO mole eq., **after 0.5 h (also at other hour intervals up to 672 h measured, lowest conc.: 4 pmol/mg wet weight after 672 h)

Guinea pig lung may contain the following mycotoxins and/or their metabolites:

TRICHOTHECENES
incidence: 6?/6, sa. const.: male guinea pigs (Hartley Strain), wt.: 450–500 g, contamination: artificial (dose: 1.04 mg T-2 toxin (labeled)/kg, injected, once; for detailed information please see the article), conc. range: ≤1,373 pmol/mg wet weight* ** (mean value), country: USA[389], *TRICHO mole eq., **after 0.5 h (also at other hour intervals up to 672 h measured, lowest conc.: 3 pmol/mg wet weight after 672 h)

Guinea pig muscle may contain the following mycotoxins and/or their metabolites:

TRICHOTHECENES
incidence: 6?/6, sa. const.: male guinea pigs (Hartley Strain), wt.: 450–500 g, contamination: artificial (dose: 1.04 mg T-2 toxin (labeled)/kg, injected, once; for detailed information please see the article), conc. range: ≤728 pmol/mg wet weight* ** (mean value), country: USA[389], *TRICHO mole eq., **after 0.5 h (also at other hour intervals up to 672 h measured, lowest conc.: 1 pmol/mg wet weight after 672 h)

Guinea pig plasma may contain the following mycotoxins and/or their metabolites:

AFLATOXIN B_1
incidence: ?/?*, sa. const.: male Hartley guniea pigs, wt.: 235–450 g, contamination: no AFB_1, conc.: nr, country: France/Japan[24], *control
incidence: ?/?, sa. const.: male Hartley guinea pigs, wt.: 235–450 g, contamination: artificial (dose: **20 µg AFB_1/kg/day**, by gavage, daily for up to 14 days), conc. range: ≈≤15 pg AFB_1-lysine eq/mg albumin* ** (mean values), country: France/Japan[24], *animals killed after 24 h of final treatment (also measured after 1, 3 and 7 days, lowest conc.: ≈4 pg AFB_1-lysine eq/mg albumin after 1 day), **AFB_1-albumin adducts
incidence: ?/?, sa. const.: male Hartley guinea pigs, wt.: 235–450 g, contamination: artificial (dose: **80 µg AFB_1/kg/day**, by gavage, daily for up to 14 days), conc. range: ≤77.6 pg AFB_1-lysine eq/mg albumin* ** (mean values), country: France/Japan[24], *animals killed after 24 h of final treatment (also measured after 1, 3 and 7 days, lowest conc.: 10 pg AFB_1-lysine eq/mg albumin after 1 day), **AFB_1-albumin adducts

TRICHOTHECENES
incidence: 6?/6, sa. const.: male guinea pigs (Hartley Strain), wt.: 450–500 g, contamination: artificial (dose: 1.04 mg T-2 toxin (labeled)/kg, injected, once; for detailed information please see the article), conc. range: ≤289 pmol/ml* ** (mean value), country: USA[389], *TRICHO mole eq., **after 12 h (also at other hour intervals up to 672 h measured, lowest conc.: 6 pmol/mg wet weight after 672 h)

Guinea pig spleen may contain the following mycotoxins and/or their metabolites:

TRICHOTHECENES
incidence: 6?/6, sa. const.: male guinea pigs (Hartley Strain), wt.: 450–500 g, contamination: artificial (dose: 1.04 mg T-2 toxin (labeled)/kg, injected, once; for detailed information please see the article), conc. range: ≤1,302 pmol/mg wet weight* ** (mean value), country: USA[389], *TRICHO mole eq., **after 0.5 h (also at other hour intervals up to 672 h measured, lowest conc.: 6 pmol/mg wet weight after 672 h)

Guinea pig testes may contain the following mycotoxins and/or their metabolites:

TRICHOTHECENES
incidence: 6?/6, sa. const.: male guinea pigs (Hartley Strain), wt.: 450–500 g, contamination: artificial (dose: 1.04 mg T-2 toxin (labeled)/kg, injected, once; for detailed information please see the article), conc. range: ≤493 pmol/mg wet weight* ** (mean value), country: USA[389], *TRICHO mole eq., **after 0.5 h (also at other hour intervals up to 672 h measured, lowest conc.: 4 pmol/mg wet weight after 168 and 336 h)

Hamster

Hamster Artificial Contamination

Hamster kidney may contain the following mycotoxins and/or their metabolites:

AFLATOXIN B_1
incidence: ?/8*, sa. const.: male Syrian Golden hamsters, wt.: 70–80 g, contamination: artificial (dose: 40 μg AFB_1 (labeled)/100 g b. wt., i.p., once; for detailed information please see the article), conc.: 1.6 AFB_1-DNA binding pmol/mg DNA** (mean value), country: USA[556], *control, **after 2 h
incidence: ?/8, sa. const.: male Syrian Golden hamsters, weight: 70–80 g, contamination: artificial (dose: 40 μg AFB_1 (labeled)/100 g b. wt., i.p., once and **pretreatment** with **L-BSO**, conc.: 4 mmol, 4 and 2 h before AFB_1 injection); for detailed information please see the article), conc.: 2.3 AFB_1-DNA binding pmol/mg DNA* (mean value), country: USA[556], *after 2 h
incidence: ?/8, sa. const.: male Syrian Golden hamsters, wt.: 70–80 g, contamination: artificial (dose: 40 μg AFB_1 (labeled)/100 g b. wt., i.p., once and **pretreatment** with **DEM** (conc.: 3.5 mmol, 4 h before AFB_1 injection) + **L-BSO**, conc.: 4 mmol, 2 h before AFB_1 injection); for detailed information please see the article), conc.: 2.8 AFB_1-DNA binding pmol/mg DNA* (mean value), country: USA[556], *after 2 h

incidence: 3?/3, sa. const.: male Golden Syrian hamster, wt.: 120–140 g, contamination: artificial (dose: 40 μg AFB_1 (labeled)/100 g, i.p., once, conc. range: ≈≤14 ng AFB_1 bound/mg rRNA* (mean value), country: UK[608], *after 6 h (also measured after 2, 24 and 48 h, lowest conc.: ≈4 ng AFB_1 bound/**mg rRNA** after 48 h)
incidence: 3?/3, sa. const.: male Golden Syrian hamster, wt.: 120–140 g, contamination: artificial (dose: 40 μg AFB_1 (labeled)/100 g, i.p., once), conc. range: ≈≤10 ng AFB_1 bound/**mg DNA*** (mean value), country: UK[608], *after 6 h (also measured after 2, 24 and 48 h, lowest conc.: ≈2 ng AFB_1 bound/mg DNA after 48 h)
incidence: 3?/3, sa. const.: male Golden Syrian hamster, wt.: 120–140 g, contamination: artificial (dose: 40 μg AFB_1 (labeled)/100 g, i.p., once), conc. range: ≈≤0.6 ng AFB_1 bound/**mg protein*** (mean value), country: UK[608], *after 6 h (also measured after 2, 24 and 48 h, lowest conc.: nd after 48 h)

Hamster liver may contain the following mycotoxins and/or their metabolites:

AFLATOXIN B_1
incidence: ?/?*, sa. const.: male Syrian Golden hamsters, wt.: 70–130 g, contamination: no AFB_1, conc.: nr, country: France/Japan[24], *control
incidence: ?/?, sa. const.: male Syrian Golden hamsters, wt.: 70–130 g, contamination: artificial (dose: 20 µg AFB_1/kg/day, by gavage, daily for up to 14 days), conc.: 0.75 pmol AFB_1-FAPy/mg DNA* ** (mean value), country: France/Japan[24], *animals killed after 24 h of final treatment, **AFB_1-DNA adducts
incidence: ?/?, sa. const.: male Syrian Golden hamsters, wt.: 70–130 g, contamination: artificial (dose: **80 µg AFB_1**/kg/day, by gavage, daily for up to 14 days), conc. range: ≤1.39 pmol AFB_1-FAPy/mg DNA* ** (mean values), country: France/Japan[24], *animals killed after 24 h of final treatment (also measured after 1, 3 and 7 days, lowest conc.: ≈0.5 pmol AFB_1-FAPy/mg DNA after 1 day), **AFB_1-DNA adducts

incidence: ?/8*, sa. const.: male Syrian Golden hamsters, wt.: 70–80 g, contamination: artificial (dose: 40 µg AFB_1 (labeled)/100 g b. wt., i.p., once; for detailed information please see the article), conc.: 5.9 AFB_1-DNA binding pmol/mg DNA** (mean value), country: USA[556], *control, **after 2 h
incidence: ?/8, sa. const.: male Syrian Golden hamsters, wt.: 70–80 g, contamination: artificial (dose: 40 µg AFB_1 (labeled)/100 g b. wt., i.p., once and **pretreatment** with **L-BSO**, conc.: 4 mmol, 4 and 2 h before AFB_1 injection); for detailed information please see the article), conc.: 6.5 pmol/mg DNA* (mean value), country: USA[556], *after 2 h
incidence: ?/8, sa. const.: male Syrian Golden hamsters, wt.: 70–80 g, contamination: artificial (dose: 40 µg AFB_1 (labeled)/100 g b. wt., i.p., once and **pretreatment** with **DEM**, conc.: 3.5 mmol, 4 h before AFB_1 injection) + **L-BSO** (conc.: 4 mmol, 2 h before AFB_1 injection), conc.: 10.3 AFB_1-DNA binding pmol/mg DNA* (mean value), country: USA[556], *after 2 h

incidence: 3?/3, sa. const.: male Golden Syrian hamster, wt.: 120–140 g, contamination: artificial (dose: 40 µg AFB_1 (labeled)/100 g, i.p., once, conc. range: ≈≤5.5 ng AFB_1 bound/mg rRNA* (mean value), country: UK[608], *after 2 h (also measured after 6, 24 and 48 h, lowest conc.: ≈2 ng AFB_1 bound/**mg rRNA** after 48 h)
incidence: 3?/3, sa. const.: male Golden Syrian hamster, wt.: 120–140 g, contamination: artificial (dose: 40 µg AFB_1 (labeled)/100 g, i.p., once, conc. range: ≈≤5 ng AFB_1 bound/mg DNA* (mean value), country: UK[608], *after 6 h (also measured after 2, 24 and 48 h, lowest conc.: ≈1 ng AFB_1 bound/**mg DNA** after 48 h)
incidence: 3?/3, sa. const.: male Golden Syrian hamster, wt.: 120–140 g, contamination: artificial (dose: 40 µg AFB_1 (labeled)/100 g, i.p., once, conc. range: ≈≤1 ng AFB_1 bound/**mg protein*** (mean value), country: UK[608], *after 2 h (also measured after 6, 24 and 48 h, lowest conc.: nd after 48 h)

Hamster plasma may contain the following mycotoxins and/or their metabolites:

AFLATOXIN B_1
incidence: ?/?*, sa. const.: male Syrian Golden hamsters, wt.: 70–130 g, contamination: no AFB_1, conc.: nr, country: France/Japan[24], *control
incidence: ?/?, sa. const.: male Syrian Golden hamsters, wt.: 70–130 g, contamination: artificial (dose: **20 µg AFB_1**/kg/day, by gavage, daily for up to 14 days), conc. range: ≈≤5 pg AFB_1-lysine eq/mg albumin* (mean values), country: France/Japan[24], *after 14 days (also measured after 1, 3 and 7 days, lowest conc.: ≈0.5 pg AFB_1-lysine eq/mg albumin after 1 day), **AFB_1-albumin adducts

incidence: ?/?, sa. const.: male Syrian Golden hamsters, wt.: 70–130 g, contamination: artificial (dose: **80 µg** AFB_1/kg/day, by gavage, daily for up to 14 days), conc. range: ≤46 pg AFB_1-lysine eq/mg albumin* (mean values), country: France/Japan[24], *after 14 days (also measured after 1, 3 and 7 days, lowest conc.: ≈2 pg AFB_1-lysine eq/mg albumin after 1 day), **AFB_1-albumin adducts

Hare

Hare Natural Contamination

Hare liver may contain the following mycotoxins and/or their metabolites:

AFLATOXIN B_1
incidence: ?/42, sa. const.: livers from free living hares of Czechoslovakia?, contamination: natural, conc. range: 0.232–1.20 µg/kg, country: Czechoslovakia[9]
incidence: ?/12, sa. const.: livers from hares of Czechoslovakia? living in boxes, contamination: natural, conc. range: ≤0.19 µg/kg, country: Czechoslovakia[9]

Heifers see Cattle

Hen

Hen Natural Contamination see also Chicken

Hen egg may contain the following mycotoxins and/or their metabolites:

AFLATOXIN B_1
incidence: ?/?, sa. const.: eggs from laying hens of the USA?, contamination: probably natural, conc.: ≈0.075 ng/g, country: USA[75]

DEOXYNIVALENOL
incidence: 17/20, sa. const.: eggs from laying hens of Belgium, contamination: natural (for detailed information please see the article), conc. range: tr–17.9 ppb (mean values), country: Belgium[576]

DEEPOXYDEOXYNIVALENOL
incidence: 4/20, sa. const.: eggs from laying hens of Belgium, contamination: natural (for detailed information please see the article), conc. range: 2.4–23.7 ppb (mean values), country: Belgium[576]

Hen Artificial Contamination

Hen may contain the following mycotoxins and/or their metabolites:

DEOXYNIVALENOL
incidence: ?/4, sa. const.: White Leghorn hens, age: 316 days, wt.: 1.3–1.7 kg, contamination artificial (dose: 2.2 mg DON (labeled)/bird, o. once; for detailed information please see the article), conc. range: ≤30.0 µg/bird* ** (mean value), country: Canada[135], *total DON eq. (DON and/or metabolites), **after 3 h (also at other hour intervals up to 96 h measured, lowest conc.: 0.1 µg/bird after 96 h)
incidence: ?/3, sa. const.: White Leghorn hens, age: 316 days, wt.: 1.3–1.7 kg, contamination artificial (dose: 2.2 mg DON (unlabeled)/day/bird, o., for days 1–6 followed by 2.2 mg DON (labeled)/day/bird, o., for days 7–12; for detailed information please see the article), conc. range: ≤16.0 µg/bird* ** (mean value), country: Canada[135], *total DON eq. (DON and/or metabolites), **after 8 days of DON-administration (also at other day intervals up to 18 days measured, lowest conc.: 1.9 µg/bird after 18 days)

Hen Adipose Tissue see Hen fat

Hen adrenal may contain the following mycotoxins and/or their metabolites:

AFLATOXINS
incidence: ?/?, sa. const.: White Leghorn hens (Cornell strain K), age: 20–22 weeks,

wt.: 1,600–1,800 g, contamination artificial (dose: 11.26 mg AFs (labeled), by stomach tube, once; for detailed information please see the article), conc. range: ≤8.05 µg/g* ** ***, country: USA[81], *AFs or their metabolites, **after 4 days (also measured after 1 and 7 days, lowest conc.: 4.00 µg/g after 1 day), ***in adrenal and thyroid

Zearalenone
incidence: ?/4, sa. const.: White Leghorn laying hens, age: 26–39 weeks, contamination: artificial (dose: 10 mg ZEA (labeled)/kg, by gavage into the crop, once), conc. range: ≤79 µg eq ZEA/100 g wet tissue* (mean value), country: USA[394], *after 24 h (also measured after 2, 4, 48 and 72 h, lowest conc.: 32 µg eq ZEA/100 g wet tissue after 4 h)

Hen Adrenal and Thyroid see Hen adrenal

Hen bile may contain the following mycotoxins and/or their metabolites:

Aflatoxins
incidence: ?/?, sa. const.: White Leghorn hens (Cornell strain K), age: 20–22 weeks, wt.: 1,600–1,800 g, contamination: artificial (dose: 11.26 mg AFs (labeled), by stomach tube, once; for detailed information please see the article), conc. range: ≤40.33 µg/g* **, country: USA[81], *AFs or their metabolites, **after 1 day (also measured after 4 and 7 days, lowest conc.: 8.17 µg/g after 7 days)

Deoxynivalenol
incidence: ?/?*, sa. const.: White Leghorn hens, age: 55 weeks, contamination: no NIV (for detailed information please see the article), conc.: nr, country: Lithuania/Sweden[375], *control
incidence: ?/?, sa. const.: White Leghorn hens, age: 55 weeks, contamination: artificial (dose: **5 mg NIV**/kg, o., for 50 days; for detailed information please see the article), conc.: 27 ng/ml* (mean value), country: Lithuania/Sweden[375], *after 50 days

Nivalenol
incidence: ?/?*, sa. const.: White Leghorn hens, age: 55 weeks, contamination: no NIV (for detailed information please see the article), conc.: nr, country: Lithuania/Sweden[375], *control
incidence: ?/?, sa. const.: White Leghorn hens, age: 55 weeks, contamination: artificial (dose: **5 mg NIV**/kg, o., for 50 days; for detailed information please see the article), conc.: 11 ng/ml* (mean value), country: Lithuania/Sweden[375], *after 50 days

Zearalenone
incidence: ?/4, sa. const.: White Leghorn laying hens, age: 26–39 weeks, contamination: artificial (dose: 10 mg ZEA (labeled)/kg, by gavage into the crop, once), conc. range: ≤56,300 µg eq ZEA/100 g wet tissue* (mean value), country: USA[394], *after 24 h (also measured after 2, 4, 48 and 72 h, lowest conc.: 3,460 µg eq ZEA/100 g wet tissue after 72 h)

Hen blood may contain the following mycotoxins and/or their metabolites:

Aflatoxin B_1
incidence: 6/6*, sa. const.: Single Comb White Leghorn hens, contamination: no AFB_1 (for detailed information please see the article), conc.: nd, country: USA[363], *control
incidence: 9/10, sa. const.: Single Comb White Leghorn hens, contamination: artificial (dose: **8 µg AFB_1**/g feed, for 7 days; for detailed information please see the article), conc. range: ≤0.07 ng/g* **, country: USA[363], *AFB_1-residues, **after 7 days

Ochratoxin A
incidence: 8?/8, sa. const.: New Hampshire-Leghorn cross hens, age: 24 weeks, contamination: artificial

(dose: **OTA 0.5 ppm** (feed weight), o., for 2 weeks), conc.: 6.9 ppb* (mean value), country: USA[316], *after 14 days
incidence: 8?/8, sa. const.: New Hampshire-Leghorn cross hens, age: 24 weeks, contamination: artificial (dose: **OTA 5.0 ppm** (feed weight), o., for 2 weeks), conc.: 7.0 ppb* (mean value), country: USA[316], *after 14 days

ZEARALENONE
incidence: ?/4*, sa. const.: White Leghorn laying hens, age: 26–39 weeks, contamination: artificial (dose: 10 mg ZEA (labeled)/kg, by gavage into the crop, once), conc. range: ≤95 µg eq ZEA/100 g wet tissue** (mean value), country: USA[394], *whole blood, **after 4 h (also measured after 2, 24, 48 and 72 h, lowest conc.: 39 µg eq ZEA/100 g wet tissue after 72 h)

Hen blood clot may contain the following mycotoxins and/or their metabolites:

AFLATOXIN B_1
incidence: 8/8*, sa. const.: White Leghorn pullets of the Shaver strain, contamination: no AFB_1 + AFB_2 (for detailed information please see the article), conc.: nr, country: USA[110], *control
incidence: ?/?, sa. const.: White Leghorn pullets of the Shaver strain, contamination: artificial (dose: **3,310 µg AFB_1/kg + 1,680 µg AFB_2/kg**, o., for 4 weeks; for detailed information please see the article), conc. range: 0.06–0.12 µg/kg* (mean values), country: USA[110], *after feeding 4 weeks an AF-contaminated diet

incidence: 8/8*, sa. const.: White Leghorn pullets of the Shaver strain, contamination: no AFB_1 + AFB_2 (for detailed information please see the article), conc.: nr, country: USA[110], *control
incidence: ?/?, sa. const.: White Leghorn pullets of the Shaver strain, contamination: artificial (dose: **3,310 µg AFB_1/kg + 1,680 µg AFB_2/kg**, o., for 4 weeks; for detailed information please see the article), conc.: 0.05 µg/kg* (mean value), country: USA[110], *2 days after feeding 4 weeks an AF-contaminated diet

AFLATOXIN B_2
incidence: 8/8*, sa. const.: White Leghorn pullets of the Shaver strain, contamination: no AFB_1 + AFB_2 (for detailed information please see the article), conc.: nr, country: USA[110], *control
incidence: ?/24, sa. const.: White Leghorn pullets of the Shaver strain, contamination: artificial (dose: **3,310 µg AFB_1/kg + 1,680 µg AFB_2/kg**, o., for 4 weeks; for detailed information please see the article), conc. range: 0.07–0.10 µg/kg* (mean values), country: USA[110], *after feeding 4 weeks an AF-contaminated diet

incidence: 8/8*, sa. const.: White Leghorn pullets of the Shaver strain, contamination: no AFB_1 + AFB_2 (for detailed information please see the article), conc.: nr, country: USA[110], *control
incidence: ?/?, sa. const.: White Leghorn pullets of the Shaver strain, contamination: artificial (dose: **3,310 µg AFB_1/kg + 1,680 µg AFB_2/kg**, o., for 4 weeks; for detailed information please see the article), conc.: 0.03 µg/kg* (mean value), country: USA[110], *2 days after feeding 4 weeks an AF-contaminated diet

AFLATOXIN B_{2a}
incidence: 8/8*, sa. const.: White Leghorn pullets of the Shaver strain, contamination: no AFB_1 + AFB_2 (for detailed information please see the article), conc.: nr, country: USA[110], *control
incidence: ?/?, sa. const.: White Leghorn pullets of the Shaver strain, contamination: artificial (dose: **3,310 µg AFB_1/kg + 1,680 µg AFB_2/kg**, o., for 4 weeks; for detailed information please see the article), conc. range: 0.02–0.04 µg/kg* (mean values), country: USA[110], *after feeding 4 weeks an AF-contaminated diet

incidence: 8/8*, sa. const.: White Leghorn pullets of the Shaver strain, contamination: no $AFB_1 + AFB_2$ (for detailed information please see the article), conc.: nr, country: USA[110], *control
incidence: ?/?, sa. const.: White Leghorn pullets of the Shaver strain, contamination artificial (dose: **3,310 µg AFB_1/kg + 1,680 µg AFB_2/kg**, o., for 4 weeks; for detailed information please see the article), conc.: tr* (mean value), country: USA[110], *2 days after feeding 4 weeks an AF-contaminated diet

Aflatoxin M_1
incidence: 8/8*, sa. const.: White Leghorn pullets of the Shaver strain, contamination: no $AFB_1 + AFB_2$ (for detailed information please see the article), conc.: nr, country: USA[110], *control
incidence: ?/?, sa. const.: White Leghorn pullets of the Shaver strain, contamination: artificial (dose: **3,310 µg AFB_1/kg + 1,680 µg AFB_2/kg**, o., for 4 weeks; for detailed information please see the article), conc. range: tr*, country: USA[110], *after feeding 4 weeks an AF-contaminated diet

incidence: 8/8*, sa. const.: White Leghorn pullets of the Shaver strain, contamination: no $AFB_1 + AFB_2$ (for detailed information please see the article), conc.: nr, country: USA[110], *control
incidence: ?/?, sa. const.: White Leghorn pullets of the Shaver strain, contamination artificial (dose: **3,310 µg AFB_1/kg + 1,680 µg AFB_2/kg**, o., for 4 weeks; for detailed information please see the article), conc.: nd*, country: USA[110], *2 days after feeding 4 weeks an AF-contaminated diet

Aflatoxin M_2
incidence: 8/8*, sa. const.: White Leghorn pullets of the Shaver strain, contamination: no $AFB_1 + AFB_2$ (for detailed information please see the article), conc.: nr, country: USA[110], *control
incidence: ?/?, sa. const.: White Leghorn pullets of the Shaver strain, contamination: artificial (dose: **3,310 µg AFB_1/kg + 1,680 µg AFB_2/kg**, o, for 4 weeks; for detailed information please see the article), conc. range: 0.01 µg/kg* (mean value), country: USA[110], *after feeding 4 weeks an AF-contaminated diet

incidence: 8/8*, sa. const.: White Leghorn pullets of the Shaver strain, contamination: no $AFB_1 + AFB_2$ (for detailed information please see the article), conc.: nr, country: USA[110], *control
incidence: ?/?, sa. const.: White Leghorn pullets of the Shaver strain, contamination artificial (dose: **3,310 µg AFB_1/kg + 1,680 µg AFB_2/kg**, o., for 4 weeks; for detailed information please see the article), conc.: nd*, country: USA[110], *2 days after feeding 4 weeks an AF-contaminated diet

Hen brain may contain the following mycotoxins and/or their metabolites:

Zearalenone
incidence: ?/4, sa. const.: White Leghorn laying hens, age: 26–39 weeks, contamination: artificial (dose: 10 mg ZEA (labeled)/kg, by gavage into the crop, once), conc. range: ≤26 µg eq ZEA/100 g wet tissue* (mean value), country: USA[394], *after 4 h (also measured after 2, 24, 48 and 72 h, lowest conc.: 11 µg eq ZEA/100 g wet tissue after 72 h)

Hen Breast see Hen muscle, breast

Hen cloaca may contain the following mycotoxins and/or their metabolites:

Ochratoxin A
incidence: 8?/8, sa. const.: New Hampshire-Leghorn cross hens, age: 24 weeks, contamination: artificial (dose: **OTA 0.5 ppm** (feed weight), o., for 2 weeks), conc.: 15.1 ppb* (mean value), country: USA[316], *after 14 days
incidence: 8?/8, sa. const.: New Hampshire-Leghorn cross hens, age:

24 weeks, contamination: artificial (dose: **OTA 5.0 ppm** (feed weight), o., for 2 weeks), conc.: 25.7 ppb* (mean value), country: USA[316], *after 14 days

Hen clutch may contain the following mycotoxins and/or their metabolites:

ZEARALENONE
incidence: ?/4, sa. const. From White Leghorn laying hens, contamination: artificial (dose: 10 mg ZEA (labeled)/kg, by gavage into the crop, once), conc. range: ≤256 μg eq ZEA/100 g wet weight* (mean value), country: USA[394], *after 21–26 h (also measured after 1–10, 43–50 and 69–72 h, lowest conc.: 19 μg eq ZEA/100 g wet weight after 1–10 h)

Hen comb may contain the following mycotoxins and/or their metabolites:

ZEARALENONE
incidence: ?/4, sa. const.: White Leghorn laying hens, age: 26–39 weeks, contamination: artificial (dose: 10 mg ZEA (labeled)/kg, by gavage into the crop, once), conc. range: ≤183 μg eq ZEA/100 g wet tissue* (mean value), country: USA[394], *after 24 h (also measured after 2, 4, 48 and 72 h, lowest conc.: 57 μg eq ZEA/100 g wet tissue after 2 h)

Hen Crop and Gizzard see Hen gizzard

Hen digestive tract may contain the following mycotoxins and/or their metabolites:

AFLATOXINS
incidence: ?/?, sa. const.: White Leghorn hens (Cornell strain K), age: 20–22 weeks, wt.: 1,600–1,800 g, contamination: artificial (dose: 11.26 mg AFs (labeled), by stomach tube, once; for detailed information please see the article), conc. range: ≤111.36 μg/g* ** ***, country: USA[81], *AFs or their metabolites, **in contents of digestive tract, ***after 4 days (also measured after 1 and 7 days, lowest conc.: 42.26 μg/g after 7 days)

Hen egg may contain the following mycotoxins and/or their metabolites:

AFLATOXICOL
incidence: 20/20*, sa. const.: eggs from laying hens, contamination: no AFB_1, conc.: nd, country: Italy[201], *control organic and aqueous phase
incidence: ?/20, sa. const.: eggs from laying hens, contamination: artificial (dose: 30 μg AFB_1/kg b. wt., by gavage, for 7 days), conc. range: ≤1.15 ng/g* (mean value), country: Italy[201], *in **organic phase**, **at day 2 after AFB_1-treatment (also measured after 7 days conc.: 0.02 ng/g)
incidence: ?/20, sa. const.: eggs from laying hens, contamination: artificial (dose: 30 μg AFB_1/kg b. wt., by gavage, for 7 days), conc.: 0.17 ng/g* (mean value), country: Italy[201], *in **aqueous phase**, **at day 7 after AFB_1-treatment (also measured after 2 days but conc.: nd)

incidence: 6/6*, sa. const.: Single Comb White Leghorn hens, contamination: no AFB_1 (for detailed information please see the article), conc.: nd, country: USA[363], *control
incidence: ?/?, sa. const.: eggs from Single Comb White Leghorn hens, contamination: artificial (dose: 8 μg AFB_1/g feed, for 7 days; for detailed information please see the article), conc. range: ≤3.3 (ng/g) × 10^{-1}* ** ***, country: USA[363], *AFL-residues, **after 6.5 days of AFB_1-administration, ***(also at other day intervals up to 14 days measured, lowest conc.: nd after 13.5 and 14 days)

AFLATOXIN B_1
incidence: 16/16*, sa. const.: eggs from White Leghorn pullets, contamination: no AFB_1 + AFB_2 (for detailed information please see the article), conc.: nr, country: USA[112], *control

incidence: ?/2–3, sa. const.: eggs from White Leghorn pullets, contamination: artificial (dose: 3,310 μg AFB_1 + 1,680 μg AFB_2/kg feed for each hen, o., for up to 4 weeks; for detailed information please see the article), conc. range: tr–0.06 μg/kg* (mean values), country: USA[112], *conc. in **eggs** laid during AF-feeding (up to 4 weeks measured)
incidence: ?/5–6, sa. const.: eggs from White Leghorn pullets, contamination: artificial (dose: 3,310 μg AFB_1 + 1,680 μg AFB_2/kg feed for each hen, o., for up to 4 weeks; for detailed information please see the article), conc. range: tr–0.06 μg/kg* (mean values), country: USA[112], *conc. in **albumen** during AF-feeding (up to 4 weeks measured)
incidence: ?/5–6, sa. const.: eggs from White Leghorn pullets, contamination: artificial (dose: 3,310 μg AFB_1 + 1,680 μg AFB_2/kg feed for each hen, o., for up to 4 weeks; for detailed information please see the article), conc. range: 0.01–0.06 μg/kg* (mean values), country: USA[112], *conc. in **yolk** during AF-feeding (up to 4 weeks measured)

incidence: 16/16*, sa. const.: eggs from White Leghorn pullets, contamination: no AFB_1 + AFB_2 (for detailed information please see the article), conc.: nr, country: USA[112], *control
incidence: ?/2–3, sa. const.: eggs from White Leghorn pullets, contamination: artificial (dose: 3,310 μg AFB_1 + 1,680 μg AFB_2/kg feed for each hen, o., for 4 weeks; for detailed information please see the article), conc. range: tr–0.03 μg/kg* (mean value), country: USA[112], *conc. in **eggs** after day 1–3 of withdrawal of the AF-containing diet (up to 4 days measured after withdrawal of AFs)
incidence: ?/?, sa. const.: eggs from White Leghorn pullets, contamination: artificial (dose: 3,310 μg AFB_1 + 1,680 μg AFB_2/kg feed for each hen, o., for 4 weeks; for detailed information please see the article), conc. range: tr–0.02 μg/kg* (mean values), country: USA[112], *conc. in **albumen** after day 1–2 of withdrawal of the AF-containing diet (up to 6 days measured after withdrawal of AFs)
incidence: ?/?, sa. const.: eggs from White Leghorn pullets, contamination: artificial (dose: 3,310 μg AFB_1 + 1,680 μg AFB_2/kg feed for each hen, o., for 4 weeks; for detailed information please see the article), conc. range: tr–0.03 μg/kg* (mean values), country: USA[112], *conc. in **yolk** after day 1–6 of withdrawal of the AF-containing diet (up to 7 days measured after withdrawal of AFs)

incidence: 10/10, sa. const.: eggs from Arbor Acre hens, contamination: no AFB_1 (for detailed information please see the article), conc.: nd, country: USA[139]
incidence: ?/10, sa. const.: eggs from Arbor Acre hens, contamination: artificial (dose: **0.1 ppm AFB_1**, o., for 10 days; for detailed information please see the article), conc. range: ≤0.35 ppb*, country: USA[139], *after 10 days on ration
incidence: ?/10, sa. const.: eggs from Arbor Acre hens, contamination: artificial (dose: **0.2 ppm AFB_1**, o., for 12 days; for detailed information please see the article), conc. range: ≤2.2 ppb*, country: USA[139], *after 11 days on ration
incidence: ?/10, sa. const.: eggs from Arbor Acre hens, contamination: artificial (dose: **0.4 ppm AFB_1**, o., for 15 days; for detailed information please see the article), conc. range: ≤3.3 ppb*, country: USA[139], *after 10 days on ration

incidence: 20/20*, sa. const.: eggs from laying hens, contamination: no AFB_1, conc.: nd, country: Italy[201], *control of organic and aqueous phase
incidence: ?/20, sa. const.: eggs from laying hens, contamination: artificial (dose: 30 μg AFB_1/kg b. wt., by gavage, for 7 days), conc. range: ≤1.70 ng/g* ** *** (mean value), country: Italy[201], *in **organic phase**, **value corrected for AFB_{2a} level, ***at day 2 after AFB_1-treatment (also measured after 7 days conc.: 0.82 ng/g)

incidence: 20/20, sa. const.: eggs from laying hens, contamination: artificial (dose: 30 µg AFB_1/kg b. wt., by gavage, for 7 days), conc.: nd* ** ***, country: Italy[201], *in **aqeous phase**, **value corrected for AFB_{2a} level, ***at day 2 and 7 after AFB_1-treatment

incidence: 39/39*, sa. const.: eggs from white HNL-type hens, contamination: no AFB_1 + AFG_1 (for detailed information please see the article), conc.: nd, country: Germany[320], *control
incidence: ?/10, sa. const.: eggs from white HNL-type hens, contamination: artificial (dose: **10,000 ppb AFB_1/13,000 ppb AFG_1**, o., for 8 weeks; for detailed information please see the article), conc. range: ≤0.26 ppb* ** (mean value), country: Germany[320], *after 8 weeks?, **in **albumen** (40 ppb AFB_1/52 ppb AFG_1, 100 ppb AFB_1/31 ppb AFG_1 or 1,000 ppb AFB_1/310 ppb AFG_1 fed = no contamination; 3,000 ppb AFB_1/930 ppb AFG_1, 5,000 ppb AFB_1/1,550 ppb AFG_1 or 10,000 ppb AFB_1/13,000 ppb AFG_1 and 4 weeks later 5,000 ppb AFB_1/6,500 ppb AFG_1 fed but lower contamination than the pr. value)
incidence: 39/39*, sa. const.: eggs from white HNL-type hens, contamination: no AFB_1 + AFG_1 (for detailed information please see the article), conc.: nd, country: Germany[320], *control
incidence: ?/10, sa. const.: eggs from white HNL-type hens, contamination: artificial (dose: **10,000 ppb AFB_1/13,000 ppb AFG_1**, o., for 8 weeks; for detailed information please see the article), conc. range: ≤0.79 ppb* ** (mean value), country: Germany[320], *after 8 weeks?, **in **yolk** (40 ppb AFB_1/52 ppb AFG_1, 100 ppb AFB_1/31 ppb AFG_1 or 1,000 ppb AFB_1/310 ppb AFG_1 fed = no contamination; 3,000 ppb AFB_1/930 ppb AFG_1, 5,000 ppb AFB_1/1,550 ppb AFG_1 or 10,000 ppb AFB_1/13,000 ppb AFG_1 and 4 weeks later 5,000 ppb AFB_1/6,500 ppb AFG_1 fed but lower contamination than the pr. value)

incidence: 39/39*, sa. const.: eggs from white HNL-type hens, contamination: no AFB_1 + AFG_1 (for detailed information please see the article), conc.: nd, country: Germany[320], *control
incidence: ?/10, sa. const.: eggs from white HNL-type hens, contamination: artificial (dose: **10,000 ppb AFB_1/13,000 ppb AFG_1**, o., for 8 weeks; for detailed information please see the article), conc. range: ≤0.43 ppb* ** (mean value), country: Germany[320], *after 8 weeks?, **in **yolk and albumen** (40 ppb AFB_1/52 ppb AFG_1, 100 ppb AFB_1/31 ppb AFG_1 or 1,000 ppb AFB_1/310 ppb AFG_1 fed = no contamination; 3,000 ppb AFB_1/930 ppb AFG_1, 5,000 ppb AFB_1/1,550 ppb AFG_1 or 10,000 ppb AFB_1/13,000 ppb AFG_1 and 4 weeks later 5,000 ppb AFB_1/6,500 ppb AFG_1 fed but lower contamination than the pr. value)

incidence: 35/35*, sa. const.: eggs from brown HNL-type hens, contamination: no AFB_1 + AFG_1 (for detailed information please see the article), conc.: nd, country: Germany[320], *control
incidence: ?/16, sa. const.: eggs from brown HNL-type hens, contamination: artificial (dose: **5,000 ppb AFB_1/1,550 ppb AFG_1**, o., for 3 weeks; for detailed information please see the article), conc. range: ≤0.16 ppb* ** (mean value), country: Germany[320], *after 3 weeks?, **in **albumen** (100 ppb AFB_1/31 ppb AFG_1 or 3,000 ppb AFB_1/930 ppb AFG_1 fed = no contamination; 10,000 ppb AFB_1/3,100 ppb AFG_1 fed but lower contamination than the pr. value)
incidence: 35/35*, sa. const.: eggs from brown HNL-type hens, contamination: no AFB_1 + AFG_1 (for detailed information please see the article), conc.: nd, country: Germany[320], *control
incidence: ?/16, sa. const.: eggs from brown HNL-type hens, contamination: artificial (dose: **5,000 ppb AFB_1/1,550 ppb AFG_1**, o., for 3 weeks; for detailed information please see the article), conc. range: ≤0.40 ppb* ** (mean value), country: Germany)[320], *after 3 weeks?, **in **yolk** (100 ppb AFB_1/31 ppb AFG_1 or 3,000 ppb AFB_1/930 ppb AFG_1 fed = no

contamination; 10,000 ppb AFB_1/3,100 ppb AFG_1 fed but lower contamination than the pr. value)

incidence: ?/?*, sa. const.: eggs from Brown Hyssex laying hens, contamination: no AF (for detailed information please see the article), conc.: nd, country: Spain[359], *control
incidence: ?/?, sa. const.: eggs from Brown Hyssex hens, contamination: artificial (dose: **2.5 mg AF**/kg feed, o., for 32 days; for detailed information please see the article), conc. range: ≤0.56 µg/kg* (mean value), country: Spain[359] (measured at 4th* day, 8th day, 16th day and 32nd day, lowest value conc.: 0.05 µg/kg after 16 days, intoxication period)
incidence: ?/?, sa. const.: eggs from Brown Hyssex hens, contamination: artificial (dose: **5.0 mg AF**/kg feed, o., for 32 days; for detailed information please see the article), conc. range: ≤0.34 µg/kg* (mean value), country: Spain[359] (measured at 4th day, 8th day, 16th* day and 32nd day, lowest value conc.: 0.05 µg/kg after 4 and 8 days, intoxication period)

incidence: ?/?*, sa. const.: eggs from Babcock hens (16 weeks old), contamination: no AFB_1 (for detailed information please see the article), conc.: nd, country: Brazil[361], *control
incidence: ?/?, sa. const.: eggs from Babcock hens (16 weeks old), contamination: artificial (dose: **500 µg AFB_1**/kg feed, for 8 weeks; for detailed information please see the article), conc. range: 0.05*–0.16** µg/kg***, Ø conc.: 0.10 µg/kg, country: Brazil[361], after 14* and 42** days of AFB_1-administration (also at other day intervals up to 54 days measured, lowest conc.: nd after 7 days), ***AFB_1-residues (100 and 300 µg AFB_1/kg feed resulted in no contamination)

incidence: 6/6*, sa. const.: Single Comb White Leghorn hens, contamination: no AFB_1 (for detailed information please see the article), conc.: nd, country: USA[363], *control
incidence: ?/?, sa. const.: eggs from Single Comb White Leghorn hens, contamination: artificial (dose: **8 µg AFB_1**/g feed, for 7 days; for detailed information please see the article), conc. range: ≤3.8 (ng/g) × 10^{-1}* ** ***, country: USA[363], *AFB_1-residues, **after 6.5 days of AFB_1-administration, ***(also at other day intervals up to 14 days measured, lowest conc.: nd in part after 13.5 and 14 days)

Aflatoxin B_2
incidence: 16/16*, sa. const.: eggs from White Leghorn pullets, contamination: no AFB_1 + AFB_2 (for detailed information please see the article), conc.: nr, country: USA[112], *control
incidence: ?/2–3, sa. const.: eggs from White Leghorn pullets, contamination: artificial (dose: 3,310 µg AFB_1 + 1,680 µg AFB_2/kg feed for each hen, o., for up to 4 weeks; for detailed information please see the article), conc. range: tr–0.05 µg/kg* (mean values), country: USA[112], *conc. in **eggs** laid during AF-feeding (up to 4 weeks measured)
incidence: ?/5–6, sa. const.: eggs from White Leghorn pullets, contamination: artificial (dose: 3,310 µg AFB_1 + 1,680 µg AFB_2/kg feed for each hen, o., for up to 4 weeks; for detailed information please see the article), conc. range: 0.01–0.04 µg/kg* (mean values), country: USA[112], *conc. in **albumen** during AF-feeding (up to 4 weeks measured)
incidence: ?/5–6, sa. const.: eggs from White Leghorn pullets, contamination: artificial (dose: 3,310 µg AFB_1 + 1,680 µg AFB_2/kg feed for each hen, o., for up to 4 weeks; for detailed information please see the article), conc. range: 0.02–0.04 µg/kg* (mean values), country: USA[112], *conc. in **yolk** during AF-feeding (up to 4 weeks measured)

incidence: 16/16*, sa. const.: eggs from White Leghorn pullets, contamination: no

AFB_1 + AFB_2 (for detailed information please see the article), conc.: nr, country: USA[112], *control
incidence: ?/2–3, sa. const.: eggs from White Leghorn pullets, contamination: artificial (dose: 3,310 μg AFB_1 + 1,680 μg AFB_2/kg feed for each hen, o., for 4 weeks; for detailed information please see the article), conc. range: ≤0.02 μg/kg* (mean value), country: USA[112], *conc. in **eggs** after day 1–3 of withdrawal of the AF-containing diet (up to 4 days measured after withdrawal of AFs)
incidence: ?/?, sa. const.: eggs from White Leghorn pullets, contamination: artificial (dose: 3,310 μg AFB_1 + 1,680 μg AFB_2/kg feed for each hen, o., for 4 weeks; for detailed information please see the article), conc. range: tr–0.01 μg/kg* (mean values), country: USA[112], *conc. in **albumen** after day 1–2 of withdrawal of the AF-containing diet (up to 6 days measured after withdrawal of AFs)
incidence: ?/?, sa. const.: eggs from White Leghorn pullets, contamination: artificial (dose: 3,310 μg AFB_1 + 1,680 μg AFB_2/kg feed for each hen, o., for 4 weeks; for detailed information please see the article), conc. range: tr–0.01 μg/kg* (mean values), country: USA[112], *conc. in **yolk** after day 1–6 withdrawal of the AF-containing diet (up to 7 days measured after withdrawal of AFs)

Aflatoxin B_{2a}
incidence: 16/16*, sa. const.: eggs from White Leghorn pullets, contamination: no AFB_1 + AFB_2 (for detailed information please see the article), conc.: nr, country: USA[112], *control
incidence: ?/2–3, sa. const.: eggs from White Leghorn pullets, contamination: artificial (dose: 3,310 μg AFB_1 + 1,680 μg AFB_2/kg feed for each hen, o., for up to 4 weeks; for detailed information please see the article), conc. range: 0.02–0.09 μg/kg* (mean values), country: USA[112], *conc. in **eggs** laid during AF-feeding (up to 4 weeks measured)
incidence: ?/5–6, sa. const.: eggs from White Leghorn pullets, contamination: artificial (dose: 3,310 μg AFB_1 + 1,680 μg AFB_2/kg feed for each hen, o., for up to 4 weeks; for detailed information please see the article), conc. range: 0.03–0.13 μg/kg* (mean values), country: USA[112], *conc. in **albumen** during AF-feeding (up to 4 weeks measured)
incidence: ?/5–6, sa. const.: eggs from White Leghorn pullets, contamination: artificial (dose: 3,310 μg AFB_1 + 1,680 μg AFB_2/kg feed for each hen, o., for up to 4 weeks; for detailed information please see the article), conc. range: 0.03–0.11 μg/kg* (mean values), country: USA[112], *conc. in **yolk** during AF-feeding (up to 4 weeks measured)

incidence: 16/16*, sa. const.: eggs from White Leghorn pullets, contamination: no AFB_1 + AFB_2 (for detailed information please see the article), conc.: nr, country: USA[112], *control
incidence: ?/2–3, sa. const.: eggs from White Leghorn pullets, contamination: artificial (dose: 3,310 μg AFB_1 + 1,680 μg AFB_2/kg feed for each hen, o., for 4 weeks; for detailed information please see the article), conc. range: tr–0.04 μg/kg* (mean values), country: USA[112], *conc. in **eggs** after day 1–3 of withdrawal of the AF-containing diet (up to 4 days measured after withdrawal of AFs)
incidence: ?/?, sa. const.: eggs from White Leghorn pullets, contamination: artificial (dose: 3,310 μg AFB_1 + 1,680 μg AFB_2/kg feed for each hen, o., for 4 weeks; for detailed information please see the article), conc. range: tr–0.02 μg/kg* (mean values), country: USA[112], *conc. in **albumen** after day 1–2 of withdrawal of the AF-containing diet (up to 6 days measured after withdrawal of AFs)
incidence: ?/?, sa. const.: eggs from White Leghorn pullets, contamination: artificial (dose: 3,310 μg AFB_1 + 1,680 μg AFB_2/kg feed for each hen, o., for 4 weeks; for detailed information please see the article), conc. range: tr–0.05 μg/kg*

(mean values), country: USA[112], *conc. in **yolk** after day 1–6 of withdrawal of the AF-containing diet (up to 7 days measured after withdrawal of AFs)

incidence: 20/20*, sa. const.: eggs from laying hens, contamination: no AFB_1, conc.: nd, country: Italy[201], *control of organic and aqueous phase
incidence: ?/20, sa. const.: eggs, contamination: artificial (dose: 30 μg AFB_1/kg b. wt., by gavage, for 7 days), conc. range: ≤0.03 ng/g* ** (mean value), country: Italy[201], *in **organic phase**, **at day 2 after AFB_1-treatment (also measured after 7 days conc.: 0.01 ng/g)
incidence: ?/20, sa. const.: eggs, contamination: artificial (dose: 30 μg AFB_1/kg b. wt., by gavage, for 7 days), conc.: 0.01 ng/g* ** (mean value), country: Italy[201], *in **aqueous phase**, **at day 7 after AFB_1-treatment (also measured after 2 days but conc.: nd)

Aflatoxin G_1
incidence: 39/39*, sa. const.: eggs from white HNL-type hens, contamination: no AFB_1 + AFG_1 (for detailed information please see the article), conc.: nd, country: Germany[320], *control
incidence: ?/10, sa. const.: eggs from white HNL-type hens, contamination: artificial (dose: **10,000 ppb AFB_1/13,000 ppb AFG_1**, o., for 8 weeks; for detailed information please see the article), conc. range: ≤0.15 ppb* ** (mean value), country: Germany[320], *after 8 weeks?, **in **albumen** (40 ppb AFB_1/52 ppb AFG_1, 100 ppb AFB_1/31 ppb AFG_1 1,000 ppb AFB_1/310 ppb AFG_1 or 3,000 ppb AFB_1/930 ppb AFG_1 fed = no contamination; 5,000 ppb AFB_1/1,550 ppb AFG_1 or 10,000 ppb AFB_1/13,000 ppb AFG_1 and 4 weeks later 5,000 ppb AFB_1/6,500 ppb AFG_1 fed but lower contamination than the pr. value)
incidence: 39/39*, sa. const.: eggs from white HNL-type hens, contamination: no AFB_1 + AFG_1 (for detailed information please see the article), conc.: nd, country: Germany[320], *control
incidence: ?/10, sa. const.: eggs from white HNL-type hens, contamination: artificial (dose: **10,000 ppb AFB_1/13,000 ppb AFG_1**, o., for 8 weeks; for detailed information please see the article), conc. range: ≤0.90 ppb* ** (mean value), country: Germany[320], *after 8 weeks?, **in **yolk** (40 ppb AFB_1/52 ppb AFG_1, 100 ppb AFB_1/31 ppb AFG_1, 1,000 ppb AFB_1/310 ppb AFG_1 or 3,000 ppb AFB_1/930 ppb AFG_1 fed = no contamination; 5,000 ppb AFB_1/1,550 ppb AFG_1 or 10,000 ppb AFB_1/13,000 ppb AFG_1 and 4 weeks later 5,000 ppb AFB_1/6,500 ppb AFG_1 fed but lower contamination than the pr. value)

incidence: 39/39*, sa. const.: eggs from white HNL-type hens, contamination: no AFB_1 + AFG_1 (for detailed information please see the article), conc.: nd, country: Germany[320], *control
incidence: ?/10, sa. const.: eggs from white HNL-type hens, contamination: artificial (dose: **10,000 ppb AFB_1/13,000 ppb AFG_1**, o., for 8 weeks; for detailed information please see the article), conc. range: ≤0.40 ppb* ** (mean value), country: Germany[320], *after 8 weeks?, **in **yolk** and albumen (40 ppb AFB_1/52 ppb AFG_1, 100 ppb AFB_1/31 ppb AFG_1, 1,000 ppb AFB_1/310 ppb AFG_1 or 3,000 ppb AFB_1/930 ppb AFG_1 fed = no contamination; 5,000 ppb AFB_1/1,550 ppb AFG_1 or 10,000 ppb AFB_1/13,000 ppb AFG_1 and 4 weeks later 5,000 ppb AFB_1/6,500 ppb AFG_1 fed but lower contamination than the pr. value)

incidence: 35/35*, sa. const.: eggs from brown HNL-type hens, contamination: no AFB_1 + AFG_1 (for detailed information please see the article), conc.: nd, country: Germany[320], *control
incidence: ?/7, sa. const.: eggs from brown HNL-type hens, contamination: artificial (dose: **10,000 ppb AFB_1/3,100 ppb AFG_1**, o., for 3 weeks; for detailed information

please see the article), conc. range: tr* ** (mean value), country: Germany[320], *after 3 weeks?, **in **albumen** (100 ppb AFB_1/31 ppb AFG_1, 3,000 ppb AFB_1/930 ppb AFG_1 or 5,000 ppb AFB_1/1,550 ppb AFG_1 fed = no contamination)
incidence: 35/35*, sa. const.: eggs from brown HNL-type hens, contamination: no AFB_1 + AFG_1 (for detailed information please see the article), conc.: nd, country: Germany[320], *control
incidence: ?/7, sa. const.: eggs from brown HNL-type hens, contamination: artificial (dose: **10,000 ppb AFB_1/3,100 ppb AFG_1**, o., for 3 weeks; for detailed information please see the article), conc. range: ≤0.14 ppb* ** (mean value), country: Germany[320], *after 3 weeks?, **in **yolk** (100 ppb AFB_1/31 ppb AFG_1 or 3,000 ppb AFB_1/930 ppb AFG_1 fed = no contamination; 5,000 ppb AFB_1/1,550 ppb AFG_1 fed but lower contamination than the pr. value)

Aflatoxin M_1
incidence: 16/16*, sa. const.: eggs from White Leghorn pullets, contamination: no AFB_1 + AFB_2 (for detailed information please see the article), conc.: nr, country: USA[112], *control
incidence: ?/2–3, sa. const.: eggs from White Leghorn pullets, contamination: artificial (dose: 3,310 μg AFB_1 + 1,680 μg AFB_2/kg feed for each hen, o., for up to 4 weeks; for detailed information please see the article), conc. range: tr–0.02 μg/kg* (mean values), country: USA[112], *conc. in **eggs** laid during AF-feeding (up to 4 weeks measured)
incidence: ?/5–6, sa. const.: eggs from White Leghorn pullets, contamination: artificial (dose: 3,310 μg AFB_1 + 1,680 μg AFB_2/kg feed for each hen, o., for up to 4 weeks; for detailed information please see the article), conc. range: tr–0.02 μg/kg* (mean values), country: USA[112], *conc. in **albumen** during AF-feeding (up to 4 weeks measured)
incidence: ?/5–6, sa. const.: eggs from White Leghorn pullets, contamination: artificial (dose: 3,310 μg AFB_1 + 1,680 μg AFB_2/kg feed for each hen, o., for up to 4 weeks; for detailed information please see the article), conc. range: tr–0.02 μg/kg* (mean values), country: USA[112], *conc. in **yolk** during AF-feeding (up to 4 weeks measured)

incidence: 16/16*, sa. const.: eggs from White Leghorn pullets, contamination: no AFB_1 + AFB_2 (for detailed information please see the article), conc.: nr, country: USA[112], *control
incidence: 2–3/2–3, sa. const.: eggs from White Leghorn pullets, contamination: artificial (dose: 3,310 μg AFB_1 + 1,680 μg AFB_2/kg feed for each hen, o., for 4 weeks; for detailed information please see the article), conc.: nd/tr* (mean value), country: USA[112], *conc. in **eggs** after day 1–2 of withdrawal of the AF-containing diet (up to 4 days measured after withdrawal of AFs)
incidence: ?/?, sa. const.: eggs from White Leghorn pullets, contamination: artificial (dose: 3,310 μg AFB_1 + 1,680 μg AFB_2/kg feed for each hen, o., for 4 weeks; for detailed information please see the article), conc.: nd*, country: USA[112], *conc. in **albumen** after withdrawal of the AF-containing diet (up to 6 days measured after withdrawal of AFs)
incidence: ?/?, sa. const.: eggs from White Leghorn pullets, contamination: artificial (dose: 3,310 μg AFB_1 + 1,680 μg AFB_2/kg feed for each hen, o., for 4 weeks; for detailed information please see the article), conc. range: tr–0.01 μg/kg* (mean values), country: USA[112], *conc. in **yolk** after day 1 of withdrawal of the AF-containing diet (up to 7 days measured after withdrawal of AFs)

incidence: 20/20*, sa. const.: eggs from laying hens, contamination: no AFB_1, conc.: nd, country: Italy[201], *control of organic and aqueous phase
incidence: ?/20, sa. const.: eggs from laying hens, contamination: artificial (dose: 30 μg AFB_1/kg b. wt., by gavage, for

7 days), conc.: 0.10 ng/g* (mean value), country: Italy[201], *in **organic phase**, **at day 7 after AFB_1-treatment (also measured after 2 days but conc.: nd)
incidence: ?/20, sa. const.: eggs from laying hens, contamination: artificial (dose: 30 µg AFB_1/kg b. wt., by gavage, for 7 days), conc.: 0.32 ng/g* (mean value), country: Italy[201], *in **aqueous phase**, **at day 7 after AFB_1-treatment (also measured after 2 days but conc.: nd)

Aflatoxin M_2
incidence: 16/16*, sa. const.: eggs from White Leghorn pullets, contamination: no AFB_1 + AFB_2 (for detailed information please see the article), conc.: nr, country: USA[112], *control
incidence: ?/2–3, sa. const.: eggs from White Leghorn pullets, contamination: artificial (dose: 3,310 µg AFB_1 + 1,680 µg AFB_2/kg feed for each hen, o., for up to 4 weeks; for detailed information please see the article), conc. range: tr–0.04 µg/kg* (mean values), country: USA[112], *conc. in **eggs** laid during AF-feeding (up to 4 weeks measured)
incidence: ?/5–6, sa. const.: eggs from White Leghorn pullets, contamination: artificial (dose: 3,310 µg AFB_1 + 1,680 µg AFB_2/kg feed for each hen, o., for up to 4 weeks; for detailed information please see the article), conc. range: 0.01–0.04 µg/kg* (mean values), country: USA[112], *conc. in **albumen** during AF-feeding (up to 4 weeks measured)
incidence: ?/5–6, sa. const.: eggs from White Leghorn pullets, contamination: artificial (dose: 3,310 µg AFB_1 + 1,680 µg AFB_2/kg feed for each hen, o., for up to 4 weeks; for detailed information please see the article), conc. range: 0.01–0.04 µg/kg* (mean values), country: USA[112], *conc. in **yolk** during AF-feeding (up to 4 weeks measured)
incidence: 16/16*, sa. const.: eggs from White Leghorn pullets, contamination: no AFB_1 + AFB_2 (for detailed information please see the article), conc.: nr, country: USA[112], *control
incidence: ?/2–3, sa. const.: eggs from White Leghorn pullets, contamination: artificial (dose: 3,310 µg AFB_1 + 1,680 µg AFB_2/kg feed for each hen, o., for 4 weeks; for detailed information please see the article), conc. range: tr–0.01 µg/kg* (mean value) (mean values), country: USA[112], *conc. in **eggs** after day 1–3 of withdrawal of the AF-containing diet (up to 4 days measured after withdrawal of AFs)
incidence: ?/?, sa. const.: eggs from White Leghorn pullets, contamination: artificial (dose: 3,310 µg AFB_1 + 1,680 µg AFB_2/kg feed for each hen, o., for 4 weeks; for detailed information please see the article), conc. range: tr–0.01 µg/kg* (mean values), country: USA[112], *conc. in **albumen** after day 1–2 of withdrawal of the AF-containing diet (up to 6 days measured after withdrawal of AFs)
incidence: ?/?, sa. const.: eggs from White Leghorn pullets, contamination: artificial (dose: 3,310 µg AFB_1 + 1,680 µg AFB_2/kg feed for each hen, o., for 4 weeks; for detailed information please see the article), conc. range: tr–0.01 µg/kg* (mean values), country: USA[112], *conc. in **yolk** after day 1–5 of withdrawal of the AF-containing diet (up to 7 days measured after withdrawal of AFs)

Aflatoxin
incidence: ?/?*, sa. const.: eggs from SCWL hens, contamination: no AFB_1 (for detailed information please see the article), conc.: nd, country: India[392], *control
incidence: ?/?, sa. const.: eggs from SCWL hens, contamination: artificial (dose: **600 ppb AFB_1** in feed, o.; for detailed information please see the article), conc. range: ≤5.9 ppb*, country: India[392], *after feeding period 3 (values after period 1 and 2 lower, period = 28 days with daily AFB_1-consumption)

incidence: ?/?, sa. const.: eggs from SCWL hens, contamination: artificial (dose: **1,250 ppb AFB_1** in feed, o.; for detailed information please see the article), conc. range: ≤6.3 ppb*, country: India[392], *after feeding period 3 (values after period 1 and 2 lower, period = 28 days with daily AFB_1-consumption)
incidence: ?/?, sa. const.: eggs from SCWL hens, contamination: artificial (dose: **2,120 ppb AFB_1** in feed, o.; for detailed information please see the article), conc. range: ≤10.0 ppb*, country: India[392], *after feeding period 3 (values after period 1 and 2 lower, period = 28 days with daily AFB_1-consumption)
incidence: ?/?, sa. const.: eggs from SCWL hens, contamination: artificial (dose: **2,850 ppb AFB_1** in feed, o.; for detailed information please see the article), conc. range: ≤11.1 ppb*, country: India[392], *after feeding period 2 (values after period 1 and 3 lower, period = 28 days with daily AFB_1-consumption)

Aflatoxins

incidence: ?/?, sa. const.: eggs from White Leghorn hens (Cornell strain K), contamination: artificial (dose: 11.26 mg AFs (labeled), by stomach tube, once; for detailed information please see the article), conc. range: ≤7.72 µg/g* ** ***, country: USA[81], *AFs or their metabolites, **in **albumen**, ***after 14 h oviposition time (also measured after 10, 24 and 48 h, lowest conc.: 4.98 µg/g after 10 h of ovulation)
incidence: ?/?, sa. const.: eggs from White Leghorn hens (Cornell strain K), contamination: artificial (dose: 11.26 mg AFs (labeled), by stomach tube, once; for detailed information please see the article), conc. range: ≤7.31 µg/g* ** ***, country: USA[81], *AFs or their metabolites, **in **yolk**, ***after 48 h oviposition time (also measured after 10, 14 and 24 h, lowest conc.:5.10 µg/g after 10 h of ovulation)
incidence: ?/?, sa. const.: eggs from White Leghorn hens (Cornell strain K), contamination: artificial (dose: 11.26 mg AFs (labeled), by stomach tube, once; for detailed information please see the article), conc. range: ≤12.47 µg/g* ** ***, country: USA[81], *AFs or their metabolites, **in **egg shell membranes**, ***after 48 h oviposition time (also measured after 10, 14 and 24 h, lowest conc.: 3.23 µg/g after 10 h of ovulation)

Citrinin

incidence: ?/?*, sa. const.: eggs from Mamourah hens, contamination: no CIT (for detailed information please see the article), conc.: nr, country: Egypt[591], *control
incidence: ?/?, sa. const.: eggs from Mamourah hens, contamination: artificial (dose: 100 ppb CIT in the diet, o., for 6 weeks; for detailed information please see the article), conc.: 10.4 ppb* **, country: Egypt[591], ***egg yolk**, **after 6 weeks
incidence: ?/?, sa. const.: eggs from Mamourah hens, contamination: artificial (dose: 100 ppb CIT in the diet, o., for 6 weeks; for detailed information please see the article), conc.: 6.16 ppb* **, country: Egypt[591], ***egg white**, **after 6 weeks

Cyclopiazonic acid

incidence: ?/?*, sa. const.: eggs from crossbred hens, contamination: no CPA (for detailed information please see the article), conc.: nd, country: Australia[367], *control
incidence: ?/?*, sa. const.: eggs from crossbred hens, contamination: artificial (dose: **2.5 mg CPA/kg**, o., daily for 9 days; for detailed information please see the article), conc. range: ≤350 ng/g ** ***, country: Australia[367], ***egg whites**, **acute study, ***after 9 days of the first dose (also at other day intervals up to 9 days measured, except for the start values lowest conc.: ≈100 ng/g after 7 days)
incidence: ?/?*, sa. const.: eggs from crossbred hens, contamination: artificial (dose: **5 mg CPA/kg**, o., daily for 9 days;

for detailed information please see the article), conc. range: ≤381 ng/g ** ***, country: Australia[367], ***egg whites**, **acute study, ***after 2 days of the first dose (also at other day intervals up to 9 days measured, except for the start value lowest conc.: ≈75 ng/g after 4 days)
incidence: ?/?*, sa. const.: eggs from crossbred hens, contamination: no CPA (for detailed information please see the article), conc.: nd, country: Australia[367], *control
incidence: ?/?*, sa. const.: eggs from crossbred hens, contamination: artificial (dose: **2.5 mg CPA**/kg, o., daily for 9 days; for detailed information please see the article), conc. range: ≤1 ng/g ** ***, country: Australia[367], ***egg yolks**, **acute study, ***after 8 days of the first dose (also at other day intervals up to 9 days measured, generally low conc.)
incidence: ?/?*, sa. const.: eggs from crossbred hens, contamination: artificial (dose: **5 mg CPA**/kg, o., daily for 9 days; for detailed information please see the article), conc. range: ≤11 ng/g ** ***, country: Australia[367], ***egg yolks**, **acute study, ***after 4 days of the first dose (also at other day intervals up to 9 days measured, generally low conc.)
incidence: ?/?*, sa. const.: eggs from crossbred hens, contamination: no CPA (for detailed information please see the article), conc.: nd?, country: Australia[367], *control
incidence: ?/?*, sa. const.: eggs from crossbred hens, contamination: artificial (dose: **1.25 mg CPA**/kg, o., daily for 28 days; for detailed information please see the article), conc. range: ≤430 ng/g**, country: Australia[367], ***egg whites**, **chronic study, ***at day 22 of CPA-administration (also at other day intervals up to 28 days measured, lowest conc.: ≈55 ng/g after 15 days)
incidence: ?/?*, sa. const.: eggs from crossbred hens, contamination: artificial (dose: **2.5 mg CPA**/kg, o., daily for 28 days; for detailed information please see the article), conc. range: ≤193 ng/g ** ***, country: Australia[367], ***egg whites**, **chronic study, ***at day 17 of CPA-administration (also at other day intervals up to 28 days measured, except for the start values lowest conc.: ≈35 ng/g after 14 days)
incidence: ?/?*, sa. const.: eggs from crossbred hens, contamination: no CPA (for detailed information please see the article), conc.: nd?, country: Australia[367], *control
incidence: ?/?*, sa. const.: eggs from crossbred hens, contamination: artificial (dose: **1.25 mg CPA**/kg, o., daily for 28 days; for detailed information please see the article), conc. range: ≤19 ng/g ** ***, country: Australia[367], ***egg yolks**, **chronic study, ***at day 10 of CPA-administration (also at other day intervals up to 28 days measured, except for the start values lowest conc.: ≈2 ng/g after 15 days)
incidence: ?/?*, sa. const.: eggs from crossbred hens, contamination: artificial (dose: **2.5 mg CPA**/kg, o., daily for 28 days; for detailed information please see the article), conc. range: ≤26 ng/g ** ***, country: Australia[367], ***egg yolks**, **chronic study, ***at day 14 of CPA-administration (also at other day intervals up to 28 days measured, except for the start value lowest conc.: ≈3 ng/g after 15 days)

Deoxynivalenol
incidence: ?/?*, sa. const.: eggs from Bovan Goldline hens, which are derived from White Leghorns with addition of other genetic material, contamination: neither DON nor ZEA (for detailed information please see the article), conc.: nd, country: Czech Republic/UK[414], *control
incidence: ?/?, sa. const.: eggs from Bovan Goldline hens, which are derived from White Leghorns with addition of other genetic material, contamination: artificial (dose: **5.0 mg DON**/kg + **137.5 μg ZEA**/kg feed, both for 21 days; for detailed information please see the article), Ø conc. range: 0.10*–0.39** μg/kg, country: Czech Republic/UK[414], after 4* and 2**

weeks of DON- and ZEA-administration (also measured after 1 and 3 weeks but values between pr. values)
incidence: ?/?, sa. const.: eggs from Bovan Goldline hens, which are derived from White Leghorns with addition of other genetic material, contamination: artificial (dose: **7.5 mg DON/kg + 206 µg ZEA/kg** feed, both for 21 days; for detailed information please see the article), Ø conc. range: 0.14*–0.54** µg/kg, country: Czech Republic/UK[414], after 4* and 2** weeks of DON- and ZEA-administration (also measured after 1 and 3 weeks but values between pr. values)
incidence: ?/?, sa. const.: eggs from Bovan Goldline hens, which are derived from White Leghorns with addition of other genetic material, contamination: artificial (dose: **10 mg DON/kg + 275 µg ZEA/kg** feed, both for 21 days; for detailed information please see the article), Ø conc. range: 0.18*–0.44** µg/kg, country: Czech Republic/UK[414], after 4* and 3** weeks of DON- and ZEA-administration (also measured after 1 and 2 weeks but values between pr. values)

incidence: ?/≤10, sa. const.: eggs from White Leghorn hens, contamination: artificial (dose: 2.2 mg DON (labeled), by intubation, once **(acute study)**; for detailed information please see the article), conc. range: ≤1.91 µg/60-g egg* ** *** (mean value), country: Canada[423], *total DON eq., ****albumen**, ***after 24 h (also measured after 48, 72 and 96 h, lowest conc.: 0.11 µg/60-g egg after 96 h)
incidence: ?/?, sa. const.: eggs from White Leghorn hens, contamination: artificial (dose: 2.2 mg DON (unlabeled), o., for 6 days and 2.2 mg DON (labeled), o. for the following 6 days and no DON for the remaining 6 days **(chronic study)**; for detailed information please see the article), conc. range: ≤4.19 µg/60-g egg* ** *** (mean value), country: Canada[423], *total DON eq., ****albumen**, ***after 13 days of the study (also at other day intervals up to 18 days measured, lowest conc.: 0.757 µg/60-g egg after 7 and conc.: 0.868 µg/60-g egg after 18 days)

incidence: ?/?*, sa. const.: eggs from White Leghorn hens, contamination: artificial (dose: no DON (for detailed information please see the article), conc.: nr, country: Canada[424], *control
incidence: ?/?, sa. const.: eggs from White Leghorn hens, contamination: artificial (dose: **5.5 ppm DON** (labeled)/bird/day, o., for 65 days; for detailed information please see the article), conc. range: ≈≤1.7 µg/60 g egg* ** (mean value), country: Canada[424], *DON and/or metabolites, **after 8 days of DON-administration (also at other day intervals up to 80 days measured, lowest conc.: almost nd after 80 days)

Ochratoxin A
incidence: ?/?, sa. const.: eggs from New Hampshire-Leghorn cross hens, contamination: artificial (dose: **0.5 ppm OTA**, o., for 2 weeks; for detailed information please see the article), conc. range: 1.9–8.4 ppb* ** (mean values), Ø conc.: 4.78 ppb* ** (mean value), country: USA[316], *conc. in **yolk**, **after 2 weeks?
incidence: ?/?, sa. const.: eggs from New Hampshire-Leghorn cross hens, contamination: artificial (dose: **0.5 ppm OTA**, o., for 2 weeks; for detailed information please see the article), conc. range: 1.7–8.7 ppb* ** (mean values), Ø conc.: 3.7 ppb* ** (mean value), country: USA[316], *conc. in **albumen**, **after 2 weeks?
incidence: ?/?, sa. const.: eggs from New Hampshire-Leghorn cross hens, contamination: artificial (dose: **5 ppm OTA**, o., for 2 weeks; for detailed information please see the article), conc. range: 3.3–7.8 ppb* ** (mean values), Ø conc.: 5.66 ppb* ** (mean value), country: USA[316], *conc. in **yolk**, **after 2 weeks?
incidence: ?/?, sa. const.: eggs from New Hampshire-Leghorn cross hens, contamination: artificial (dose: **5 ppm**

OTA, o., for 2 weeks; for detailed information please see the article), conc. range: 1.4–2.3 ppb* ** (mean values), Ø conc.: 1.88 ppb* ** (mean value), country: USA[316], *conc. in **albumen**, **after 2 weeks?

incidence: 5/5*, sa. const.: eggs from White Leghorn hens, contamination: no OTA (for detailed information please see the article), conc.: nd, country: Germany[374], *control
incidence: ?/5, sa. const.: eggs from White Leghorn hens, contamination: artificial (dose: **1.3 mg OTA**/kg feed, o., for 28 days; for detailed information please see the article), conc.: 1.6 μg/kg* ** (mean value), country: Germany[374], *in **yolk**, **after feeding OTA for 28 days
incidence: ?/5, sa. const.: eggs from White Leghorn hens, contamination: artificial (dose: **2.6 mg OTA**/kg feed, o., for 28 days; for detailed information please see the article), conc.: 2.5 μg/kg* ** (mean value), country: Germany[374], *in **yolk**, **after feeding OTA for 28 days
incidence: ?/5, sa. const.: eggs from White Leghorn hens, contamination: artificial (dose: **5.2 mg OTA**/kg feed, o., for 28 days; for detailed information please see the article), conc.: 4.0 μg/kg* ** (mean value), country: Germany[374], *in **yolk**, **after feeding OTA for 28 days
incidence: 5/5*, sa. const.: eggs from White Leghorn hens, contamination: no OTA (for detailed information please see the article), conc.: nd, country: Germany[374], *control
incidence: ?/5, sa. const.: eggs from White Leghorn hens, contamination: artificial (dose: **1.3 mg OTA**/kg feed, o., for 28 days; for detailed information please see the article), conc.: 0.1 μg/kg* ** (mean value), country: Germany[374], *in **albumen**, **after feeding OTA for 28 days
incidence: ?/5, sa. const.: eggs from White Leghorn hens, contamination: artificial (dose: **2.6 mg OTA**/kg feed, o., for 28 days; for detailed information please see the article), conc.: 0.2 μg/kg* ** (mean value), country: Germany[374], *in **albumen**, **after feeding OTA for 28 days
incidence: ?/5, sa. const.: eggs from White Leghorn hens, contamination: artificial (dose: **5.2 mg OTA**/kg feed, o., for 28 days; for detailed information please see the article), conc.: 0.1 μg/kg* ** (mean value), country: Germany[374], *in **albumen**, **after feeding OTA for 28 days

incidence: ?/?*, sa. const.: eggs from Rhode Island Red hens, contamination: no OTA (for detailed information please see the article), conc.: nr, country: Poland[590], *control
incidence: ?/?, sa. const.: eggs from Rhode Island Red hens, contamination: artificial (dose: **2.1 ppm OTA** in the diet, o., for 5 weeks; for detailed information please see the article), conc.: 4.1 ppb*, country: Poland[590], *collection of eggs began after 14 days of OTA-administration
incidence: ?/?, sa. const.: eggs from Rhode Island Red hens, contamination: artificial (dose: **4.1 ppm OTA** in the diet, o., for 5 weeks; for detailed information please see the article), conc.: 7.9 ppb*, country: Poland[590], *collection of eggs began after 14 days of OTA-administration

Zearalenone
incidence: ?/7, sa. const.: eggs from White Leghorn hens, contamination: artificial (dose: 10 mg ZEA (labeled)/kg, by gavage into the crop, once), conc. range: ≤22 μg eq ZEA/100 g wet weight* ** (mean value), country: USA[394], *in **white egg**, **after 43–50 h (also measured after 1–10, 21–26 and 69–72 h, lowest conc.: 7 μg eq ZEA/100 g wet weight after 1–10 h)
incidence: ?/4, sa. const.: eggs from White Leghorn hens, contamination: artificial (dose: 10 mg ZEA (labeled)/kg, by gavage into the crop, once), conc. range: ≤195 μg eq ZEA/100 g wet weight* ** (mean value), country: USA[394], *in **yolk egg**, **after 69–72 h (also measured after

1–10, 21–26 and 43–50 h, lowest conc.: 23 μg eq ZEA/100 g wet weight after 1–10 h)

Hen excreta may contain the following mycotoxins and/or their metabolites:

Aflatoxin B_1
incidence: 2/2, sa. const.: brown hens (HNL-type), contamination: artificial (dose: AFB_1 + AFG_1 addition; for detailed information please see the article), conc. range: ≤131.8 ppb*, country: Germany[320], *after 24 h (also measured after 48, 72, 96 and 120 h, lowest conc.: tr after 120 h)

Aflatoxins
incidence: ?/?, sa. const.: White Leghorn hens (Cornell strain K), age: 20–22 weeks, wt.: 1,600–1,800 g, contamination: artificial (dose: 11.26 mg AFs (labeled), by stomach tube, once; for detailed information please see the article), conc.: 0.3976 mg/g (total)* ** ***, country: USA[81], *AFs or their metabolites, **in urinary-fecal excretion, ***mean value from 0 to 168 h (up to 168 h measured)

Deoxynivalenol
incidence: 10/10*, sa. const.: Single Comb White Leghorn hens, age: 26 weeks, contamination: no DAS, DON and ZEA (for detailed information please see the article), conc.: nd, country: Canada[377], *control
incidence: ?/10, sa. const.: Single Comb White Leghorn hens, age: 26 weeks, contamination: artificial (dose: **82.80 mg DON/kg + 10.70 mg ZEA/kg** diet (DAS can be neglected), o., for 27 days; for detailed information please see the article), conc.: 8.7 ppm* (mean value), country: Canada[377], *after 27 days

Fusarenon-X
incidence: ?/?*, sa. const.: White Leghorn hens, age: 55 weeks, contamination: no NIV (for detailed information please see the article), conc.: nr, country: Lithuania/Sweden[375], *control
incidence: ?/?, sa. const.: White Leghorn hens, age: 55 weeks, contamination: artificial (dose: **5 mg NIV**/kg, o., for 50 days; for detailed information please see the article), conc.: tr*, country: Lithuania/Sweden[375], *during the investigation

Nivalenol
incidence: ?/?*, sa. const.: White Leghorn hens, age: 55 weeks, contamination: no NIV (for detailed information please see the article), conc.: nr, country: Lithuania/Sweden[375], *control
incidence: ?/?, sa. const.: White Leghorn hens, age: 55 weeks, contamination: artificial (dose: **5 mg NIV**/kg, o., for 50 days; for detailed information please see the article), conc. range: ≤174 ng/g* (mean value), country: Lithuania/Sweden[375], *after 4–6 days feeding contaminated diet (also measured after 1–3, 19–21 and 48–50 days, lowest conc.: 29 ng/g after 1–3 days)

Deepoxynivalenol
incidence: ?/?*, sa. const.: White Leghorn hens, age: 55 weeks, contamination: no NIV (for detailed information please see the article), conc.: nr, country: Lithuania/Sweden[375], *control
incidence: ?/?, sa. const.: White Leghorn hens, age: 55 weeks, contamination: artificial (dose: **5 mg NIV**/kg, o., for 50 days; for detailed information please see the article), conc. range: ≤292 ng/g* (mean value), country: Lithuania/Sweden[375], *after 4–6 days feeding contaminated diet (also measured after 1–3, 19–21 and 48–50 days, lowest conc.: 13 ng/g after 1–3 days)

Hen fat may contain the following mycotoxins and/or their metabolites:

Aflatoxins
incidence: ?/?*, sa. const.: White Leghorn hens (Cornell strain K), age: 20–22 weeks,

wt.: 1,600–1,800 g, contamination: artificial (dose: 11.26 mg AFs (labeled), by stomach tube, once; for detailed information please see the article), conc. range: ≤6.14 µg/g** ***, country: USA[81], *adipose tissue, **AFs or their metabolites, ***after 4 days (also measured after 1 and 7 days, lowest conc.: 4.14 µg/g after 1 day)

Deoxynivalenol
incidence: ?/?, sa. const.: White Leghorn hens, age: 187 days, wt.: 1.3–1.7 kg, contamination: artificial (dose: 2.2 mg DON (labeled)/bird, o., once; for detailed information please see the article), conc.: 25 ng eq DON* **, country: Canada[135], *DON (or metabolites) in cutaneous fat, **after 96 h?

Ochratoxin A
incidence: 8?/8, sa. const.: New Hampshire-Leghorn cross hens, age: 24 weeks, contamination: artificial (dose: **OTA 0.5 ppm** (feed weight), o., for 2 weeks), conc.: 11.9 ppb* (mean value), country: USA[316], *after 14 days
incidence: 8?/8, sa. const.: New Hampshire-Leghorn cross hens, age: 24 weeks, contamination: artificial (dose: **OTA 5.0 ppm** (feed weight), o., for 2 weeks), conc.: 37.2 ppb* (mean value), country: USA[316], *after 14 days

Zearalenone
incidence: ?/4, sa. const.: White Leghorn laying hens, age: 26–39 weeks, contamination: artificial (dose: 10 mg ZEA (labeled)/kg, by gavage into the crop, once), conc. range: ≤41 µg eq ZEA/100 g wet tissue* (mean value), country: USA[394], *after 72 h (also measured after 2, 4, 24 and 48 h, lowest conc.: 26 µg eq ZEA/100 g wet tissue after 2 h)

Hen Feces see Hen excreta

Hen gallbladder may contain the following mycotoxins and/or their metabolites:

Zearalenone
incidence: ?/4, sa. const.: White Leghorn laying hens, age: 26–39 weeks, contamination: artificial (dose: 10 mg ZEA (labeled)/kg, by gavage into the crop, once), conc. range: ≤4,080 µg eq ZEA/100 g wet tissue* (mean value), country: USA[394], *after 24 h (also measured after 2, 4, 48 and 72 h, lowest conc.: 386 µg eq ZEA/100 g wet tissue after 72 h)

Hen gizzard may contain the following mycotoxins and/or their metabolites:

Aflatoxin B_1
incidence: 8/8*, sa. const.: White Leghorn pullets of the Shaver strain, contamination no AFB_1 + AFB_2 (for detailed information please see the article), conc.: nr, country: USA[110], *control
incidence: 8/8, sa. const.: White Leghorn pullets of the Shaver strain, contamination artificial (dose: **3,310 µg AFB_1/kg + 1,680 µg AFB_2/kg**, o., for 4 weeks; for detailed information please see the article), conc. range: 0.34–1.29 µg/kg*, Ø conc.: 0.67 µg/kg*, country: USA[110], *after feeding 4 weeks an AF-contaminated diet

incidence: 8/8*, sa. const.: White Leghorn pullets of the Shaver strain, contamination no AFB_1 + AFB_2 (for detailed information please see the article), conc.: nr, country: USA[110], *control
incidence: 3/8, sa. const.: White Leghorn pullets of the Shaver strain, contamination artificial (dose: **3,310 µg AFB_1/kg + 1,680 µg AFB_2/kg**, o., for 4 weeks; for detailed information please see the article), conc. range: tr–0.04 µg/kg*,

country: USA[110], *2 days after feeding 4 weeks an AF-contaminated diet

incidence: 16?/16?*, sa. const.: Brown Hyssex laying hens, age: 27 days, contamination: no AF (for detailed information please see the article), conc.: nd, country: Spain[359], *control
incidence: 16?/16?, sa. const.: Brown Hyssex laying hens, age: 27 weeks, contamination: artificial (dose: **2.5 mg AF/kg** feed, o., for 32 days; for detailed information please see the article), conc. range: ≤7.41 μg/kg* (mean value), country: Spain[359] (measured at 4th, 8th*, 16th and 32nd day, lowest conc.: 0.84 μg/kg after 16 days, intoxication period)
incidence: 16?/16?, sa. const.: Brown Hyssex laying hens, age: 27 weeks, contamination: artificial (dose: **5.0 mg AF/kg** feed, o., for 32 days; for detailed information please see the article), conc. range: ≤22.5 μg/kg* (mean value), country: Spain[359] (measured at 4th*, 8th, 16th and 32nd day, lowest conc.: 3.58 μg/kg after 16 days, intoxication period)

Aflatoxin B_2
incidence: 8/8*, sa. const.: White Leghorn pullets of the Shaver strain, contamination no AFB_1 + AFB_2 (for detailed information please see the article), conc.: nr, country: USA[110], *control
incidence: 8/8, sa. const.: White Leghorn pullets of the Shaver strain, contamination artificial (dose: **3,310 μg AFB_1/kg + 1,680 μg AFB_2/kg**, o., for 4 weeks; for detailed information please see the article), conc. range: 0.16–0.73 μg/kg*, Ø conc.: 0.38 μg/kg*, country: USA[110], *after feeding 4 weeks an AF-contaminated diet

incidence: 8/8*, sa. const.: White Leghorn pullets of the Shaver strain, contamination no AFB_1 + AFB_2 (for detailed information please see the article), conc.: nr, country: USA[110], *control
incidence: 6/8, sa. const.: White Leghorn pullets of the Shaver strain, contamination artificial (dose: **3,310 μg AFB_1/kg + 1,680 μg AFB_2/kg**, o., for 4 weeks; for detailed information please see the article), conc. range: tr–0.05 μg/kg*, country: USA[110], *2 days after feeding 4 weeks an AF-contaminated diet

incidence: 16?/16?*, sa. const.: Brown Hyssex laying hens, age: 27 days, contamination: no AF (for detailed information please see the article), conc.: nd, country: Spain[359], *control
incidence: 16?/16?, sa. const.: Brown Hyssex laying hens, age: 27 weeks, contamination: artificial (dose: **2.5 mg AF/kg** feed, o., for 32 days; for detailed information please see the article), conc. range: ≤0.19 μg/kg* (mean value), country: Spain[359] (measured at 4th, 8th*, 16th and 32nd day, lowest conc.: 0.04 μg/kg after 16 and 32 days, intoxication period)
incidence: 16?/16?, sa. const.: Brown Hyssex laying hens, age: 27 weeks, contamination: artificial (dose: **5.0 mg AF/kg** feed, o., for 32 days; for detailed information please see the article), conc. range: ≤0.67 μg/kg* (mean value), country: Spain[359] (measured at 4th*, 8th, 16th and 32nd day, lowest conc.: 0.08 μg/kg after 16 days, intoxication period)

Aflatoxin B_{2a}
incidence: 8/8*, sa. const.: White Leghorn pullets of the Shaver strain, contamination: no AFB_1 + AFB_2 (for detailed information please see the article), conc.: nr, country: USA[110], *control
incidence: 7/8, sa. const.: White Leghorn pullets of the Shaver strain, contamination: artificial (dose: **3,310 μg AFB_1/kg + 1,680 μg AFB_2/kg**, o., for 4 weeks; for detailed information please see

the article), conc. range: tr–0.10 µg/kg*, country: USA[110], *after feeding 4 weeks an AF-contaminated diet

incidence: 8/8*, sa. const.: White Leghorn pullets of the Shaver strain, contamination: no AFB_1 + AFB_2 (for detailed information please see the article), conc.: nr, country: USA[110], *control
incidence: 2/8, sa. const.: White Leghorn pullets of the Shaver strain, contamination: artificial (dose: **3,310 µg AFB_1/kg + 1,680 µg AFB_2/kg**, o., for 4 weeks; for detailed information please see the article), conc. range: tr*, country: USA[110], *2 days after feeding 4 weeks an AF-contaminated diet

Aflatoxin G_1
incidence: 16?/16?*, sa. const.: Brown Hyssex laying hens, age: 27 days, contamination: no AF (for detailed information please see the article), conc.: nd, country: Spain[359], *control
incidence: 16?/16?, sa. const.: Brown Hyssex laying hens, age: 27 weeks, contamination: artificial (dose: **2.5 mg AF/kg** feed, o., for 32 days; for detailed information please see the article), conc. range: ≤1.62 µg/kg* (mean value), country: Spain[359] (measured at 4th, 8th*, 16th and 32nd day, lowest conc.: 0.11 µg/kg after 32 days)
incidence: 16?/16?, sa. const.: Brown Hyssex laying hens, age: 27 weeks, contamination: artificial (dose: **5.0 mg AF/kg** feed, o., for 32 days; for detailed information please see the article), conc. range: ≤7.32 µg/kg* (mean value), country: Spain[359] (measured at 4th*, 8th, 16th and 32nd day, lowest conc.: 0.94 µg/kg after 16 days)

Aflatoxin G_2
incidence: 16?/16?*, sa. const.: Brown Hyssex laying hens, age: 27 days, contamination: no AF (for detailed information please see the article), conc.: nd, country: Spain[359], *control
incidence: 16?/16?, sa. const.: Brown Hyssex laying hens, age: 27 weeks, contamination: artificial (dose: **2.5 mg AF/kg** feed, o., for 32 days; for detailed information please see the article), conc.: nd, country: Spain[359] (measured at 4th, 8th, 16th and 32nd day, intoxication period)
incidence: 2/16?, sa. const.: Brown Hyssex laying hens, age: 27 weeks, contamination: artificial (dose: **5.0 mg AF/kg** feed, o., for 32 days; for detailed information please see the article), conc. range: tr*, country: Spain[359] (measured at 4th*, 8th, 16th and 32nd day, intoxication period)

Aflatoxin M_1
incidence: 8/8*, sa. const.: White Leghorn pullets of the Shaver strain, contamination: no AFB_1 + AFB_2 (for detailed information please see the article), conc.: nr, country: USA[110], *control
incidence: 4/8, sa. const.: White Leghorn pullets of the Shaver strain, contamination: artificial (dose: **3,310 µg AFB_1/kg + 1,680 µg AFB_2/kg**, o., for 4 weeks; for detailed information please see the article), conc. range: tr–0.01 µg/kg*, country: USA[110], *after feeding 4 weeks an AF-contaminated diet

incidence: 8/8*, sa. const.: White Leghorn pullets of the Shaver strain, contamination: no AFB_1 + AFB_2 (for detailed information please see the article), conc.: nr, country: USA[110], *control
incidence: 8/8, sa. const.: White Leghorn pullets of the Shaver strain, contamination: artificial (dose: **3,310 µg AFB_1/kg + 1,680 µg AFB_2/kg**, o., for 4 weeks; for detailed information please see the article), conc.: nd* and NR, country: USA[110], *2 days after feeding 4 weeks an AF-contaminated diet

Aflatoxin M_2
incidence: 8/8*, sa. const.: White Leghorn pullets of the Shaver strain, contamination: no AFB_1 + AFB_2 (for

detailed information please see the article), conc.: nr, country: USA[110], *control
incidence: 5/8, sa. const.: White Leghorn pullets of the Shaver strain, contamination: artificial (dose: **3,310 µg AFB$_1$/kg + 1,680 µg AFB$_2$/kg**, o., for 4 weeks; for detailed information please see the article), conc. range: tr–0.02 µg/kg*, country: USA[110], *after feeding 4 weeks an AF-contaminated diet

incidence: 8/8*, sa. const.: White Leghorn pullets of the Shaver strain, contamination: no AFB$_1$ + AFB$_2$ (for detailed information please see the article), conc.: nr, country: USA[110], *control
incidence: 16/16, sa. const.: White Leghorn pullets of the Shaver strain, contamination: artificial (dose: **3,310 µg AFB$_1$/kg + 1,680 µg AFB$_2$/kg**, o., for 4 weeks; for detailed information please see the article), conc.: nd* and NR, country: USA[110], *2 days after feeding 4 weeks an AF-contaminated diet

Aflatoxins
incidence: ?/?, sa. const.: White Leghorn hens (Cornell strain K), age: 20–22 weeks, wt.: 1,600–1,800 g, contamination: artificial (dose: 11.26 mg AFs (labeled), by stomach tube, once; for detailed information please see the article), conc. range: ≤8.65 µg/g* **, country: USA[81], *AFs or their metabolites, **after 1 day (also measured after 4 and 7 days, lowest conc.: 4.48 µg/g after 4 days)
incidence: ?/?, sa. const.: White Leghorn hens (Cornell strain K), age: 20–22 weeks, wt.: 1,600–1,800 g, contamination artificial (dose: 11.26 mg AFs (labeled), by stomach tube, once; for detailed information please see the article), conc. range: ≤103.10 µg/g* ** ***, country: USA[81], *AFs or their metabolites, **in **contents of crop and gizzard**, ***after 4 days (also measured after 1 and 7 days, lowest conc.: 31.33 µg/g after 7 days)

Deoxynivalenol
incidence: 10/10*, sa. const.: Single Comb White Leghorn hens, age: 26 weeks, contamination: no DAS, DON and ZEA (for detailed information please see the article), conc.: nd, country: Canada[377], *control
incidence: ?/10, sa. const.: Single Comb White Leghorn hens, age: 26 weeks, contamination: artificial (dose: **82.80 mg DON/kg + 10.70 mg ZEA**/kg diet (DAS can be neglected), o., for 27 days; for detailed information please see the article), conc.: ≈20 ppm* (mean value), country: Canada[377], *after 27 days

Ochratoxin A
incidence: 8?/8, sa. const.: New Hampshire-Leghorn cross hens, age: 24 weeks, contamination: artificial (dose: **OTA 0.5 ppm** (feed weight), o., for 2 weeks), conc.: 7.3 ppb* (mean value), country: USA[316], *after 14 days
incidence: 8?/8, sa. const.: New Hampshire-Leghorn cross hens, age: 24 weeks, contamination: artificial (dose: **OTA 5.0 ppm** (feed weight), o., for 2 weeks), conc.: 73.7 ppb* (mean value), country: USA[316], *after 14 days

Hen heart may contain the following mycotoxins and/or their metabolites:

Aflatoxin B$_1$
incidence: 8/8*, sa. const.: White Leghorn pullets of the Shaver strain, contamination: no AFB$_1$ + AFB$_2$ (for detailed information please see the article), conc.: nr, country: USA[110], *control
incidence: ?/?, sa. const.: White Leghorn pullets of the Shaver strain, contamination: artificial (dose: **3,310 µg AFB$_1$/kg + 1,680 µg AFB$_2$/kg**, o., for 4 weeks; for detailed information please see the article), conc. range: 0.08–0.18 µg/kg* (mean values), country: USA[110], *after feeding 4 weeks an AF-contaminated diet

incidence: 8/8*, sa. const.: White Leghorn pullets of the Shaver strain, contamination: no AFB_1 + AFB_2 (for detailed information please see the article), conc.: nr, country: USA[110], *control
incidence: 16/16, sa. const.: White Leghorn pullets of the Shaver strain, contamination: artificial (dose: **3,310 μg AFB_1/kg + 1,680 μg AFB_2/kg**, o., for 4 weeks; for detailed information please see the article), conc.: nd*, country: USA[110], *2 days after feeding 4 weeks an AF-contaminated diet

Aflatoxin B_2
incidence: 8/8*, sa. const.: White Leghorn pullets of the Shaver strain, contamination: no AFB_1 + AFB_2 (for detailed information please see the article), conc.: nr, country: USA[110], *control
incidence: ?/?, sa. const.: White Leghorn pullets of the Shaver strain, contamination: artificial (dose: **3,310 μg AFB_1/kg + 1,680 μg AFB_2/kg**, o., for 4 weeks; for detailed information please see the article), conc. range: 0.05 μg/kg* (mean value), country: USA[110], *after feeding 4 weeks an AF-contaminated diet

incidence: 8/8*, sa. const.: White Leghorn pullets of the Shaver strain, contamination: no AFB_1 + AFB_2 (for detailed information please see the article), conc.: nr, country: USA[110], *control
incidence: 16/16, sa. const.: White Leghorn pullets of the Shaver strain, contamination: artificial (dose: **3,310 μg AFB_1/kg + 1,680 μg AFB_2/kg**, o., for 4 weeks; for detailed information please see the article), conc.: nd*, country: USA[110], *2 days after feeding 4 weeks an AF-contaminated diet

Aflatoxin B_{2a}
incidence: 8/8*, sa. const.: White Leghorn pullets of the Shaver strain, contamination: no AFB_1 + AFB_2 (for detailed information please see the article), conc.: nr, country: USA[110], *control
incidence: ?/?, sa. const.: White Leghorn pullets of the Shaver strain, contamination: artificial (dose: **3,310 μg AFB_1/kg + 1,680 μg AFB_2/kg**, o., for 4 weeks; for detailed information please see the article), conc. range: 0.05–0.06 μg/kg* (mean values), country: USA[110], *after feeding 4 weeks an AF-contaminated diet

incidence: 8/8*, sa. const.: White Leghorn pullets of the Shaver strain, contamination: no AFB_1 + AFB_2 (for detailed information please see the article), conc.: nr, country: USA[110], *control
incidence: 16/16, sa. const.: White Leghorn pullets of the Shaver strain, contamination: artificial (dose: **3,310 μg AFB_1/kg + 1,680 μg AFB_2/kg**, o., for 4 weeks; for detailed information please see the article), conc.: nd*, country: USA[110], *2 days after feeding 4 weeks an AF-contaminated diet

Aflatoxin M_1
incidence: 8/8*, sa. const.: White Leghorn pullets of the Shaver strain, contamination: no AFB_1 + AFB_2 (for detailed information please see the article), conc.: nr, country: USA[110], *control
incidence: ?/?, sa. const.: White Leghorn pullets of the Shaver strain, contamination: artificial (dose: **3,310 μg AFB_1/kg + 1,680 μg AFB_2/kg**, o., for 4 weeks; for detailed information please see the article), conc.: tr* (mean value), country: USA[110], *after feeding 4 weeks an AF-contaminated diet

incidence: 8/8*, sa. const.: White Leghorn pullets of the Shaver strain, contamination: no AFB_1 + AFB_2 (for detailed information please see the article), conc.: nr, country: USA[110], *control
incidence: 16/16, sa. const.: White Leghorn pullets of the Shaver strain,

contamination: artificial (dose: **3,310 μg AFB_1/kg + 1,680 μg AFB_2/kg**, o., for 4 weeks; for detailed information please see the article), conc.: nd*, country: USA[110], *2 days after feeding 4 weeks an AF-contaminated diet

Aflatoxin M_2

incidence: 8/8*, sa. const.: White Leghorn pullets of the Shaver strain, contamination: no AFB_1 + AFB_2 (for detailed information please see the article), conc.: nr, country: USA[110], *control
incidence: ?/?, sa. const.: White Leghorn pullets of the Shaver strain, contamination: artificial (dose: **3,310 μg AFB_1/kg + 1,680 μg AFB_2/kg**, o., for 4 weeks; for detailed information please see the article), conc.: 0.02 μg/kg* (mean value), country: USA[110], *after feeding 4 weeks an AF-contaminated diet
incidence: 8/8*, sa. const.: White Leghorn pullets of the Shaver strain, contamination: no AFB_1 + AFB_2 (for detailed information please see the article), conc.: nr, country: USA[110], *control
incidence: 16/16, sa. const.: White Leghorn pullets of the Shaver strain, contamination: artificial (dose: **3,310 μg AFB_1/kg + 1,680 μg AFB_2/kg**, o., for 4 weeks; for detailed information please see the article), conc.: nd*, country: USA[110], *2 days after feeding 4 weeks an AF-contaminated diet

Aflatoxins

incidence: ?/?, sa. const.: White Leghorn hens (Cornell strain K), age: 20–22 weeks, wt.: 1,600–1,800 g, contamination: artificial (11.26 mg AFs (labeled), by stomach tube, once; for detailed information please see the article), conc. range: ≤16.59 μg/g* **, country: USA[81], *AFs or their metabolites, **after 7 days (also measured after 1 and 4 days, lowest conc.: 7.07 μg/g after 4 days)

Ochratoxin A

incidence: 8?/8, sa. const.: New Hampshire-Leghorn cross hens, age: 24 weeks, contamination: artificial (dose: **OTA 0.5 ppm** (feed weight), o., for 2 weeks), conc.: 12.0 ppb* (mean value), country: USA[316], *after 14 days
incidence: 8?/8, sa. const.: New Hampshire-Leghorn cross hens, age: 24 weeks, contamination: artificial (dose: **OTA 5.0 ppm** (feed weight), o., for 2 weeks), conc.: 43.8 ppb* (mean value), country: USA[316], *after 14 days

incidence: 24/24*, sa. const.: Rhode Island Red hens, age: 54 weeks, contamination: no OTA (for detailed information please see the article), conc.: nd, country: Poland[590], *control
incidence: ?/22, sa. const.: Rhode Island Red **hens**, age: 54 weeks, contamination: artificial (dose: **2.1 ppm OTA** in the diet, o., for 5 weeks; for detailed information please see the article), conc.: 2.1 ppb*, country: Poland[590], *after 5 weeks?
incidence: ?/2, sa. const.: Rhode Island Red **cockerels**, age: 54 weeks, contamination: artificial (dose: **2.1 ppm OTA** in the diet, o., for 5 weeks; for detailed information please see the article), conc.: 0.07 ppb*, country: Poland[590], *after 5 weeks?
incidence: ?/22, sa. const.: Rhode Island Red **hens**, age: 54 weeks, contamination: artificial (dose: **4.1 ppm OTA** in the diet, o., for 5 weeks; for detailed information please see the article), conc.: 18.6 ppb*, country: Poland[590], *after 5 weeks?
incidence: ?/2, sa. const.: Rhode Island Red **cockerels**, age: 54 weeks, contamination: artificial (dose: **4.1 ppm OTA** in the diet, o., for 5 weeks; for detailed information please see the article), conc.: nd*, country: Poland[590], *after 5 weeks?

Zearalenone

incidence: ?/4, sa. const.: White Leghorn laying hens, age: 26–39 weeks, contamination: artificial (dose: 10 mg ZEA (labeled)/kg, by gavage into the crop, once), conc. range: ≤57 μg eq ZEA/100 g wet tissue (mean value), country: USA[394], *after 2 h (also measured after 4, 24, 48 and 72 h, lowest conc.: 25 µg eq ZEA/100 g wet tissue after 48 h)

Hen intestine may contain the following mycotoxins and/or their metabolites:

Ochratoxin A
incidence: 8?/8, sa. const.: New Hampshire-Leghorn cross hens, age: 24 weeks, contamination: artificial (dose: **OTA 0.5 ppm** (feed weight), o., for 2 weeks), conc.: 21.3 ppb* (mean value), country: USA[316], *after 14 days
incidence: 8?/8, sa. const.: New Hampshire-Leghorn cross hens, age: 24 weeks, contamination: artificial (dose: **OTA 5.0 ppm** (feed weight), o., for 2 weeks), conc.: 33.8 ppb* (mean value), country: USA[316], *after 14 days

Hen kidney may contain the following mycotoxins and/or their metabolites:

Aflatoxicol
incidence: ?/?*, sa. const.: Hubbard strain laying hens, age: 14 days, contamination: no AFB_1 (for detailed information please see the article), conc.: nr, country: Italy[232], *control
incidence: 3?/3, sa. const.: Hubbard strain laying hens, age: 14 days, contamination: artificial (dose: **50 µg AFB_1/kg** feed, o., for 169 days; for detailed information please see the article), conc.: 0.04 µg/kg* (mean value), country: Italy[232], *after 169 days

incidence: 6/6*, sa. const.: Single Comb White Leghorn hens, contamination: no AFB_1 (for detailed information please see the article), conc.: nd, country: USA[363], *control
incidence: 9/10, sa. const.: Single Comb White Leghorns hens, contamination: artificial (dose: **8 µg AFB_1**/g feed, for 7 days; for detailed information please see the article), conc. range: ≤0.15 ng/g* **, country: USA[363], *AFL-residues, **after 7 days

incidence: ?/?*, sa. const.: Hubbard strain laying hens, age: 14 days, contamination: no AFB_1 + OTA (for detailed information please see the article), conc.: nr, country: Italy[515], *control
incidence: 3?/3, sa. const.: Hubbard strain laying hens, age: 14 days, contamination: artificial (dose: **50 µg AFB_1/kg** feed + **50 µg OTA/kg** feed, o., for 169 days; for detailed information please see the article), conc.: 0.40 µg/kg* (mean value), country: Italy[515], *after 169 days

Aflatoxin B_1
incidence: 8/8*, sa. const.: White Leghorn pullets of the Shaver strain, contamination: no AFB_1 + AFB_2 (for detailed information please see the article), conc.: nr, country: USA[110], *control
incidence: ?/?, sa. const.: White Leghorn pullets of the Shaver strain, contamination: artificial (dose: **3,310 µg AFB_1/kg + 1,680 µg AFB_2/kg**, o., for 4 weeks; for detailed information please see the article), conc. range: 0.27–0.87 µg/kg* (mean values), country: USA[110], *after feeding 4 weeks an AF-contaminated diet

incidence: 8/8*, sa. const.: White Leghorn pullets of the Shaver strain, contamination: no AFB_1 + AFB_2 (for detailed information please see the article), conc.: nr, country: USA[110], *control
incidence: ?/?, sa. const.: White Leghorn pullets of the Shaver strain, contamination: artificial (dose: **3,310 µg AFB_1/kg + 1,680 µg AFB_2/kg**, o., for 4 weeks; for detailed information please see the article), conc.: 0.02 µg/kg* (mean value), country: USA[110], *2 days after feeding 4 weeks an AF-contaminated diet

incidence: ?/?*, sa. const.: Hubbard strain laying hens, age: 14 days, contamination: no AFB_1 (for detailed information please see the article), conc.: nr, country: Italy[232], *control incidence: 3?/3, sa. const.: Hubbard strain laying hens, age: 14 days, contamination: artificial (dose: **50 µg AFB_1/kg** feed, o., for 169 days; for detailed information please see the article), conc.: 0.08 µg/kg* (mean value), country: Italy[232], *after 169 days

incidence: 16?/16?*, sa. const.: Brown Hyssex laying hens, age: 27 days, contamination: no AF (for detailed information please see the article), conc.: nd, country: Spain[359], *control
incidence: 16?/16?, sa. const.: Brown Hyssex laying hens, age: 27 weeks, contamination: artificial (dose: **2.5 mg AF/kg** feed, o., for 32 days; for detailed information please see the article), conc. range: ≤0.25 µg/kg* (mean value), country: Spain[359] (measured at 4th, 8th*, 16th and 32nd day, lowest conc.: 0.07 µg/kg after 32 days, intoxication period)
incidence: 16?/16?, sa. const.: Brown Hyssex laying hens, age: 27 weeks, contamination: artificial (dose: **5.0 mg AF/kg** feed, o., for 32 days; for detailed information please see the article), conc. range: ≤0.37 µg/kg* (mean value), country: Spain[359] (measured at 4th, 8th*, 16th and 32nd* day, lowest conc.: 0.1 µg/kg after 32 days, intoxication period)

incidence: 6/6*, sa. const.: Single Comb White Leghorn hens, contamination: no AFB_1 (for detailed information please see the article), conc.: nd, country: USA[363], *control
incidence: 7/10, sa. const.: Single Comb White Leghorns hens, contamination: artificial (dose: **8 µg AFB_1/g** feed, for 7 days; for detailed information please see the article), conc. range: ≤0.62 ng/g* **, country: USA[363], *AFB_1-residues, **after 7 days

incidence: ?/?*, sa. const.: Hubbard strain laying hens, age: 14 days, contamination: no AFB_1 + OTA (for detailed information please see the article), conc.: nr, country: Italy[515], *control
incidence: 3?/3, sa. const.: Hubbard strain laying hens, age: 14 days, contamination: artificial (dose: **50 µg AFB_1/kg** feed + **50 µg OTA/kg** feed, o., for 169 days; for detailed information please see the article), conc.: 0.32 µg/kg* (mean value), country: Italy[515], *after 169 days

Aflatoxin B_2
incidence: 8/8*, sa. const.: White Leghorn pullets of the Shaver strain, contamination: no AFB_1 + AFB_2 (for detailed information please see the article), conc.: nr, country: USA[110], *control
incidence: ?/?, sa. const.: White Leghorn pullets of the Shaver strain, contamination: artificial (dose: **3,310 µg AFB_1/kg + 1,680 µg AFB_2/kg**, o., for 4 weeks; for detailed information please see the article), conc. range: 0.12–0.30 µg/kg* (mean values), country: USA[110], *after feeding 4 weeks an AF-contaminated diet

incidence: 8/8*, sa. const.: White Leghorn pullets of the Shaver strain, contamination: no AFB_1 + AFB_2 (for detailed information please see the article), conc.: nr, country: USA[110], *control
incidence: ?/?, sa. const.: White Leghorn pullets of the Shaver strain, contamination: artificial (dose: **3,310 µg AFB_1/kg + 1,680 µg AFB_2/kg**, o., for 4 weeks; for detailed information please see the article), conc. range: 0.01 µg/kg* (mean value), Ø conc.: 0.01 µg/kg**, country: USA[110], *2 days after feeding 4 weeks an AF-contaminated diet

Aflatoxin B_{2a}
incidence: 8/8*, sa. const.: White Leghorn pullets of the Shaver strain, contamination: no AFB_1 + AFB_2 (for detailed information please see the article), conc.: nr, country: USA[110], *control
incidence: ?/?, sa. const.: White Leghorn pullets of the Shaver strain, contamination: artificial (dose: **3,310 µg AFB_1/kg + 1,680 µg AFB_2/kg**, o., for 4 weeks; for detailed information please see the article), conc. range: 1.93–2.30 µg/kg* (mean values), country: USA[110], *after feeding 4 weeks an AF-contaminated diet

incidence: 8/8*, sa. const.: White Leghorn pullets of the Shaver strain, contamination: no AFB_1 + AFB_2 (for detailed information please see the article), conc.: nr, country: USA[110], *control
incidence: ?/?, sa. const.: White Leghorn pullets of the Shaver strain, contamination: artificial (dose: **3,310 µg AFB_1/kg + 1,680 µg AFB_2/kg**, o., for 4 weeks; for detailed information please see the article), conc. range: 0.03 µg/kg* (mean value), Ø conc.: 0.03 µg/kg*, country: USA[110], *2 days after feeding 4 weeks an AF-contaminated diet

Aflatoxin M_1
incidence: 8/8*, sa. const.: White Leghorn pullets of the Shaver strain, contamination: no AFB_1 + AFB_2 (for detailed information please see the article), conc.: nr, country: USA[110], *control
incidence: ?/?, sa. const.: White Leghorn pullets of the Shaver strain, contamination: artificial (dose: **3,310 µg AFB_1/kg + 1,680 µg AFB_2/kg**, o., for 4 weeks; for detailed information please see the article), conc. range: 0.07–0.11 µg/kg* (mean values), country: USA[110], *after feeding 4 weeks an AF-contaminated diet

incidence: 8/8*, sa. const.: White Leghorn pullets of the Shaver strain, contamination: no AFB_1 + AFB_2 (for detailed information please see the article), conc.: nr, country: USA[110], *control
incidence: 16/16, sa. const.: White Leghorn pullets of the Shaver strain, contamination: artificial (dose: **3,310 µg AFB_1/kg + 1,680 µg AFB_2/kg**, o., for 4 weeks; for detailed information please see the article), conc.: nd*, country: USA[110], *2 days after feeding 4 weeks an AF-contaminated diet

incidence: ?/?*, sa. const.: Hubbard strain laying hens, age: 14 days, contamination: no AFB_1 (for detailed information please see the article), conc.: nr, country: Italy[232], *control
incidence: 3?/3, sa. const.: Hubbard strain laying hens, age: 14 days, contamination: artificial (dose: **50 µg AFB_1/kg** feed, o., for 169 days; for detailed information please see the article), conc.: 0.01 µg/kg* (mean value), country: Italy[232], *after 169 days

incidence: 6/6*, sa. const.: Single Comb White Leghorn hens, contamination: no AFB_1 (for detailed information please see the article), conc.: nd, country: USA[363], *control
incidence: 9/10, sa. const.: Single Comb White Leghorns hens, contamination: artificial (dose: **8 µg AFB_1/g** feed, for 7 days; for detailed information please see the article), conc. range: ≤0.10 ng/g* **, country: USA[363], *AFM_1-residues, **after 7 days

incidence: ?/?*, sa. const.: Hubbard strain laying hens, age: 14 days, contamination: no AFB_1 + OTA (for detailed information please see the article), conc.: nr, country: Italy[515], *control
incidence: 3?/3, sa. const.: Hubbard strain laying hens, age: 14 days, contamination: artificial (dose: **50 µg AFB_1 + 50 µg OTA**/kg feed, o., for 169 days; for detailed information please see the article), conc.: 0.01 µg/kg* (mean value), country: Italy[515], *after 169 days

Aflatoxin M_2
incidence: 8/8*, sa. const.: White Leghorn pullets of the Shaver strain, contamination: no AFB_1 + AFB_2 (for detailed information please see the article), conc.: nr, country: USA[110], *control
incidence: ?/?, sa. const.: White Leghorn pullets of the Shaver strain, contamination: artificial (dose: **3,310 µg AFB_1/kg + 1,680 µg AFB_2/kg**, o., for 4 weeks; for detailed information please see the article), conc. range: 0.03–0.08 µg/kg* (mean values), country: USA[110], *after feeding 4 weeks an AF-contaminated diet

incidence: 8/8*, sa. const.: White Leghorn pullets of the Shaver strain, contamination: no AFB_1 + AFB_2 (for detailed information please see the article), conc.: nr, country: USA[110], *control
incidence: 16/16, sa. const.: White Leghorn pullets of the Shaver strain, contamination: artificial (dose: **3,310 µg AFB_1/kg + 1,680 µg AFB_2/kg**, o., for 4 weeks; for detailed information please see the article), conc.: nd*, country: USA[110], *2 days after feeding 4 weeks an AF-contaminated diet

Deoxynivalenol
incidence: ?/?, sa. const.: White Leghorn hens, age: 316 days, wt.: 1.3–1.7 kg, contamination: artificial (dose: 2.2 mg DON (unlabeled)/day/bird, o., for 6 days followed by 2.2 mg DON (labeled)/day/bird, o., for 6 days; for detailed information please see the article), conc. range: ≈≤60 ng/g* (mean value), country: Canada[135], *after 8 days of DON-administration (also measured after 10, 12, 14, 16 and 18 days, lowest conc.: nd after 96 h?)
incidence: ?/?, sa. const.: White Leghorn hens, age: 187 days, wt.: 1.3–1.7 kg, contamination: artificial (dose: 2.2 mg DON (labeled)/bird, o., once; for detailed information please see the article), conc.: 25 ng eq DON* **, country: Canada[135], *DON (or metabolites), **after 96 h? (lowest conc.: significantly lower than the above indicated value after 18 days)

Ochratoxin A
incidence: 8?/8, sa. const.: New Hampshire-Leghorn cross hens, age: 24 weeks, contamination: artificial (dose: **OTA 0.5 ppm** (feed weight), o., for 2 weeks), conc.: 124.1 ppb* (mean value), country: USA[316], *after 14 days
incidence: 8?/8, sa. const.: New Hampshire-Leghorn cross hens, age: 24 weeks, contamination: artificial (dose: **OTA 5.0 ppm** (feed weight), o., for 2 weeks), conc.: 124.0 ppb* (mean value), country: USA[316], *after 14 days

incidence: ?/?*, sa. const.: Hubbard strain laying hens, age: 14 days, contamination: no AFB_1 + OTA (for detailed information please see the article), conc.: nr, country: Italy[515], *control
incidence: 3?/3, sa. const.: Hubbard strain laying hens, age: 14 days, contamination: artificial (dose: **50 µg AFB_1 + 50 µg OTA/kg** feed, o., for 169 days; for detailed information please see the article), conc.: 3.1 µg/kg* (mean value), country: Italy[515], *after 169 days
incidence: ?/?*, sa. const.: Hubbard strain laying hens, age: 14 days, contamination: no AFB_1 + OTA (for detailed information please see the article), conc.: nr, country: Italy[515], *control
incidence: 3?/3, sa. const.: Hubbard strain laying hens, age: 14 days, contamination: artificial (dose: **50 µg AFB_1 + 50 µg OTA/kg** feed, o., for 87 days; for detailed information please see the article), conc. range: 0.5 µg/kg* (mean value), country: Italy[515], *33 and 82 days after withdrawal from treatment

incidence: 12/12*, sa. const.: H and N White Leghorn hens, age: 26 weeks, contamination: no OTA (for detailed information please see the article), conc.: nd, country: Canada[525], *control
incidence: ?/8, sa. const.: H and N White Leghorn hens, age: 26 weeks, contamination: artificial (dose: **0.5 ppm OTA** in feed, o., for 42 days; for detailed information please see the article), conc.: 36.8 ppm* (mean value), country: Canada[525], *directly after withdrawal from treated feed for 6 weeks
incidence: ?/12, sa. const.: H and N White Leghorn hens, age: 26 weeks, contamination: artificial (dose: **1.0 ppm OTA** in feed, o., for 42 days; for detailed information please see the article), conc.: 77.0 ppm* (mean value), country:

Canada[525], *directly after withdrawal from treated feed for 6 weeks
incidence: ?/16, sa. const.: H and N White Leghorn hens, age: 26 weeks, contamination: artificial (dose: **4.0 ppm OTA** in feed, o., for 42 days; for detailed information please see the article), conc.: 106.9 ppm* (mean value), country: Canada[525], *directly after withdrawal from treated feed for 6 weeks
incidence: ?/16, sa. const.: H and N White Leghorn hens, age: 26 weeks, contamination: artificial (dose: **4.0 ppm OTA** in feed, o., for 42 days; for detailed information please see the article), conc.: 31.0 ppm* (mean value), country: Canada[525], ***after 24 h off** treated feed for 6 weeks
incidence: ?/16, sa. const.: H and N White Leghorn hens, age: 26 weeks, contamination: artificial (dose: **4.0 ppm OTA** in feed, o., for 42 days; for detailed information please see the article), conc.: 13.1 ppm* (mean value), country: Canada[525], ***after 48 h off** treated feed for 6 weeks

incidence: 24/24*, sa. const.: Rhode Island Red hens, age: 54 weeks, contamination: no OTA (for detailed information please see the article), conc.: nd, country: Poland[590], *control
incidence: ?/22, sa. const.: Rhode Island Red **hens**, age: 54 weeks, contamination: artificial (dose: **2.1 ppm OTA** in the diet, o., for 5 weeks; for detailed information please see the article), conc.: 8.1 ppb*, country: Poland[590], *after 5 weeks?
incidence: ?/2, sa. const.: Rhode Island Red **cockerels**, age: 54 weeks, contamination: artificial (dose: **2.1 ppm OTA** in the diet, o., for 5 weeks; for detailed information please see the article), conc.: 2.9 ppb*, country: Poland[590], *after 5 weeks?
incidence: ?/22, sa. const.: Rhode Island Red **hens**, age: 54 weeks, contamination: artificial (dose: **4.1 ppm OTA** in the diet, o., for 5 weeks; for detailed information please see the article), conc.: 9.0 ppb*, country: Poland[590], *after 5 weeks?
incidence: ?/2, sa. const.: Rhode Island Red **cockerels**, age: 54 weeks, contamination: artificial (dose: **4.1 ppm OTA** in the diet, o., for 5 weeks; for detailed information please see the article), conc.: 5.9 ppb*, country: Poland[590], *after 5 weeks?

ZEARALENONE
incidence: ?/4, sa. const.: White Leghorn laying hens, age: 26–39 weeks, contamination: artificial (dose: 10 mg ZEA (labeled)/kg, by gavage into the crop, once), conc. range: ≤144 µg eq ZEA/100 g wet tissue* (mean value), country: USA[394], *after 24 h (also measured after 2, 4, 48 and 72 h, lowest conc.: 31 µg eq ZEA/100 g wet tissue after 72 h)

Hen Leg see Hen muscle, leg

Hen liver may contain the following mycotoxins and/or their metabolites:

AFLATOXICOL
incidence: ?/?*, sa. const.: Hubbard strain laying hens, age: 14 days, contamination: no AFB_1 (for detailed information please see the article), conc.: nr, country: Italy[232], *control
incidence: 3?/3, sa. const.: Hubbard strain laying hens, age: 14 days, contamination: artificial (dose: **50 µg AFB_1/kg** feed, o., for 169 days; for detailed information please see the article), conc.: 0.60 µg/kg* (mean value), country: Italy[232], *after 169 days

incidence: 6/6*, sa. const.: Single Comb White Leghorn hens, contamination: no AFB_1 (for detailed information please see the article), conc.: nd, country: USA[363], *control
incidence: 10/10, sa. const.: Single Comb White Leghorns hens, contamination: artificial (dose: **8 µg AFB_1/g** feed, for

7 days; for detailed information please see the article), conc. range: ≤0.32 ng/g* **, country: USA[363], *AFL-residues, **after 7 days

incidence: ?/?*, sa. const.: Hubbard strain laying hens, age: 14 days, contamination: no AFB_1 + OTA (for detailed information please see the article), conc.: nr, country: Italy[515], *control
incidence: 3?/3, sa. const.: Hubbard strain laying hens, age: 14 days, contamination: artificial (dose: **50 μg AFB_1/kg** feed + **50 μg OTA/kg** feed, o., for 169 days; for detailed information please see the article), conc.: 1.80 μg/kg* (mean value), country: Italy[515], *after 169 days

Aflatoxin B_1
incidence: 8/8*, sa. const.: White Leghorn pullets of the Shaver strain, contamination: no AFB_1 + AFB_2 (for detailed information please see the article), conc.: nr, country: USA[110], *control
incidence: 8/8, sa. const.: White Leghorn pullets of the Shaver strain, contamination: artificial (dose: **3,310 μg AFB_1/kg** + **1,680 μg AFB_2/kg**, o., for 4 weeks; for detailed information please see the article), conc. range: 0.07–0.44 μg/kg*, Ø conc.: 0.2 μg/kg*, country: USA[110], *after feeding 4 weeks an AF-contaminated diet

incidence: 8/8*, sa. const.: White Leghorn pullets of the Shaver strain, contamination: no AFB_1 + AFB_2 (for detailed information please see the article), conc.: nr, country: USA[110], *control
incidence: 1/8, sa. const.: White Leghorn pullets of the Shaver strain, contamination: artificial (dose: **3,310 μg AFB_1/kg** + **1,680 μg AFB_2/kg**, o., for 4 weeks; for detailed information please see the article), conc.: tr*, country: USA[110], *2 days after feeding 4 weeks an AF-contaminated diet

incidence: ?/?*, sa. const.: Hubbard strain laying hens, age: 14 days, contamination: no AFB_1 (for detailed information please see the article), conc.: nr, country: Italy[232], *control
incidence: 3?/3, sa. const.: Hubbard strain laying hens, age: 14 days, contamination: artificial (dose: **50 μg AFB_1/kg** feed, o., for 169 days; for detailed information please see the article), conc.: 0.10 μg/kg* (mean value), country: Italy[232], *after 169 days

incidence: ?/?*, sa. const.: Warren laying hens, age: 11 months, wt.: ≈2.2 kg, contamination: neither AFB_1 nor CPL (for detailed information please see the article), conc.: nd, country: Italy[280], *control
incidence: ?/?, sa. const.: Warren laying hens, age: 11 months, wt.: ≈2.2 kg, contamination: no AFB_1 but **CPL** (2%) in the diet for 4 weeks; for detailed information please see the article), conc.: nd*, country: Italy[280], *after 4 weeks
incidence: ?/?, sa. const.: Warren laying hens, age: 11 months, wt.: ≈2.2 kg, contamination: artificial (dose: **2.5 ppm AFB_1**, o. and diet without CPL for 4 weeks; for detailed information please see the article), conc.: 2.21 ng/g* (mean value), country: Italy[280], *after 4 weeks
incidence: ?/?, sa. const.: Warren laying hens, age: 11 months, wt.: ≈2.2 kg, contamination: artificial (dose: **2.5 ppm AFB_1**, o. and diet with **CPL** (2%) for 4 weeks; for detailed information please see the article), conc.: 0.98 ng/g* (mean value), country: Italy[280], *after 4 weeks

incidence: 1/?, sa. const.: brown hens (HNL-type), contamination: artificial (dose: AFB_1 + AFG_1 fed; for detailed information please see the article), conc. range: ≤3.83 ppb, country: Germany[320]

incidence: 16?/16?*, sa. const.: Brown Hyssex laying hens, age: 27 days, contamination: no AF (for detailed information please see the article), conc.: nd, country: Spain[359], *control

incidence: 16?/16?, sa. const.: Brown Hyssex laying hens, age: 27 weeks, contamination: artificial (dose: **2.5 mg AF/kg** feed, o., for 32 days; for detailed information please see the article), conc. range: ≤0.35 μg/kg* (mean value), country: Spain[359] (measured at 4th, 8th*, 16th and 32nd day, lowest conc.: 0.04 μg/kg after 32 days, intoxication period)
incidence: 16?/16?, sa. const.: Brown Hyssex laying hens, age: 27 weeks, contamination: artificial (dose: **5.0 mg AF/kg** feed, o., for 32 days; for detailed information please see the article), conc. range: ≤0.23 μg/kg* (mean value), country: Spain[359] (measured at 4th*, 8th, 16th and 32nd day, lowest conc.: 0.07 μg/kg after 32 days, intoxication period)

incidence: 6/6*, sa. const.: Single Comb White Leghorn hens, contamination: no AFB_1 (for detailed information please see the article), conc.: nd, country: USA[363], *control
incidence: 10/10, sa. const.: Single Comb White Leghorns hens, contamination: artificial (dose: **8 μg AFB_1/g** feed, for 7 days; for detailed information please see the article), conc. range: ≤0.83 ng/g* **, country: USA[363], *AFB_1-residues, **after 7 days

incidence: ?/?*, sa. const.: eggs from SCWL hens, contamination: no AFB_1 (for detailed information please see the article), conc.: nd, country: India[392], *control
incidence: ?/?, sa. const.: eggs from SCWL hens, contamination: artificial (dose: **600 ppb AFB_1** in feed; o.; for detailed information please see the article), conc.: 569 ppb*, country: India[392], *after feeding period 3 (84 days) (period = 28 days with daily AFB_1-consumption)
incidence: ?/?, sa. const.: eggs from SCWL hens, contamination: artificial (dose: **1,250 ppb AFB_1** in feed; o.; for detailed information please see the article), conc.: 1,244 ppb*, country: India[392], *after feeding period 3 (84 days) (period = 28 days with daily AFB_1-consumption)
incidence: ?/?, sa. const.: eggs from SCWL hens, contamination: artificial (dose: **2,120 ppb AFB_1** in feed; o.; for detailed information please see the article), conc.: 2,245 ppb*, country: India[392], *after feeding period 3 (84 days) (period = 28 days with daily AFB_1-consumption)
incidence: ?/?, sa. const.: eggs from SCWL hens, contamination: artificial (dose: **2,850 ppb AFB_1** in feed; o.; for detailed information please see the article), conc.: 2,186 ppb*, country: India[392], *after feeding period 3 (84 days) (period = 28 days with daily AFB_1-consumption)

incidence: ?/?*, sa. const.: Hubbard strain laying hens, age: 14 days, contamination: no AFB_1 + OTA (for detailed information please see the article), conc.: nr, country: Italy[515], *control
incidence: 3?/3, sa. const.: Hubbard strain laying hens, age: 14 days, contamination: artificial (dose: **50 μg AFB_1/kg** feed + **50 μg OTA/kg** feed, o., for 169 days; for detailed information please see the article), conc.: 0.20 μg/kg* (mean value), country: Italy[515], *after 169 days

incidence: 24/24*, sa. const.: Warren laying hens, age: 44 weeks, wt.: Ø 2.2 kg, contamination: no AFB_1 (for detailed information please see the article), conc.: nd, country: Italy[630], *control
incidence: ?/24, sa. const.: Warren laying hens, age: 44 weeks, wt.: Ø 2.2 kg, contamination: **2.5 ppm AFB_1**, o., for 4 weeks (for detailed information please see the article), conc.: 4.13 ppb*, country: Italy[630], *after 4 weeks
incidence: 24/24, sa. const.: Warren laying hens, age: 44 weeks, wt.: Ø 2.2 kg, contamination: no AFB_1 but **MOS-diet** (0.11%) for 4 weeks (for detailed information please see the article), conc.: nd*, country: Italy[630], *after 4 weeks
incidence: ?/24, sa. const.: Warren laying hens, age: 44 weeks, wt.: Ø 2.2 kg,

contamination: **2.5 ppm AFB$_1$**, o. together with **MOS-diet** (0.11%) for 4 weeks (for detailed information please see the article), conc.: 2.21 ppb*, country: Italy[630], *after 4 weeks

Aflatoxin B$_2$

incidence: 8/8*, sa. const.: White Leghorn pullets of the Shaver strain, contamination: no AFB$_1$ + AFB$_2$ (for detailed information please see the article), conc.: nr, country: USA[110], *control

incidence: 8/8, sa. const.: White Leghorn pullets of the Shaver strain, contamination: artificial (dose: **3,310 μg AFB$_1$/kg + 1,680 μg AFB$_2$/kg**, o., for 4 weeks; for detailed information please see the article), conc. range: 0.03–0.26 μg/kg*, Ø conc.: 0.13 μg/kg*, country: USA[110], *after feeding 4 weeks an AF-contaminated diet

incidence: 8/8*, sa. const.: White Leghorn pullets of the Shaver strain, contamination: no AFB$_1$ + AFB$_2$ (for detailed information please see the article), conc.: nr, country: USA[110], *control

incidence: 2/8, sa. const.: White Leghorn pullets of the Shaver strain, contamination: artificial (dose: **3,310 μg AFB$_1$/kg + 1,680 μg AFB$_2$/kg**, o., for 4 weeks; for detailed information please see the article), conc. range: 0.01–0.02 μg/kg*, Ø conc.: 0.015 μg/kg*, country: USA[110], *2 days after feeding 4 weeks an AF-contaminated diet

Aflatoxin B$_{2a}$

incidence: 8/8*, sa. const.: White Leghorn pullets of the Shaver strain, contamination: no AFB$_1$ + AFB$_2$ (for detailed information please see the article), conc.: nr, country: USA[110], *control

incidence: 8/8, sa. const.: White Leghorn pullets of the Shaver strain, contamination: artificial (dose: **3,310 μg AFB$_1$/kg + 1,680 μg AFB$_2$/kg**, o, for 4 weeks; for detailed information please see the article), conc. range: 0.35–3.87 μg/kg*, Ø conc.: 1.52 μg/kg*, country: USA[110], *after feeding 4 weeks an AF-contaminated diet

incidence: 8/8*, sa. const.: White Leghorn pullets of the Shaver strain, contamination: no AFB$_1$ + AFB$_2$ (for detailed information please see the article), conc.: nr, country: USA[110], *control

incidence: 7/8, sa. const.: White Leghorn pullets of the Shaver strain, contamination: artificial (dose: 3,310 μg AFB$_1$/kg + 1,680 μg AFB$_2$/kg, o., for 4 weeks; for detailed information please see the article), conc. range: tr–0.09 μg/kg*, country: USA[110], *2 days after feeding 4 weeks an AF-contaminated diet

incidence: 16/16?*, sa. const.: Brown Hyssex laying hens, age: 27 days, contamination: no AF (for detailed information please see the article), conc.: nd, country: Spain[359], *control

incidence: 16?/16?, sa. const.: Brown Hyssex laying hens, age: 27 weeks, contamination: artificial (dose: **2.5 mg AF/kg** feed, o., for 32 days; for detailed information please see the article), conc. range: ≤0.41 μg/kg* (mean value), country: Spain[359] (measured at 4th, 8th*, 16th and 32nd day, lowest conc.: nd after 16 days, intoxication period)

incidence: 16?/16?, sa. const.: Brown Hyssex laying hens, age: 27 weeks, contamination: artificial (dose: **5.0 mg AF/kg** feed, o., for 32 days; for detailed information please see the article), conc. range: ≤0.67 μg/kg* (mean value), country: Spain[359] (measured at 4th*, 8th, 16th and 32nd day, lowest conc.: 0.18 μg/kg after 16 days, intoxication period)

Aflatoxin G$_1$

incidence: 1/?, sa. const.: brown hens (HNL-type), contamination: artificial (dose: AFB$_1$ addition; for detailed information please see the article), conc.: 0.28 ppb, country: Germany[320]

AFLATOXIN M_1

incidence: 8/8*, sa. const.: White Leghorn pullets of the Shaver strain, contamination: no AFB_1 + AFB_2 (for detailed information please see the article), conc.: nr, country: USA[110], *control
incidence: 4/8, sa. const.: White Leghorn pullets of the Shaver strain, contamination: artificial (dose: **3,310 μg AFB_1/kg + 1,680 μg AFB_2/kg**, o., for 4 weeks; for detailed information please see the article), conc. range: tr–0.05 μg/kg*, country: USA[110], *after feeding 4 weeks an AF-contaminated diet

incidence: 8/8*, sa. const.: White Leghorn pullets of the Shaver strain, contamination: no AFB_1 + AFB_2 (for detailed information please see the article), conc.: nr, country: USA[110], *control
incidence: 16/16, sa. const.: White Leghorn pullets of the Shaver strain, contamination: artificial (dose: **3,310 μg AFB_1/kg + 1,680 μg AFB_2/kg**, o., for 4 weeks; for detailed information please see the article), conc.: nd* and NR, country: USA[110], *2 days after feeding 4 weeks an AF-contaminated diet

incidence: 16?/16?*, sa. const.: Brown Hyssex laying hens, age: 27 days, contamination: no AF (for detailed information please see the article), conc.: nd, country: Spain[359], *control
incidence: 16?/16?, sa. const.: Brown Hyssex laying hens, age: 27 weeks, contamination: artificial (dose: **2.5 mg AF/kg** feed, o., for 32 days; for detailed information please see the article), conc. range: ≤0.32 μg/kg* (mean value), country: Spain[359] (measured at 4th, 8th, 16th and 32nd* day, lowest conc.: nd after 4 and 16 days, intoxication period)
incidence: 16?/16?, sa. const.: Brown Hyssex laying hens, age: 27 weeks, contamination: artificial (dose: **5.0 mg AF/kg** feed, o., for 32 days; for detailed information please see the article), conc. range: ≤0.11 μg/kg* (mean value), country: Spain[359] (measured at 4th, 8th, 16th* and 32nd day, lowest conc.: nd after 32 days, intoxication period)

incidence: ?/?*, sa. const.: Hubbard strain laying hens, age: 14 days, contamination: no AFB_1 + OTA (for detailed information please see the article), conc.: nr, country: Italy[515], *control
incidence: 3?/3, sa. const.: Hubbard strain laying hens, age: 14 days, contamination: artificial (dose: **50 μg AFB_1/kg** feed + **50 μg OTA/kg** feed, o., for 169 days; for detailed information please see the article), conc.: <0.01 μg/kg* (mean value), country: Italy[515], *after 169 days

AFLATOXIN M_2

incidence: 8/8*, sa. const.: White Leghorn pullets of the Shaver strain, contamination: no AFB_1 + AFB_2 (for detailed information please see the article), conc.: nr, country: USA[110], *control
incidence: 7/8, sa. const.: White Leghorn pullets of the Shaver strain, contamination: artificial (dose: **3,310 μg AFB_1/kg + 1,680 μg AFB_2/kg**, o., for 4 weeks; for detailed information please see the article), conc. range: tr–0.10 μg/kg*, country: USA[110], *after feeding 4 weeks an AF-contaminated diet

incidence: 8/8*, sa. const.: White Leghorn pullets of the Shaver strain, contamination: no AFB_1 + AFB_2 (for detailed information please see the article), conc.: nr, country: USA[110], *control
incidence: 1/8, sa. const.: White Leghorn pullets of the Shaver strain, contamination: artificial (dose: **3,310 μg AFB_1/kg + 1,680 μg AFB_2/kg**, o., for 4 weeks; for detailed information please see the article), conc.: tr*, country: USA[110],

*2 days after feeding 4 weeks an AF-contaminated diet

Aflatoxins
incidence: ?/?, sa. const.: White Leghorn hens (Cornell strain K), age: 20–22 weeks, wt.: 1,600–1,800 g, contamination: artificial (dose: 11.26 mg AFs (labeled), by stomach tube, once; for detailed information please see the article), conc. range: ≤18.69 μg/g* **, country: USA[81], *AFs or their metabolites, **after 4 days (also measured after 1 and 7 days, lowest conc.: 10.65 μg/g after 7 days)

Deoxynivalenol
incidence: ?/?*, sa. const.: White Leghorn hens, age: 55 weeks, contamination: no NIV (for detailed information please see the article), conc.: nr, country: Lithuania/Sweden[375], *control
incidence: ?/?, sa. const.: White Leghorn hens, age: 55 weeks, contamination: artificial (dose: **5 mg NIV**/kg, o., for 50 days; for detailed information please see the article), conc.: tr*, country: Lithuania/Sweden[375], *after 50 days

Nivalenol
incidence: ?/?*, sa. const.: White Leghorn hens, age: 55 weeks, contamination: no NIV (for detailed information please see the article), conc.: nr, country: Lithuania/Sweden[375], *control
incidence: ?/?, sa. const.: White Leghorn hens, age: 55 weeks, contamination: artificial (dose: **5 mg NIV**/kg, o., for 50 days; for detailed information please see the article), conc.: tr*, country: Lithuania/ weden[375], *after 50 days

Ochratoxin A
incidence: 8?/8, sa. const.: New Hampshire-Leghorn cross hens, age: 24 weeks, contamination: artificial (dose: **OTA 0.5 ppm** (feed weight), o., for 2 weeks), conc.: 12.5 ppb* (mean value), country: USA[316], *after 14 days
incidence: 8?/8, sa. const.: New Hampshire-Leghorn cross hens, age: 24 weeks, contamination: artificial (dose: **OTA 5.0 ppm** (feed weight), o., for 2 weeks), conc.: 80.2 ppb* (mean value), country: USA[316], *after 14 days

incidence: 5/5*, sa. const.: White Leghorn hens, contamination: no OTA (for detailed information please see the article), conc.: nd, country: Germany[374], *control
incidence: ?/5, sa. const.: White Leghorn hens, contamination: artificial (dose: **1.3 mg OTA**/kg feed, o., for 28 days; for detailed information please see the article), conc.: 9.1 μg/kg* (mean value), country: Germany[374], **after feeding OTA for 28 days
incidence: ?/5, sa. const.: White Leghorn hens, contamination: artificial (dose: **2.6 mg OTA**/kg feed, o., for 28 days; for detailed information please see the article), conc.: 17.9 μg/kg* (mean value), country: Germany[374], **after feeding OTA for 28 days
incidence: ?/5, sa. const.: eggs from White Leghorn hens, contamination: artificial (dose: **5.2 mg OTA**/kg feed, o., for 28 days; for detailed information please see the article), conc.: 18.0 μg/kg* (mean value), country: Germany[374], **after feeding OTA for 28 days

incidence: ?/?*, sa. const.: Hubbard strain laying hens, age: 14 days, contamination: no AFB_1 + OTA (for detailed information please see the article), conc.: nr, country: Italy[515], *control
incidence: 3?/3, sa. const.: Hubbard strain laying hens, age: 14 days, contamination: artificial (dose: **50 μg AFB_1**/kg feed + **50 μg OTA**/kg feed, o., for **169 days**; for detailed information please see the article), conc.: 2.0 μg/kg* (mean value), country: Italy[515], *after 169 days
incidence: ?/?*, sa. const.: Hubbard strain laying hens, age: 14 days, contamination: no AFB_1 + OTA (for detailed information please see the article), conc.: nr, country: Italy[515], *control

incidence: 3?/3, sa. const.: Hubbard strain laying hens, age: 14 days, contamination: artificial (dose: **50 µg AFB_1/kg** feed + **50 µg OTA**/kg feed, o., for **87 days**; for detailed information please see the article), conc. range: 0.5*–0.9** µg/kg (mean value), country: Italy[515], 33** and 82* days after withdrawal from treatment

incidence: 12/12*, sa. const.: H and N White Leghorn hens, age: 26 weeks, contamination: no OTA (for detailed information please see the article), conc.: nd, country: Canada[525], *control
incidence: ?/8, sa. const.: H and N White Leghorn hens, age: 26 weeks, contamination: artificial (dose: **0.5 ppm OTA** in feed, o., for 42 days; for detailed information please see the article), conc.: 26.3 ppm* (mean value), country: Canada[525], *directly after withdrawal from treated feed for 6 weeks
incidence: ?/12, sa. const.: H and N White Leghorn hens, age: 26 weeks, contamination: artificial (dose: **1.0 ppm OTA** in feed, o., for 42 days; for detailed information please see the article), conc.: 57.6 ppm* (mean value), country: Canada[525], *directly after withdrawal from treated feed for 6 weeks
incidence: ?/16, sa. const.: H and N White Leghorn hens, age: 26 weeks, contamination: artificial (dose: **4.0 ppm OTA** in feed, o., for 42 days; for detailed information please see the article), conc.: 72.6 ppm* (mean value), country: Canada[525], *directly after withdrawal from treated feed for 6 weeks
incidence: ?/16, sa. const.: H and N White Leghorn hens, age: 26 weeks, contamination: artificial (dose: **4.0 ppm OTA** in feed, o., for 42 days; for detailed information please see the article), conc.: 26.3 ppm* (mean value), country: Canada[525], ***after 24 h off** treated feed for 6 weeks
incidence: ?/16, sa. const.: H and N White Leghorn hens, age: 26 weeks, contamination: artificial (dose: **4.0 ppm OTA** in feed, o., for 42 days; for detailed information please see the article), conc.: 8.7 ppm* (mean value), country: Canada[525], *after 48 h off treated feed for 6 weeks

incidence: 24/24*, sa. const.: Rhode Island Red **hens**, age: 54 weeks, contamination: no OTA (for detailed information please see the article), conc.: nd, country: Poland[590], *control
incidence: ?/22, sa. const.: Rhode Island Red hens, age: 54 weeks, contamination: artificial (dose: **2.1 ppm OTA** in the diet, o., for 5 weeks; for detailed information please see the article), conc.: 3.3 ppb*, country: Poland[590], *after 5 weeks?
incidence: ?/2, sa. const.: Rhode Island Red **cockerels**, age: 54 weeks, contamination: artificial (dose: **2.1 ppm OTA** in the diet, o., for 5 weeks; for detailed information please see the article), conc.: 1.3 ppb*, country: Poland[590], *after 5 weeks?
incidence: ?/22, sa. const.: Rhode Island Red **hens**, age: 54 weeks, contamination: artificial (dose: **4.1 ppm OTA** in the diet, o., for 5 weeks; for detailed information please see the article), conc.: 11.9 ppb*, country: Poland[590], *after 5 weeks?
incidence: ?/2, sa. const.: Rhode Island Red **cockerels**, age: 54 weeks, contamination: artificial (dose: **4.1 ppm OTA** in the diet, o., for 5 weeks; for detailed information please see the article), conc.: 1.7 ppb*, country: Poland[590], *after 5 weeks?

Zearalenone
incidence: ?/4, sa. const.: White Leghorn laying hens, age: 26–39 weeks, contamination: artificial (dose: 10 mg ZEA (labeled)/kg, by gavage into the crop, once), conc. range: ≤397 µg eq ZEA/100 g wet tissue* (mean value), country: USA[394], *after 4 h (also measured after 2, 24, 48 and 72 h, lowest conc.: 50 µg eq ZEA/100 g wet tissue after 72 h)

Hen lung may contain the following mycotoxins and/or their metabolites:

AFLATOXINS
incidence: ?/?, sa. const.: White Leghorn hens (Cornell strain K), age: 20–22 weeks, wt.: 1,600–1,800 g, contamination: artificial (dose: 11.26 mg AFs, by stomach tube, once; for detailed information please see the article), conc. range: ≤8.10 µg/g* ** ***, country: USA[81], *AFs or their metabolites, **after 7 days (also measured after 1 day and 4 days, lowest conc.: 3.20 µg/g after 1 day), ***in lung and trachea

OCHRATOXIN A
incidence: 8?/8, sa. const.: New Hampshire-Leghorn cross hens, age: 24 weeks, contamination: artificial (dose: **OTA 0.5 ppm** (feed weight), o., for 2 weeks), conc.: 17.3 ppb* (mean value), country: USA[316], *after 14 days
incidence: 8?/8, sa. const.: New Hampshire-Leghorn cross hens, age: 24 weeks, contamination: artificial (dose: **OTA 5.0 ppm** (feed weight), o., for 2 weeks), conc.: 87.0 ppb* (mean value), country: USA[316], *after 14 days

ZEARALENONE
incidence: ?/4, sa. const.: White Leghorn laying hens, age: 26–39 weeks, contamination: artificial (dose: 10 mg ZEA (labeled)/kg, by gavage into the crop, once), conc. range: ≤63 µg eq ZEA/100 g wet tissue* (mean value), country: USA[394], *after 48 h (also measured after 2, 4, 24 and 72 h, lowest conc.: 26 µg eq ZEA/100 g wet tissue after 72 h)

Hen Lungs and Trachea see Hen lung

Hen muscle may contain the following mycotoxins and/or their metabolites:

AFLATOXIN B_1
incidence: 16?/16?*, sa. const.: Brown Hyssex laying hens, age: 27 days, contamination: no AF (for detailed information please see the article), conc.: nd, country: Spain[359], *control
incidence: 16?/16?, sa. const.: Brown Hyssex laying hens, age: 27 weeks, contamination: artificial (dose: **2.5 mg AF/kg** feed, o., for 32 days; for detailed information please see the article), conc. range: ≤0.2 µg/kg* (mean value), country: Spain[359] (measured at 4th*, 8th, 16th and 32nd* day, lowest conc.: 0.02 µg/kg after 8 and 16 days, intoxication period)
incidence: 16?/16?, sa. const.: Brown Hyssex laying hens, age: 27 weeks, contamination: artificial (dose: **5.0 mg AF/kg** feed, o., for 32 days; for detailed information please see the article), conc. range: ≤0.1 µg/kg* (mean value), country: Spain[359] (measured at 4th*, 8th, 16th and 32nd day, lowest conc.: 0.02 µg/kg after 32 days, intoxication period)

Hen muscle, breast may contain the following mycotoxins and/or their metabolites:

AFLATOXICOL
incidence: 6/6*, sa. const.: Single Comb White Leghorn hens, contamination: no AFB_1 (for detailed information please see the article), conc.: nd, country: USA[363], *control
incidence: 10/10, sa. const.: Single Comb White Leghorns hens, contamination: artificial (dose: **8 µg AFB_1/g** feed, for 7 days; for detailed information please see the article), conc. range: ≤0.11 ng/g* **, country: USA[363], *AFL-residues, **after 7 days

AFLATOXIN B_1
incidence: 8/8*, sa. const.: White Leghorn pullets of the Shaver strain, contamination no AFB_1 + AFB_2 (for detailed information please see the article), conc.: nr, country: USA[110], *control

incidence: 2/8, sa. const.: White Leghorn pullets of the Shaver strain, contamination: artificial (dose: **3,310 µg AFB_1/kg + 1,680 µg AFB_2/kg**, o., for 4 weeks; for detailed information please see the article), conc. range: tr–0.05 µg/kg*, country: USA[110], *after feeding 4 weeks an AF-contaminated diet

incidence: 8/8*, sa. const.: White Leghorn pullets of the Shaver strain, contamination: no AFB_1 + AFB_2 (for detailed information please see the article), conc.: nr, country: USA[110], *control
incidence: 4/8, sa. const.: White Leghorn pullets of the Shaver strain, contamination: artificial (dose: **3,310 µg AFB_1/kg + 1,680 µg AFB_2/kg**, o., for 4 weeks; for detailed information please see the article), conc. range: tr–0.01 µg/kg*, country: USA[110], *2 days after feeding 4 weeks an AF-contaminated diet

Aflatoxin B_2
incidence: 8/8*, sa. const.: White Leghorn pullets of the Shaver strain, contamination: no AFB_1 + AFB_2 (for detailed information please see the article), conc.: nr, country: USA[110], *control
incidence: 4/8, sa. const.: White Leghorn pullets of the Shaver strain, contamination: artificial (dose: **3,310 µg AFB_1/kg + 1,680 µg AFB_2/kg**, o., for 4 weeks; for detailed information please see the article), conc. range: tr–0.05 µg/kg*, country: USA[110], *after feeding 4 weeks an AF-contaminated diet

incidence: 8/8*, sa. const.: White Leghorn pullets of the Shaver strain, contamination: no AFB_1 + AFB_2 (for detailed information please see the article), conc.: nr, country: USA[110], *control
incidence: 6/8, sa. const.: White Leghorn pullets of the Shaver strain, contamination: artificial (dose: **3,310 µg AFB_1/kg + 1,680 µg AFB_2/kg**, o., for 4 weeks; for detailed information please see the article), conc. range: tr–0.02 µg/kg*, country: USA[110], *2 days after feeding 4 weeks an AF-contaminated diet

Aflatoxin B_{2a}
incidence: 8/8*, sa. const.: White Leghorn pullets of the Shaver strain, contamination: no AFB_1 + AFB_2 (for detailed information please see the article), conc.: nr, country: USA[110], *control
incidence: 7/8, sa. const.: White Leghorn pullets of the Shaver strain, contamination: artificial (dose: **3,310 µg AFB_1/kg + 1,680 µg AFB_2/kg**, o., for 4 weeks; for detailed information please see the article), conc. range: tr–0.03 µg/kg*, country: USA[110], *after feeding 4 weeks an AF-contaminated diet

incidence: 8/8*, sa. const.: White Leghorn pullets of the Shaver strain, contamination: no AFB_1 + AFB_2 (for detailed information please see the article), conc.: nr, country: USA[110], *control
incidence: 6/8, sa. const.: White Leghorn pullets of the Shaver strain, contamination: artificial (dose: **3,310 µg AFB_1/kg + 1,680 µg AFB_2/kg**, o., for 4 weeks; for detailed information please see the article), conc. range: tr*, country: USA[110], *2 days after feeding 4 weeks an AF-contaminated diet

Aflatoxin M_1
incidence: 8/8*, sa. const.: White Leghorn pullets of the Shaver strain, contamination: no AFB_1 + AFB_2 (for detailed information please see the article), conc.: nr, country: USA[110], *control
incidence: 1/16, sa. const.: White Leghorn pullets of the Shaver strain, contamination: artificial (dose: **3,310 µg AFB_1/kg + 1,680 µg AFB_2/kg**, o., for 4 weeks; for detailed information please see the article), conc.: tr*, country: USA[110], *after feeding 4 weeks an AF-contaminated diet

incidence: 8/8*, sa. const.: White Leghorn pullets of the Shaver strain,

contamination: no AFB_1 + AFB_2 (for detailed information please see the article), conc.: nr, country: USA[110], *control
incidence: 16/16, sa. const.: White Leghorn pullets of the Shaver strain, contamination: artificial (dose: **3,310 µg AFB_1/kg + 1,680 µg AFB_2/kg**, o., for 4 weeks; for detailed information please see the article), conc.: nd*, country: USA[110], *2 days after feeding 4 weeks an AF-contaminated diet

Aflatoxin M_2
incidence: 8/8*, sa. const.: White Leghorn pullets of the Shaver strain, contamination: no AFB_1 + AFB_2 (for detailed information please see the article), conc.: nr, country: USA[110], *control
incidence: 4/8, sa. const.: White Leghorn pullets of the Shaver strain, contamination: artificial (dose: **3,310 µg AFB_1/kg + 1,680 µg AFB_2/kg**, o., for 4 weeks; for detailed information please see the article), conc. range: tr–0.02 µg/kg*, country: USA[110], *after feeding 4 weeks an AF-contaminated diet

incidence: 8/8*, sa. const.: White Leghorn pullets of the Shaver strain, contamination: no AFB_1 + AFB_2 (for detailed information please see the article), conc.: nr, country: USA[110], *control
incidence: 16/16, sa. const.: White Leghorn pullets of the Shaver strain, contamination: artificial (dose: 3,310 µg AFB_1/kg + 1,680 µg AFB_2/kg, o., for 4 weeks; for detailed information please see the article), conc.: nd* and CPR, country: USA[110], *2 days after feeding 4 weeks an AF-contaminated diet

Aflatoxins
incidence: ?/?, sa. const.: White Leghorn hens (Cornell strain K), age: 20–22 weeks, wt.: 1,600–1,800 g, contamination: artificial (dose: 11.26 mg AFs (labeled), by stomach tube, once; for detailed information please see the article), conc. range: ≤9.9 µg/g* **, country: USA[81], *AFs or their metabolites, **after 7 days (also measured after 1 and 4 days, lowest conc.: 6.46 µg/g after 4 days)

Ochratoxin A
incidence: 8?/8, sa. const.: New Hampshire-Leghorn cross hens, age: 24 weeks, contamination: artificial (dose: **OTA 0.5 ppm** (feed weight), o., for 2 weeks), conc.: 3.3 ppb* (mean value), country: USA[316], *after 14 days
incidence: 8?/8, sa. const.: New Hampshire-Leghorn cross hens, age: 24 weeks, contamination: artificial (dose: **OTA 5.0 ppm** (feed weight), o., for 2 weeks), conc.: 8.4 ppb* (mean value), country: USA[316], *after 14 days

Zearalenone
incidence: ?/4, sa. const.: White Leghorn laying hens, age: 26–39 weeks, contamination: artificial (dose: 10 mg ZEA (labeled)/kg, by gavage into the crop, once), conc. range: ≤12 µg eq ZEA/100 g wet tissue* (mean value), country: USA[394], *after 72 h (also measured after 2, 4, 24 and 48 h, lowest conc.: 7 µg eq ZEA/100 g wet tissue after 2 h)

Hen muscle, leg may contain the following mycotoxins and/or their metabolites:

Aflatoxin B_1
incidence: 8/8*, sa. const.: White Leghorn pullets of the Shaver strain, contamination: no AFB_1 + AFB_2 (for detailed information please see the article), conc.: nr, country: USA[110], *control
incidence: 8/8, sa. const.: White Leghorn pullets of the Shaver strain, contamination: artificial (dose: **3,310 µg AFB_1/kg + 1,680 µg AFB_2/kg**, o., for 4 weeks; for detailed information please see the article), conc. range: tr–0.11 µg/kg*, country: USA[110], *after feeding 4 weeks an AF-contaminated diet

incidence: 8/8*, sa. const.: White Leghorn pullets of the Shaver strain, contamination: no AFB_1 + AFB_2 (for detailed information please see the article), conc.: nr, country: USA[110], *control
incidence: 5/8, sa. const.: White Leghorn pullets of the Shaver strain, contamination: artificial (dose: **3,310 µg AFB_1/kg + 1,680 µg AFB_2/kg**, o., for 4 weeks; for detailed information please see the article), conc. range: tr–0.01 µg/kg*, country: USA[110], *2 days after feeding 4 weeks an AF-contaminated diet

Aflatoxin B_2
incidence: 8/8*, sa. const.: White Leghorn pullets of the Shaver strain, contamination: no AFB_1 + AFB_2 (for detailed information please see the article), conc.: nr, country: USA[110], *control
incidence: 8/8, sa. const.: White Leghorn pullets of the Shaver strain, contamination: artificial (dose: **3,310 µg AFB_1/kg + 1,680 µg AFB_2/kg**, o., for 4 weeks; for detailed information please see the article), conc. range: 0.02–0.22 µg/kg*, Ø conc.: 0.056 µg/kg*, country: USA[110], *after feeding 4 weeks an AF-contaminated diet

incidence: 8/8*, sa. const.: White Leghorn pullets of the Shaver strain, contamination: no AFB_1 + AFB_2 (for detailed information please see the article), conc.: nr, country: USA[110], *control
incidence: 6/8, sa. const.: White Leghorn pullets of the Shaver strain, contamination: artificial (dose: **3,310 µg AFB_1/kg + 1,680 µg AFB_2/kg**, o., for 4 weeks; for detailed information please see the article), conc. range: tr–0.03 µg/kg*, country: USA[110], *2 days after feeding 4 weeks an AF-contaminated diet
please like in the way of the other mycotoxins

Aflatoxin B_{2a}
incidence: 8/8*, sa. const.: White Leghorn pullets of the Shaver strain, contamination: no AFB_1 + AFB_2 (for detailed information please see the article), conc.: nr, country: USA[110], *control
incidence: 8/8, sa. const.: White Leghorn pullets of the Shaver strain, contamination: artificial (dose: **3,310 µg AFB_1/kg + 1,680 µg AFB_2/kg**, o., for 4 weeks; for detailed information please see the article), conc. range: 0.02–0.04 µg/kg*, Ø conc.: 0.03 µg/kg*, country: USA[110], *after feeding 4 weeks an AF-contaminated diet

incidence: 8/8*, sa. const.: White Leghorn pullets of the Shaver strain, contamination: no AFB_1 + AFB_2 (for detailed information please see the article), conc.: nr, country: USA[110], *control
incidence: 6/8, sa. const.: White Leghorn pullets of the Shaver strain, contamination: artificial (dose: **3,310 µg AFB_1/kg + 1,680 µg AFB_2/kg**, o., for 4 weeks; for detailed information please see the article), conc. range: tr–0.02 µg/kg*, country: USA[110], *2 days after feeding 4 weeks an AF-contaminated diet

Aflatoxin M_1
incidence: 8/8*, sa. const.: White Leghorn pullets of the Shaver strain, contamination: no AFB_1 + AFB_2 (for detailed information please see the article), conc.: nr, country: USA[110], *control
incidence: 2/8, sa. const.: White Leghorn pullets of the Shaver strain, contamination: artificial (dose: **3,310 µg AFB_1/kg + 1,680 µg AFB_2/kg**, o., for 4 weeks; for detailed information please see the article), conc. range: tr–0.01 µg/kg*, country: USA[110], *after feeding 4 weeks an AF-contaminated diet

incidence: 8/8*, sa. const.: White Leghorn pullets of the Shaver strain, contamination: no AFB_1 + AFB_2 (for detailed information please see the article), conc.: nr, country: USA[110], *control

incidence: 16/16, sa. const.: White Leghorn pullets of the Shaver strain, contamination: artificial (dose: **3,310 µg AFB_1/kg + 1,680 µg AFB_2/kg**, o., for 4 weeks; for detailed information please see the article), conc.: nd*, country: USA[110], *2 days after feeding 4 weeks an AF-contaminated diet

Aflatoxin M_2

incidence: 8/8*, sa. const.: White Leghorn pullets of the Shaver strain, contamination: no AFB_1 + AFB_2 (for detailed information please see the article), conc.: nr, country: USA[110], *control
incidence: 8/8, sa. const.: White Leghorn pullets of the Shaver strain, contamination: artificial (dose: **3,310 µg AFB_1/kg + 1,680 µg AFB_2/kg**, o., for 4 weeks; for detailed information please see the article), conc. range: tr–0.02 µg/kg*, country: USA[110], *after feeding 4 weeks an AF-contaminated diet

incidence: 8/8*, sa. const.: White Leghorn pullets of the Shaver strain, contamination: no AFB_1 + AFB_2 (for detailed information please see the article), conc.: nr, country: USA[110], *control
incidence: 2/8, sa. const.: White Leghorn pullets of the Shaver strain, contamination: artificial (dose: **3,310 µg AFB_1/kg + 1,680 µg AFB_2/kg**, o., for 4 weeks; for detailed information please see the article), conc. range: tr*, country: USA[110], *2 days after feeding 4 weeks an AF-contaminated diet

Aflatoxins

incidence: ?/?, sa. const.: White Leghorn hens (Cornell strain K), age: 20–22 weeks, wt.: 1,600–1,800 g, contamination: artificial (dose: 11.26 mg AFs (labeled), by stomach tube, once; for detailed information please see the article), conc. range: ≤7.89 µg/g* **, country: USA[81], *AFs or their metabolites, **after 4 days (also measured after 1 and 7 days, lowest conc.: 5.70 µg/g after 7 days)

Ochratoxin A

incidence: 8?/8, sa. const.: New Hampshire-Leghorn cross hens, age: 24 weeks, contamination: artificial (dose: **OTA 0.5 ppm** (feed weight), o., for 2 weeks), conc.: 4.2 ppb* (mean value), country: USA[316], *after 14 days
incidence: 8?/8, sa. const.: New Hampshire-Leghorn cross hens, age: 24 weeks, contamination: artificial (dose: **OTA 5.0 ppm** (feed weight), o., for 2 weeks), conc.: 7.2 ppb* (mean value), country: USA[316], *after 14 days

Zearalenone

incidence: ?/4, sa. const.: White Leghorn laying hens, age: 26–39 weeks, contamination: artificial (dose: 10 mg ZEA (labeled)/kg, by gavage into the crop, once), conc. range: ≤18 µg eq ZEA/100 g wet tissue* (mean value), country: USA[394], *after 48 h (also measured after 2, 4, 24 and 72 h, lowest conc.: 9 µg eq ZEA/100 g wet tissue after 2 h)

Hen muscle, red may contain the following mycotoxins and/or their metabolites:

Citrinin

incidence: 9/9*, sa. const.: sex mature Mamourah hens, age: ≈13 months, b. wt.: Ø 1,998 g, contamination: no CIT (for detailed information please see the article), conc.: nr, country: Egypt[591], *control
incidence: ?/9, sa. const.: sex mature Mamourah hens, age: ≈13 months, b. wt.: Ø 1,998 g, contamination: artificial (dose: **100 ppb CIT** in the diet, o., for 6 weeks; for detailed information please see the article), conc.: 9.84 ppb*, country: Egypt[591], *after 6 weeks

Ochratoxin A

incidence: 12/12*, sa. const.: H and N White Leghorn hens, age: 26 weeks, contamination: no OTA (for detailed information please see the article), conc.: nd, country: Canada[525], *control

incidence: ?/8, sa. const.: H and N White Leghorn hens, age: 26 weeks, contamination: artificial (dose: **0.5 ppm OTA** in feed, o., for 42 days; for detailed information please see the article), conc.: 8.0 ppm* (mean value), country: Canada[525], *directly after withdrawal from treated feed for 6 weeks
incidence: ?/12, sa. const.: H and N White Leghorn hens, age: 26 weeks, contamination: artificial (dose: **1.0 ppm OTA** in feed, o., for 42 days; for detailed information please see the article), conc.: 12.6 ppm* (mean value), country: Canada[525], *directly after withdrawal from treated feed for 6 weeks
incidence: ?/16, sa. const.: H and N White Leghorn hens, age: 26 weeks, contamination: artificial (dose: **4.0 ppm OTA** in feed, o., for 42 days; for detailed information please see the article), conc.: 20.8 ppm* (mean value), country: Canada[525], *directly after withdrawal from treated feed for 6 weeks
incidence: ?/16, sa. const.: H and N White Leghorn hens, age: 26 weeks, contamination: artificial (dose: **4.0 ppm OTA** in feed, o., for 42 days; for detailed information please see the article), conc.: ndr, country: Canada[525], ***after 24 h off** treated feed for 6 weeks
incidence: ?/16, sa. const.: H and N White Leghorn hens, age: 26 weeks, contamination: artificial (dose: **4.0 ppm OTA** in feed, o., for 42 days; for detailed information please see the article), conc.: ndr, country: Canada[525], ***after 48 h off** treated feed for 6 weeks

Hen muscle, thigh may contain the following mycotoxins and/or their metabolites:

Aflatoxicol
incidence: ?/?*, sa. const.: Hubbard strain laying hens, age: 14 days, contamination: no AFB_1 + OTA (for detailed information please see the article), conc.: nr, country: Italy[515], *control
incidence: 3?/3, sa. const.: Hubbard strain laying hens, age: 14 days, contamination: artificial (dose: **50 µg AFB_1/kg** feed + **50 µg OTA/kg** feed, o., for 169 days; for detailed information please see the article), conc.: <0.04 µg/kg* (mean value), country: Italy[515], *after 169 days

Aflatoxin B_1
incidence: ?/?*, sa. const.: Hubbard strain laying hens, age: 14 days, contamination: no AFB_1 (for detailed information please see the article), conc.: nr, country: Italy[232], *control
incidence: 3?/3, sa. const.: Hubbard strain laying hens, age: 14 days, contamination: artificial (dose: **50 µg AFB_1/kg** feed, o., for 169 days; for detailed information please see the article), conc.: 0.07 µg/kg* (mean value), country: Italy[232], *after 169 days

incidence: ?/?*, sa. const.: Hubbard strain laying hens, age: 14 days, contamination: no AFB_1 + OTA (for detailed information please see the article), conc.: nr, country: Italy[515], *control
incidence: 3?/3, sa. const.: Hubbard strain laying hens, age: 14 days, contamination: artificial (dose: **50 µg AFB_1/kg** feed + **50 µg OTA/kg** feed, o., for 169 days; for detailed information please see the article), conc.: 0.06 µg/kg* (mean value), country: Italy[515], *after 169 days

Aflatoxin M_1
incidence: ?/?*, sa. const.: Hubbard strain laying hens, age: 14 days, contamination: no AFB_1 (for detailed information please see the article), conc.: nr, country: Italy[232], *control
incidence: 3?/3, sa. const.: Hubbard strain laying hens, age: 14 days, contamination: artificial (dose: **50 µg AFB_1/kg** feed, o., for 169 days; for detailed information please see the article), conc.: 0.02 µg/kg* (mean value), country: Italy[232], *after 169 days

incidence: ?/?*, sa. const.: Hubbard strain laying hens, age: 14 days, contamination:

no AFB_1 + OTA (for detailed information please see the article), conc.: nr, country: Italy[515], *control
incidence: 3?/3, sa. const.: Hubbard strain laying hens, age: 14 days, contamination: artificial (dose: **50 µg AFB_1/kg feed + 50 µg OTA/kg** feed, o., for 169 days; for detailed information please see the article), conc.: 0.10 µg/kg* (mean value), country: Italy[515], *after 169 days

Ochratoxin A
incidence: ?/?*, sa. const.: Hubbard strain laying hens, age: 14 days, contamination: no AFB_1 + OTA (for detailed information please see the article), conc.: nr, country: Italy[515], *control
incidence: 3?/3, sa. const.: Hubbard strain laying hens, age: 14 days, contamination: artificial (dose: **50 µg AFB_1/kg** feed + **50 µg OTA/kg** feed, o., for 169 days; for detailed information please see the article), conc.: 1.0 µg/kg* (mean value), country: Italy[515], *after 169 days
incidence: ?/?*, sa. const.: Hubbard strain laying hens, age: 14 days, contamination: no AFB_1 + OTA (for detailed information please see the article), conc.: nr, country: Italy[515], *control
incidence: 3?/3, sa. const.: Hubbard strain laying hens, age: 14 days, contamination: artificial (dose: **50 µg AFB_1/kg** feed + **50 µg OTA/kg** feed, o., for 87 days; for detailed information please see the article), conc. range: 0.5 µg/kg* (mean value), country: Italy[515], *33 and 82 days after withdrawal from treatment

Hen muscle, white may contain the following mycotoxins and/or their metabolites:

Citrinin
incidence: 9/9*, sa. const.: sex mature Mamourah hens, age: ≈13 months, Ø wt.: 1,998 g, contamination: no CIT (for detailed information please see the article), conc.: nr, country: Egypt[591], *control
incidence: ?/9, sa. const.: sex mature Mamourah hens, age: ≈13 months, Ø wt.: 1,998 g, contamination: artificial (dose: **100 ppb CIT** in the diet, o., for 6 weeks; for detailed information please see the article), conc.: 10.3 ppb*, country: Egypt[591], *after 6 weeks

Ochratoxin A
incidence: 12/12*, sa. const.: H and N White Leghorn hens, age: 26 weeks, contamination: no OTA (for detailed information please see the article), conc.: nd, country: Canada[525], *control
incidence: ?/12, sa. const.: H and N White Leghorn hens, age: 26 weeks, contamination: artificial (dose: **0.5 ppm OTA** in feed, o., for 42 days; for detailed information please see the article), conc.: nd, country: Canada[525], *directly after withdrawal from treated feed for 6 weeks
incidence: ?/12, sa. const.: H and N White Leghorn hens, age: 26 weeks, contamination: artificial (dose: **1.0 ppm OTA** in feed, o., for 42 days; for detailed information please see the article), conc.: 4.6 ppm* (mean value), country: Canada[525], *directly after withdrawal from treated feed for 6 weeks
incidence: ?/16, sa. const.: H and N White Leghorn hens, age: 26 weeks, contamination: artificial (dose: **4.0 ppm OTA** in feed, o., for 42 days; for detailed information please see the article), conc.: 15.9 ppm* (mean value), country: Canada[525], *directly after withdrawal from treated feed for 6 weeks
incidence: ?/16, sa. const.: H and N White Leghorn hens, age: 26 weeks, contamination: artificial (dose: **4.0 ppm OTA** in feed, o., for 42 days; for detailed information please see the article), conc.: nd, country: Canada[525], ***after 24 h off** treated feed for 6 weeks
incidence: ?/16, sa. const.: H and N White Leghorn hens, age: 26 weeks, contamination: artificial (dose: **4.0 ppm OTA** in feed, o., for 42 days; for detailed information please see the article), conc.: nd, country: Canada[525], ***after 48 h off** treated feed for 6 weeks

Hen muscle, wing may contain the following mycotoxins and/or their metabolites:

AFLATOXINS
incidence: ?/?, sa. const.: White Leghorn hens (Cornell strain K), age: 20–22 weeks, wt.: 1,600–1,800 g, contamination: artificial (dose: 11.26 mg AFs (labeled), by stomach tube, once; for detailed information please see the article), conc. range: ≤8.18 µg/g* **, country: USA[81], *AFs or their metabolites, **after 1 day (also measured after 4 and 7 days, lowest conc.: 6.29 µg/g after 7 days)

ZEARALENONE
incidence: ?/4, sa. const.: White Leghorn laying hens, age: 26–39 weeks, contamination: artificial (dose: 10 mg ZEA (labeled)/kg, by gavage into the crop, once), conc. range: ≤12 µg eq ZEA/100 g wet tissue* (mean value), country: USA[394], *after 24 h (also measured after 2, 4, 48 and 72 h, lowest conc.: 8 µg eq ZEA/100 g wet tissue after 2 h)

Hen ova may contain the following mycotoxins and/or their metabolites:

AFLATOXICOL
incidence: 6/6*, sa. const.: Single Comb White Leghorn hens, contamination: no AFB_1 (for detailed information please see the article), conc.: nd, country: USA[363], *control
incidence: 10/10, sa. const.: Single Comb White Leghorns hens, contamination: artificial (dose: **8 µg AFB_1/g** feed, for 7 days; for detailed information please see the article), conc. range: ≤0.42 ng/g* **, country: USA[363], *AFL-residues, **after 7 days

AFLATOXIN B_1
incidence: 8/8*, sa. const.: White Leghorn pullets of the Shaver strain, contamination no AFB_1 + AFB_2 (for detailed information please see the article), conc.: nr, country: USA[110], *control
incidence: 6/8, sa. const.: White Leghorn pullets of the Shaver strain, contamination: artificial (dose: **3,310 µg AFB_1/kg + 1,680 µg AFB_2/kg**, o., for 4 weeks; for detailed information please see the article), conc. range: 0.03–0.15 µg/kg*, Ø conc.: 0.07 µg/kg*, country: USA[110], *after feeding 4 weeks an AF-contaminated diet

incidence: 8/8*, sa. const.: White Legorn pullets of the Shaver strain, contamination no AFB_1 + AFB_2 (for detailed information please see the article), conc.: nr, country: USA[110], *control
incidence: 8/8, sa. const.: White Leghorn pullets of the Shaver strain, contamination: artificial (dose: **3,310 µg AFB_1/kg + 1,680 µg AFB_2/kg**, o., for 4 weeks; for detailed information please see the article), conc. range: tr–0.08 µg/kg*, country: USA[110], *2 days after feeding 4 weeks an AF-contaminated diet

incidence: 6/6*, sa. const.: Single Comb White Leghorn hens, contamination: no AFB_1 (for detailed information please see the article), conc.: nd, country: USA[363], *control
incidence: 10/10, sa. const.: Single Comb White Leghorns hens, contamination: artificial (dose: **8 µg AFB_1/g** feed, for 7 days; for detailed information please see the article), conc. range: ≤0.37 ng/g* **, country: USA[363], *AFB_1-residues, **after 7 days

AFLATOXIN B_2
incidence: 8/8*, sa. const.: White Leghorn pullets of the Shaver strain, contamination no AFB_1 + AFB_2 (for detailed information please see the article), conc.: nr, country: USA[110], *control
incidence: 6/8, sa. const.: White Leghorn pullets of the Shaver strain, contamination: artificial (dose: **3,310 µg AFB_1/kg + 1,680 µg AFB_2/kg**, o., for 4 weeks; for detailed information please see

the article), conc. range: 0.02–0.04 μg/kg*, Ø conc.: 0.03 μg/kg*, country: USA[110], *after feeding 4 weeks an AF-contaminated diet

incidence: 8/8*, sa. const.: White Leghorn pullets of the Shaver strain, contamination no AFB_1 + AFB_2 (for detailed information please see the article), conc.: nr, country: USA[110], *control
incidence: 8/8, sa. const.: White Leghorn pullets of the Shaver strain, contamination: artificial (dose: **3,310 μg AFB_1/kg + 1,680 μg AFB_2/kg**, o., for 4 weeks; for detailed information please see the article), conc. range: tr–0.03 μg/kg*, country: USA[110], *2 days after feeding 4 weeks an AF-contaminated diet

Aflatoxin B_{2a}
incidence: 8/8*, sa. const.: White Leghorn pullets of the Shaver strain, contamination no AFB_1 + AFB_2 (for detailed information please see the article), conc.: nr, country: USA[110], *control
incidence: 6/8, sa. const.: White Leghorn pullets of the Shaver strain, contamination: artificial (dose: **3,310 μg AFB_1/kg + 1,680 μg AFB_2/kg**, o, for 4 weeks; for detailed information please see the article), conc. range: 0.02–0.04 μg/kg*, Ø conc.: 0.03 μg/kg*, country: USA[110], *after feeding 4 weeks an AF-contaminated diet

incidence: 8/8*, sa. const.: White Leghorn pullets of the Shaver strain, contamination no AFB_1 + AFB_2 (for detailed information please see the article), conc.: nr, country: USA[110], *control
incidence: 7/8, sa. const.: White Leghorn pullets of the Shaver strain, contamination: artificial (dose: **3,310 μg AFB_1/kg + 1,680 μg AFB_2/kg**, o., for 4 weeks; for detailed information please see the article), conc. range: tr–0.05 μg/kg*, country: USA[110], *2 days after feeding 4 weeks an AF-contaminated diet

Aflatoxin M_1
incidence: 8/8*, sa. const.: White Leghorn pullets of the Shaver strain, contamination no AFB_1 + AFB_2 (for detailed information please see the article), conc.: nr, country: USA[110], *control
incidence: 5/8, sa. const.: White Leghorn pullets of the Shaver strain, contamination: artificial (dose: **3,310 μg AFB_1/kg + 1,680 μg AFB_2/kg**, o., for 4 weeks; for detailed information please see the article), conc. range: tr–0.02 μg/kg*, country: USA[110], *after feeding 4 weeks an AF-contaminated diet

incidence: 8/8*, sa. const.: White Leghorn pullets of the Shaver strain, contamination no AFB_1 + AFB_2 (for detailed information please see the article), conc.: nr, country: USA[110], *control
incidence: 1/8, sa. const.: White Leghorn pullets of the Shaver strain, contamination: artificial (dose: **3,310 μg AFB_1/kg + 1,680 μg AFB_2/kg**, o., for 4 weeks; for detailed information please see the article), conc.: 0.01 μg/kg*, country: USA[110], *2 days after feeding 4 weeks an AF-contaminated diet

Aflatoxin M_2
incidence: 8/8*, sa. const.: White Leghorn pullets of the Shaver strain, contamination no AFB_1 + AFB_2 (for detailed information please see the article), conc.: nr, country: USA[110], *control
incidence: 6/8, sa. const.: White Leghorn pullets of the Shaver strain, contamination: artificial (dose: **3,310 μg AFB_1/kg + 1,680 μg AFB_2/kg**, o., for 4 weeks; for detailed information please see the article), conc. range: tr–0.02 μg/kg*, country: USA[110], *after feeding 4 weeks an AF-contaminated diet

incidence: 8/8*, sa. const.: White Leghorn pullets of the Shaver strain, contamination no AFB_1 + AFB_2 (for detailed information please see the article), conc.: nr, country: USA[110], *control incidence: 4/8, sa. const.: White Leghorn pullets of the Shaver strain, contamination: artificial (dose: **3,310 µg AFB_1/kg + 1,680 µg AFB_2/kg**, o., for 4 weeks; for detailed information please see the article), conc. range: tr–0.02 µg/kg*, country: USA[110], *2 days after feeding 4 weeks an AF-contaminated diet

AFLATOXINS
incidence: ?/?, sa. const.: White Leghorn hens (Cornell strain K), age: 20–22 weeks, wt.: 1,600–1,800 g, contamination: artificial (dose: 11.26 mg AFs (labeled), by stomach tube, once; for detailed information please see the article), conc. range: ≤8.29 µg/g* ** ***, country: USA[81], *AFs or their metabolites, **in large ova (>10 mm), ***after 4 days (also measured after 1 and 7 days, lowest conc.: 6.75 µg/g after 1 day)
incidence: ?/?, sa. const.: White Leghorn hens (Cornell strain K), age: 20–22 weeks, wt.: 1,600–1,800 g, contamination: artificial (dose: 11.26 mg AFs (labeled), by stomach tube, once; for detailed information please see the article), conc. range: ≤13.72 µg/g* ** ***, country: USA[81], *AFs or their metabolites, **in small ova (<10 mm), ***after 7 days (also measured after 1 and 4 days, lowest conc.: 7.09 µg/g after 4 days)

Hen Ovaries see Hen ova

Hen oviduct may contain the following mycotoxins and/or their metabolites:

ZEARALENONE
incidence: ?/4, sa. const.: White Leghorn laying hens, age: 26–39 weeks, contamination: artificial (dose: 10 mg ZEA (labeled)/kg, by gavage into the crop, once), conc. range: ≤68 µg eq ZEA/100 g wet tissue* (mean value), country: USA[394], *after 24 h (also measured after 2, 4, 48 and 72 h, lowest conc.: 39 µg eq ZEA/100 g wet tissue after 2 h)

Hen pancreas may contain the following mycotoxins and/or their metabolites:

AFLATOXINS
incidence: ?/?, sa. const.: White Leghorn hens (Cornell strain K), age: 20–22 weeks, wt.: 1,600–1,800 g, contamination: artificial (dose: 11.26 mg AFs (labeled), stomach tube, once; for detailed information please see the article), conc. range: ≤12.54 µg/g* **, country: USA[81], *AFs or their metabolites, **after 7 days (also measured after 1 and 4 days, lowest conc.: 5.10 µg/g after 1 day)

ZEARALENONE
incidence: ?/20, sa. const.: White Leghorn laying hens, age: 26–39 weeks, contamination: artificial (dose: 10 mg ZEA (labeled)/kg, by gavage into the crop, once), conc. range: ≤38 µg eq zearalenone/100 g wet tissue* (mean value), country: USA[394], *after 2 h (also measured after 4, 24, 48 and 72 h, lowest conc.: 15 µg eq ZEA/100 g wet tissue after 72 h)

Hen plasma may contain the following mycotoxins and/or their metabolites:

DEOXYNIVALENOL
incidence: ?/4, sa. const.: White Leghorn hens, age: 316 days, wt.: 1.3–1.7 kg, contamination: artificial (dose: 2.2 mg DON (unlabeled)/day/bird, o., for 6 days followed by 2.2 mg DON (labeled)/day/bird, o., for 6 days; for detailed information please see the article), conc. range: ≈≤38–54 ng/ml* (mean value), country: Canada[135], *after 8–12 days exposure to DON (also measured after 14, 16 and 18 days, lowest conc.: significantly lower than the above indicated value after 18 days)

FUMONISIN B_1
incidence: 6/6, sa. const.: White Leghorn laying hens, age: 30 weeks, wt.: 1.30–1.68 kg, contamination: artificial (dose: 2.0 mg FB_1 (labeled and unlabeled)/kg b.

wt., o., once; for detailed information please see the article), conc. range: 28–103 ng/ml* **, country: Canada[129], *FB_1 and/or metabolites, **(measured at other hour intervals up to 28 h, lowest conc.: nd after 24 h)

Ochratoxin A
incidence: ?/24*, sa. const.: Rhode Island Red hens, age: 54 weeks, contamination: no OTA (for detailed information please see the article), conc.: nd, country: Poland[590], *control
incidence: ?/22, sa. const.: Rhode Island Red **hens**, age: 54 weeks, contamination: artificial (dose: **2.1 ppm OTA** in the diet, o., for 5 weeks; for detailed information please see the article), conc.: 13.3 ppb*, country: Poland[590], *after 4 weeks of OTA-administration
incidence: ?/2, sa. const.: Rhode Island Red **cockerels**, age: 54 weeks, contamination: artificial (dose: **2.1 ppm OTA** in the diet, o., for 5 weeks; for detailed information please see the article), conc.: 5.7 ppb*, country: Poland[590], *after 4 weeks of OTA-administration
incidence: ?/22, sa. const.: Rhode Island Red **hens**, age: 54 weeks, contamination: artificial (dose: **4.1 ppm OTA** in the diet, o., for 5 weeks; for detailed information please see the article), conc.: 37.0 ppb*, country: Poland[590], *after 4 weeks of OTA-administration
incidence: ?/2, sa. const.: Rhode Island Red **cockerels**, age: 54 weeks, contamination: artificial (dose: **4.1 ppm OTA** in the diet, o., for 5 weeks; for detailed information please see the article), conc.: 16.0 ppb*, country: Poland[590], *after 4 weeks of OTA-administration

Zearalenone
incidence: ?/4, sa. const.: White Leghorn laying hens, age: 26–39 weeks, contamination: artificial (dose: 10 mg ZEA (labeled)/kg, by gavage into the crop, once), conc. range: ≤82 µg eq ZEA/100 g wet tissue* (mean value), country: USA[394], *after 4 h (also measured after 2, 24, 48 and 72 h, lowest conc.: 12 µg eq ZEA/100 g wet tissue after 72 h)

Hen red blood cells may contain the following mycotoxins and/or their metabolites:

Zearalenone
incidence: ?/4, sa. const.: White Leghorn laying hens, age: 26–39 weeks, contamination: artificial (dose: 10 mg ZEA (labeled)/kg, by gavage into the crop, once), conc. range: ≤269 µg eq ZEA/100 g wet tissue* (mean value), country: USA[394], *after 48 h (also measured after 2, 4, 24 and 72 h, lowest conc.: 125 µg eq ZEA/100 g wet tissue after 72 h)

Hen reproductive organs may contain the following mycotoxins and/or their metabolites:

Aflatoxins
incidence: ?/?, sa. const.: White Leghorn hens (Cornell strain K), age: 20–22 weeks, wt.: 1,600–1,800 g, contamination: artificial (dose: 11.26 mg AFs (labeled), stomach tube, once; for detailed information please see the article), conc. range: ≤13.48 µg/g* **, country: USA[81], *AFs or their metabolites, **after 1 day (also measured after 4 and 7 days, lowest conc.: 4.49 µg/g after 4 days)

Hen serum may contain the following mycotoxins and/or their metabolites:

Aflatoxin B_1
incidence: 8/8*, sa. const.: White Leghorn pullets of the Shaver strain, contamination no AFB_1 + AFB_2 (for detailed information please see the article), conc.: nr, country: USA[110], *control
incidence: ?/?, sa. const.: White Leghorn pullets of the Shaver strain, contamination: artificial (dose: **3,310 µg AFB_1/kg + 1,680 µg AFB_2/kg**, o, for 4 weeks; for detailed information please

see the article), conc. range: 0.02–0.03 µg/kg* (mean values), country: USA[110], *after feeding 4 weeks an AF-contaminated diet
incidence: 8/8*, sa. const.: White Leghorn pullets of the Shaver strain, contamination no AFB_1 + AFB_2 (for detailed information please see the article), conc.: nr, country: USA[110], *control
incidence: ?/?, sa. const.: White Leghorn pullets of the Shaver strain, contamination: artificial (dose: **3,310 µg AFB_1/kg + 1,680 µg AFB_2**/kg, o, for 4 weeks; for detailed information please see the article), conc.: 0.01 µg/kg* (mean value), country: USA[110], *2 days after feeding 4 weeks an AF-contaminated diet

Aflatoxin B_2
incidence: 8/8*, sa. const.: White Leghorn pullets of the Shaver strain, contamination no AFB_1 + AFB_2 (for detailed information please see the article), conc.: nr, country: USA[110], *control
incidence: ?/?, sa. const.: White Leghorn pullets of the Shaver strain, contamination: artificial (dose: **3,310 µg AFB_1/kg + 1,680 µg AFB_2**/kg, o, for 4 weeks; for detailed information please see the article), conc. range: 0.04–0.05 µg/kg* (mean values), country: USA[110], *after feeding 4 weeks an AF-contaminated diet

incidence: 8/8*, sa. const.: White Leghorn pullets of the Shaver strain, contamination no AFB_1 + AFB_2 (for detailed information please see the article), conc.: nr, country: USA[110], *control
incidence: ?/?, sa. const.: White Leghorn pullets of the Shaver strain, contamination: artificial (dose: **3,310 µg AFB_1/kg + 1,680 µg AFB_2**/kg, o., for 4 weeks; for detailed information please see the article), conc.: 0.01 µg/kg * (mean value), country: USA[110], *2 days after feeding 4 weeks an AF-contaminated diet

Aflatoxin B_{2a}
incidence: 8/8*, sa. const.: White Leghorn pullets of the Shaver strain, contamination no AFB_1 + AFB_2 (for detailed information please see the article), conc.: nr, country: USA[110], *control
incidence: ?/?, sa. const.: White Leghorn pullets of the Shaver strain, contamination: artificial (dose: **3,310 µg AFB_1/kg + 1,680 µg AFB_2**/kg, o, for 4 weeks; for detailed information please see the article), conc. range: 0.01–0.02 µg/kg* (mean values), country: USA[110], *after feeding 4 weeks an AF-contaminated diet

incidence: 8/8*, sa. const.: White Leghorn pullets of the Shaver strain, contamination no AFB_1 + AFB_2 (for detailed information please see the article), conc.: nr, country: USA[110], *control
incidence: ?/?, sa. const.: White Leghorn pullets of the Shaver strain, contamination: artificial (dose: **3,310 µg AFB_1/kg + 1,680 µg AFB_2**/kg, o., for 4 weeks; for detailed information please see the article), conc.: tr* (mean value), country: USA[110], *2 days after feeding 4 weeks an AF-contaminated diet

Aflatoxin M_1
incidence: 8/8*, sa. const.: White Leghorn pullets of the Shaver strain, contamination no AFB_1 + AFB_2 (for detailed information please see the article), conc.: nr, country: USA[110], *control
incidence: 16/24, sa. const.: White Leghorn pullets of the Shaver strain, contamination: artificial (dose: **3,310 µg AFB_1/kg + 1,680 µg AFB_2**/kg, o, for 4 weeks; for detailed information please see the article), conc.: nd*, country: USA[110], *after feeding 4 weeks an AF-contaminated diet

incidence: 8/8*, sa. const.: White Leghorn pullets of the Shaver strain, contamination no AFB_1 + AFB_2 (for detailed information please see the article), conc.: nr, country: USA[110], *control
incidence: ?/?, sa. const.: White Leghorn pullets of the Shaver strain,

contamination: artificial (dose: **3,310 µg AFB_1/kg + 1,680 µg AFB_2/kg**, o., for 4 weeks; for detailed information please see the article), conc.: nd*, country: USA[110], *2 days after feeding 4 weeks an AF-contaminated diet

Aflatoxin M_2

incidence: 8/8*, sa. const.: White Leghorn pullets of the Shaver strain, contamination no AFB_1 + AFB_2 (for detailed information please see the article), conc.: nr, country: USA[110], *control

incidence: ?/?, sa. const.: White Leghorn pullets of the Shaver strain, contamination: artificial (dose: **3,310 µg AFB_1/kg + 1,680 µg AFB_2/kg**, o., for 4 weeks; for detailed information please see the article), conc. range: tr* (mean value), country: USA[110], *after feeding 4 weeks an AF-contaminated diet

incidence: 8/8*, sa. const.: White Leghorn pullets of the Shaver strain, contamination no AFB_1 + AFB_2 (for detailed information please see the article), conc.: nr, country: USA[110], *control

incidence: ?/?, sa. const.: White Leghorn pullets of the Shaver strain, contamination: artificial (dose: **3,310 µg AFB_1/kg + 1,680 µg AFB_2/kg**, o., for 4 weeks; for detailed information please see the article), conc.: nd*, country: USA[110], *2 days after feeding 4 weeks an AF-contaminated diet

Ochratoxin A

incidence: 5/5*, sa. const.: White Leghorn hens, contamination: no OTA (for detailed information please see the article), conc.: nd, country: Germany[374], *control

incidence: ?/5, sa. const.: White Leghorn hens, contamination: artificial (dose: **1.3 mg OTA/kg** feed, o., for 28 days; for detailed information please see the article), conc.: 4.7 µg/l* (mean value), country: Germany[374], **after feeding OTA for 28 days

incidence: ?/5, sa. const.: White Leghorn hens, contamination: artificial (dose: **2.6 mg OTA/kg** feed, o., for 28 days; for detailed information please see the article), conc.: 14.1 µg/l* (mean value), country: Germany[374], **after feeding OTA for 28 days

incidence: ?/4, sa. const.: eggs from White Leghorn hens, contamination: artificial (dose: **5.2 mg OTA/kg** feed, o., for 28 days; for detailed information please see the article), conc.: 11.7 µg/l* (mean value), country: Germany[374], **after feeding OTA for 28 days

Hen skin may contain the following mycotoxins and/or their metabolites:

Aflatoxicol

incidence: ?/?*, sa. const.: Hubbard strain laying hens, age: 14 days, contamination: no AFB_1 + OTA (for detailed information please see the article), conc.: nr, country: Italy[515], *control

incidence: 3?/3, sa. const.: Hubbard strain laying hens, age: 14 days, contamination: artificial (dose: **50 µg AFB_1/kg** feed + **50 µg OTA/kg** feed, o., for 169 days; for detailed information please see the article), conc.: <0.04 µg/kg* (mean value), country: Italy[515], *after 169 days

Aflatoxin B_1

incidence: ?/?*, sa. const.: Hubbard strain laying hens, age: 14 days, contamination: no AFB_1 (for detailed information please see the article), conc.: nr, country: Italy[232], *control

incidence: 3?/3, sa. const.: Hubbard strain laying hens, age: 14 days, contamination: artificial (dose: **50 µg AFB_1/kg** feed, o., for 169 days; for detailed information please see the article), conc.: 0.12 µg/kg* (mean value), country: Italy[232], *after 169 days

incidence: ?/?*, sa. const.: Hubbard strain laying hens, age: 14 days, contamination: no AFB_1 + OTA (for detailed information please see the article), conc.: nr, country: Italy[515], *control incidence: 3?/3, sa. const.: Hubbard strain laying hens, age: 14 days, contamination: artificial (dose: **50 µg AFB_1/kg feed + 50 µg OTA/kg** feed, o., for 169 days; for detailed information please see the article), conc.: 0.04 µg/kg* (mean value), country: Italy[515], *after 169 days

Aflatoxin M_1
incidence: ?/?*, sa. const.: Hubbard strain laying hens, age: 14 days, contamination: no AFB_1 + OTA (for detailed information please see the article), conc.: nr, country: Italy[515], *control
incidence: 3?/3, sa. const.: Hubbard strain laying hens, age: 14 days, contamination: artificial (dose: **50 µg AFB_1/kg feed + 50 µg OTA/kg** feed, o., for 169 days; for detailed information please see the article), conc.: <0.01 µg/kg* (mean value), country: Italy[515], *after 169 days

Aflatoxins
incidence: ?/?, sa. const.: White Leghorn hens (Cornell strain K), age: 20–22 weeks, wt.: 1,600–1,800 g, contamination: artificial (dose: 11.26 mg AFs (labeled), by stomach tube, once; for detailed information please see the article), conc. range: ≤4.71 µg/g* **, country: USA[81], *AFs or their metabolites, **after 1 day (also measured after 4 and 7 days, lowest conc.: 3.25 µg/g after 4 days)

Ochratoxin A
incidence: ?/?*, sa. const.: Hubbard strain laying hens, age: 14 days, contamination: no AFB_1 + OTA (for detailed information please see the article), conc.: nr, country: Italy[515], *control
incidence: 3?/3, sa. const.: Hubbard strain laying hens, age: 14 days, contamination: artificial (dose: **50 µg AFB_1/kg feed + 50 µg OTA/kg** feed, o., for 169 days; for detailed information please see the article), conc.: 2.0 µg/kg* (mean value), country: Italy[515], *after 169 days
incidence: ?/?*, sa. const.: Hubbard strain laying hens, age: 14 days, contamination: no AFB_1 + OTA (for detailed information please see the article), conc.: nr, country: Italy[515], *control
incidence: 3?/3, sa. const.: Hubbard strain laying hens, age: 14 days, contamination: artificial (dose: **50 µg AFB_1/kg feed + 50 µg OTA/kg** feed, o., for 87 days; for detailed information please see the article), conc. range: <0.5 µg/kg* (mean value), country: Italy[515], *33 and 82 days after withdrawal from treatment

Hen spleen may contain the following mycotoxins and/or their metabolites:

Aflatoxins
incidence: ?/?, sa. const.: White Leghorn hens (Cornell strain K), age: 20–22 weeks, wt.: 1,600–1,800 g, contamination: artificial (dose: 11.26 mg AFs (labeled), by stomach tube, once; for detailed information please see the article), conc. range: ≤12.47 µg/g* ** ***, country: USA[81], *AFs or their metabolites, **after 7 days (also measured after 1 and 4 days, lowest conc.: 7.14 µg/g after 4 days), ***in spleen and kidney

Zearalenone
incidence: ?/4, sa. const.: White Leghorn laying hens, age: 26–39 weeks, contamination: artificial (dose: 10 mg ZEA (labeled)/kg, by gavage into the crop, once), conc. range: ≤64 µg eq ZEA/100 g wet tissue* (mean value), country: USA[394], *after 24 h (also measured after 2, 4, 48 and 72 h, lowest conc.: 28 µg eq ZEA/100 g wet tissue after 72 h)

Hen Spleen and Kidney see Hen spleen

Hen stomach may contain the following mycotoxins and/or their metabolites:

Ochratoxin A
incidence: 8?/8, sa. const.: New Hampshire-Leghorn cross hens, age: 24 weeks, contamination: artificial (dose: **OTA 0.5 ppm** (feed weight), o.,

for 2 weeks), conc.: 19.8 ppb* (mean value), country: USA[316], *after 14 days
incidence: 8?/8, sa. const.: New Hampshire-Leghorn cross hens, age: 24 weeks, contamination: artificial (dose: **OTA 5.0 ppm** (feed weight), o., for 2 weeks), conc.: 94.3 ppb* (mean value), country: USA[316], *after 14 days

Horse

Horse Natural Contamination

Horse liver may contain the following mycotoxins and/or their metabolites:

Aflatoxin B_1
incidence: 2?/?, sa. const.: livers from horses of Thailand and the USA, contamination: natural, conc. range: 9–18 ppb, Ø conc.: 13.5 ppb, country: Thailand/USA[141]

Horse urine may contain the following mycotoxins and/or their metabolites:

Zearalanols
incidence: 47/76*, sa. const.: urine from horses of New Zealand, contamination: natural, conc. range: ≤18.8 ng/ml**, country: New Zealand[230], *export animals, **most probable of *Fusarium* origin

Zearalenols
incidence: 47/76*, sa. const.: urine from horses of New Zealand, contamination: natural, conc. range: ≤2,157 ng/ml, country: New Zealand[230], *export animals

Lamb

Lamb Natural Contamination

Lamb urine may contain the following mycotoxins and/or their metabolites:

Zearalanols
incidence: 38/90*, sa. const.: urine from lambs of New Zealand, contamination: natural, conc. range: ≤0.77 ng/ml**, country: New Zealand[230], *export animals, **most probable of *Fusarium* origin

Zearalenols
incidence: 38/90*, sa. const.: urine from lambs of New Zealand, contamination: natural, conc. range: ≤34 ng/ml, country: New Zealand[230], *export animals

Lamb Artificial Contamination

Lamb feces may contain the following mycotoxins and/or their metabolites:

Aflatoxin B_1
incidence: 6/6*, sa. const.: male Rasa Aragonesa lambs, age: 40–45 days, wt.: ≈15.3 kg, contamination: no AF, conc.: nd, country: Spain[358], *control
incidence: ?/12?, sa. const.: male Rasa Aragonesa lambs, age: 40–45 days, wt.: ≈15.3 kg, contamination: artificial (dose: 2.5 mg AF, o., for 21 days), conc. range: 6.86–36.29 µg/kg* (mean values), Ø conc.: 17.25 µg/kg* (mean value), country: Spain[358], *during **intoxication period** over 21 days (measured after 7, 14* and 21 days) (for overall information please see the article)
incidence: 6/6*, sa. const.: male Rasa Aragonesa lambs, age: 40–45 days, wt.: ≈15.3 kg, contamination: no AF, conc.: nd, country: Spain[358], *control
incidence: ?/12?, sa. const.: male Rasa Aragonesa lambs, age: 40–45 days, wt.: ≈15.3 kg, contamination: artificial (dose: 2.5 mg AF, o., for 21 days), conc. range: 7.95–19.66 µg/kg*, Ø conc.: 13.33 µg/kg*, country: Spain[358], *during **clearance period** (at the 1st, 2nd* and 4th day contamination, at the 8th day no contamination anymore) (for overall information please see the article)

Aflatoxin G_1
incidence: 6/6*, sa. const.: male Rasa Aragonesa lambs, age: 40–45 days, wt.: ≈15.3 kg, contamination: no AF, conc.: nd, country: Spain[358], *control

incidence: ?/?, sa. const.: male Rasa Aragonesa lambs, age: 40–45 days, wt.: ≈15.3 kg, contamination: artificial (dose: 2.5 mg AF, o., for 21 days), conc. range: 1.3*–27.47** µg/kg*** (mean values), Ø conc.: 10.96 µg/kg*** (mean value), country: Spain[358], ***during **intoxication period** over 21 days (measured after 7**, 14 and 21* days) (for overall information please see the article)
incidence: 6/6*, sa. const.: male Rasa Aragonesa lambs, age: 40–45 days, wt.: ≈15.3 kg, contamination: no AF, conc.: nd, country: Spain[358], *control
incidence: ?/?, sa. const.: male Rasa Aragonesa lambs, age: 40–45 days, wt.: ≈15.3 kg, contamination: artificial (dose: 2.5 mg AF, o., for 21 days), conc. range: 1.22–1.4 µg/kg*, Ø conc.: 1.31 µg/kg*, country: Spain[358], *during **clearance period** (at the 1st* day still contamination, at the 2nd, 4th and 8th day no contamination anymore) (for overall information please see the article)

Aflatoxin M_1
incidence: 6/6*, sa. const.: male Rasa Aragonesa lambs, age: 40–45 days, wt.: ≈15.3 kg, contamination: no AF, conc.: nd, country: Spain[358], *control
incidence: ?/?, sa. const.: male Rasa Aragonesa lambs, age: 40–45 days, wt.: ≈15.3 kg, contamination: artificial (dose: 2.5 mg AF, o., for 21 days), conc. range: 8.16*–61.82** µg/kg*** (mean value), Ø conc.: 27.2 µg/kg*** (mean value), country: Spain[358], ***during **intoxication period** over 21 days (measured after 7**, 14* and 21 days) (for overall information please see the article)
incidence: 6/6*, sa. const.: male Rasa Aragonesa lambs, age: 40–45 days, wt.: ≈15.3 kg, contamination: no AF, conc.: nd, country: Spain[358], *control
incidence: ?/?, sa. const.: male Rasa Aragonesa lambs, age: 40–45 days, wt.: ≈15.3 kg, contamination: artificial (dose: 2.5 mg AF, o., for 21 days), conc. range: 0.23*–13.94** µg/kg***, Ø conc.: 4.01 µg/kg***, country: Spain[358], ***during **clearance period** (at the 1st**, 2nd 4th and 8th* day contamination) (for overall information please see the article)

Lamb kidney may contain the following mycotoxins and/or their metabolites:

Aflatoxicol
incidence: 6/6*, sa. const.: male Rasa Aragonesa lambs, age: 40–45 days, wt.: ≈15.3 kg, contamination: no AF, conc.: nd, country: Spain[358], *control
incidence: 2/12?, sa. const.: male Rasa Aragonesa lambs, age: 40–45 days, wt.: ≈15.3 kg, contamination: artificial (dose: 2.5 mg AF, o., for 21 days), conc. range: tr?*, country: Spain[358], ***clearance period** (8 days after removal of the contaminated diet)

Aflatoxin B_1
incidence: 6/6*, sa. const.: male Rasa Aragonesa lambs, age: 40–45 days, wt.: ≈15.3 kg, contamination: no AF, conc.: nd, country: Spain[358], *control
incidence: 12/12, sa. const.: male Rasa Aragonesa lambs, age: 40–45 days, wt.: ≈15.3 kg, contamination: artificial (dose: 2.5 mg AF, o., for 21 days), conc. range: 0.38–2.93 µg/kg*, Ø conc.: 1.29 µg/kg*, country: Spain[358], ***intoxication period** (after 21 days of intoxication)
incidence: 6/6*, sa. const.: male Rasa Aragonesa lambs, age: 40–45 days, wt.: ≈15.3 kg, contamination: no AF, conc.: nd, country: Spain[358], *control
incidence: 7/12, sa. const.: male Rasa Aragonesa lambs, age: 40–45 days, wt.: ≈15.3 kg, contamination: artificial (dose: 2.5 mg AF, o., for 21 days), conc. range: 0.02–0.71 µg/kg*, Ø conc.: 0.20 µg/kg*, country: Spain[358], ***clearance period** (8 days after removal of the contaminated diet)

Aflatoxin B_{2a}
incidence: 6/6*, sa. const.: male Rasa Aragonesa lambs, age: 40–45 days, wt.:

≈15.3 kg, contamination: no AF, conc.: nd, country: Spain[358], *control
incidence: 3/12, sa. const.: male Rasa Aragonesa lambs, age: 40–45 days, wt.: ≈15.3 kg, contamination: artificial (dose: 2.5 mg AF, o., for 21 days), conc. range: 0.02–0.09 µg/kg, Ø conc.: 0.05 µg/kg*, country: Spain[358], ***intoxication period** (after 21 days of intoxication)
incidence: 6/6*, sa. const.: male Rasa Aragonesa lambs, age: 40–45 days, wt.: ≈15.3 kg, contamination: no AF, conc.: nd, country: Spain[358], *control
incidence: 5/12, sa. const.: male Rasa Aragonesa lambs, age: 40–45 days, wt.: ≈15.3 kg, contamination: artificial (dose: 2.5 mg AF, o., for 21 days), conc. range: 0.03–0.69 µg/kg*, Ø conc.: 0.19 µg/kg*, country: Spain[358], ***clearance period** (8 days after removal of the contaminated diet)

Aflatoxin G_1
incidence: 6/6*, sa. const.: male Rasa Aragonesa lambs, age: 40–45 days, wt.: ≈15.3 kg, contamination: no AF, conc.: nd, country: Spain[358], *control
incidence: 12/12, sa. const.: male Rasa Aragonesa lambs, age: 40–45 days, wt.: ≈15.3 kg, contamination: artificial (dose: 2.5 mg AF, o., for 21 days), conc. range: 0.03–0.52 µg/kg*, Ø conc.: 0.31 µg/kg*, country: Spain[358], ***intoxication period** (after 21 days of intoxication)
incidence: 6/6*, sa. const.: male Rasa Aragonesa lambs, age: 40–45 days, wt.: ≈15.3 kg, contamination: no AF, conc.: nd, country: Spain[358], *control
incidence: 3/12, sa. const.: male Rasa Aragonesa lambs, age: 40–45 days, wt.: ≈15.3 kg, contamination: artificial (dose: 2.5 mg AF, o., for 21 days), conc. range: 0.01–0.04 µg/kg*, Ø conc.: 0.03 µg/kg*, country: Spain[358], ***clearance period** (8 days after removal of the contaminated diet)

Aflatoxin M_1
incidence: 6/6*, sa. const.: male Rasa Aragonesa lambs, age: 40–45 days, wt.: ≈15.3 kg, contamination: no AF, conc.: nd, country: Spain[358], *control
incidence: 12/12, sa. const.: male Rasa Aragonesa lambs, age: 40–45 days, wt.: ≈15.3 kg, contamination: artificial (dose: 2.5 mg AF, o., for 21 days), conc. range: 0.98–15.46 µg/kg*, Ø conc.: 5.45 µg/kg*, country: Spain[358], ***intoxication period** (after 21 days of intoxication)
incidence: 6/6*, sa. const.: male Rasa Aragonesa lambs, age: 40–45 days, wt.: ≈15.3 kg, contamination: no AF, conc.: nd, country: Spain[358], *control
incidence: 9/12, sa. const.: male Rasa Aragonesa lambs, age: 40–45 days, wt.: ≈15.3 kg, contamination: artificial (dose: 2.5 mg AF, o., for 21 days), conc. range: 0.05–1.27 µg/kg*, Ø conc.: 0.35 µg/kg*, country: Spain[358], ***clearance period** (8 days after removal of the contaminated diet)

Lamb liver may contain the following mycotoxins and/or their metabolites:

Aflatoxin B_1
incidence: 6/6*, sa. const.: male Rasa Aragonesa lambs, age: 40–45 days, wt.: ≈15.3 kg, contamination: no AF, conc.: nd, country: Spain[358], *control
incidence: 12/12, sa. const.: male Rasa Aragonesa lambs, age: 40–45 days, wt.: ≈15.3 kg, contamination: artificial (dose: 2.5 mg AF, o., for 21 days), conc. range: 0.34–4.05 µg/kg, Ø conc.: 1.94 µg/kg*, country: Spain[358], ***intoxication period** (after 21 days of intoxication)
incidence: 6/6*, sa. const.: male Rasa Aragonesa lambs, age: 40–45 days, wt.: ≈15.3 kg, contamination: no AF, conc.: nd, country: Spain[358], *control
incidence: 9/12, sa. const.: male Rasa Aragonesa lambs, age: 40–45 days, wt.: ≈15.3 kg, contamination: artificial (dose: 2.5 mg AF, o., for 21 days), conc. range: 0.14–1.03 µg/kg*, Ø conc.: 0.44 µg/kg*, country: Spain[358], ***clearance period** (8 days after removal of the contaminated diet)

Aflatoxin G_1
incidence: 6/6*, sa. const.: male Rasa Aragonesa lambs, age: 40–45 days, wt.: ≈15.3 kg, contamination: no AF, conc.: nd, country: Spain[358], *control

incidence: 11/12, sa. const.: male Rasa Aragonesa lambs, age: 40–45 days, wt: ≈15.3 kg, contamination: artificial (dose: 2.5 mg AF, o., for 21 days), conc. range: 0.03–2.94 μg/kg*, Ø conc.: 1.3 μg/kg*, country: Spain[358], ***intoxication period** (after 21 days of intoxication)
incidence: 6/6*, sa. const.: male Rasa Aragonesa lambs, age: 40–45 days, wt.: ≈15.3 kg, contamination: no AF, conc.: nd, country: Spain[358], *control
incidence: 2/12, sa. const.: male Rasa Aragonesa lambs, age: 40–45 days, wt.: ≈15.3 kg, contamination: artificial (dose: 2.5 mg AF, o., for 21 days), conc. range: 0.02–0.03 μg/kg*, Ø conc.: 0.025 μg/kg*, country: Spain[358], ***clearance period** (8 days after removal of the contaminated diet)

Aflatoxin M_1
incidence: 6/6*, sa. const.: male Rasa Aragonesa lambs, age: 40–45 days, wt.: ≈15.3 kg, contamination: no AF, conc.: nd, country: Spain[358], *control
incidence: 12/12, sa. const.: male Rasa Aragonesa lambs, age: 40–45 days, wt.: ≈15.3 kg, contamination: artificial (dose: 2.5 mg AF, o., for 21 days), conc. range: 0.10–0.72 μg/kg*, Ø conc.: 0.35 μg/kg*, country: Spain[358], ***intoxication period** (after 21 days of intoxication)
incidence: 6/6*, sa. const.: male Rasa Aragonesa lambs, age: 40–45 days, wt.: ≈15.3 kg, contamination: no AF, conc.: nd, country: Spain[358], *control
incidence: 7/12, sa. const.: male Rasa Aragonesa lambs, age: 40–45 days, wt.: ≈15.3 kg, contamination: artificial (dose: 2.5 mg AF, o., for 21 days), conc. range: 0.07–0.16 μg/kg*, Ø conc.: 0.12 μg/kg*, country: Spain[358], ***clearance period** (8 days after removal of the contaminated diet)

Lamb urine may contain the following mycotoxins and/or their metabolites:

Aflatoxin B_1
incidence: 6/6*, sa. const.: male Rasa Aragonesa lambs, age: 40–45 days, wt.: ≈15.3 kg, contamination: no AF, conc.: nd, country: Spain[358], *control
incidence: 12/12, sa. const.: male Rasa Aragonesa lambs, age: 40–45 days, wt.: ≈15.3 kg, contamination: artificial (dose: 2.5 mg AF, o., for 21 days), conc. range: 0.30–5.16 μg/l*, Ø conc.: 1.78 μg/l*, country: Spain[358], ***intoxication period** (after 21 days of intoxication)
incidence: 6/6*, sa. const.: male Rasa Aragonesa lambs, age: 40–45 days, wt.: ≈15.3 kg, contamination: no AF, conc.: nd, country: Spain[358], *control
incidence: 3/12, sa. const.: male Rasa Aragonesa lambs, age: 40–45 days, wt.: ≈15.3 kg, contamination: artificial (dose: 2.5 mg AF, o., for 21 days), conc. range: 0.07–0.19 μg/l*, Ø conc.: 0.14 μg/l*, country: Spain[358], ***clearance period** (8 days after removal of the contaminated diet)

Aflatoxin B_{2a}
incidence: 6/6*, sa. const.: male Rasa Aragonesa lambs, age: 40–45 days, wt.: ≈15.3 kg, contamination: no AF, conc.: nd, country: Spain[358], *control
incidence: 12/12, sa. const.: male Rasa Aragonesa lambs, age: 40–45 days, wt.: ≈15.3 kg, contamination: artificial (dose: 2.5 mg AF, o., for 21 days), conc. range: 0.01–0.98 μg/l, Ø conc.: 0.35 μg/l*, country: Spain[358], ***intoxication period** (after 21 days of intoxication)
incidence: 6/6*, sa. const.: male Rasa Aragonesa lambs, age: 40–45 days, wt.: ≈15.3 kg, contamination: no AF, conc.: nd, country: Spain[358], *control
incidence: 10/12, sa. const.: male Rasa Aragonesa lambs, age: 40–45 days, wt.: ≈15.3 kg, contamination: artificial (dose: 2.5 mg AF, o., for 21 days), conc. range: 0.09–1.26 μg/l*, Ø conc.: 0.29 μg/l*, country: Spain[358], ***clearance period** (8 days after removal of the contaminated diet)

Aflatoxin G_1
incidence: 6/6*, sa. const.: male Rasa Aragonesa lambs, age: 40–45 days, wt.: ≈15.3 kg, contamination: no AF, conc.: nd, country: Spain[358], *control

incidence: 12/12, sa. const.: male Rasa Aragonesa lambs, age: 40–45 days, wt.: ≈15.3 kg, contamination: artificial (dose: 2.5 mg AF, o., for 21 days), conc. range: 0.46–10.52 μg/l*, Ø conc.: 2.69 μg/l*, country: Spain[358], ***intoxication period** (after 21 days of intoxication)
incidence: 6/6*, sa. const.: male Rasa Aragonesa lambs, age: 40–45 days, wt.: ≈15.3 kg, contamination: no AF, conc.: nd, country: Spain[358], *control
incidence: 1/12, sa. const.: male Rasa Aragonesa lambs, age: 40–45 days, wt.: ≈15.3 kg, contamination: artificial (dose: 2.5 mg AF, o., for 21 days), conc.: 1.67 μg/l*, country: Spain[358], ***clearance period** (8 days after removal of the contaminated diet)

AFLATOXIN M_1
incidence: 6/6*, sa. const.: male Rasa Aragonesa lambs, age: 40–45 days, wt.: ≈15.3 kg, contamination: no AF, conc.: nd, country: Spain[358], *control
incidence: 12/12, sa. const.: male Rasa Aragonesa lambs, age: 40–45 days, wt.: ≈15.3 kg, contamination: artificial (dose: 2.5 mg AF, o., for 21 days), conc. range: 1.90–27.84 μg/l*, Ø conc.: 7.37 μg/l*, country: Spain[358], ***intoxication period** (after 21 days of intoxication)
incidence: 6/6*, sa. const.: male Rasa Aragonesa lambs, age: 40–45 days, wt.: ≈15.3 kg, contamination: no AF, conc.: nd, country: Spain[358], *control
incidence: 10/12, sa. const.: male Rasa Aragonesa lambs, age: 40–45 days, wt.: ≈15.3 kg, contamination: artificial (dose: 2.5 mg AF, o., for 21 days), conc. range: 0.22–1.13 μg/l*, Ø conc.: 0.54 μg/l*, country: Spain[358], ***clearance period** (8 days after removal of the contaminated diet)

Monkey

Monkey Artificial Contamination

Monkey, Macaque

Monkey, macaque bile may contain the following mycotoxins and/or their metabolites:

AFLATOXIN B_1
incidence: 3/4*, sa. const.: female macaque monkeys, age: 38–44 months, wt.: 1.2–2.4 kg, contamination: no AFB_1 (for detailed information please see the article), conc. range: tr, country: USA/Thailand[98], *control
incidence: 15/16, sa. const.: female macaques monkeys, age: 38–44 months, wt.: 1.2–2.4 kg, contamination: artificial (dose: **0.5*, 1.5*, 4.5*, 13.5* or 40.5* mg crystalline AFB_1**/kg b. wt., o., once; for detailed information please see the article), conc. range: tr–163 μg/ml* **, country: USA/Thailand[98], **7 days after AFB_1-administration

Monkey, macaque blood may contain the following mycotoxins and/or their metabolites:

AFLATOXIN B_1
incidence: 4/4, sa. const.: female macaque monkeys, age: 38–44 months, wt.: 1.2–2.4 kg, contamination: no AFB_1 (for detailed information please see the article), conc.: no specimen?, country: USA/Thailand[98]
incidence: 1/?*, sa. const.: female macaques monkeys, age: 38–44 months, wt.: 1.2–2.4 kg, contamination: artificial (dose: **0.5, 1.5, 4.5, 13.5 or 40.5** mg AFB_1**/kg b. wt., o., once; for detailed information please see the article), conc.: 37 μg/ml** ***, country: USA/Thailand[98], *heart blood, ***7 days after AFB_1-administration

Monkey, macaque brain may contain the following mycotoxins and/or their metabolites:

AFLATOXIN B_1
incidence: 4/4, sa. const.: female macaque monkeys, age: 38–44 months, wt.:

1.2–2.4 kg, contamination: no AFB_1 (for detailed information please see the article), conc.: nd, country: USA/Thailand[98]
incidence: 4/19, sa. const.: female macaque monkeys, age: 38–44 months, wt.: 1.2–2.4 kg, contamination: artificial (dose: **0.5, 1.5, 4.5, 13.5* or 40.5* mg AFB_1**/kg b. wt., o., once; for detailed information please see the article), conc. range: tr–30 µg/kg* **, country: USA/Thailand[98], **7 days after AFB_1-administration

Monkey, macaque heart may contain the following mycotoxins and/or their metabolites:

AFLATOXIN B_1
incidence: 4/4, sa. const.: female macaque monkeys, age: 38–44 months, wt.: 1.2–2.4 kg, contamination: no AFB_1 (for detailed information please see the article), conc.: no specimen, country: USA/Thailand[98]
incidence: 6/8, sa. const.: female macaque monkeys, age: 38–44 months, wt.: 1.2–2.4 kg, contamination: artificial (dose: **0.5*, 1.5*, 4.5, 13.5* or 40.5 mg AFB_1**/kg b. wt., o., once; for detailed information please see the article), conc. range: tr–176 µg/kg* **, country: USA/Thailand[98], **7 days after AFB_1-administration

Monkey, Macaque Heart Blood see Monkey blood

Monkey, macaque kidney may contain the following mycotoxins and/or their metabolites:

AFLATOXIN B_1
incidence: 4/4, sa. const.: female macaque monkeys, age: 38–44 months, wt.: 1.2–2.4 kg, contamination: no AFB_1 (for detailed information please see the article), conc. range: tr, country: USA/Thailand[98]
incidence: 4/16, sa. const.: female macaques monkeys, age: 38–44 months, wt.: 1.2–2.4 kg, contamination: artificial (dose: **0.5*, 1.5*, 4.5*, 13.5* or 40.5* mg AFB_1**/kg b. wt., o., once; for detailed information please see the article), conc. range: tr–162 µg/kg* **, country: USA/Thailand[98], **7 days after AFB_1-administration

Monkey, macaque liver may contain the following mycotoxins and/or their metabolites:

AFLATOXIN B_1
incidence: 4/4, sa. const.: female macaque monkeys, age: 38–44 months, wt.: 1.2–2.4 kg, contamination: no AFB_1 (for detailed information please see the article), conc.: nd, country: USA/Thailand[98]
incidence: 4/19, sa. const.: female macaque monkeys, age: 38–44 months, wt.: 1.2–2.4 kg, contamination: artificial (dose: **0.5, 1.5, 4.5, 13.5* or 40.5* mg AFB_1**/kg b. wt., o., once; for detailed information please see the article), conc. range: tr–163 µg/kg* **, country: USA/Thailand[98], **7 days after AFB_1-administration

Monkey, macaque lung may contain the following mycotoxins and/or their metabolites:

AFLATOXIN B_1
incidence: 4/4, sa. const.: female macaque monkeys, age: 38–44 months, wt.: 1.2–2.4 kg, contamination: no AFB_1 (for detailed information please see the article), conc.: nd?, country: USA/Thailand[98]
incidence: 2/?, sa. const.: female macaque monkeys, age: 38–44 months, wt.: 1.2–2.4 kg, contamination: artificial (dose: **0.5, 1.5, 4.5*, 13.5* or 40.5 mg AFB_1**/kg b. wt., o., once; for detailed information please see the article), conc. range: tr* **, country: USA/Thailand[98], **7 days after AFB_1-administration

Monkey, macaque pancreas may contain the following mycotoxins and/or their metabolites:

Aflatoxin B_1
incidence: 4/4, sa. const.: female macaque monkeys, age: 38–44 months, wt.: 1.2–2.4 kg, contamination: no AFB_1 (for detailed information please see the article), conc.: nd?, country: USA/Thailand[98]
incidence: 1/?, sa. const.: female macaque monkeys, age: 38–44 months, wt.: 1.2–2.4 kg, contamination: artificial (dose: **0.5, 1.5, 4.5, 13.5* or 40.5 mg AFB_1/kg** b. wt., o., once; for detailed information please see the article), conc.: tr* **, country: USA/Thailand[98], **7 days after AFB_1-administration

Monkey, macaque plasma may contain the following mycotoxins and/or their metabolites:

Ochratoxin A
incidence: 1/1, sa. const.: female macaque monkeys, age: adult, wt.: 4 kg, contamination: artificial (dose: 50 ng OTA ng/g b. wt., **o.**, once), conc.: 500 ng/ml, country: Sweden/Yugoslavia[193] (at different min, hour and day intervals up to 38 days measured)
incidence: 1/1, sa. const.: male macaque monkey, age: adult, wt.: 5 kg, contamination: artificial (dose: 50 ng OTA ng/g b. wt., **i.v.**, once), conc.: 980 ng/ml, country: Sweden/Yugoslavia[193] (at different min, hour and day intervals up to 38 days measured)

Monkey, macaque spleen may contain the following mycotoxins and/or their metabolites:

Aflatoxin B_1
incidence: 4/4, sa. const.: female macaque monkeys, age: 38–44 months, wt.: 1.2–2.4 kg, contamination: no AFB_1 (for detailed information please see the article), conc.: nd?, country: USA/Thailand[98]
incidence: 2/?, sa. const.: female macaque monkeys, age: 38–44 months, wt.: 1.2–2.4 kg, contamination: artificial (dose: **0.5, 1.5, 4.5*, 13.5* or 40.5 mg AFB_1/kg** b. wt., o., once; for detailed information please see the article), conc. range: tr* **, country: USA/Thailand[98], **7 days after AFB_1-administration

Monkey, Marmoset

Monkey, marmoset liver may contain the following mycotoxins and/or their metabolites:

Aflatoxin B_1
incidence: 4/4, sa. const.: male marmosets, age: 3–9 years, wt.: 250–406 g, contamination: artificial (dose: 100 µg AFB_1 (labeled)/kg, by gavage, twice; for detailed information please see the article), conc. range: 0.48–1.44 pmol [^{3}H]-AFB eq/mg DNA* ** ***, country: USA[430], *control, **AFB_1-DNA adducts, ***on day 28
incidence: 4/4, sa. const.: male marmosets, age: 3–9 years, wt.: 250–406 g, contamination: artificial (dose: 100 µg AFB_1 (labeled)/kg, by gavage, twice; for detailed information please see the article), conc. range: 0.23–0.45 pmol [^{3}H]-AFB eq/mg DNA* ** ***, country: USA[430], ***oltipraz-diet** (18 mg/kg/day) for 13 days (day 16–28) at day 0 and day 26 AFB_1 was administered, **AFB_1-DNA adducts, ***on day 28
incidence: 3/3, sa. const.: male marmosets, age: 3–9 years, wt.: 250–406 g, contamination: artificial (dose: 100 µg AFB_1 (labeled)/kg, by gavage, twice; for detailed information please see the article), conc. range: 0.18–0.84 pmol [^{3}H]-AFB eq/mg DNA* ** ***, country: USA[430], ***EQ-diet** (30 mg/kg/day) for 13 days (day 16–28) at day 0 and day 26 AFB_1 was administered, **AFB_1-DNA adducts, ***on day 28

incidence: 1/1, sa. const.: female marmoset, age: adult, wt.: 300–350 g, contamination: artificial (dose: 2 µg AFB_1

(labeled)/kg, i.v., once; for detailed information please see the article), conc.: ≤122 pmol AFB_1 × 10^3/mg protein*, country: Sweden[614], *after 15 min (also measured after 2 and 6 h, lowest conc.: 2 pmol AFB_1 × 10^3/mg protein after 6 h)
incidence: 1/1, sa. const.: female marmoset, age: adult, wt.: 300–350 g, contamination: artificial (dose: 2 µg AFB_1 (labeled)/kg, i.v., once; for detailed information please see the article), conc.: ≤26 pmol AFB_1-metabolites × 10^3/mg protein* **, country: Sweden[614], *after 15 min (also measured after 2 and 6 h, lowest conc.: 5 pmol AFB_1-metabolites × 10^3/mg protein after 6 h), ****chloroform-soluble AFB_1-metabolites**
incidence: 1/1, sa. const.: female marmoset, age: adult, wt.: 300–350 g, contamination: artificial (dose: 2 µg AFB_1 (labeled)/kg, i.v., once; for detailed information please see the article), conc.: ≤92 pmol AFB_1-metabolites × 10^3/mg protein* **, country: Sweden[614], *after 2 h (also measured after 15 min and 6 h, lowest conc.: 38 pmol AFB_1-metabolites × 10^3/mg protein after 6 h), **water-soluble AFB_1-metabolites
incidence: 1/1, sa. const.: female marmoset, age: adult, wt.: 300–350 g, contamination: artificial (dose: 2 µg AFB_1 (labeled)/kg, i.v., once; for detailed information please see the article), conc.: ≤66 pmol AFB_1-metabolites × 10^3/mg protein* **, country: Sweden[614], *after 2 h (also measured after 15 min and 6 h, lowest conc.: 27 pmol AFB_1-metabolites × 10^3/mg protein after 15 min), ****tissue-bound AFB_1-metabolites**

Monkey, Vervet

Monkey, vervet plasma may contain the following mycotoxins and/or their metabolites:

Fumonisin B_1
incidence: 1/1, sa. const.: male vervet monkey, age: 20–26 months, wt.: 2.10–2.32 kg, contamination: artificial (dose: 6.42 mg FB_1 (labeled)/kg b. wt., by gavage, once), conc. range: <210 ng/ml*, country: South Africa/Canada[130], ***1–2 h** after dosing (at 30 min and also at other hour intervals up to 7 h measured, lowest conc.: ≈38 ng/ml after 7 h).
incidence: 1/1, sa. const.: male vervet monkey, age: 20–26 months, wt.: 2.10–2.32 kg, contamination: artificial (dose: 6.42 mg FB_1 (labeled)/kg b. wt., by gavage, once), conc. range: 85 ng/ml*, country: South Africa/Canada[130], ***2 h** after dosing (at 30 min and also at other hour intervals up to 7 h measured, lowest conc.: ≈30 ng/ml after 0.5 and 4 h)

incidence: 2/2, sa. const.: female vervet monkeys, age: 20–43 months, weight: 2.1–3.1 kg, contamination: artificial (1.6 mg FB_1/kg b. wt., i.v., once), conc. range: ≤8 µg/ml*, country: South Africa[535], *after ≈10 min (also at other min intervals up to 180 min measured, lowest conc.: <0.2 µg/ml after 180 min)

Fumonisin B_2
incidence: 1/2, sa. const.: male vervet monkeys, age: ≈11 and 13 years, wt.: 5.23 and 4.76 kg, contamination: artificial (7.5 mg FB_2/kg b. wt., by gavage, once), conc.: 25–40 ng/ml*, country: South Africa[127], *3–5 h after dosing

Mothers' Breast Milk see Human breast milk

Mouse

Mouse Artificial Contamination

Mouse amniotic fluid may contain the following mycotoxins and/or their metabolites:

Ochratoxin A
incidence: ?/?*, sa. const.: embryos of Slc:ICR mice, contamination: no OTA (for detailed information please see the article), conc.: nr, country: Japan[287],* control

incidence: ?/4, sa. const.: embryos of Slc:ICR mice, contamination: artificial (dose: **5 mg crystalline OTA**/kg, i.p., once at 13th day of gestation; for detailed information please see the article), conc. range: ≤409 ng/ml* ** (mean value), country: Japan[287], *in 13-day-pregnant mice, **after 24 h of OTA-administration (also at other hour intervals up to 72 h measured, lowest conc.: ≈20 ng/ml after ≈2 h)

Mouse blood may contain the following mycotoxins and/or their metabolites:

T-2 Toxin
incidence: 2?/2, sa. const.: ddYS male mice, age: 6 weeks, wt.: 30 g, contamination: artificial (dose: 1 mg T-2 toxin (labeled)/kg, o., once; for detailed information please see the article), conc.: 0.7 μg/ml, country: Japan[385], *after ≈1 h (also measured after 3, 24, 48 and 72 h)

Mouse brain may contain the following mycotoxins and/or their metabolites:

Deoxynivalenol
incidence: ?/?, sa. const.: male B6C3F1 (C57B1/6J × C3H/HeJ) mice, age: 7 weeks, contamination: no DON (for detailed information please see the article), conc.: nr, country: USA[494]
incidence: ?/?, sa. const.: male B6C3F1 (C57B1/6J × C3H/HeJ) mice, age: 7 weeks, contamination: artificial (dose: **25 mg DON**/kg b. wt., by o. gavage, once; for detailed information please see the article), conc. range: 0.8 μg/g* (mean value), country: USA[494], *after 5 min (also at other hour intervals up to 25 h measured)

Penitrem A
incidence: 1/1, sa. const.: male C57BL/6 mouse, age: ≈20 weeks, wt.: 30 g, contamination: artificial (dose: 8 mg PNA/kg, o., once), conc.: ≤19.2 nmol/g*, country: Norway/New Zealand[631], *after 60 min (also measured after 30 and 120 min)

Mouse embryo may contain the following mycotoxins and/or their metabolites:

Ochratoxin A
incidence: ?/?*, sa. const.: embryos of Slc:ICR mice, contamination: no OTA (for detailed information please see the article), conc.: nr**, country: Japan[287],* control, **in 11-day and 13-day embryos
incidence: ?/8–13, sa. const.: embryos of Slc:ICR mice, contamination: artificial (dose: 5 mg crystalline OTA/kg, i.p., once at 11th day of gestation; for detailed information please see the article), conc. range: ≤360 ng/g* ** (mean value), country: Japan[287], *in **11-day embryos**, **after 48 h of OTA-administration (also measured at other hour intervals up to 72 h, lowest conc.: almost nd after 2 h)
incidence: ?/5, sa. const.: embryos of Slc:ICR mice, contamination: artificial (dose: 5 mg crystalline OTA/kg, i.p., once at 13th day of gestation; for detailed information please see the article), conc. range: ≤492 ng/g* ** (mean value), country: Japan[287], *in **13-day embryos**, **after 30 h of OTA-administration (also measured at other hour intervals up to 72 h, lowest conc.: 30 ng/g after 2 h)

Mouse gastrointestinal tract may contain the following mycotoxins and/or their metabolites:

Penitrem A
incidence: 1/1, sa. const.: male C57BL/6 mouse, age: ≈20 weeks, wt.: 30 g, contamination: artificial (dose: 8 mg PNA/kg, o., once), conc.: ≤710 nmol/g*, country: Norway/New Zealand[631], *after 60 min (also measured after 30 and 120 min)

Mouse heart may contain the following mycotoxins and/or their metabolites:

Deoxynivalenol
incidence: ?/?, sa. const.: male B6C3F1 (C57B1/6J × C3H/HeJ) mice, age: 7 weeks, contamination: no DON (for detailed information please see the article), conc.: nr, country: USA[494]
incidence: ?/?, sa. const.: sa. const.: male B6C3F1 (C57B1/6J × C3H/HeJ) mice, age: 7 weeks, contamination: artificial (dose: **25 mg DON/kg** b. wt., by o. gavage, once; for detailed information please see the article), conc. range: ≤6.8 µg/g* (mean value), country: USA[494], *after 15 min (also at other hour intervals up to 25 h measured)

Mouse intestine may contain the following mycotoxins and/or their metabolites:

Deoxynivalenol
incidence: ?/?*, sa. const.: male B6C3F1 [C57BL/6(H-2^b) × C3H/HeN(H-2^k) mice, age: 8–10 weeks, contamination: no DON (for detailed information please see the article), conc.: nr, country: USA[370], *control
incidence: 3?/3, sa. const.: male B6C3F1 [C57BL/6(H-2^b) × C3H/HeN(H-2^k) mice, age: 8–10 weeks, contamination: artificial (dose: **5 mg DON** (labeled)/kg b. wt., by o. gavage, once; for detailed information please see the article), conc. range: ≤4.38 pmol [^{3}H]DON/mg wet weight* ** (mean value), country: USA[370], *in **small intestine**, **after 0.5 h (also at other hour intervals up to 24 h measured, lowest conc.: 0.23 pmol [^{3}H]DON/mg wet weight after 24 h)

incidence: 3?/3, sa. const.: male B6C3F1 [C57BL/6(H-2^b) × C3H/HeN(H-2^k) mice, age: 8–10 weeks, contamination: artificial (dose: **25 mg DON** (labeled)/kg b. wt., by o. gavage, once; for detailed information please see the article), conc. range: ≤13.9 pmol [^{3}H]DON/mg wet weight* ** (mean value), country: USA[370], *in **small intestine**, **after 1 h (also at other hour intervals up to 24 h measured, lowest conc.: 2.50 pmol [^{3}H]DON/mg wet weight after 24 h)

incidence: ?/?*, sa. const.: male B6C3F1 [C57BL/6(H-2^b) × C3H/HeN(H-2^k) mice, age: 8–10 weeks, contamination: no DON (for detailed information please see the article), conc.: nr, country: USA[370], *control
incidence: 3?/3, sa. const.: male B6C3F1 [C57BL/6(H-2^b) × C3H/HeN(H-2^k) mice, age: 8–10 weeks, contamination: artificial (dose: **5 mg DON** (labeled)/kg b. wt., by o. gavage, once; for detailed information please see the article), conc. range: ≤2.25 pmol [^{3}H]DON/mg wet weight* ** (mean value), country: USA[370], *in **large intestine**, **after 0.5 h (also at other hour intervals up to 24 h measured, lowest conc.: 0.78 pmol [^{3}H]DON/mg wet weight after 24 h)
incidence: 3?/3, sa. const.: male B6C3F1 [C57BL/6(H-2^b) × C3H/HeN(H-2^k) mice, age: 8–10 weeks, contamination: artificial (dose: **25 mg DON** (labeled)/kg b. wt., by o. gavage, once; for detailed information please see the article), conc. range: ≤12.4 pmol [^{3}H]DON/mg wet weight* ** (mean value), country: USA[370], *in **large intestine**, **after 0.5 h (also at other hour intervals up to 24 h measured, lowest conc.: 1.49 pmol [^{3}H]DON/mg wet weight after 4 h)

Mouse kidney may contain the following mycotoxins and/or their metabolites:

Aflatoxin B_1
incidence: 3?/3, sa. const.: BALB/c mice, age: 6–8 weeks, contamination: artificial (dose: 6 mg AFB_1 (labeled)/kg, i.p., once), conc.: 0.2 µmol/mol DNA* ** *** (mean value), country: USA/Taiwan, Republic of China[49], *after 4 h, **AFB_1-DNA adducts, ***measured by **scintillation counting**
incidence: 3/3, sa. const.: BALB/c mice, age: 6–8 weeks, contamination: artificial (dose: 6 mg AFB_1 (labeled)/kg, i.p., once),

conc.: nd, country: USA/Taiwan, Republic of China[49], *after 4 h, **AFB_1-FAPy adducts, ***measured by **ELISA**

incidence: 3?/3, sa. const.: BALB/c mice, age: 6–8 weeks, contamination: artificial (dose: 12 mg AFB_1 (labeled)/kg, i.p., once), conc.: 2.3 µmol/mol DNA* ** *** (mean value), country: USA/Taiwan, Republic of China[49], *after 4 h, **AFB_1-DNA adducts, ***measured by **scintillation counting**

incidence: 3/3, sa. const.: BALB/c mice, age: 6–8 weeks, contamination: artificial (dose: 12 mg AFB_1 (labeled)/kg, i.p., once), conc.: nd, country: USA/Taiwan, Republic of China[49], *after 4 h, **AFB_1-FAPy adducts, ***measured by **ELISA**

Deoxynivalenol

incidence: ?/?*, sa. const.: male B6C3F1 [C57BL/6(H-2^b) × C3H/HeN(H-2^k) mice, age: 8–10 weeks, contamination: no DON (for detailed information please see the article), conc.: nr, country: USA[370], *control

incidence: 3?/3, sa. const.: male B6C3F1 [C57BL/6(H-2^b) × C3H/HeN(H-2^k) mice, age: 8–10 weeks, contamination: artificial (dose: **5 mg DON** (labeled)/kg b. wt., by o. gavage, once; for detailed information please see the article), conc. range: ≤5.61 pmol [^{3}H]DON/mg wet weight* (mean value), country: USA[370], *after 0.5 h (also at other hour intervals up to 24 h measured, lowest conc.: 0.17 pmol [^{3}H]DON/mg wet weight after 24 h)

incidence: 3?/3, sa. const.: male B6C3F1 [C57BL/6(H-2^b) × C3H/HeN(H-2^k) mice, age: 8–10 weeks, contamination: artificial (dose: **25 mg DON** (labeled)/kg b. wt., by o. gavage, once; for detailed information please see the article), conc. range: ≤19.8 pmol [^{3}H]DON/mg wet weight* (mean value), country: USA[370], *after 0.5 h (also at other hour intervals up to 24 h measured, lowest conc.: 4.13 pmol [^{3}H] DON/mg wet weight after 24 h)

incidence: ?/?, sa. const.: male B6C3F1 (C57B1/6J × C3H/HeJ) mice, age: 7 weeks, contamination: no DON (for detailed information please see the article), conc.: nr, country: USA[494]

incidence: ?/?, sa. const.: sa. const.: male B6C3F1 (C57B1/6J × C3H/HeJ) mice, age: 7 weeks, contamination: artificial (dose: **25 mg DON**/kg b. wt., by o. gavage, once; for detailed information please see the article), conc. range: 9.0 µg/g* (mean value), country: USA[494], *after 15 min (also at other hour intervals up to 25 h measured)

incidence: ?/?, sa. const.: pathogen-free female B6C3F1 mice, age: 9–10 weeks, contamination: artificial (dose: 5 mg DON/kg b. wt., **i.n.**, once; for detailed information please see the article), conc. range: ≤3.73 µg/g* ** (mean value), country: USA[534], *after 15 min (also measured after 30, 60 and 120 min, lowest conc.: ≈0.8 µg/g** after 120 min), **DON eq.

incidence: ?/?, sa. const.: pathogen-free female B6C3F1 mice, age: 9–10 weeks, contamination: artificial (dose: 5 mg DON/kg b. wt., by **o.** gavage, once; for detailed information please see the article), conc. range: ≤1.76 µg/g* ** (mean value), country: USA[534], *after 15 min (also measured after 30, 60 and 120 min, lowest conc.: ≈0.6 µg/g** after 120 min), **DON eq.

Ochratoxin A

incidence: ?/9, sa. const.: pregnant Slc:ICR mice, contamination: artificial (dose: 5 mg crystalline OTA/kg, i.p., once at **11th day of gestation**; for detailed information please see the article), conc. range: ≈≤9 µg/g* (mean value), country: Japan[287], *after 2 h of OTA-administration (also at other hour intervals up to 72 h measured, lowest conc.: ≈0.25 µg/g after 72 h)

incidence: ?/?*, sa. const.: embryos of Slc:ICR mice, contamination: no OTA (for detailed information please see the article), conc.: nr**, country: Japan[287],* control, **on day 11 of gestation

incidence: ?/5, sa. const.: pregnant Slc:ICR mice, contamination: artificial (dose: 5 mg crystalline OTA/kg, i.p., once at **13th day of gestation**; for detailed information please see the article), conc. range: ≈≤6 μg/g* (mean value), country: Japan[287], *after 2 h of OTA-administration (also at other hour intervals up to 72 h measured, lowest conc.: ≈0.2 μg/g after 72 h)
incidence: ?/?*, sa. const.: embryos of Slc:ICR mice, contamination: no OTA (for detailed information please see the article), conc.: nr**, country: Japan[287],* control, **on day 13 of gestation

Penitrem A
incidence: 1/1, sa. const.: male C57BL/6 mouse, age: ≈20 weeks, wt.: 30 g, contamination: artificial (dose: 8 mg PNA/kg, o., once), conc.: ≤35.1 nmol/g*, country: Norway/New Zealand[631], *after 60 min (also measured after 30 and 120 min)

Mouse liver may contain the following mycotoxins and/or their metabolites:

Aflatoxin B_1
incidence: ?/?*, sa. const.: male C57BL mice, weight: 16–20 g, contamination: artificial no AFB_1, conc.: nr, country: France/Japan[24], *control
incidence: ?/?, sa. const.: male C57BL mice, wt.: 16–20 g, contamination: artificial (dose: **20 μg AFB_1**/kg/day, by gavage, daily for up to 14 days), conc.: 0.44 pmol AFB_1-FAPy/mg DNA* ** (mean value), country: France/Japan[24], *animals killed after 24 h of final treatment, **AFB_1-DNA adduct
incidence: ?/?, sa. const.: male C57BL mice, wt.: 16–20 g, contamination: artificial (dose: **80 μg AFB_1**/kg/day, by gavage, daily for up to 14 days), conc.: ≈1 pmol AFB_1-FAPy/mg DNA* ** (mean value), country: France/Japan[24], *after 1 and 14 days of AFB_1-administration, **AFB_1-DNA adducts

incidence: 3?/3, sa. const.: BALB/c mice, age: 6–8 weeks, contamination: artificial (dose: 6 mg AFB_1 (labeled)/kg, i.p., once), conc.: 1 μmol/mol DNA* ** *** (mean value), country: USA/Taiwan, Republic of China[49], *after 4 h, **AFB_1-DNA adducts, ***measured by **scintillation counting**
incidence: 3/3, sa. const.: BALB/c mice, age: 6–8 weeks, contamination: artificial (dose: 6 mg AFB_1 (labeled)/kg, i.p., once), conc.: nd, country: USA/Taiwan, Republic of China[49], *after 4 h, **AFB_1-FAPy adducts, ***measured by **ELISA**

incidence: 3?/3, sa. const.: BALB/c mice, age: 6–8 weeks, contamination: artificial (dose: 12 mg AFB_1 (labeled)/kg, i.p., once), conc.: 65.1 μmol/mol DNA* ** *** (mean value), country: USA/Taiwan, Republic of China[49], *after 4 h, **AFB_1-DNA adducts, ***measured by **scintillation counting**
incidence: 3?/3, sa. const.: BALB/c mice, age: 6–8 weeks, contamination: artificial (dose: 12 mg AFB_1 (labeled)/kg, i.p., once), conc.: 41.7 μmol/mol DNA* ** *** (mean value), country: USA/Taiwan, Republic of China[49], *after 4 h, **AFB_1-FAPy adducts, ***measured by **ELISA**

incidence: ?/6*, sa. const.: male Swiss-Webster mice, wt.: 28–32 g, contamination: artificial (dose: 0.25 mg AFB_1 (labeled)/kg, i.p., once; for detailed information please see the article), conc.: 6.24 adducts/10^7 nucleotides** *** (mean value), country: USA[511], *control, **after 2 h, ***AFB-DNA adducts
incidence: ?/6*, sa. const.: male Swiss-Webster mice, wt.: 28–32 g, contamination: artificial (dose: 0.25 mg AFB_1 (labeled)/kg, i.p., once; for detailed information please see the article), conc. range: 184 adducts/10^7 nucleotides** *** (mean value), country: USA[511], *control, on day 14 **BSO** (0.6 g/kg i.p.) and (**DEM** 0.75 ml/kg i.p.) at 2 and 15 h prior to AFB_1-treatment, **after 2 h, ***AFB-DNA adducts
incidence: ?/7*, sa. const.: male Swiss-Webster mice, wt.: 28–32 g, contamination:

artificial (dose: 0.25 mg AFB_1 (labeled)/kg, i.p., once; for detailed information please see the article), conc. range: 3.39 adducts/10^7 nucleotides** *** (mean value), country: USA[511], ***BHA-diet** (0.75%) for 14 days prior to AFB_1-treatment, **after 2 h, ***AFB-DNA adducts

incidence: ?/7*, sa. const.: male Swiss-Webster mice, wt.: 28–32 g, contamination: artificial (dose: 0.25 mg AFB_1 (labeled)/kg, i.p., once; for detailed information please see the article), conc. range: 5.51 adducts/10^7 nucleotides** *** (mean value), country: USA[511], *BHA-diet (0.75%) for 14 days + on day 14 **BSO** (0.6 g/kg i.p.) and (**DEM** 0.75 ml/kg i.p.) at 2 and 15 h prior to AFB_1-treatment, **after 2 h, ***AFB-DNA adducts

incidence: 3?/3*, sa. const.: female CD-1 mice, wt.: 25–30 g, contamination: artificial (dose: 0.25 mg AFB_1 (labeled)/kg, i.p., once; for detailed information please see the article), conc.: ≈4 AFB adduct/10^7 DNA nucleotides** (mean value), country: USA[616], *control, **after 2 h

incidence: 3?/3, sa. const.: female CD-1 mice, wt.: 25–30 g, contamination: artificial (dose: 0.25 mg AFB_1 (labeled)/kg, i.p., once; for detailed information please see the article), conc.: ≈1 AFB adduct/10^7 DNA nucleotides* ** (mean value), country: USA[616], ***BHA-diet** (0.75%) for 10 days prior to AFB_1-treatment, **after 2 h

Deoxynivalenol

incidence: ?/?*, sa. const.: male B6C3F1 [C57BL/6(H-2^b) × C3H/HeN(H-2^k) mice, age: 8–10 weeks, contamination: no DON (for detailed information please see the article), conc.: nr, country: USA[370], *control

incidence: 3?/3, sa. const.: male B6C3F1 [C57BL/6(H-2^b) × C3H/HeN(H-2^k) mice, age: 8–10 weeks, contamination: artificial (dose: **5 mg DON** (labeled)/kg b. wt., by o. gavage, once; for detailed information please see the article), conc. range: ≤4.88 pmol [^{3}H]DON/mg wet weight* (mean value), country: USA[370], *after 0.5 h (also at other hour intervals up to 24 h measured, lowest conc.: 0.30 pmol [^{3}H] DON/mg wet weight after 24 h)

incidence: 3?/3, sa. const.: male B6C3F1 [C57BL/6(H-2^b) × C3H/HeN(H-2^k) mice, age: 8–10 weeks, contamination: artificial (dose: **25 mg DON** (labeled)/kg b. wt., by o. gavage, once; for detailed information please see the article), conc. range: ≤18.4 pmol [^{3}H]DON/mg wet weight* (mean value), country: USA[370], *after 0.5 h (also at other hour intervals up to 24 h measured, lowest conc.: 2.29 pmol [^{3}H] DON/mg wet weight after 24 h)

incidence: ?/?, sa. const.: male B6C3F1 (C57B1/6J × C3H/HeJ) mice, age: 7 weeks, contamination: no DON; for detailed information please see the article), conc.: nr, country: USA[494]

incidence: ?/?, sa. const.: male B6C3F1 (C57B1/6J × C3H/HeJ) mice, age: 7 weeks, contamination: artificial (dose: **25 mg DON**/kg b. wt., by o. gavage, once; for detailed information please see the article), conc. range: 19.6 µg/g* (mean value), country: USA[494], *after 5 min (also at other hour intervals up to 25 h measured)

incidence: ?/?, sa. const.: pathogen-free female B6C3F1 mice, age: 9–10 weeks, contamination: artificial (dose: 5 mg DON/kg b. wt., **i.n.**, once; for detailed information please see the article), conc. range: ≤2.37 µg/g* ** (mean value), country: USA[534], *after 30 min (also measured after 15, 60 and 120 min, lowest conc.: ≈0.6 µg/g** after 120 min), **DON eq.

incidence: ?/?, sa. const.: pathogen-free female B6C3F1 mice, age: 9–10 weeks, contamination: artificial (dose: 5 mg DON/kg b. wt., by **o.** gavage, once; for detailed information please see the article), conc. range: ≤1.10 µg/g* ** (mean value), country: USA[534], *after 15 min (also measured after 30, 60 and 120 min, lowest conc.: ≈0.4 µg/g** after 120 min), **DON eq.

Ochratoxin A

incidence: ?/9, sa. const.: pregnant Slc:ICR mice, contamination: artificial (dose: 5 mg crystalline OTA/kg, i.p., once at **11th day of gestation;** for detailed information please see the article), conc. range: ≈≤10.5 µg/g* (mean value), country: Japan[287], *after 2 h of OTA-administration (also at other hour intervals up to 72 h measured, lowest conc.: ≈0.35 µg/g after 48 h)
incidence: ?/?*, sa. const.: embryos of Slc:ICR mice, contamination: no OTA (for detailed information please see the article), conc.: nr**, country: Japan[287], *control, **on day 11 of gestation
incidence: ?/5, sa. const.: pregnant Slc:ICR mice, contamination: artificial (dose: 5 mg crystalline OTA/kg, i.p., once at **13th day of gestation;** for detailed information please see the article), conc. range: ≈≤8.5 µg/g* (mean value), country: Japan[287], *after 2 h of OTA-administration (also at other hour intervals up to 72 h measured, lowest conc.: ≈0.5 µg/g after 72 h)
incidence: ?/?*, sa. const.: embryos of Slc:ICR mice, contamination: no OTA (for detailed information please see the article), conc.: nr**, country: Japan[287],* control, **on day 13 of gestation

Penitrem A

incidence: 1/1, sa. const.: male C57BL/6 mouse, age: ≈20 weeks, wt.: 30 g, contamination: artificial (dose: 8 mg PNA/kg, o., once), conc.: ≤117 nmol/g*, country: Norway/New Zealand[631], *after 60 min (also measured after 30 and 120 min)

Mouse lung

may contain the following mycotoxins and/or their metabolites:

Deoxynivalenol

incidence: ?/?, sa. const.: pathogen-free female B6C3F1 mice, age: 9–10 weeks, contamination: artificial (dose: 5 mg DON/kg b. wt., **i.n.,** once; for detailed information please see the article), conc. range: ≤2.20 µg/g* ** (mean value), country: USA[534], *after 30 min (also measured after 15, 60 and 120 min, lowest conc.: ≈0.5 µg/g** after 120 min), **DON eq.
incidence: ?/?, sa. const.: pathogen-free female B6C3F1 mice, age: 9–10 weeks, contamination: artificial (dose: 5 mg DON/kg b. wt., by **o.** gavage, once; for detailed information please see the article), conc. range: ≤0.95 µg/g* ** (mean value), country: USA[534], *after 15 min (also measured after 30, 60 and 120 min, lowest conc.: ≈0.3 µg/g** after 120 min), **DON eq.

Gliotoxin

incidence: ?/10*, sa. const.: female Swiss-webster mice, age: 38–44 months, wt.: 20–25 g, contamination: no *Aspergillus fumigatus* (for detailed information please see the article), conc.: 3,976 ng/g (mean value), country: USA[78], *control
incidence: ?/10*, sa. const.: female Swiss-webster mice, age: 38–44 months, wt.: 20–25 g, contamination: artificial (dose: *Aspergillus fumigatus* addition; for detailed information please see the article), conc.: ≈1,150 ng/g (mean value), country: USA[78], ***AMB deoxycholate (0.25 mg/kg/day)** i.p. daily until 96 h after *A. fumigatus* inoculation
incidence: ?/20*, sa. const.: female Swiss-webster mice, age: 38–44 months, wt.: 20–25 g, contamination: artificial (dose: *Aspergillus fumigatus* addition; for detailed information please see the article), conc.: ≈1,800 ng/g (mean value), country: USA[78], ***AMB deoxycholate (0.5 mg/kg/day)** i.p. daily until 96 h after *A. fumigatus* inoculation
incidence: ?/15*, sa. const.: female Swiss-webster mice, age: 38–44 months, wt.: 20–25 g, contamination: artificial (dose: *Aspergillus fumigatus* addition; for

detailed information please see the article), conc.: ≈1,900 ng/g (mean value), country: USA[78], ***AMB deoxycholate (1 mg/kg/day)** i.p. daily until 96 h after *A. fumigatus* inoculation

Mouse milk may contain the following mycotoxins and/or their metabolites:

AFLATOXIN B_1
incidence: ?/?, sa. const.: lactating mice (strain-HA/ICR), Ø wt.: 50 g, contamination: artificial (dose: 5 μg AFB_1 (labeled and unlabeled)/mouse, i.p., once; for detailed information please see the article), conc. range: ≤7 ng/ml (mean value), country: USA[124], *after 1 h (also measured after 3 and 10 h, lowest conc.: 3.1 ng/ml after 10 h)

AFLATOXIN M_1
incidence: ?/4–5, sa. const.: lactating mice (strain-HA/ICR), Ø wt.: 50 g, contamination: artificial (dose: **5 μg AFB_1** (labeled and unlabeled)/mouse, i.p., once), conc. range: ≤7.0 ng/ml* (mean value), country: USA[124], *after 0.5 h (also at other hour intervals up to 24 h measured, lowest conc.: 0.6 ng/ml after 24 h)
incidence: ?/4–5, sa. const.: lactating mice (strain-HA/ICR), Ø wt.: 50 g, contamination: artificial (dose: **50 μg AFB_1**/mouse, i.p., once), conc. range: ≤63.3 ng/ml* (mean value), country: USA[124], *after 0.5 h (also at other hour intervals up to 24 h measured, lowest conc.: 1.6 ng/ml after 24 h)

Mouse Peyer's patches may contain the following mycotoxins and/or their metabolites:

DEOXYNIVALENOL
incidence: ?/?*, sa. const.: male B6C3F1 [C57BL/6(H-2^b) × C3H/HeN(H-2^k) mice, age: 8–10 weeks, contamination: no DON (for detailed information please see the article), conc.: nr, country: USA[370], *control
incidence: 3?/3, sa. const.: male B6C3F1 [C57BL/6(H-2^b) × C3H/HeN(H-2^k) mice, age: 8–10 weeks, contamination: artificial (dose: **5 mg DON** (labeled)/kg b. wt., by o. gavage, once; for detailed information please see the article), conc. range: ≤1.89 pmol [^{3}H]DON/mg wet weight* (mean value), country: USA[370], *after 0.5 h (also at other hour intervals up to 24 h measured, lowest conc.: 0.13 pmol [^{3}H]DON/mg wet weight after 24 h)
incidence: 3?/3, sa. const.: male B6C3F1 [C57BL/6(H-2^b) × C3H/HeN(H-2^k) mice, age: 8–10 weeks, contamination: artificial (dose: **25 mg DON** (labeled)/kg b.w., by o. gavage, once; for detailed information please see the article), conc. range: ≤1.95 pmol [^{3}H]DON/mg wet weight* (mean value), country: USA[370], *after 0.5 h (also at other hour intervals up to 24 h measured, lowest conc.: 0.82 pmol [^{3}H]DON/mg wet weight after 24 h)

Mouse placenta may contain the following mycotoxins and/or their metabolites:

OCHRATOXIN A
incidence: ?/9, sa. const.: pregnant Slc:ICR mice, contamination: artificial (dose: 5 mg crystalline OTA/kg, i.p., once at **11th day of gestation;** for detailed information please see the article), conc. range: ≈≤4.6 μg/g* (mean value), country: Japan[287], *after 2 h of OTA-administration (also at other hour intervals up to 72 h measured, lowest conc.: ≈1 μg/g after 72 h)
incidence: ?/?*, sa. const.: embryos of Slc:ICR mice, contamination: no OTA (for detailed information please see the article), conc.: nr**, country: Japan[287], *control, **on day 11 of gestation
incidence: ?/5, sa. const.: pregnant Slc:ICR mice, contamination: artificial (dose: 5 mg crystalline OTA/kg, i.p., once at **13th day of gestation;** for detailed information please see the article), conc. range: ≈≤3.1 μg/g* (mean value), country: Japan[287], *after 2 h of OTA-administration (also at other hour intervals up to 72 h measured, lowest conc.: ≈0.7 μg/g after 72 h)

incidence: ?/?*, sa. const.: embryos of Slc:ICR mice, contamination: no OTA (for detailed information please see the article), conc.: nr**, country: Japan[287], *control, **on day 13 of gestation

Mouse plasma may contain the following mycotoxins and/or their metabolites:

Aflatoxin B_1
incidence:?/?*, sa. const.: male C57BL mice, weight: 16–20 g, contamination: artificial no AFB_1, conc.: nr, country: France/Japan[24], *control
incidence:?/?, sa. const.: male C57BL mice, weight: 16–20 g, contamination: artificial (dose: **20 µg AFB_1**/kg/day, by gavage, daily for up to 14 days), conc.: ndr* (for detailed information please see the article), country: France/Japan[24], *AFB_1-albumin adducts
incidence:?/?, sa. const.: male C57BL mice, weight: 16–20 g, contamination: artificial (dose: **80 µg AFB_1**/kg/day, by gavage, daily for up to 14 days), conc. range: ≤13.4 pg AFB_1-lysine eq/mg albumin* ** (mean values), country: France/Japan[24], *animals killed after 24 h of final treatment (also measured after 1, 3 and 7 days, lowest conc.: ≈nd after 1 day), **AFB_1-albumin adducts

Deoxynivalenol
incidence: ?/?*, sa. const.: male B6C3F1 [C57BL/6(H-2^b) × C3H/HeN(H-2^k) mice, age: 8–10 weeks, contamination: no DON (for detailed information please see the article), conc.: nr, country: USA[370], *control
incidence: 3?/3, sa. const.: male B6C3F1 [C57BL/6(H-2^b) × C3H/HeN(H-2^k) mice, age: 8–10 weeks, contamination: artificial (dose: **5 mg DON** (labeled)/kg b. wt., by o. gavage, once; for detailed information please see the article), conc. range: ≤4.42 pmol [^{3}H]DON/µl* (mean value), country: USA[370], *after 0.5 h (also at other hour intervals up to 24 h measured, lowest conc.: 0.29 pmol [^{3}H]DON/µl after 24 h)
incidence: 3?/3, sa. const.: male B6C3F1 [C57BL/6(H-2^b) × C3H/HeN(H-2^k) mice, age: 8–10 weeks, contamination: artificial (dose: **25 mg DON** (labeled)/kg b. wt., by o. gavage, once; for detailed information please see the article), conc. range: ≤22.3 pmol [^{3}H]DON/µl* (mean value), country: USA[370], *after 0.5 h (also at other hour intervals up to 24 h measured, lowest conc.: 2.59 pmol [^{3}H]DON/µl after 24 h)

incidence: ?/?, sa. const.: male B6C3F1 (C57Bl/6J × C3H/HeJ) mice, age: 7 weeks, contamination: no DON (for detailed information please see the article), conc.: nr, country: USA[494]
incidence: ?/?, sa. const.: male B6C3F1 (C57Bl/6J × C3H/HeJ) mice, age: 7 weeks, contamination: artificial (dose: **25 mg DON**/kg b. wt., by o. gavage, once; for detailed information please see the article), conc. range: ≤12.1 µg/ml (mean value), country: USA[494], *after 5 min (also at other hour intervals up to 25 h measured)

incidence: 3/3, sa. const.: male B6C3F1 (C57Bl/6J × C3H/HeJ) mice, age: 7 weeks, contamination: no DON (for detailed information please see the article), conc.: nd, country: USA[494]
incidence: ?/3, sa. const.: male B6C3F1 (C57Bl/6J × C3H/HeJ) mice, age: 7 weeks, contamination: artificial (dose: **2 mg DON**/kg diet, o., for 4 weeks (subchronic dietary exposure); for detailed information please see the article), conc.: 20 ng/ml (mean value), country: USA[494], *after 4 weeks
incidence: ?/3, sa. const.: male B6C3F1 (C57Bl/6J × C3H/HeJ) mice, age: 7 weeks, contamination: artificial (dose: **5 mg DON**/kg diet, o., for 4 weeks (subchronic dietary exposure); for detailed information please see the article), conc.: ≈33 ng/ml (mean value), country: USA[494], *after 4 weeks
incidence: ?/3, sa. const.: male B6C3F1 (C57Bl/6J × C3H/HeJ) mice, age: 7 weeks, contamination: artificial (dose: **10 mg DON**/kg diet, o., for 4 weeks (subchronic dietary exposure); for detailed

information please see the article), conc.: ≈58 ng/ml (mean value), country: USA[494], *after 4 weeks
incidence: ?/3, sa. const.: male B6C3F1 (C57B1/6J × C3H/HeJ) mice, age: 7 weeks, contamination: artificial (dose: **20 mg DON/kg** diet, o., for 4 weeks (subchronic dietary exposure); for detailed information please see the article), conc.: 100 ng/ml (mean value), country: USA[494], *after 4 weeks

incidence: ?/?, sa. const.: pathogen-free female B6C3F1 mice, age: 9–10 weeks, contamination: artificial (dose: 5 mg DON/kg b. wt., **i.n.**, once; for detailed information please see the article), conc. range: ≈≤3.4 μg/ml* ** (mean value), country: USA[534], *after 30 min (also measured after 15, 60 and 120 min, lowest conc.: ≈0.5 μg/ml** after 120 min), **DON eq.
incidence: ?/?, sa. const.: pathogen-free female B6C3F1 mice, age: 9–10 weeks, contamination: artificial (dose: 5 mg DON/kg b. wt., by **o.** gavage, once; for detailed information please see the article), conc. range: ≈≤1.0 μg/ml* ** (mean value), country: USA[534], *after 15 min (also measured after 30, 60 and 120 min, lowest conc.: ≈0.2 μg/ml** after 120 min), **DON eq.

Ochratoxin A
incidence: ?/?, sa. const.: white male NIH-Bethesda mice, wt.: ≈20 g, contamination: artificial (dose: 50 ng OTA ng/g b. wt., **o.**, once), conc.: 370 ng/ml (mean value), country: Sweden/Yugoslavia[193] (at different min, hour and day intervals up to 7 days measured)
incidence: ?/?, sa. const.: white male NIH-Bethesda mice, wt.: ≈20 g, contamination: artificial (dose: 50 ng OTA ng/g b. wt., **i.v.**, once), conc.: 370 ng/ml (mean value), country: Sweden/Yugoslavia[193] (at different min, hour and day intervals up to 7 days measured)

Mouse serum may contain the following mycotoxins and/or their metabolites:

Aflatoxin B_1
incidence: ?/11*, sa. const.: female HBV-transgenic mice of the linea 50–4 mated with nontransgenic C57 male mice, age: 3 months, contamination: artificial (dose: 1.0 mg AFB_1 (labeled)/kg b. wt., once?; for detailed information please see the article), conc.: 4.84 ng AFB_1/mg albumin** *** **** (mean value), country: France/USA[16], ***HBV-positive,** **after 24 h, ***AFB_1-albumin adducts, ****P450 nd
incidence: ?/11*, sa. const.: female HBV-transgenic mice of the linea 50–4 mated with nontransgenic C57 male mice, age: 3 months, contamination: artificial (dose: 1.0 mg AFB_1 (labeled)/kg b.w., once?; for detailed information please see the article), conc.: 4.01 ng AFB_1/mg albumin** *** **** (mean value), country: France/USA[16], ***HBV-negative,** **after 24 h, ***AFB_1-albumin adducts, ****P450 nd
incidence: ?/7*, sa. const.: female HBV-transgenic mice of the linea 50–4 mated with nontransgenic C57 male mice, age: 9 months, contamination: no AFB_1 (for detailed information please see the article), conc.: nd** *** (mean value), country: France/USA[16] ***HBV-positive,** **after 24 h, ***together with 0.56 nmol P450/mg microsomal protein measured
incidence: ?/8*, sa. const.: female HBV-transgenic mice of the linea 50–4 mated with nontransgenic C57 male mice, age: 9 months, contamination: no AFB_1 (for detailed information please see the article), conc.: nd** *** (mean value), country: France/USA[16], ***HBV-negative,** **after 24 h, ***together with 0.58 nmol P450/mg microsomal protein measured
incidence: ?/10*, sa. const.: female HBV-transgenic mice of the linea 50–4 mated with nontransgenic C57 male mice, age: 12 months, contamination: artificial

(dose: 1.0 mg AFB_1 (labeled)/kg b. wt., once?; for detailed information please see the article), conc.: 5.22 ng AFB_1/mg albumin** *** **** (mean value), country: France/USA[16], ***HBV-positive,** **after 24 h, ***AFB_1-albumin adducts, ****together with 0.50 nmol P450/mg microsomal protein measured
incidence: ?/9*, sa. const.: female HBV-transgenic mice of the linea 50–4 mated with nontransgenic C57 male mice, age: 12 months, contamination: artificial (dose: 1.0 mg AFB_1 (labeled)/kg b. wt., once?; for detailed information please see the article), conc.: 7.33 ng AFB_1/mg albumin** *** **** (mean value), country: France/USA[16], ***HBV-negative,** **after 24 h, ***AFB_1-albumin adducts, ****together with 0.7 nmol P450/mg microsomal protein measured

Gliotoxin
incidence: ?/10*, sa. const.: female Swiss-Webster mice, age: 38–44 months, wt.: 20–25 g, contamination: no *Aspergillus fumigatus* (for detailed information please see the article), conc.: 37 ng/ml (mean value), country: USA[78], *control
incidence: ?/10*, sa. const.: female Swiss-webster mice, age: 38–44 months, wt.: 20–25 g, contamination: artificial (dose: *Aspergillus fumigatus* addition; for detailed information please see the article), conc.: ≈11 ng/mg (mean value), country: USA[78], ***AMB deoxycholate (0.25 mg/kg/day)** i.p. daily until 96 h after *A. fumigatus* inoculation
incidence: ?/20*, sa. const.: female Swiss-webster mice, age: 38–44 months, wt.: 20–25 g, contamination: artificial (dose: *Aspergillus fumigatus* addition; for detailed information please see the article), conc.: ≈10.5 ng/mg (mean value), country: USA[78], ***AMB deoxycholate (0.5 mg/kg/day)** i.p. daily until 96 h after *A. fumigatus* inoculation
incidence: ?/15*, sa. const.: female Swiss-webster mice, age: 38–44 months, wt.: 20–25 g, contamination: artificial (dose: *Aspergillus fumigatus* addition; for detailed information please see the article), conc.: ≈3.5 ng/mg (mean value), country: USA[78], ***AMB deoxycholate (1 mg/kg/day)** i.p. daily until 96 h after *A. fumigatus* inoculation

Ochratoxin A
incidence: ?/?*, sa. const.: pregnant Slc:ICR mice, contamination: artificial (dose: 5 mg crystalline OTA/kg, i.p., once at **11th day of gestation;** for detailed information please see the article), conc. range: ≈≤49 μg/ml** (mean value), country: Japan[287], *maternal serum, **after 2 h of OTA-administration (also at other hour intervals up to 72 h measured, lowest conc.: ≈5 μg/g after 72 h)
incidence: ?/?*, sa. const.: embryos of Slc:ICR mice, contamination: no OTA (for detailed information please see the article), conc.: nr**, country: Japan[287], *control, **on day 11 of gestation
incidence: ?/5*, sa. const.: pregnant Slc:ICR mice, contamination: artificial (dose: 5 mg crystalline OTA/kg, i.p., once at **13th day of gestation;** for detailed information please see the article), conc. range: ≈≤49 μg/ml** (mean value), country: Japan[287], *maternal serum, **after 2 h of OTA-administration (also at other hour intervals up to 72 h measured, lowest conc.: ≈3 μg/g after 72 h)
incidence: ?/?*, sa. const.: embryos of Slc:ICR mice, contamination: no OTA (for detailed information please see the article), conc.: nr**, country: Japan[287], *control, **on day 13 of gestation

Mouse spleen may contain the following mycotoxins and/or their metabolites:

Deoxynivalenol
incidence: ?/?*, sa. const.: male B6C3F1 [C57BL/6(H-2^b) × C3H/HeN(H-2^k) mice, age: 8–10 weeks, contamination: no DON (for detailed information please see the article), conc.: nr, country: USA[370], *control

incidence: 3?/3, sa. const.: male B6C3F1 [C57BL/6(H-2^b) × C3H/HeN(H-2^k) mice, age: 8–10 weeks, contamination: artificial (dose: **5 mg DON** (labeled)/kg b. wt., by o. gavage, once; for detailed information please see the article), conc. range: ≤2.29 pmol [^{3}H]DON/mg wet weight* (mean value), country: USA[370], *after 0.5 h (also at other hour intervals up to 24 h measured, lowest conc.: 0.08 pmol [^{3}H] DON/mg wet weight after 24 h)
incidence: 3?/3, sa. const.: male B6C3F1 [C57BL/6(H-2^b) × C3H/HeN(H-2^k) mice, age: 8–10 weeks, contamination: artificial (dose: **25 mg DON** (labeled)/kg b. wt., by o. gavage, once; for detailed information please see the article), conc. range: ≤6.58 pmol [^{3}H]DON/mg wet weight* (mean value), country: USA[370], *after 0.5 h (also at other hour intervals up to 24 h measured, lowest conc.: 0.87 pmol [^{3}H] DON/mg wet weight after 24 h)

incidence: ?/?, sa. const.: male B6C3F1 (C57B1/6J × C3H/HeJ) mice, age: 7 weeks, contamination: no DON; for detailed information please see the article), conc.: nr, country: USA[494]
incidence: ?/?, sa. const.: male B6C3F1 (C57B1/6J × C3H/HeJ) mice, age: 7 weeks, contamination: artificial (dose: **25 mg DON**/kg b. wt., by o. gavage, once; for detailed information please see the article), conc. range: ≤7.9 µg/g* (mean value), country: USA[494], *after 15 min (also at other hour intervals up to 25 h measured)

incidence: ?/?, sa. const.: pathogen-free female B6C3F1 mice, age: 9–10 weeks, contamination: artificial (dose: 5 mg DON/kg b. wt., **i.n.**, once; for detailed information please see the article), conc. range: ≤1.87 µg/g* ** (mean value), country: USA[534], *after 15 min (also measured after 30, 60 and 120 min, lowest conc.: ≈0.4 µg/g** after 120 min), **DON eq.
incidence: ?/?, sa. const.: pathogen-free female B6C3F1 mice, age: 9–10 weeks, contamination: artificial (dose: 5 mg DON/kg b. wt., by **o.** gavage, once; for detailed information please see the article), conc. range: ≤0.77 µg/g* ** (mean value), country: USA[534], *after 15 min (also measured after 30, 60 and 120 min, lowest conc.: ≈0.2 µg/g** after 120 min), **DON eq.

Pig

Pig Natural Contamination

Pig bile may contain the following mycotoxins and/or their metabolites:

DEOXYNIVALENOL
incidence: 6?/6*, sa. const.: Pietrain × DL growing pigs, wt.: ≈22.2 kg, contamination: natural (for detailed information please see the article), conc. range: ≤21.2 ng/ml** (mean value), country: Germany[475], *fed **organic diet (wheat variety: *Contur*)**, **in the year 2001, further data of the years 1999 and 2000 in the article
incidence: 6?/6*, sa. const.: Pietrain × DL growing pigs, wt.: ≈22.2 kg, contamination: natural (for detailed information please see the article), conc. range: ≤34.8 ng/ml** (mean value), country: Germany[475], *fed **conventional diet (wheat variety: *Contur*)**, **in the year 2001, further data of the years 1999 and 2000 in the article
incidence: 6?/6*, sa. const.: Pietrain × DL growing pigs, wt.: ≈22.2 kg, contamination: natural (for detailed information please see the article), conc. range: ≤19.2 ng/ml** (mean value), country: Germany[475], *fed **organic diet (wheat variety: *Batis*)**, **in the year 2000, further data of the years 1999 and 2001 in the article
incidence: 6?/6*, sa. const.: Pietrain × DL growing pigs, wt.: ≈22.2 kg, contamination: natural (for detailed information please see the article), conc.: ±15.6 ng/ml** (mean value), country: Germany[475], *fed

conventional **diet (wheat variety: *Batis*)**, **in the year 2000, further data of the years 1999 and 2001 in the article

incidence: 6?/6*, sa. const.: Pietrain × DL growing pigs, wt.: ≈22.2 kg, contamination: natural (for detailed information please see the article), conc. range: ≤6.9 ng/ml** (mean value), country: Germany[475], *fed **organic diet (wheat variety: *Petrus*)**, **in the year 2001, further data of the year 2000 in the article

incidence: 6?/6*, sa. const.: Pietrain × DL growing pigs, wt.: ≈22.2 kg, contamination: natural (for detailed information please see the article), conc. range: ≤49.2 ng/ml** (mean value), country: Germany[475], *fed **conventional diet (wheat variety: *Petrus*)**, **in the year 2001, further data of the year 2000 in the article

Zearalenone

incidence: 764/794, sa. const.: breeding sows, contamination: natural, conc. range: ≤220.4 ng/g*, country: Germany[404], *ZEA and/or ZEA derivatives

incidence: 6?/6*, sa. const.: Pietrain × DL growing pigs, wt.: ≈22.2 kg, contamination: natural (for detailed information please see the article), conc. range: ≤54.0 ng/ml** (mean value), country: Germany[475], *fed **organic diet (wheat variety: *Contur*)**, **sum of ZEA, α- and β-ZEAOL, **in the year 2001, further data of the years 1999 and 2000 in the article

incidence: 6?/6*, sa. const.: Pietrain × DL growing pigs, wt.: ≈22.2 kg, contamination: natural (for detailed information please see the article), conc.: ±94.7 ng/ml** (mean value), country: Germany[475], *fed **conventional diet (wheat variety: *Contur*)**, **sum of ZEA, α- and β-ZEAOL, **in the year 2001, further data of the years 1999 and 2000 in the article

incidence: 6?/6*, sa. const.: Pietrain × DL growing pigs, wt.: ≈22.2 kg, contamination: natural (for detailed information please see the article), conc. range: ≤81.0 ng/ml** (mean value), country: Germany[475], *fed **organic diet (wheat variety: *Batis*)**, **sum of ZEA, α- and β-ZEAOL, **in the year 2001, further data of the years 1999 and 2000 in the article

incidence: 6?/6*, sa. const.: Pietrain × DL growing pigs, wt.: ≈22.2 kg, contamination: natural (for detailed information please see the article), conc.: ±92.4 ng/ml** (mean value), country: Germany[475], *fed **conventional diet (wheat variety: *Batis*)**, **sum of ZEA, α- and β-ZEAOL, **in the year 2001, further data of the years 1999 and 2000 in the article

incidence: 6?/6*, sa. const.: Pietrain × DL growing pigs, wt.: ≈22.2 kg, contamination: natural (for detailed information please see the article), conc. range: ≤44.5 ng/ml** (mean value), country: Germany[475], *fed **organic diet (wheat variety: *Petrus*)**, **sum of ZEA, α- and β-ZEAOL, **in the year 2000, further data of the year 2001 in the article

incidence: 6?/6*, sa. const.: Pietrain × DL growing pigs, wt.: ≈22.2 kg, contamination: natural (for detailed information please see the article), conc. range: ≤80.6 ng/ml** (mean value), country: Germany[475], *fed **conventional diet (wheat variety: *Petrus*)**, **sum of ZEA, α- and β-ZEAOL, **in the year 2000, further data of the year 2001 in the article

incidence: 52/52, sa. const.: gilts with reproductive problems, contamination: natural, conc. range: ≤40.0 ng/ml, country: Germany[526]

α-Zearalenol

incidence: 52/52, sa. const.: gilts with reproductive problems, contamination: natural, conc. range: ≤66.1 ng/ml, country: Germany[526]

Pig blood may contain the following mycotoxins and/or their metabolites:

Ochratoxin A

incidence: 40/261, sa. const.: blood from pigs of Germany, contamination: natural, conc. range: ≤2.05 ppb, country: Germany[264]

incidence: 147/255, sa. const.: blood from pigs of Czechoslovakia, contamination: natural, conc. range: 0.1–1 μg/l (98 sa), 1–5 μg/l (44 sa), 5–20 μg/l (5 sa), Ø conc.: 1.9 μg/l, country: Czechoslovakia[281]

incidence: 36/195, sa. const.: blood from pigs of Poland, contamination: natural, conc. range: 3–270 ng/ml, country: Poland/Sweden[304]

incidence: 47/279, sa. const.: blood from slaughter pigs of Sweden, contamination: natural, conc. range: 2–187 ng/ml, Ø conc.: 15.74 ng/ml, country: Sweden[306]

incidence: 26/45*, sa. const.: blood from pigs of Poland, contamination: natural, conc. range: ≤69.5 ng/ml, country: Poland[326], *blood serum

incidence: 63/105*, sa. const.: blood from pigs of Poland, contamination: natural, conc. range: ≤122 ng/ml, country: Poland[327], *blood serum

incidence: 910/1200, sa. const.: blood from pigs of Canada, contamination: natural, conc. range: <10 ng/ml (774 sa), 10–20 ng/ml (87 sa), 20–50 ng/ml (36 sa), 50–100 ng/ml (8 sa), 100–150 ng/ml (2 sa), 150–200 ng/ml (2 sa), 229 ng/ml (1 sa), country: Canada[334]

incidence: 26/122, sa. const.: blood from slaughter pigs of Sweden, contamination: natural, conc. range: 2.0–62 ng/ml, Ø conc.: 8.69 ng/ml, country: Sweden[371]

incidence: 6/38, sa. const.: blood from pigs of Yugoslavia, contamination: natural, conc. range: 36–77 μg/l, country: Crotia/Yugoslavia[393]

incidence: 179/359*, sa. const.: blood from pigs of Sweden, contamination: natural, conc. range: ≥2 ng/ml (136 sa), ≥5 ng/ml (29 sa), ≥10 ng/ml (14 sa), Ø conc.: 8.2 ng/ml, country: Sweden[411], *fed with **short stored grain**

incidence: 94/174*, sa. const.: blood from pigs of Sweden, contamination: natural, conc. range: ≥2 ng/ml (49 sa), ≥5 ng/ml (29 sa), ≥10 ng/ml (16 sa), Ø conc.: 13.2 ng/ml, country: Sweden[411], *fed with **long stored grain**

Pig digesta may contain the following mycotoxins and/or their metabolites:

Deoxynivalenol

incidence: 1/1, sa. const.: digesta from a pig of Germany?, contamination: natural, conc.: 145 ng/g, country: Germany[364]

Deepoxydeoxynivalenol

incidence: 1/1, sa. const.: digesta from a pig of Germany?, contamination: natural, conc.: 274 ng/g, country: Germany[364]

Pig kidney may contain the following mycotoxins and/or their metabolites:

Aflatoxin M_1

incidence: 1/40, sa. const.: kidneys from pigs of Brazil, contamination: natural, conc.: tr, country: Brazil[249]

Citrinin

incidence: 9/125, sa. const.: kidneys from sows of the UK, contamination: natural, conc. range: 0.1–<1 μg/kg (1 sa), 1–<5 μg/kg (4 sa), 5–<10 μg/kg (2 sa), >10 μg/kg (2 sa), country: UK[289]

Ochratoxin A

incidence: 41/300, sa. const.: kidneys from pigs of Germany, contamination: natural, conc. range: 0.5–10 μg/kg, country: Germany[263]

incidence: 6/54, sa. const.: kidneys from pigs of Germany, contamination: natural, conc. range: ≤0.75 ppb, country: Germany[264]

incidence: 1/63*, sa. const.: kidneys from pigs of Czechoslovakia, contamination: natural, conc.: 2.8 μg/kg, country: Czechoslovakia[281], *healthy

incidence: 20/96*, sa. const.: kidneys from pigs of Czechoslovakia, contamination:

natural, conc. range: 1–5 μg/kg (18 sa), 5–20 μg/kg (2 sa), country: Czechoslovakia[281], *macroscopic lesions

incidence: 136/378, sa. const.: kidneys from sows of the UK, contamination: natural, conc. range: 0.5–<1 μg/kg (53 sa), 1–<5 μg/kg (68 sa), 5–<10 μg/kg (10 sa), 10–<25 μg/kg (4 sa), >25 μg/kg (1 sa), country: UK[289]

incidence: 22/104*, sa. const.: kidneys from pigs of Germany, contamination: natural, conc. range: 0.1–1.8 μg/kg, Ø conc.: 0.45 μg/kg, country: Germany[291], *suspected

incidence: 20/20*, sa. const.: kidneys from pigs of Denmark, contamination: natural, conc. range: 0.2–195.5 μg/kg, Ø conc.: 34.24 μg/kg, country: Germany [291], *suspected

incidence: 12/36, sa. const.: kidneys from pigs of Switzerland, contamination: natural, conc. range: 0.1–0.2 μg/kg (11 sa), 0.3 μg/kg (1 sa), country: Switzerland[292]

incidence: 4498/7639, sa. const.: kidneys from pigs of Denmark, contamination: natural, conc. range: >25 μg/kg (4293 sa), >150 μg/kg (205 sa), country: Denmark[295]

incidence: 238/10, sa. const.: kidneys from pigs of France, contamination: natural, conc. range: tr (184 sa), 0.5-5 μg/kg (54 sa), country: France297

incidence: 21/71*, sa. const.: kidneys from pigs of Czechoslovakia, contamination: natural, conc. range: 1–5 μg/kg (18 sa), 5–20 μg/kg (3 sa), country: Czechoslovakia[300], *suspected

incidence: 27/113*, sa. const.: kidneys from pigs of Poland, contamination: natural, conc. range: tr–23 ng/g, country: Poland/Sweden[304], *suspected

incidence: 52/122*, sa. const.: kidneys from pigs of Poland, contamination: natural, conc. range: 1≤x<2 ng/g (27 sa), 2≤x<10 ng/g (25 sa), country: Poland/Sweden[305], *suspected

incidence: 24/90, sa. const.: kidneys from pigs of Sweden, contamination: natural, conc. range: >2–88 μg/kg, country: Sweden[315]

incidence: 284/300, sa. const.: kidneys from pigs of Denmark, contamination: natural, conc. range: 0.02–0.06 μg/kg (54 sa), 0.06–0.09 μg/kg (27 sa), 0.09–0.50 μg/kg (140 sa), 0.5–1.00 μg/kg (39 sa), ≤15 μg/kg (24 sa), Ø conc.: 0.50 μg/kg, country: Denmark[324]

incidence: 81/191, sa. const.: kidneys from slaughter pigs of Austria, contamination: natural, conc. range: <10 ppb (69 sa), 10–25 ppb (7 sa), ≤88.89 ppb (5 sa), country: Austria[325]

incidence: 35/85, sa. const.: kidneys from pigs of Poland, contamination: natural, conc. range: ≤3.1 ng/g, country: Poland[327]

incidence: 21/60*, sa. const.: kidneys from pigs of Denmark, contamination: natural, conc. range: ≤68 μg/kg, country: Denmark[328], *condemned

incidence: 15/104, sa. const.: kidneys from pigs of the UK or imported, contamination: natural, conc. range: 1–5 μg/kg (12 sa), ≤9.3 μg/kg (3 sa), country: UK[331]

incidence: 112/303*, sa. const.: kidneys from pigs of the UK, contamination: natural, conc. range: 0.5–<1 ng/g (51 sa), 1–<2 ng/g (39 sa), 2–<5 ng/g (14 sa), 5–<10 ng/g (6 sa), ≤12.4 ng/g (2 sa), country: UK[335], *rejected as unsuitable for human consumption

incidence: 2/131, sa. const.: kidneys from pigs of Norway?, contamination: natural, conc. range: 7–10 μg/kg, Ø conc.: 8.5 μg/kg, country: Norway[337]

incidence: 10/10, sa. const.: kidneys from pigs of Denmark, contamination: natural, conc. range: 4–112.7 ng/g, Ø conc.: 46.05 ng/g, country: Belgium/Scotland, UK[342]

incidence: 28/95, sa. const.: kidneys from pigs of Belgium, contamination: natural,

conc. range: 0.2–0.99 ng/g (6 sa), 1–4.99 ng/g (19 sa), 5–9.99 ng/g (3 sa), country: Belgium[343]
incidence: 6/13, sa. const.: kidneys from pigs of Belgium, contamination: natural, conc. range: 1.0–1.6 ng/g, Ø conc.: 1.22 ng/g, country: Belgium[343]
incidence: 3/4, sa. const.: kidneys from piglets of Belgium, contamination: natural, conc. range: 0.5–1.8 ng/g, Ø conc.: 1.33 ng/g, country: Belgium[343]

incidence: 32/129, sa. const.: kidneys from pigs of Sweden, contamination: natural, conc. range: ≥2–<5 ppb (25 sa), ≥5–<10 ppb (2 sa), ≥10–≤104 ppb (5 sa), country: Sweden[344]

incidence: 10/193*, sa. const.: kidneys from pigs of Finland, contamination: natural, conc. range: <1–5 μg/kg, country: Finland[348], *33 suspected

incidence: 82/104, sa. const.: kidneys from pigs of Romania, contamination: natural, conc. range: ≤3.18 ng/g, Ø conc.: 0.54 ng/g, country: Romania/Germany[350]

incidence: 7/25, sa. const.: kidneys from slaughter pigs of Germany, contamination: natural, conc. range: 0.4–5.1 μg/kg, Ø conc.: 1.16 μg/kg, country: Germany[379]

incidence: 3/38, sa. const.: kidneys from pigs of Croatia, contamination: natural, conc. range: 26–76 μg/kg, country: Croatia/Yugoslavia[393]

incidence: 1/1, sa. const.: kidneys from pigs of Italy, contamination: natural, conc.: 1.9 ng/g, country: Italy[431]

incidence: 42/54*, sa. const.: kidneys from pigs of different countries, contamination: natural, conc. range: 0.26–3.05 ng/g, country: Italy[441], *contaminated sa. from Belgium (9 sa), Germany (13 sa), Italy (9 sa) and The Netherlands (11 sa)

incidence: 16/60, sa. const.: kidneys from pigs of Spain, contamination: natural, conc. range: 0.5–1 ng/g (12 sa), 1–3 ng/g (4 sa), country: Spain[527]

incidence: 187/250, sa. const. kidneys from pigs of the UK, contamination: natural, conc. range: 0.2–0.5 μg/kg (151 sa), 0.51–1.0 μg/kg (29 sa), 1.01–1.5 μg/kg (4 sa), 1.51–2.0 μg/kg (2 sa), 2.3 μg/kg (1 sa), country: UK[553]

Pig liver may contain the following mycotoxins and/or their metabolites:

Aflatoxin B_1
incidence: 1/3, sa. const.: livers from feeder pigs of the USA, contamination: natural, conc.: 0.012 ng/g, country: USA[113]

incidence: 1/43, sa. const.: livers from pigs of Brazil, contamination: natural, conc.: 27 ng/g, country: Brazil[249]

incidence: 8/160, sa. const.: livers from hogs of the USA, contamination: natural, conc. range: 0.04–0.24 ppb, country: USA[519]

Aflatoxin M_1
incidence: 4/160, sa. const.: livers from hogs of the USA, contamination: natural, conc. range: 0.20–0.44 ppb, country: USA[519]

Deoxynivalenol
incidence: 6?/6*, sa. const.: livers from Pietrain × DL growing pigs, wt.: ≈22.2 kg, contamination: natural (for detailed information please see the article), conc. range: ≤10.9 ng/mg** (mean value), country: Germany[475], *fed **organic diet (wheat variety: *Contur*)**, **in the year 2001, further data of the years 1999 and 2000 in the article
incidence: 6?/6*, sa. const.: livers from Pietrain × DL growing pigs, wt.: ≈22.2 kg, contamination: natural (for detailed information please see the article), conc. range: ≤10.8 ng/mg** (mean value), country: Germany[475], *fed **conventional diet (wheat variety: *Contur*)**, **in the year 2001, further data of the years 1999 and 2000 in the article

incidence: 6?/6*, sa. const.: livers from Pietrain × DL growing pigs, wt.: ≈22.2 kg, contamination: natural (for detailed information please see the article), conc. range: ≤9.0 ng/mg** (mean value), country: Germany[475], *fed **organic diet (wheat variety: *Batis*),** **in the year 2001, further data of the years 1999 and 2000 in the article
incidence: 6?/6*, sa. const.: livers from Pietrain × DL growing pigs, wt.: ≈22.2 kg, contamination: natural (for detailed information please see the article), conc. range: ≤15.7 ng/mg** (mean value), country: Germany[475], *fed **conventional diet (wheat variety: *Batis*),** **in the year 2001, further data of the years 1999 and 2000 in the article
incidence: 6?/6*, sa. const.: livers from Pietrain × DL growing pigs, wt.: ≈22.2 kg, contamination: natural (for detailed information please see the article), conc. range: ≤5.2 ng/mg** (mean value), country: Germany[475], *fed **organic diet (wheat variety: *Petrus*),** **in the year 2001, further data of the year 2000 in the article
incidence: 6?/6*, sa. const.: livers from Pietrain × DL growing pigs, wt.: ≈22.2 kg, contamination: natural (for detailed information please see the article), conc. range: 8.3 ng/mg** (mean value), country: Germany[475], *fed conventional diet (wheat variety: *Petrus*), **in the year 2001, further data of the year 2000 in the article

Ochratoxin A
incidence: 73/191, sa. const.: livers from slaughter pigs of Austria, contamination: natural, conc. range: <10 ppb (61 sa), 10–25 ppb (8 sa), ≤97.33 ppb (4 sa), country: Austria[325]
incidence: 39/52, sa. const.: livers from pigs of Romania, contamination: natural, conc. range: ≤0.61 ng/g, Ø conc.: 0.16 ng/g, country: Romania/Germany[350]

Zearalenone
incidence: 6?/6*, sa. const.: livers from Pietrain × DL growing pigs, wt.: ≈22.2 kg, contamination: natural (for detailed information please see the article), conc. range: ≤5.6 ng/mg** (mean value), country: Germany[475], *fed **organic diet (wheat variety: *Contur*),** **sum of ZEA, α- and β-ZEAOL, **in the year 2001, further data of the years 1999 and 2000 in the article
incidence: 6?/6*, sa. const.: livers from Pietrain × DL growing pigs, wt.: ≈22.2 kg, contamination: natural (for detailed information please see the article), conc. range: ≤22.0 ng/mg** (mean value), country: Germany[475], *fed **conventional diet (wheat variety: *Contur*),** **sum of ZEA, α- and β-ZEAOL, **in the year 2001, further data for the years 1999 and 2000 in the article
incidence: 6?/6*, sa. const.: livers from Pietrain × DL growing pigs, wt.: ≈22.2 kg, contamination: natural (for detailed information please see the article), conc. range: ≤18.5 ng/mg** (mean value), country: Germany[475], *fed **organic diet (wheat variety: *Batis*),** **sum of ZEA, α- and β-ZEAOL, **in the year 2001, further data of the years 1999 and 2000 in the article
incidence: 6?/6*, sa. const.: livers from Pietrain × DL growing pigs, wt.: ≈22.2 kg, contamination: natural (for detailed information please see the article), conc. range: ≤5.9 ng/mg** (mean value), country: Germany[475], *fed **conventional diet (wheat variety: *Batis*),** **sum of ZEA, α- and β-ZEAOL, **in the year 2001, further data of the years 1999 and 2000 in the article
incidence: 6?/6*, sa. const.: livers from Pietrain × DL growing pigs, wt.: ≈22.2 kg, contamination: natural (for detailed information please see the article), conc. range: ≤7.4 ng/mg** (mean value), country: Germany[475], *fed **organic diet (wheat variety: *Petrus*),** **sum of ZEA, α- and β-ZEAOL, **in the year 2001, further data of the year 2000 in the article
incidence: 6?/6*, sa. const.: livers from Pietrain × DL growing pigs, wt.: ≈22.2 kg, contamination: natural (for detailed information please see the article), conc. range: 6.0 ng/mg** (mean value), country: Germany[475], *fed **conventional diet (wheat variety: *Petrus*),** **sum of ZEA, α- and

β-ZEAOL, **in the year 2001, further data of the year 2000 in the article

Pig meat may contain the following mycotoxins and/or their metabolites:

OCHRATOXIN A
incidence: 228/300, sa. const.: meat from pigs of Denmark, contamination: natural, conc. range: 0.03–0.06 μg/kg (134 sa), 0.06–0.09 μg/kg (27 sa), 0.09–0.50 μg/kg (55 sa), 0.50–1.00 μg/kg (3 sa), ≤2.9 μg/kg (9 sa), Ø conc.: 0.12 μg/kg, country: Denmark[324]

Pig muscle may contain the following mycotoxins and/or their metabolites:

DEOXYNIVALENOL
incidence: 6?/6*, sa. const.: muscles from Pietrain × DL growing pigs, wt.: ≈22.2 kg, contamination: natural (for detailed information please see the article), conc. range: ≤2.1 ng/mg** (mean value), country: Germany[475], *fed **organic diet (wheat variety: *Contur*)**, **in the year 2001, further data of the years 1999 and 2000 in the article
incidence: 6?/6*, sa. const.: muscles from Pietrain × DL growing pigs, wt.: ≈22.2 kg, contamination: natural (for detailed information please see the article), conc. range: ≤8.0 ng/mg** (mean value), country: Germany[475], *fed **conventional diet (wheat variety: *Contur*)**, **in the year 2001, further data of the years 1999 and 2000 in the article
incidence: 6?/6*, sa. const.: muscles from Pietrain × DL growing pigs, wt.: ≈22.2 kg, contamination: natural (for detailed information please see the article), conc. range: ≤1.4 ng/mg** (mean value), country: Germany[475], *fed **organic diet (wheat variety: *Batis*)**, **in the year 1999, further data of the years 2000 and 2001 in the article
incidence: 6?/6*, sa. const.: muscles from Pietrain × DL growing pigs, wt.: ≈22.2 kg, contamination: natural (for detailed information please see the article), conc.: ±0.4 ng/mg** (mean value), country: Germany[475], *fed **conventional diet (wheat variety: *Batis*)**, **in the year 1999, further data of the years 2000 and 2001 in the article
incidence: 6?/6*, sa. const.: muscles from Pietrain × DL growing pigs, wt.: ≈22.2 kg, contamination: natural (for detailed information please see the article), conc. range: ≤1.6 ng/mg** (mean value), country: Germany[475], *fed **organic diet (wheat variety: *Petrus*)**, **in the year 2000, further data of the year 2001 in the article
incidence: 6?/6*, sa. const.: muscles from Pietrain × DL growing pigs, wt.: ≈22.2 kg, contamination: natural (for detailed information please see the article), conc. range: 2.7 ng/mg** (mean value), country: Germany[475], *fed **conventional diet (wheat variety: *Petrus*)**, **in the year 2000, further data of the year 2001 in the article

OCHRATOXIN A
incidence: 2/22*, sa. const.: muscles from pigs of Italy, contamination: natural, conc. range: ≤0.06 μg/kg, Ø conc.: 0.05 μg/kg, country: Italy[339], *for ham
incidence: 9/52, sa. const.: muscles from pigs of Romania, contamination: natural, conc. range: ≤0.53 ng/g, Ø conc.: 0.15 ng/g, country: Romania/Germany[350]

ZEARALENONE
incidence: 6?/6*, sa. const.: muscles from Pietrain × DL growing pigs, wt.: ≈22.2 kg, contamination: natural (for detailed information please see the article), conc. range: ≤6.1 ng/mg** (mean value), country: Germany[475], *fed **organic diet (wheat variety: *Contur*)**, **sum of ZEA, α- and β-ZEAOL, **in the year 1999, further data of the years 2000 and 2001 in the article
incidence: 6?/6*, sa. const.: muscles from Pietrain × DL growing pigs, wt.: ≈22.2 kg, contamination: natural (for detailed information please see the article), conc.: ±0.9 ng/mg** (mean value), country: Germany[475], *fed **conventional diet (wheat variety: *Contur*)**, **sum of ZEA, α- and β-ZEAOL, **in the year 1999, further data of the years 2000 and 2001 in the article
incidence: 6?/6*, sa. const.: muscles from Pietrain × DL growing pigs, wt.: ≈22.2 kg,

contamination: natural (for detailed information please see the article), conc. range: ≤6.2 ng/mg** (mean value), country: Germany[475], *fed **organic diet (wheat variety: *Batis*)**,**sum of ZEA, α- and β-ZEAOL, **in the year 2001, further data of the years 1999 and 2000 in the article
incidence: 6?/6*, sa. const.: muscles from Pietrain × DL growing pigs, wt.: ≈22.2 kg, contamination: natural (for detailed information please see the article), conc. range: ≤1.2 ng/mg** (mean value), country: Germany[475], *fed **conventional diet (wheat variety: *Batis*)**, **sum of ZEA, α- and β-ZEAOL, **in the year 2001, further data of the years 1999 and 2000 in the article
incidence: 6?/6*, sa. const.: muscles from Pietrain × DL growing pigs, wt.: ≈22.2 kg, contamination: natural (for detailed information please see the article), conc. range: ≤16.1 ng/mg** (mean value), country: Germany[475], *fed **organic diet (wheat variety: *Petrus*)**, **sum of ZEA, α- and β-ZEAOL, **in the year 2001, further data of the year 2000 in the article
incidence: 6?/6*, sa. const.: muscles from Pietrain × DL growing pigs, wt.: ≈22.2 kg, contamination: natural (for detailed information please see the article), conc. range: 14.3 ng/mg** (mean value), country: Germany[475], *fed **conventional diet (wheat variety: *Petrus*)**, **sum of ZEA, α- and β-ZEAOL, **in the year 2001, further data of the year 2000 in the article

Pig plasma may contain the following mycotoxins and/or their metabolites:

Aflatoxin B_1
incidence: 3/9, sa. const.: plasma from feeder pigs of the USA, contamination: natural, conc. range: 5.1–36.7 ng/ml, country: USA[113]

Ochratoxin A
incidence: 191/216, sa. const.: plasma from pigs of Norway, contamination: natural, conc. range: ≥0.1 ng/ml (178 sa), ≥1 ng/ml (11 sa), ≤12.5 ng/ml (2 sa), Ø conc.: 0.5 ng/ml, country: Norway[307]

Pig serum may contain the following mycotoxins and/or their metabolites:

Citrinin
incidence: 7/10, sa. const.: serum from pigs with nephropathy problems of Bulgaria, contamination: natural, conc.: 1.3 μg/l* (mean value), country: Bulgaria/South Africa[581], *in the year 2006
incidence: 8/10, sa. const.: serum from pigs with nephropathy problems of Bulgaria, contamination: natural, conc.: 1.6 μg/l* (mean value), country: Bulgaria/South Africa[581], *in the year 2007

Deoxynivalenol
incidence: 1/1, sa. const.: serum from a pig of Germany?, contamination: natural, conc.: 33 ng/ml, country: Germany[364]
incidence: 1/10, sa. const.: serum from pigs with nephropathy problems of Bulgaria, contamination: natural, conc.: 7.6 μg/l*, country: Bulgaria/South Africa[581], *in the year 2006
incidence: 10/10, sa. const.: serum from pigs with nephropathy problems of Bulgaria, contamination: natural, conc.: nd*, country: Bulgaria/South Africa[581], *in the year 2007

Ochratoxin A
incidence: 93/191, sa. const.: serum from slaughter pigs of Germany, contamination: natural, conc. range: 0.1–67.3 μg/kg, Ø conc.: 5.8 μg/kg, country: Germany[291]
incidence: 52/200, sa. const.: serum from pigs of Austria, contamination: natural, conc. range: <1 ppb (50 sa), ≤1.24 ppb (2 sa), country: Austria[296]
incidence: 66/287, sa. const.: serum from pigs of Austria, contamination: natural, conc. range: <1 ppb (59 sa), >1 ppb (5 sa), ≤30.36 ppb (2 sa), country: Austria[296]
incidence: 3/300, sa. const.: serum from pigs of France, contamination: natural, conc. range: 0.40–1.4 μg/kg, Ø conc.: 0.787 μg/kg, country: France[297]
incidence: 6/100, sa. const.: serum from pigs of France, contamination: natural,

conc. range: 0.16–0.48 μg/kg, country: France[297]

incidence: 148/388, sa. const.: serum from pigs of Poland, contamination: natural, conc. range: 1–520 ng/ml, country: Poland[305]

incidence: 45/85, sa. const.: serum from pigs of Germany, contamination: natural, conc. range: 0.29–17.6 μg/kg (mean values), country: Germany[333]

incidence: 572/1588, sa. const.: serum from pigs of Canada, contamination: natural, conc. range: 0.3–211 ng/ml, Ø conc.: 14.1 ng/ml, country: Canada[338]

incidence: 2/4, sa. const.: serum from sows of Belgium, contamination: natural, conc. range: 3.1–3.7 ng/ml, Ø conc.: 3.4 ng/ml, country: Belgium[343]
incidence: 4/4, sa. const.: serum from sows of Belgium, contamination: natural, conc. range: 2.3–3.7 ng/ml, Ø conc.: 2.95 ng/ml, country: Belgium[343]

incidence: 51/52, sa. const.: serum from pigs of Romania, contamination: natural, conc. range: ≤13.4 ng/ml, Ø conc.: 2.43 ng/ml, country: Romania/Germany[350]

incidence: 3/25, sa. const.: serum from slaughter pigs of Germany, contamination: natural, conc. range: 2.8–12.9 μg/kg, Ø conc.: 6.77 μg/kg, country: Germany[379]

incidence: 3/4, sa. const.: serum from pigs of Belgium, contamination: natural, conc. range: 209.4–363.1 ng/ml, Ø conc.: 285.06 ng/ml, country: Belgium[410]

incidence: 8/10, sa. const.: serum from pigs with nephropathy problems of Bulgaria, contamination: natural, conc.: 28.8 μg/l* (mean value), country: Bulgaria/South Africa[581], *in the year 2006
incidence: 9/10, sa. const.: serum from pigs with nephropathy problems of Bulgaria, contamination: natural, conc.: 6.3 μg/l* (mean value), country: Bulgaria/South Africa[581], *in the year 2007

Penicillic Acid
incidence: 8/10, sa. const.: serum from pigs with nephropathy problems of Bulgaria, contamination: natural, conc.: 23.3 μg/l* (mean value), country: Bulgaria/South Africa[581], *in the year 2006
incidence: 9/10, sa. const.: serum from pigs with nephropathy problems of Bulgaria, contamination: natural, conc.: 22.9 μg/l* (mean value), country: Bulgaria/South Africa[581], *in the year 2007

Penitrem A
incidence: 3/10, sa. const.: serum from pigs with nephropathy problems of Bulgaria, contamination: natural, conc.: 64.0 μg/l* (mean value), country: Bulgaria/South Africa[581], *in the year 2006
incidence: 3/10, sa. const.: serum from pigs with nephropathy problems of Bulgaria, contamination: natural, conc.: 45.6 μg/l* (mean value), country: Bulgaria/South Africa[581], *in the year 2007

Zearalenone
incidence: 9/52, sa. const.: serum from pigs of romania, contamination: natural, conc. range: ≤0.96 ng/ml, Ø conc.: 0.8 ng/ml, country: Romania/Germany[350]

incidence: 5/10, sa. const.: serum from pigs with nephropathy problems of Bulgaria, contamination: natural, conc.: 0.24 μg/l* (mean value), country: Bulgaria/South Africa[581], *in the year 2006
incidence: 5/10, sa. const.: serum from pigs with nephropathy problems of Bulgaria, contamination: natural, conc.: 0.33 μg/l* (mean value), country: Bulgaria/South Africa[581], *in the year 2007

Pig urine may contain the following mycotoxins and/or their metabolites:

Citrinin
incidence: 10/10, sa. const.: urine from pigs with nephropathy problems of Bulgaria, contamination: natural, conc.:

1.7 μg/l* (mean value), country: Bulgaria/South Africa[581], *in the year 2006
incidence: 10/10, sa. const.: urine from pigs with nephropathy problems of Bulgaria, contamination: natural, conc.: 1.8 μg/l* (mean value), country: Bulgaria/South Africa[581], *in the year 2007

Deoxynivalenol
incidence: 1/10, sa. const.: urine from pigs with nephropathy problems of Bulgaria, contamination: natural, conc.: 5.1 μg/l*, country: Bulgaria/South Africa[581], *in the year 2006
incidence: 10/10, sa. const.: urine from pigs with nephropathy problems of Bulgaria, contamination: natural, conc.: nd*, country: Bulgaria/South Africa[581], *in the year 2007

Ochratoxin A
incidence: 10/10, sa. const.: urine from pigs with nephropathy problems of Bulgaria, contamination: natural, conc.: 3.5 μg/l* (mean value), country: Bulgaria/South Africa[581], *in the year 2006
incidence: 10/10, sa. const.: urine from pigs with nephropathy problems of Bulgaria, contamination: natural, conc.: 6.2 μg/l* (mean value), country: Bulgaria/South Africa[581], *in the year 2007

Penicillic Acid
incidence: 6/10, sa. const.: urine from pigs with nephropathy problems of Bulgaria, contamination: natural, conc.: 1.6 μg/l* (mean value), country: Bulgaria/South Africa[581], *in the year 2006
incidence: 6/10, sa. const.: urine from pigs with nephropathy problems of Bulgaria, contamination: natural, conc.: 1.7 μg/l* (mean value), country: Bulgaria/South Africa[581], *in the year 2007

Penitrem A
incidence: 10/10, sa. const.: urine from pigs with nephropathy problems of Bulgaria, contamination: natural, conc.: nd*, country: Bulgaria/South Africa[581], *in the year 2006
incidence: 10/10, sa. const.: urine from pigs with nephropathy problems of Bulgaria, contamination: natural, conc.: nd*, country: Bulgaria/South Africa[581], *in the year 2007

Zearalenone
incidence: 10/10, sa. const.: urine from pigs with nephropathy problems of Bulgaria, contamination: natural, conc.: 9.4 μg/l* (mean value), country: Bulgaria/South Africa[581], *in the year 2006
incidence: 10/10, sa. const.: urine from pigs with nephropathy problems of Bulgaria, contamination: natural, conc.: 13.1 μg/l* (mean value), country: Bulgaria/South Africa[581], *in the year 2007

Pig Artificial Contamination

Pig Adipose Tissue see Pig fat

Pig adrenals may contain the following mycotoxins and/or their metabolites:

Deoxynivalenol
incidence: 5/5*, sa. const.: healthy Yorkshire barrows, age: ≈11–15 weeks, wt.: 17–22 kg, contamination: no DON, conc.: nd, country: Canada[407], *control
incidence: ?/4, sa. const.: healthy Yorkshire barrows, age: ≈11–15 weeks, wt.: 17–22 kg, contamination: artificial (dose: 1.0 mg DON/kg b. wt., i.v., once), conc. range: ≤242.2 ng/g* (mean value), country: Canada[407], *after 0.33 h (also measured after 1, 3, 8, and 24 h, lowest conc.: nd after 24 h)

Pig bile may contain the following mycotoxins and/or their metabolites:

Aflatoxin B_1
incidence: 1/1, sa. const.: female Hampshire × Deutsches Edelschwein piglets, wt.: 15 kg, contamination: artificial (dose: 3.1 μg AFB_1 (labeled)/kg b. wt., o., once; for detailed information please see the article), conc.: 1.3 ppb* **, country: Switzerland[66], *AFB_1 eq., **after 24 h

incidence: 1/1, sa. const.: female Hampshire × Deutsches Edelschwein piglets, wt.: 15 kg, contamination: artificial (dose: 3.1 µg AFB_1 (labeled)/kg b. wt., o., once; for detailed information please see the article), conc.: 2.9 ppb* **, country: Switzerland[66], *AFB_1 eq., **after 48 h

Deoxynivalenol

incidence: 5/5*, sa. const.: healthy Yorkshire barrows, age: ≈11–15 weeks, wt.: 17–22 kg, contamination: no DON, conc.: nd, country: Canada[407], *control
incidence: ?/4, sa. const.: healthy Yorkshire barrows, age: ≈11–15 weeks, wt.: 17–22 kg, contamination: artificial (dose: **1.0 mg DON**/kg b. wt., i.v., once), conc. range: ≤10,962.5 ng/g* (mean value), country: Canada[407], *after 8 h (also measured after 0.33, 1, 3, and 24 h, lowest conc.: 1,283.0 ng/g after 24 h)

incidence: 5?/5, sa. const.: castrated Large White × German Landrace, db Classic crossbred pigs, wt.: ≈24.6 kg, contamination: artificial (dose: 0.05, 0.57, or 1.23 mg DON/kg mash or 0.07, 0.55, or 1.13* mg DON/kg pellets*, o., for 11 weeks; for detailed information please see the article), conc. range: ≤144.0 ng/ml* ** *** (mean value), country: Germany[483], **after 78/79 days (thereof 11 weeks of DON-administration), ***values of the other DON-treatments lower

incidence: 9?/9, sa. const.: German Landrace gilts, age: 180 days, wt.: ≈103 kg, contamination: artificial (dose: DON/ZEA in wheat in different conc., o., for 35 days; for detailed information please see the article), conc. range: ≤223.8 ng/ml* ** (mean value), country: Germany[537], *9.57 mg DON and 0.358 mg ZEA/kg diet fed (both fed in highest conc.), **after 36 days (thereof 35 days of DON- and ZEA-administration)

Deepoxydeoxynivalenol

incidence: 5?/5, sa. const.: castrated Large White × German Landrace, db Classic crossbred pigs, weight: ≈24.6 kg, contamination: artificial (dose: 0.05, 0.57, or 1.23* mg DON/kg mash* or 0.07, 0.55, or 1.13 mg DON/kg pellets, o., for 11 weeks; for detailed information please see the article), conc. range: ≤44.0 ng/ml* ** *** (mean value), country: Germany[483], **after 78/79 days (thereof 11 weeks of DON-administration), ***values of the other DON-treatments lower

incidence: 9?/9, sa. const.: German Landrace gilts, age: 180 days, wt.: ≈103 kg, contamination: artificial (dose: DON/ZEA in wheat in different conc., o., for 35 days; for detailed information please see the article), conc. range: ≤103.2 ng/ml* ** (mean value), country: Germany[537], *9.57 mg DON and 0.358 mg ZEA/kg diet fed (both fed in highest conc.), **after 36 days (thereof 35 days of DON- and ZEA-administration)

Fumonisin B_1

incidence: 5/5*, sa. const.: weaned barrows of the same genotype, wt.: 12–14 kg, contamination: no FB_1, FB_2 + FB_3 (for detailed information please see the article), conc.: nd, country: Hungary/Germany[87], *control
incidence: 1/10, sa. const.: weaned barrows of the same genotype, wt.: 12–14 kg, contamination: artificial (dose: **50 mg FB_1, 20 mg FB_2 + 5 mg FB_3/animal**, o., for 22 days; for detailed information please see the article), conc.: 322.8 ng/g*, country: Hungary/Germany[87], *after 22 days toxin feeding period

Fumonisin B_2

incidence: 5/5*, sa. const.: weaned barrows of the same genotype, wt.: 12–14 kg, contamination: no FB_1, FB_2 + FB_3 (for detailed information please see the article), conc.: nd, country: Hungary/Germany[87], *control
incidence: 10/10, sa. const.: weaned barrows of the same genotype, wt.: 12–14 kg, contamination: artificial (dose: **50 mg FB_1, 20 mg FB_2 + 5 mg FB_3/animal**,

o., for 22 days; for detailed information please see the article), conc.: nd*, country: Hungary/Germany[87], *after 22 days toxin feeding period

HT-2 Toxin
incidence: 2/2, sa. const.: female crossbred Yorkshire × Hampshire swines, wt.: 20 kg, contamination: artificial (dose: 0.15 mg T-2 toxin (labeled and unlabeled)/kg b. wt., i.v., once), conc. range: 2,336–4,698 ng/ml* **, Ø conc.: 3,517 ng/ml* **, country: USA[308], *after 4 h, **free and conjugated HT-2 toxin

3′-Hydroxy HT-2 Toxin
incidence: 2/2, sa. const.: female crossbred Yorkshire × Hampshire swines, wt.: 20 kg, contamination: artificial (dose: 0.15 mg T-2 toxin (labeled and unlabeled)/kg b. wt., i.v., once), conc. range: 1,712–2,097 ng/ml* **, Ø conc.: 1,904.5 ng/ml* **, country: USA[308], *after 4 h, **free and conjugated 3′-OH HT-2 toxin

Neosolaniol
incidence: 2/2, sa. const.: female crossbred Yorkshire × Hampshire swines, wt.: 20 kg, contamination: artificial (dose: 0.15 mg T-2 toxin (labeled and unlabeled)/kg b. wt., i.v., once), conc. range: 93–232 ng/ml* **, Ø conc.: 162.5 ng/ml* **, country: USA[308], *after 4 h, **free and conjugated neosolaniol

4-Deacetylneosolaniol
incidence: 2/2, sa. const.: female crossbred Yorkshire × Hampshire swines, wt.: 20 kg, contamination: artificial (dose: 0.15 mg T-2 toxin (labeled and unlabeled)/kg b. wt., i.v., once), conc. range: 160–324 ng/ml* **, Ø conc.: 242 ng/ml* **, country: USA[308], *after 4 h, **free and conjugated 4-deacetylneosolaniol

Ochratoxin A
incidence: 4/4*, sa. const.: male and females pigs, wt.: ≈70 kg, contamination: no OTA (for detailed information please see the article), conc.: nd, country: Germany[365], *control
incidence: 8?/8, sa. const.: male and females pigs, wt.: ≈70 kg, contamination: artificial (dose: **0.09 mg natural** OTA/kg diet/day, in the morning and evening **half of OTA-ration**, o., for 28 days; for detailed information please see the article), conc.: 17.03 ng/ml* (mean value), country: Germany[365], *after 28 days
incidence: 8?/8, sa. const.: male and females pigs, wt.: ≈70 kg, contamination: artificial (dose: **0.09 mg crystalline OTA**/kg diet/day, in the morning and evening **half of OTA-ration**, o., for 28 days; for detailed information please see the article), conc.: 5.13 ng/ml* (mean value), country: Germany[365], *after 28 days
incidence: 8?/8, sa. const.: male and females pigs, wt.: ≈70 kg, contamination: artificial (dose: **0.09 mg crystalline OTA**/kg diet/day, in the morning **total OTA-ration**, o., for 28 days; for detailed information please see the article), conc.: 4.78 ng/ml* (mean value), country: Germany[365], *after 28 days

incidence: 8?/8*, sa. const.: pigs, contamination: no OTA (for detailed information please see the article), conc.: <0.59 ng/ml (mean value), country: Germany[366], *control
incidence: 8?/8, sa. const.: pigs, contamination: artificial (dose: **22.11 mg OTA** (in total), o., for 90 days; for detailed information please see the article), conc.: no value*, country: Germany[366], *after 90 days
incidence: 8?/8, sa. const.: pigs, contamination: artificial (dose: **88.44 mg OTA** (in total), o., for 90 days; for detailed information please see the article), conc.: 77.6 ng/ml* (mean value), country: Germany[366], *after 90 days

T-2 Toxin
incidence: 2/2, sa. const.: female crossbred Yorkshire × Hampshire swines, wt.: 20 kg, contamination: artificial (dose: 0.15 mg T-2 toxin (labeled and unlabeled)/kg b. wt., i.v., once), conc. range: 3,634–11,831 ng/ml* **, Ø conc.: 7,732.5 ng/ml*

**, country: USA[308], *after 4 h, **free and conjugated T-2 toxin

incidence: 2/2*, sa. const.: female swines of mixed breeding, wt.: 26–66 kg, contamination: no T-2 toxin, conc.: nr, country: USA[403], *control
incidence: 1/4, sa. const.: female swines of mixed breeding, wt.: 26–66 kg, contamination: artificial (dose: **1.2 mg T-2 toxin**/kg, i.a., once), conc.: <40 ppb*, country: USA[403], *after 130 min

3′-Hydroxy T-2 Toxin
incidence: 2/2, sa. const.: female crossbred Yorkshire × Hampshire swines, wt.: 20 kg, contamination: artificial (dose: 0.15 mg T-2 toxin (labeled and unlabeled)/kg b. wt., i.v., once), conc. range: 432–1,202 ng/ml* **, Ø conc.: 817 ng/ml* **, country: USA[308], *after 4 h, **free and conjugated 3′-OH T-2 toxin

T-2 Tetraol
incidence: 2/2, sa. const.: female crossbred Yorkshire × Hampshire swines, wt.: 20 kg, contamination: artificial (dose: 0.15 mg T-2 toxin (labeled and unlabeled)/kg b. wt., i.v., once), conc. range: 322–366 ng/ml* **, Ø conc.: 344 ng/ml* **, country: USA[308], *after 4 h, **free and conjugated T-2 tetraol

T-2 Triol
incidence: 2/2, sa. const.: female crossbred Yorkshire × Hampshire swines, wt.: 20 kg, contamination: artificial (dose: 0.15 mg T-2 toxin (labeled and unlabeled)/kg b. wt., i.v., once), conc. range: 801–1,008 ng/ml* **, Ø conc.: 904.5 ng/ml* **, country: USA[308], *after 4 h, **free and conjugated T-2 triol

Zearalenone
incidence: 9?/9, sa. const.: German Landrace gilts, age: 180 days, wt.: ≈103 kg, contamination: artificial (dose: DON/ZEA in wheat in different conc., o., for 35 days; for detailed information please see the article), conc. range: ≤268.3 ng/g* ** (mean value), country: Germany[537], *9.57 mg DON and 0.358 mg ZEA/kg diet fed (both fed in highest conc.), **after 36 days (thereof 35 days of DON- and ZEA-administration)

α-Zearalenol
incidence: 9?/9, sa. const.: German Landrace gilts, age: 180 days, wt.: ≈103 kg, contamination: artificial (dose: DON/ZEA in wheat in different conc., o., for 35 days; for detailed information please see the article), conc. range: ≤309.5 ng/g* ** (mean value), country: Germany[537], *9.57 mg DON and 0.358 mg ZEA/kg diet fed (both fed in highest conc.), **after 36 days (thereof 35 days of DON- and ZEA-administration)

β-Zearalenol
incidence: 9?/9, sa. const.: German Landrace gilts, age: 180 days, wt.: ≈103 kg, contamination: artificial (dose: DON/ZEA in wheat in different conc., o., for 35 days; for detailed information please see the article), conc. range: ≤17.8 ng/g* ** (mean value), country: Germany[537], *9.57 mg DON and 0.358 mg ZEA/kg diet fed (both fed in highest conc.), **after 36 days (thereof 35 days of DON- and ZEA-administration)

Pig blood may contain the following mycotoxins and/or their metabolites:

Aflatoxin B_1
incidence: 4/4*, sa. const.: male and female feeder pigs, wt.: 54.2–71.6 kg, contamination: no AFB_1, conc.: na, country: USA[136], *control
incidence: 4/4, sa. const.: male and female feeder pigs, wt.: 54.2–71.6 kg, contamination: artificial (dose: **100 µg AFB_1**/kg diet, o., for 4 weeks), conc. range: 0.03–0.30 µg/kg*, Ø conc.: 0.17 µg/kg*, country: USA[136], *after 4 weeks
incidence: 4/4, sa. const.: male and female feeder pigs, wt.: 54.2–71.6 kg, contamination: artificial (dose: **200 µg AFB_1**/kg diet, o., for 4 weeks), conc. range: 0.21–0.47 µg/kg*, Ø conc.: 0.33 µg/kg*, country: USA[136], *after 4 weeks

incidence: 4/4, sa. const.: male and female feeder pigs, wt.: 54.2–71.6 kg, contamination: artificial (dose: **400 µg AFB$_1$/kg** diet, o., for 4 weeks), conc. range: 0.32–3.33 µg/kg*, Ø conc.: 1.15 µg/kg*, country: USA[136], *after 4 weeks

Aflatoxin M$_1$
incidence: 4/4*, sa. const.: male and female feeder pigs, wt.: 54.2–71.6 kg, contamination: no AFB$_1$, conc.: na, country: USA[136], *control
incidence: 4/4, sa. const.: male and female feeder pigs, wt.: 54.2–71.6 kg, contamination: artificial (dose: **100 µg AFB$_1$/kg** diet, o., for 4 weeks), conc. range: 0.04–0.11 µg/kg*, Ø conc.: 0.06 µg/kg*, country: USA[136], *after 4 weeks
incidence: 4/4, sa. const.: male and female feeder pigs, wt.: 54.2–71.6 kg, contamination: artificial (dose: **200 µg AFB$_1$/kg** diet, o., for 4 weeks), conc. range: 0.07–0.18 µg/kg*, Ø conc.: 0.12 µg/kg*, country: USA[136], *after 4 weeks
incidence: 4/4, sa. const.: male and female feeder pigs, wt.: 54.2–71.6 kg, contamination: artificial (dose: **400 µg AFB$_1$/kg** diet, o., for 4 weeks), conc. range: 0.06–0.23 µg/kg*, Ø conc.: 0.14 µg/kg*, country: USA[136], *after 4 weeks

Nivalenol
incidence: 3/3, sa. const.: male castrated Swedish Landrace × Yorkshire pigs, wt.: 37–63 kg, contamination: artificial (dose: 0.05 mg NIV/kg b. wt., o., twice daily for 3 days (**trial 1**); for detailed information please see the article), conc. range: ≤4.4 ng/ml* ** (mean value), country: Sweden[501], *on 3rd day, **systemic concentration
incidence: 3/3, sa. const.: male castrated Swedish Landrace × Yorkshire pigs, wt.: 37–63 kg, contamination: artificial (dose: 0.05 mg NIV/kg b. wt., o., twice daily for 3 days (**trial 2** = NIV-administration also 16 h before first blood sa taken); for detailed information please see the article), conc. range: ≤4.8 ng/ml* (mean value), country: Sweden[501], *on 3rd day

Ochratoxin A
incidence: 4?/4*, sa. const.: male and females pigs, wt.: ≈70 kg, contamination: no OTA (for detailed information please see the article), conc. range: ≤0.32 ng/ml** *** (mean value), country: Germany[365], *control, **in full blood, ***after the start of the experiment (also at other day intervals up to 28 days measured, lowest conc.: 0.25 ng/ml after 28 days)
incidence: 8?/8, sa. const.: male and females pigs, wt.: ≈70 kg, contamination: artificial (dose: **0.09 mg natural OTA**/kg diet/day, in the morning and evening **half of OTA-ration**, o., for 28 days; for detailed information please see the article), conc. range: ≤50.76 ng/ml* ** *** (mean value), country: Germany[365], **in full blood, ***after 22 days of OTA-administration (also at other day intervals up to 28 days measured, lowest conc.: 0.71 ng/ml after the start of the experiment)
incidence: 8?/8, sa. const.: male and females pigs, wt.: ≈70 kg, contamination: artificial (dose: **0.09 mg crystalline OTA**/kg diet/day, in the morning and evening **half of OTA-ration**, o., for 28 days; for detailed information please see the article), conc. range: ≤14.67 ng/ml* ** *** (mean value), country: Germany[365], **in full blood, ***after 22 days of OTA-administration (also at other day intervals up to 28 days measured, lowest conc.: 0.70 ng/ml after the start of the experiment)
incidence: 8?/8, sa. const.: male and females pigs, wt.: ≈70 kg, contamination: artificial (dose: **0.09 mg crystalline OTA**/kg diet/day, in the morning **total OTA-ration**, o., for 28 days; for detailed information please see the article), conc. range: ≤17.26 ng/ml* ** *** (mean value), country: Germany[365], **in full blood, ***after 22 days of OTA-administration (also at other day intervals up to 28 days

measured, lowest conc.: 0.90 ng/ml after the start of the experiment)

incidence: 8?/8*, sa. const.: pigs, contamination: no OTA (for detailed information please see the article), conc.: 0.85 ng/ml** (mean value), country: Germany[366], *control, **after 90 days
incidence: 8?/8, sa. const.: pigs, contamination: artificial (dose: **22.11 mg OTA** (in total), o., for 90 days; for detailed information please see the article), conc.: 20.0 ng/ml* ** (mean value), country: Germany[366], *in full blood, **after 90 days
incidence: 8?/8, sa. const.: pigs, contamination: artificial (dose: **88.44 mg OTA** (in total), o., for 90 days; for detailed information please see the article), conc.: 77.0 ng/ml* ** (mean value), country: Germany[366], *in full blood, **after 90 days

incidence: 9/9*, sa. const.: Deutsches Landschwein, Deutsches Edelschwein × Piétrain, Belgische Landrasse, wt.: 25 kg, contamination: no OTA, conc.: nd, country: Germany[409], *control
incidence: 2*/9, sa. const.: Deutsches Landschwein, Deutsches Edelschwein × Piétrain, Belgische Landrasse, wt.: 25 kg, contamination: artificial (dose: **0.15 mg natural OTA**/kg feed, o., daily for 28 days), conc. range: ≤12.0 μg/l* **, country: Germany[409], *at day 14 (also measured after 28 days, lowest conc.: <3 μg/l after 14 and 28 days (several animals)), **in full blood
incidence: 8/8*, sa. const.: Deutsches Landschwein, Deutsches Edelschwein × Piétrain, Belgische Landrasse, wt.: 25 kg, contamination: no OTA, conc.: nd**, country: Germany[409], *control, **in full blood
incidence: 6/6, sa. const.: Deutsches Landschwein, Deutsches Edelschwein × Piétrain, Belgische Landrasse, wt.: 25 kg, contamination: artificial (dose: **0.58 mg natural OTA**/kg feed, o., daily for 28 days), conc. range: ≤43.2 μg/l* **, country: Germany[409], *at day 28 (also measured after 14 days, lowest conc.: 18.4 μg/l after 14 days), **in full blood

T-2 Toxin
incidence: 1/1, sa. const.: female weanling crossbred Yorkshire × Duroc × Hampshire swine, wt.: 7.5 kg, contamination: artificial (dose: 0.1 mg T-2 toxin (labeled)/kg b. wt., intubated, once), conc.: 4.0 ppb* ** ***, country: USA[318], *calculated residue level, **T-2 toxin and/ or metabolites, ***after 18 h

Zearalenone
incidence: 8/8*, sa. const.: female hybrid Large Polish × Polish White gilts, age: 120–125 days, wt.: ≈49.2 kg, contamination: no ZEA (for detailed information please see the article), conc.: nd, country: Poland[418], *control
incidence: 8?/8, sa. const.: female hybrid Large Polish × Polish White gilts, age: 120–125 days, wt.: ≈49.2 kg, contamination: artificial (dose: **200 μg ZEA**/kg b. wt., o., for 7 days; for detailed information please see the article), conc. range: ≤8.07 ng/ml* (mean value), country: Poland[418], *after 5.5 h on day 1 of ZEA-administration (also measured after 2 and 7.5 h, lowest conc.: 0.93 ng/ml after 7.5 h)
incidence: 8?/8, sa. const.: female hybrid Large Polish × Polish White gilts, age: 120–125 days, wt.: ≈49.2 kg, contamination: artificial (dose: **400 μg ZEA**/kg b. wt., o., for 7 days; for detailed information please see the article), conc. range: ≤7.22 ng/ml* (mean value), country: Poland[418], *after 5.5 h on day 1 of ZEA-administration (also measured after 2 and 7.5 h, lowest conc.: 2.50 ng/ml after 7.5 h)

incidence: 8/8*, sa. const.: female hybrid Large Polish × Polish White gilts, age: 120–125 days, wt.: ≈49.2 kg, contamination: no ZEA (for detailed

information please see the article), conc.: nd, country: Poland[418], *control
incidence: 8?/8, sa. const.: female hybrid Large Polish × Polish White gilts, age: 120–125 days, wt.: ≈49.2 kg, contamination: artificial (dose: **200 μg ZEA**/kg b. wt., o., for 7 days; for detailed information please see the article), conc. range: ≤9.92 ng/ml* (mean value), country: Poland[418], *after 5 days of ZEA-administration (also after other day intervals up to 7 days measured, lowest conc.: 0.10 ng/ml after 4 days)
incidence: 8?/8, sa. const.: female hybrid Large Polish × Polish White gilts, age: 120–125 days, wt.: ≈49.2 kg, contamination: artificial (dose: **400 μg ZEA**/kg b. wt., o., for 7 days; for detailed information please see the article), conc. range: ≤14.45 ng/ml* (mean value), country: Poland[418], *after 7 days of ZEA-administration (also after other day intervals up to 7 days measured, lowest conc.: 2.34 ng/ml after 4 days)

α-Zearalenol
incidence: 8/8*, sa. const.: female hybrid Large Polish × Polish White gilts, age: 120–125 days, wt.: ≈49.2 kg, contamination: no ZEA (for detailed information please see the article), conc.: nd, country: Poland[418], *control
incidence: 8?/8, sa. const.: female hybrid Large Polish × Polish White gilts, age: 120–125 days, wt.: ≈49.2 kg, contamination: artificial (dose: **200 μg ZEA**/kg b. wt., o., for 7 days; for detailed information please see the article), conc. range: ≤9.43 ng/ml* (mean value), country: Poland[418], *after 5.5 h on day 1 of ZEA-administration (also measured after 2 and 7.5 h, lowest conc.: 6.97 ng/ml after 7.5 h)
incidence: 8?/8, sa. const.: female hybrid Large Polish × Polish White gilts, age: 120–125 days, wt.: ≈49.2 kg, contamination: artificial (dose: **400 μg ZEA**/kg b. wt., o., for 7 days; for detailed information please see the article), conc. range: ≤24.13 ng/ml* (mean value), country: Poland[418], *after 2 h on day 1 of ZEA-administration (also measured after 5.5 and 7.5 h, lowest conc.: 12.84 ng/ml after 7.5 h)

incidence: 8/8*, sa. const.: female hybrid Large Polish × Polish White gilts, age: 120–125 days, wt.: ≈49.2 kg, contamination: no ZEA (for detailed information please see the article), conc.: nd, country: Poland[418], *control
incidence: 8?/8, sa. const.: female hybrid Large Polish × Polish White gilts, age: 120–125 days, wt.: ≈49.2 kg, contamination: artificial (dose: **200 μg ZEA**/kg b. wt., o., for 7 days; for detailed information please see the article), conc. range: ≤8.11 ng/ml* (mean value), country: Poland[418], *after 1 day of ZEA-administration (also at other day intervals up to 7 days measured, lowest conc.: 2.86 ng/ml after 7 days)
incidence: 8?/8, sa. const.: female hybrid Large Polish × Polish White gilts, age: 120–125 days, wt: ≈49.2 kg, contamination: artificial (dose: **400 μg ZEA**/kg b. wt., o., for 7 days; for detailed information please see the article), conc. range: ≤17.79 ng/ml* (mean value), country: Poland[418], *after 1 day of ZEA-administration (also at other day intervals up to 7 days measured, lowest conc.: 4.48 ng/ml after 7 days)

Pig bone marrow may contain the following mycotoxins and/or their metabolites:

HT-2 Toxin
incidence: 2/2, sa. const.: female crossbred Yorkshire × Hampshire swines, wt.: 20 kg, contamination: artificial (dose: 0.15 mg T-2 toxin (labeled and unlabeled)/kg b. wt., i.vs., once), conc. range: 1.16–1.35 ng/g*, Ø conc.: 1.255 ng/g*, country: USA[425], *after 4 h

3′-Hydroxy HT-2 Toxin
incidence: 2/2, sa. const.: female crossbred Yorkshire × Hampshire swines, wt.: 20 kg, contamination: artificial (dose: 0.15 mg T-2 toxin (labeled and unlabeled)/kg b. wt., i.vs.,

once), conc. range: 1.67–2.51 ng/g*, Ø conc.: 2.09 ng/g*, country: USA[425], *after 4 h

T-2 Toxin
incidence: 2/2, sa. const.: female crossbred Yorkshire × Hampshire swines, wt.: 20 kg, contamination: artificial (dose: 0.15 mg T-2 toxin (labeled and unlabeled)/kg b. wt., i.vs., once), conc. range: 0.38–0.58 ng/g*, Ø conc.: 0.48 ng/g*, country: USA[425], *after 4 h
incidence: 2/2, sa. const.: female crossbred Yorkshire × Hampshire swines, wt.: 20 kg, contamination: artificial (dose: 0.15 mg T-2 toxin (labeled and unlabeled)/kg b. wt., i.vs., once), conc. range: 18–25 ng/g*, Ø conc.: 21.50 ng/g* **, country: USA[425], *after 4 h, ****total metabolites**

3′-Hydroxy T-2 Toxin
incidence: 2/2, sa. const.: female crossbred Yorkshire × Hampshire swines, wt.: 20 kg, contamination: artificial (dose: 0.15 mg T-2 toxin (labeled and unlabeled)/kg b. wt., i.vs., once), conc. range: 0.40–1.32 ng/g*, Ø conc.: 0.86 ng/g*, country: USA[425], *after 4 h

T-2 Tetraol
incidence: 1/2, sa. const.: female crossbred Yorkshire × Hampshire swines, wt.: 20 kg, contamination: artificial (dose: 0.15 mg T-2 toxin (labeled and unlabeled)/kg b. wt., i.vs., once), conc.: 0.04 ng/g*, country: USA[425], *after 4 h

Deepoxy T-2 Tetraol
incidence: 2/2, sa. const.: female crossbred Yorkshire × Hampshire swines, wt.: 20 kg, contamination: artificial (dose: 0.15 mg T-2 toxin (labeled and unlabeled)/kg b. wt., i.vs., once), conc. range: 0.66–0.74 ng/g*, Ø conc.: 0.70 ng/g*, country: USA[425], *after 4 h

T-2 Triol
incidence: 2/2, sa. const.: female crossbred Yorkshire × Hampshire swines, wt.: 20 kg, contamination: artificial (dose: 0.15 mg T-2 toxin (labeled and unlabeled)/kg b. wt., i. vs., once), conc. range: 0.07–0.15 ng/g*, Ø conc.: 0.11 ng/g*, country: USA[425], *after 4 h

Pig brain may contain the following mycotoxins:

Aflatoxin B_1
incidence: 1/1, sa. const.: female Hampshire × Deutsches Edelschwein piglet, wt.: 15 kg, contamination: artificial (dose: 3.1 μg AFB_1 (labeled)/kg b. wt., o., once; for detailed information please see the article), conc.: <0.2 ppb* **, country: Switzerland[66], *AFB_1 eq., **after **24 h**
incidence: 1/1, sa. const.: female Hampshire × Deutsches Edelschwein piglet, wt.: 15 kg, contamination: artificial (dose: 3.1 μg AFB_1 (labeled)/kg b. wt., o., once; for detailed information please see the article), conc.: <0.2 ppb*, country: Switzerland[66], *AFB_1 eq., **after **48 h**

Deoxynivalenol
incidence: 5/5*, sa. const.: healthy Yorkshire barrows, age: ≈11–15 weeks, wt.: 17–22 kg, contamination: no DON, conc.: nd, country: Canada[407], *control
incidence: ?/4, sa. const.: healthy Yorkshire barrows, age: ≈11–15 weeks, wt.: 17–22 kg, contamination: artificial (dose: **1.0 mg DON**/kg b. wt., i.v., once), conc. range: ≤54.9 ng/g* (mean value), country: Canada[407], *after 0.33 h (also measured after 1, 3, 8 and 24 h, lowest conc.: nd after 24 h)

Fumonisin B_1
incidence: 6/6*, sa. const.: castrated pigs of identical genotype, wt.: ≈12–14 kg, contamination: no FB_1 (for detailed information please see the article), conc.: nd, country: Germany/Hungary[109], *control
incidence: 8/13, sa. const.: castrated pigs of identical genotype, wt.: ≈12–14 kg, contamination: artificial (dose: **100 mg FB_1** daily, o., for 5–11 days; for detailed information please see the article), conc. range: 1–1,860 μg/kg*, Ø conc.: 240.5 μg/kg*, country: Germany/Hungary[109], *on 6th day of the experiment

HT-2 Toxin
incidence: 2/2, sa. const.: female crossbred Yorkshire × Hampshire swines, wt.: 20 kg,

contamination: artificial (dose: 0.15 mg T-2 toxin (labeled and unlabeled)/kg b. wt., i.vs., once), conc. range: 0.10–0.13 ng/g*, Ø conc.: 0.115 ng/g*, country: USA[425], *after 4 h

3′-Hydroxy HT-2 Toxin
incidence: 2/2, sa. const.: female crossbred Yorkshire × Hampshire swines, wt.: 20 kg, contamination: artificial (dose: 0.15 mg T-2 toxin (labeled and unlabeled)/kg b. wt., i.vs., once), conc. range: 0.07–0.22 ng/g*, Ø conc.: 0.145 ng/g*, country: USA[425], *after 4 h

Ochratoxin A
incidence: 2/2*, sa. const.: pigs, contamination: no OTA and/or DON (for detailed information please see the article), conc.: nd, country: Germany[378], *control
incidence: 5/6, sa. const.: pigs, contamination: artificial (dose: **0.1 ppm crystalline OTA + 1.0 ppm crystalline DON**, o., twice daily for 90 days; for detailed information please see the article), conc. range: 0.14–0.34 ng/g*, Ø conc.: 0.224 ng/g*, country: Germany[378], *after 90 days
incidence: 2/3, sa. const.: pigs, contamination: artificial (dose: **0.1 ppm crystalline OTA**, o., twice daily for 90 days; for detailed information please see the article), conc. range: 0.14–0.20 ng/g*, Ø conc.: 0.17 ng/g*, country: Germany[378], *after 90 days
incidence: 6/6, sa. const.: pigs, contamination: artificial (dose: **1.0 ppm crystalline DON**, o., twice daily for 90 days; for detailed information please see the article), conc.: nd*, country: Germany[378], *after 90 days

T-2 Toxin
incidence: 2/2, sa. const.: female crossbred Yorkshire × Hampshire swines, wt.: 20 kg, contamination: artificial (dose: 0.15 mg T-2 toxin (labeled and unlabeled)/kg b. wt., i.vs., once), conc. range: 0.02–0.05 ng/g*, Ø conc.: 0.035 ng/g*, country: USA[425], *after 4 h
incidence: 2/2, sa. const.: female crossbred Yorkshire × Hampshire swines, wt.: 20 kg, contamination: artificial (dose: 0.15 mg T-2 toxin (labeled and unlabeled)/kg b. wt., i.vs., once), conc. range: 8–13 ng/g*, Ø conc.: 10.5 ng/g* **, country: USA[425], *after 4 h, **__total metabolites__

3′-Hydroxy T-2 Toxin
incidence: 2/2, sa. const.: female crossbred Yorkshire × Hampshire swines, wt.: 20 kg, contamination: artificial (dose: 0.15 mg T-2 toxin (labeled and unlabeled)/kg b. wt., i. vs., once), conc. range: 0.07–0.10 ng/g*, Ø conc.: 0.085 ng/g*, country: USA[425], *after 4 h

Deepoxy T-2 Tetraol
incidence: 2/2, sa. const.: female crossbred Yorkshire × Hampshire swines, wt.: 20 kg, contamination: artificial (dose: 0.15 mg T-2 toxin (labeled and unlabeled)/kg b. wt., i. vs., once), conc. range: 0.28–0.32 ng/g*, Ø conc.: 0.30 ng/g*, country: USA[425], *after 4 h

T-2 Triol
incidence: 2/2, sa. const.: female crossbred Yorkshire × Hampshire swines, wt.: 20 kg, contamination: artificial (dose: 0.15 mg T-2 toxin (labeled and unlabeled)/kg b. wt., i. vs., once), conc. range: 0.03–0.05 ng/g*, Ø conc.: 0.04 ng/g*, country: USA[425], *after 4 h

Pig cecum may contain the following mycotoxins and/or their metabolites:

Deoxynivalenol + Deepoxydeoxynivalenol
incidence: ?/11, sa. const.: castrated male pigs, Ø wt.: 88.1 kg, contamination: artificial (dose: 4.2 mg DON/kg, o., for 7 days), conc. range: ≈≤660 μg/g* (mean value), country: Germany[413], *≈5 h after final DON-administration (also at other hour intervals up to 24 h measured, lowest conc.: ≈65 μg/g after 24 h)

Pig colon may contain the following mycotoxins and/or their metabolites:

Deoxynivalenol
incidence: 9/9, sa. const.: male and female crossbred piglets, age: 5 weeks, contamination: artificial (dose: 5.8 ppm DON in the diet, for 1–5 weeks; for detailed

information please see the article), conc. range: tr–327 ppb* **, country: Canada/ USA[70], *in colon contents (pigs fed 0.7 or 3.1 ppm DON = no data), **measured at 1st, 4th, and 5th week (pr. residue values are each lowest and highest value of 1st to 5th week measurement)

Deoxynivalenol + Deepoxydeoxynivalenol
incidence: ?/11, sa. const.: castrated male pigs, Ø wt.: 88.1 kg, contamination: artificial (dose: 4.2 mg DON/kg, o., for 7 days), conc. range: ≈≤1,020 μg/g* (mean value), country: Germany[413], *≈5 h after final DON-administration (also at other hour intervals up to 24 h measured, lowest conc.: ≈150 μg/g after 24 h)

T-2 Toxin
incidence: 2?/2*, sa. const.: female swines of mixed breeding, wt.: 26–66 kg, contamination: no T-2 toxin (for detailed information please see the article), conc.: nr**, country: USA[403], *control, **in spiral colon contents
incidence: 1/?, sa. const.: female swines of mixed breeding, wt.: 26–66 kg, contamination: artificial (dose: **2.4 mg T-2 toxin**/kg, i.a., once), conc.: 66 ppb* **, country: USA[403], *in spiral colon contents, **after ≈19.5 h

Pig duodenum may contain the following mycotoxins and/or their metabolites:

HT-2 Toxin
incidence: 2/2, sa. const.: female crossbred Yorkshire × Hampshire swines, wt.: 20 kg, contamination: artificial (dose: 0.15 mg T-2 toxin (labeled and unlabeled)/kg b. wt., i. vs., once), conc. range: 1.80–2.84 ng/g*, Ø conc.: 2.32 ng/g*, country: USA[425], *after 4 h
incidence: 2/2, sa. const.: female crossbred Yorkshire × Hampshire swines, wt.: 20 kg, contamination: artificial (dose: 0.15 mg T-2 toxin (labeled and unlabeled)/kg b. wt., i. vs., once), conc. range: 1.76–3.75 ng/g* **, Ø conc.: 2.755 ng/g* **, country: USA[425], *in **duodenum contents**, **after 4 h

3′-Hydroxy HT-2 Toxin
incidence: 2/2, sa. const.: female crossbred Yorkshire × Hampshire swines, wt.: 20 kg, contamination: artificial (dose: 0.15 mg T-2 toxin (labeled and unlabeled)/kg b. wt., i. vs., once), conc. range: 2.87–5.83 ng/g*, Ø conc.: 4.35 ng/g*, country: USA[425], *after 4 h
incidence: 2/2, sa. const.: female crossbred Yorkshire × Hampshire swines, wt.: 20 kg, contamination: artificial (dose: 0.15 mg T-2 toxin (labeled and unlabeled)/kg b. wt., i. vs., once), conc. range: 4.08–7.76 ng/g* **, Ø conc.: 5.92 ng/g* **, country: USA[425], *in **duodenum contents**, **after 4 h

T-2 Toxin
incidence: 2/2, sa. const.: female crossbred Yorkshire × Hampshire swines, wt.: 20 kg, contamination: artificial (dose: 0.15 mg T-2 toxin (labeled and unlabeled)/kg b. wt., ivs., once), conc. range: 0.30–0.61 ng/g*, Ø conc.: 0.455 ng/g*, country: USA[425], *after 4 h
incidence: 2/2, sa. const.: female crossbred Yorkshire × Hampshire swines, wt.: 20 kg, contamination: artificial (dose: 0.15 mg T-2 toxin (labeled and unlabeled)/kg b. wt., i.vs., once), conc. range: 32–101 ng/g*, Ø conc.: 66.5 ng/g* **, country: USA[425], *after 4 h, ****total metabolites**
incidence: 2/2, sa. const.: female crossbred Yorkshire × Hampshire swines, wt.: 20 kg, contamination: artificial (dose: 0.15 mg T-2 toxin (labeled and unlabeled)/kg b. wt., i.vs., once), conc. range: 0.22–0.68 ng/g* **, Ø conc.: 0.45 ng/g* **, country: USA[425], *in **duodenum contents**, **after 4 h
incidence: 2/2, sa. const.: female crossbred Yorkshire × Hampshire swines, wt.: 20 kg, contamination: artificial (dose: 0.15 mg T-2 toxin (labeled and unlabeled)/kg b. wt., i.vs., once), conc. range: 55–144 ng/g* **, Ø conc.: 99.5 ng/g* ** ***, country: USA[425], *in **duodenum contents**, **after 4 h, *****total metabolites**

3′-Hydroxy T-2 Toxin
incidence: 2/2, sa. const.: female crossbred Yorkshire × Hampshire swines, wt.: 20 kg, contamination: artificial (dose: 0.15 mg T-2

toxin (labeled and unlabeled)/kg b. wt., i.vs., once), conc. range: 0.51–1.42 ng/g*, Ø conc.: 0.965 ng/g*, country: USA[425], *after 4 h
incidence: 2/2, sa. const.: female crossbred Yorkshire × Hampshire swines, wt.: 20 kg, contamination: artificial (dose: 0.15 mg T-2 toxin (labeled and unlabeled)/kg b. wt., i. vs., once), conc. range: 0.55–2.10 ng/g* **, Ø conc.: 1.325 ng/g* **, country: USA[425], *in **duodenum contents,** **after 4 h

T-2 Tetraol
incidence: 2/2, sa. const.: female crossbred Yorkshire × Hampshire swines, wt.: 20 kg, contamination: artificial (dose: 0.15 mg T-2 toxin (labeled and unlabeled)/kg b. wt., i. vs., once), conc. range: 0.19–0.69 ng/g*, Ø conc.: 0.44 ng/g*, country: USA[425], *after 4 h
incidence: 2/2, sa. const.: female crossbred Yorkshire × Hampshire swines, wt.: 20 kg, contamination: artificial (dose: 0.15 mg T-2 toxin (labeled and unlabeled)/kg b. wt., i.vs., once), conc. range: 0.09–0.13 ng/g* **, Ø conc.: 0.11 ng/g* **, country: USA[425], *in **duodenum contents,** **after 4 h

Deepoxy T-2 Tetraol
incidence: 2/2, sa. const.: female crossbred Yorkshire × Hampshire swines, wt.: 20 kg, contamination: artificial (dose: 0.15 mg T-2 toxin (labeled and unlabeled)/kg b. wt., i.vs., once), conc. range: 0.62–1.59 ng/g*, Ø conc.: 1.105 ng/g*, country: USA[425], *after 4 h
incidence: 2/2, sa. const.: female crossbred Yorkshire × Hampshire swines, wt.: 20 kg, contamination: artificial (dose: 0.15 mg T-2 toxin (labeled and unlabeled)/kg b. wt., i. vs., once), conc. range: 0.95–3.27 ng/g* **, Ø conc.: 2.11 ng/g* **, country: USA[425], *in **duodenum contents,** **after 4 h

T-2 Triol
incidence: 1/2, sa. const.: female crossbred Yorkshire × Hampshire swines, wt: 20 kg, contamination: artificial (dose: 0.15 mg T-2 toxin (labeled and unlabeled)/kg b. wt., i.vs., once), conc.: 0.40 ng/g*, country: USA[425], *after 4 h
incidence: 2/2, sa. const.: female crossbred Yorkshire × Hampshire swines, wt.: 20 kg, contamination: artificial (dose: 0.15 mg T-2 toxin (labeled and unlabeled)/kg b. wt., i. vs., once), conc. range: 0.13–0.33 ng/g* **, Ø conc.: 0.23 ng/g* **, country: USA[425], *in **duodenum contents,** **after 4 h

Deepoxy T-2 Triol
incidence: 2/2, sa. const.: female crossbred Yorkshire × Hampshire swines, wt.: 20 kg, contamination: artificial (dose: 0.15 mg T-2 toxin (labeled and unlabeled)/kg b. wt., i.vs., once), conc.: nd*, country: USA[425], *after 4 h
incidence: 1/2, sa. const.: female crossbred Yorkshire × Hampshire swines, wt.: 20 kg, contamination: artificial (dose: 0.15 mg T-2 toxin (labeled and unlabeled)/kg b. wt., i. vs., once), conc.: 0.18 ng/g* **, country: USA[425], *in **duodenum contents,** **after 4 h

Pig eye may contain the following mycotoxins and/or their metabolites:

Fumonisin B_1
incidence: 6/6*, sa. const.: castrated pigs of identical genotype, weight: ≈12–14 kg, contamination: no FB_1 (for detailed information please see the article), conc.: nd, country: Germany/Hungary[109], *control
incidence: 12/13, sa. const.: castrated pigs of identical genotype, wt.: ≈12–14 kg, contamination: artificial (dose: **100 mg FB_1** daily, o., for 5–11 days; for detailed information please see the article), conc. range: 8–226 µg/kg*, Ø conc.: 57.31 µg/kg*, country: Germany/Hungary[109], *on 6th day of the experiment

Pig fat may contain the following mycotoxins and/or their metabolites:

Aflatoxin B_1
incidence: 1/1, sa. const.: female Hampshire × Deutsches Edelschwein piglet, wt.: 15 kg, contamination: artificial (dose: 3.1 µg AFB_1 (labeled)/kg b. wt., o., once; for detailed information please see the article), conc.: 0.5 ppb* **, country: Switzerland[66], *AFB_1 eq., **after **24 h**
incidence: 1/1, sa. const.: female Hampshire × Deutsches Edelschwein piglet, wt.: 15 kg, contamination: artificial

(dose: 3.1 μg AFB_1 (labeled)/kg b. wt., o., once; for detailed information please see the article), conc.: 0.2 ppb* **, country: Switzerland[66], *AFB_1 eq., **after **48 h**

incidence: 20/20*, sa. const.: male and female Danish Landrace pigs, age: ≈8 weeks, wt.: 20 kg up to slaughter at 90 kg, contamination: no AFB_1 + AFB_2 (for detailed information please see the article), conc.: nd, country: Denmark/USA[101], *control
incidence: 7/18*, sa. const.: male and female Danish Landrace pigs, age: ≈8 weeks, wt.: 20 kg up to slaughter at 90 kg, contamination: artificial (dose: AFB_1 + AFB_2 (overall **300 ppb AFs**), o., for **120–164 days****; for detailed information please see the article), conc. range: tr** ***, country: Denmark/USA[101], *livers of some of these pigs rejected at meat inspection, **after 120–164 days, ***in adipose tissue
incidence: 10/18*, sa. const.: male and female Danish Landrace pigs, age: ≈8 weeks, wt.: 20 kg up to slaughter at 90 kg, contamination: artificial (dose: AFB_1 + AFB_2 addition (overall **500 ppb AFs**), o., for **135–216 days****; for detailed information please see the article), conc. range: tr** ***, country: Denmark/USA[101] *livers of some of these pigs rejected at meat inspection, **after 135–216 days, ***in adipose tissue

incidence: 5/5*, sa. const.: cross-bred pigs, contamination: (dose: 9 ng/g AFB_1 + AFB_2, o., for 35 days; for detailed information please see the article), conc.: nd, country: USA[597], *control
incidence: 5?/5, sa. const.: cross-bred pigs, contamination: artificial (dose: **524 ng/g AFB_1 + AFB_2**, o., for 35 days; for detailed information please see the article), conc.: 0.030 ng/g* (mean value), country: USA[597], *after **35 days**
incidence: 5?/5, sa. const.: cross-bred pigs, contamination: artificial (dose: **524 ng/g AFB_1 + AFB_2 + HSCAS** (0.5%), o., for 35 days; for detailed information please see the article), conc.: 0.010 ng/g* (mean value), country: USA[597], *after **35 days**
incidence: 5/5*, sa. const.: male miniature pigs, contamination: (dose: 1 ng/g AFB_1 + AFB_2, o., for 15 days; for detailed information please see the article), conc.: nd, country: USA[597], *control
incidence: 5?/5, sa. const.: male miniature pigs, contamination: artificial (dose: **590 ng/g AFB_1 + AFB_2**, o., for 15 days; for detailed information please see the article), conc.: 0.060 ng/g* (mean value), country: USA[597], *after **15 days**
incidence: 5?/5, sa. const.: male miniature pigs, contamination: artificial (dose: **590 ng/g AFB_1 + AFB_2 + HSCAS** (0.5%), o., for 15 days; for detailed information please see the article), conc.: 0.020 ng/g* (mean value), country: USA[597], *after **15 days**
incidence: 5/5, sa. const.: male miniature pigs, contamination: artificial (dose: **590 ng/g AFB_1 + AFB_2**, o., for 15 days followed by **2 weeks control diet**; for detailed information please see the article), conc.: 0.010 ng/g* (mean value), country: USA[597], *after **29 days** (thereof 15 days of AFB_1- and AFB_2-administration)

Aflatoxin B_2
incidence: 20/20*, sa. const.: male and female Danish Landrace pigs, age: ≈8 weeks, wt.: 20 kg up to slaughter at 90 kg, contamination: no AFB_1 + AFB_2 (for detailed information please see the article), conc.: nd, country: Denmark/USA[101], *control
incidence: 7/18*, sa. const.: male and female Danish Landrace pigs, age: ≈8 weeks, wt.: 20 kg up to slaughter at 90 kg, contamination: artificial (dose: AFB_1 + AFB_2 addition (overall **300 ppb AFs**), o., for 120–231 days**; for detailed information please see the article), conc.: nd** ***, country: Denmark/USA[101], *livers of some of these pigs rejected at meat inspection, **after 120–231 days, ***in adipose tissue

incidence: 10/18*, sa. const.: male and female Danish Landrace pigs, age: ≈8 weeks, wt.: 20 kg up to slaughter at 90 kg, contamination: artificial (dose: AFB_1 + AFB_2 addition (overall **500 ppb AFs**), o., for 120–231 days**; for detailed information please see the article), conc.: nd** ***, country: Denmark/USA[101] *livers of some of these pigs rejected at meat inspection, **after 120–231 days, ***in adipose tissue

incidence: 5/5*, sa. const.: cross-bred pigs, contamination: (dose: 9 ng/g AFB_1 + AFB_2, o., for 35 days; for detailed information please see the article), conc.: nd, country: USA[597], *control
incidence: 5/5, sa. const.: cross-bred pigs, contamination: artificial (dose: **524 ng/g AFB_1 + AFB_2**, o., for 35 days; for detailed information please see the article), conc.: nd*, country: USA[597], *after **35 days**
incidence: 5/5, sa. const.: cross-bred pigs, contamination: artificial (dose: **524 ng/g AFB_1 + AFB_2 + HSCAS** (0.5%), o., for 35 days; for detailed information please see the article), conc.: nd*, country: USA[597], *after **35 days**

incidence: 5/5*, sa. const.: male miniature pigs, contamination: (dose: 1 ng/g AFB_1 + AFB_2, o., for 15 days; for detailed information please see the article), conc.: nd, country: USA[597], *control
incidence: 5/5, sa. const.: male miniature pigs, contamination: artificial (dose: **590 ng/g AFB_1 + AFB_2**, o., for 15 days; for detailed information please see the article), conc.: nd*, country: USA[597], *after **15 days**
incidence: 5/5, sa. const.: male miniature pigs, contamination: artificial (dose: **590 ng/g AFB_1 + AFB_2 + HSCAS** (0.5%), o., for 15 days; for detailed information please see the article), conc.: nd*, country: USA[597], *after **15 days**
incidence: 5/5, sa. const.: male miniature pigs, contamination: artificial (dose: **590 ng/g AFB_1 + AFB_2**, o., for 15 days followed by **2 weeks control diet**; for detailed information please see the article), conc.: nd*, country: USA[597], *after **29 days** (thereof 15 days of AFB_1- and AFB_2-administration)

Aflatoxin M_1
incidence: 5/5*, sa. const.: cross-bred pigs, contamination: (dose: 9 ng/g AFB_1 + AFB_2, o., for 35 days; for detailed information please see the article), conc.: nd, country: USA[597], *control

incidence: 5?/5, sa. const.: cross-bred pigs, contamination: artificial (dose: **524 ng/g AFB_1 + AFB_2**, o., for 35 days; for detailed information please see the article), conc.: 0.010 ng/g* (mean value), country: USA[597], *after **35 days**
incidence: 5/5, sa. const.: cross-bred pigs, contamination: artificial (dose: **524 ng/g AFB_1 + AFB_2 + HSCAS** (0.5%), o., for 35 days; for detailed information please see the article), conc.: nd*, country: USA[597], *after **35 days**

incidence: 5/5*, sa. const.: male miniature pigs, contamination: (dose: 1 ng/g AFB_1 + AFB_2, o., for 15 days; for detailed information please see the article), conc.: nd, country: USA[597], *control
incidence: 5?/5, sa. const.: male miniature pigs, contamination: artificial (dose: **590 ng/g AFB_1 + AFB_2**, o., for 15 days; for detailed information please see the article), conc.: 0.070 ng/g* (mean value), country: USA[597], *after **15 days**
incidence: 5/5, sa. const.: male miniature pigs, contamination: artificial (dose: **590 ng/g AFB_1 + AFB_2 + HSCAS** (0.5%), o., for 15 days; for detailed information please see the article), conc.: nd*, country: USA[597], *after **15 days**
incidence: 5/5, sa. const.: male miniature pigs, contamination: artificial (dose: **590 ng/g AFB_1 + AFB_2**, o., for 15 days followed by **2 weeks control diet**; for detailed information please see the article), conc.: nd*, country: USA[597], *after **29 days** (thereof 15 days of AFB_1- and AFB_2-administration)

Aflatoxin M

incidence: 20/20*, sa. const.: male and female Danish Landrace pigs, age: ≈8 weeks, wt.: 20 kg up to slaughter at 90 kg, contamination: no $AFB_1 + AFB_2$ (for detailed information please see the article), conc.: nd, country: Denmark/USA[101], *control

incidence: 7/18*, sa. const.: male and female Danish Landrace pigs, age: ≈8 weeks, wt.: 20 kg up to slaughter at 90 kg, contamination: artificial (dose: $AFB_1 + AFB_2$ addition (overall **300 ppb AFs**), o., for 120–231 days**; for detailed information please see the article), conc.: nd** ***, country: Denmark/USA[101], *livers of some of these pigs rejected at meat inspection, **after 120–231 days, ***in adipose tissue

incidence: 10/18*, sa. const.: male and female Danish Landrace pigs, age: ≈8 weeks, wt.: 20 kg up to slaughter at 90 kg, contamination: artificial (dose: $AFB_1 + AFB_2$ addition (overall **500 ppb AFs**), o., for 120–231 days**; for detailed information please see the article), conc.: nd** ***, country: Denmark/USA[101] *livers of some of these pigs rejected at meat inspection, **after 120–231 days, ***in adipose tissue

Deoxynivalenol

incidence: 6/6*, sa. const.: barrows (Yorkshire), wt.: ≈25 kg, contamination: no DON (for detailed information please see the article), conc.: nd, country: Canada[406], *control

incidence: 50?/50, sa. const.: barrows (Yorkshire), wt.: ≈25 kg, contamination: artificial (dose: **6.0 mg natural DON**/kg dry weight, o., for **3 weeks** (feeding trial 1); for detailed information please see the article), conc. range: ≤51.3 ng/g* **, country: Canada[406], *in back fat, **after 3 weeks

incidence: 6/6*, sa. const.: barrows (Yorkshire), wt.: ≈25 kg, contamination: no DON (for detailed information please see the article), conc.: nd, country: Canada[406], *control

incidence: 6?/6, sa. const.: barrows (Yorkshire), wt.: ≈25 kg, contamination: artificial (dose: **6.0 mg crystalline DON/** kg dry weight, o., for **4 weeks** (feeding trial 2); for detailed information please see the article), conc. range: ≤22.7 ng/g* **, country: Canada[406], *in back fat, **after 4 weeks

incidence: 6/6*, sa. const.: barrows (Yorkshire), wt.: ≈25 kg, contamination: no DON (for detailed information please see the article), conc.: nd, country: Canada[406], *control

incidence: 6?/6, sa. const.: barrows (Yorkshire), wt.: ≈25 kg, contamination: artificial (dose: **7.6 mg natural DON**/kg dry weight, o., for **7 weeks** (feeding trial 3); for detailed information please see the article), conc. range: ≤19.5 ng/g* **, country: Canada[406], *in back fat, **after 7 weeks

incidence: 5/5*, sa. const.: healthy Yorkshire barrows, age: ≈11–15 weeks, wt.: 17–22 kg, contamination: no DON, conc.: nd, country: Canada[407], *control

incidence: ?/4, sa. const.: healthy Yorkshire barrows, age: ≈11–15 weeks, wt.: 17–22 kg, contamination: artificial (dose: **1.0 mg DON**/kg b. wt., i.v., once), conc. range: ≤491.6 ng/g* ** (mean value), country: Canada[407], *after 1 h (also measured after 0.33, 3, 8 and 24 h, lowest conc.: 3.4 ng/g after 24 h), **in **abdominal fat**

incidence: ?/25, sa. const.: healthy Yorkshire barrows, age: ≈11–15 weeks, wt.: 17–22 kg, contamination: artificial (dose: **1.0 mg DON**/kg b. wt., i.v., once), conc. range: ≤294.7 ng/g* ** (mean value), country: Canada[407], *after 1 h (also measured after 0.33, 3, 8 and 24 h, lowest conc.: 12.4 ng/g after 24 h), **in **back fat**

Fumonisin B_1

incidence: 5/5*, sa. const.: weaned barrows of the same genotype, wt.: 12–14 kg, contamination: no FB_1, $FB_2 + FB_3$ (for detailed information please see the article), conc.: nd**, country: Hungary/Germany[87], *control, **neither in abdominal nor in subcutaneous fat

incidence: 9/10, sa. const.: weaned barrows of the same genotype, wt.: 12–14 kg, contamination: artificial (dose: **50 mg FB_1, 20 mg FB_2 + 5 mg FB_3**/animal, o., for 22 days; for detailed information please see the article), conc. range: 0.8–111.2 ng/g* **, Ø conc.: 16.27 ng/g* **, country: Hungary/Germany[87], *in **abdominal fat**, **after 22 days toxin feeding period
incidence: 8/10, sa. const.: weaned barrows of the same genotype, wt.: 12–14 kg, contamination: artificial (dose: **50 mg FB_1, 20 mg FB_2 + 5 mg FB_3**/animal, o., for 22 days; for detailed information please see the article), conc. range: 0.8–8.8 ng/g* **, Ø conc.: 3.4 ng/g* **, country: Hungary/Germany[87], *in **subcutaneous fat**, **after 22 days toxin feeding period

incidence: 6/6*, sa. const.: castrated pigs of identical genotype, wt.: ≈12–14 kg, contamination: no FB_1 (for detailed information please see the article), conc.: nd, country: Germany/Hungary[109], *control
incidence: 6/13, sa. const.: castrated pigs of identical genotype, wt.: ≈12–14 kg, contamination: artificial (dose: **100 mg FB_1** daily, o., for 5–11 days; for detailed information please see the article), conc. range: 1–11 µg/kg*, Ø conc.: 4.83 µg/kg*, country: Germany/Hungary[109], *on 6th day of the experiment

Fumonisin B_2
incidence: 5/5*, sa. const.: weaned barrows of the same genotype, wt.: 12–14 kg, contamination: no FB_1, FB_2 + FB_3 (for detailed information please see the article), conc.: nd**, country: Hungary/Germany[87], *control, **neither in abdominal nor in subcutaneous fat
incidence: 5/10, sa. const.: weaned barrows of the same genotype, wt.: 12–14 kg, contamination: artificial (dose: **50 mg FB_1, 20 mg FB_2 + 5 mg FB_3**/animal, o., for 22 days; for detailed information please see the article), conc. range: 1.6–16 ng/g* **, Ø conc.: 5.28 ng/g* **, country: Hungary/Germany[87], *in **abdominal fat**, **after 22 days toxin feeding period
incidence: 7/10, sa. const.: weaned barrows of the same genotype, wt.: 12–14 kg, contamination: artificial (dose: **50 mg FB_1, 20 mg FB_2 + 5 mg FB_3**/animal, o., for 22 days; for detailed information please see the article), conc. range: 0.8–3.2 ng/g* **, Ø conc.: 1.71 ng/g* **, country: Hungary/Germany[87], *in **subcutaneous fat**, **after 22 days toxin feeding period

Ochratoxin A
incidence: 5/5*, sa. const.: female pigs of Danish Landrace, age: ≈8 weeks, wt.: ≈20 kg, contamination: no OTA (for detailed information please see the article), conc.: nd, country: Denmark/USA[102], *control
incidence: ?/5, sa. const.: female pigs of Danish Landrace, age: ≈8 weeks, wt.: ≈20 kg, contamination: artificial (dose: 1 ppm crystalline OTA, o., once daily for 1 month, afterwards toxin-free diets for various intervals; for detailed information please see the article), conc.: 5.95 µg/kg* (mean value), country: Denmark/USA[102], ***1 day after termination** of OTA-exposure
incidence: ?/5, sa. const.: female pigs of Danish Landrace, age: ≈8 weeks, wt.: ≈20 kg, contamination: artificial (dose: 1 ppm crystalline OTA, o., once daily for 1 month, afterwards toxin-free diets for various intervals; for detailed information please see the article), conc.: 2.54 µg/kg* (mean value), country: Denmark/USA[102], ***8 days after termination of OTA-exposure**
incidence: 5/5, sa. const.: female pigs of Danish Landrace, age: ≈8 weeks, wt.: ≈20 kg, contamination: artificial (dose: 1 ppm crystalline OTA, o., once daily for 1 month, afterwards toxin-free diets for various intervals; for detailed information please see the article), conc.: nd* (mean value), country: Denmark/USA[102], ***15 days after termination** of OTA-exposure

incidence: 5/5, sa. const.: female pigs of Danish Landrace, age: ≈8 weeks, wt.: ≈20 kg, contamination: artificial (dose: 1 ppm crystalline OTA, o., once daily for 1 month, afterwards toxin-free diets for various intervals; for detailed information please see the article), conc.: nd* (mean value), country: Denmark/USA[102], ***29 days after termination** of OTA-exposure

incidence: ?/?, sa. const.: weaners (specific pathogen free), wt.: 14–18 kg, contamination: artificial (dose: ? μg OTA addition; for detailed information please see the article), conc. range: ≤17 μg/kg*, country: Denmark[204], *calculated value based on the amount of OTA in blood after 24 h on toxin-free diet

incidence: 13/13*, sa. const.: castrated pigs, weight: 20 kg, contamination: no OTA and/or CIT (for detailed information please see the article), conc.: nd, country: Denmark/USA[330], *control
incidence: 1/1?, sa. const.: castrated pigs, weight: 20 kg, contamination: artificial (dose: **1,400 μg crystalline OTA**/kg feed, o., for 6 weeks; for detailed information please see the article), conc.: 4 μg/kg* ** (mean value), country: Denmark/USA[330], *in leaf fat, **after 6 weeks
incidence: 1/1?, sa. const.: castrated pigs, weight: 20 kg, contamination: artificial (dose: **650 μg crystalline CIT**/kg feed, o., for 6 weeks; for detailed information please see the article), conc.: nd* **, country: Denmark/USA[330], *in leaf fat, **after 6 weeks
incidence: 1/1?, sa. const.: castrated pigs, weight: 20 kg, contamination: artificial (dose: **1,400 μg crystalline OTA + 650 μg crystalline CIT**/kg feed, o., for 6 weeks; for detailed information please see the article), conc.: 3 μg/kg* **, (mean value), country: Denmark/USA[330], *in leaf fat, **after 6 weeks
incidence: 1/1?, sa. const.: castrated pigs, weight: 20 kg, contamination: artificial (dose: **1,400 μg natural OTA + 650 μg natural CIT**/kg feed, o., for 6 weeks; for detailed information please see the article), conc.: 8 μg/kg* ** (mean value), country: Denmark/USA[330], *in leaf fat, **after 6 weeks

incidence: 4/4*, sa. const.: male and females pigs, wt.: ≈70 kg, contamination: no OTA (for detailed information please see the article), conc.: nd, country: Germany[365], *control
incidence: 8?/8, sa. const.: male and females pigs, wt.: ≈70 kg, contamination: artificial (dose: **0.09 mg natural OTA**/kg diet/day, in the morning and evening **half of OTA-ration**, o., for 28 days; for detailed information please see the article), conc.: 6.56 ng/g* ** (mean value), country: Germany[365], *in **leaf fat**, **after 28 days
incidence: 8?/8, sa. const.: male and females pigs, wt.: ≈70 kg, contamination: artificial (dose: **0.09 mg crystalline OTA**/kg diet/day, in the morning and evening **half of OTA-ration**, o., for 28 days; for detailed information please see the article), conc.: 1.81 ng/g* ** (mean value), country: Germany[365], *in **leaf fat**, **after 28 days
incidence: 8?/8, sa. const.: male and females pigs, wt.: ≈70 kg, contamination: artificial (dose: **0.09 mg crystalline OTA**/kg diet/day, in the morning **total OTA-ration**, o., for 28 days; for detailed information please see the article), conc.: 2.49 ng/g* ** (mean value), country: Germany[365], *in **leaf fat**, **after 28 days

incidence: 4/4*, sa. const.: male and females pigs, wt.: ≈70 kg, contamination: no OTA (for detailed information please see the article), conc.: nd, country: Germany[365], *control
incidence: 8?/8, sa. const.: male and females pigs, wt.: ≈70 kg, contamination: artificial (dose: **0.09 mg natural OTA**/kg diet/day, in the morning and evening **half of OTA-ration**, o., for 28 days; for detailed information please see the article), conc.: 5.58 ng/g* ** (mean value), country: Germany[365], *in **neck fat**, **after 28 days

incidence: 8?/8, sa. const.: male and females pigs, wt.: ≈70 kg, contamination: artificial (dose: **0.09 mg crystalline OTA/kg** diet/day, in the morning and evening **half of OTA-ration**, o., for 28 days; for detailed information please see the article), conc.: 1.54 ng/g* ** (mean value), country: Germany[365], *in **neck fat**, **after 28 days
incidence: 8?/8, sa. const.: male and females pigs, wt.: ≈70 kg, contamination: artificial (dose: **0.09 mg crystalline OTA/** kg diet/day, in the morning **total OTA-ration**, o., for 28 days; for detailed information please see the article), conc.: 1.86 ng/g* ** (mean value), country: Germany[365], *in **neck fat**, **after 28 days

incidence: 2/2*, sa. const.: pigs, contamination: no OTA and/or DON (for detailed information please see the article), conc.: nd, country: Germany[378], *control
incidence: 6/6, sa. const.: pigs, contamination: artificial (dose: **0.1 ppm crystalline OTA + 1.0 ppm crystalline DON**, o., twice daily for 90 days; for detailed information please see the article), conc. range: 0.31–1.11 ng/g*, Ø conc.: 0.828 ng/g*, country: Germany[378], *after 90 days
incidence: 3/3, sa. const.: pigs, contamination: artificial (dose: **0.1 ppm crystalline OTA**, o., twice daily for 90 days; for detailed information please see the article), conc. range: 0.56–0.82 ng/g*, Ø conc.: 0.696 ng/g*, country: Germany[378], *after 90 days
incidence: 6/6, sa. const.: pigs, contamination: artificial (dose: **1.0 ppm crystalline DON**, o., twice daily for 90 days; for detailed information please see the article), conc. range: nd*, country: Germany[378], *after 90 days

incidence: 2?/2*, sa. const.: pigs, contamination: no OTA and/or ZEA (for detailed information please see the article), conc.: nr, country: Germany[380], *control
incidence: 6/6, sa. const.: pigs, contamination: artificial (dose: **0.1 ppm OTA + 0.25 ppm ZEA**, o., twice daily for 90 days; for detailed information please see the article), conc. range: ≤2.31 ng/g*, country: Germany[380], *after 91 days (thereof 90 days of OTA- and ZEA-administration)

incidence: 2/3, sa. const.: pigs, contamination: artificial (dose: **0.1 ppm OTA**, o., twice daily for 90 days; for detailed information please see the article), conc. range: ≤1.41 ng/g*, country: Germany[380], *after 91 days (thereof 90 days of OTA-administration)

incidence: 3/3*, sa. const.: female pigs of Danish Landrace, age: 8–10 weeks, wt.: ≈20 kg, contamination: no OTA (for detailed information please see the article), conc.: nd, country: Denmark/USA/Sweden[383], *control
incidence: 4?/4, sa. const.: female pigs of Danish Landrace, age: 8–10 weeks, wt.: ≈20 kg, contamination: artificial (dose: **1 mg crystalline OTA/kg** feed, o., for 3 months; for detailed information please see the article), conc.: ≈2.5 µg/kg (mean value), country: Denmark/USA/Sweden[383], *after ≈3 months

T-2 Toxin
incidence: 1/1, sa. const.: female weanling crossbred Yorkshire × Duroc × Hampshire swine, wt.: 7.5 kg, contamination: artificial (dose: 0.1 mg T-2 toxin (labeled)/kg b. wt., intubated, once), conc.: 4.9 ppb* ** ***, country: USA[318], *calculated residue level, **T-2 toxin and/or metabolites, ***after 18 h

α-Zearalenol
incidence: 2?/2?*, sa. const.: male KAHYP pigs, wt.: ?, contamination: no ZEA, conc.: nd, country: Hungary[632], *control
incidence: 2?/2, sa. const.: male KAHYP pigs, wt.: 60 kg, contamination: artificial (dose: 15 ppm ZEA, o., for 14 days, conc.: 10 µg/kg* **, country: Hungary[632], *after 14 days, **in adipose tissues

β-Zearalenol
incidence: 2?/2?*, sa. const.: male KAHYP pigs, wt.: ?, contamination: no ZEA, conc.: nd, country: Hungary[632], *control
incidence: 2?/2, sa. const.: male KAHYP pigs, wt.: 60 kg, contamination: artificial (dose: **15 ppm ZEA**, o., for 14 days, conc.: pr* **, country: Hungary[632], *after 14 days, **in adipose tissues

Pig Fat Around Kidney see Pig kidney

Pig feces may contain the following mycotoxins and/or their metabolites:

Deoxynivalenol
incidence: 5?/5, sa. const.: castrated male Swedish Landrace pigs, wt.: ≈20 kg, contamination: artificial (dose: 2.5 mg 3-aDON/kg feed, o., 5 times in 2.5 days), conc.: pr, country: Sweden[416]

incidence: 16/16*, sa. const.: barrows, wt.: 28 kg, contamination: artificial (dose: 0.09 mg DON/kg wheat (but DON not intended; wheat roportion in the diet 0%); for detailed information please see the article), conc.: nd, country: Germany[484], *control
incidence: ?/16, sa. const.: barrows, wt.: 28 kg, contamination: artificial (dose: **2.64 mg natural DON**/kg wheat (**wheat proportion** in the diet **17.5%**), o., for 70 days?; for detailed information please see the article), conc.: 0.009 mg/kg* (mean value), country: Germany[484], *after 13 weeks
incidence: ?/16, sa. const.: barrows, wt.: 28 kg, contamination: artificial (dose: **4.41 mg natural DON**/kg wheat (**wheat proportion** in the diet **35%**), o., for 70 days?; for detailed information please see the article), conc.: 0.020 mg/kg* (mean value), country: Germany[484], *after 13 weeks

incidence: 5/5*, sa. const.: male castrated and female fattening pigs (Deutsches Bundeshybridzuchtprogramm), wt.: 34.1–103.7 kg, contamination: artificial (dose: **0.15 mg DON**/kg wheat, contaminated (proportion in the diet 0%), o., for 7 days?; for detailed information please see the article), conc.: nd, country: Austria/Germany[530], *control
incidence: ?/5, sa. const.: male castrated and female fattening pigs (Deutsches Bundeshybridzuchtprogramm), wt.: 34.1–103.7 kg, contamination: artificial (dose: **3.86 mg DON**/kg wheat, contaminated (proportion in the diet 40%), o., for 7 days?; for detailed information please see the article), conc. range: ≤0.018 mg/kg* **(mean value), country: Austria/Germany[530], *in freeze dried feces, **at the end of the collection period

incidence: 7?/7, sa. const.: Yorkshire barrows, wt.: ≈35 kg, contamination: artificial (dose: **0.19 mg DON**/kg diet dry matter, o., for 2 weeks (diet C); for detailed information please see the article), conc.: 0.05 mg* (mean value), country: Canada[606], *in a 5-days collection period

incidence: 7?/7, sa. const.: Yorkshire barrows, wt.: ≈35 kg, contamination: artificial (dose: **4.66 mg DON**/kg diet dry matter, o., for 1 week (diet C in the first week); for detailed information please see the article), conc.: 0.25 mg* (mean value), country: Canada[606], *in a 5-days collection period

Deepoxydeoxynivalenol
incidence: 5?/5, sa. const.: castrated male Swedish Landrace pigs, wt.: ≈20 kg, contamination: artificial (dose: 2.5 mg 3-aDON/kg feed, o., 5 times in 2.5 days), conc.: pr, country: Sweden[416]

incidence: 5?/5*, sa. const.: male castrated and female fattening pigs (Deutsches Bundeshybridzuchtprogramm), wt.: 34.1–103.7 kg, contamination: artificial (dose: **0.15 mg DON**/kg wheat, contaminated (proportion in the diet 0%), o., for 7 days?; for detailed information please see the article), conc. range: ≤0.020 mg/kg** *** (mean value), country: Austria/Germany[530], *control, **in freeze dried feces, ***at the end of the collection period

incidence: 5?/5, sa. const.: male castrated and female fattening pigs (Deutsches Bundeshybridzuchtprogramm), wt.: 34.1–103.7 kg, contamination: artificial (dose: **3.86 mg DON/kg** wheat, contaminated (proportion in the diet 40%), o., for 7 days?; for detailed information please see the article), conc. range: ≤0.459 mg/kg* *** (mean value), country: Austria/Germany[530], *in freeze dried feces, **at the end of the collection period

incidence: 7?/7, sa. const.: Yorkshire barrows, wt.: ≈35 kg, contamination: artificial (dose: **0.19 mg DON**/kg diet dry matter, o., for 2 weeks (diet C); for detailed information please see the article), conc.: nd*, country: Canada[606], *daily
incidence: 7?/7, sa. const.: Yorkshire barrows, wt.: ≈35 kg, contamination: artificial (dose: **4.66 mg DON**/kg diet dry matter, o., for 1 week (diet C in the first week); for detailed information please see the article), conc.: 0.068 mg* (mean value), country: Canada[606], *daily

Fumonisin B_1
incidence: 5/5*, sa. const.: weaned barrows of the same genotype, wt.: 12–14 kg, contamination: no FB_1, FB_2 + FB_3 (for detailed information please see the article), conc.: nd, country: Hungary/Germany[87], *control
incidence: 10/10, sa. const.: weaned barrows of the same genotype, wt.: 12–14 kg, contamination: artificial (dose: **50 mg FB_1, 20 mg FB_2 + 5 mg FB_3**/animal, o., for 22 days; for detailed information please see the article), conc. range: 0.3–76.1 mg*, Ø conc.: 28.21 mg*, country: Hungary/Germany[87], *excretion between days 13 and 17

Fumonisin B_2
incidence: 5/5*, sa. const.: weaned barrows of the same genotype, wt.: 12–14 kg, contamination: no FB_1, FB_2 + FB_3 (for detailed information please see the article), conc.: nd, country: Hungary/Germany[87], *control
incidence: 10/10, sa. const.: weaned barrows of the same genotype, wt.: 12–14 kg, contamination: artificial (dose: **50 mg FB_1, 20 mg FB_2 + 5 mg FB_3**/animal, o., for 22 days; for detailed information please see the article), conc. range: 0.4–3.5 mg*, Ø conc.: 2.02 mg*, country: Hungary/Germany[87], *excretion between days 13 and 17

Nivalenol
incidence: 3/3, sa. const.: male castrated Swedish Landrace × Yorkshire pigs, wt.: 37–63 kg, contamination: artificial (dose: 0.05 mg NIV/kg b. wt., o., twice daily for 3 days (NIV-administration also 16 h before first blood sa. taken); for detailed information please see the article), conc. range: ≤3,034 ng/g* (mean value), country: Sweden[501], *on 3rd day

Ochratoxin A
incidence: 1/1*, sa. const.: pregnant gilt, contamination: neither OTA nor OTB, conc.: nd, country: UK[266], *control
incidence: 2/2, sa. const.: pregnant gilts, contamination: artificial (dose: **0.38 mg OTA + 0.13 mg OTB**/kg b. wt., o., for 8 days during early pregnancy), conc. range: 1.3–8.1 µg/g* ** dry matter, Ø conc.: 4.7 µg/g* ** dry matter, country: UK[266], *very approximately, **measured during dosing

Ochratoxin α
incidence: 1/1*, sa. const.: pregnant gilt, contamination: neither OTA nor OTB, conc.: nd, country: UK[266], *control
incidence: 2/2, sa. const.: pregnant gilts, contamination: artificial (dose: **0.38 mg OTA + 0.13 mg OTB**/kg b. wt., o., for 8 days during early pregnancy), conc. range: 100.4–143.2 µg/g* dry matter, Ø conc.: 121.8 µg/g* dry matter, country: UK[266], *measured during dosing

Ochratoxin ß
incidence: 1/1*, sa. const.: pregnant gilt, contamination: neither OTA nor OTB, conc.: nd, country: UK[266], *control

incidence: 2/2, sa. const.: pregnant gilts, contamination: artificial (dose: **0.38 mg OTA + 0.13 mg OTB**/kg b. wt., o., for 8 days during early pregnancy), conc. range: 200.2–201.5 µg/g* dry matter, Ø conc.: 200.85 µg/g* dry matter, country: UK[266], *measured during dosing

Zearalenone
incidence: 2?/2?*, sa. const.: male KAHYP pigs, wt.: ?, contamination: no ZEA, conc.: nd, country: Hungary[632], *control
incidence: 2?/2, sa. const.: male KAHYP pigs, wt.: 60 kg, contamination: artificial (dose: **15 ppm ZEA**, o., for 14 days, conc.: 3,710 µg/kg*, country: Hungary[632], *after 13/14 days)

α-Zearalenol
incidence: 2?/2?*, sa. const.: male KAHYP pigs, wt.: ?, contamination: no ZEA, conc.: nd, country: Hungary[632], *control
incidence: 2?/2, sa. const.: male KAHYP pigs, wt.: 60 kg, contamination: artificial (dose: **15 ppm ZEA**, o., for 14 days, conc.: 15,750 µg/kg*, country: Hungary[632], *after 13/14 days)

β-Zearalenol
incidence: 2?/2?*, sa. const.: male KAHYP pigs, wt.: ?, contamination: no ZEA, conc.: nd, country: Hungary[632], *control
incidence: 2?/2, sa. const.: male KAHYP pigs, wt.: 60 kg, contamination: artificial (dose: **15 ppm ZEA**, o., for 14 days, conc.: 4,140 µg/kg*, country: Hungary[632], *after 13/14 days)

Pig gallbladder may contain the following mycotoxins and/or their metabolites:

Aflatoxin B_1
incidence: 2?/2, sa. const.: adult swines, contamination: artificial (dose: 1.08–1.09 mg AFB_1 (besides other AF), o., daily for 33 days; for detailed information please see the article), conc.: 0.04 ppb* **, country: France[88], *value expressed for total bile contents, **after 33 days

Aflatoxin B_2
incidence: 2?/2, sa. const.: adult swines, contamination: artificial (dose: 1.08–1.09 mg AFB_1 (besides other AF), o., daily for 33 days; for detailed information please see the article), conc.: 0.02 ppb* **, country: France[88], *value expressed for total bile contents, **after 33 days

incidence: ?/?*, sa. const.: Large White growing female pigs, age: 79 days, wt.: 20 kg, contamination: no AFs (for detailed information please see the article), conc.: nd, country: France[314], *control
incidence: 1/1, sa. const.: Large White growing female pigs, age: 79 days, wt.: 20 kg, contamination: artificial (dose: natural AFs addition (treatment 1: mixed feeding, avg. daily intake of **870 µg AFB_1**) for 26 days; for detailed information please see the article), conc.: nd*, country: France[314], *final wt. of the animal 109 kg
incidence: 1/1, sa. const.: Large White growing female pigs, age: 79 days, wt.: 20 kg, contamination: artificial (dose: natural AFs addition (treatment 2: separate feeding = peanut oil meal 40% and corn gluten meal 30%, avg. daily intake of **1,566 µg AFB_1**) for 19 days; for detailed information please see the article), conc.: 0.143 µg/kg*, country: France[314], *final wt. of the animal 97 kg
incidence: 1/1, sa. const.: Large White growing female pigs, age: 79 days, wt.: 20 kg, contamination: artificial (dose: natural AFs addition (treatment 3: separate feeding = peanut oil meal 0% and corn gluten meal 0%, avg. daily intake of **642 µg AFB_1**) for 26 days; for detailed information please see the article), conc.: nd*, country: France[314], *final wt. of the animal 104 kg

Aflatoxin M_1
incidence: 2?/2, sa. const.: adult swines, contamination: artificial (dose: 1.08–1.09 mg AFB_1 (besides other AF), o., daily for 33 days; for detailed information please see the article), conc.: 0.02 ppb* **, country: France[88], *value

expressed for total bile contents, *after 33 days

Aflatoxin M

incidence: ?/?*, sa. const.: Large White growing female pigs, age: 79 days, wt.: 20 kg, contamination: no AFs (for detailed information please see the article), conc.: nd, country: France[314], *control
incidence: 1/1, sa. const.: Large White growing female pigs, age: 79 days, wt.: 20 kg, contamination: artificial (dose: natural AFs addition (treatment 1: mixed feeding, avg. daily intake of **870 μg AFB_1**) for 26 days; for detailed information please see the article), conc.: nd*, country: France[314], *final wt. of the animal 109 kg
incidence: 1/1, sa. const.: Large White growing female pigs, age: 79 days, wt.: 20 kg, contamination: artificial (dose: natural AFs addition (treatment 2: separate feeding = peanut oil meal 40% and corn gluten meal 30%, avg. daily intake of **1,566 μg AFB_1**) for 19 days; for detailed information please see the article), conc.: 0.176 μg/kg*, country: France[314], *final wt. of the animal 97 kg
incidence: 1/1, sa. const.: Large White growing female pigs, age: 79 days, wt.: 20 kg, contamination: artificial (dose: natural AFs addition (treatment 3: separate feeding = peanut oil meal 0% and corn gluten meal 0%, avg. daily intake of **642 μg AFB_1**) for 26 days; for detailed information please see the article), conc.: nd*, country: France[314], *final wt. of the animal 104 kg

Pig heart may contain the following mycotoxins and/or their metabolites:

Aflatoxin B_1

incidence: 1/1, sa. const.: female Hampshire × Deutsches Edelschwein piglet, wt.: 15 kg, contamination: artificial (dose: 3.1 μg AFB_1 (labeled)/kg b. wt., o., once; for detailed information please see the article), conc.: 0.5 ppb* **, country: Switzerland[66], *AFB_1 eq., **after **24 h**

incidence: 1/1, sa. const.: female Hampshire × Deutsches Edelschwein piglets, wt.: 15 kg, contamination: artificial (dose: 3.1 μg AFB_1 (labeled)/kg b. wt., o., once; for detailed information please see the article), conc.: 0.5 ppb* **, country: Switzerland[66], *AFB_1 eq., **after **48 h**

incidence: 1/2, sa. const.: adult swines, contamination: artificial (dose: 1.08–1.09 mg AFB_1 (besides other AF), o., daily for 33 days; for detailed information please see the article), conc.: 0.92 ppb*, country: France[88], *after 33 days

incidence: 20/20*, sa. const.: male and female Danish Landrace pigs, age: ≈8 weeks, wt.: 20 kg up to slaughter at 90 kg, contamination: no AFB_1 + AFB_2 (for detailed information please see the article), conc.: nd, country: Denmark/USA[101], *control
incidence: 1/16*, sa. const.: male and female Danish Landrace pigs, age: ≈8 weeks, wt.: 20 kg up to slaughter at 90 kg, contamination: artificial (dose: AFB_1 + AFB_2 addition (overall **300 ppb AFs**), o., for **141 days****; for detailed information please see the article), conc.: tr**, country: Denmark/USA[101], *livers of some of these pigs were rejected at meat inspection, **after 141 days
incidence: 4/17*, sa. const.: male and female Danish Landrace pigs, age: ≈8 weeks, wt.: 20 kg up to slaughter at 90 kg, contamination: artificial (dose: AFB_1 + AFB_2 addition (overall **500 ppb AFs**), o., for **141–216 days****; for detailed information please see the article), conc. range: tr**, country: Denmark/USA[101], *livers of some of these pigs were rejected at meat inspection, after 141–216 days**

incidence: 4/4*, sa. const.: crossbred (Duroc × Yorkshire) barrows, wt.: 24.5–26.3 kg, contamination: no AFs, conc.: nd, country: USA[138], *control
incidence: 4?/4, sa. const.: crossbred (Duroc × Yorkshire) barrows, wt.: 24.5–26.3 kg, contamination: artificial

(dose: **662 µg AFB_1/kg** diet, **273 µg AFB_2/kg** diet, **300 µg AFG_1/kg** diet and **285 µg AFG_2/kg** diet, o., for 21 days), conc. range: 0.05–1.41 µg/kg*, country: USA[138], *after ≈22 days (thereof 21 days of AFs-administration)

incidence: ?/?*, sa. const.: Large White growing female pigs, age: 79 days, wt.: 20 kg, contamination: no AFs (for detailed information please see the article), conc.: nd, country: France[314], *control
incidence: 1/1, sa. const.: Large White growing female pigs, age: 79 days, wt.: 20 kg, contamination: artificial (dose: natural AFs addition (treatment 1: mixed feeding, avg. daily intake of **870 µg AFB_1**) for 26 days; for detailed information please see the article), conc.: nd*, country: France[314], *final wt. of the animal 109 kg
incidence: 1/1, sa. const.: Large White growing female pigs, age: 79 days, wt.: 20 kg, contamination: artificial (dose: natural AFs addition (treatment 2: separate feeding = peanut oil meal 40% and corn gluten meal 30%, avg. daily intake of **1,566 µg AFB_1**) for 19 days; for detailed information please see the article), conc.: 1.17 µg/kg*, country: France[314], *final wt. of the animal 97 kg
incidence: 1/1, sa. const.: Large White growing female pigs, age: 79 days, wt.: 20 kg, contamination: artificial (dose: natural AFs addition (treatment 3: separate feeding = peanut oil meal 0% and corn gluten meal 0%, avg. daily intake of **642 µg AFB_1**) for 26 days; for detailed information please see the article), conc.: nd*, country: France[314], *final wt. of the animal 104 kg

Aflatoxin B_2
incidence: 1/2, sa. const.: adult swines, contamination: artificial (dose: 1.08–1.09 mg AFB_1 (besides other AF), o., daily for 33 days; for detailed information please see the article), conc.: 0.34 ppb*, country: France[88], *after 33 days

incidence: 20/20*, sa. const.: male and female Danish Landrace pigs, age: ≈8 weeks, wt.: 20 kg up to slaughter at 90 kg, contamination: no AFB_1 + AFB_2 (for detailed information please see the article), conc.: nd, country: Denmark/USA[101], *control
incidence: 16/16*, sa. const.: male and female Danish Landrace pigs, age: ≈8 weeks, wt.: 20 kg up to slaughter at 90 kg, contamination: artificial (dose: AFB_1 + AFB_2 addition (overall **300 ppb AFs**), o., for **120–231 days****; for detailed information please see the article), conc.: nd**, country: Denmark/USA[101], *livers of some of these pigs were rejected at meat inspection, **after 120–231 days
incidence: 2/17*, sa. const.: male and female Danish Landrace pigs, age: ≈8 weeks, wt.: 20 kg up to slaughter at 90 kg, contamination: artificial (dose: AFB_1 + AFB_2 addition (overall **500 ppb AFs**), o., for **150 or 186 days****; for detailed information please see the article), conc. range: tr**, country: Denmark/USA[101], *livers of some of these pigs were rejected at meat inspection, **after 150 or 186 days

incidence: 4/4*, sa. const.: crossbred (Duroc × Yorkshire) barrows, wt.: 24.5–26.3 kg, contamination: no AFs, conc.: nd, country: USA[138], *control
incidence: 4?/4, sa. const.: crossbred (Duroc × Yorkshire) barrows, wt.: 24.5–26.3 kg, contamination: artificial (dose: **662 µg AFB_1/kg** diet, **273 µg AFB_2/kg** diet, **300 µg AFG_1/kg** diet and **285 µg AFG_2/kg** diet, o., for 21 days), conc. range: tr–0.14 µg/kg*, country: USA[138], *after ≈22 days (thereof 21 days of AF-administration)

incidence: ?/?*, sa. const.: Large White growing female pigs, age: 79 days, wt.: 20 kg, contamination: no AFs (for detailed information please see the article), conc.: nd, country: France[314], *control
incidence: 1/1, sa. const.: Large White growing female pigs, age: 79 days, wt.: 20 kg, contamination: artificial (dose: natural AFs addition (treatment 1: mixed feeding, avg. daily intake of **870 µg AFB_1**)

for 26 days; for detailed information please see the article), conc.: nd*, country: France[314], *final wt. of the animal 109 kg
incidence: 1/1, sa. const.: Large White growing female pigs, age: 79 days, wt.: 20 kg, contamination: artificial (dose: natural AFs addition (treatment 2: separate feeding = peanut oil meal 40% and corn gluten meal 30%, avg. daily intake of **1,566 µg AFB_1**) for 19 days; for detailed information please see the article), conc.: 1.17 µg/kg*, country: France[314], *final wt. of the animal 97 kg
incidence: 1/1, sa. const.: Large White growing female pigs, age: 79 days, wt.: 20 kg, contamination: artificial (dose: natural AFs addition (treatment 3: separate feeding = peanut oil meal 0% and corn gluten meal 0%, avg. daily intake of **642 µg AFB_1**) for 26 days; for detailed information please see the article), conc.: nd*, country: France[314], *final wt. of the animal 104 kg

Aflatoxin B_{2a}
incidence: 4/4*, sa. const.: crossbred (Duroc × Yorkshire) barrows, weight: 24.5–26.3 kg, contamination: no AFs fed, conc.: nd, country: USA[138], *control
incidence: 4?/4, sa. const.: crossbred (Duroc × Yorkshire) barrows, wt.: 24.5–26.3 kg, contamination: artificial (**662 µg AFB_1**/kg diet, **273 µg AFB_2**/kg diet, **300 µg AFG_1**/kg diet and **285 µg AFG_2**/kg diet, o., for 21 days), conc. range: appreciable amounts*, country: USA[138], *after ≈22 days (thereof 21 days of AF-administration)

Aflatoxin M_1
incidence: 1/2, sa. const.: adult swines, contamination: artificial (dose: 1.08–1.09 mg AFB_1 (besides other AF), o., daily for 33 days; for detailed information please see the article), conc.: 0.18 ppb*, country: France[88], *after 33 days

incidence: 4/4*, sa. const.: crossbred (Duroc × Yorkshire) barrows, wt.: 24.5–26.3 kg, contamination: no AFs, conc.: nd, country: USA[138], *control
incidence: 4?/4, sa. const.: crossbred (Duroc × Yorkshire) barrows, wt.: 24.5–26.3 kg, contamination: artificial (dose: **662 µg AFB_1**/kg diet, **273 µg AFB_2**/kg diet, **300 µg AFG_1**/kg diet and **285 µg AFG_2**/kg diet, o., for 21 days), conc. range: 0.05–0.54 µg/kg*, country: USA[138], *after ≈22 days (thereof 21 days of AF-administration)

Aflatoxin M
incidence: 20/20*, sa. const.: male and female Danish Landrace pigs, age: ≈8 weeks, wt.: 20 kg up to slaughter at 90 kg, contamination: no AFB_1 + AFB_2 (for detailed information please see the article), conc.: nd, country: Denmark/USA[101], *control
incidence: 16/16*, sa. const.: male and female Danish Landrace pigs, age: ≈8 weeks, wt.: 20 kg up to slaughter at 90 kg, contamination: artificial (dose: AFB_1 + AFB_2 addition (overall **300 ppb AFs**), o., for **120–231 days****; for detailed information please see the article), conc.: nd**, country: Denmark/USA[101], *livers of some of these pigs were rejected at meat inspection, **after 120–231 days
incidence: 1/17*, sa. const.: male and female Danish Landrace pigs, age: ≈8 weeks, wt.: 20 kg up to slaughter at 90 kg, contamination: artificial (dose: AFB_1 + AFB_2 addition (overall **500 ppb AFs**), o., for **150 days****; for detailed information please see the article), conc.: tr**, country: Denmark/USA[101], *livers of some of these pigs were rejected at meat inspection, **after 150 days

incidence: ?/?*, sa. const.: Large White growing female pigs, age: 79 days, wt.: 20 kg, contamination: no AFs (for detailed information please see the article), conc.: nd, country: France[314], *control
incidence: 1/1, sa. const.: Large White growing female pigs, age: 79 days, wt.: 20 kg, contamination: artificial (dose: natural AFs addition (treatment 1: mixed feeding, avg. daily intake of **870 µg AFB_1**) for 26 days; for detailed information

please see the article), conc.: nd*, country: France[314], *final wt. of the animal 109 kg
incidence: 1/1, sa. const.: Large White growing female pigs, age: 79 days, wt.: 20 kg, contamination: artificial (dose: natural AFs addition (treatment 2: separate feeding = peanut oil meal 40% and corn gluten meal 30%, avg. daily intake of **1,566 µg AFB_1**) for 19 days; for detailed information please see the article), conc.: 0.15 µg/kg*, country: France[314], *final wt. of the animal 97 kg
incidence: 1/1, sa. const.: Large White growing female pigs, age: 79 days, wt.: 20 kg, contamination: artificial (dose: natural AFs addition (treatment 3: separate feeding = peanut oil meal 0% and corn gluten meal 0%, avg. daily intake of **642 µg AFB_1**) for 26 days; for detailed information please see the article), conc.: nd*, country: France[314], *final wt. of the animal 104 kg

Deoxynivalenol
incidence: 4/4*, sa. const.: barrows and gilts, Ø wt.: 7.7 kg, contamination: no DON (for detailed information please see the article), conc.: nd, country: USA[64], *control
incidence: ?/4, sa. const.: barrows and gilts, Ø wt.: 7.7 kg, contamination: artificial (dose: **0.9 ppm DON** (analyzed value) in the diet, o., for 3 weeks; for detailed information please see the article), conc.: 18 ppb* (mean value), country: USA[64], *after 3 weeks post-treatment
incidence: ?/4, sa. const.: barrows and gilts, Ø wt.: 7.7 kg, contamination: artificial (dose: **2.0 ppm DON** (analyzed value) in the diet, o., for 3 weeks; for detailed information please see the article), conc.: 9 ppb* (mean value), country: USA[64], *after 3 weeks post-treatment
incidence: ?/4, sa. const.: barrows and gilts, Ø wt.: 7.7 kg, contamination: artificial (dose: **2.8 ppm DON** (analyzed value) in the diet, o., for 3 weeks; for detailed information please see the article), conc.: 10 ppb* (mean value), country: USA[64], *after 3 weeks post-treatment

incidence: 5/5*, sa. const.: healthy Yorkshire barrows, age: ≈11–15 weeks, wt.: 17–22 kg, contamination: no DON, conc.: nd, country: Canada[407], *control
incidence: ?/4, sa. const.: healthy Yorkshire barrows, age: ≈11–15 weeks, wt.: 17–22 kg, contamination: artificial (dose: **1.0 mg DON/** kg b. wt., i.v., once), conc. range: ≤48.0 ng/g* (mean value), country: Canada[407], *after 1 h (also measured after 0.33, 3, 8, and 24 h, lowest conc.: nd after 24 h)

HT-2 Toxin
incidence: 2/2, sa. const.: female crossbred Yorkshire × Hampshire swines, wt.: 20 kg, contamination: artificial (dose: 0.15 mg T-2 toxin (labeled and unlabeled)/kg b. wt., i. vs., once), conc. range: 0.48–0.68 ng/g*, Ø conc.: 0.58 ng/g*, country: USA[425], *after 4 h

3′-Hydroxy HT-2 Toxin
incidence: 2/2, sa. const.: female crossbred Yorkshire × Hampshire swines, wt.: 20 kg, contamination: artificial (dose: 0.15 mg T-2 toxin (labeled and unlabeled)/kg b. wt., i.vs., once), conc. range: 0.80–1.23 ng/g*, Ø conc.: 1.015 ng/g*, country: USA[425], *after 4 h

Ochratoxin A
incidence: 4/4*, sa. const.: male and females pigs, wt.: ≈70 kg, contamination: no OTA (for detailed information please see the article), conc.: nd, country: Germany[365], *control
incidence: 8?/8, sa. const.: male and females pigs, wt.: ≈70 kg, contamination: artificial (dose: **0.09 mg natural OTA**/kg diet/day, in the morning and evening **half of OTA-ration**, o., for 28 days; for detailed information please see the article), conc.: 19.29 ng/g* (mean value), country: Germany[365], *after 28 days
incidence: 8?/8, sa. const.: male and females pigs, wt.: ≈70 kg, contamination: artificial (dose: **0.09 mg crystalline OTA**/kg diet/day, in the morning and

evening **half of OTA-ration**, o., for 28 days; for detailed information please see the article), conc.: 5.05 ng/g* (mean value), country: Germany[365], *after 28 days
incidence: 8?/8, sa. const.: male and females pigs, wt.: ≈70 kg, contamination: artificial (dose: **0.09 mg crystalline OTA**/kg diet/day, in the morning **total OTA-ration**, o., for 28 days; for detailed information please see the article), conc.: 6.28 ng/g* (mean value), country: Germany[365], *after 28 days

incidence: 8?/8*, sa. const.: pigs, contamination: no OTA (for detailed information please see the article), conc.: <0.39 μg/kg (mean value), country: Germany[366], *control
incidence: ?/8, sa. const.: pigs, contamination: artificial (dose: **22.11 mg OTA** (in total), o., for 90 days; for detailed information please see the article), conc.: 12.9 μg/kg (mean value), country: Germany[366], *after 90 days
incidence: ?/8, sa. const.: pigs, contamination: artificial (dose: **88.44 mg OTA** (in total), o., for 90 days; for detailed information please see the article), conc.: 50.1 μg/kg (mean value), country: Germany[366], *after 90 days

T-2 Toxin
incidence: 1/1, sa. const.: female weanling crossbred Yorkshire × Duroc × Hampshire swine, wt.: 7.5 kg, contamination: artificial (dose: 0.1 mg T-2 toxin (labeled)/kg b. wt., intubated, once), conc.: 3.9 ppb* ** ***, country: USA[318], *calculated residue level, **T-2 toxin and/or metabolites, ***after 18 h

incidence: 2/2, sa. const.: female crossbred Yorkshire × Hampshire swines, wt: 20 kg, contamination: artificial (dose: 0.15 mg T-2 toxin (labeled and unlabeled)/kg b. wt., i.vs., once), conc. range: 0.06–0.32 ng/g*, Ø conc.: 0.19 ng/g*, country: USA[425], *after 4 h

incidence: 2/2, sa. const.: female crossbred Yorkshire × Hampshire swines, wt.: 20 kg, contamination: artificial (dose: 0.15 mg T-2 toxin (labeled and unlabeled)/kg b. wt., i.vs., once), conc. range: 23–26 ng/g*, Ø conc.: 24.5 ng/g* **, country: USA[425], *after 4 h, ****total metabolites**

3′-Hydroxy T-2 Toxin
incidence: 2/2, sa. const.: female crossbred Yorkshire × Hampshire swines, wt.: 20 kg, contamination: artificial (dose: 0.15 mg T-2 toxin (labeled and unlabeled)/kg b. wt., i.vs., once), conc. range: 0.19–0.58 ng/g*, Ø conc.: 0.385 ng/g*, country: USA[425], *after 4 h

T-2 Tetraol
incidence: 2/2, sa. const.: female crossbred Yorkshire × Hampshire swines, wt.: 20 kg, contamination: artificial (dose: 0.15 mg T-2 toxin (labeled and unlabeled)/kg b. wt., i. vs., once), conc. range: 0.04 ng/g*, Ø conc.: 0.04 ng/g*, country: USA[425], *after 4 h

Deepoxy T-2 Tetraol
incidence: 2/2, sa. const.: female crossbred Yorkshire × Hampshire swines, wt.: 20 kg, contamination: artificial (dose: 0.15 mg T-2 toxin (labeled and unlabeled)/kg b. wt., i.vs., once), conc. range: 0.88–1.38 ng/g*, Ø conc.: 1.13 ng/g*, country: USA[425], *after 4 h

T-2 Triol
incidence: 2/2, sa. const.: female crossbred Yorkshire × Hampshire swines, wt.: 20 kg, contamination: artificial (dose: 0.15 mg T-2 toxin (labeled and unlabeled)/kg b. wt., i. vs., once), conc. range: 0.04–0.14 ng/g*, Ø conc.: 0.09 ng/g*, country: USA[425], *after 4 h

α-Zearalenol
incidence: 2?/2?*, sa. const.: male KAHYP pigs, wt.: ?, contamination: no ZEA, conc.: nd, country Hungary[632], *control
incidence: 2?/2, sa. const.: male KAHYP pigs, wt.: 60 kg, contamination: artificial (dose: **15 ppm ZEA**, o., for 14 days, conc.: <10 μg/kg*, country: Hungary[632], *after 14 days)

Pig ileum may contain the following mycotoxins and/or their metabolites:

HT-2 Toxin
incidence: 2/2, sa. const.: female crossbred Yorkshire × Hampshire swines, wt.: 20 kg, contamination: artificial (dose: 0.15 mg T-2 toxin (labeled and unlabeled)/kg b. wt., i. vs., once), conc. range: 7.09–9.75 ng/g*, Ø conc.: 8.42 ng/g*, country: USA[425], *after 4 h
incidence: 2/2, sa. const.: female crossbred Yorkshire × Hampshire swines, wt.: 20 kg, contamination: artificial (dose: 0.15 mg T-2 toxin (labeled and unlabeled)/kg b. wt., i. vs., once), conc. range: 6.77–17.69 ng/g* **, Ø conc.: 12.23 ng/g* **, country: USA[425], *in **ileum contents**, **after 4 h

Deepoxy HT-2 Toxin
incidence: 1/2, sa. const.: female crossbred Yorkshire × Hampshire swines, wt.: 20 kg, contamination: artificial (dose: 0.15 mg T-2 toxin (labeled and unlabeled)/kg b. wt., i.vs., once), conc.: 1.72 ng/g*, country: USA[425], *after 4 h
incidence: 2/2, sa. const.: female crossbred Yorkshire × Hampshire swines, wt.: 20 kg, contamination: artificial (dose: 0.15 mg T-2 toxin (labeled and unlabeled)/kg b. wt., i. vs., once), conc. range: 2.06–3.12 ng/g* **, Ø conc.: 2.59 ng/g* **, country: USA[425], *in **ileum contents**, **after 4 h

3′-Hydroxy HT-2 Toxin
incidence: 2/2, sa. const.: female crossbred Yorkshire × Hampshire swines, wt.: 20 kg, contamination: artificial (dose: 0.15 mg T-2 toxin (labeled and unlabeled)/kg b. wt., i.vs., once), conc. range: 24.83–29.27 ng/g*, Ø conc.: 27.05 ng/g*, country: USA[425], *after 4 h
incidence: 2/2, sa. const.: female crossbred Yorkshire × Hampshire swines, wt.: 20 kg, contamination: artificial (dose: 0.15 mg T-2 toxin (labeled and unlabeled)/kg b. wt., i.vs., once), conc. range: 66.70–67.56 ng/g* **, Ø conc.: 67.13 ng/g* **, country: USA[425], *in **ileum contents**, **after 4 h

T-2 Toxin
incidence: 2/2, sa. const.: female crossbred Yorkshire × Hampshire swines, wt.: 20 kg, contamination: artificial (dose: 0.15 mg T-2 toxin (labeled and unlabeled)/kg b. wt., i.vs., once), conc. range: 5.02–7.32 ng/g*, Ø conc.: 6.17 ng/g*, country: USA[425], *after 4 h
incidence: 2/2, sa. const.: female crossbred Yorkshire × Hampshire swines, wt.: 20 kg, contamination: artificial (dose: 0.15 mg T-2 toxin (labeled and unlabeled)/kg b. wt., i.vs., once), conc. range: 498–554 ng/g*, Ø conc.: 526 ng/g* **, country: USA[425], *after 4 h, ****total metabolites**
incidence: 2/2, sa. const.: female crossbred Yorkshire × Hampshire swines, wt.: 20 kg, contamination: artificial (dose: 0.15 mg T-2 toxin (labeled and unlabeled)/kg b. wt., i.vs., once), conc. range: 10.64–12.61 ng/g* **, Ø conc.: 11.625 ng/g* **, country: USA[425], *in **ileum contents**, **after 4 h
incidence: 2/2, sa. const.: female crossbred Yorkshire × Hampshire swines, wt.: 20 kg, contamination: artificial (dose: 0.15 mg T-2 toxin (labeled and unlabeled)/kg b. wt., i.vs., once), conc. range: 1,406–1,644 ng/g* **, Ø conc.: 1,525 ng/g* ** ***, country: USA[425], *in **ileum contents**, **after 4 h, *****total metabolites**

3′-Hydroxy T-2 Toxin
incidence: 2/2, sa. const.: female crossbred Yorkshire × Hampshire swines, wt.: 20 kg, contamination: artificial (dose: 0.15 mg T-2 toxin (labeled and unlabeled)/kg b. wt., i.vs., once), conc. range: 1.41–3.02 ng/g*, Ø conc.: 2.22 ng/g*, country: USA[425], *after 4 h
incidence: 2/2, sa. const.: female crossbred Yorkshire × Hampshire swines, wt.: 20 kg, contamination: artificial (dose: 0.15 mg T-2 toxin (labeled and unlabeled)/kg b. wt., i.vs., once), conc. range: 5.43–6.64 ng/g* **, Ø conc.: 6.035 ng/g* **, country: USA[425], *in **ileum contents**, **after 4 h

T-2 Tetraol
incidence: 2/2, sa. const.: female crossbred Yorkshire × Hampshire swines, wt.: 20 kg, contamination: artificial (dose: 0.15 mg T-2 toxin (labeled and unlabeled)/kg b. wt., i.vs., once), conc. range: 4.29–13.36 ng/g*, Ø conc.: 8.825 ng/g*, country: USA[425], *after 4 h
incidence: 2/2, sa. const.: female crossbred Yorkshire × Hampshire swines, wt.: 20 kg, contamination: artificial (dose: 0.15 mg T-2 toxin (labeled and unlabeled)/kg b. wt., i.vs., once), conc. range: 13.66–41.75 ng/g* **, Ø conc.: 27.705 ng/g* **, country: USA[425], *in **ileum contents**, **after 4 h

Deepoxy T-2 Tetraol
incidence: 2/2, sa. const.: female crossbred Yorkshire × Hampshire swines, wt.: 20 kg, contamination: artificial (dose: 0.15 mg T-2 toxin (labeled and unlabeled)/kg b. wt., i.vs., once), conc. range: 1.76–5.98 ng/g*, Ø conc.: 3.87 ng/g*, country: USA[425], *after 4 h
incidence: 2/2, sa. const.: female crossbred Yorkshire × Hampshire swines, wt.: 20 kg, contamination: artificial (dose: 0.15 mg T-2 toxin (labeled and unlabeled)/kg b. wt., i.vs., once), conc. range: 13.57–14.99 ng/g* **, Ø conc.: 14.28 ng/g* **, country: USA[425], *in **ileum contents**, **after 4 h

T-2 Triol
incidence: 2/2, sa. const.: female crossbred Yorkshire × Hampshire swines, wt.: 20 kg, contamination: artificial (dose: 0.15 mg T-2 toxin (labeled and unlabeled)/kg b. wt., i.vs., once), conc. range: 0.67–0.86 ng/g*, Ø conc.: 0.765 ng/g*, country: USA[425], *after 4 h
incidence: 2/2, sa. const.: female crossbred Yorkshire × Hampshire swines, wt.: 20 kg, contamination: artificial (dose: 0.15 mg T-2 toxin (labeled and unlabeled)/kg b. wt., i.vs., once), conc. range: 3.63–6.24 ng/g* **, Ø conc.: 4.935 ng/g* **, country: USA[425], *in **ileum contents**, **after 4 h

Deepoxy T-2 Triol
incidence: 2/2, sa. const.: female crossbred Yorkshire × Hampshire swines, wt.: 20 kg, contamination: artificial (dose: 0.15 mg T-2 toxin (labeled and unlabeled)/kg b. wt., i.vs., once), conc. range: 0.55–0.75 ng/g*, Ø conc.: 0.65 ng/g*, country: USA[425], *after 4 h
incidence: 1/2, sa. const.: female crossbred Yorkshire × Hampshire swines, wt.: 20 kg, contamination: artificial (dose: 0.15 mg T-2 toxin (labeled and unlabeled)/kg b. wt., i.vs., once), conc.: 1.97 ng/g* **, country: USA[425], *in **ileum contents**, **after 4 h

Pig intestine may contain the following mycotoxins and/or their metabolites:

Deoxynivalenol
incidence: 5/5*, sa. const.: healthy Yorkshire barrows, age: ≈11–15 weeks, wt.: 17–22 kg, contamination: no DON, conc.: nd, country: Canada[407], *control
incidence: ?/4, sa. const.: healthy Yorkshire barrows, age: ≈11–15 weeks, wt.: 17–22 kg, contamination: artificial (dose: **1.0 mg DON**/kg b. wt., i.v., once), conc. range: ≤20.4 ng/g* (mean value), country: Canada[407], *after 0.33 h (also measured after 1, 3, 8, and 24 h, lowest conc.: nd after 24 h)

incidence: ?/11, sa. const.: castrated male pigs, Ø wt.: 88.1 kg, contamination: artificial (dose: 4.2 mg DON/kg, o., for 7 days), conc. range: ≈≤1,300 ng/g* ** (mean value), country: Germany[413], *in small intestine, **4 h after final DON-administration (also at other hour intervals up to 24 h measured, lowest conc.: ≈20 ng/g after 24 h)

HT-2 Toxin
incidence: 2/2, sa. const.: female crossbred Yorkshire × Hampshire swines, wt.: 20 kg, contamination: artificial (dose: 0.15 mg T-2 toxin (labeled and

unlabeled)/kg b. wt., i.vs., once), conc. range: 1.50–7.20 ng/g* **, Ø conc.: 4.35 ng/g* **, country: USA[425], *in **large intestine,** **after 4 h
incidence: 2/2, sa. const.: female crossbred Yorkshire × Hampshire swines, wt.: 20 kg, contamination: artificial (dose: 0.15 mg T-2 toxin (labeled and unlabeled)/kg b. wt., i.vs., once), conc. range: 1.32–8.91 ng/g* **, Ø conc.: 5.115 ng/g* **, country: USA[425], *in **large intestine contents,** **after 4 h

Deepoxy HT-2 Toxin
incidence: 1/2, sa. const.: female crossbred Yorkshire × Hampshire swines, wt.: 20 kg, contamination: artificial (dose: 0.15 mg T-2 toxin (labeled and unlabeled)/kg b. wt., i.vs., once), conc.: 0.17 ng/g* **, country: USA[425], *in **large intestine,** **after 4 h
incidence: 12, sa. const.: female crossbred Yorkshire × Hampshire swines, wt.: 20 kg, contamination: artificial (dose: 0.15 mg T-2 toxin (labeled and unlabeled)/kg b. wt., i.vs., once), conc.: 0.31 ng/g* **, country: USA[425], *in **large intestine contents,** **after 4 h

3′-Hydroxy HT-2 Toxin
incidence: 2/2, sa. const.: female crossbred Yorkshire × Hampshire swines, wt.: 20 kg, contamination: artificial (dose: 0.15 mg T-2 toxin (labeled and unlabeled)/kg b. wt., i.vs., once), conc. range: 3.56–13.99 ng/g* **, Ø conc.: 8.775 ng/g* **, country: USA[425], *in **large intestine,** **after 4 h
incidence: 2/2, sa. const.: female crossbred Yorkshire × Hampshire swines, wt.: 20 kg, contamination: artificial (dose: 0.15 mg T-2 toxin (labeled and unlabeled)/kg b. wt., i.vs., once), conc. range: 7.02–13.49 ng/g* **, Ø conc.: 10.255 ng/g* **, country: USA[425], *in **large intestine contents,** **after 4 h

Ochratoxin A
incidence: 4/4*, sa. const.: male and females pigs, wt.: ≈70 kg, contamination: no OTA (for detailed information please see the article), conc.: nd, country: Germany[365], *control
incidence: 8?/8, sa. const.: male and females pigs, wt.: ≈70 kg, contamination: artificial (dose: **0.09 mg natural OTA**/kg diet/day, in the morning and evening **half of OTA-ration,** o., for 28 days; for detailed information please see the article), conc.: 15.68 ng/g* (mean value), country: Germany[365], *after 28 days
incidence: 8?/8, sa. const.: male and females pigs, wt.: ≈70 kg, contamination: artificial (dose: **0.09 mg crystalline OTA**/kg diet/day, in the morning and evening **half of OTA-ration,** o., for 28 days; for detailed information please see the article), conc.: 4.37 ng/g* (mean value), country: Germany[365], *after 28 days
incidence: 8?/8, sa. const.: male and females pigs, wt.: ≈70 kg, contamination: artificial (dose: **0.09 mg crystalline OTA**/kg diet/day, in the morning **total OTA-ration,** o., for 28 days; for detailed information please see the article), conc.: 5.21 ng/g* (mean value), country: Germany[365], *after 28 days

T-2 Toxin
incidence: 2/2, sa. const.: female crossbred Yorkshire × Hampshire swines, wt.: 20 kg, contamination: artificial (dose: 0.15 mg T-2 toxin (labeled and unlabeled)/kg b. wt., i.vs., once), conc. range: 0.42–5.82 ng/g* **, Ø conc.: 3.12 ng/g* **, country: USA[425], *in **large intestine,** **after 4 h
incidence: 2/2, sa. const.: female crossbred Yorkshire × Hampshire swines, wt.: 20 kg, contamination: artificial (dose: 0.15 mg T-2 toxin (labeled and unlabeled)/kg b. wt., i.vs., once), conc. range: 48–142 ng/g* **, Ø conc.: 95 ng/g* ** ***, country: USA[425], *in **large intestine,** **after 4 h, *****total metabolites**
incidence: 2/2, sa. const.: female crossbred Yorkshire × Hampshire swines, wt.: 20 kg,

contamination: artificial (dose: 0.15 mg T-2 toxin (labeled and unlabeled)/kg b. wt., i.vs., once), conc. range: 0.31–9.87 ng/g* **, Ø conc.: 5.09 ng/g* **, country: USA[425], *in **large intestine** contents, **after 4 h
incidence: 2/2, sa. const.: female crossbred Yorkshire × Hampshire swines, wt.: 20 kg, contamination: artificial (dose: 0.15 mg T-2 toxin (labeled and unlabeled)/kg b. wt., i.vs., once), conc. range: 68–181 ng/g* **, Ø conc.: 124.5 ng/g* ** ***, country: USA[425], *in **large intestine contents**, **after 4 h, ***__total metabolites__

3′-Hydroxy T-2 Toxin
incidence: 2/2, sa. const.: female crossbred Yorkshire × Hampshire swines, wt.: 20 kg, contamination: artificial (dose: 0.15 mg T-2 toxin (labeled and unlabeled)/kg b. wt., i.vs., once), conc. range: 1.45–3.65 ng/g* **, Ø conc.: 2.55 ng/g* **, country: USA[425], *in **large intestine**, **after 4 h
incidence: 1/2, sa. const.: female crossbred Yorkshire × Hampshire swines, wt.: 20 kg, contamination: artificial (dose: 0.15 mg T-2 toxin (labeled and unlabeled)/kg b. wt., i.vs., once), conc.: 4.71 ng/g* **, country: USA[425], *in **large intestine contents**, **after 4 h

Deepoxy T-2 Tetraol
incidence: 2/2, sa. const.: female crossbred Yorkshire × Hampshire swines, wt.: 20 kg, contamination: artificial (dose: 0.15 mg T-2 toxin (labeled and unlabeled)/kg b. wt., i.vs., once), conc. range: 1.02–3.06 ng/g* **, Ø conc.: 2.04 ng/g* **, country: USA[425], *in **large intestine**, **after 4 h
incidence: 2/2, sa. const.: female crossbred Yorkshire × Hampshire swines, wt.: 20 kg, contamination: artificial (dose: 0.15 mg T-2 toxin (labeled and unlabeled)/kg b. wt., i.vs., once), conc. range: 1.10–2.60 ng/g* **, Ø conc.: 1.85 ng/g* **, country: USA[425], *in **large intestine contents**, **after 4 h

T-2 Triol
incidence: 2/2, sa. const.: female crossbred Yorkshire × Hampshire swines, wt.: 20 kg, contamination: artificial (dose: 0.15 mg T-2 toxin (labeled and unlabeled)/kg b. wt., i.vs., once), conc. range: 0.33–1.38 ng/g* **, Ø conc.: 0.855 ng/g* **, country: USA[425], *in **large intestine**, **after 4 h
incidence: 2/2, sa. const.: female crossbred Yorkshire × Hampshire swines, wt.: 20 kg, contamination: artificial (dose: 0.15 mg T-2 toxin (labeled and unlabeled)/kg b. wt., i.vs., once), conc. range: 0.55–2.17 ng/g* **, Ø conc.: 1.36 ng/g* **, country: USA[425], *in **large intestine contents**, **after 4 h

Deepoxy T-2 Triol
incidence: 2/2, sa. const.: female crossbred Yorkshire × Hampshire swines, wt.: 20 kg, contamination: artificial (dose: 0.15 mg T-2 toxin (labeled and unlabeled)/kg b. wt., i.vs., once), conc.: nd* **, country: USA[425], *in **large intestine**, **after 4 h
incidence: 1/2, sa. const.: female crossbred Yorkshire × Hampshire swines, wt.: 20 kg, contamination: artificial (dose: 0.15 mg T-2 toxin (labeled and unlabeled)/kg b. wt., i.vs., once), conc.: 0.76 ng/g* **, country: USA[425], *in **large intestine contents**, **after 4 h

Pig jejunum may contain the following mycotoxins and/or their metabolites:

HT-2 Toxin
incidence: 2/2, sa. const.: female crossbred Yorkshire × Hampshire swines, wt.: 20 kg, contamination: artificial (dose: 0.15 mg T-2 toxin (labeled and unlabeled)/kg b. wt., i.vs., once), conc. range: 3.54–3.58 ng/g*, Ø conc.: 3.56 ng/g*, country: USA[425], *after 4 h
incidence: 2/2, sa. const.: female crossbred Yorkshire × Hampshire swines, wt.: 20 kg, contamination: artificial (dose: 0.15 mg T-2 toxin (labeled and unlabeled)/kg b. wt., i.vs., once), conc. range: 4.22–4.25 ng/g* **,

Ø conc.: 4.235 ng/g* **, country: USA[425], *in **jejunum contents**, **after 4 h

DEEPOXY HT-2 TOXIN
incidence: 1/2, sa. const.: female crossbred Yorkshire × Hampshire swines, wt.: 20 kg, contamination: artificial (dose: 0.15 mg T-2 toxin (labeled and unlabeled)/kg b. wt., i.vs., once), conc.: 0.40 ng/g*, country: USA[425], *after 4 h
incidence: 2/2, sa. const.: female crossbred Yorkshire × Hampshire swines, wt.: 20 kg, contamination: artificial (dose: 0.15 mg T-2 toxin (labeled and unlabeled)/kg b. wt., i.vs., once), conc.: nd* **, country: USA[425], *in **jejunum contents**, **after 4 h

3′-HYDROXY HT-2 TOXIN
incidence: 2/2, sa. const.: female crossbred Yorkshire × Hampshire swines, wt.: 20 kg, contamination: artificial (dose: 0.15 mg T-2 toxin (labeled and unlabeled)/kg b. wt., i.vs., once), conc. range: 7.36–7.38 ng/g*, Ø conc.: 7.37 ng/g*, country: USA[425], *after 4 h

incidence: 2/2, sa. const.: female crossbred Yorkshire × Hampshire swines, wt.: 20 kg, contamination: artificial (dose: 0.15 mg T-2 toxin (labeled and unlabeled)/kg b. wt., i.vs., once), conc. range: 12.23–15.85 ng/g* **, Ø conc.: 14.04 ng/g* **, country: USA[425], *in **jejunum contents**, **after 4 h

T-2 TOXIN
incidence: 2?/2*, sa. const.: female swines of mixed breeding, wt.: 26–66 kg, contamination: no T-2 toxin (for detailed information please see the article), conc.: nr, country: USA[403], *control
incidence: 1/?, sa. const.: female swines of mixed breeding, wt.: 26–66 kg, contamination: artificial (dose: 2.4 mg T-2 toxin/kg, i.a., once), conc.: 374 ppb* **, country: USA[403], *in **jejunum contents**, **after ≈19.5 h

incidence: 2/2, sa. const.: female crossbred Yorkshire × Hampshire swines, wt.: 20 kg, contamination: artificial (dose: 0.15 mg T-2 toxin (labeled and unlabeled)/kg b. wt., i.vs., once), conc. range: 1.04–1.32 ng/g*, Ø conc.: 1.18 ng/g*, country: USA[425], *after 4 h
incidence: 2/2, sa. const.: female crossbred Yorkshire × Hampshire swines, wt.: 20 kg, contamination: artificial (dose: 0.15 mg T-2 toxin (labeled and unlabeled)/kg b. wt., i.vs., once), conc. range: 149–180 ng/g*, Ø conc.: 164.5 ng/g* **, country: USA[425], *after 4 h, ****total metabolites**
incidence: 2/2, sa. const.: female crossbred Yorkshire × Hampshire swines, wt.: 20 kg, contamination: artificial (dose: 0.15 mg T-2 toxin (labeled and unlabeled)/kg b. wt., i.vs., once), conc. range: 1.17–1.40 ng/g* **, Ø conc.: 1.285 ng/g* **, country: USA[425], *in **jejunum contents**, **after 4 h
incidence: 2/2, sa. const.: female crossbred Yorkshire × Hampshire swines, wt.: 20 kg, contamination: artificial (dose: 0.15 mg T-2 toxin (labeled and unlabeled)/kg b. wt., i.vs., once), conc. range: 303–316 ng/g* **, Ø conc.: 309.5 ng/g* ** ***, country: USA[425], *in **jejunum contents**, **after 4 h, *****total metabolites**

3′-HYDROXY T-2 TOXIN
incidence: 2/2, sa. const.: female crossbred Yorkshire × Hampshire swines, wt.: 20 kg, contamination: artificial (dose: 0.15 mg T-2 toxin (labeled and unlabeled)/kg b. wt., i.vs., once), conc. range: 0.99–1.71 ng/g*, Ø conc.: 1.35 ng/g*, country: USA[425], *after 4 h
incidence: 2/2, sa. const.: female crossbred Yorkshire × Hampshire swines, wt.: 20 kg, contamination: artificial (dose: 0.15 mg T-2 toxin (labeled and unlabeled)/kg b. wt., i.vs., once), conc. range: 0.64–2.42 ng/g* **, Ø conc.: 1.53 ng/g* **, country: USA[425], *in **jejunum contents**, **after 4 h

T-2 TETRAOL
incidence: 2/2, sa. const.: female crossbred Yorkshire × Hampshire swines, wt.: 20 kg, contamination: artificial (dose: 0.15 mg T-2 toxin (labeled and unlabeled)/kg b. wt., i.vs., once), conc. range: 2.67–3.78 ng/g*,

Ø conc.: 3.225 ng/g*, country: USA[425], *after 4 h
incidence: 2/2, sa. const.: female crossbred Yorkshire × Hampshire swines, wt.: 20 kg, contamination: artificial (dose: 0.15 mg T-2 toxin (labeled and unlabeled)/kg b. wt., i.vs., once), conc. range: 3.37–7.52 ng/g* **, Ø conc.: 5.445 ng/g* **, country: USA[425], *in **jejunum contents**, **after 4 h

DEEPOXY T-2 TETRAOL
incidence: 2/2, sa. const.: female crossbred Yorkshire × Hampshire swines, wt.: 20 kg, contamination: artificial (dose: 0.15 mg T-2 toxin (labeled and unlabeled)/kg b. wt., i.vs., once), conc. range: 1.70–1.82 ng/g*, Ø conc.: 1.76 ng/g*, country: USA[425], *after 4 h
incidence: 2/2, sa. const.: female crossbred Yorkshire × Hampshire swines, wt.: 20 kg, contamination: artificial (dose: 0.15 mg T-2 toxin (labeled and unlabeled)/kg b. wt., i.vs., once), conc. range: 3.40–4.57 ng/g* **, Ø conc.: 3.985 ng/g* **, country: USA[425], *in **jejunum contents**, **after 4 h

T-2 TRIOL
incidence: 2/2, sa. const.: female crossbred Yorkshire × Hampshire swines, wt.: 20 kg, contamination: artificial (dose: 0.15 mg T-2 toxin (labeled and unlabeled)/kg b. wt., i.vs., once), conc. range: 0.36–0.39 ng/g*, Ø conc.: 0.375 ng/g*, country: USA[425], *after 4 h
incidence: 2/2, sa. const.: female crossbred Yorkshire × Hampshire swines, wt.: 20 kg, contamination: artificial (dose: 0.15 mg T-2 toxin (labeled and unlabeled)/kg b. wt., i.vs., once), conc. range: 0.40–0.54 ng/g* **, Ø conc.: 0.47 ng/g* **, country: USA[425], *in **jejunum contents**, **after 4 h

DEEPOXY T-2 TRIOL
incidence: 2/2, sa. const.: female crossbred Yorkshire × Hampshire swines, wt.: 20 kg, contamination: artificial (dose: 0.15 mg T-2 toxin (labeled and unlabeled)/kg b. wt., i.vs., once), conc.: nd*, country: USA[425], *after 4 h
incidence: 1/2, sa. const.: female crossbred Yorkshire × Hampshire swines, wt.: 20 kg, contamination: artificial (dose: 0.15 mg T-2 toxin (labeled and unlabeled)/kg b. wt., i. vs., once), conc.: 0.44 ng/g* **, country: USA[425], *in **jejunum contents**, **after 4 h

Pig kidney may contain the following mycotoxins and/or their metabolites:

AFLATOXICOL
incidence: 3/3, sa. const.: castrated male pigs of mixed breed, wt.: 9–11 kg, contamination: artificial (acute study, dose: 1 mg AFB_1/kg b. wt., o., once), conc. range: ≤4.47 ng/g*, country: USA[362], *in pig that died after 22 h (2 other pigs sacrificed 24 and 72 h after treatment showed lower mycotoxin values)

AFLATOXIN B_1
incidence: 2/2*, sa. const.: castrated male Yorkshire-Hampshire-Duroc tricross pigs, age: 3–4 weeks, contamination: artificial (dose: **41 ng AFB_1**/g diet, o., for 3 weeks (control); for detailed information please see the article), conc.: nd**, country: USA[60], *control, **sacrificed on withdrawal day 0 (animals sacrificed later showed no AFB_1 contamination)
incidence: 2/2, sa. const.: castrated male Yorkshire-Hampshire-Duroc tricross pigs, age: 3–4 weeks, contamination: artificial (dose: **341 ng AFB_1**/g diet, o., for 3 weeks; for detailed information please see the article), conc.: nd*, country: USA[60], *sacrificed on withdrawal day 0 (animals sacrificed later showed no AFB_1 contamination)
incidence: 2/2, sa. const.: castrated male Yorkshire-Hampshire-Duroc tricross pigs, age: 3–4 weeks, contamination: artificial (dose: **866 ng AFB_1**/g diet, o., for 3 weeks; for detailed information please see the article), conc.: nd*, country: USA[60], *sacrificed on withdrawal day 0 (animals sacrificed later showed no AFB_1 contamination)
incidence: 2/2, sa. const.: castrated male Yorkshire-Hampshire-Duroc tricross pigs, age: 3–4 weeks, contamination: artificial

(dose: **1,253 ng AFB_1/g** diet, o., for 3 weeks; for detailed information please see the article), conc. range: 0.10–0.22 ng/g tissue*, Ø conc.: 0.16 ng/g tissue*, country: USA[60], *sacrificed on withdrawal day 0 (animals sacrificed later showed no AFB_1 contamination)

incidence: 1/1, sa. const.: female Hampshire × Deutsches Edelschwein piglet, wt.: 15 kg, contamination: artificial (dose: 3.1 μg AFB_1 (labeled)/kg b. wt., o., once; for detailed information please see the article), conc.: 2.4 ppb* **, country: Switzerland[66], *AFB_1 eq., **after **24 h**
incidence: 1/1, sa. const.: female Hampshire × Deutsches Edelschwein piglet, wt.: 15 kg, contamination: artificial (dose: 3.1 μg AFB_1 (labeled)/kg b. wt., o., once; for detailed information please see the article), conc.: 1.3 ppb* **, country: Switzerland[66], *AFB_1 eq., **after **48 h**

incidence: 2/2, sa. const.: adult swines, contamination: artificial (dose: 1.08–1.09 mg AFB_1 (besides other AFs), o., daily for 33 days; for detailed information please see the article), conc. range: 0.08–2.47 ppb*, Ø conc.: 1.28 ppb*, country: France[88], *after 33 days

incidence: 20/20*, sa. const.: male and female Danish Landrace pigs, age: ≈8 weeks, wt.: 20 kg up to slaughter at 90 kg, contamination: no AFB_1 + AFB_2 (for detailed information please see the article), conc.: nd, country: Denmark/USA[101], *control
incidence: 11/16*, sa. const.: male and female Danish Landrace pigs, age: ≈8 weeks, wt.: 20 kg up to slaughter at 90 kg, contamination: artificial (dose: AFB_1 + AFB_2 addition (overall **300 ppb AFs**), o., for **150** or 186*** days**; for detailed information please see the article), conc. range: tr**–10*** ppb, country: Denmark/USA[101], *livers of some of these pigs were rejected at meat inspection, after 150** or 186*** days
incidence: 11/18*, sa. const.: male and female Danish Landrace pigs, age: ≈8 weeks, wt.: 20 kg up to slaughter at 90 kg, contamination: artificial (dose: AFB_1 + AFB_2 addition (overall **500 ppb AFs**), o., for **135** or 159*** days**; for detailed information please see the article), conc. range: tr***–50** ppb, country: Denmark/USA[101], *livers of some of these pigs were rejected at meat inspection, after 135** or 159*** days

incidence: 3/4*, sa. const.: male and female feeder pigs, wt.: 54.2–71.6 kg, contamination: no AFB_1, conc.: <0.12 μg/kg, country: USA[136], *control
incidence: 4/4, sa. const.: male and female feeder pigs, wt.: 54.2–71.6 kg, contamination: artificial (dose: **100 μg AFB_1/kg** diet, o., for 4 weeks), conc. range: 0.10–0.37 μg/kg*, Ø conc.: 0.23 μg/kg*, country: USA[136], *after 4 weeks
incidence: 4/4, sa. const.: male and female feeder pigs, wt.: 54.2–71.6 kg, contamination: artificial (dose: **200 μg AFB_1/kg** diet, o., for 4 weeks), conc. range: 0.23–1.50 μg/kg*, Ø conc.: 0.70 μg/kg*, country: USA[136], *after 4 weeks
incidence: 4/4, sa. const.: male and female feeder pigs, wt.: 54.2–71.6 kg, contamination: artificial (dose: **400 μg AFB_1/kg** diet, o., for 4 weeks), conc. range: 0.63–10.0 μg/kg, Ø conc.: 4.44 μg/kg, country: USA[136], *after 4 weeks

incidence: 4/4*, sa. const.: crossbred (Duroc × Yorkshire) barrows, wt.: 24.5–26.3 kg, contamination: no AFs, conc.: nd, country: USA[138], *control
incidence: 4?/4, sa. const.: crossbred (Duroc × Yorkshire) barrows, wt.: 24.5–26.3 kg, contamination: artificial (dose: **662 μg AFB_1/kg** diet, **273 μg AFB_2/kg** diet, **300 μg AFG_1/kg** diet and **285 μg AFG_2/kg** diet, o., for 21 days), conc. range: 0.05–0.75 μg/kg*, country: USA[138], *after ≈22 days (thereof 21 days of AF-administration)

incidence: 10/10*, sa. const.: mixed breed feeder pigs, contamination: chronic study, no AFs (for detailed information please see the article), conc.: nd, country: USA[182], *control
incidence: 10?/10, sa. const.: mixed breed feeder pigs, contamination: artificial (chronic study, dose: **400 ng natural AFs/g** diet, for 10 weeks; for detailed information please see the article), conc.: 0.20 ng/g* (wet matter basis) (mean value), country: USA[182], *after 10 weeks
incidence: 10?/10, sa. const.: mixed breed feeder pigs, contamination: artificial (chronic study, dose: **800 ng natural AFs/g** diet, for 10 weeks; for detailed information please see the article), conc.: 0.25 ng/g* (wet matter basis) (mean value), country: USA[182], *after 10 weeks

incidence: 8/8*, sa. const.: mixed breed feeder pigs, contamination: acute study, no AFs, conc.: nd, country: USA[182], *control
incidence: 1/1(8)*, sa. const.: mixed breed feeder pigs, contamination: artificial (acute study, dose: **1.2 mg total AFs** (AFB_1+AFB_2+AFG_1+AFG_2)/kg b. wt., o., once), conc. range: ≤3.80 ng/g** (wet matter basis), country: USA[182], *for overall information please see the article, **12 h post-dosage (also measured 24, 48 and 72 h post-dosage, lowest conc.: nd after 72 h)

incidence: ?/?*, sa. const.: Large White growing female pigs, age: 79 days, wt.: 20 kg, contamination: no AFs (for detailed information please see the article), conc.: nd, country: France[314], *control
incidence: 1/1, sa. const.: Large White growing female pigs, age: 79 days, wt.: 20 kg, contamination: artificial (dose: natural AFs addition (treatment 1: mixed feeding, avg. daily intake of **870 µg AFB_1**) for 26 days; for detailed information please see the article), conc.: nd*, country: France[314], *final wt. of the animal 109 kg
incidence: 2/2, sa. const.: Large White growing female pigs, age: 79 days, wt.: 20 kg, contamination: artificial (dose: natural AFs addition (treatment 2: separate feeding = peanut oil meal 40% and corn gluten meal 30%, avg. daily intake of **1,566 µg AFB_1**) for 19* or 26** days; for detailed information please see the article), conc. range: 0.26*–0.244** µg/kg, country: France[314], final wt. of the animals 97* and 105** kg
incidence: 1/1, sa. const.: Large White growing female pigs, age: 79 days, wt.: 20 kg, contamination: artificial (dose: natural AFs addition (treatment 3: separate feeding = peanut oil meal 0% and corn gluten meal 0%, avg. daily intake of **642 µg AFB_1**) for 26 days; for detailed information please see the article), conc.: 0.73 µg/kg*, country: France[314], *final wt. of the animal 104 kg

incidence: 3/3, sa. const.: castrated male pigs of mixed breed, wt.: 9–11 kg, contamination: artificial (**acute study**, dose: 1 mg AFB_1/kg b.w., o., once), conc. range: ≤23.6 ng/g*, country: USA[362], *in pig that died after 22 h (2 other pigs sacrificed 24 and 72 h after treatment showed lower mycotoxin values)
incidence: 2/5, sa. const.: market-weight pigs, wt.: ≈92 kg, contamination: artificial (**subacute study**, dose: ≈15 µg AFB_1 as well as AFB_2, AFG_1 + AFG_2 (all natural)/kg b. wt., o., for 14 days; for detailed information please see the article), conc. range: 0.13–0.6 ng/g*, Ø conc.: 0.365 ng/g*, country: USA[362], *after 14 days

incidence: 5?/5*, sa. const.: cross-bred pigs, contamination: (dose: 9 ng/g AFB_1 + AFB_2, o., for 35 days; for detailed information please see the article), conc.: 0.027 ng/g** (mean value), country: USA[597], *control, **after 35 days
incidence: 5?/5, sa. const.: cross-bred pigs, contamination: artificial (dose: **524 ng/g AFB_1 + AFB_2**, o., for 35 days; for detailed information please see the article), conc.: 0.681 ng/g* (mean value), country: USA[597], *after 35 days
incidence: 5?/5, sa. const.: cross-bred pigs, contamination: artificial (dose: **524 ng/g AFB_1 + AFB_2 + 0.5% HSCAS**, o.,

for 35 days; for detailed information please see the article), conc.: 0.410 ng/g* (mean value), country: USA[597], *after 35 days

incidence: 5/5*, sa. const.: male miniature pigs, contamination: (dose: 1 ng/g AFB_1 + AFB_2, o., for 15 days; for detailed information please see the article), conc.: nd, country: USA[597], *control
incidence: 5?/5, sa. const.: male miniature pigs, contamination: artificial (dose: **590 ng/g AFB_1 + AFB_2**, o., for 15 days; for detailed information please see the article), conc.: 0.410 ng/g* (mean value), country: USA[597], *after 15 days
incidence: 5?/5, sa. const.: male miniature pigs, contamination: artificial (dose: **590 ng/g AFB_1 + AFB_2 + HSCAS (0.5%)**, o., for 15 days; for detailed information please see the article), conc.: 0.250 ng/g* (mean value), country: USA[597], *after 15 days
incidence: 5?/5, sa. const.: male miniature pigs, contamination: artificial (dose: **590 ng/g AFB_1 + AFB_2**, o., for 15 days followed by **2 weeks control diet**; for detailed information please see the article), conc.: nd*, country: USA[597], *after 29 days (thereof 15 days of AFB_1- and AFB_2-administration)

Aflatoxin B_2
incidence: 2/2, sa. const.: adult swines, contamination: artificial (dose: 1.08–1.09 mg AFB_1 (besides other AFs), o., daily for 33 days; for detailed information please see the article), conc. range: tr–1.01 ppb*, country: France[88], *after 33 days

incidence: 20/20*, sa. const.: male and female Danish Landrace pigs, age: ≈8 weeks, wt.: 20 kg up to slaughter at 90 kg, contamination: no AFB_1 + AFB_2 (for detailed information please see the article), conc.: nd, country: Denmark/USA[101], *control
incidence: 6/16*, sa. const.: male and female Danish Landrace pigs, age: ≈8 weeks, wt.: 20 kg up to slaughter at 90 kg, contamination: artificial (dose: AFB_1 + AFB_2 addition (overall **300 ppb AFs**), o., for **164** or 180*** days**; for detailed information please see the article), conc. range: tr**–10*** ppb, country: Denmark/USA[101], *livers of some of these pigs were rejected at meat inspection, after 164** or 180*** days
incidence: 7/17*, sa. const.: male and female Danish Landrace pigs, age: ≈8 weeks, wt.: 20 kg up to slaughter at 90 kg, contamination: artificial (dose: AFB_1 + AFB_2 addition (overall **500 ppb AFs**), o., for **141** or 216*** days**; for detailed information please see the article), conc. range: tr***–10 ppb**, country: Denmark/USA[101], *livers of some of these pigs were rejected at meat inspection, after 141** or 216*** days

incidence: 4/4*, sa. const.: crossbred (Duroc × Yorkshire) barrows, wt.: 24.5–26.3 kg, contamination: no AFs, conc.: nd, country: USA[138], *control
incidence: 4?/4, sa. const.: crossbred (Duroc × Yorkshire) barrows, wt.: 24.5–26.3 kg, contamination: artificial (dose: **662 μg AFB_1/kg** diet, **273 μg AFB_2/kg** diet, **300 μg AFG_1/kg** diet and **285 μg AFG_2/kg** diet, o., for 21 days), conc. range: tr–0.55 μg/kg*, country: USA[138], *after ≈22 days (thereof 21 days of AF-administration)

incidence: 10/10*, sa. const.: mixed breed feeder pigs, contamination: chronic study, no AFs (for detailed information please see the article), conc.: nd, country: USA[182], *control
incidence: 10?/10, sa. const.: mixed breed feeder pigs, contamination: artificial (chronic study, dose: **400 ng natural AFs/g** diet, for 10 weeks; for detailed information please see the article), conc.: 0.02 ng/g* (wet matter basis) (mean value), country: USA[182], *after 10 weeks
incidence: 10?/10, sa. const.: mixed breed feeder pigs, contamination: artificial (chronic study, dose: **800 ng natural AFs/g** diet, for 10 weeks; for detailed

information please see the article), conc.: 0.05 ng/g* (wet matter basis) (mean value), country: USA[182], *after 10 weeks

incidence: 8/8*, sa. const.: mixed breed feeder pigs, contamination: acute study, no AFs, conc.: nd, country: USA[182], *control
incidence: 1/1(8)*, sa. const.: mixed breed feeder pigs, contamination: artificial (acute study, dose: **1.2 mg total AFs** (AFB_1+AFB_2+AFG_1+AFG_2)/kg b. wt., o., once, conc. range: ≤1.52 ng/g** (wet matter basis), country: USA[182], *for overall information please see the article, **24 h post-dosage (also measured 12, 48 and 72 h post-dosage, lowest conc.: nd after 72 h)

incidence: ?/?*, sa. const.: Large White growing female pigs, age: 79 days, wt.: 20 kg, contamination: no AFs (for detailed information please see the article), conc.: nd, country: France[314], *control
incidence: 1/1, sa. const.: Large White growing female pigs, age: 79 days, wt.: 20 kg, contamination: artificial (dose: natural AFs addition (treatment 1: mixed feeding, avg. daily intake of **870 µg AFB_1**) for 26 days; for detailed information please see the article), conc.: nd*, country: France[314], *final wt. of the animal 109 kg
incidence: 1/1, sa. const.: Large White growing female pigs, age: 79 days, wt.: 20 kg, contamination: artificial (dose: natural AFs addition (treatment 2: separate feeding = peanut oil meal 40% and corn gluten meal 30%, avg. daily intake of **1,566 µg AFB_1**) for 19 days; for detailed information please see the article), conc.: 0.109 µg/kg*, country: France[314], *final wt. of the animal 97 kg
incidence: 1/1, sa. const.: Large White growing female pigs, age: 79 days, wt.: 20 kg, contamination: artificial (dose: natural AFs addition (treatment 3: separate feeding = peanut oil meal 0% and corn gluten meal 0%, avg. daily intake of **642 µg AFB_1**) for 26 days; for detailed information please see the article), conc.: 1.60 µg/kg*, country: France[314], *final wt. of the animal 104 kg

incidence: 5/5*, sa. const.: cross-bred pigs, contamination: (dose: 9 ng/g AFB_1 + AFB_2, o., for 35 days; for detailed information please see the article), conc.: nd, country: USA[597], *control
incidence: 5?/5, sa. const.: cross-bred pigs, contamination: artificial (dose: **524 ng/g AFB_1 + AFB_2**, o., for 35 days; for detailed information please see the article), conc.: 0.138 ng/g* (mean value), country: USA[597], *after 35 days
incidence: 5?/5, sa. const.: cross-bred pigs, contamination: artificial (dose: **524 ng/g AFB_1 + AFB_2 + HSCAS** (0.5%), o., for 35 days; for detailed information please see the article), conc.: 0.002 ng/g* (mean value), country: USA[597], *after 35 days

incidence: 5/5*, sa. const.: male miniature pigs, contamination: (dose: 1 ng/g AFB_1 + AFB_2, o., for 15 days; for detailed information please see the article), conc.: nd, country: USA[597], *control
incidence: 5?/5, sa. const.: male miniature pigs, contamination: artificial (dose: **590 ng/g AFB_1 + AFB_2**, o., for 15 days; for detailed information please see the article), conc.: 0.080 ng/g* (mean value), country: USA[597], *after 15 days
incidence: 5?/5, sa. const.: male miniature pigs, contamination: artificial (dose: **590 ng/g AFB_1 + AFB_2 + HSCAS** (0.5%), o., for 15 days; for detailed information please see the article), conc.: 0.050 ng/g* (mean value), country: USA[597], *after 15 days
incidence: 5?/5, sa. const.: male miniature pigs, contamination: artificial (dose: **590 ng/g AFB_1 + AFB_2**, o., for 15 days followed by **2 weeks control diet**; for detailed information please see the article), conc.: nd*, country: USA[597], *after 29 days (thereof 15 days of AFB_1- and AFB_2-administration)

Aflatoxin B_{2a}
incidence: 4/4*, sa. const.: crossbred (Duroc × Yorkshire) barrows, wt.: 24.5–26.3 kg, contamination: no AFs, conc.: nd, country: USA[138], *control

incidence: 4?/4, sa. const.: crossbred (Duroc × Yorkshire) barrows, wt.: 24.5–26.3 kg, contamination: artificial (dose: **662 µg AFB_1/kg** diet, **273 µg AFB_2/** kg diet, **300 µg AFG_1/kg** diet and **285 µg AFG_2/kg** diet, o., for 21 days), conc. range: appreciable amounts*, country: USA[138], *after ≈22 days (thereof 21 days of AF-administration)

Aflatoxin G_1
incidence: 8/8*, sa. const.: mixed breed feeder pigs, contamination: acute study, no AFs, conc.: nd, country: USA[182], *control
incidence: 1/1(8)*, sa. const.: mixed breed feeder pigs, contamination: artificial (acute study, dose: **1.2 mg total AFs** (AFB_1+AFB_2+AFG_1+AFG_2)/kg b. wt., o., once, conc. range: ≤0.60 ng/g** (wet matter basis), country: USA[182], *for overall information please see the article, **12 h post-dosage (also measured 24, 48 and 72 h post-dosage, lowest conc.: nd after 24, 48 and 72 h (several animals))

Aflatoxin G_2
incidence: 8/8*, sa. const.: mixed breed feeder pigs, contamination: acute study, no AFs, conc.: nd, country: USA[182], *control
incidence: 1/1(8)*, sa. const.: mixed breed feeder pigs, contamination: artificial (acute study, dose: **1.2 mg total AFs** (AFB_1+AFB_2+AFG_1+AFG_2)/kg b. wt., o., once, conc. range: ≤0.07 ng/g** (wet matter basis), country: USA[182], *for overall information please see the article, **12 h post-dosage (also measured 24, 48 and 72 h post-dosage, lowest conc.: nd after 24, 48 and 72 h (several animals))

Aflatoxin M_1
incidence: 1/2*, sa. const.: castrated male Yorkshire-Hampshire-Duroc tricross pigs, age: 3–4 weeks, contamination: artificial (dose: **41 ng AFB_1/g** diet, o., for 3 weeks (control); for detailed information please see the article), conc.: tr**, country: USA[60], *control, **sacrificed on withdrawal day 0 (animals sacrificed later showed no AFM_1-contamination)
incidence: 2/2, sa. const.: castrated male Yorkshire-Hampshire-Duroc tricross pigs, age: 3–4 weeks, contamination: artificial (dose: **341 ng AFB_1/g** diet, o., for 3 weeks; for detailed information please see the article), conc. range: 0.18–0.34 ng/g tissue*, Ø conc.: 0.26 ng/g tissue*, country: USA[60], *sacrificed on withdrawal day 0 (animals sacrificed later showed no AFM_1-contamination)
incidence: 2/2, sa. const.: castrated male Yorkshire-Hampshire-Duroc tricross pigs, age: 3–4 weeks, contamination: artificial (dose: **866 ng AFB_1/g** diet, o., for 3 weeks; for detailed information please see the article), conc. range: 0.67–1.02 ng/g tissue*, Ø conc.: 0.845 ng/g tissue*, country: USA[60], *sacrificed on withdrawal day 0 (animals sacrificed later showed lower or no residue values)
incidence: 2/2, sa. const.: castrated male Yorkshire-Hampshire-Duroc tricross pigs, age: 3–4 weeks, contamination: artificial (dose: **1,253 ng AFB_1/g** diet, o., for 3 weeks; for detailed information please see the article), conc. range: 0.093–1.30 ng/g tissue*, Ø conc.: 0.604 ng/g tissue*, country: USA[60], *sacrificed on withdrawal day 0 (animals sacrificed later showed lower or no residue values)

incidence: 2/2, sa. const.: adult swines, contamination: artificial (dose: 1.08–1.09 mg AFB_1 (besides other AFs), o., daily for 33 days; for detailed information please see the article), conc. range: 2.03–10.00 ppb*, Ø conc.: 6.02 ppb*, country: France[88], *after 33 days

incidence: 3/4*, sa. const.: male and female feeder pigs, wt.: 54.2–71.6 kg, contamination: no AFB_1, conc.: <0.07 µg/kg, country: USA[136], *control
incidence: 4/4, sa. const.: male and female feeder pigs, wt.: 54.2–71.6 kg, contamination: artificial (dose: **100 µg AFB_1/kg** diet, o., for 4 weeks), conc. range: 0.09–0.23 µg/kg*, Ø conc.: 0.18 µg/kg*, country: USA[136], *after 4 weeks

incidence: 4/4, sa. const.: male and female feeder pigs, wt.: 54.2–71.6 kg, contamination: artificial (dose: **200 µg AFB$_1$/kg** diet, o., for 4 weeks), conc. range: 0.29–1.29 µg/kg*, Ø conc.: 0.75 µg/kg*, country: USA[136], *after 4 weeks
incidence: 4/4, sa. const.: male and female feeder pigs, wt.: 54.2–71.6 kg, contamination: artificial (dose: **400 µg AFB$_1$/kg** diet, o., for 4 weeks), conc. range: 0.18–0.68 µg/kg*, Ø conc.: 0.38 µg/kg*, country: USA[136], *after 4 weeks

incidence: 4/4*, sa. const.: crossbred (Duroc × Yorkshire) barrows, wt.: 24.5–26.3 kg, contamination: no AFs, conc.: nd, country: USA[138], *control
incidence: 4?/4, sa. const.: crossbred (Duroc × Yorkshire) barrows, wt.: 24.5–26.3 kg, contamination: artificial (dose: **662 µg AFB$_1$/kg** diet, **273 µg AFB$_2$/**kg diet, **300 µg AFG$_1$/kg** diet and **285 µg AFG$_2$/kg** diet, o., for 21 days), conc. range: probably tr*, country: USA[138], *after ≈22 days (thereof 21 days of AF-administration)

incidence: 10/10*, sa. const.: mixed breed feeder pigs, contamination: chronic study, no AFs (for detailed information please see the article), conc.: nd, country: USA[182], *control
incidence: 10?/10, sa. const.: mixed breed feeder pigs, contamination: artificial (chronic study, dose: **400 ng natural AFs/g** diet, for 10 weeks; for detailed information please see the article), conc.: 0.61 ng/g* (wet matter basis) (mean value), country: USA[182], *after 10 weeks
incidence: 10?/10, sa. const.: mixed breed feeder pigs, contamination: artificial (chronic study, dose: **800 ng natural AFs/g** diet, for 10 weeks; for detailed information please see the article), conc.: 0.91 ng/g* (wet matter basis) (mean value), country: USA[182], *after 10 weeks

incidence: 8/8*, sa. const.: mixed breed feeder pigs, contamination: acute study, no AFs, conc.: nd, country: USA[182], *control

incidence: 1/1(8)*, sa. const.: mixed breed feeder pigs, contamination: artificial (acute study, dose: 1.2 mg total AFs (AFB$_1$+AFB$_2$+AFG$_1$+AFG$_2$)/kg b. wt., o., once, conc. range: ≤4.10 ng/g** (wet matter basis), country: USA[182], *for overall information please see the article, **24 h post-dosage (also measured 12, 48 and 72 h post-dosage, lowest conc.: nd after 72 h)

incidence: 3/3, sa. const.: castrated male pigs of mixed breed, weight: 9–11 kg, contamination: artificial (**acute study**, dose: 1 mg AFB$_1$/kg b. wt., o., once), conc. range: ≤11.2 ng/g*, country: USA[362], *in pig that died after 22 h (2 other pigs sacrificed 24 and 72 h after treatment showed lower mycotoxin values)
incidence: 5/5, sa. const.: market-weight pigs, weight: ≈92 kg, contamination: artificial (**subacute study**, dose: ≈15 µg AFB$_1$ as well as AFB$_2$, AFG$_1$ + AFG$_2$ (all natural)/kg b. wt., o., for 14 days; for detailed information please see the article), conc. range: 1.10–2.63 ng/g*, Ø conc.: 1.46 ng/g*, country: USA[362], *after 14 days

incidence: 5?/5*, sa. const.: cross-bred pigs, contamination: (dose: 9 ng/g AFB$_1$ + AFB$_2$, o., for 35 days; for detailed information please see the article), conc.: 0.166 ng/g** (mean value), country: USA[597], *control, **after **35 days**
incidence: 5?/5, sa. const.: cross-bred pigs, contamination: artificial (dose: **524 ng/g AFB$_1$ + AFB$_2$**, o., for 35 days; for detailed information please see the article), conc.: 3.132 ng/g* (mean value), country: USA[597], *after **35 days**
incidence: 5?/5, sa. const.: cross-bred pigs, contamination: artificial (dose: **524 ng/g AFB$_1$ + AFB$_2$ + HSCAS** (0.5%), o., for 35 days; for detailed information please see the article), conc.: 0.804 ng/g* (mean value), country: USA[597], *after **35 days**

incidence: 5?/5*, sa. const.: male miniature pigs, contamination: (dose: 1 ng/g AFB$_1$ + AFB$_2$, o., for 15 days; for detailed

information please see the article), conc.: 0.030 ng/g** (mean value), country: USA[597], *control, **after **15 days**
incidence: 5?/5, sa. const.: male miniature pigs, contamination: artificial (dose: **590 ng/g AFB_1 + AFB_2**, o., for 15 days; for detailed information please see the article), conc.: 5.980 ng/g* (mean value), country: USA[597], *after **15 days**
incidence: 5?/5, sa. const.: male miniature pigs, contamination: artificial (dose: **590 ng/g AFB_1 + AFB_2 + HSCAS** (0.5%), o., for 15 days; for detailed information please see the article), conc.: 1.690 ng/g* (mean value), country: USA[597], *after **15 days**
incidence: 5?/5, sa. const.: male miniature pigs, contamination: artificial (dose: 590 ng/g AFB_1 + AFB_2, o., for 15 days followed by 2 weeks control diet; for detailed information please see the article), conc.: 0.050 ng/g* (mean value), country: USA[597], *after **29 days** (thereof 15 days of AFB_1- and AFB_2-administration)

Aflatoxin M
incidence: 20/20*, sa. const.: male and female Danish Landrace pigs, age: ≈8 weeks, wt.: 20 kg up to slaughter at 90 kg, contamination: no AFB_1 + AFB_2 (for detailed information please see the article), conc.: nd, country: Denmark/USA[101], *control
incidence: 2/16*, sa. const.: male and female Danish Landrace pigs, age: ≈8 weeks, wt.: 20 kg up to slaughter at 90 kg, contamination: artificial (dose: AFB_1 + AFB_2 addition (overall **300 ppb AFs**), o., for **120** or 164*** days**; for detailed information please see the article), conc. range: tr***–3** ppb, country: Denmark/USA[101], *livers of some of these pigs were rejected at meat inspection, after 120** or 164*** days
incidence: 4/17*, sa. const.: male and female Danish Landrace pigs, age: ≈8 weeks, wt.: 20 kg up to slaughter at 90 kg, contamination: artificial (dose: AFB_1 + AFB_2 addition (overall **500 ppb AFs**), o., for **135** or 186*** days**; for detailed information please see the article), conc. range: tr***–3** ppb, country: Denmark/USA[101], *livers of some of these pigs were rejected at meat inspection, after 135** or 186*** days

incidence: ?/?*, sa. const.: Large White growing female pigs, age: 79 days, wt.: 20 kg, contamination: no AFs (for detailed information please see the article), conc.: nd, country: France[314], *control
incidence: 1/1, sa. const.: Large White growing female pigs, age: 79 days, wt.: 20 kg, contamination: artificial (dose: natural AFs addition (treatment 1: mixed feeding, avg. daily intake of **870 µg AFB_1**) for 26 days; for detailed information please see the article), conc.: nd*, country: France[314], *final wt. of the animal 109 kg
incidence: 1/1, sa. const.: Large White growing female pigs, age: 79 days, wt.: 20 kg, contamination: artificial (dose: natural AFs addition (treatment 2: separate feeding = peanut oil meal 40% and corn gluten meal 30%, avg. daily intake of **1,566 µg AFB_1**) for 19 days; for detailed information please see the article), conc.: 1.32 µg/kg*, country: France[314], *final wt. of the animal 97 kg
incidence: 1/1, sa. const.: Large White growing female pigs, age: 79 days, wt.: 20 kg, contamination: artificial (dose: natural AFs addition (treatment 3: separate feeding = peanut oil meal 0% and corn gluten meal 0%, avg. daily intake of **642 µg AFB_1**) for 26 days; for detailed information please see the article), conc.: 2.50 µg/kg*, country: France[314], *final wt. of the animal 104 kg

Citrinin
incidence: 13/13*, sa. const.: castrated pigs, wt.: 20 kg, contamination: no OTA and/or CIT (for detailed information please see the article), conc.: nd?, country: Denmark/USA[330], *control

incidence: 13/13, sa. const.: castrated pigs, wt.: 20 kg, contamination: artificial (dose: **1,400 µg crystalline OTA**/kg feed, o., for 6 weeks; for detailed information please see the article), conc.: nd*, country: Denmark/USA[330], *after 6 weeks
incidence: 13?/13, sa. const.: castrated pigs, wt.: 20 kg, contamination: artificial (dose: **650 µg crystalline CIT**/kg feed, o., for 6 weeks; for detailed information please see the article), conc.: pr*, country: Denmark/USA[330], *after 6 weeks
incidence: 13?/13, sa. const.: castrated pigs, wt.: 20 kg, contamination: artificial (dose: **1,400 µg crystalline OTA + 650 µg crystalline CIT**/kg feed, o., for 6 weeks; for detailed information please see the article), conc.: pr*, country: Denmark/USA[330], *after 6 weeks
incidence: 13?/13, sa. const.: castrated pigs, wt.: 20 kg, contamination: artificial (dose: **1,400 µg natural OTA + 650 µg natural CIT**/kg feed, o., for 6 weeks; for detailed information please see the article), conc.: pr*, country: Denmark/USA[330], *after 6 weeks

Deoxynivalenol
incidence: 4/4*, sa. const.: barrows and gilts, Ø wt.: 7.7 kg, contamination: no DON (for detailed information please see the article), conc.: nd, country: USA[64], *control
incidence: 4?/4, sa. const.: barrows and gilts, Ø wt.: 7.7 kg, contamination: artificial (dose: **0.9 ppm DON** (analyzed value) in the diet, o., for 3 weeks; for detailed information please see the article), conc.: 19 ppb*, country: USA[64], *after 3 weeks post-treatment
incidence: 4?/4, sa. const.: barrows and gilts, Ø wt.: 7.7 kg, contamination: artificial (dose: **2.0 ppm DON** (analyzed value) in the diet, o., for 3 weeks; for detailed information please see the article), conc.: 19 ppb* (mean value), country: USA[64], *after 3 weeks post-treatment
incidence: 4?/4, sa. const.: barrows and gilts, Ø wt.: 7.7 kg, contamination: artificial (dose: **2.8 ppm DON** (analyzed value) in the diet, o., for 3 weeks; for detailed information please see the article), conc.: 23 ppb* (mean value), country: USA[64], *after 3 weeks post-treatment

incidence: 6/6*, sa. const.: barrows (Yorkshire), wt.: ≈25 kg, contamination: no DON (for detailed information please see the article), conc.: nd, country: Canada[406], *control
incidence: 50?/50, sa. const.: barrows (Yorkshire), wt.: ≈25 kg, contamination: artificial (dose: **6.0 mg natural DON**/kg dry weight, o., for **3 weeks**; for detailed information please see the article), conc. range: ≤52.5 ng/g*, country: Canada[406], *after 3 weeks
incidence: 6/6*, sa. const.: barrows (Yorkshire), wt.: ≈25 kg, contamination: no DON (for detailed information please see the article), conc.: nd, country: Canada[406], *control
incidence: 6?/6, sa. const.: barrows (Yorkshire), wt.: ≈25 kg, contamination: artificial (dose: **6.0 mg crystalline DON**/kg dry weight, o., for **4 weeks**; for detailed information please see the article), conc. range: ≤15.7 ng/g*, country: Canada[406], *after 4 weeks
incidence: 6/6*, sa. const.: barrows (Yorkshire), wt.: ≈25 kg, contamination: no DON (for detailed information please see the article), conc.: nd, country: Canada[406], *control
incidence: 6?/6, sa. const.: barrows (Yorkshire), wt.: ≈25 kg, contamination: artificial (dose: **7.6 mg natural DON**/kg dry weight, o., for **7 weeks**; for detailed information please see the article), conc. range: ≤6.5 ng/g*, country: Canada[406], *after 7 weeks

incidence: 5/5*, sa. const.: healthy Yorkshire barrows, age: ≈11–15 weeks, wt.: 17–22 kg, contamination: no DON, conc.: nd, country: Canada[407], *control
incidence: ?/4, sa. const.: healthy Yorkshire barrows, age: ≈11–15 weeks, wt.: 17–22 kg, contamination: artificial (dose: **1.0 mg**

DON/kg b. wt., i.v., once), conc. range: ≤1,985.3 ng/g* (mean value), country: Canada[407], *after 1 h (also measured after 0.33, 3, 8 and 24 h, lowest conc.: 10.0 ng/g after 24 h)

incidence: 5?/5, sa. const.: castrated Large White × German Landrace, db Classic crossbred pigs, wt.: ≈24.6 kg, contamination: artificial (dose: 0.05, 0.57 or 1.23* mg DON/kg mash* or 0.07, 0.55 or 1.13 mg DON/kg pellets, o., for 11 weeks; for detailed information please see the article), conc. range: ≤19.3 ng/g* ** *** (mean value), country: Germany[483], **after 78/79 days (thereof 11 weeks of DON-administration), ***values of the other DON-treatments lower

Deepoxydeoxynivalenol
incidence: 5?/5, sa. const.: castrated Large White × German Landrace, db Classic crossbred pigs, wt.: ≈24.6 kg, contamination: artificial (dose: 0.05, 0.57 or 1.23 mg DON/kg mash or 0.07, 0.55 or 1.13* mg DON/kg pellets*, o., for 11 weeks; for detailed information please see the article), conc. range: ≤2.4 ng/g* ** *** (mean value), country: Germany[483], **after 78/79 days (thereof 11 weeks of DON-administration), ***values of the other DON-treatments lower

Fumonisin B_1
incidence: 5/5*, sa. const.: weaned barrows of the same genotype, wt.: 12–14 kg, contamination: no FB_1, FB_2 + FB_3 (for detailed information please see the article), conc.: nd, country: Hungary/Germany[87], *control
incidence: 10/10, sa. const.: weaned barrows of the same genotype, wt.: 12–14 kg, contamination: artificial (dose: **50 mg FB_1, 20 mg FB_2 + 5 mg FB_3**/animal, o., for 22 days; for detailed information please see the article), conc. range: 22.4–47.2 ng/g*, Ø conc.: 30.56 ng/g*, country: Hungary/Germany[87], *after 22 days toxin feeding period

incidence: 6?/6*, sa. const.: Yorkshire barrows, age: 6–8 weeks, wt.: 9–13 kg, contamination: no FB_1 (for detailed information please see the article), conc.: nr**, country: Canada[108], *control
incidence: ?/2, sa. const.: Yorkshire barrows, age: 6–8 weeks, wt.: 9–13 kg, contamination: artificial (dose: **3.0 mg FB_1** (labeled)/kg feed, o., for 12 days and 2.0 mg FB_1 (labeled)/kg feed, o., for another 12 days; for detailed information please see the article), conc.: ≈65 ng/g tissue* ** (mean value), country: Canada[108], *after 24 days (also at other day intervals up to 33 days measured), **FB_1 and/or metabolites

incidence: 6/6*, sa. const.: castrated pigs of identical genotype, wt.: ≈12–14 kg, contamination: no FB_1 (for detailed information please see the article), conc.: nd, country: Germany/Hungary[109], *control
incidence: 13/13, sa. const.: castrated pigs of identical genotype, wt.: ≈12–14 kg, contamination: artificial (dose: **100 mg FB_1** daily, o., for 5–11 days; for detailed information please see the article), conc. range: 82–4,760 μg/kg*, Ø conc.: 772.43 μg/kg*, country: Germany/Hungary[109], *on 6th day of the experiment

Fumonisin B_2
incidence: 5/5*, sa. const.: weaned barrows of the same genotype, wt.: 12–14 kg, contamination: no FB_1, FB_2 + FB_3 (for detailed information please see the article), conc.: nd, country: Hungary/Germany[87], *control
incidence: 10/10, sa. const.: weaned barrows of the same genotype, wt.: 12–14 kg, contamination: artificial (dose: **50 mg FB_1, 20 mg FB_2 + 5 mg FB_3**/animal, o., for 22 days; for detailed information please see the article), conc.: nd*, country: Hungary/Germany[87], *after 22 days toxin feeding period

HT-2 Toxin
incidence: 2/2, sa. const.: female crossbred Yorkshire × Hampshire swines, wt.: 20 kg,

contamination: artificial (dose: 0.15 mg T-2 toxin (labeled and unlabeled)/kg b. wt., i.vs., once), conc. range: 1.08–1.45 ng/g*, Ø conc.: 1.265 ng/g*, country: USA[425], *after 4 h

Deepoxy HT-2 toxin
incidence: 1/2, sa. const.: female crossbred Yorkshire × Hampshire swines, wt.: 20 kg, contamination: artificial (dose: 0.15 mg T-2 toxin (labeled and unlabeled)/kg b. wt., i.vs., once), conc.: 0.08 ng/g*, country: USA[425], *after 4 h

3′-Hydroxy HT-2 Toxin
incidence: 2/2, sa. const.: female crossbred Yorkshire × Hampshire swines, wt.: 20 kg, contamination: artificial (dose: 0.15 mg T-2 toxin (labeled and unlabeled)/kg b. wt., i.vs., once), conc. range: 4.96–5.41 ng/g*, Ø conc.: 5.185 ng/g*, country: USA[425], *after 4 h

Ochratoxin A
incidence: 5/5*, sa. const.: female pigs of Danish Landrace, age: ≈8 weeks, wt.: ≈20 kg, contamination: no OTA (for detailed information please see the article), conc.: nd, country: Denmark/USA[102], *control
incidence: 5?/5, sa. const.: female pigs of Danish Landrace, age: ≈8 weeks, wt.: ≈20 kg, contamination: artificial (dose: 1 ppm crystalline OTA, o., once daily for 1 month, afterwards toxin-free diets for various intervals fed; for detailed information please see the article), conc.: 25.70 µg/kg* (mean value), country: Denmark/USA[102], ***1 day after termination** of OTA-exposure
incidence: 5?/5, sa. const.: female pigs of Danish Landrace, age: ≈8 weeks, wt.: ≈20 kg, contamination: artificial (dose: 1 ppm crystalline OTA, o., once daily for 1 month, afterwards toxin-free diets for various intervals fed; for detailed information please see the article), conc.: 7.66 µg/kg* (mean value), country: Denmark/USA[102], ***8 days after termination** of OTA-exposure
incidence: 5?/5, sa. const.: female pigs of Danish Landrace, age: ≈8 weeks, wt.: ≈20 kg, contamination: artificial (dose: 1 ppm crystalline OTA, o., once daily for 1 month, afterwards toxin-free diets for various intervals fed; for detailed information please see the article), conc.: 3.02 µg/kg* (mean value), country: Denmark/USA[102], ***15 days after termination** of OTA-exposure
incidence: 5?/5, sa. const.: female pigs of Danish Landrace, age: ≈8 weeks, wt.: ≈20 kg, contamination: artificial (dose: 1 ppm crystalline OTA, o., once daily for 1 month, afterwards toxin-free diets for various intervalls fed; for detailed information please see the article), conc.: 0.54 µg/kg* (mean value), country: Denmark/USA[102], ***29 days after termination** of OTA-exposure

incidence: ?/?, sa. const.: weaners (specific pathogen free), wt.: 14–18 kg, contamination: artificial (dose: ? µg OTA addition; for detailed information please see the article), conc. range: ≤61 µg/kg*, country: Denmark[204], *calculated value based on the amount of OTA in blood after 24 h on toxin free diet

incidence: 1/1*, sa. const.: pregnant gilt, contamination: neither OTA nor OTB, conc.: nd, country: UK[266], *control
incidence: 2/2, sa. const.: pregnant gilts, contamination: artificial (dose: **0.38 mg OTA + 0.13 mg OTB**/kg b. wt., o., for 8 days during early pregnancy), conc. range: 0.43–0.70 µg/g*, Ø conc.: 0.565 µg/g*, country: UK[266], *measured on day 30 of pregnancy

incidence: 13/13*, sa. const.: castrated pigs, wt.: 20 kg, contamination: no OTA and/or CIT (for detailed information please see the article), conc.: nd, country: Denmark/USA[330], *control
incidence: 13?/13, sa. const.: castrated pigs, wt.: 20 kg, contamination: artificial (dose: **1,400 µg crystalline OTA**/kg feed, o., for 6 weeks; for detailed information please see the article), conc.: 17 µg/kg*

(mean value), country: Denmark/USA[330], *after 6 weeks
incidence: 13?/13, sa. const.: castrated pigs, wt.: 20 kg, contamination: artificial (dose: **650 µg crystalline CIT**/kg feed, o., for 6 weeks; for detailed information please see the article), conc.: nd*, country: Denmark/USA[330], *after 6 weeks
incidence: 13?/13, sa. const.: castrated pigs, wt.: 20 kg, contamination: artificial (dose: **1,400 µg crystalline OTA + 650 µg crystalline CIT**/kg feed, o., for 6 weeks; for detailed information please see the article), conc.: 17 µg/kg* (mean value), country: Denmark/USA[330], *after 6 weeks
incidence: 13?/13, sa. const.: castrated pigs, wt.: 20 kg, contamination: artificial (dose: **1,400 µg natural OTA + 650 µg natural CIT**/kg feed, o., for 6 weeks; for detailed information please see the article), conc.: 67 µg/kg* (mean value), country: Denmark/USA[330], *after 6 weeks

incidence: 8?/8*, sa. const.: Danish Landrace piglets, age: 8 weeks, contamination: artificial (dose: **18.6 µg OTA**/kg liveweight/day, o., over a period of 6 weeks (treatment 1); for detailed information please see the article), conc.: 16 µg/kg* (mean value), country: Denmark[336], *sa. taken at weaning (pigs 8 weeks old)
incidence: 6?/6*, sa. const.: Danish Landrace piglets, age: 8 weeks, contamination: artificial (dose: **8.0 µg OTA**/kg liveweight/day, o., over a period of 6 weeks (treatment 2); for detailed information please see the article), conc.: 16 µg/kg* (mean value), country: Denmark[336], *sa. taken at weaning (pigs 8 weeks old)
incidence: 8?/8*, sa. const.: Danish Landrace piglets, age: 8 weeks, contamination: artificial (dose: **19.7 µg OTA**/kg liveweight/day, o., over a period of 6 weeks (treatment 3); for detailed information please see the article), conc.: 23 µg/kg* (mean value), country: Denmark[336], *sa. taken at weaning (pigs 8 weeks old)

incidence: 4/4*, sa. const.: male and females pigs, wt.: ≈70 kg, contamination: no OTA (for detailed information please see the article), conc.: nd, country: Germany[365], *control
incidence: 8?/8, sa. const.: male and females pigs, wt.: ≈70 kg, contamination: artificial (dose: **0.09 mg natural OTA**/kg diet/day, in the morning and evening **half of OTA-ration**, o., for 28 days; for detailed information please see the article), conc.: 20.73 ng/g* (mean value), country: Germany[365], *after 28 days
incidence: 8?/8, sa. const.: male and females pigs, wt.: ≈70 kg, contamination: artificial (dose: **0.09 mg crystalline OTA**/kg diet/day, in the morning and evening **half of OTA-ration**, o., for 28 days; for detailed information please see the article), conc.: 8.69 ng/g* (mean value), country: Germany[365], *after 28 days
incidence: 8?/8, sa. const.: male and females pigs, wt.: ≈70 kg, contamination: artificial (dose: **0.09 mg crystalline OTA**/kg diet/day, in the morning **total OTA-ration**, o., for 28 days; for detailed information please see the article), conc.: 10.54 ng/g* (mean value), country: Germany[365], *after 28 days

incidence: 8?/8*, sa. const.: pigs, contamination: no OTA (for detailed information please see the article), conc.: <0.95 µg/kg (mean value), country: Germany[366], *control
incidence: 8?/8, sa. const.: pigs, contamination: artificial (dose: **22.11 mg OTA** (in total) fed for 90 days; for detailed information please see the article), conc.: 16.2 µg/kg (mean value), country: Germany[366], *after 90 days
incidence: 8?/8, sa. const.: pigs, contamination: artificial (dose: **88.44 mg OTA** (in total) fed for 90 days; for detailed information please see the article), conc.: 43.6 µg/kg (mean value), country: Germany[366], *after 90 days

incidence: 2/2*, sa. const.: pigs, contamination: no OTA and/or DON (for detailed information please see the

article), conc. range: 0.16–0.35 ng/g, Ø conc.: 0.255 ng/g, country: Germany[378], *control
incidence: 6/6, sa. const.: pigs, contamination: artificial (dose: **0.1 ppm crystalline OTA + 1.0 ppm crystalline DON**, o., twice daily for 90 days; for detailed information please see the article), conc. range: 3.35–8.25 ng/g*, Ø conc.: 6.05 ng/g*, country: Germany[378], *after 90 days
incidence: 3/3, sa. const.: pigs, contamination: artificial (dose: **0.1 ppm crystalline OTA**, o., twice daily for 90 days; for detailed information please see the article), conc. range: 3.80–4.58 ng/g*, Ø conc.: 4.09 ng/g*, country: Germany[378], *after 90 days
incidence: 6/6, sa. const.: pigs, contamination: artificial (dose: **1.0 ppm crystalline DON!**, o., twice daily for 90 days; for detailed information please see the article), conc. range: 0.42–0.81 ng/g*, Ø conc.: 0.61 ng/g*, country: Germany[378], *after 90 days

incidence: 2/2*, sa. const.: pigs, contamination: no OTA and/or ZEA (for detailed information please see the article), conc.: nr, country: Germany[380], *control
incidence: 6/6, sa. const.: pigs, contamination: artificial (dose: **0.1 ppm OTA + 0.25 ppm ZEA**, o., twice daily for 90 days; for detailed information please see the article), conc. range: ≤9.64 ng/g*, country: Germany[380], *after 91 days (thereof 90 days of OTA- and ZEA-administration)
incidence: 3/3, sa. const.: pigs, contamination: artificial (dose: **0.1 ppm OTA**, o., twice daily for 90 days; for detailed information please see the article), conc. range: ≤4.63 ng/g*, country: Germany[380], *after 91 days (thereof 90 days of OTA-administration)

incidence: 3/3*, sa. const.: female pigs of Danish Landrace, age: 8–10 weeks, wt.: ≈20 kg, contamination: no OTA (for detailed information please see the article), conc.: nd, country: Denmark/USA/Sweden[383], *control
incidence: 4?/4, sa. const.: female pigs of Danish Landrace, age: 8–10 weeks, wt.: ≈20 kg, contamination: artificial (dose: **1 mg crystalline OTA**/kg feed, o., for 3 months; for detailed information please see the article), conc.: ≈26.5 μg/kg* (mean value), country: Denmark/USA/Sweden[383], *after ≈3 months

incidence: 9/9*, sa. const.: Deutsches Landschwein, Deutsches Edelschwein × Piétrain, Belgische Landrasse, wt.: 25 kg, contamination: no OTA, conc.: nd, country: Germany[409], *control
incidence: 2/9, sa. const.: Deutsches Landschwein, Deutsches Edelschwein × Piétrain, Belgische Landrasse, wt.: 25 kg, contamination: artificial (dose: **0.15 mg natural OTA**/kg feed, o., daily for 28 days), conc. range: ≤4.0 μg/kg*, country: Germany[409], *after 28 days
incidence: 8/8*, sa. const.: Deutsches Landschwein, Deutsches Edelschwein × Piétrain, Belgische Landrasse, wt.: 25 kg, contamination: no OTA, conc.: nd, country: Germany[409], *control
incidence: 6/6, sa. const.: Deutsches Landschwein, Deutsches Edelschwein × Piétrain, Belgische Landrasse, wt.: 25 kg, contamination: artificial (dose: **0.58 mg natural OTA**/kg feed, o., daily for 28 days), conc. range: ≤3.6 μg/kg*, country: Germany[409], *after 28 days

T-2 Toxin
incidence: 1/1, sa. const.: female weanling crossbred Yorkshire × Duroc × Hampshire swines, wt.: 7.5 kg, contamination: artificial (dose: **0.1 mg T-2 toxin** (labeled)/kg b. wt., intubated, once), conc.: 15.9 ppb* ** ***, country: USA[318], *calculated residue level, **T-2 toxin and/or metabolites, ***after 18 h
incidence: 1/1, sa. const.: female weanling crossbred Yorkshire × Duroc × Hampshire

swine, wt.: 9.5 kg, contamination: artificial (dose: **0.4 mg T-2 toxin** (labeled)/kg b. wt., intubated, once), conc.: 61.4 ppb* ** ***, country: USA[318], *calculated residue level, **T-2 toxin and/or metabolites, ***after 18 h

incidence: 2/2*, sa. const.: female swines of mixed breeding, wt.: 26–66 kg, contamination: no T-2 toxin, conc.: nr, country: USA[403], *control
incidence: 2/4, sa. const.: female swines of mixed breeding, wt.: 26–66 kg, contamination: artificial (dose: **1.2 mg T-2 toxin**/kg, i.a., once), conc. range: ≈≤30 ppb*, country: USA[403], *after ≈2.1 h (also measured after ≈1 and 3 h, lowest value conc.: under limit of reliable quantitation after 3 h)

incidence: 2/2, sa. const.: female crossbred Yorkshire × Hampshire swines, wt.: 20 kg, contamination: artificial (dose: 0.15 mg T-2 toxin (labeled and unlabeled)/kg b. wt., i.vs., once), conc. range: 0.05–0.08 ng/g*, Ø conc.: 0.065 ng/g*, country: USA[425], *after 4 h
incidence: 2/2, sa. const.: female crossbred Yorkshire × Hampshire swines, wt.: 20 kg, contamination: artificial (dose: 0.15 mg T-2 toxin (labeled and unlabeled)/kg b. wt., i.vs., once), conc. range: 68–74 ng/g* **, Ø conc.: 71 ng/g* **, country: USA[425], *after 4 h, ****total metabolites**

3′-Hydroxy T-2 Toxin
incidence: 2/2, sa. const.: female crossbred Yorkshire × Hampshire swines, wt.: 20 kg, contamination: artificial (dose: 0.15 mg T-2 toxin (labeled and unlabeled)/kg b. wt., i.vs., once), conc. range: 0.47–0.81 ng/g*, Ø conc.: 0.64 ng/g*, country: USA[425], *after 4 h

T-2 Tetraol
incidence: 2/2, sa. const.: female crossbred Yorkshire × Hampshire swines, wt.: 20 kg, contamination: artificial (dose: 0.15 mg T-2 toxin (labeled and unlabeled)/kg b. wt., i.vs., once), conc. range: tr–0.26 ng/g*, country: USA[425], *after 4 h

Deepoxy T-2 Tetraol
incidence: 2/2, sa. const.: female crossbred Yorkshire × Hampshire swines, wt.: 20 kg, contamination: artificial (dose: 0.15 mg T-2 toxin (labeled and unlabeled)/kg b. wt., i.vs., once), conc. range: 4.03–4.30 ng/g*, Ø conc.: 4.165 ng/g*, country: USA[425], *after 4 h

T-2 Triol
incidence: 2/2, sa. const.: female crossbred Yorkshire × Hampshire swines, wt.: 20 kg, contamination: artificial (dose: 0.15 mg T-2 toxin (labeled and unlabeled)/kg b. wt., i.vs., once), conc. range: 0.14–0.16 ng/g*, Ø conc.: 0.15 ng/g*, country: USA[425], *after 4 h

Deepoxy T-2 Triol
incidence: 1/2, sa. const.: female crossbred Yorkshire × Hampshire swines, wt.: 20 kg, contamination: artificial (dose: 0.15 mg T-2 toxin (labeled and unlabeled)/kg b. wt., i.vs., once), conc.: 0.06 ng/g*, country: USA[425], *after 4 h

Zearalenone
incidence: 2?/2?*, sa. const.: male KAHYP pigs, wt.: ?, contamination: no ZEA, conc.: nd, country: Hungary[632], *control
incidence: 2?/2, sa. const.: male KAHYP pigs, wt.: 60 kg, contamination: artificial (dose: **15 ppm ZEA**, o., for 14 days, conc.: 30 µg/kg*, country: Hungary[632], *after 14 days)
incidence: 2?/2?*, sa. const.: male KAHYP pigs, wt.: ?, contamination: no ZEA, conc.: nd, country: Hungary[632], *control
incidence: 2?/2, sa. const.: male KAHYP pigs, wt.: 60 kg, contamination: artificial (dose: **15 ppm ZEA**, o., for 14 days, conc.: <10 µg/kg* **, country: Hungary[632], *after 14 days, **in **fat around kidney**)

α-Zearalenol
incidence: 2?/2*, sa. const.: pigs, contamination: no OTA and/or ZEA (for detailed information please see the article), conc.: nr, country: Germany[380], *control
incidence: 6/6, sa. const.: pigs, contamination: artificial (dose: **0.1 ppm**

OTA + 0.25 ppm ZEA, o., twice daily for 90 days; for detailed information please see the article), conc. range: ≤4.30 ng/g*, country: Germany[380], *after 91 days (thereof 90 days of OTA- and ZEA-administration)

incidence: 2?/2?*, sa. const.: male KAHYP pigs, wt.: ?, contamination: no ZEA, conc.: nd, country: Hungary[632], *control
incidence: 2?/2, sa. const.: male KAHYP pigs, wt.: 60 kg, contamination: artificial (dose: **15 ppm ZEA**, o., for 14 days, conc.: 220 μg/kg*, country: Hungary[632], *after 14 days)
incidence: 2?/2?*, sa. const.: male KAHYP pigs, wt.: ?, contamination: no ZEA, conc.: nd, country: Hungary[632], *control
incidence: 2?/2, sa. const.: male KAHYP pigs, wt.: 60 kg, contamination: artificial (dose: **15 ppm ZEA**, o., for 14 days, conc.: 15 μg/kg* **, country: Hungary[632], *after 14 days, **in **fat around kidney**)

β-Zearalenol
incidence: 2?/2?*, sa. const.: male KAHYP pigs, wt.: ?, contamination: no ZEA, conc.: nd, country: Hungaryy[632], *control
incidence: 2?/2, sa. const.: male KAHYP pigs, wt.: 60 kg, contamination: artificial (dose: **15 ppm ZEA**, o., for 14 days, conc.: 85 μg/kg*, country: Hungary[632], *after 14 days)
incidence: 2?/2?*, sa. const.: male KAHYP pigs, wt.: ?, contamination: no ZEA, conc.: nd, country: Hungary[632], *control
incidence: 2?/2, sa. const.: male KAHYP pigs, wt.: 60 kg, contamination: artificial (dose: **15 ppm ZEA**, o., for 14 days, conc.: <10 μg/kg* **, country: Hungary[632], *after 14 days, **in **fat around kidney**)

Pig Leaf Fat see Pig fat

Pig Lean see Pig muscle

Pig liver may contain the following mycotoxins and/or their metabolites:

Aflatoxicol
incidence: 3/3, sa. const.: castrated male pigs of mixed breed, wt.t: 9–11 kg, contamination: artificial (**acute study**, dose: 1 mg AFB_1/kg b. wt., o., once), conc. range: ≤1.68 ng/g*, country: USA[362], *in pig that died after 22 h (2 other pigs sacrificed 24 and 72 h after treatment showed lower mycotoxin values)
incidence: 4/5, sa. const.: market-weight pigs, weight: ≈92 kg, contamination: artificial (**subacute study**, dose: ≈15 μg AFB_1 as well as AFB_2, AFG_1 + AFG_2 (all natural)/kg b. wt., o., for 14 days; for detailed information please see the article), conc. range: 0.01–0.02 ng/g*, Ø conc.: 0.015 ng/g*, country: USA[362], *after 14 days

Aflatoxin B_1
incidence: 2/2*, sa. const.: castrated male Yorkshire-Hampshire-Duroc tricross pigs, age: 3–4 weeks, contamination: artificial (dose: **41 ng AFB_1**/g diet, o., for 3 weeks (control); for detailed information please see the article), conc.: nd**, country: USA[60], *control, **sacrificed on withdrawal day 0 (animals sacrificed later showed no AFB_1-contamination)
incidence: 2/2, sa. const.: castrated male Yorkshire-Hampshire-Duroc tricross pigs, age: 3–4 weeks, contamination: artificial (dose: **341 ng AFB_1**/g diet, o., for 3 weeks; for detailed information please see the article), conc.: nd*, country: USA[60], *sacrificed on withdrawal day 0 (animals sacrificed later showed no AFB_1-contamination)
incidence: 2/2, sa. const.: castrated male Yorkshire-Hampshire-Duroc tricross pigs, age: 3–4 weeks, contamination: artificial (dose: **866 ng AFB_1**/g diet, o., for 3 weeks; for detailed information please see the article), conc. range: 0.17–0.18 ng/g tissue*, Ø conc.: 0.175 ng/g tissue*, country: USA[60], *sacrificed on withdrawal day 0 (animals sacrificed later showed no AFB_1-contamination)
incidence: 2/2, sa. const.: castrated male Yorkshire-Hampshire-Duroc tricross pigs, age: 3–4 weeks, contamination: artificial (dose: **1,253 ng AFB_1**/g diet, o., for 3 weeks; for detailed information please

see the article), conc. range: 0.36–0.43 ng/g tissue*, Ø conc.: 0.395 ng/g tissue*, country: USA[60], *sacrificed on withdrawal day 0 (animals sacrificed later showed no AFB_1-contamination)

incidence: 1/1, sa. const.: female Hampshire × Deutsches Edelschwein piglets, wt.: 15 kg, contamination: artificial (dose: 3.1 µg AFB_1 (labeled)/kg b.w., o., once; for detailed information please see the article), conc.: 17.3 ppb* ** ***, country: Switzerland[66], *AFB_1 eq., **total activity, ***after **24 h**
incidence: 1/1, sa. const.: female Hampshire × Deutsches Edelschwein piglets, wt.: 15 kg, contamination: artificial (dose: 3.1 µg AFB_1 (labeled)/kg b.w., o., once; for detailed information please see the article), conc.: 17.3 ppb* ** ***, country: Switzerland[66], *AFB_1 eq., **total activity, ***after **48 h**

incidence: 1/1, sa. const.: pig, contamination: artificial (dose: 1,200 ng AFB_1/g feed, o., for 3 weeks), conc.: 0.27 ng/g*, country: USA[75], *after 3 weeks?

incidence: 2/2, sa. const.: adult swines, contamination: artificial (dose: 1.08–1.09 mg AFB_1 (besides other AFs), o., daily for 33 days; for detailed information please see the article), conc. range: 0.08–2.50 ppb*, Ø conc.: 1.29 ppb*, country: France[88], *after 33 days

incidence: 20/20*, sa. const.: male and female Danish Landrace pigs, age: ≈8 weeks, wt.: 20 kg up to slaughter at 90 kg, contamination: no AFB_1 + AFB_2 (for detailed information please see the article), conc.: nd, country: Denmark/USA[101], *control
incidence: 14/18*, sa. const.: male and female Danish Landrace pigs, age: ≈8 weeks, wt.: 20 kg up to slaughter at 90 kg, contamination: artificial (dose: AFB_1 + AFB_2 addition (overall **300 ppb AFs**), o., for **141** or 150*** days**; for detailed information please see the article), conc. range: tr***–92** ppb, country: Denmark/USA[101], *included are liver sa. rejected at meat inspection, after 141** or 150*** days
incidence: 14/18*, sa. const.: male and female Danish Landrace pigs, age: ≈8 weeks, wt.: 20 kg up to slaughter at 90 kg, contamination: artificial (dose: AFB_1 + AFB_2 addition (overall **500 ppb AFs**), o., for **157** or 231*** days**; for detailed information please see the article), conc. range: tr***–51 ppb**, country: Denmark/USA[101], *included are liver sa. rejected at meat inspection, after 157** or 231*** days

incidence: 2/4*, sa. const.: male and female feeder pigs, wt.: 54.2–71.6 kg, contamination: no AFB_1, conc.: <0.12 µg/kg, country: USA[136], *control
incidence: 4/4, sa. const.: male and female feeder pigs, wt.: 54.2–71.6 kg, contamination: artificial (dose: **100 µg AFB_1/kg** diet, o., for 4 weeks), conc. range: 0.18–0.25 µg/kg*, Ø conc.: 0.23 µg/kg*, country: USA[136], *after 4 weeks
incidence: 4/4, sa. const.: male and female feeder pigs, wt.: 54.2–71.6 kg, contamination: artificial (dose: **200 µg AFB_1/kg** diet, o., for 4 weeks), conc. range: 0.18–0.75 µg/kg*, Ø conc.: 0.48 µg/kg*, country: USA[136], *after 4 weeks
incidence: 4/4, sa. const.: male and female feeder pigs, wt. 54.2–71.6 kg, contamination: artificial (dose: **400 µg AFB_1/kg** diet, o., for 4 weeks), conc. range: 0.70–2.66 µg/kg*, Ø conc.: 1.51 µg/kg*, country: USA[136], *after 4 weeks

incidence: 4/4*, sa. const.: crossbred (Duroc × Yorkshire) barrows, wt.: 24.5–26.3 kg, contamination: no AFs, conc.: nd, country: USA[138], *control
incidence: 4?/4, sa. const.: crossbred (Duroc × Yorkshire) barrows, wt.: 24.5–26.3 kg, contamination: artificial (dose: **662 µg AFB_1**/kg diet, **273 µg AFB_2**/kg diet, **300 µg AFG_1**/kg diet and **285 µg AFG_2**/kg diet, o., for 21 days), conc. range: 0.05–0.10 µg/kg*, country: USA[138], *after ≈22 days (thereof 21 days of AFs-administration)

incidence: 10/10*, sa. const.: mixed breed feeder pigs, contamination: chronic study, no AFs (for detailed information please see the article), conc.: nd, country: USA[182], *control
incidence: 10?/10, sa. const.: mixed breed feeder pigs, contamination: artificial (chronic study, dose: **400 ng natural AFs/g** diet, o., for 10 weeks; for detailed information please see the article), conc.: 0.51 ng/g* (wet matter basis) (mean value), country: USA[182], *after 10 weeks
incidence: 10?/10, sa. const.: mixed breed feeder pigs, contamination: artificial (chronic study, dose: **800 ng natural AFs/g** diet, o., for 10 weeks; for detailed information please see the article), conc.: 1.57 ng/g* (wet matter basis) (mean value), country: USA[182], *after 10 weeks

incidence: 8/8*, sa. const.: mixed breed feeder pigs, contamination: acute study, no AFs, conc.: nd, country: USA[182], *control
incidence: 1/1(8)*, sa. const.: mixed breed feeder pigs, contamination: artificial (acute study, dose: **1.2 mg total AFs** (AFB_1+AFB_2+AFG_1+AFG_2)/kg b. wt., o., once, conc. range: ≤9.00 ng/g** (wet matter basis), country: USA[182], *for overall information please see the article, **12 h post-dosage (also measured 24, 48 and 72 h post-dosage, lowest conc.: nd after 72 h)

incidence: ?/?*, sa. const.: Large White growing female pigs, age: 79 days, wt.: 20 kg, contamination: no AFs (for detailed information please see the article), conc.: nd, country: France[314], *control
incidence: 1/1, sa. const.: Large White growing female pigs, age: 79 days, wt.: 20 kg, contamination: artificial (dose: natural AFs addition (treatment 1: mixed feeding, avg. daily intake of **870 µg AFB_1**) for 26 days; for detailed information please see the article), conc.: nd*, country: France[314], *final wt. of the animal 109 kg
incidence: 2/2, sa. const.: Large White growing female pigs, age: 79 days, wt.: 20 kg, contamination: artificial (dose: natural AFs addition (treatment 2: separate feeding = peanut oil meal 40% and corn gluten meal 30%, avg. daily intake of **1,566 µg AFB_1**) for 26* or 19** days; for detailed information please see the article), conc. range: 0.45*–6.10** µg/kg, country: France[314], final wt. of the animal 105* and 97** kg
incidence: 1/1, sa. const.: Large White growing female pigs, age: 79 days, wt.: 20 kg, contamination: artificial (dose: natural AFs addition (treatment 3: separate feeding = peanut oil meal 0% and corn gluten meal 0%, avg. daily intake of **642 µg AFB_1**) for 26 days; for detailed information please see the article), conc.: 0.403 µg/kg*, country: France[314], *final wt. of the animal 104 kg

incidence: 3/3, sa. const.: castrated male pigs of mixed breed, wt.: 9–11 kg, contamination: artificial (**acute study**, dose: 1 mg AFB_1/kg b. wt., o., once), conc. range: ≤36.5 ng/g*, country: USA[362], *in 1 pig that died after 22 h (2 other pigs sacrificed 24 and 72 h after treatment showed lower mycotoxin values)
incidence: 5/5, sa. const.: market-weight pigs, wt.: ≈92 kg, contamination: artificial (**subacute study**, dose: ≈15 µg AFB_1 as well as AFB_2, AFG_1 and AFG_2 (all natural)/kg b. wt., o., for 14 days; for detailed information please see the article), conc. range: 0.15–0.68 ng/g*, Ø conc.: 0.344 ng/g*, country: USA[362], *after 14 days

incidence: 5?/5*, sa. const.: cross-bred pigs, contamination: (dose: 9 ng/g AFB_1 + AFB_2, o., for 35 days; for detailed information please see the article), conc.: 0.002 ng/g** (mean value), country: USA[597], *control, **after 35 days
incidence: 5?/5, sa. const.: cross-bred pigs, contamination: artificial (dose: **524 ng/g AFB_1 + AFB_2**, o., for 35 days; for detailed information please see the article), conc.: 0.484 ng/g* (mean value), country: USA[597], *after 35 days
incidence: 5?/5, sa. const.: cross-bred pigs, contamination: artificial (dose: **524 ng/g**

AFB_1 + AFB_2 + HSCAS (0.5%), o., for 35 days; for detailed information please see the article), conc.: 0.500 ng/g* (mean value), country: USA[597], *after 35 days

incidence: 5/5*, sa. const.: male miniature pigs, contamination: (dose: 1 ng/g AFB_1 + AFB_2, o., for 15 days; for detailed information please see the article), conc.: nd, country: USA[597], *control
incidence: 5?/5, sa. const.: male miniature pigs, contamination: artificial (dose: **590 ng/g AFB_1 + AFB_2**, o., for 15 days; for detailed information please see the article), conc.: 0.310 ng/g* (mean value), country: USA[597], *after 15 days
incidence: 5?/5, sa. const.: male miniature pigs, contamination: artificial (dose: **590 ng/g AFB_1 + AFB_2 + HSCAS** (0.5%), o., for 15 days; for detailed information please see the article), conc.: 0.240 ng/g* (mean value), country: USA[597], *after 15 days
incidence: 5/5, sa. const.: male miniature pigs, contamination: artificial (dose: **590 ng/g AFB_1 + AFB_2**, o., for 15 days followed by **2 weeks control diet**; for detailed information please see the article), conc.: nd*, country: USA[597], *after 29 days (thereof 15 days of AFB_1- and AFB_2-administration)

Aflatoxin B_2
incidence: 2/2, sa. const.: adult swines, contamination: artificial (dose: 1.08–1.09 mg AFB_1 (besides other AFs), o., daily for 33 days; for detailed information please see the article), conc. range: 0.05–0.90 ppb*, Ø conc.: 0.48 ppb*, country: France[88], *after 33 days

incidence: 20/20*, sa. const.: male and female Danish Landrace pigs, age: ≈8 weeks, wt.: 20 kg up to slaughter at 90 kg, contamination: no AFB_1 + AFB_2 (for detailed information please see the article), conc.: nd, country: Denmark/USA[101], *control
incidence: 7/18*, sa. const.: male and female Danish Landrace pigs, age: ≈8 weeks, wt.: 20 kg up to slaughter at 90 kg, contamination: artificial (dose: AFB_1 + AFB_2-addition (overall **300 ppb AFs**), o., for **141** or 150*** days**; for detailed information please see the article), conc. range: tr***–45** ppb, country: Denmark/USA[101], *included are liver sa. rejected at meat inspection, after 141** or 150*** days
incidence: 10/18*, sa. const.: male and female Danish Landrace pigs, age: ≈8 weeks, wt.: 20 kg up to slaughter at 90 kg, contamination: artificial (dose: AFB_1 + AFB_2-addition (overall **500 ppb AFs**), o., for **141** or 231*** days**; for detailed information please see the article), conc. range: tr***–15** ppb, country: Denmark/USA[101], *included are liver sa. rejected at meat inspection, after 141** or 231*** days

incidence: 4/4*, sa. const.: crossbred (Duroc × Yorkshire) barrows, weight: 24.5–26.3 kg, contamination: no AFs, conc.: nd, country: USA[138], *control
incidence: 4?/4, sa. const.: crossbred (Duroc × Yorkshire) barrows, wt.: 24.5–26.3 kg, contamination: artificial (dose: **662 µg AFB_1/kg** diet, **273 µg AFB_2/kg** diet, **300 µg AFG_1/kg** diet and **285 µg AFG_2/kg** diet, o., for 21 days), conc. range: tr–0.06 µg/kg*, country: USA[138], *after ≈22 days (thereof 21 days of AF-administration)

incidence: 10/10*, sa. const.: mixed breed feeder pigs, contamination: no AFs, chronic study, (for detailed information please see the article), conc.: nd, country: USA[182], *control
incidence: 10?/10, sa. const.: mixed breed feeder pigs, contamination: artificial (chronic study, dose: **400 ng natural AFs**/g diet, o., for 10 weeks; for detailed information please see the article), conc.: 0.03 ng/g* (wet matter basis) (mean value), country: USA[182], *after 10 weeks
incidence: 10?/10, sa. const.: mixed breed feeder pigs, contamination: artificial (chronic study, dose: **800 ng natural AFs**/g diet, o., for 10 weeks; for detailed

information please see the article), conc.: 0.17 ng/g* (wet matter basis) (mean value), country: USA[182], *after 10 weeks

incidence: 8/8*, sa. const.: mixed breed feeder pigs, contamination: acute study, no AFs, conc.: nd, country: USA[182], *control
incidence: 1/1(8)*, sa. const.: mixed breed feeder pigs, contamination: artificial (acute study, dose: **1.2 mg total AFs** (AFB_1+AFB_2+AFG_1+AFG_2)/kg b. wt., o., once, conc. range: ≤0.75 ng/g** (wet matter basis), country: USA[182], *for overall information please see the article, **48 h post-dosage (also measured 12, 24 and 72 h post-dosage, lowest conc.: nd after 72 h)

incidence: ?/?*, sa. const.: Large White growing female pigs, age: 79 days, wt.: 20 kg, contamination: no AFs (for detailed information please see the article), conc.: nd, country: France[314], *control
incidence: 1/1, sa. const.: Large White growing female pigs, age: 79 days, wt.: 20 kg, contamination: artificial (dose: natural AFs addition (treatment 1: mixed feeding, avg. daily intake of **870 µg AFB_1**), for 26 days; for detailed information please see the article), conc.: nd*, country: France[314], *final wt. of the animal 109 kg
incidence: 2/2, sa. const.: Large White growing female pigs, age: 79 days, wt.: 20 kg, contamination: artificial (dose: natural AFs addition (treatment 2: separate feeding = peanut oil meal 40% and corn gluten meal 30%, avg. daily intake of **1,566 µg AFB_1**), for 26* or 19** days; for detailed information please see the article), conc. range: 0.45*–1.05** µg/kg, country: France[314], final wt. of the animal 105* and 97** kg
incidence: 1/1, sa. const.: Large White growing female pigs, age: 79 days, wt.: 20 kg, contamination: artificial (dose: natural AFs addition (treatment 3: separate feeding = peanut oil meal 0% and corn gluten meal 0%, avg. daily intake of **642 µg AFB_1**), for 26 days; for detailed information please see the article), conc.: 0.403 µg/kg*, country: France[314], *final wt. of the animal 104 kg

incidence: 5/5*, sa. const.: cross-bred pigs, contamination: (dose: 9 ng/g AFB_1 + AFB_2, o., for 35 days; for detailed information please see the article), conc.: nd, country: USA[597], *control
incidence: 5?/5, sa. const.: cross-bred pigs, contamination: artificial (dose: **524 ng/g AFB_1 + AFB_2**, o., for 35 days; for detailed information please see the article), conc.: 0.053 ng/g* (mean value), country: USA[597], *after **35 days**
incidence: 5?/5, sa. const.: cross-bred pigs, contamination: artificial (dose: **524 ng/g AFB_1 + AFB_2 + HSCAS** (0.5%), o., for 35 days; for detailed information please see the article), conc.: 0.015 ng/g* (mean value), country: USA[597], *after **35 days**

incidence: 5/5*, sa. const.: male miniature pigs, contamination: (dose: 1 ng/g AFB_1 + AFB_2, o., for 15 days; for detailed information please see the article), conc.: nd, country: USA[597], *control
incidence: 5?/5, sa. const.: male miniature pigs, contamination: artificial (dose: **590 ng/g AFB_1 + AFB_2**, o., for 15 days; for detailed information please see the article), conc.: 0.010 ng/g* (mean value), country: USA[597], *after **15 days**
incidence: 5?/5, sa. const.: male miniature pigs, contamination: artificial (dose: **590 ng/g AFB_1 + AFB_2 + HSCAS** (0.5%), o., for 15 days; for detailed information please see the article), conc.: 0.020 ng/g* (mean value), country: USA[597], *after **15 days**
incidence: 5/5, sa. const.: male miniature pigs, contamination: artificial (dose: **590 ng/g AFB_1 + AFB_2**, o., for 15 days followed by **2 weeks control diet**; for detailed information please see the article), conc.: nd*, country: USA[597], *after **29 days** (thereof 15 days of AFB_1- and AFB_2-administration)

Aflatoxin B_{2a}

incidence: 4/4*, sa. const.: crossbred (Duroc × Yorkshire) barrows, wt.: 24.5–26.3 kg, contamination: no AFs, conc.: nd, country: USA[138], *control
incidence: 4?/4, sa. const.: crossbred (Duroc × Yorkshire) barrows, wt.: 24.5–26.3 kg, contamination: artificial (dose: **662 µg AFB_1/kg diet, 273 µg AFB_2/kg diet, 300 µg AFG_1/kg** diet and **285 µg AFG_2/kg** diet, o., for 21 days), conc. range: appreciable amounts*, country: USA[138], *after ≈22 days (thereof 21 days of AF-administration)

Aflatoxin B

incidence: 5?/5*, sa. const.: cross-bred pigs, contamination: (dose: 9 ng/g AFB_1 + AFB_2, o., for 35 days; for detailed information please see the article), conc.: 6.893 pmol/mg DNA** *** (mean value), country: USA[597], *control, **AFB N^7 formamido-pyrimidine DNA adducts, ***after 35 days
incidence: 5?/5, sa. const.: cross-bred pigs, contamination: artificial (dose: **524 ng/g AFB_1 + AFB_2**, o., for 35 days; for detailed information please see the article), conc.: 32.473 pmol/mg DNA* ** (mean value), country: USA[597], *after **35 days**, ** AFB N^7 formamido-pyrimidine DNA adducts
incidence: 5?/5, sa. const.: cross-bred pigs, contamination: artificial (dose: **524 ng/g AFB_1 + AFB_2 + HSCAS** (0.5%), o., for 35 days; for detailed information please see the article), conc.: 10.197 pmol/mg DNA* ** (mean value), country: USA[597], *after **35 days**, ** AFB N^7 formamido-pyrimidine DNA adducts

incidence: 5/5*, sa. const.: male miniature pigs, contamination: (dose: 1 ng/g AFB_1 + AFB_2, o., for 15 days; for detailed information please see the article), conc.: nd, country: USA[597], *control
incidence: 5?/5, sa. const.: male miniature pigs, contamination: artificial (dose: **590 ng/g AFB_1 + AFB_2**, o., for 15 days; for detailed information please see the article), conc.: 7.400 pmol/mg DNA* ** (mean value), country: USA[597], *AFB N^7 formamido-pyrimidine DNA adducts, **after **15 days**
incidence: 5?/5, sa. const.: male miniature pigs, contamination: artificial (dose: **590 ng/g AFB_1 + AFB_2 + HSCAS** (0.5%), o., for 15 days; for detailed information please see the article), conc.: 4.670 pmol/mg DNA* ** (mean value), country: USA[597], *AFB N^7 formamido-pyrimidine DNA adducts, **after **15 days**
incidence: 5/5, sa. const.: male miniature pigs, contamination: artificial (dose: **590 ng/g AFB_1 + AFB_2**, o., for 15 days followed by **2 weeks control diet**; for detailed information please see the article), conc.: 7.750 pmol/mg DNA* ** (mean value), country: USA[597], *AFB N^7 formamido-pyrimidine DNA adducts, **after **29 days** (thereof 15 days of AFB_1- and AFB_2-administration)

Aflatoxin G_1

incidence: 1/1, sa. const.: pig, contamination: artificial (dose: 1,200 ng AFB_1/g feed, o., for 3 weeks), conc.: 0.1 ng/g*, country: USA[75], *after 3 weeks?

incidence: 10/10*, sa. const.: mixed breed feeder pigs, contamination: no AFs, chronic study, (for detailed information please see the article), conc.: nd, country: USA[182], *control
incidence: 10?/10, sa. const.: mixed breed feeder pigs, contamination: artificial (chronic study, dose: **400 ng natural AFs**/g diet, o., for 10 weeks; for detailed information please see the article), conc.: 0.31 ng/g* (wet matter basis) (mean value), country: USA[182], *after 10 weeks
incidence: 10?/10, sa. const.: mixed breed feeder pigs, contamination: artificial (chronic study, dose: **800 ng natural AFs**/g diet, o., for 10 weeks; for detailed information please see the article), conc.: nd*, country: USA[182], *after 10 weeks

incidence: 8/8*, sa. const.: mixed breed feeder pigs, contamination: no AFs, acute study, conc.: nd, country: USA[182], *control
incidence: 1/1(8)*, sa. const.: mixed breed feeder pigs, contamination: artificial (acute study, dose: **1.2 mg total AFs** (AFB_1+AFB_2+AFG_1+AFG_2)/kg b. wt., o., once, conc. range: ≤0.53 ng/g** (wet matter basis), country: USA[182], *for overall information please see the article, **24 h post-dosage (also measured 12, 48 and 72 h post-dosage, lowest conc.: nd after 72 h)

Aflatoxin G_2
incidence: 8/8*, sa. const.: mixed breed feeder pigs, contamination: no AFs, acute study, conc.: nd, country: USA[182], *control
incidence: 1/1(8)*, sa. const.: mixed breed feeder pigs, contamination: artificial (acute study, dose: **1.2 mg total AFs** (AFB_1+AFB_2+AFG_1+AFG_2)/kg b. wt., o., once, conc. range: ≤0.08 ng/g** (wet matter basis), country: USA[182], *for overall information please see the article, **12 h post-dosage (also measured 24, 48 and 72 h post-dosage, lowest conc.: nd after 48 and 72 h)

Aflatoxin M_1
incidence: 2/2*, sa. const.: castrated male Yorkshire-Hampshire-Duroc tricross pigs, age: 3–4 weeks, contamination: artificial (dose: **41 ng AFB_1**/g diet, o., for 3 weeks; for detailed information please see the article), conc.: nd**, country: USA[60], *control, **sacrificed on withdrawal day 0 (animals sacrificed later showed no AFM_1-contamination)
incidence: 2/2, sa. const.: castrated male Yorkshire-Hampshire-Duroc tricross pigs, age: 3–4 weeks, contamination: artificial (dose: **341 ng AFB_1**/g diet, o., for 3 weeks; for detailed information please see the article), conc. range: 0.32–0.39 ng/g tissue*, Ø conc.: 0.355 ng/g tissue*, country: USA[60], *sacrificed on withdrawal day 0 (animals sacrificed later showed no AFM_1-contamination)
incidence: 2/2, sa. const.: castrated male Yorkshire-Hampshire-Duroc tricross pigs, age: 3–4 weeks, contamination: artificial (dose: **866 ng AFB_1**/g diet, o., for 3 weeks; for detailed information please see the article), conc. range: 0.28–0.30 ng/g tissue*, Ø conc.: 0.29 ng/g tissue*, country: USA[60], *sacrificed on withdrawal day 0 (animals sacrificed later showed lower or no residue values)
incidence: 2/2, sa. const.: castrated male Yorkshire-Hampshire-Duroc tricross pigs, age: 3–4 weeks, contamination: artificial (dose: **1,253 ng AFB_1**/g diet, o., for 3 weeks; for detailed information please see the article), conc. range: 0.59–0.77 ng/g tissue*, Ø conc.: 0.68 ng/g tissue*, country: USA[60], *sacrificed on withdrawal day 0 (animals sacrificed later showed lower or no residue values)

incidence: ?/?, sa. const.: pigs, contamination: artificial (dose: 1,200 ng AFB_1/g feed, o., for 3 weeks), conc.: 0.76 ng/g*, country: USA[75], *after 3 weeks

incidence: 1/2, sa. const.: adult swines, contamination: artificial (dose: 1.08–1.09 mg AFB_1 (besides other AFs), o., daily for 33 days; for detailed information please see the article), conc.: tr*, country: France[88], *after 33 days

incidence: 2/4*, sa. const.: male and female feeder pigs, wt.: 54.2–71.6 kg, contamination: no AFB_1, conc.: <0.03 µg/kg**, country: USA[136], *control, **after 4 weeks
incidence: 4/4, sa. const.: male and female feeder pigs, wt.: 54.2–71.6 kg, contamination: artificial (dose: **100 µg AFB_1**/kg diet, o., for 4 weeks), conc. range: 0.05–0.23 µg/kg*, Ø conc.: 0.14 µg/kg*, country: USA[136], *after 4 weeks
incidence: 4/4, sa. const.: male and female feeder pigs, wt.: 54.2–71.6 kg, contamination: artificial (dose: **200 µg AFB_1**/kg diet, o., for 4 weeks), conc. range: 0.11–1.50 µg/kg*, Ø conc.: 0.58 µg/kg*, country: USA[136], *after 4 weeks

incidence: 4/4, sa. const.: male and female feeder pigs, wt.: 54.2–71.6 kg, contamination: artificial (dose: **400 μg AFB_1/kg** diet, o., for 4 weeks), conc. range: 1.02–2.00 μg/kg*, Ø conc.: 1.43 μg/kg*, country: USA[136], *after 4 weeks

incidence: 4/4*, sa. const.: crossbred (Duroc × Yorkshire) barrows, wt.: 24.5–26.3 kg, contamination: no AFs, conc.: nd, country: USA[138], *control
incidence: 4?/4, sa. const.: crossbred (Duroc × Yorkshire) barrows, wt.: 24.5–26.3 kg, contamination: artificial (dose: **662 μg AFB_1**/kg diet, **273 μg AFB_2**/kg diet, **300 μg AFG_1**/kg diet and **285 μg AFG_2**/kg diet, o., for 21 days), conc. range: tr–0.20 μg/kg*, country: USA[138], *after ≈22 days (thereof 21 days of AF-administration)

incidence: 10/10*, sa. const.: mixed breed feeder pigs, contamination: no AFs, chronic study (for detailed information please see the article), conc.: nd, country: USA[182], *control
incidence: 10?/10, sa. const.: mixed breed feeder pigs, contamination: artificial (chronic study, dose: **400 ng natural AFs**/g diet, o., for 10 weeks; for detailed information please see the article), conc.: 0.58 ng/g* (wet matter basis) (mean value), country: USA[182], *after 10 weeks
incidence: 10?/10, sa. const.: mixed breed feeder pigs, contamination: artificial (chronic study, dose: **800 ng natural AFs**/g diet, o., for 10 weeks; for detailed information please see the article), conc.: 1.07 ng/g* (wet matter basis) (mean value), country: USA[182], *after 10 weeks

incidence: 8/8*, sa. const.: mixed breed feeder pigs, contamination: no AFs, acute study (for detailed information please see the article), conc.: nd, country: USA[182], *control
incidence: 1/1(8)*, sa. const.: mixed breed feeder pigs, contamination: artificial (acute study, dose: **1.2 mg total AFs** (AFB_1+AFB_2+AFG_1+AFG_2)/kg b. wt., o., once, conc. range: ≤16.80 ng/g** (wet matter basis), country: USA[182], *for overall information please see the article, **24 h post-dosage (also measured 12, 48 and 72 h post-dosage, lowest conc.: nd after 72 h)

incidence: 3/3, sa. const.: castrated male pigs of mixed breed, wt.: 9–11 kg, contamination: artificial (**acute study**, dose: 1 mg AFB_1/kg b. wt., o., once), conc. range: ≤6.0 ng/g*, country: USA[362], *in pig that died after 22 h (2 other pigs sacrificed 24 and 72 h after treatment showed lower mycotoxin values)
incidence: 5/5, sa. const.: market-weight pigs, wt.: ≈92 kg, contamination: artificial (**subacute study**, dose: ≈15 μg AFB_1 as well as AFB_2, AFG_1 and AFG_2 (all natural)/kg b. wt., o., for 14 days; for detailed information please see the article), conc. range: 0.51–1.70 ng/g*, Ø conc.: 0.886 ng/g*, country: USA[362], *after 14 days

incidence: 5?/5*, sa. const.: cross-bred pigs, contamination: (dose: 9 ng/g AFB_1 + AFB_2, o., for 35 days; for detailed information please see the article), conc.: 0.071 ng/g** (mean value), country: USA[597], *control, **after 35 days
incidence: 5?/5, sa. const.: cross-bred pigs, contamination: artificial (dose: **524 ng/g AFB_1 + AFB_2**, o., for 35 days; for detailed information please see the article), conc.: 1.479 ng/g* (mean value), country: USA[597], *after **35 days**
incidence: 5?/5, sa. const.: cross-bred pigs, contamination: artificial (dose: **524 ng/g AFB_1 + AFB_2 + HSCAS** (0.5%), o., for 35 days; for detailed information please see the article), conc.: 0.547 ng/g* (mean value), country: USA[597], *after **35 days**

incidence: 5/5*, sa. const.: male miniature pigs, contamination: (dose: 1 ng/g AFB_1 + AFB_2, o., for 15 days; for detailed

information please see the article), conc.: nd, country: USA[597], *control
incidence: 5?/5, sa. const.: male miniature pigs, contamination: artificial (dose: **590 ng/g AFB_1 + AFB_2**, o., for 15 days; for detailed information please see the article), conc.: 2.850 ng/g* (mean value), country: USA[597], *after **15 days**
incidence: 5?/5, sa. const.: male miniature pigs, contamination: artificial (dose: **590 ng/g AFB_1 + AFB_2 + HSCAS** (0.5%), o., for 15 days; for detailed information please see the article), conc.: 0.780 ng/g* (mean value), country: USA[597], *after **15 days**
incidence: 5?/5, sa. const.: male miniature pigs, contamination: artificial (dose: **590 ng/g AFB_1 + AFB_2**, o., for 15 days followed by **2 weeks control diet**; for detailed information please see the article), conc.: 0.050 ng/g* (mean value), country: USA[597], *after **29 days** (thereof 15 days of AFB_1- and AFB_2-administration)

Aflatoxin M
incidence: 1/20*, sa. const.: male and female Danish Landrace pigs, age: ≈8 weeks, wt.: 20 kg up to slaughter at 90 kg, contamination: no AFB_1 + AFB_2 (for detailed information please see the article), conc.: pr**, country: Denmark/USA[101], *control, **estimated
incidence: 3/18*, sa. const.: male and female Danish Landrace pigs, age: ≈8 weeks, wt.: 20 kg up to slaughter at 90 kg, contamination: artificial (dose: AFB_1 + AFB_2 addition (overall **300 ppb AFs**), o., for **120** or 141*** days**; for detailed information please see the article), conc. range: tr***–3** ppb*, country: Denmark/USA[101], *included are liver sa. rejected at meat inspection, after 120** or 141*** days
incidence: 3/18*, sa. const.: male and female Danish Landrace pigs, age: ≈8 weeks, wt.: 20 kg up to slaughter at 90 kg, contamination: artificial (dose: AFB_1 + AFB_2 addition (overall **500 ppb AFs**), o., for **135** or 186*** days**; for detailed information please see the article), conc. range: tr***–3** ppb, country: Denmark/USA[101], *included are liver sa. rejected at meat inspection, after 135** or 186*** days

incidence: ?/?*, sa. const.: Large White growing female pigs, age: 79 days, wt.: 20 kg, contamination: no AFs (for detailed information please see the article), conc.: nd, country: France[314], *control
incidence: 1/1, sa. const.: Large White growing female pigs, age: 79 days, wt.: 20 kg, contamination: artificial (dose: natural AFs addition (treatment 1: mixed feeding, avg. daily intake of **870 µg AFB_1**) for 26 days; for detailed information please see the article), conc.: nd*, country: France[314], *final wt. of the animal 109 kg
incidence: 1/1, sa. const.: Large White growing female pigs, age: 79 days, wt.: 20 kg, contamination: artificial (dose: natural AFs addition (treatment 2: separate feeding = peanut oil meal 40% and corn gluten meal 30%, avg. daily intake of **1,566 µg AFB_1**) for 19 days; for detailed information please see the article), conc.: 2.82 µg/kg*, country: France[314], *final wt. of the animal 97 kg
incidence: 1/1, sa. const.: Large White growing female pigs, age: 79 days, wt.: 20 kg, contamination: artificial (dose: natural AFs addition (treatment 3: separate feeding = peanut oil meal 0% and corn gluten meal 0%, avg. daily intake of **642 µg AFB_1**) for 26 days; for detailed information please see the article), conc.: 0.250 µg/kg*, country: France[314], *final wt. of the animal 104 kg

Deoxynivalenol
incidence: 4/4*, sa. const.: barrows and gilts, Ø wt.: 7.7 kg, contamination: no DON (for detailed information please see the article), conc.: nd, country: USA[64], *control
incidence: ?/4, sa. const.: barrows and gilts, Ø wt.: 7.7 kg, contamination:

artificial (dose: **0.9 ppm DON** (analyzed value) in the diet, o., for 3 weeks; for detailed information please see the article), conc.: 5 ppb*, country: USA[64], *after 3 weeks post-treatment
incidence: ?/4, sa. const.: barrows and gilts, Ø wt.: 7.7 kg, contamination: artificial (dose: **2.0 ppm DON** (analyzed value) in the diet, o., for 3 weeks; for detailed information please see the article), conc.: 10 ppb* (mean value), country: USA[64], *after 3 weeks post-treatment
incidence: ?/4, sa. const.: barrows and gilts, Ø wt.: 7.7 kg, contamination: artificial (dose: **2.8 ppm DON** (analyzed value) in the diet, o., for 3 weeks; for detailed information please see the article), conc.: 12 ppb* (mean value), country: USA[64], *after 3 weeks post-treatment

incidence: 6/6*, sa. const.: barrows (Yorkshire), wt.: ≈25 kg, contamination: no DON (for detailed information please see the article), conc.: nd, country: Canada[406], *control
incidence: 50?/50, sa. const.: barrows (Yorkshire), wt.: ≈25 kg, contamination: artificial (dose: **6.0 mg natural DON**/kg dry weight, o., for **3 weeks** (feeding trial 1); for detailed information please see the article), conc. range: ≤14.6 ng/g*, country: Canada[406], *after 3 weeks
incidence: 6/6*, sa. const.: barrows (Yorkshire), wt.: ≈25 kg, contamination: no DON (for detailed information please see the article), conc.: nd, country: Canada[406], *control
incidence: 6?/6, sa. const.: barrows (Yorkshire), wt.: ≈25 kg, contamination: artificial (dose: **6.0 mg crystalline DON**/kg dry weight, o., for **4 weeks** (feeding trial 2); for detailed information please see the article), conc. range: ≤8.7 ng/g*, country: Canada[406], *after 4 weeks
incidence: 6/6*, sa. const.: barrows (Yorkshire), wt.: ≈25 kg, contamination: no DON (for detailed information please see the article), conc.: nd, country: Canada[406], *control
incidence: 6?/6, sa. const.: barrows (Yorkshire), wt.: ≈25 kg, contamination: artificial (dose: **7.6 mg natural DON**/kg dry weight, o., for **7 weeks** (feeding trial 3); for detailed information please see the article), conc. range: ≤7.8 ng/g*, country: Canada[406], *after 7 weeks

incidence: 5/5*, sa. const.: healthy Yorkshire barrows, age: ≈11–15 weeks, wt.: 17–22 kg, contamination: no DON, conc.: nd, country: Canada[407], *control
incidence: ?/4, sa. const.: healthy Yorkshire barrows, age: ≈11–15 weeks, wt.: 17–22 kg, contamination: artificial (dose: **1.0 mg DON**/kg b. wt., i.v., once), conc. range: ≤1,114.3 ng/g* (mean value), country: Canada[407], *after 0.33 h (also measured after 1, 3, 8 and 24 h, lowest conc.: 8.2 ng/g after 24 h)

incidence: 5?/5, sa. const.: castrated Large White × German Landrace, db Classic crossbred pigs, wt.: ≈24.6 kg, contamination: artificial (dose: 0.05, 0.57, or 1.23 mg DON/kg mash or 0.07, 0.55 or 1.13* mg DON/kg pellets*, o., for 11 weeks; for detailed information please see the article), conc. range: ≤4.8 ng/g* ** *** (mean value), country: Germany[483], **after 78/79 days (thereof 11 weeks of DON-administration), ***values of the other DON-treatments were lower

incidence: 9?/9, sa. const.: German Landrace gilts, age: 180 days, wt.: 103 kg, contamination: artificial (dose: DON/ZEA in wheat in different conc., o., for 35 days; for detailed information please see the article), conc. range: ≤8.2 ng/g* ** (mean value), country: Germany[537], *9.57 mg DON and 0.358 mg ZEA/kg diet fed (both fed in highest conc.), **after 36 days (thereof 35 days of DON- and ZEA-administration)

Deepoxydeoxynivalenol
incidence: 5/5, sa. const.: castrated Large White × German Landrace, db Classic crossbred pigs, wt.: ≈24.6 kg, contamination: artificial (dose: 0.05, 0.57, or 1.23 mg DON/kg mash* or 0.07, 0.55 or 1.13 mg DON/kg

pellets*, o., for 11 weeks; for detailed information please see the article), conc.: nd*, country: Germany[483], *after 78/79 days (thereof 11 weeks of DON-administration)

incidence: 9?/9, sa. const.: German Landrace gilts, age: 180 days, weight: ≈103 kg, contamination: artificial (dose: DON/ZEA in wheat in different conc., o., for 35 days; for detailed information please see the article), conc. range: ≤4.8 ng/g* ** (mean value), country: Germany[537], *6.1 mg DON and 0.235 mg ZEA/kg diet fed (both fed in second highest conc.), **after 36 days (thereof 35 days of DON- and ZEA-administration)

Fumonisin B_1
incidence: 5/5*, sa. const.: weaned barrows of the same genotype, wt.: 12–14 kg, contamination: no FB_1, FB_2 + FB_3 (for detailed information please see the article), conc.: nd, country: Hungary/Germany[87], *control
incidence: 10/10, sa. const.: weaned barrows of the same genotype, wt.: 12–14 kg, contamination: artificial (dose: **50 mg FB_1, 20 mg FB_2 + 5 mg FB_3**/animal, o., for 22 days; for detailed information please see the article), conc. range: 38.4–158.4 ng/g*, Ø conc.: 99.36 ng/g*, country: Hungary/Germany[87], *after 22 days toxin feeding period

incidence: 6/6*, sa. const.: Yorkshire barrows, age: 6–8 weeks, wt.: 9–13 kg, contamination: no FB_1 (for detailed information please see the article), conc.: nr, country: Canada[108], *control
incidence: ?/2, sa. const.: Yorkshire barrows, age: 6–8 weeks, wt.: 9–13 kg, contamination: artificial (dose: **3.0 mg FB_1** (labeled)/kg feed, o., for 12 days and 2.0 mg FB_1 (labeled)/kg feed, o., for another 12 days; for detailed information please see the article), conc.: ≈160 ng/g tissue* ** (mean value), country: Canada[108], *after 24 days (also at other day intervals up to 33 days measured), **FB_1 and/or metabolites

incidence: 6/6*, sa. const.: castrated pigs of identical genotype, wt.: ≈12–14 kg, contamination: no FB_1 (for detailed information please see the article), conc.: nd, country: Germany/Hungary[109], *control
incidence: 13/13, sa. const.: castrated pigs of identical genotype, wt.: ≈12–14 kg, contamination: artificial (dose: **100 mg FB_1** daily, o., for 5–11 days; for detailed information please see the article), conc. range: 74–710 µg/kg*, Ø conc.: 214.57 µg/kg*, country: Germany/Hungary[109], *on 6th day of the experiment

Fumonisin B_2
incidence: 5/5*, sa. const.: weaned barrows of the same genotype, wt.: 12–14 kg, contamination: no FB_1, FB_2 + FB_3 (for detailed information please see the article), conc.: nd, country: Hungary/Germany[87], *control
incidence: 5/10, sa. const.: weaned barrows of the same genotype, wt.: 12–14 kg, contamination: artificial (dose: **50 mg FB_1, 20 mg FB_2 + 5 mg FB_3**/animal, o., for 22 days; for detailed information please see the article), conc. range: 0.8–7.2 ng/g*, Ø conc.: 2.88 ng/g*, country: Hungary/Germany[87], *after 22 days toxin feeding period

HT-2 Toxin
incidence: 2/2, sa. const.: female crossbred Yorkshire × Hampshire swines, wt.: 20 kg, contamination: artificial (dose: 0.15 mg T-2 toxin (labeled and unlabeled)/kg b. wt., i.vs., once), conc. range: 1.47–2.66 ng/g*, Ø conc.: 2.065 ng/g*, country: USA[425], *after 4 h

Deepoxy HT-2 Toxin
incidence: 2/2, sa. const.: female crossbred Yorkshire × Hampshire swines, wt.t: 20 kg, contamination: artificial (dose: 0.15 mg T-2 toxin/kg b. wt., i.vs., once), conc. range: 0.14–0.26 ng/g*, Ø conc.: 0.20 ng/g*, country: USA[425], *after 4 h

3′-Hydroxy HT-2 Toxin
incidence: 2/2, sa. const.: female crossbred Yorkshire × Hampshire swines, wt.: 20 kg,

contamination: artificial (dose: 0.15 mg T-2 toxin (labeled and unlabeled)/kg b. wt., i.vs., once), conc. range: 2.42–4.63 ng/g*, Ø conc.: 3.525 ng/g*, country: USA[425], *after 4 h

Ochratoxin A

incidence: 5/5*, sa. const.: female pigs of Danish Landrace, age: ≈8 weeks, wt.: ≈20 kg, contamination: no OTA (for detailed information please see the article), conc.: nd, country: Denmark/USA[102], *control
incidence: ?/5, sa. const.: female pigs of Danish Landrace, age: ≈8 weeks, wt.: ≈20 kg, contamination: artificial (dose: 1 ppm crystalline OTA, o., once daily for 1 month, afterwards toxin-free diets for various intervals fed; for detailed information please see the article), conc.: 17.80 μg/kg* (mean value), country: Denmark/USA[102], ***1 day after termination** of OTA-exposure
incidence: ?/5, sa. const.: female pigs of Danish Landrace, age: ≈8 weeks, wt.: ≈20 kg, contamination: artificial (dose: 1 ppm crystalline OTA, o., once daily for 1 month, afterwards toxin-free diets for various intervals fed; for detailed information please see the article), conc.: 5.14 μg/kg* (mean value), country: Denmark/USA[102], ***8 days after termination** of OTA-exposure
incidence: ?/5, sa. const.: female pigs of Danish Landrace, age: ≈8 weeks, wt.: ≈20 kg, contamination: artificial (dose: 1 ppm crystalline OTA, o., once daily for 1 month, afterwards toxin-free diets for various intervals fed; for detailed information please see the article), conc.: 0.58 μg/kg* (mean value), country: Denmark/USA[102], ***15 days after termination** of OTA-exposure
incidence: ?/5, sa. const.: female pigs of Danish Landrace, age: ≈8 weeks, wt.: ≈20 kg, contamination: artificial (dose: 1 ppm crystalline OTA, o., once daily for 1 month, afterwards toxin-free diets for various intervals fed; for detailed information please see the article), conc.: nd* (mean value), country: Denmark/USA[102], ***29 days after termination** of OTA-exposure
incidence: ?/?, sa. const.: weaners (specific pathogen free), wt.: 14–18 kg, contamination: artificial (dose: ? μg OTA addition; for detailed information please see the article), conc. range: ≤25 μg/kg*, country: Denmark[204], *calculated value based on the amount of OTA in blood after 24 h on toxin free diet

incidence: 1/1*, sa. const.: pregnant gilts, contamination: neither OTA nor OTB, conc.: nd, country: UK[266], *control
incidence: 2/2, sa. const.: pregnant gilts, contamination: artificial (dose: **0.38 mg OTA + 0.13 mg OTB**/kg b. wt., o., for 8 days during early pregnancy), conc. range: 0.30–0.34 μg/g*, Ø conc.: 0.32 μg/g*, country: UK[266], *measured on day 30 of pregnancy

incidence: 13/13*, sa. const.: castrated pigs, wt.: 20 kg, contamination: no OTA and/or CIT (for detailed information please see the article), conc.: nd, country: Denmark/USA[330], *control
incidence: ?/13, sa. const.: castrated pigs, wt.: 20 kg, contamination: artificial (dose: **1,400 μg crystalline OTA**/kg feed, o., for 6 weeks; for detailed information please see the article), conc.: 6 μg/kg* (mean value), country: Denmark/USA[330], *after 6 weeks
incidence: ?/13, sa. const.: castrated pigs, wt.: 20 kg, contamination: artificial (dose: **650 μg crystalline CIT**/kg feed, o., for 6 weeks; for detailed information please see the article), conc.: nd*, country: Denmark/USA[330], *after 6 weeks
incidence: ?/13, sa. const.: castrated pigs, wt.: 20 kg, contamination: artificial (dose: **1,400 μg crystalline OTA + 650 μg crystalline CIT**/kg feed, o., for 6 weeks; for detailed information please see the article), conc.: 7 μg/kg* (mean value), country: Denmark/USA[330], *after 6 weeks
incidence: ?/13, sa. const.: castrated pigs, wt.: 20 kg, contamination: artificial (dose: **1,400 μg natural OTA + 650 μg natural CIT**/kg feed, o., for 6 weeks; for detailed

information please see the article), conc.: 30 μg/kg* (mean value), country: Denmark/USA[330], *after 6 weeks

incidence: 8?/8*, sa. const.: Danish Landrace piglets, age: 8 weeks, contamination: artificial (dose: **18.6 μg OTA/kg** liveweight/day, o., over a period of 6 weeks (treatment 1); for detailed information please see the article), conc.: 11 μg/kg* (mean value), country: Denmark[336], *sa. taken at weaning (pigs 8 weeks old)
incidence: 6?/6*, sa. const.: Danish Landrace piglets, age: 8 weeks, contamination: artificial (dose: **8.0 μg OTA/kg** liveweight/day, o., over a period of 6 weeks (treatment 2); for detailed information please see the article), conc.: 5 μg/kg* (mean value), country: Denmark[336], *sa. taken at weaning (pigs 8 weeks old)
incidence: 8?/8*, sa. const.: Danish Landrace piglets, age: 8 weeks, contamination: artificial (dose: **19.7 μg OTA/kg** liveweight/day, o., over a period of 6 weeks (treatment 3); for detailed information please see the article), conc.: 9 μg/kg* (mean value), country: Denmark[336], *sa. taken at weaning (pigs 8 weeks old)

incidence: 4/4*, sa. const.: male and females pigs, wt.: ≈70 kg, contamination: no OTA (for detailed information please see the article), conc.: nd, country: Germany[365], *control
incidence: 8?/8, sa. const.: male and females pigs, wt.: ≈70 kg, contamination: artificial (dose: **0.09 mg natural OTA**/kg diet/day, in the morning and evening **half of OTA-ration**, o., for 28 days; for detailed information please see the article), conc.: 12.35 ng/g* (mean value), country: Germany[365], *after 28 days
incidence: 8?/8, sa. const.: male and females pigs, wt.: ≈70 kg, contamination: artificial (dose: **0.09 mg crystalline OTA**/kg diet/day, in the morning and evening **half of OTA-ration**, o., for 28 days; for detailed information please see the article), conc.: 2.21 ng/g* (mean value), country: Germany[365], *after 28 days
incidence: 8?/8, sa. const.: male and females pigs, wt.: ≈70 kg, contamination: artificial (dose: **0.09 mg crystalline OTA**/kg diet/day, in the morning **total OTA-ration**, o., for 28 days; for detailed information please see the article), conc.: 3.19 ng/g* (mean value), country: Germany[365], *after 28 days

incidence: 8?/8*, sa. const.: pigs, contamination: no OTA (for detailed information please see the article), conc.: <0.66 μg/kg (mean value), country: Germany[366], *control
incidence: 8?/8, sa. const.: pigs, contamination: artificial (dose: **22.11 mg OTA** (in total), o., for 90 days; for detailed information please see the article), conc.: 7.9 μg/kg (mean value), country: Germany[366], *after 90 days
incidence: 8?/8, sa. const.: pigs, contamination: artificial (dose: **88.44 mg OTA** (in total), o., for 90 days; for detailed information please see the article), conc.: 36.7 μg/kg* (mean value), country: Germany[366], *after 90 days

incidence: 2/2*, sa. const.: pigs, contamination: no OTA and/or DON (for detailed information please see the article), conc.: nd, country: Germany[378], *control
incidence: 6/6, sa. const.: pigs, contamination: artificial (dose: **0.1 ppm crystalline OTA + 1.0 ppm crystalline DON**, o., twice daily for 90 days; for detailed information please see the article), conc. range: 0.98–1.74 ng/g*, Ø conc.: 1.33 ng/g*, country: Germany[378], *after 90 days
incidence: 3/3, sa. const.: pigs, contamination: artificial (dose: **0.1 ppm crystalline OTA**, o., twice daily for 90 days; for detailed information please see the article), conc. range: 0.88–1.67 ng/g*, Ø conc.: 1.26 ng/g*, country: Germany[378], *after 90 days
incidence: 3/3, sa. const.: pigs, contamination: artificial (dose: **1.0 ppm**

crystalline DON, o., twice daily for 90 days; for detailed information please see the article), conc.: nd*, country: Germany[378], *after 90 days

incidence: 2?/2*, sa. const.: pigs, contamination: no OTA and/or ZEA (for detailed information please see the article), conc.: nr, country: Germany[380], *control
incidence: 6/6, sa. const.: pigs, contamination: artificial (dose: **0.1 ppm OTA + 0.25 ppm ZEA**, o., twice daily for 90 days; for detailed information please see the article), conc. range: ≤4.18 ng/g*, country: Germany[380], *after 91 days (thereof 90 days of OTA- and ZEA-administration)
incidence: 3/3, sa. const.: pigs, contamination: artificial (dose: **0.1 ppm OTA**, o., twice daily for 90 days; for detailed information please see the article), conc. range: ≤4.21 ng/g*, country: Germany[380], *after 91 days (thereof 90 days of OTA-administration)

incidence: 3/3*, sa. const.: female pigs of Danish Landrace, age: 8–10 weeks, wt.: ≈20 kg, contamination: no OTA (for detailed information please see the article), conc.: nd, country: Denmark/USA/Sweden[383], *control
incidence: 4?/4, sa. const.: female pigs of Danish Landrace, age: 8–10 weeks, wt.: ≈20 kg, contamination: artificial (dose: **1 mg crystalline OTA**/kg feed, o., for 3 months; for detailed information please see the article), conc.: ≈11.5 μg/kg (mean value), country: Denmark/USA/Sweden[383], *after ≈3 months

incidence: 9/9*, sa. const.: Deutsches Landschwein, Deutsches Edelschwein × Piétrain, Belgische Landrasse, wt.: 25 kg, contamination: no OTA, conc.: nd, country: Germany[409], *control
incidence: 9/9, sa. const.: Deutsches Landschwein, Deutsches Edelschwein × Piétrain, Belgische Landrasse, wt.: 25 kg, contamination: artificial (dose: **0.15 mg natural OTA**/kg feed, o., daily for 28 days), conc.: nd*, country: Germany[409], *after 28 days

incidence: 8/8*, sa. const.: Deutsches Landschwein, Deutsches Edelschwein × Piétrain, Belgische Landrasse, wt.: 25 kg, contamination: no OTA, conc.: nd, country: Germany[409], *control
incidence: 6/6, sa. const.: Deutsches Landschwein, Deutsches Edelschwein × Piétrain, Belgische Landrasse, wt.: 25 kg, contamination: artificial (dose: **0.58 mg natural OTA**/kg feed, o., daily for 28 days), conc. range: ≤6.4 μg/kg*, country: Germany[409], *after 28 days

T-2 Toxin
incidence: 1/1, sa. const.: female weanling crossbred Yorkshire × Duroc × Hampshire swine, wt.: 7.5 kg, contamination: artificial (dose: **0.1 mg T-2 toxin** (labeled)/kg b. wt., intubated, once), conc.: 13.8 ppb* ** ***, country: USA[318], *calculated residue level, **T-2 toxin and/or metabolites, ***after 18 h
incidence: 1/1, sa. const.: female weanling crossbred Yorkshire × Duroc × Hampshire swine, wt.: 9.5 kg, contamination: artificial (dose: **0.4 mg T-2 toxin** (labeled)/kg b. wt., intubated, once), conc.: 37.7 ppb* ** ***, country: USA[318], *calculated residue level, **T-2 toxin and/or metabolites, ***after 18 h

incidence: 2/2, sa. const.: female crossbred Yorkshire × Hampshire swines, wt.: 20 kg, contamination: artificial (dose: 0.15 mg T-2 toxin (labeled and unlabeled)/kg b. wt., i.vs., once), conc. range: 0.27–0.71 ng/g*, Ø conc.: 0.49 ng/g*, country: USA[425], *after 4 h
incidence: 2/2, sa. const.: female crossbred Yorkshire × Hampshire swines, wt.: 20 kg, contamination: artificial (dose: 0.15 mg T-2 toxin (labeled and

unlabeled)/kg b. wt., i.vs., once), conc. range: 39–107 ng/g* **, Ø conc.: 73 ng/g* **, country: USA[425], *after 4 h, ****total metabolites**

3′-Hydroxy T-2 Toxin
incidence: 2/2, sa. const.: female crossbred Yorkshire × Hampshire swines, wt.: 20 kg, contamination: artificial (dose: 0.15 mg T-2 toxin (labeled and unlabeled)/kg b. wt., i.vs., once), conc. range: 0.67–0.78 ng/g*, Ø conc.: 0.725 ng/g*, country: USA[425], *after 4 h

T-2 Tetraol
incidence: 2/2, sa. const.: female crossbred Yorkshire × Hampshire swines, wt.: 20 kg, contamination: artificial (dose: 0.15 mg T-2 toxin (labeled and unlabeled)/kg b. wt., i.vs., once), conc. range: 0.71–1.73 ng/g*, Ø conc.: 1.22 ng/g*, country: USA[425], *after 4 h

Deepoxy T-2 Tetraol
incidence: 2/2, sa. const.: female crossbred Yorkshire × Hampshire swines, wt.: 20 kg, contamination: artificial (dose: 0.15 mg T-2 toxin (labeled and unlabeled)/kg b. wt., i.vs., once), conc. range: 1.66–1.72 ng/g*, Ø conc.: 1.69 ng/g*, country: USA[425], *after 4 h

T-2 Triol
incidence: 2/2, sa. const.: female crossbred Yorkshire × Hampshire swines, wt.: 20 kg, contamination: artificial (dose: 0.15 mg T-2 toxin (labeled and unlabeled)/kg b. wt., i.vs., once), conc. range: 0.16–0.35 ng/g*, Ø conc.: 0.255 ng/g*, country: USA[425], *after 4 h

Deepoxy T-2 Triol
incidence: 2/2, sa. const.: female crossbred Yorkshire × Hampshire swines, wt.: 20 kg, contamination: artificial (dose: 0.15 mg T-2 toxin (labeled and unlabeled)/kg b. wt., i.vs., once), conc. range: 0.07–0.23 ng/g*, Ø conc.: 0.15 ng/g*, country: USA[425], *after 4 h

Zearalenone
incidence: 3?/3, sa. const.: Yorkshire gilts, wt.: 8–11 kg, contamination: artificial (dose: 40 µg ZEA/g feed additionally to **0% alfalfa** in the diet, for 4 weeks; for detailed information please see the article), conc.: 0.64 µg/5 g liver* (mean value), country: Canada[84], *after 28 days
incidence: 3?/3, sa. const.: Yorkshire gilts, wt.: 8–11 kg, contamination: artificial (dose: 40 µg ZEA/g feed additionally to **15% alfalfa** in the diet, for 4 weeks; for detailed information please see the article), conc.: 0.39 µg/5 g liver* (mean value), country: Canada[84], *after 28 days
incidence: 3?/3, sa. const.: Yorkshire gilts, wt.: 8–11 kg, contamination: artificial (dose: 40 µg ZEA/g feed additionally to **25% alfalfa** in the diet, for 4 weeks; for detailed information please see the article), conc.: 0.39 µg/5 g liver* (mean value), country: Canada[84], *after 28 days

incidence: 7/7*, sa. const.: hybrids of *Deutsches Edelschwein* and *Pietrain* (female pigs), age: 3 months, Ø wt.: 58.6 kg, contamination: (dose: 60 µg ZEA/animal besides other *Fusarium* mycotoxins, o., daily for 18 days; for detailed information please see the article), conc.: nr, country: Austria[600], *control
incidence: 7/7, sa. const.: hybrids of *Deutsches Edelschwein* and *Pietrain* (female pigs), age: 3 months, Ø wt.: 58.6 kg, contamination: artificial (dose: **1.1 mg ZEA**/animal besides other *Fusarium* mycotoxins, o., daily for 18 days; for detailed information please see the article), conc. range: tr–3.1 µg/kg*, country: Austria[600], *after 18 days

incidence: 2?/2?*, sa. const.: male KAHYP pigs, wt.: ?, contamination: no ZEA, conc.: nd, country: Hungary[632], *control
incidence: 2?/2, sa. const.: male KAHYP pigs, wt.: 60 kg, contamination: artificial (dose: **15 ppm ZEA**, o., for 14 days, conc.:

70 μg/kg*, country: Hungary[632], *after 14 days)

α-Zearalenol

incidence: 3?/3, sa. const.: Yorkshire gilts, wt.: 8–11 kg, contamination: artificial (dose: 40 μg ZEA/g feed additionally to **0% of alfalfa** in the diet, for 4 weeks; for detailed information please see the article), conc.: 0.47 μg/5 g liver* (mean value), country: Canada[84], *after 28 days
incidence: 3?/3, sa. const.: Yorkshire gilts, wt.: 8–11 kg, contamination: artificial (dose: 40 μg ZEA/g feed additionally to **15% of alfalfa** in the diet, for 4 weeks; for detailed information please see the article), conc.: 0.72 μg/5 g liver* (mean value), country: Canada[84], *after 28 days
incidence: 3?/3, sa. const.: Yorkshire gilts, wt.: 8–11 kg, contamination: artificial (dose: 40 μg ZEA/g feed additionally to **25% of alfalfa** in the diet, for 4 weeks; for detailed information please see the article), conc.: 1.55 μg/5 g liver* (mean value), country: Canada[84], *after 28 days

incidence: 2?/2*, sa. const.: pigs, contamination: no OTA and/or ZEA (for detailed information please see the article), conc.: nr, country: Germany[380], *control
incidence: 6/6, sa. const.: pigs, contamination: artificial (dose: **0.1 ppm OTA + 0.25 ppm ZEA**, o., twice daily for 90 days; for detailed information please see the article), conc. range: ≤4.91 ng/g*, country: Germany[380], *after 91 days (thereof 90 days of OTA- and ZEA-administration)

incidence: 7?/7*, sa. const.: hybrids of *Deutsches Edelschwein* and *Pietrain* (female pigs), age: 3 months, Ø wt.: 58.6 kg, contamination: (dose: 60 μg ZEA/animal besides other *Fusarium* mycotoxins, o., daily for 18 days; for detailed information please see the article), conc.: nr, country: Austria[600], *control
incidence: 7/7, sa. const.: hybrids of *Deutsches Edelschwein* and *Pietrain* (female pigs), age: 3 months, Ø wt.: 58.6 kg, contamination: artificial (dose: **1.1 mg ZEA**/animal besides other *Fusarium* mycotoxins, o., daily for 18 days; for detailed information please see the article), conc. range: 3.6–12.0 μg/kg*, country: Austria[600], *after 18 days

incidence: 2?/2?*, sa. const.: male KAHYP pigs, wt.: ?, contamination: no ZEA, conc.: nd, country: Hungary[632], *control
incidence: 2?/2, sa. const.: male KAHYP pigs, wt.: 60 kg, contamination: artificial (dose: **15 ppm ZEA**, o., for 14 days, conc.: 160 μg/kg*, country: Hungary[632], *after 14 days)

β-Zearalenol

incidence: 7?/7*, sa. const.: hybrids of *Deutsches Edelschwein* and *Pietrain* (female pigs), age: 3 months, Ø wt.: 58.6 kg, contamination: (dose: 60 μg ZEA/animal besides other *Fusarium* mycotoxins, o., daily for 18 days; for detailed information please see the article), conc.: nr, country: Austria[600], *control
incidence: 7/7, sa. const.: hybrids of *Deutsches Edelschwein* and *Pietrain* (female pigs), age: 3 months, Ø wt.: 58.6 kg, contamination: artificial (dose: **1.1 mg ZEA**/animal besides other *Fusarium* mycotoxins, o., daily for 18 days; for detailed information please see the article), conc. range: 1.9–4.8 μg/kg*, country: Austria[600], *after 18 days

incidence: 2?/2?*, sa. const.: male KAHYP pigs, wt.: ?, contamination: no ZEA, conc.: nd, country: Hungary[632], *control
incidence: 2?/2, sa. const.: male KAHYP pigs, wt.: 60 kg, contamination: artificial (dose: 15 ppm ZEA, o., for 14 days, conc.: 28 μg/kg*, country: Hungary[632], *after 14 days)

Pig lung may contain the following mycotoxins and/or their metabolites:

Aflatoxin B_1
incidence: 1/1, sa. const.: female Hampshire × Deutsches Edelschwein piglet, wt.: 15 kg, contamination: artificial (dose: 3.1 μg AFB_1 (labeled)/kg b. wt., o., once; for detailed information please see the article), conc.: 1 ppb* **, country: Switzerland[66], *AFB_1 eq., **after **24 h**
incidence: 1/1, sa. const.: female Hampshire × Deutsches Edelschwein piglets, wt.: 15 kg, contamination: artificial (dose: 3.1 μg AFB_1 (labeled)/kg b. wt., o., once; for detailed information please see the article), conc.: 0.7 ppb* **, country: Switzerland[66], *AFB_1 eq., **after **48 h**

Deoxynivalenol
incidence: 5/5*, sa. const.: healthy Yorkshire barrows, age: ≈11–15 weeks, wt.: 17–22 kg, contamination: no DON, conc.: nd, country: Canada[407], *control
incidence: ?/4, sa. const.: healthy Yorkshire barrows, age: ≈11–15 weeks, wt.: 17–22 kg, contamination: artificial (dose: **1.0 mg DON**/kg b. wt., i.v., once), conc. range: ≤265.9 ng/g* (mean value), country: Canada[407], *after 0.33 h (also measured after 1, 3, 8 and 24 h, lowest conc.: 1.0 ng/g after 24 h)

Fumonisin B_1
incidence: 5/5*, sa. const.: weaned barrows of the same genotype, wt.: 12–14 kg, contamination: no FB_1, FB_2 + FB_3 (for detailed information please see the article), conc.: nd, country: Hungary/Germany[87], *control
incidence: 10/10, sa. const.: weaned barrows of the same genotype, wt.: 12–14 kg, contamination: artificial (dose: **50 mg FB_1, 20 mg FB_2 + 5 mg FB_3**/animal, o., for 22 days; for detailed information please see the article), conc. range: 1.6–5.6 ng/g*, Ø conc.: 2.72 ng/g*, country: Hungary/Germany[87], *after 22 days toxin feeding period

incidence: 6/6*, sa. const.: castrated pigs of identical genotype, wt.: ≈12–14 kg, contamination: no FB_1 (for detailed information please see the article), conc.: nd, country: Germany/Hungary[109], *control
incidence: 13/13, sa. const.: castrated pigs of identical genotype, wt.: ≈12–14 kg, contamination: artificial (dose: **100 mg FB_1** daily, o., for 5–11 days; for detailed information please see the article), conc. range: 6–1,150 μg/kg*, Ø conc.: 158.14 μg/kg*, country: Germany/Hungary[109], *on 6th day of the experiment

Fumonisin B_2
incidence: 5/5*, sa. const.: weaned barrows of the same genotype, wt.: 12–14 kg, contamination: no FB_1, FB_2 + FB_3 (for detailed information please see the article), conc.: nd, country: Hungary/Germany[87], *control
incidence: 4/10, sa. const.: weaned barrows of the same genotype, wt.: 12–14 kg, contamination: artificial (dose: **50 mg FB_1, 20 mg FB_2 + 5 mg FB_3**/animal, o., for 22 days; for detailed information please see the article), conc. range: 0.8–2.4 ng/g*, Ø conc.: 1.6 ng/g*, country: Hungary/Germany[87], *after 22 days toxin feeding period

HT-2 Toxin
incidence: 2/2, sa. const.: female crossbred Yorkshire × Hampshire swines, wt.: 20 kg, contamination: artificial (dose: 0.15 mg T-2 toxin (labeled and unlabeled)/kg b. wt., i.vs., once), conc. range: 0.85–1.69 ng/g*, Ø conc.: 1.27 ng/g*, country: USA[425], *after 4 h

Deepoxy HT-2 Toxin
incidence: 1/2, sa. const.: female crossbred Yorkshire × Hampshire swines, wt.: 20 kg, contamination: artificial (dose: 0.15 mg T-2 toxin (labeled and unlabeled)/kg b. wt., i.vs., once), conc.: 0.13 ng/g*, country: USA[425], *after 4 h

3′-Hydroxy HT-2 Toxin
incidence: 2/2, sa. const.: female crossbred Yorkshire × Hampshire swines, wt.: 20 kg,

contamination: artificial (dose: 0.15 mg T-2 toxin (labeled and unlabeled)/kg b. wt., i.vs., once), conc. range: 1.24–2.15 ng/g*, Ø conc.: 1.695 ng/g*, country: USA[425], *after 4 h

T-2 Toxin
incidence: 2/2, sa. const.: female crossbred Yorkshire × Hampshire swines, wt.: 20 kg, contamination: artificial (dose: 0.15 mg T-2 toxin (labeled and unlabeled)/kg b. wt., i.vs., once), conc. range: 0.13–0.85 ng/g*, Ø conc.: 0.49 ng/g*, country: USA[425], *after 4 h
incidence: 2/2, sa. const.: female crossbred Yorkshire × Hampshire swines, wt.: 20 kg, contamination: artificial (dose: 0.15 mg T-2 toxin (labeled and unlabeled)/kg b. wt., i.vs., once), conc. range: 21–24 ng/g* **, Ø conc.: 22.5 ng/g* **, country: USA[425], *after 4 h, ****total metabolites**

3′-Hydroxy T-2 Toxin
incidence: 2/2, sa. const.: female crossbred Yorkshire × Hampshire swines, wt.: 20 kg, contamination: artificial (dose: 0.15 mg T-2 toxin (labeled and unlabeled)/kg b. wt., i.vs., once), conc. range: 0.27–0.87 ng/g*, Ø conc.: 0.57 ng/g*, country: USA[425], *after 4 h

T-2 Tetraol
incidence: 1/2, sa. const.: female crossbred Yorkshire × Hampshire swines, wt.: 20 kg, contamination: artificial (dose: 0.15 mg T-2 toxin (labeled and unlabeled)/kg b. wt., i.vs., once), conc.: 0.06 ng/*, country: USA[425], *after 4 h

Deepoxy T-2 Tetraol
incidence: 2/2, sa. const.: female crossbred Yorkshire × Hampshire swines, wt.: 20 kg, contamination: artificial (dose: 0.15 mg T-2 toxin (labeled and unlabeled)/kg b. wt., i.vs., once), conc. range: 1.03–1.28 ng/g*, Ø conc.: 1.155 ng/g*, country: USA[425], *after 4 h

T-2 Triol
incidence: 2/2, sa. const.: female crossbred Yorkshire × Hampshire swines, wt.: 20 kg, contamination: artificial (dose: 0.15 mg T-2 toxin (labeled and unlabeled)/kg b. wt., i.vs., once), conc. range: 0.11–0.17 ng/g*, Ø conc.: 0.14 ng/g*, country: USA[425], *after 4 h

Pig Lymphatic Gland see Pig lymph

Pig lymph may contain the following mycotoxins and/or their metabolites:

Deoxynivalenol
incidence: 5/5*, sa. const.: healthy Yorkshire barrows, age: ≈11–15 weeks, wt.: 17–22 kg, contamination: no DON, conc.: nd, country: Canada[407], *control
incidence: ?/4, sa. const.: healthy Yorkshire barrows, age: ≈11–15 weeks, wt.: 17–22 kg, contamination: artificial (dose: **1.0 mg DON**/kg b. wt., i.v., once), conc. range: ≤292.3 ng/g* (mean value), country: Canada[407], *after 1 h (also measured after 0.33, 3, 8 and 24 h, lowest conc.: 0.8 ng/g after 24 h)

HT-2 Toxin
incidence: 2/2, sa. const.: female crossbred Yorkshire × Hampshire swines, wt.: 20 kg, contamination: artificial (dose: 0.15 mg T-2 toxin (labeled and unlabeled)/kg b. wt., i.vs., once), conc. range: 3.76–5.21 ng/g* **, Ø conc.: 4.485 ng/g* **, country: USA[425], *in mesenteric lymph nodes, **after 4 h

3′-Hydroxy HT-2 Toxin
incidence: 2/2, sa. const.: female crossbred Yorkshire × Hampshire swines, wt.: 20 kg, contamination: artificial (dose: 0.15 mg T-2 toxin (labeled and unlabeled)/kg b. wt., i.vs., once), conc. range: 3.03–4.02 ng/g* **, Ø conc.: 3.525 ng/g* **, country: USA[425], *in mesenteric lymph nodes, **after 4 h

T-2 Toxin
incidence: 2/2*, sa. const.: female swines of mixed breeding, wt.: 26–66 kg, contamination: no T-2 toxin, conc.: nr, country: USA[403], *control
incidence: 1/1, sa. const.: female swine of mixed breeding, wt.: 26–66 kg, contamination: artificial (dose: **1.2 mg**

T-2 toxin, i.a., once), conc.: ≈116 ppb**, country: USA[403], *after 3 h, **in mesenteric lymph nodes

incidence: 2/2, sa. const.: female crossbred Yorkshire × Hampshire swines, wt.: 20 kg, contamination: artificial (dose: 0.15 mg T-2 toxin (labeled and unlabeled)/kg b. wt., i.vs., once), conc. range: 1.37–1.81 ng/g* **, Ø conc.: 1.59 ng/g* **, country: USA[425], *in mesenteric lymph nodes, **after 4 h
incidence: 2/2, sa. const.: female crossbred Yorkshire × Hampshire swines, wt.: 20 kg, contamination: artificial (dose: 0.15 mg T-2 toxin (labeled and unlabeled)/kg b. wt., i.vs., once), conc. range: 33–46 ng/g* ** ***, Ø conc.: 39.5 ng/g* ** ***, country: USA[425], *in mesenteric lymph nodes, **after 4 h, ***__total metabolites__

3′-Hydroxy T-2 Toxin
incidence: 2/2, sa. const.: female crossbred Yorkshire × Hampshire swines, wt.: 20 kg, contamination: artificial (dose: 0.15 mg T-2 toxin (labeled and unlabeled)/kg b. wt., i.vs., once), conc. range: 0.63–1.69 ng/g* **, Ø conc.: 1.16 ng/g* **, country: USA[425], *in mesenteric lymph nodes, **after 4 h

T-2 Tetraol
incidence: 1/2, sa. const.: female crossbred Yorkshire × Hampshire swines, wt.: 20 kg, contamination: artificial (dose: 0.15 mg T-2 toxin (labeled and unlabeled)/kg b. wt., i.vs., once), conc.: 0.09 ng/g* **, country: USA[425], *in mesenteric lymph nodes, **after 4 h

Deepoxy T-2 Tetraol
incidence: 2/2, sa. const.: female crossbred Yorkshire × Hampshire swines, wt.: 20 kg, contamination: artificial (dose: 0.15 mg T-2 toxin (labeled and unlabeled)/kg b. wt., i.vs., once), conc. range: 1.05–1.51 ng/g* **, Ø conc.: 1.28 ng/g* **, country: USA[425], *in mesenteric lymph nodes, **after 4 h

T-2 Triol
incidence: 2/2, sa. const.: female crossbred Yorkshire × Hampshire swines, wt.: 20 kg, contamination: artificial (dose: 0.15 mg T-2 toxin (labeled and unlabeled)/kg b. wt., i.vs., once), conc. range: 0.22–0.27 ng/g* **, Ø conc.: 0.245 ng/g* **, country: USA[425], *in mesenteric lymph nodes, **after 4 h

α-Zearalenol
incidence: 2?/2?*, sa. const.: male KAHYP pigs, wt.: ?, contamination: no ZEA, conc.: nd, country: Hungary[632], *control
incidence: 2?/2, sa. const.: male KAHYP pigs, wt.: 60 kg, contamination: artificial (dose: 15 ppm ZEA, o., for 14 days, conc.: 10 μg/kg*, country: Hungary[632], *after 14 days

Pig Mesenteric Lymph Nodes see Pig lymph

Pig muscle may contain the following mycotoxins and/or their metabolites:

Aflatoxicol
incidence: 2/2, sa. const.: castrated male pigs of mixed breed, wt.: 9–11 kg, contamination: artificial (acute study, dose: 1 mg AFB_1/kg b. wt., o., once), conc. range: ≤0.06 ng/g*, country: USA[362], *sacrificed after 24 h (1 other pig sacrificed 72 h after treatment showed a lower mycotoxin value)

Aflatoxin B_1
incidence: 1/1, sa. const.: female Hampshire × Deutsches Edelschwein piglets, wt.: 15 kg, contamination: artificial (dose: 3.1 μg AFB_1 (labeled)/kg b. wt., o., once; for detailed information please see the article), conc.: <0.2 ppb* **, country: Switzerland[66], *AFB_1 eq., **after 24 h
incidence: 1/1, sa. const.: female Hampshire × Deutsches Edelschwein piglets, wt.: 15 kg, contamination: artificial (dose: 3.1 μg AFB_1 (labeled)/kg b. wt., o., once; for detailed information please see the article), conc.: <0.2 ppb* **, country: Switzerland[66], *AFB_1 eq., **after 48 h

incidence: 1/2, sa. const.: adult swines, contamination: artificial (dose:

1.08–1.09 mg AFB_1 (besides other AF), o., daily for 33 days; for detailed information please see the article), conc.: 0.85 ppb*, country: France[88], *after 33 days

incidence: 20/20*, sa. const.: male and female Danish Landrace pigs, age: ≈8 weeks, wt.: 20 kg up to slaughter at 90 kg, contamination: no AFB_1 + AFB_2 (for detailed information please see the article), conc.: nd, country: Denmark/USA[101], *control
incidence: 2/17*, sa. const.: male and female Danish Landrace pigs, age: ≈8 weeks, wt.: 20 kg up to slaughter at 90 kg, contamination: artificial (dose: AFB_1 + AFB_2 addition (overall **300 ppb AFs**), o., for **120 or 180 days****; for detailed information please see the article), conc. range: tr**, country: Denmark/USA[101], *livers of some of these pigs were rejected at meat inspection, **after 120 or 180 days
incidence: 10/18*, sa. const.: male and female Danish Landrace pigs, age: ≈8 weeks, wt.: 20 kg up to slaughter at 90 kg, contamination: artificial (dose: AFB_1 + AFB_2 addition (overall **500 ppb AFs**), o., for **135–231 days****; for detailed information please see the article), conc. range: tr**, country: Denmark/USA[101], *livers of some of these pigs were rejected at meat inspection, **after 135–231 days

incidence: 4/4*, sa. const.: male and female feeder pigs, wt.: 54.2–71.6 kg, contamination: no AFB_1, conc.: na, country: USA[136], *control
incidence: 4/4, sa. const.: male and female feeder pigs, wt.: 54.2–71.6 kg, contamination: artificial (dose: **100 μg AFB_1**/kg diet, o., for 4 weeks), conc. range: 0.13–0.23 μg/kg*, Ø conc.: 0.19 μg/kg*, country: USA[136], *after 4 weeks
incidence: 4/4, sa. const.: male and female feeder pigs, wt.: 54.2–71.6 kg, contamination: artificial (dose: **200 μg AFB_1**/kg diet, o., for 4 weeks), conc. range: 0.19–0.69 μg/kg*, Ø conc.: 0.46 μg/kg*, country: USA[136], *after 4 weeks
incidence: 4/4, sa. const.: male and female feeder pigs, wt.: 54.2–71.6 kg, contamination: artificial (dose: **400 μg AFB_1**/kg diet, o., for 4 weeks), conc. range: 0.36–2.22 μg/kg*, Ø conc.: 1.04 μg/kg*, country: USA[136], *after 4 weeks
incidence: 4/4*, sa. const.: crossbred (Duroc × Yorkshire) barrows, wt.: 24.5–26.3 kg, contamination: no AFs, conc.: nd, country: USA[138], *control

incidence: 4?/4, sa. const.: crossbred (Duroc × Yorkshire) barrows, wt.: 24.5–26.3 kg, contamination: artificial (dose: **662 μg AFB_1**/kg diet, **273 μg AFB_2**/kg diet, **300 μg AFG_1**/kg diet and **285 μg AFG_2**/kg diet, o., for 21 days), conc. range: tr–0.10 μg/kg*, country: USA[138], *after ≈22 days (thereof 21 days of AF-administration)

incidence: 10/10*, sa. const.: mixed breed feeder pigs, contamination: chronic study, no AFs (for detailed information please see the article), conc.: nd, country: USA[182], *control
incidence: 10?/10, sa. const.: mixed breed feeder pigs, contamination: artificial (chronic study, dose: **400 ng natural AFs**/g diet, o., for 10 weeks; for detailed information please see the article), conc.: nd*, country: USA[182], *after 10 weeks
incidence: 10?/10, sa. const.: mixed breed feeder pigs, contamination: artificial (chronic study, dose: **800 ng natural AFs**/g diet, o., for 10 weeks; for detailed information please see the article), conc.: 0.19 ng/g* (wet matter basis) (mean value), country: USA[182], *after 10 weeks

incidence: 8/8*, sa. const.: mixed breed feeder pigs, contamination: acute study, no AFs, conc.: nd, country: USA[182], *control
incidence: 1/1(8)*, sa. const.: mixed breed feeder pigs, contamination: artificial (acute study, dose: 1.2 mg total AFs (AFB_1+AFB_2+AFG_1+AFG_2)/kg b. wt., o., once, conc. range: ≤1.57 ng/g** (wet matter basis), country: USA[182], *for overall information please see the article, **48 h post-dosage (also measured

12, 24 and 72 h post-dosage, lowest conc.: nd after 12 and 72 h)

incidence: ?/?*, sa. const.: Large White growing female pigs, age: 79 days, wt.: 20 kg, contamination: no AFs (for detailed information please see the article), conc.: nd, country: France[314], *control
incidence: 1/1, sa. const.: Large White growing female pigs, age: 79 days, wt.: 20 kg, contamination: artificial (dose: natural AFs addition (treatment 1: mixed feeding, avg. daily intake of **870 µg AFB_1**) for 26 days; for detailed information please see the article), conc.: nd*, country: France[314], *final wt. of the animal 109 kg
incidence: 1/1, sa. const.: Large White growing female pigs, age: 79 days, wt.: 20 kg, contamination: artificial (dose: natural AFs addition (treatment 2: separate feeding = peanut oil meal 40% and corn gluten meal 30%, avg. daily intake of **1,566 µg AFB_1**) for 19 days; for detailed information please see the article), conc.: 0.1 µg/kg*, country: France[314], *final wt. of the animal 97 kg
incidence: 1/1, sa. const.: Large White growing female pig, age: 79 days, wt.: 20 kg, contamination: artificial (dose: natural AFs addition (treatment 3: separate feeding = peanut oil meal 0% and corn gluten meal 0%, avg. daily intake of **642 µg AFB_1**) for 26 days; for detailed information please see the article), conc.: nd*, country: France[314], *final wt. of the animal 104 kg

incidence: 2/2, sa. const.: castrated male pigs of mixed breed, wt.: 9–11 kg, contamination: artificial (**acute study**, dose: 1 mg AFB_1/kg b. wt., o., once), conc. range: ≤2.9 ng/g*, country: USA[362], *sacrificed after 24 h (1 other pig sacrificed 72 h after treatment showed a lower mycotoxin value)
incidence: 2/5, sa. const.: market-weight pigs, wt.: ≈92 kg, contamination: artificial (**subacute study**, dose: ≈15 µg AFB_1 as well as AFB_2, AFG_1 and AFG_2 (all natural)/kg b. wt., o., for 14 days; for detailed information please see the article), conc. range: 0.04 ng/g* **, Ø conc.: 0.04 ng/g* **, country: USA[362], *in loin muscle, **after 14 days

incidence: 5?/5*, sa. const.: cross-bred pigs, contamination: (dose: 9 ng/g AFB_1 + AFB_2, o., for 35 days; for detailed information please see the article), conc.: 0.010 ng/g** (mean value), country: USA[597], *control, **after 35 days
incidence: 5?/5, sa. const.: cross-bred pigs, contamination: artificial (dose: **524 ng/g AFB_1 + AFB_2**, o., for 35 days; for detailed information please see the article), conc.: 0.210 ng/g* (mean value), country: USA[597], *after **35 days**
incidence: 5?/5, sa. const.: cross-bred pigs, contamination: artificial (dose: **524 ng/g AFB_1 + AFB_2 + HSCAS** (0.5%), o., for 35 days; for detailed information please see the article), conc.: 0.008 ng/g* (mean value), country: USA[597], *after **35 days**

incidence: 5/5*, sa. const.: male miniature pigs, contamination: (dose: 1 ng/g AFB_1 + AFB_2, o., for 15 days; for detailed information please see the article), conc.: nd, country: USA[597], *control
incidence: 5?/5, sa. const.: male miniature pigs, contamination: artificial (dose: **590 ng/g AFB_1 + AFB_2**, o., for 15 days; for detailed information please see the article), conc.: 0.130 ng/g* (mean value), country: USA[597], *after **15 days**
incidence: 5?/5, sa. const.: male miniature pigs, contamination: artificial (dose: **590 ng/g AFB_1 + AFB_2 + HSCAS** (0.5%), o., for 15 days; for detailed information please see the article), conc.: 0.070 ng/g* (mean value), country: USA[597], *after **15 days**
incidence: 5?/5, sa. const.: male miniature pigs, contamination: artificial (dose: **590 ng/g AFB_1 + AFB_2**, o., for 15 days followed by **2 weeks control diet**; for detailed information please see the article), conc.: 0.010 ng/g* (mean value), country: USA[597], *after **29 days** (thereof 15 days of AFB_1- and AFB_2-administration)

Aflatoxin B_2

incidence: 1/2, sa. const.: adult swines, contamination: artificial (dose: 1.08–1.09 mg AFB_1 (besides other AFs), o., daily for 33 days; for detailed information please see the article), conc.: 0.46 ppb*, country: France[88], *after 33 days

incidence: 20/20*, sa. const.: male and female Danish Landrace pigs, age: ≈8 weeks, wt.: 20 kg up to slaughter at 90 kg, contamination: no AFB_1 + AFB_2 (for detailed information please see the article), conc.: nd, country: Denmark/USA[101], *control
incidence: 18/18*, sa. const.: male and female Danish Landrace pigs, age: ≈8 weeks, wt.: 20 kg up to slaughter at 90 kg, contamination: artificial (dose: AFB_1 + AFB_2 addition (overall **300 ppb AFs**), o., for **120–231 days****; for detailed information please see the article), conc.: nd**, country: Denmark/USA[101], *livers of some of these pigs were rejected at meat inspection, **after 120–231 days
incidence: 2/18*, sa. const.: male and female Danish Landrace pigs, age: ≈8 weeks, wt.: 20 kg up to slaughter at 90 kg, contamination: artificial (dose: AFB_1 + AFB_2 addition (overall **500 ppb AFs**), o., for **150 or 159 days****; for detailed information please see the article), conc. range: tr**, country: Denmark/USA[101], *livers of some of these pigs were rejected at meat inspection, **after 150 or 159 days

incidence: 4/4*, sa. const.: crossbred (Duroc × Yorkshire) barrows, wt.: 24.5–26.3 kg, contamination: no AFs, conc.: nd, country: USA[138], *control
incidence: 4?/4, sa. const.: crossbred (Duroc × Yorkshire) barrows, wt.: 24.5–26.3 kg, contamination: artificial (dose: **662 μg AFB_1**/kg diet, **273 μg AFB_2**/kg diet, **300 μg AFG_1**/kg diet and **285 μg AFG_2**/kg diet, o., for 21 days), conc. range: tr–0.05 μg/kg*, country: USA[138], *after ≈22 days (thereof 21 days of AF-administration)

incidence: 8/8*, sa. const.: mixed breed feeder pigs, contamination: acute study, no AFs, conc.: nd, country: USA[182], *control
incincidence: 1/1(8)*, sa. const.: mixed breed feeder pigs, contamination: artificial (acute study, dose: **1.2 mg total AFs** (AFB_1+AFB_2+AFG_1+AFG_2)/kg b. wt., o., once, conc. range: ≤0.45 ng/g** (wet matter basis), country: USA[182], *for overall information please see the article, **48 h post-dosage (also measured 12, 24 and 72 h post-dosage, lowest conc.: nd after 12 and 72 h)

incidence: ?/?*, sa. const.: Large White growing female pigs, age: 79 days, wt.: 20 kg, contamination: no AFs (for detailed information please see the article), conc.: nd, country: France[314], *control
incidence: 1/1, sa. const.: Large White growing female pigs, age: 79 days, wt.: 20 kg, contamination: artificial (dose: natural AFs addition (treatment 1: mixed feeding, avg. daily intake of **870 μg AFB_1**) for 26 days; for detailed information please see the article), conc.: nd*, country: France[314], *final wt. of the animal 109 kg
incidence: 1/1, sa. const.: Large White growing female pigs, age: 79 days, wt.: 20 kg, contamination: artificial (dose: natural AFs addition (treatment 2: separate feeding = peanut oil meal 40% and corn gluten meal 30%, avg. daily intake of **1,566 μg AFB_1**) for 19 days; for detailed information please see the article), conc.: 0.05 μg/kg*, country: France[314], *final wt. of the animal 97 kg
incidence: 1/1, sa. const.: Large White growing female pigs, age: 79 days, wt.: 20 kg, contamination: artificial (dose: natural AFs addition (treatment 3: separate feeding = peanut oil meal 0% and corn gluten meal 0%, avg. daily intake of **642 μg AFB_1**) for 26 days; for detailed information please see the article), conc.: nd*, country: France[314], *final wt. of the animal 104 kg

incidence: 5?/5*, sa. const.: cross-bred pigs, contamination: (dose: 9 ng/g AFB_1 + AFB_2, o., for 35 days; for detailed information please see the article), conc.: 0.001 ng/g** (mean value), country: USA[597], *control, **after 35 days
incidence: 5?/5, sa. const.: cross-bred pigs, contamination: artificial (dose: **524 ng/g AFB_1 + AFB_2**, o., for 35 days; for detailed information please see the article), conc.: 0.027 ng/g* (mean value), country: USA[597], *after **35 days**
incidence: 5?/5, sa. const.: cross-bred pigs, contamination: artificial (dose: **524 ng/g AFB_1 + AFB_2 + HSCAS** (0.5%), o., for 35 days; for detailed information please see the article), conc.: 0.003 ng/g* (mean value), country: USA[597], *after **35 days**

incidence: 5/5*, sa. const.: male miniature pigs, contamination: (dose: 1 ng/g AFB_1 + AFB_2, o., for 15 days; for detailed information please see the article), conc.: nd, country: USA[597], *control
incidence: 5?/5, sa. const.: male miniature pigs, contamination: artificial (dose: **590 ng/g AFB_1 + AFB_2**, o., for 15 days; for detailed information please see the article), conc.: 0.010 ng/g* (mean value), country: USA[597], *after **15 days**
incidence: 5?/5, sa. const.: male miniature pigs, contamination: artificial (dose: **590 ng/g AFB_1 + AFB_2 + HSCAS** (0.5%), o., for 15 days; for detailed information please see the article), conc.: 0.010 ng/g* (mean value), country: USA[597], *after **15 days**
incidence: 5/5, sa. const.: male miniature pigs, contamination: artificial (dose: **590 ng/g AFB_1 + AFB_2**, o., for 15 days followed by **2 weeks control diet**; for detailed information please see the article), conc.: nd*, country: USA[597], *after **29 days** (thereof 15 days of AFB_1- and AFB_2-administration)

Aflatoxin B_{2a}
incidence: 4/4*, sa. const.: crossbred (Duroc × Yorkshire) barrows, wt.: 24.5–26.3 kg, contamination: no AFs, conc.: nd, country: USA[138], *control
incidence: 4?/4, sa. const.: crossbred (Duroc × Yorkshire) barrows, wt.: 24.5–26.3 kg, contamination: artificial (dose: **662 µg AFB_1**/kg diet, **273 µg AFB_2**/kg diet, **300 µg AFG_1**/kg diet and **285 µg AFG_2**/kg diet, o., for 21 days), conc. range: appreciable amounts*, country: USA[138], *after ≈22 days (thereof 21 days of AF-administration)

Aflatoxin M_1
incidence: 1/2, sa. const.: adult swines, contamination: artificial (dose: 1.08–1.09 mg AFB_1 (besides other AFs), o., daily for 33 days; for detailed information please see the article), conc.: 0.09 ppb*, country: France[88], *after 33 days

incidence: 4/4*, sa. const.: male and female feeder pigs, wt.: 54.2–71.6 kg, contamination: no AFB_1, conc.: na, country: USA[136], *control
incidence: 3/4, sa. const.: male and female feeder pigs, wt.: 54.2–71.6 kg, contamination: artificial (dose: **100 µg AFB_1**/kg diet, o., for 4 weeks), conc. range: 0.03–0.04 µg/kg*, Ø conc.: 0.036 µg/kg*, country: USA[136], *after 4 weeks
incidence: 4/4, sa. const.: male and female feeder pigs, wt.: 54.2–71.6 kg, contamination: artificial (dose: **200 µg AFB_1**/kg diet, o., for 4 weeks), conc. range: 0.04–0.09 µg/kg*, Ø conc.: 0.07 µg/kg*, country: USA[136], *after 4 weeks
incidence: 3/4, sa. const.: male and female feeder pigs, wt.: 54.2–71.6 kg, contamination: artificial (dose: **400 µg AFB_1**/kg diet, o., for 4 weeks), conc. range: 0.17–0.35 µg/kg*, Ø conc.: 0.28 µg/kg*, country: USA[136], *after 4 weeks

incidence: 4/4*, sa. const.: crossbred (Duroc × Yorkshire) barrows, wt.: 24.5–26.3 kg, contamination: no AFs, conc.: nd, country: USA[138], *control incidence: 4?/4, sa. const.: crossbred (Duroc × Yorkshire) barrows, wt.: 24.5–26.3 kg, contamination: artificial (dose: **662 µg AFB$_1$/kg** diet, **273 µg AFB$_2$**/kg diet, **300 µg AFG$_1$**/kg diet and **285 µg AFG$_2$**/kg diet, o., for 21 days), conc. range: tr–0.24 µg/kg*, country: USA[138], *after ≈22 days (thereof 21 days of AF-administration)

incidence: 10/10*, sa. const.: mixed breed feeder pigs, contamination: chronic study, no AFs (for detailed information please see the article), conc.: nd, country: USA[182], *control
incidence: 10/10, sa. const.: mixed breed feeder pigs, contamination: artificial (chronic study, dose: **400 ng natural AFs/g** diet, o., for 10 weeks; for detailed information please see the article), conc.: nd*, country: USA[182], *after 10 weeks
incidence: 10?/10, sa. const.: mixed breed feeder pigs, contamination: artificial (chronic study, dose: **800 ng natural AFs/g** diet, o., for 10 weeks; for detailed information please see the article), conc.: 0.45 ng/g* (wet matter basis) (mean value), country: USA[182], *after 10 weeks

incidence: 8/8*, sa. const.: mixed breed feeder pigs, contamination: acute study, no AFs, conc.: nd, country: USA[182], *control
incidence: 1/1(8)*, sa. const.: mixed breed feeder pigs, contamination: artificial (acute study, dose: **1.2 mg total AFs** (AFB$_1$+AFB$_2$+AFG$_1$+AFG$_2$)/kg b. wt., o., once, conc. range: ≤2.18 ng/g** (wet matter basis), country: USA[182], *for overall information please see the article, **24 h post-dosage (also measured 12, 48 and 72 h post-dosage, lowest conc.: nd after 12 and 72 h)

incidence: 2/2, sa. const.: castrated male pigs of mixed breed, wt.: 9–11 kg, contamination: artificial (acute study, dose: 1 mg AFB$_1$/kg b. wt., o., once), conc. range: ≤1.3 ng/g*, country: USA[362], *sacrificed after 24 h (1 other pig sacrificed 72 h after treatment showed a lower mycotoxin value)

incidence: 5/5*, sa. const.: cross-bred pigs, contamination: (dose: 9 ng/g AFB$_1$ + AFB$_2$, o., for 35 days; for detailed information please see the article), conc.: nd, country: USA[597], *control
incidence: 5?/5, sa. const.: cross-bred pigs, contamination: artificial (dose: **524 ng/g AFB$_1$ + AFB$_2$**, o., for 35 days; for detailed information please see the article), conc.: 0.206 ng/g* (mean value), country: USA[597], *after **35 days**
incidence: 5?/5, sa. const.: cross-bred pigs, contamination: artificial (dose: **524 ng/g AFB$_1$ + AFB$_2$ + HSCAS** (0.5%), o., for 35 days; for detailed information please see the article), conc.: 0.075 ng/g* (mean value), country: USA[597], *after **35 days**

incidence: 5?/5*, sa. const.: male miniature pigs, contamination: (dose: 1 ng/g AFB$_1$ + AFB$_2$, o., for 15 days; for detailed information please see the article), conc.: 0.010 ng/g** (mean value), country: USA[597], *control, **after 15 days
incidence: 5?/5, sa. const.: male miniature pigs, contamination: artificial (dose: **590 ng/g AFB$_1$ + AFB$_2$**, o., for 15 days; for detailed information please see the article), conc.: 0.380 ng/g* (mean value), country: USA[597], *after **15 days**
incidence: 5?/5, sa. const.: male miniature pigs, contamination: artificial (dose: **590 ng/g AFB$_1$ + AFB$_2$ + HSCAS** (0.5%), o., for 15 days; for detailed information please see the article), conc.: 0.110 ng/g* (mean value), country: USA[597], *after **15 days**
incidence: 5?/5, sa. const.: male miniature pigs, contamination: artificial (dose: **590 ng/g AFB$_1$ + AFB$_2$**, o., for 15 days followed by **2 weeks control diet**; for detailed information please see the article), conc.: 0.010 ng/g* (mean value), country: USA[597], *after **29 days** (thereof 15 days of AFB$_1$- and AFB$_2$-administration)

Aflatoxin M

incidence: ?/?*, sa. const.: Large White growing female pigs, age: 79 days, wt.: 20 kg, contamination: no AFs (for detailed information please see the article), conc.: nd, country: France[314], *control
incidence: 1/1, sa. const.: Large White growing female pigs, age: 79 days, wt.: 20 kg, contamination: artificial (dose: natural AFs addition (treatment 1: mixed feeding, avg. daily intake of **870 µg AFB_1**) for 26 days; for detailed information please see the article), conc.: nd*, country: France[314], *final wt. of the animal 109 kg
incidence: 1/1, sa. const.: Large White growing female pigs, age: 79 days, wt.: 20 kg, contamination: artificial (dose: natural AFs addition (treatment 2: separate feeding = peanut oil meal 40% and corn gluten meal 30%, avg. daily intake of **1,566 µg AFB_1**) for 19 days; for detailed information please see the article), conc.: 0.05 µg/kg*, country: France[314], *final wt. of the animal 97 kg
incidence: 1/1, sa. const.: Large White growing female pigs, age: 79 days, wt.: 20 kg, contamination: artificial (dose: natural AFs addition (treatment 3: separate feeding = peanut oil meal 0% and corn gluten meal 0%, avg. daily intake of **642 µg AFB_1**) for 26 days; for detailed information please see the article), conc.: nd*, country: France[314], *final wt. of the animal 104 kg

Cyclopiazonic Acid

incidence: 3?/3, sa. const.: Piétran crossbred pigs, wt.: ≈97 kg, contamination: artificial (dose: 10 mg CPA/kg diet, o., for 6 days; for detailed information please see the article), conc.: 393 ng/g* **(mean value), country: USA[572], *in skeletal muscle (***longissimus***), **at day 6 (also measured at day 3, 4 and 5 but conc.: nd)
incidence: 3?/3, sa. const.: Piétran crossbred pigs, wt.: ≈97 kg, contamination: artificial (dose: 10 mg CPA/kg diet, o., for 6 days; for detailed information please see the article), conc.: 545 ng/g* ** (mean value), country: USA[572], *in skeletal muscle (***semitendinosus***), **at day 6 (also measured at day 3, 4 and 5 but conc.: nd)

Deoxynivalenol

incidence: 5/5*, sa. const.: healthy Yorkshire barrows, age: ≈11–15 weeks, wt.: 17–22 kg, contamination: no DON, conc.: nd, country: Canada[407], *control
incidence: ?/4, sa. const.: healthy Yorkshire barrows, age: ≈11–15 weeks, wt.: 17–22 kg, contamination: artificial (dose: **1.0 mg DON**/kg b. wt., i.v., once), conc. range: ≤33.1 ng/g* (mean value), country: Canada[407], *after 1 h (also measured after 0.33, 3, 8, and 24 h, lowest conc.: nd after 24 h)

incidence: 5?/5, sa. const.: castrated Large White × German Landrace, db Classic crossbred pigs, wt.: ≈24.6 kg, contamination: artificial (dose: 0.05, 0.57, or 1.23* mg DON/kg mash* or 0.07, 0.55, or 1.13 mg DON/kg pellets, o., for 11 weeks; for detailed information please see the article), conc. range: ≤5.2 ng/g* ** *** (mean value), country: Germany[483], **after 78/79 days (thereof 11 weeks of DON-administration), ***values of the other DON-treatments were lower

Deepoxydeoxynivalenol

incidence: 5/5, sa. const.: castrated Large White × German Landrace, db Classic crossbred pigs, wt.: ≈24.6 kg, contamination: artificial (dose: 0.05, 0.57, or 1.23 mg DON/kg mash* or 0.07, 0.55, or 1.13 mg DON/kg pellets*, o., for 11 weeks; for detailed information please see the article), conc.: nd*, country: Germany[483], *after 78/79 days (thereof 11 weeks of DON-administration)

Fumonisin B_1

incidence: 5/5*, sa. const.: weaned barrows of the same genotype, wt.: 12–14 kg, contamination: no FB_1, FB_2 + FB_3 (for detailed information please see the article), conc.: nd, country: Hungary/Germany[87], *control

incidence: 3/10, sa. const.: weaned barrows of the same genotype, wt.: 12–14 kg, contamination: artificial (dose: **50 mg FB_1, 20 mg FB_2 + 5 mg FB_3**/animal, o., for 22 days; for detailed information please see the article), conc. range: 0.8–6 ng/g* **, Ø conc.: 2.67 ng/g* **, country: Hungary/Germany[87], *in muscle ***psoas major***, **after 22 days toxin feeding period
incidence: 3/10, sa. const.: weaned barrows of the same genotype, wt.: 12–14 kg, contamination: artificial (dose: **50 mg FB_1, 20 mg FB_2 + 5 mg FB_3**/animal, o., for 22 days; for detailed information please see the article), conc. range: 1.2–1.6 ng/g* **, Ø conc.: 1.47 ng/g* **, country: Hungary/Germany[87], *in muscle ***longissimus dorsi***, **after 22 days toxin feeding period

incidence: 6/6*, sa. const.: castrated pigs of identical genotype, wt.: ≈12–14 kg, contamination: no FB_1 (for detailed information please see the article), conc.: nd, country: Germany/Hungary[109], *control
incidence: 13/13, sa. const.: castrated pigs of identical genotype, wt.: ≈12–14 kg, contamination: artificial (dose: **100 mg FB_1** daily, o., for 5–11 days; for detailed information please see the article), conc. range: 3–50 μg/kg* **, Ø conc.: 20.14 μg/kg* **, country: Germany/Hungary[109], *in muscle ***biceps femoris***, **on 6th day of the experiment
incidence: 13/13, sa. const.: castrated pigs of identical genotype, wt.: ≈12–14 kg, contamination: artificial (dose: **100 mg FB_1** daily, o., for 5–11 days; for detailed information please see the article), conc. range: 3–68 μg/kg* **, Ø conc.: 15.57 μg/kg* **, country: Germany/Hungary[109], *in muscle ***longissimus dorsi***, **on 6th day of the experiment
incidence: 12/13, sa. const.: castrated pigs of identical genotype, wt.: ≈12–14 kg, contamination: artificial (dose: **100 mg FB_1** daily, o., for 5–11 days; for detailed information please see the article), conc. range: 2–256 μg/kg* **, Ø conc.: 37.46 μg/kg* **, country: Germany/Hungary[109], *in muscle ***psoas major***, **on 6th day of the experiment

Fumonisin B_2
incidence: 5/5*, sa. const.: weaned barrows of the same genotype, wt.: 12–14 kg, contamination: no FB_1, FB_2 and FB_3 (for detailed information please see the article), conc.: nd, country: Hungary/Germany[87], *control
incidence: 1/10, sa. const. weaned barrows of the same genotype, wt.: 12–14 kg, contamination: artificial (dose: **50 mg FB_1, 20 mg FB_2 + 5 mg FB_3**/animaly, o., for 22 days; for detailed information please see the article), conc.: 1.6 ng/g* **, country: Hungary/Germany[87], *in muscle ***psoas major***, **after 22 days toxin feeding period
incidence: 10/10, sa. const.: weaned barrows of the same genotype, wt.: 12–14 kg, contamination: artificial (dose: **50 mg FB_1, 20 mg FB_2 + 5 mg FB_3**/animal, o., for 22 days; for detailed information please see the article), conc.: nd* **, country: Hungary/Germany[87], *in muscle ***longissimus dorsi***, **after 22 days toxin feeding period

HT-2 Toxin
incidence: 2/2, sa. const.: female crossbred Yorkshire × Hampshire swines, wt.: 20 kg, contamination: artificial (dose: 0.15 mg T-2 toxin (labeled and unlabeled)/kg b. wt., i.vs., once), conc. range: 1.69–2.06 ng/g*, Ø conc.: 1.875 ng/g*, country: USA[425], *after 4 h

3′-Hydroxy HT-2 Toxin
incidence: 2/2, sa. const.: female crossbred Yorkshire × Hampshire swines, wt.: 20 kg, contamination: artificial (dose: 0.15 mg T-2 toxin (labeled and unlabeled)/kg b. wt., i.vs., once), conc. range: 1.10–1.90 ng/g*, Ø conc.: 1.50 ng/g*, country: USA[425], *after 4 h

Ochratoxin A
incidence: 5/5*, sa. const.: female pigs of Danish Landrace, age: ≈8 weeks, wt.: ≈20 kg, contamination: no OTA

(for detailed information please see the article), conc.: nd, country: Denmark/USA[102], *control
incidence: ?/5, sa. const.: female pigs of Danish Landrace, age: ≈8 weeks, wt.: ≈20 kg, contamination: artificial (dose: 1 ppm crystalline OTA, o., once daily for 1 month, afterwards toxin-free diets for various intervals fed; for detailed information please see the article), conc.: 11.54 μg/kg* (mean value), country: Denmark/USA[102], ***1 day after termination** of OTA-exposure
incidence: ?/5, sa. const.: female pigs of Danish Landrace, age: ≈8 weeks, wt.: ≈20 kg, contamination: artificial (dose: 1 ppm crystalline OTA, o., once daily for 1 month, afterwards toxin-free diets for various intervals fed; for detailed information please see the article), conc.: 2.22 μg/kg* (mean value), country: Denmark/USA[102], ***8 days after termination** of OTA-exposure
incidence: ?/5, sa. const.: female pigs of Danish Landrace, age: ≈8 weeks, wt.: ≈20 kg, contamination: artificial (dose: 1 ppm crystalline OTA, o., once daily for 1 month, afterwards toxin-free diets for various intervals fed; for detailed information please see the article), conc.: nd*, country: Denmark/USA[102], ***15 days after termination** of OTA-exposure
incidence: ?/5, sa. const.: female pigs of Danish Landrace, age: ≈8 weeks, wt.: ≈20 kg, contamination: artificial (dose: 1 ppm crystalline OTA, o., once daily for 1 month, afterwards toxin-free diets for various intervals fed; for detailed information please see the article), conc.: nd*, country: Denmark/USA[102], ***29 days after termination** of OTA-exposure

incidence: ?/?, sa. const.: weaners (specific pathogen free), wt.: 14–18 kg, contamination: artificial (dose: ? μg OTA addition; for detailed information please see the article), conc. range: ≤32 μg/kg* **, country: Denmark[204], *calculated value based on the amount of OTA in blood after 24 h on toxin-free diet, **in muscle *psoas major*

incidence: 1/1*, sa. const.: pregnant gilts, contamination: neither OTA nor OTB, conc.: nd, country: UK[266], *control
incidence: 2/2, sa. const.: pregnant gilts, contamination: artificial (dose: **0.38 mg OTA + 0.13 mg OTB**/kg b. wt., o., for 8 days during early pregnancy), conc. range: 0.13–0.15 μg/g*, Ø conc.: 0.14 μg/g*, country: UK[266], *measured on day 30 of pregnancy

incidence: 13/13*, sa. const.: castrated pigs, wt.: 20 kg, contamination: no OTA and/or CIT (for detailed information please see the article), conc.: nd, country: Denmark/USA[330], *control
incidence: ?/13, sa. const.: castrated pigs, wt.: 20 kg, contamination: artificial (dose: **1,400 μg crystalline OTA**/kg feed, o., for 6 weeks; for detailed information please see the article), conc.: 8 μg/kg* ** (mean value), country: Denmark/USA[330], *in muscle *psoas major*, **after 6 weeks
incidence: ?/13, sa. const.: castrated pigs, wt.: 20 kg, contamination: artificial (dose: **650 μg crystalline CIT**/kg feed, o., for 6 weeks; for detailed information please see the article), conc.: nd* **, country: Denmark/USA[330], *in muscle *psoas major*, **after 6 weeks
incidence: ?/13, sa. const.: castrated pigs, wt.: 20 kg, contamination: artificial (dose: **1,400 μg crystalline OTA + 650 μg crystalline CIT**/kg feed, o., for 6 weeks; for detailed information please see the article), conc.: 8 μg/kg* ** (mean value), country: Denmark/USA[330], *in muscle *psoas major*, **after 6 weeks
incidence: ?/13, sa. const.: castrated pigs, wt.: 20 kg, contamination: artificial (dose: **1,400 μg natural OTA + 650 μg natural CIT**/kg feed, o., for 6 weeks; for detailed information please see the article), conc.: 37 μg/kg* ** (mean value), country: Denmark/USA[330], *in muscle *psoas major*, **after 6 weeks

incidence: 8?/8*, sa. const.: Danish Landrace piglets, age: 8 weeks, contamination: artificial (dose: **18.6 µg OTA**/kg liveweight/day, o., over a period of 6 weeks (treatment 1); for detailed information please see the article), conc.: 13 µg/kg* ** (mean value), country: Denmark[336], *sa. taken at weaning (pigs 8 weeks old), **in muscle *psoas major*
incidence: 6?/6*, sa. const.: Danish Landrace piglets, age: 8 weeks, contamination: artificial (dose: **8.0 µg OTA**/kg liveweight/day, o., over a period of 6 weeks (treatment 2); for detailed information please see the article), conc.: nr* **, country: Denmark[336], *sa. taken at weaning (pigs 8 weeks old), **in muscle *psoas major*
incidence: 8?/8*, sa. const.: Danish Landrace piglets, age: 8 weeks, contamination: artificial (dose: **19.7 µg OTA**/kg liveweight/day, o., over a period of 6 weeks (treatment 3); for detailed information please see the article), conc.: 18 µg/kg* ** (mean value), country: Denmark[336], *sa. taken at weaning (pigs 8 weeks old), **in muscle *psoas major*

incidence: 4/4*, sa. const.: male and females pigs, wt.: ≈70 kg, contamination: no OTA (for detailed information please see the article), conc.: nd, country: Germany[365], *control
incidence: 8?/8, sa. const.: male and females pigs, wt.: ≈70 kg, contamination: artificial (dose: **0.09 mg natural OTA**/kg diet/day, in the morning and evening **half of OTA-ration**, o., for 28 days; for detailed information please see the article), conc.: 4.23 ng/g* (mean value), country: Germany[365], *after 28 days
incidence: 8?/8, sa. const.: male and females pigs, wt.: ≈70 kg, contamination: artificial (dose: **0.09 mg crystalline OTA/** kg diet/day, in the morning and evening **half of OTA-ration**, o., for 28 days; for detailed information please see the article), conc.: 1.44 ng/g* (mean value), country: Germany[365], *after 28 days
incidence: 8?/8, sa. const.: male and females pigs, wt.: ≈70 kg, contamination: artificial (dose: **0.09 mg crystalline OTA**/kg diet/day, in the morning **total OTA-ration**, o., for 28 days; for detailed information please see the article), conc.: 1.57 ng/g* (mean value), country: Germany[365], *after 28 days

incidence: 8?/8*, sa. const.: pigs, contamination: no OTA (for detailed information please see the article), conc.: <0.39 µg/kg (mean value), country: Germany[366], *control
incidence: 8?/8, sa. const.: pigs, contamination: artificial (dose: **22.11 mg OTA** (in total), o., for 90 days; for detailed information please see the article), conc.: 2.7 µg/kg* (mean value), country: Germany[366], *after 90 days
incidence: 8?/8, sa. const.: pigs, contamination: artificial (dose: **88.44 mg OTA** (in total), o., for 90 days; for detailed information please see the article), conc.: 10.3 µg/kg* (mean value), country: Germany[366], *after 90 days

incidence: 2/2*, sa. const.: pigs, contamination: no OTA and/or DON (for detailed information please see the article), conc.: nd, country: Germany[378], *control
incidence: 6/6, sa. const.: pigs, contamination: artificial (dose: **0.1 ppm crystalline OTA + 1.0 ppm crystalline DON**, o., twice daily for 90 days; for detailed information please see the article), conc. range: 0.42–0.89 ng/g*, Ø conc.: 0.63 ng/g*, country: Germany[378], *after 90 days
incidence: 3/3, sa. const.: pigs, contamination: artificial (dose: **0.1 ppm crystalline OTA**, o., twice daily for 90 days; for detailed information please see the article), conc. range: 0.27–0.56 ng/g*, Ø conc.: 0.383 ng/g*, country: Germany[378], *after 90 days
incidence: 6/6, sa. const.: pigs, contamination: artificial (dose: **1.0 ppm crystalline DON**, o., twice daily for 90 days; for detailed information please

see the article), conc.: nd*, country: Germany[378], *after 90 days

incidence: 2?/2*, sa. const.: pigs, contamination: no OTA and/or ZEA (for detailed information please see the article), conc.: nr, country: Germany[380], *control
incidence: 6/6, sa. const.: pigs, contamination: artificial (dose: **0.1 ppm OTA + 0.25 ppm ZEA**, o., twice daily for 90 days; for detailed information please see the article), conc. range: ≤2.44 ng/g*, country: Germany[380], *after 91 days (thereof 90 days of OTA- and ZEA-administration)
incidence: 3/3, sa. const.: pigs, contamination: artificial (dose: **0.1 ppm OTA**, o., twice daily for 90 days; for detailed information please see the article), conc. range: ≤2.13 ng/g*, country: Germany[380], *after 91 days (thereof 90 days of OTA-administration)

incidence: 3/3*, sa. const.: female pigs of Danish Landrace, age: 8–10 weeks, wt.: ≈20 kg, contamination: no OTA (for detailed information please see the article), conc.: nd, country: Denmark/USA/Sweden[383], *control
incidence: 4?/4, sa. const.: female pigs of Danish Landrace, age: 8–10 weeks, wt.: ≈20 kg, contamination: artificial (dose: **1 mg crystalline OTA**/kg feed, o., for 3 months; for detailed information please see the article), conc.: ≈9.5 µg/kg* (mean value), country: Denmark/USA/Sweden[383], *after ≈3 months

T-2 Toxin
incidence: 1/1, sa. const.: female weanling crossbred Yorkshire × Duroc × Hampshire swine, wt.: 7.5 kg, contamination: artificial (dose: **0.1 mg T-2 toxin** (labeled)/kg b. wt., intubated, once), conc.: 3.1 ppb* ** ***, country: USA[318], *calculated residue level, **T-2 toxin and/or metabolites, ***after 18 h
incidence: 1/1, sa. const.: female weanling crossbred Yorkshire × Duroc × Hampshire swine, wt.: 9.5 kg, contamination: artificial (dose: **0.4 mg T-2 toxin** (labeled)/kg b. wt., intubated, once), conc.: 11.5 ppb* ** ***, country: USA[318], *calculated residue level, **T-2 toxin and/or metabolites, ***after 18 h

incidence: 2/2*, sa. const.: female swines of mixed breeding, wt.: 26–66 kg, contamination: no T-2 toxin, conc.: nr, country: USA[403], *control
incidence: 4/4, sa. const.: female swines of mixed breeding, wt.: 26–66 kg, contamination: artificial (dose: **1.2 mg T-2 toxin**, i.a., once), conc. range: ≈≤57 ppb*, country: USA[403], *after ≈1 h (also measured after ≈2.1, 3 and 4 h, lowest value conc.: under limit of reliable quantitation after 4 h)

incidence: 2/2, sa. const.: female crossbred Yorkshire × Hampshire swines, wt.: 20 kg, contamination: artificial (dose: 0.15 mg T-2 toxin (labeled and unlabeled)/kg b. wt., i.vs., once), conc. range: 0.64–0.85 ng/g*, Ø conc.: 0.745 ng/g*, country: USA[425], *after 4 h
incidence: 2/2, sa. const.: female crossbred Yorkshire × Hampshire swines, wt.: 20 kg, contamination: artificial (dose: 0.15 mg T-2 toxin (labeled and unlabeled)/kg b. wt., i.vs., once), conc. range: 18–19 ng/g* **, Ø conc.: 18.5 ng/g* **, country: USA[425], *after 4 h, ****total metabolites**

3′-Hydroxy T-2 Toxin
incidence: 2/2, sa. const.: female crossbred Yorkshire × Hampshire swines, wt.: 20 kg, contamination: artificial (dose: 0.15 mg T-2 toxin (labeled and unlabeled)/kg b. wt., i.vs., once), conc. range: 0.48–1.07 ng/g*, Ø conc.: 0.775 ng/g*, country: USA[425], *after 4 h

Deepoxy T-2 Tetraol
incidence: 2/2, sa. const.: female crossbred Yorkshire × Hampshire swines, wt.: 20 kg, contamination: artificial (dose: 0.15 mg T-2 toxin (labeled and unlabeled)/kg b. wt., i.vs., once), conc. range: 0.98–1.13 ng/g*, Ø conc.: 1.055 ng/g*, country: USA[425], *after 4 h

T-2 Triol
incidence: 2/2, sa. const.: female crossbred Yorkshire × Hampshire swines, wt.: 20 kg, contamination: artificial (dose: 0.15 mg T-2 toxin (labeled and unlabeled)/kg b. wt., i.vs., once), conc. range: 0.14–0.18 ng/g*, Ø conc.: 0.16 ng/g*, country: USA[425], *after 4 h

Zearalenone
incidence: 2?/2?*, sa. const.: male KAHYP pigs, wt.: ?, contamination: no ZEA, conc.: nd, country: Hungary[632], *control
incidence: 2/2, sa. const.: male KAHYP pigs, wt.: 60 kg, contamination: artificial (dose: **15 ppm ZEA**, o., for 14 days), conc.: nd* **, country: Hungary[632], *after 14 days, **in **femoral muscle**
incidence: 2/2, sa. const.: male KAHYP pigs, wt.: 60 kg, contamination: artificial (dose: **15 ppm ZEA**, o., for 14 days), conc.: nd* **, country: Hungary[632], *after 14 days, **in **shoulder blade muscle**

α-Zearalenol
incidence: 2?/2?*, sa. const.: male KAHYP pigs, wt.: ?, contamination: no ZEA, conc.: nd, country: Hungary[632], *control
incidence: 2?/2, sa. const.: male KAHYP pigs, wt.: 60 kg, contamination: artificial (dose: **15 ppm ZEA**, o., for 14 days), conc.: <10 μg/kg* **, country: Hungary[632], *after 14 days, **in **femoral muscle**
incidence: 2?/2, sa. const.: male KAHYP pigs, wt.: 60 kg, contamination: artificial (dose: **15 ppm ZEA**, o., for 14 days), conc.: <10 μg/kg* **, country: Hungary[632], *after 14 days, **in **shoulder blade muscle**

β-Zearalenol
incidence: 2?/2?*, sa. const.: male KAHYP pigs, wt.: ?, contamination: no ZEA, conc.: nd, country: Hungary[632], *control
incidence: 2?/2, sa. const.: male KAHYP pigs, wt.: 60 kg, contamination: artificial (dose: **15 ppm ZEA**, o., for 14 days), conc.: pr* **, country: Hungary[632], *after 14 days, **in **femoral muscle**
incidence: 2?/2, sa. const.: male KAHYP pigs, wt.: 60 kg, contamination: artificial (dose: **15 ppm ZEA**, o., for 14 days), conc.: pr* **, country: Hungary[632], *after 14 days, **in **shoulder blade muscle**

Pig myocardium may contain the following mycotoxins and/or their metabolites:

Fumonisin B_1
incidence: 5/5*, sa. const.: weaned barrows of the same genotype, wt.: 12–14 kg, contamination: no FB_1, FB_2 + FB_3 (for detailed information please see the article), conc.: nd, country: Hungary/Germany[87], *control
incidence: 7/10, sa. const.: weaned barrows of the same genotype, wt.: 12–14 kg, contamination: artificial (dose: **50 mg FB_1, 20 mg FB_2 + 5 mg FB_3**/animal, o., for 22 days; for detailed information please see the article), conc. range: 7.2–35.2 ng/g*, Ø conc.: 13.26 ng/g*, country: Hungary/Germany[87], *after 22 days toxin feeding period

incidence: 6/6*, sa. const.: castrated pigs of identical genotype, wt.: ≈12–14 kg, contamination: no FB_1 (for detailed information please see the article), conc.: nd, country: Germany/Hungary[109], *control
incidence: 13/13, sa. const.: castrated pigs of identical genotype, wt.: ≈12–14 kg, contamination: artificial (dose: **100 mg FB_1** daily, o., for 5–11 days; for detailed information please see the article), conc. range: 3–838 μg/kg*, Ø conc.: 77.43 μg/kg*, country: Germany/Hungary[109], *on 6th day of the experiment

Fumonisin B_2
incidence: 5/5*, sa. const.: weaned barrows of the same genotype, wt.: 12–14 kg, contamination: no FB_1, FB_2 + FB_3 (for detailed information please see the article), conc.: nd, country: Hungary/Germany[87], *control
incidence: 10/10, sa. const.: weaned barrows of the same genotype, wt.:

12–14 kg, contamination: artificial (dose: **50 mg FB_1, 20 mg FB_2 + 5 mg FB_3/** animal, o., for 22 days; for detailed information please see the article), conc.: nd*, country: Hungary/Germany[87], *after 22 days toxin feeding period

Pig pancreas may contain the following mycotoxins and/or their metabolites:

DEOXYNIVALENOL
incidence: 5/5*, sa. const.: healthy Yorkshire barrows, age: ≈11–15 weeks, wt.: 17–22 kg, contamination: no DON, conc.: nd, country: Canada[407], *control
incidence: ?/4, sa. const.: healthy Yorkshire barrows, age: ≈11–15 weeks, wt.: 17–22 kg, contamination: artificial (dose: **1.0 mg DON**/kg b. wt., iv., once), conc. range: ≤18.7 ng/g* (mean value), country: Canada[407], *after 1 h (also measured after 0.33, 3, 8 and 24 h, lowest conc.: nd after 8 and 24 h)

FUMONISIN B_1
incidence: 6/6*, sa. const.: castrated pigs of identical genotype, wt.: ≈12–14 kg, contamination: no FB_1 (for detailed information please see the article), conc.: nd, country: Germany/Hungary[109], *control
incidence: 12/12, sa. const.: castrated pigs of identical genotype, wt.: ≈12–14 kg, contamination: artificial (dose: **100 mg FB_1** daily, o., for 5–11 days; for detailed information please see the article), conc. range: 24–464 μg/kg*, Ø conc.: 126.85 μg/kg*, country: Germany/Hungary[109], *on 6th day of the experiment

HT-2 TOXIN
incidence: 1/1, sa. const.: female crossbred Yorkshire × Hampshire swine, wt.: 20 kg, contamination: artificial (dose: 0.15 mg T-2 toxin (labeled and unlabeled)/kg b. wt., i.vs., once), conc.: 11.86 ng/g*, country: USA[425], *after 4 h

3′-HYDROXY HT-2 TOXIN
incidence: 1/1, sa. const.: female crossbred Yorkshire × Hampshire swine, wt.: 20 kg, contamination: artificial (dose: 0.15 mg T-2 toxin (labeled and unlabeled)/kg b. wt., i. vs., once), conc.: 17.79 ng/g*, country: USA[425], *after 4 h

T-2 TOXIN
incidence: 1/1, sa. const.: female crossbred Yorkshire × Hampshire swines, wt.: 20 kg, contamination: artificial (dose: 0.15 mg T-2 toxin (labeled and unlabeled)/kg b. wt., i.vs., once), conc.: 0.52 ng/g*, country: USA[425], *after 4 h
incidence: 1/1, sa. const.: female crossbred Yorkshire × Hampshire swines, wt.: 20 kg, contamination: artificial (dose: 0.15 mg T-2 toxin (labeled and unlabeled)/kg b. wt., i.vs., once), conc.: 159 ng/g* **, country: USA[425], *after 4 h, ****total metabolites**

3′-HYDROXY T-2 TOXIN
incidence: 1/1, sa. const.: female crossbred Yorkshire × Hampshire swine, wt.: 20 kg, contamination: artificial (dose: 0.15 mg T-2 toxin (labeled and unlabeled)/kg b. wt., i.vs., once), conc.: 3.13 ng/g*, country: USA[425], *after 4 h

T-2 TETRAOL
incidence: 1/1, sa. const.: female crossbred Yorkshire × Hampshire swine, wt.: 20 kg, contamination: artificial (dose: 0.15 mg T-2 toxin (labeled and unlabeled)/kg b. wt., i.vs., once), conc.: 0.82 ng/g*, country: USA[425], *after 4 h

DEEPOXY T-2 TETRAOL
incidence: 1/1, sa. const.: female crossbred Yorkshire × Hampshire swine, wt.: 20 kg, contamination: artificial (dose: 0.15 mg T-2 toxin (labeled and unlabeled)/kg b. wt., i.vs., once), conc.: 3.07 ng/g*, country: USA[425], *after 4 h

T-2 TRIOL
incidence: 1/1, sa. const.: female crossbred Yorkshire × Hampshire swine, wt.: 20 kg,

contamination: artificial (dose: 0.15 mg T-2 toxin (labeled and unlabeled)/kg b. wt., i.vs., once), conc.: 0.65 ng/g*, country: USA[425], *after 4 h

α-Zearalenol
incidence: 2?/2?*, sa. const.: male KAHYP pigs, wt.: ?, contamination: no ZEA, conc.: nd, country: Hungary[632], *control
incidence: 2?/2, sa. const.: male KAHYP pigs, wt.: 60 kg, contamination: artificial (dose: **15 ppm ZEA**, o., for 14 days, conc.: 10 μg/kg*, country: Hungary[632], *after 14 days)

Pig placenta may contain the following mycotoxins and/or their metabolites:

Ochratoxin A
incidence: 1/1*, sa. const.: pregnant gilts, contamination: neither OTA nor OTB, conc.: nd, country: UK[266], *control
incidence: 2/2, sa. const.: pregnant gilts, contamination: artificial (dose: **0.38 mg OTA + 0.13 mg OTB**/kg b. wt., o., for 8 days during early pregnancy), conc. range: 0.04–0.06 μg/g*, Ø conc.: 0.05 μg/g*, country: UK[266], *measured on day 30 of pregnancy

Pig plasma may contain the following mycotoxins and/or their metabolites:

Cyclopiazonic Acid
incidence: 3?/3, sa. const.: Piétran crossbred pigs, wt.: ≈97 kg, contamination: artificial (dose: 10 mg CPA/kg diet, o., for 6 days; for detailed information please see the article), conc. range: ≤464 ng/ml* (mean value), country: USA[572], *at day 6 (also measured at day 3, 4 and 5, lowest conc.: 370 ng/ml at day 3)

Deoxynivalenol
incidence: 1/10, sa. const.: male and female crossbred piglets, age: 5 weeks, contamination: artificial (dose: **0.7 ppm DON** in the diet, for 1–5 weeks; for detailed information please see the article), conc.: 53 ppb*, country: Canada/USA[70], *at 4 weeks (also measured at 1st and 5th week but conc.: nd)
incidence: 4/8, sa. const.: male and female crossbred piglets, age: 5 weeks, contamination: artificial (dose: **3.1 ppm DON** in the diet, for 1–5 weeks; for detailed information please see the article), conc. range: 79–100 ppb*, country: Canada/USA[70], *measured at 1st, 4th, and 5th week (pr. residue values are each lowest and highest value of 1st to 5th week measurement)
incidence: 2/10, sa. const.: male and female crossbred piglets, age: 5 weeks, contamination: artificial (dose: **5.8 ppm DON** in the diet, for 1–5 weeks; for detailed information please see the article), conc. range: 32–64 ppb*, country: Canada/USA[70], *measured at 1st, 4th, and 5th week (pr. residue values are each lowest and highest value of 4th to 5th week measurement, 1st week conc.: nd)

incidence: 5/5*, sa. const.: healthy Yorkshire barrows, age: ≈11–15 weeks, wt.: 17–22 kg, contamination: no DON, conc.: nd, country: Canada[407], *control
incidence: ?/4, sa. const.: healthy Yorkshire barrows, age: ≈11–15 weeks, wt.: 17–22 kg, contamination: artificial (dose: **1.0 mg DON**/kg b. wt., i.v., once), conc. range: ≤1,337.3 ng/g* (mean value), country: Canada[407], *after 0.33 h (also measured after 1, 3, 8 and 24 h, lowest conc.: 17.8 ng/g after 24 h)

incidence: 5?/5, sa. const.: castrated male Swedish Landrace pigs, wt.: ≈20 kg, contamination: artificial (dose: 2.5 mg 3-aDON/kg feed, o., 5 times in 2.5 days), conc. range: ≤64.4 ng/ml*, country: Sweden[416], *after **1 day** (also measured after 3 days)
incidence: 5?/5, sa. const.: castrated male Swedish Landrace pigs, wt.: ≈20 kg, contamination: artificial (dose: 2.5 mg 3-aDON/kg feed, o., 5 times in 2.5 days), conc. range: ≤62.0 ng/ml*, country:

Sweden[416], *after **3 days** (also measured after 1 day)

Ochratoxin A

incidence: ?/4, sa. const.: Large White castrated pigs, wt.: ≈35 kg, contamination: artificial (dose: 0.5 mg OTA/kg, **o.**, once; for detailed information please see the article), conc. range: ≤1.74 µg/ml* (mean value), country: France[172], *after 10 h (also at other hour intervals up to 72 h measured, lowest conc.: ≈1 µg/ml after 72 h)
incidence: ?/4, sa. const.: large White castrated pigs, wt.: ≈35 kg, contamination: artificial (dose: 0.5 mg OTA/kg, **i.v.**, once; for detailed information please see the article), conc. range: ≈≤10 µg/ml* (mean value), country: France[172], *after 1 h? (also at other hour up to 72 h measured, lowest conc.: ≈2 µg/ml after 72 h)

incidence: 4/4*, sa. const.: sexually matured boars Hungarian Large White and Dutch Landrace), wt.: 250 kg, contamination: no OTA (for detailed information please see the article), conc.: nr**, country: Hungary[395], *control (low- and high-dosed boars), **in seminal plasma
incidence: 1/1*, sa. const.: sexually matured boar Hungarian Large White and Dutch Landrace), wt.: 250 kg, contamination: artificial (dose: 20 µg OTA, o., daily for 5 weeks; for detailed information please see the article), conc. range: ≤0.501 ng/ml** ***, country: Hungary[395], ***low-dosed boar**, after 63** days (thereof 5 weeks of OTA-administration) (also at other day intervals up to 77 days measured, except for the start values lowest conc.: ≈0.2 ng/ml after 77 days), ***in seminal plasma
incidence: 1/1*, sa. const.: sexually matured boar Hungarian Large White and Dutch Landrace), wt.: 250 kg, contamination: artificial (dose: 20 µg OTA, o., daily for 5 weeks; for detailed information please see the article), conc. range: ≤0.579 ng/ml** ***, country: Hungary[395], ***low-dosed boar**, after 84** days (thereof 5 weeks of OTA-administration) (also at other day intervals up to 91 days measured, except for the start values lowest conc.: ≈0.1 ng/ml after 14 and 70 days), ***in seminal plasma
incidence: 1/1*, sa. const.: sexually matured boar (Hungarian Large White and Dutch Landrace), wt.: 250 kg, contamination: artificial (dose: 40 µg OTA, o., daily for 5 weeks; for detailed information please see the article), conc. range: ≤0.709 ng/ml** ***, country: Hungary[395], ***high-dosed boar**, **after 28 days (within 5 weeks of OTA-administration) (also at other day intervals up to 77 days measured, except for the start values lowest conc.: ≈0.25 ng/ml after 49 days), ***in seminal plasma
incidence: 1/1*, sa. const.: sexually matured boar (Hungarian Large White and Dutch Landrace), wt.: 250 kg, contamination: artificial (dose: 40 µg OTA, o., daily for 5 weeks; for detailed information please see the article), conc. range: ≤0.673 ng/ml** ***, country: Hungary[395], ***high-dosed boar**, **after 28 days (within 5 weeks of OTA-administration) (also at other day intervals up to 91 days measured, except for the start values lowest conc.: ≈0.1 ng/ml after 91 days), ***in seminal plasma

Scirpentriol

incidence: 4/4*, sa. const.: male and female crossbred feeder pigs, wt.: 22.2–49.2 kg, contamination: no DAS (for detailed information please see the article), conc.: nd, country: USA[512], *control
incidence: 3/4, sa. const.: male and female crossbred feeder pigs, wt.: 22.2–49.2 kg, contamination: artificial (dose: **0.5 or 1 mg DAS**/kg b. wt., injection, once; for

detailed information please see the article), conc. range: ≤240 μg/ml*, country: USA[512], *collected after 8 h

T-2 Toxin

incidence: 2/2*, sa. const.: female swine of mixed breeding, wt.: 26–66 kg, contamination: no T-2 toxin, conc.: nd?, country: USA[403], *control
incidence: 2?/2, sa. const.: female swines of mixed breeding, wt.: 26–66 kg, contamination: artificial (dose: **0.3 mg T-2 toxin**/kg, i.a., once), conc. range: ≈≤550 ppb* (mean value), country: USA[403], *after 0 min (also at other min intervals up to 20 min measured, lowest value conc.: ≈55 ppb after 15 min)
incidence: 2?/2, sa. const.: female swines of mixed breeding, wt.: 26–66 kg, contamination: artificial (dose: **0.6 mg T-2 toxin**/kg, i.a., once), conc. range: ≈≤2,700 ppb* (mean value), country: USA[403], *after 0 min (also at other min intervals up to 20 min measured, lowest value conc.: ≈63 ppb after 20 min)
incidence: 6?/6, sa. const.: female swines of mixed breeding, wt.: 26–66 kg, contamination: artificial (dose: **1.2 mg T-2 toxin**/kg, i.a., once), conc. range: ≈≤3,650 ppb* (mean value), country: USA[403], *after 0 min (also at other min intervals up to 30 min measured, lowest value conc.: ≈125 ppb after 30 min)

Zearalenone

incidence: ?/?, sa. const.: female Yorkshire pigs, age: 10–14 weeks, wt.: 15–25 kg, contamination: artificial (dose: 5 mg ZEA (labeled)/kg b. wt., i.g., once; for detailed information please see the article), conc. range: ≤2.30 nmol/ml* **, country: Canada[128], *[^{3}H]ZEA-eq., **after 2–3 h (also at other minute, hour and day intervals up to 4 days measured)

incidence: 1/1, sa. const.: prepubertal gilt of Swedish Landrace × Yorkshire breed, age: 5 months, wt.: 60 kg, contamination: artificial (dose: 192 μg ZEA/kg b. wt./day, o., for 4 days), conc.: ≈2.4 ng/ml*, country: Sweden[309], *after ≈1.4 days of ZEA-administration (also at other day intervals up to 10 days measured, lowest conc.: nd after 10 days)

incidence: 3/3*, sa. const.: uncastrated **male** weanling Yorkshire pigs, wt.: 8.9–13.0 kg, contamination: no ZEA (for detailed information please see the article), conc.: nd, country: Canada[422], *control
incidence: 3?/3, sa. const.: uncastrated **male** weanling Yorkshire pigs, wt.: 8.9–13.0 kg, contamination: artificial (dose: **5 mg crystalline ZEA**/kg b. wt., o., 4 times; for detailed information please see the article), conc.: 216 ng/ml* (avg. maximum conc.), country: Canada[422], *sa. period over 7 h
incidence: 3?/3, sa. const.: uncastrated **male** weanling Yorkshire pigs, wt.: 8.9–13.0 kg, contamination: artificial (dose: **10 mg crystalline ZEA**/kg b. wt., o., 4 times; for detailed information please see the article), conc.: 327 ng/ml* (avg. maximum conc.), country: Canada[422], *sa. period over 7 h
incidence: 3?/3, sa. const.: uncastrated **male** weanling Yorkshire pigs, wt.: 8.9–13.0 kg, contamination: artificial (dose: **15 mg crystalline ZEA**/kg b. wt., o., 4 times; for detailed information please see the article), conc.: 298 ng/ml* (avg. maximum conc.), country: Canada[422], *sa. period over 7 h
incidence: 3/3*, sa. const.: **female** weanling Yorkshire pigs, wt.: 7.6–20.0 kg, contamination: no ZEA (for detailed information please see the article), conc.: nd, country: Canada[422], *control
incidence: 3?/3, sa. const.: **female** weanling Yorkshire pigs, wt.: 7.6–20.0 kg, contamination: artificial (dose: **5 mg crystalline ZEA**/kg b. wt., o., 4 times; for detailed information please see the article), conc.: 317 ng/ml* (avg. maximum conc.), country: Canada[422], *sa. period over 7 h
incidence: 3?/3, sa. const.: **female** weanling Yorkshire pigs, wt.: 7.6–20.0 kg, contamination: artificial (dose: **10 mg**

crystalline ZEA/kg b. wt., o., 4 times; for detailed information please see the article), conc.: 290 ng/ml* (avg. maximum conc.), country: Canada[422], *sa. period over 7 h
incidence: 3?/3, sa. const.: **female** weanling Yorkshire pigs, wt.: 7.6–20.0 kg, contamination: artificial (dose: **15 mg crystalline ZEA**/kg b. wt., o., 4 times; for detailed information please see the article), conc.: 139 ng/ml* (avg. maximum conc.), country: Canada[422], *sa. period over 7 h

α-Zearalenol
incidence: 1/1, sa. const.: prepubertal gilt of Swedish Landrace × Yorkshire breed, age: 5 months, wt.: 60 kg, contamination: artificial (dose: 192 μg ZEA/kg b. wt./day, o., for 4 days), conc.: ≈8.8 ng/ml*, country: Sweden[309], *after ≈3.5 days of ZEA-administration (also at other day intervals up to 10 days measured, lowest conc.: nd after 10 days)

incidence: 3/3*, sa. const.: uncastrated **male** weanling Yorkshire pigs, wt.: 8.9–13.0 kg, contamination: no ZEA (for detailed information please see the article), conc.: nd, country: Canada[422], *control
incidence: 3/3, sa. const.: uncastrated **male** weanling Yorkshire pigs, wt.: 8.9–13.0 kg, contamination: (dose: **5 mg crystalline ZEA**/kg b. wt., o., 4 times; for detailed information please see the article), conc.: 65 ng/ml* (avg. maximum conc.), country: Canada[422], *sa. period over 7 h
incidence: 3?/3, sa. const.: uncastrated **male** weanling Yorkshire pigs, wt.: 8.9–13.0 kg, contamination: artificial (dose: **10 mg crystalline ZEA**/kg b. wt., o., 4 times; for detailed information please see the article), conc.: 233 ng/ml* (avg. maximum conc.), country: Canada[422], *sa. period over 7 h
incidence: 3?/3, sa. const.: uncastrated **male** weanling Yorkshire pigs, wt.: 8.9–13.0 kg, contamination: artificial (dose: **15 mg crystalline ZEA**/kg b. wt., o., 4 times; for detailed information please see the article), conc.: 199 ng/ml* (avg. maximum conc.), country: Canada[422], *sa. period over 7 h
incidence: 3/3*, sa. const.: **female** weanling Yorkshire pigs, wt.t: 7.6–20.0 kg, contamination: no ZEA; (for detailed information please see the article), conc.: nd, country: Canada[422], *control
incidence: 3?/3, sa. const.: **female** weanling Yorkshire pigs, wt.: 7.6–20.0 kg, contamination: artificial (dose: **5 mg crystalline ZEA**/kg b. wt., o., 4 times; for detailed information please see the article), conc.: 319 ng/ml* (avg. maximum conc.), country: Canada[422], *sa. period over 7 h
incidence: 3?/3, sa. const.: **female** weanling Yorkshire pigs, wt.: 7.6–20.0 kg, contamination: artificial (dose: **10 mg crystalline ZEA**/kg b. wt., o., 4 times; for detailed information please see the article), conc.: 190 ng/ml* (avg. maximum conc.), country: Canada[422], *sa. period over 7 h
incidence: 3?/3, sa. const.: **female** weanling Yorkshire pigs, wt.: 7.6–20.0 kg, contamination: artificial (dose: **15 mg crystalline ZEA**/kg b. wt., o., 4 times; for detailed information please see the article), conc.: 72 ng/ml* (avg. maximum conc.), country: Canada[422], *sa. period over 7 h

Pig rectum may contain the following mycotoxins and/or their metabolites:

Deoxynivalenol + Deepoxydeoxynivalenol
incidence: ?/11, sa. const.: castrated male pigs, Ø wt.: 88.1 kg, contamination: artificial (dose: 4.2 mg DON/kg, o., for 7 days), conc. range: ≈≤610 μg/g* (mean value), country: Germany[413], *≈5 h after final DON-administration (also at other hour intervals up to 24 h measured, lowest conc.: ≈90 μg/g after 3 h)

Pig serum may contain the following mycotoxins and/or their metabolites:

Deoxynivalenol

incidence: ?/11, sa. const.: castrated male pigs, Ø wt.: 88.1 kg, contamination: artificial (dose: 4.2 mg DON/kg, o., for 7 days), conc. range: ≈≤14 ng/ml*, country: Germany[413], *6 h after final DON-administration (also at other hour intervals up to 24 h measured, lowest conc.: nd after 15 and 24 h)

incidence: 5?/5, sa. const.: castrated Large White × German Landrace, db Classic crossbred pigs, wt.: ≈24.6 kg, contamination: artificial (dose: 0.05, 0.57*, or 1.23 mg DON/kg mash* or 0.07, 0.55, or 1.13 mg DON/kg pellets, o., for 11 weeks; for detailed information please see the article), conc. range: ≤12.2 ng/ml* ** (mean value), country: Germany[483], **after 78/79 days (thereof 11 weeks of DON-administration), ***values of the other DON-treatments lower

incidence: 9?/9, sa. const.: German Landrace gilts, age: 180 days, wt.: ≈103 kg, contamination: artificial (dose: DON/ZEA in wheat in different conc., o., for 35 days; for detailed information please see the article), conc. range: ≤21.6 ng/ml* ** (mean value), country: Germany[537], *9.57 mg DON and 0.358 mg ZEA/kg diet fed (both fed in highest conc.), **after 36 days (thereof 35 days of DON- and ZEA-administration)

Deepoxydeoxynivalenol

incidence: 5/5, sa. const.: castrated Large White × German Landrace, db Classic crossbred pigs, wt.: ≈24.6 kg, contamination: artificial (dose: 0.05, 0.57, or 1.23 mg DON/kg mash* or 0.07, 0.55, or 1.13 mg DON/kg pellets*, o., for 11 weeks; for detailed information please see the article), conc.: nd*, country: Germany[483], *after 78/79 days (thereof 11 weeks of DON-administration)
incidence: 9?/9, sa. const.: German Landrace gilts, age: 180 days, wt.: ≈103 kg, contamination: artificial (dose: DON/ZEA in wheat in different conc., o., for 35 days; for detailed information please see the article), conc. range: ≤4.1 ng/ml* ** (mean value), country: Germany[537], *9.57 DON and 0.358 ZEA (mg/kg diet) fed (both fed in highest conc.), **after 36 days (thereof 35 days of DON- and ZEA-administration)

Ochratoxin A

incidence: ?/?, sa. const.: weaners (specific pathogen free), wt.: 14–18 kg, contamination: artificial (dose: ? µg OTA addition; for detailed iwnformation please see the article), conc. range: ≤838 µg/l*, country: Denmark[204], *blood sa. taken after 24 h on toxin free diet

incidence: 1/1*, sa. const.: pregnant gilts, contamination: neither OTA nor OTB, conc.: nd, country: UK[266], *control
incidence: 1/1, sa. const.: pregnant gilts, contamination: artificial (dose: 0.38 mg OTA + 0.13 mg OTB/kg b. wt., o., for 8 days during early pregnancy), conc.: 6.99 µg/ml*, country: UK[266], *measured on day 30 of pregnancy

incidence: 8?/8*, sa. const.: Danish Landrace piglets, age: 8 weeks, contamination: artificial (dose: **18.6 µg OTA**/kg liveweight/day, o., over a period of 6 weeks (treatment 1); for detailed information please see the article), conc.: 316 µg/l* (mean value), country: Denmark[336], *sa. taken at weaning (pigs 8 weeks old)
incidence: 6?/6*, sa. const.: Danish Landrace piglets, age: 8 weeks, contamination: artificial (dose: **8.0 µg OTA**/kg liveweight/ day, o., over a period of 6 weeks (treatment 2); for detailed information please see the article), conc.: 258 µg/l* (mean value), country: Denmark[336], *sa. taken at weaning (pigs 8 weeks old)
incidence: 8?/8*, sa. const.: Danish Landrace piglets, age: 8 weeks, contamination: artificial (dose: **19.7 µg OTA**/kg liveweight/day, o., over a period of 6 weeks (treatment 3); for detailed information please see the article), conc.:

386 µg/l* (mean value), country: Denmark[336], *sa. taken at weaning (pigs 8 weeks old)

incidence: 2?/2*, sa. const.: pigs, contamination: no OTA and/or DON (for detailed information please see the article), conc. range: ≤1.93 ng/g**, country: Germany[378], *control, **after 90 days (also at other day intervals up to 90 days measured, lowest conc.: 0.65 ng/g directly after the start of the experiment)
incidence: 6?/6, sa. const.: pigs, contamination: artificial (dose: **0.1 ppm crystalline OTA + 1.0 ppm crystalline DON**, o., twice daily for 90 days; for detailed information please see the article), conc. range: ≤59.19 ng/g* (mean value), country: Germany[378], *after 10 days of OTA- and DON-administration (also at other day intervals up to 90 days measured, lowest conc.: 0.77 ng/g directly after the start of the experiment)
incidence: 3?/3, sa. const.: pigs, contamination: artificial (dose: **0.1 ppm crystalline OTA**, o., twice daily for 90 days; for detailed information please see the article), conc. range: ≤32.42 ng/g* (mean value), country: Germany[378], *after 90 days of OTA-administration (also at other day intervals up to 90 days measured, lowest conc.: 0.68 ng/g directly after the start of the experiment)
incidence: 6?/6, sa. const.: pigs, contamination: artificial (dose: **1.0 ppm crystalline DON!**, o., twice daily for 90 days; for detailed information please see the article), conc. range: ≤2.76 ng/g* (mean value), country: Germany[378], *after 90 days of DON-administration (also at other day intervals up to 90 days measured, lowest conc.: 0.93 ng/g directly after the start of the experiment)

incidence: 2?/2*, sa. const.: pigs, contamination: no OTA and/or ZEA (for detailed information please see the article), conc. range: ≤1.65 ng/g** (mean value), country: Germany[380], *control, **after 90 days after the start of the experiment (also at other day intervals up to 90 days measured, lowest conc.: 0.62 ng/g directly after the start of the experiment)
incidence: 6?/6, sa. const.: pigs, contamination: artificial (dose: 0.1 **ppm OTA + 0.25 ppm ZEA**, o., twice daily for 90 days; for detailed information please see the article), conc. range: ≤70.53 ng/g* (mean value), country: Germany[380], *after 56 days of OTA- and ZEA-administration (also at other day intervals up to 90 days measured, lowest conc.: 0.71 ng/g directly after the start of the experiment)
incidence: 3?/3, sa. const.: pigs, contamination: artificial (dose: **0.1 ppm OTA**, o., twice daily for 90 days; for detailed information please see the article), conc. range: ≤65.26 ng/g* (mean value), country: Germany[380], *after 56 days of OTA-administration (also at other days up to 90 day intervals measured, lowest conc.: 0.72 ng/g directly after the start of the experiment)
incidence: 6?/6, sa. const.: pigs, contamination: artificial (dose: **0.25 ppm ZEA!**, o., twice daily for 90 days; for detailed information please see the article), conc. range: ≤4.29 ng/g* (mean value), country: Germany[380], *after 90 days of ZEA-administration (also at other days up to 90 day intervals measured, lowest conc.: 1.07 ng/g directly after the start of the experiment)

incidence: ?/4*, sa. const.: sexually matured boars Hungarian Large White and Dutch Landrace), wt.: 250 kg, contamination: no OTA (for detailed information please see the article), conc.: nr, country: Hungary[395], *control (low- and high-dosed boars)
incidence: 1/1*, sa. const.: sexually matured boar (Hungarian Large White and Dutch Landrace), wt.: 250 kg, contamination: artificial (dose: 20 µg OTA, o., daily for 5 weeks; for detailed information please see the article), conc. range: ≤0.794 ng/ml**, country: Hungary[395], ***low-dosed boar**, **after

28 days (within 5 weeks of OTA-administration) (also at other day intervals up to 77 days measured, except for the start values lowest conc.: ≈0.1 ng/ml after 63, 70 and 77 days)
incidence: 1/1*, sa. const.: sexually matured boar (Hungarian Large White and Dutch Landrace), wt.: 250 kg, contamination: artificial (dose: 20 μg OTA, o., daily for 5 weeks; for detailed information please see the article), conc. range: ≤2.06 ng/ml**, country: Hungary[395], ***low-dosed boar**, **after 35 days (thereof 5 weeks of OTA-administration) (also at other day intervals up to 84 days measured, except for the start values lowest conc.: ≈0.1 ng/ml after 58, 63 and 77 days)
incidence: 1/1*, sa. const.: sexually matured boar (Hungarian Large White and Dutch Landrace), wt.: 250 kg, contamination: artificial (dose: 40 μg OTA, o., daily for 5 weeks; for detailed information please see the article), conc. range: ≤2.56 ng/ml**, country: Hungary[395], ***high-dosed boar**, **after 42 days (thereof 5 weeks of OTA-administration) (also at other day intervals up to 77 days measured, except for the start values lowest conc.: ≈1.05 ng/ml after 21 days)
incidence: 1/1*, sa. const.: sexually matured boar (Hungarian Large White and Dutch Landrace), wt.: 250 kg, contamination: artificial (dose: 40 μg OTA, o., daily for 5 weeks; for detailed information please see the article), conc. range: ≤1.462 ng/ml**, country: Hungary[395], ***high-dosed boar**, **after 42 days (thereof 5 weeks of OTA-administration) (also at other day intervals up to 84 days measured, except for the start values lowest conc.: ≈0.1 ng/ml after 70 and 84 days)

incidence: 6/6*, sa. const.: pathogen-free male and female pigs, age: ≈8 weeks, wt.: 10–12 kg, contamination: no OTA (for detailed information please see the article), conc.: nd, country: Bulgaria/France[419], *control
incidence: 6?/6, sa. const.: pathogen-free male and female pigs, age: ≈8 weeks, wt.: 10–12 kg, contamination: artificial (dose: **90 ng OTA/g** feed, o., for 3 months; for detailed information please see the article), conc.: 11.47 ng/ml* (mean value), country: Bulgaria/France[419], *at the end of 3-month feeding period
incidence: 6?/6, sa. const.: pathogen-free male and female pigs, age: ≈8 weeks, wt.: 10–12 kg, contamination: artificial (dose: **130 ng OTA/g** feed, o., for 3 months; for detailed information please see the article), conc.: 20.92 ng/ml* (mean value), country: Bulgaria/France[419], *at the end of 3-month feeding period
incidence: 6?/6, sa. const.: pathogen-free male and female pigs, age: ≈8 weeks, wt. 10–12 kg, contamination: artificial (dose: **180 ng OTA/g** feed, o., for 3 months; for detailed information please see the article), conc.: 32.31 ng/ml* (mean value), country: Bulgaria/France[419], *at the end of 3-month feeding period

incidence: 6?/6, sa. const.: pathogen-free male and female pigs, age: ≈8 weeks, wt.: 10–12 kg, contamination: artificial (dose: 130, 305 or 790 ng OTA/g feed, o., for 5 months; for detailed information please see the article), conc.: 18.2 ng/ml* (mean and residue value of the 3 given conc. together), country: Bulgaria/France[419], ***at the end of 5-month feeding period**
incidence: 3?/3, sa. const.: pathogen-free male and female pigs, age: ≈8 weeks, wt.: 10–12 kg, contamination: artificial (dose: 130, 305 or 790 ng OTA/g feed, o., for 5 months; for detailed information please see the article), conc.: tr (mean and residue value of the 3 given conc. together), country: Bulgaria/France[419], ***at the end of 5-month feeding period and 1 week on OTA-free diet**
incidence: 3/3, sa. const.: pathogen-free male and female pigs, age: ≈8 weeks, wt.: 10–12 kg, contamination: artificial (dose: 130, 305 or 790 ng OTA/g feed, o., for 5 months; for detailed information please see the article), conc.: nd (residue value of the 3 given conc. together), country:

Bulgaria/France[419], ***at the end of 5-month feeding period and 1 month on OTA-free diet**

Ochratoxin B

incidence: 1/1*, sa. const.: pregnant gilts, contamination: neither OTA nor OTB, conc.: nd, country: UK[266], *control

incidence: 1/1, sa. const.: pregnant gilts, contamination: artificial (dose: **0.38 mg OTA + 0.13 mg OTB**/kg b. wt., o., for 8 days during early pregnancy), conc.: 1.71 μg/ml*, country: UK[266], *measured on day 30 of pregnancy

Diacetoxyscirpenol

incidence: 1/1*, sa. const.: female pig, wt.: ≈20 kg, contamination: no DAS, conc.: nd, country: Germany[421], *control

incidence: 4/4, sa. const.: female pigs, wt.: ≈20 kg, contamination: artificial (dose: **2 mg DAS**/kg b. wt., by intubation, once), conc. range: ≈9.2 ng/ml* (mean value), country: Germany[421], *after 0.5 h (also at other hour intervals up to 96 h measured, lowest conc.: nd after 48 h)

Monoacetoxyscirpenol

incidence: 1/1*, sa. const.: female pig, wt.: ≈20 kg, contamination: no DAS, conc.: nd, country: Germany[421], *control

incidence: 4/4, sa. const.: female pigs, wt.: ≈20 kg, contamination: artificial (dose: **2 mg DAS**/kg b. wt., by intubation, once), conc. range: ≈4.2 ng/ml*, country: Germany[421], *after 1 h (also at other hour intervals up to 96 h measured, lowest conc.: nd after 48 h)

Scirpentriol

incidence: 1/1*, sa. const.: female pig, wt.: ≈20 kg, contamination: no DAS, conc.: nd, country: Germany[421], *control

incidence: 4/4, sa. const.: female pigs, wt.: ≈20 kg, contamination: artificial (dose: **2 mg DAS**/kg b. wt., by intubation, once), conc. range: ≈4.3 ng/ml**, country: Germany[421], *after 1 h (also at other hour intervals up to 96 h measured, lowest conc.: nd after 48 h)

Zearalenone

incidence: 2?/2?*, sa. const.: male KAHYP pigs, wt.: ?, contamination: no ZEA, conc.: nd, country: Hungary[632], *control

incidence: 2?/2, sa. const.: male KAHYP pigs, wt.: 60 kg, contamination: artificial (dose: **15 ppm ZEA**, o., for 14 days), conc.: <10 μg/kg*, country: Hungary[632], *after 14 days

α-Zearalenol

incidence: 2?/2?*, sa. const.: male KAHYP pigs, wt.: ?, contamination: no ZEA, conc.: nd, country: Hungary[632], *control

incidence: 2?/2, sa. const.: male KAHYP pigs, wt.: 60 kg, contamination: artificial (dose: **15 ppm ZEA**, o., for 14 days), conc.: 10 μg/kg*, country: Hungary[632], *after 14 days

β-Zearalenol

incidence: 2?/2?*, sa. const.: male KAHYP pigs, wt.: ?, contamination: no ZEA, conc.: nd, country: Hungary[632], *control

incidence: 2?/2, sa. const.: male KAHYP pigs, wt.: 60 kg, contamination: artificial (dose: **15 ppm ZEA**, o., for 14 days), conc.: <10 μg/kg*, country: Hungary[632], *after 14 days

Pig skin may contain the following mycotoxins and/or their metabolites:

Deoxynivalenol

incidence: 5/5*, sa. const.: healthy Yorkshire barrows, age: ≈11–15 weeks, wt.: 17–22 kg, contamination: no DON, conc.: nd, country: Canada[407], *control

incidence: ?/4, sa. const.: healthy Yorkshire barrows, age: ≈11–15 weeks, wt.: 17–22 kg, contamination: artificial (dose: **1.0 mg DON**/kg b. wt., i.v., once), conc. range: ≤31.0 ng/g* (mean value), country: Canada[407], *after 0.33 h (also measured after 1, 3, 8 and 24 h, lowest conc.: nd after 24 h)

HT-2 Toxin

incidence: 8/8*, sa. const.: specific-pathogen-free derived crossbred female pigs, age: 11–12 weeks, wt.: 31–43 kg, contamination: no T-2 toxin (for detailed

information please see the article), conc.: nd, country: USA[499], *control
incidence: 4?/4, sa. const.: specific-pathogen-free derived crossbred female pigs, age: 11–12 weeks, wt.: 31–43 kg, contamination: artificial (dose: **15 mg T-2 toxin**/kg, t., once; for detailed information please see the article), conc. range: ≤75.00 ppm* (mean value), country: USA[499], *after 7 days (also measured after 1, 3 and 14 days, lowest conc.: 2.20 ppm after 1 day)

Neosolaniol
incidence: 8/8*, sa. const.: specific-pathogen-free derived crossbred female pigs, age: 11–12 weeks, wt.: 31–43 kg, contamination: no T-2 toxin (for detailed information please see the article), conc.: nd, country: USA[499], *control
incidence: 4?/4, sa. const.: specific-pathogen-free derived crossbred female pigs, age: 11–12 weeks, wt.: 31–43 kg, contamination: artificial (dose: **15 mg T-2 toxin**/kg, t., once; for detailed information please see the article), conc. range: ≤12.00 ppm* (mean value), country: USA[499], *after 7 days (also measured after 1, 3 and 14 days, lowest conc.: 0.09 ppm after 1 day)

4-Deacetylneosolaniol
incidence: 8/8*, sa. const.: specific-pathogen-free derived crossbred female pigs, age: 11–12 weeks, wt.: 31–43 kg, contamination: no T-2 toxin (for detailed information please see the article), conc.: nd, country: USA[499], *control
incidence: 4?/4, sa. const.: specific-pathogen-free derived crossbred female pigs, age: 11–12 weeks, wt.: 31–43 kg, contamination: artificial (dose: **15 mg T-2 toxin**/kg, t., once; for detailed information please see the article), conc. range: ≤24.00 ppm* (mean value), country: USA[499], *after 7 days (also measured after 1, 3 and 14 days, lowest conc.: 0.73 ppm after 1 day)

T-2 Toxin
incidence: 8/8*, sa. const.: specific-pathogen-free derived crossbred female pigs, age: 11–12 weeks, wt.: 31–43 kg, contamination: no T-2 toxin (for detailed information please see the article), conc.: nd, country: USA[499], *control
incidence: ?/4, sa. const.: specific-pathogen-free derived crossbred female pigs, age: 11–12 weeks, wt.: 31–43 kg, contamination: artificial (dose: **15 mg T-2 toxin**/kg, t., once; for detailed information please see the article), conc. range: ≤247.00 ppm* (mean value), country: USA[499], *after 3 days (also measured after 1, 7 and 14 days, lowest conc.: 40.00 ppm after 14 days)

T-2 Tetraol
incidence: 8/8*, sa. const.: specific-pathogen-free derived crossbred female pigs, age: 11–12 weeks, wt.: 31–43 kg, contamination: no T-2 toxin (for detailed information please see the article), conc.: nd, country: USA[499], *control
incidence: 5?/5, sa. const.: specific-pathogen-free derived crossbred female pigs, age: 11–12 weeks, wt.: 31–43 kg, contamination: artificial (dose: **15 mg T-2 toxin**/kg, t., once; for detailed information please see the article), conc. range: ≤5.60 ppm* (mean value), country: USA[499], *after 14 days (also measured after 1, 3 and 7 days, lowest conc.: nd after 1 day)

T-2 Triol
incidence: 8/8*, sa. const.: specific-pathogen-free derived crossbred female pigs, age: 11–12 weeks, wt.: 31–43 kg, contamination: no T-2 toxin (for detailed information please see the article), conc.: nd, country: USA[499], *control
incidence: 5?/5, sa. const.: specific-pathogen-free derived crossbred female pigs, age: 11–12 weeks, wt.:

31–43 kg, contamination: artificial (dose: **15 mg T-2 toxin**/kg, t., once; for detailed information please see the article), conc. range: ≤8.70 ppm* (mean value), country: USA[499], *after 14 days (also measured after 1, 3 and 7 days, lowest conc.: 0.21 ppm after 3 days)

Pig spleen may contain the following mycotoxins and/or their metabolites:

AFLATOXIN B_1
incidence: 1/1, sa. const.: female Hampshire × Deutsches Edelschwein piglet, wt.: 15 kg, contamination: artificial (dose: 3.1 μg AFB_1 (labeled)/kg b. wt., o., once; for detailed information please see the article), conc.: 0.4 ppb* **, country: Switzerland[66], *AFB_1 eq., **after **24 h**
incidence: 1/1, sa. const.: female Hampshire × Deutsches Edelschwein piglet, wt.: 15 kg, contamination: artificial (dose: 3.1 μg AFB_1 (labeled)/kg b. wt., o., once; for detailed information please see the article), conc.: 0.4 ppb* **, country: Switzerland[66], *AFB_1 eq., **after **48 h**

incidence: 1/2, sa. const.: adult swines, contamination: artificial (dose: 1.08–1.09 mg AFB_1 (besides other AFs), o., daily for 33 days; for detailed information please see the article), conc.: 2.60 ppb*, country: France[88], *after 33 days

incidence: 4/4*, sa. const.: crossbred (Duroc × Yorkshire) barrows, wt.: 24.5–26.3 kg, contamination: no AFs, conc.: nd, country: USA[138], *control
incidence: 4?/4, sa. const.: crossbred (Duroc × Yorkshire) barrows, wt.: 24.5–26.3 kg, contamination: artificial (dose: **662 μg AFB_1**/kg diet, **273 μg AFB_2**/kg diet, **300 μg AFG_1**/kg diet and **285 μg AFG_2**/kg diet, o., for 21 days), conc. range: tr–0.15 μg/kg*, country: USA[138], *after ≈22 days (thereof 21 days of AF-administration)
incidence: ?/?*, sa. const.: Large White growing female pigs, age: 79 days, wt.: 20 kg, contamination: no AFs (for detailed information please see the article), conc.: nd, country: France[314], *control
incidence: 1/1, sa. const.: Large White growing female pigs, age: 79 days, wt.: 20 kg, contamination: artificial (dose: natural AFs addition (treatment 1: mixed feeding, avg. daily intake of **870 μg AFB_1**) for 26 days; for detailed information please see the article), conc.: nd*, country: France[314], *final wt. of the animal 109 kg
incidence: 1/1, sa. const.: Large White growing female pigs, age: 79 days, wt.: 20 kg, contamination: artificial (dose: natural AFs addition (treatment 2: separate feeding = peanut oil meal 40% and corn gluten meal 30%, avg. daily intake of **1,566 μg AFB_1**) for 19 days; for detailed information please see the article), conc.: 3.3 μg/kg*, country: France[314], *final wt. of the animal 97 kg
incidence: 1/1, sa. const.: Large White growing female pigs, age: 79 days, wt.: 20 kg, contamination: artificial (dose: natural AFs addition (treatment 3: separate feeding = peanut oil meal 0% and corn gluten meal 0%, avg. daily intake of **642 μg AFB_1**) for 26 days; for detailed information please see the article), conc.: nd*, country: France[314], *final wt. of the animal 104 kg

AFLATOXIN B_2
incidence: 1/2, sa. const.: adult swines, contamination: artificial (dose: 1.08–1.09 mg AFB_1 (besides other AFs), o., daily for 33 days; for detailed information please see the article), conc.: 2.70 ppb*, country: France[88], *after 33 days

incidence: 4/4*, sa. const.: crossbred (Duroc × Yorkshire) barrows, wt.: 24.5–26.3 kg, contamination: no AFs, conc.: nd, country: USA[138], *control
incidence: 4?/4, sa. const.: crossbred (Duroc × Yorkshire) barrows, wt.: 24.5–26.3 kg, contamination: artificial (dose: **662 μg AFB_1**/kg diet, **273 μg AFB_2**/

kg diet, **300 μg AFG_1/kg** diet and **285 μg AFG_2/kg** diet, o., for 21 days), conc. range: tr–0.05 μg/kg*, country: USA[138], *after ≈22 days (thereof 21 days of AF-administration)

incidence: ?/?*, sa. const.: Large White growing female pigs, age: 79 days, wt.: 20 kg, contamination: no AFs (for detailed information please see the article), conc.: nd, country: France[314], *control
incidence: 1/1, sa. const.: Large White growing female pigs, age: 79 days, wt.: 20 kg, contamination: artificial (dose: natural AFs addition (treatment 1: mixed feeding, avg. daily intake of **870 μg AFB_1**), for 26 days; for detailed information please see the article), conc.: nd*, country: France[314], *final wt. of the animal 109 kg
incidence: 1/1, sa. const.: Large White growing female pigs, age: 79 days, wt.: 20 kg, contamination: artificial (dose: natural AFs addition (treatment 2: separate feeding = peanut oil meal 40% and corn gluten meal 30%, avg. daily intake of **1,566 μg AFB_1**), for 19 days; for detailed information please see the article), conc.: 1.0 μg/kg*, country: France[314], *final wt. of the animal 97 kg
incidence: 1/1, sa. const.: Large White growing female pigs, age: 79 days, wt.: 20 kg, contamination: artificial (dose: natural AFs addition (treatment 3: separate feeding = peanut oil meal 0% and corn gluten meal 0%, avg. daily intake of **642 μg AFB_1**), for 26 days; for detailed information please see the article), conc.: nd*, country: France[314], *final wt. of the animal 104 kg

Aflatoxin B_{2a}
incidence: 4/4*, sa. const.: crossbred (Duroc × Yorkshire) barrows, wt.: 24.5–26.3 kg, contamination: no AFs, conc.: nd, country: USA[138], *control
incidence: 4?/4, sa. const.: crossbred (Duroc × Yorkshire) barrows, wt.: 24.5–26.3 kg, contamination: artificial (dose: **662 μg AFB_1/kg** diet, **273 μg AFB_2/kg** diet, **300 μg AFG_1/kg** diet and **285 μg AFG_2/kg** diet, o., for 21 days), conc. range: appreciable amounts*, country: USA[138], *after ≈22 days (thereof 21 days of AF-administration)

Aflatoxin M_1
incidence: 1/2, sa. const.: adult swines, contamination: artificial (dose: 1.08–1.09 mg AFB_1 (besides other AFs), o., daily for 33 days; for detailed information please see the article), conc.: 0.90 ppb*, country: France[88], *after 33 days

incidence: 4/4*, sa. const.: crossbred (Duroc × Yorkshire) barrows, wt.: 24.5–26.3 kg, contamination: no AFs, conc.: nd, country: USA[138], *control
incidence: 4?/4, sa. const.: crossbred (Duroc × Yorkshire) barrows, wt.: 24.5–26.3 kg, contamination: artificial (dose: **662 μg AFB_1/kg** diet, **273 μg AFB_2/kg** diet, **300 μg AFG_1/kg** diet and **285 μg AFG_2/kg** diet, o., for 21 days), conc. range: tr*, country: USA[138], *after ≈22 days (thereof 21 days of AF-administration)

Aflatoxin M
incidence: ?/?*, sa. const.: Large White growing female pigs, age: 79 days, wt.: 20 kg, contamination: no AFs (for detailed information please see the article), conc.: nd, country: France[314], *control
incidence: 1/1, sa. const.: Large White growing female pigs, age: 79 days, wt.: 20 kg, contamination: artificial (dose: natural AFs addition (treatment 1: mixed feeding, avg. daily intake of **870 μg AFB_1**), for 26 days; for detailed information please see the article), conc.: nd*, country: France[314], *final wt. of the animal 109 kg
incidence: 1/1, sa. const.: Large White growing female pigs, age: 79 days, wt.: 20 kg, contamination: artificial (dose: natural AFs addition (treatment 2: separate feeding = peanut oil meal 40% and corn gluten meal 30%, avg. daily intake of **1,566 μg AFB_1**), for 19 days; for detailed information please see the article), conc.: 0.4 μg/kg*, country: France[314], *final wt. of the animal 97 kg

incidence: 1/1, sa. const.: Large White growing female pigs, age: 79 days, wt.: 20 kg, contamination: artificial (dose: natural AFs addition (treatment 3: separate feeding = peanut oil meal 0% and corn gluten meal 0%, avg. daily intake of **642 µg AFB_1**), for 26 days; for detailed information please see the article), conc.: nd*, country: France[314], *final wt. of the animal 104 kg

Deoxynivalenol
incidence: 4/4*, sa. const.: barrows and gilts, Ø wt.: 7.7 kg, contamination: no DON addition (for detailed information please see the article), conc.: nd, country: USA[64], *control
incidence: ?/4, sa. const.: barrows and gilts, Ø wt.: 7.7 kg, contamination: artificial (dose: **0.9 ppm DON** (analyzed value) in the diet, o., for 3 weeks; for detailed information please see the article), conc.: 19 ppb* (mean value), country: USA[64], *after 3 weeks post-treatment
incidence: ?/4, sa. const.: barrows and gilts, Ø wt.: 7.7 kg, contamination: artificial (dose: **2.0 ppm DON** (analyzed value) in the diet, o., for 3 weeks; for detailed information please see the article), conc.: 9 ppb* (mean value), country: USA[64], *after 3 weeks post-treatment
incidence: ?/4, sa. const.: barrows and gilts, Ø wt.: 7.7 kg, contamination: artificial (dose: **2.8 ppm DON** (analyzed value) in the diet, o., for 3 weeks; for detailed information please see the article), conc.: 16 ppb* (mean value), country: USA[64], *after 3 weeks post-treatment

incidence: 5/5*, sa. const.: healthy Yorkshire barrows, age: ≈11–15 weeks, wt.: 17–22 kg, contamination: no DON, conc.: nd, country: Canada[407], *control
incidence: ?/4, sa. const.: healthy Yorkshire barrows, age: ≈11–15 weeks, wt.: 17–22 kg, contamination: artificial (dose: **1.0 mg DON/kg** b. wt., iv., once), conc. range: ≤165.0 ng/g* (mean value), country: Canada[407], *after 0.33 h (also measured after 1, 3, 8 and 24 h, lowest conc.: nd after 24 h)

Fumonisin B_1
incidence: 5/5*, sa. const.: weaned barrows of the same genotype, wt.: 12–14 kg, contamination: no FB_1, FB_2 + FB_3 (for detailed information please see the article), conc.: nd, country: Hungary/Germany[87], *control
incidence: 8/10, sa. const.: weaned barrows of the same genotype, wt.: 12–14 kg, contamination: artificial (dose: **50 mg FB_1, 20 mg FB_2** + **5 mg FB_3**/animal, o., for 22 days; for detailed information please see the article), conc. range: 2.4-31.2 ng/g*, Ø conc.: 9.2 ng/g*, country: Hungary/Germany[87], *after 22 days toxin feeding period

incidence: 6/6*, sa. const.: castrated pigs of identical genotype, wt.: ≈12–14 kg, contamination: no FB_1 (for detailed information please see the article), conc.: nd, country: Germany/Hungary[109], *control
incidence: 13/13, sa. const.: castrated pigs of identical genotype, wt.: ≈12–14 kg, contamination: artificial (dose: **100 mg FB_1** daily, o., for 5–11 days; for detailed information please see the article), conc. range: 26–7,980 µg/kg*, Ø conc.: 792.64 µg/kg*, country: Germany/Hungary[109], *on 6th day of the experiment

Fumonisin B_2
incidence: 5/5*, sa. const.: weaned barrows of the same genotype, wt.: 12–14 kg, contamination: no FB_1, FB_2 and FB_3 (for detailed information please see the article), conc.: nd, country: Hungary/Germany[87], *control
incidence: 10/10, sa. const.: weaned barrows of the same genotype, wt.: 12–14 kg, contamination: artificial (dose: **50 mg FB_1, 20 mg FB_2** + **5 mg FB_3**/animal, o., for 22 days; for detailed information please see the article), conc.: nd*, country:

Hungary/Germany[87], *after 22 days toxin feeding period

HT-2 Toxin
incidence: 2/2, sa. const.: female crossbred Yorkshire × Hampshire swines, wt.: 20 kg, contamination: artificial (dose: 0.15 mg T-2 toxin (labeled and unlabeled)/kg b. wt., i.vs., once), conc. range: 2.16–2.25 ng/g*, Ø conc.: 2.205 ng/g*, country: USA[425], *after 4 h

3′-Hydroxy HT-2 Toxin
incidence: 2/2, sa. const.: female crossbred Yorkshire × Hampshire swines, wt.: 20 kg, contamination: artificial (dose: 0.15 mg T-2 toxin (labeled and unlabeled)/kg b. wt., i.vs., once), conc. range: 2.74–3.44 ng/g*, Ø conc.: 3.09 ng/g*, country: USA[425], *after 4 h

Ochratoxin A
incidence: 4/4*, sa. const.: male and females pigs, wt.: ≈70 kg, contamination: no OTA (for detailed information please see the article), conc.: nd, country: Germany[365], *control
incidence: 8?/8, sa. const.: male and females pigs, wt.: ≈70 kg, contamination: artificial (dose: **0.09 mg natural OTA**/kg diet/day, in the morning and evening **half of OTA-ration**, o., for 28 days; for detailed information please see the article), conc.: 9.57 ng/g* (mean value), country: Germany[365], *after 28 days
incidence: 8?/8, sa. const.: male and females pigs, wt.: ≈70 kg, contamination: artificial (dose: **0.09 mg crystalline OTA**/kg diet/day, in the morning and evening **half of OTA-ration**, o., for 28 days; for detailed information please see the article), conc.: 2.57 ng/g* (mean value), country: Germany[365], *after 28 days
incidence: 8?/8, sa. const.: male and females pigs, wt.: ≈70 kg, contamination: artificial (dose: **0.09 mg crystalline OTA/** kg diet/day, in the morning **total OTA-ration**, o., for 28 days; for detailed information please see the article), conc.: 2.94 ng/g* (mean value), country: Germany[365], *after 28 days

T-2 Toxin
incidence: 2/2*, sa. const.: female swines of mixed breeding, wt.: 26–66 kg, contamination: no T-2 toxin, conc.: nr, country: USA[403], *control
incidence: 4/4, sa. const.: female swines of mixed breeding, wt.: 26–66 kg, contamination: artificial (dose: **1.2 mg T-2 toxin**/kg, i.a., once), conc. range: ≈≤161 ppb*, country: USA[403], *after ≈1 h (also measured after ≈2.1, 3 and 4 h, lowest value conc.: under limit of reliable quantitation after 4 h)

incidence: 2/2, sa. const.: female crossbred Yorkshire × Hampshire swines, wt.: 20 kg, contamination: artificial (dose: 0.15 mg T-2 toxin (labeled and unlabeled)/kg b. wt., i.vs., once), conc. range: 1.04–2.73 ng/g*, Ø conc.: 1.885 ng/g*, country: USA[425], *after 4 h
incidence: 2/2, sa. const.: female crossbred Yorkshire × Hampshire swines, wt.: 20 kg, contamination: artificial (dose: 0.15 mg T-2 toxin (labeled and unlabeled)/kg b. wt., i.vs., once), conc. range: 29–42 ng/g* **, Ø conc.: 35.5 ng/g* **, country: USA[425], *after 4 h, ****total metabolites**

3′-Hydroxy T-2 Toxin
incidence: 2/2, sa. const.: female crossbred Yorkshire × Hampshire swines, wt.: 20 kg, contamination: artificial (dose: 0.15 mg T-2 toxin (labeled and unlabeled)/kg b. wt., i.vs., once), conc. range: 0.54–1.78 ng/g*, Ø conc.: 1.16 ng/g*, country: USA[425], *after 4 h

T-2 Tetraol
incidence: 2/2, sa. const.: female crossbred Yorkshire × Hampshire swines, wt.: 20 kg, contamination: artificial (dose: 0.15 mg T-2 toxin (labeled and unlabeled)/kg b. wt., i.vs., once), conc. range: 0.04–0.05 ng/g*, Ø conc.: 0.045 ng/g*, country: USA[425], *after 4 h

Deepoxy T-2 Tetraol
incidence: 2/2, sa. const.: female crossbred Yorkshire × Hampshire swines, wt.: 20 kg, contamination: artificial (dose: 0.15 mg

T-2 toxin (labeled and unlabeled)/kg b. wt., i.vs., once), conc. range: 0.92–1.52 ng/g*, Ø conc.: 1.22 ng/g*, country: USA[425], *after 4 h

T-2 TRIOL
incidence: 2/2, sa. const.: female crossbred Yorkshire × Hampshire swines, wt.: 20 kg, contamination: artificial (dose: 0.15 mg T-2 toxin (labeled and unlabeled)/kg b. wt., i.vs., once), conc. range: 0.11–0.26 ng/g*, Ø conc.: 0.185 ng/g*, country: USA[425], *after 4 h

DEEPOXY T-2 TRIOL
incidence: 1/2, sa. const.: female crossbred Yorkshire × Hampshire swines, wt.: 20 kg, contamination: artificial (dose: 0.15 mg T-2 toxin (labeled and unlabeled)/kg b. wt., i.vs., once), conc.: 0.03 ng/g*, country: USA[425], *after 4 h

ZEARALENONE
incidence: 2?/2?*, sa. const.: male KAHYP pigs, wt.: ?, contamination: no ZEA, conc.: nd, country: Hungary[632], *control
incidence: 2?/2, sa. const.: male KAHYP pigs, wt.: 60 kg, contamination: artificial (dose: **15 ppm ZEA**, o., for 14 days), conc.: ≈10 μg/kg*, country: Hungary[632], *after 14 days

α-ZEARALENOL
incidence: 2?/2?*, sa. const.: male KAHYP pigs, wt.: ?, contamination: no ZEA, conc.: nd, country: Hungary[632], *control
incidence: 2?/2, sa. const.: male KAHYP pigs, wt.: 60 kg, contamination: artificial (dose: **15 ppm ZEA**, o., for 14 days), conc.: 25 μg/kg*, country: Hungary[632], *after 14 days

β-ZEARALENOL
incidence: 2?/2?*, sa. const.: male KAHYP pigs, wt.: ?, contamination: no ZEA, conc.: nd, country: Hungary[632], *control
incidence: 2?/2, sa. const.: male KAHYP pigs, wt.: 60 kg, contamination: artificial (dose: **15 ppm ZEA**, o., for 14 days), conc.: ≈10 μg/kg*, country: Hungary[632], *after 14 days

Pig stomach may contain the following mycotoxins and/or their metabolites:

DEOXYNIVALENOL
incidence: 8/8, sa. const.: male and female crossbred piglets, age: 5 weeks, contamination: artificial (dose: 5.8 ppm DON in the diet, for 1–5 weeks; for detailed information please see the article), conc. range: tr–1,600 ppb* **, country: Canada/USA[70], *in stomach contents (pigs fed **0.7 or 3.1 ppm DON** diets = no data), **measured at 1st, 4th, and 5th week (pr. residue values are each lowest and highest value of 1st to 5th week measurement)

HT-2 TOXIN
incidence: 2/2, sa. const.: female crossbred Yorkshire × Hampshire swines, wt.: 20 kg, contamination: artificial (dose: 0.15 mg T-2 toxin (labeled and unlabeled)/kg b. wt., i.vs., once), conc. range: 2.86–3.21 ng/g*, Ø conc.: 3.035 ng/g*, country: USA[425], *after 4 h
incidence: 2/2, sa. const.: female crossbred Yorkshire × Hampshire swines, wt.: 20 kg, contamination: artificial (dose: 0.15 mg T-2 toxin (labeled and unlabeled)/kg b. wt., i.vs., once), conc. range: 0.76–2.80 ng/g* **, Ø conc.: 1.78 ng/g* **, country: USA[425], *in **stomach contents**, **after 4 h

DEEPOXY HT-2 TOXIN
incidence: 2/2, sa. const.: female crossbred Yorkshire × Hampshire swines, wt.: 20 kg, contamination: artificial (dose: 0.15 mg T-2 toxin (labeled and unlabeled)/kg b. wt., i.vs., once), conc. range: 0.28–0.62 ng/g*, Ø conc.: 0.45 ng/g*, country: USA[425], *after 4 h
incidence: 2/2, sa. const.: female crossbred Yorkshire × Hampshire swines, wt.: 20 kg, contamination: artificial (dose: 0.15 mg T-2 toxin (labeled and unlabeled)/kg b. wt., i.vs., once), conc. range: 0.51–1.15 ng/g* **, Ø conc.: 0.83 ng/g* **, country: USA[425], *in **stomach contents**, **after 4 h

3′-Hydroxy HT-2 Toxin

incidence: 2/2, sa. const.: female crossbred Yorkshire × Hampshire swines, wt.: 20 kg, contamination: artificial (dose: 0.15 mg T-2 toxin (labeled and unlabeled)/kg b. wt., i.vs., once), conc. range: 2.53–3.45 ng/g*, Ø conc.: 2.99 ng/g*, country: USA[425], *after 4 h
incidence: 2/2, sa. const.: female crossbred Yorkshire × Hampshire swines, wt.: 20 kg, contamination: artificial (dose: 0.15 mg T-2 toxin (labeled and unlabeled)/kg b. wt., i.vs., once), conc. range: 1.47–8.68 ng/g* **, Ø conc.: 5.075 ng/g* **, country: USA[425], *in **stomach contents**, **after 4 h

Ochratoxin A

incidence: 4/4*, sa. const.: male and females pigs, wt.: ≈70 kg, contamination: no OTA (for detailed information please see the article), conc.: nd, country: Germany[365], *control
incidence: 8?/8, sa. const.: male and females pigs, wt.: ≈70 kg, contamination: artificial (dose: **0.09 mg natural OTA**/kg diet/day, in the morning and evening **half of OTA-ration**, o., for 28 days; for detailed information please see the article), conc.: 14.61 ng/g* (mean value), country: Germany[365], *after 28 days
incidence: 8?/8, sa. const.: male and females pigs, wt.: ≈70 kg, contamination: artificial (dose: **0.09 mg crystalline OTA**/kg diet/day, in the morning and evening **half of OTA-ration**, o., for 28 days; for detailed information please see the article), conc.: 4.24 ng/g* (mean value), country: Germany[365], *after 28 days
incidence: 8?/8, sa. const.: male and females pigs, wt.: ≈70 kg, contamination: artificial (dose: **0.09 mg crystalline OTA/** kg diet/day, in the morning **total OTA-ration**, o., for 28 days; for detailed information please see the article), conc.: 5.13 ng/g* (mean value), country: Germany[365], *after 28 days

T-2 Toxin

incidence: 2/2*, sa. const.: female swines of mixed breeding, wt.: 26–66 kg, contamination: no T-2 toxin (for detailed information please see the article), conc.: nr, country: USA[403], *control
incidence: 3/?, sa. const.: female swines of mixed breeding, wt.: 26–66 kg, contamination: artificial (dose: **2.4 mg T-2 toxin**/kg, i.a., once), conc. range: tr–16.4 ppm* **, country: USA[403], *in stomach contents, **after ≈19.5 h

incidence: 2/2, sa. const.: female crossbred Yorkshire × Hampshire swines, wt.: 20 kg, contamination: artificial (dose: 0.15 mg T-2 toxin (labeled and unlabeled)/kg b. wt., i.vs., once), conc. range: 0.66–3.34 ng/g*, Ø conc.: 2.00 ng/g*, country: USA[425], *after 4 h
incidence: 2/2, sa. const.: female crossbred Yorkshire × Hampshire swines, wt.: 20 kg, contamination: artificial (dose: 0.15 mg T-2 toxin (labeled and unlabeled)/kg b. wt., i.vs., once), conc. range: 86–91 ng/g* **, Ø conc.: 88.5 ng/g* **, country: USA[425], *after 4 h, ****total metabolites**
incidence: 2/2, sa. const.: female crossbred Yorkshire × Hampshire swines, wt.: 20 kg, contamination: artificial (dose: 0.15 mg T-2 toxin (labeled and unlabeled)/kg b. wt., i.vs., once), conc. range: 0.46–3.02 ng/g* **, Ø conc.: 1.74 ng/g* **, country: USA[425], *in **stomach contents**, **after 4 h
incidence: 2/2, sa. const.: female crossbred Yorkshire × Hampshire swines, wt.: 20 kg, contamination: artificial (dose: 0.15 mg T-2 toxin (labeled and unlabeled)/kg b. wt., i.vs., once), conc. range: 79–399 ng/g* ** ***, Ø conc.: 239 ng/g* ** ***, country: USA[425], *in **stomach contents**, **after 4 h, *****total metabolites**

3′-Hydroxy T-2 Toxin

incidence: 2/2, sa. const.: female crossbred Yorkshire × Hampshire swines, wt.: 20 kg, contamination: artificial (dose: 0.15 mg T-2 toxin (labeled and unlabeled)/kg b. wt., i.vs., once), conc. range: 1.11–2.09 ng/g*,

Ø conc.: 1.6 ng/g*, country: USA[425], *after 4 h
incidence: 2/2, sa. const.: female crossbred Yorkshire × Hampshire swines, wt.: 20 kg, contamination: artificial (dose: 0.15 mg T-2 toxin (labeled and unlabeled)/kg b. wt., i.vs., once), conc. range: 1.62–2.77 ng/g* **, Ø conc.: 2.195 ng/g* **, country: USA[425], *in **stomach contents**, **after 4 h

T-2 Tetraol
incidence: 2/2, sa. const.: female crossbred Yorkshire × Hampshire swines, wt.: 20 kg, contamination: artificial (dose: 0.15 mg T-2 toxin (labeled and unlabeled)/kg b. wt., i.vs., once), conc. range: tr-2.09 ng/g*, country: USA[425], *after 4 h
incidence: 2/2, sa. const.: female crossbred Yorkshire × Hampshire swines, wt.: 20 kg, contamination: artificial (dose: 0.15 mg T-2 toxin (labeled and unlabeled)/kg b. wt., i. vs., once), conc. range: 0.11–11.86 ng/g* **, Ø conc.: 5.985 ng/g* **, country: USA[425], *in **stomach contents**, **after 4 h

Deepoxy T-2 Tetraol
incidence: 2/2, sa. const.: female crossbred Yorkshire × Hampshire swines, wt.: 20 kg, contamination: artificial (dose: 0.15 mg T-2 toxin (labeled and unlabeled)/kg b. wt., i.vs., once), conc. range: 1.23–1.36 ng/g*, Ø conc.: 1.295 ng/g*, country: USA[425], *after 4 h
incidence: 2/2, sa. const.: female crossbred Yorkshire × Hampshire swines, wt.: 20 kg, contamination: artificial (dose: 0.15 mg T-2 toxin (labeled and unlabeled)/kg b. wt., i.vs., once), conc. range: 0.64–1.87 ng/g* **, Ø conc.: 1.255 ng/g* **, country: USA[425], *in **stomach contents**, **after 4 h

T-2 Triol
incidence: 2/2, sa. const.: female crossbred Yorkshire × Hampshire swines, wt.: 20 kg, contamination: artificial (dose: 0.15 mg T-2 toxin (labeled and unlabeled)/kg b. wt., i.vs., once), conc. range: 0.20–0.28 ng/g*, Ø conc.: 0.24 ng/g*, country: USA[425], *after 4 h
incidence: 2/2, sa. const.: female crossbred Yorkshire × Hampshire swines, wt.: 20 kg, contamination: artificial (dose: 0.15 mg T-2 toxin (labeled and unlabeled)/kg b. wt., i.vs., once), conc. range: 0.29–1.17 ng/g* **, Ø conc.: 0.73 ng/g* **, country: USA[425], *in **stomach contents**, **after 4 h

Deepoxy T-2 Triol
incidence: 2/2, sa. const.: female crossbred Yorkshire × Hampshire swines, wt.: 20 kg, contamination: artificial (dose: 0.15 mg T-2 toxin (labeled and unlabeled)/kg b. wt., i.vs., once), conc. range: 0.09–0.10 ng/g*, Ø conc.: 0.095 ng/g*, country: USA[425], *after 4 h
incidence: 1/2, sa. const.: female crossbred Yorkshire × Hampshire swines, wt.: 20 kg, contamination: artificial (dose: 0.15 mg T-2 toxin (labeled and unlabeled)/kg b. wt., i.vs., once), conc.: 0.50 ng/g* **, country: USA[425], *in **stomach contents**, **after 4 h

Pig testes may contain the following mycotoxins and/or their metabolites:

Deoxynivalenol
incidence: 5/5*, sa. const.: healthy Yorkshire barrows, age: ≈11–15 weeks, wt.: 17–22 kg, contamination: no DON, conc.: nd, country: Canada[407], *control
incidence: ?/4, sa. const.: healthy Yorkshire barrows, age: ≈11–15 weeks, wt.: 17–22 kg, contamination: artificial (dose: **1.0 mg DON**/kg b. wt., i.v., once), conc. range: ≤123.8 ng/g* (mean value), country: Canada[407], *after 1 h (also measured after 0.33, 3, 8, and 24 h, lowest conc.: nd after 24 h)

Pig tissue may contain the following mycotoxins and/or their metabolites:

T-2 Toxin
incidence: 8/8*, sa. const.: specific-pathogen-free derived crossbred female pigs, age: 11–12 weeks, wt.: 31–43 kg, contamination: no T-2 toxin (for detailed information please see the article), conc.: nd, country: USA[499], *control
incidence: 3?/3, sa. const.: specific-pathogen-free derived crossbred female

pigs, age: 11–12 weeks, wt.: 31–43 kg, contamination: artificial (dose: **15 mg T-2 toxin**/kg, t., once; for detailed information please see the article), conc. range: ≤34.00 ppm* ** (mean value), country: USA[499], *in subcutaneous tissue, **after 1 day (also measured after 3, 7, and 14 days, lowest conc.: 3.00 ppm after 14 days)

Pig urine may contain the following mycotoxins and/or their metabolites:

Aflatoxin B_1

incidence: 4/4*, sa. const.: male and female feeder pigs, wt.: 54.2–71.6 kg, contamination: no AFB_1, conc.: nr, country: USA[136], *control

incidence: 4/4, sa. const.: male and female feeder pigs, wt.: 54.2–71.6 kg, contamination: artificial (dose: **400 µg AFB_1/kg** diet, o., for 4 weeks), conc. range: 0.22–1.17 µg/kg*, Ø conc.: 0.565 µg/kg*, country: USA[136], * after 3 weeks (within 4 weeks of AFB_1-administration)

Aflatoxin M_1

incidence: 4/4*, sa. const.: male and female feeder pigs, wt.: 54.2–71.6 kg, contamination: no AFB_1, conc.: nr, country: USA[136], *control

incidence: 4/4, sa. const.: male and female feeder pigs, wt.: 54.2–71.6 kg, contamination: artificial (dose: **400 µg AFB_1/kg** diet, o., for 4 weeks), conc. range: 1.8–8.16 µg/kg*, Ø conc.: 3.68 µg/kg*, country: USA[136], * after 3 weeks (within 4 weeks of AFB_1-administration)

Deoxynivalenol

incidence: 1/1, sa. const.: male and female crossbred piglets, age: 5 weeks, contamination: artificial (dose: **0.7 ppm DON** in the diet, for 4 weeks; for detailed information please see the article), conc.: 200 ppb*, country: Canada/USA[70], *at 4 weeks

incidence: 3/3, sa. const.: male and female crossbred piglets, age: 5 weeks, contamination: artificial (dose: **3.1 ppm DON** in the diet, for 4 weeks; for detailed information please see the article), conc. range: 150–570 ppb*, Ø conc.: 340 ppb*, country: Canada/USA[70], *at 4 weeks

incidence: 4/4, sa. const.: male and female crossbred piglets, age: 5 weeks, contamination: artificial (dose: **5.8 ppm DON** in the diet, for 1–5 weeks; for detailed information please see the article), conc. range: 40–4,320 ppb*, country: Canada/USA[70], *measured at 1st, 4th, and 5th week (pr. residue values are each lowest and highest value of 1st to 5th week measurement)

incidence: 7/7*, sa. const.: pigs, contamination: no DON, conc.: nd or tr**, country: Austria[369], *control, **measured over 16 days

incidence: ?/7, sa. const.: pigs, contamination: artificial (dose: **0.23 mg DON**/kg b. wt./day, o., for 16 days), Ø conc.: 580 µg/l* (mean value), country: Austria[369], *measured over 16 days

incidence: 5/5*, sa. const.: healthy Yorkshire barrows, age: ≈11–15 weeks, wt.: 17–22 kg, contamination: no DON, conc.: nd, country: Canada[407], *control

incidence: ?/4, sa. const.: healthy Yorkshire barrows, age: ≈11–15 weeks, wt.: 17–22 kg, contamination: artificial (dose: **1.0 mg DON**/kg b. wt., i.v., once), conc. range: ≤139,885.0 ng/g* (mean value), country: Canada[407], *after 1 h (also measured after 0.33, 3, 8, and 24 h, lowest conc.: 477.5 ng/g after 24 h)

incidence: 5?/5, sa. const.: castrated male Swedish Landrace pigs, wt.: ≈20 kg, contamination: artificial (dose: 2.5 mg 3-aDON/kg feed, o., 5 times in 2.5 days), conc. range: 58–131 ng/ml*, country: Sweden[416], *total DON after feeding

incidence: ?/16*, sa. const.: barrows, wt.: 28 kg, contamination: artificial (dose: 0.09 mg DON/kg wheat (but DON not intended; wheat proportion in the diet

0%); for detailed information please see the article), conc.: 0.035 mg/kg (mean value), country: Germany[484], *control
incidence: ?/16, sa. const.: barrows, wt.: 28 kg, contamination: artificial (dose: **2.64 mg natural DON**/kg wheat (**wheat proportion** in the diet **17.5%**), o., for 70 days?; for detailed information please see the article), conc.: 0.681 mg/kg* (mean value), country: Germany[484], *after 13 weeks
incidence: ?/16, sa. const.: barrows, wt.: 28 kg, contamination: artificial (dose: **4.41 mg natural DON**/kg wheat (**wheat proportion** in the diet **35%**), o., for 70 days?; for detailed information please see the article), conc.: 1.167 mg/kg* (mean value), country: Germany[484], *after 13 weeks

incidence: ?/5*, sa. const.: male castrated and female fattening pigs (Deutsches Bundeshybridzuchtprogramm), wt.: 34.1–103.7 kg, contamination: artificial (dose: 0.15 mg DON/kg wheat, contaminated (proportion in the diet 0%), o., for 7 days?; for detailed information please see the article), conc. range: 0.057–0.119 mg/l (mean value), country: Austria/Germany[530], *control
incidence: ?/5, sa. const.: male castrated and female fattening pigs (Deutsches Bundeshybridzuchtprogramm), wt.: 34.1–103.7 kg, contamination: artificial (dose: **3.86 mg DON**/kg wheat, contaminated (proportion in the diet 40%), o., for 7 days?; for detailed information please see the article), conc. range: 1.270–1.678 mg/l (mean value), country: Austria/Germany[530]

incidence: 9?/9, sa. const.: German Landrace gilts, age: 180 days, wt.: ≈103 kg, contamination: artificial (dose: DON/ZEA in wheat in different conc., o., for 35 days; for detailed information please see the article), conc. range: ≤1,572 ng/ml* ** (mean value), country: Germany[537], *9.57 mg DON and 0.358 mg ZEA/kg diet fed (both fed in highest conc.), **after 36 days (thereof 35 days of DON- and ZEA-administration)

incidence: 7?/7, sa. const.: Yorkshire barrows, wt.: ≈35 kg, contamination: artificial (dose: **0.19 mg DON**/kg diet dry matter, o., for 2 weeks (diet C); for detailed information please see the article), conc.: 0.07 mg* (mean value), country: Canada[606], *in a 5-days collection period
incidence: 7?/7, sa. const.: Yorkshire barrows, wt.: ≈35 kg, contamination: artificial (dose: **4.66 mg DON**/kg diet dry matter, o., for 1 week (diet C in the first week); for detailed information please see the article), conc.: 2.21 mg* (mean value), country: Canada[606], *in a 5-days collection period

DEEPOXYDEOXYNIVALENOL

incidence: 7/7*, sa. const.: pigs, contamination: no DON, conc.: nd or tr**, country: Austria[369], *control, **measured over 16 days
incidence: ?/7, sa. const.: pigs, contamination: artificial (dose: **0.23 mg DON**/kg b. wt./day, o., for 16 days), Ø conc.: 32 µg/l (mean value), country: Austria[369], *measured over 16 days

incidence: 16/16*, sa. const.: barrows, wt.: 28 kg, contamination: artificial (dose: 0.09 mg DON/kg wheat (but DON not intended; wheat proportion in the diet 0%); for detailed information please see the article), conc.: nd, country: Germany[484], *control
incidence: ?/16, sa. const.: barrows, wt.: 28 kg, contamination: artificial (dose: **2.64 mg natural DON**/kg wheat (**wheat proportion** in the diet **17.5%**), o., for 70 days?; for detailed information please see the article), conc.: 0.025 mg/kg* (mean value), country: Germany[484], *after 13 weeks
incidence: ?/16, sa. const.: barrows, wt.: 28 kg, contamination: artificial (dose: **4.41 mg natural DON**/kg wheat (**wheat proportion** in the diet **35%**), o.,

for 70 days?; for detailed information please see the article), conc.: 0.048 mg/kg* (mean value), country: Germany[484], *after 13 weeks

incidence: 5/5*, sa. const.: male castrated and female fattening pigs (Deutsches Bundeshybridzuchtprogramm), wt.: 34.1–103.7 kg, contamination: 0.15 mg DON/kg wheat, contaminated (proportion in the diet 0%), o., for 7 days?; (for detailed information please see the article), conc.: nd, country: Austria/ Germany[530], *control
incidence: ?/5, sa. const.: male castrated and female fattening pigs (Deutsches Bundeshybridzuchtprogramm), wt.: 34.1–103.7 kg, contamination: artificial (dose: **3.86 mg DON**/kg wheat, contaminated (proportion in the diet 40%), o., for 7 days?; for detailed information please see the article), conc. range: 0.060–0.095 mg/l (mean value), country: Austria/ Germany[530]

incidence: 9?/9, sa. const.: German Landrace gilts, age: 180 days, wt.: ≈103 kg, contamination: artificial (dose: DON/ZEA in wheat in different conc., o., for 35 days; for detailed information please see the article), conc. range: ≤289 ng/ml* ** (mean value), country: Germany[537], *9.57 mg DON and 0.358 mg ZEA/kg diet fed (both fed in highest conc.), **after 36 days (thereof 35 days of DON- and ZEA-administration)

incidence: 7?/7, sa. const.: Yorkshire barrows, wt.: ≈35 kg, contamination: artificial (dose: **0.19 mg DON**/kg diet dry matter, o., for 2 weeks (diet C); for detailed information please see the article), conc.: nd*, country: Canada[606], *daily
incidence: 7?/7, sa. const.: Yorkshire barrows, wt.: ≈35 kg, contamination: artificial (dose: **4.66 mg DON**/kg diet dry matter, o., for 1 week (diet C in the first week); for detailed information please see the article), conc.: 0.149 mg* (mean value), country: Canada[606], *daily

Fumonisin B_1
incidence: 5/5*, sa. const.: weaned barrows of the same genotype, wt.: 12–14 kg, contamination: no FB_1, FB_2 + FB_3 (for detailed information please see the article), conc.: nd, country: Hungary/Germany[87], *control
incidence: 10/10, sa. const.: weaned barrows of the same genotype, wt.: 12–14 kg, contamination: artificial (dose: **50 mg FB_1, 20 mg FB_2 + 5 mg FB_3/** animal, o., for 22 days; for detailed information please see the article), conc. range: 0.9–11.5 mg*, Ø conc.: 4.48 mg*, country: Hungary/Germany[87], *excreted between days 13 and 17

Fumonisin B_2
incidence: 5/5*, sa. const.: weaned barrows of the same genotype, wt.: 12–14 kg, contamination: no FB_1, FB_2 + FB_3 (for detailed information please see the article), conc.: nd, country: Hungary/Germany[87], *control
incidence: 10/10, sa. const.: weaned barrows of the same genotype, wt.: 12–14 kg, contamination: artificial (dose: 50 mg FB_1, 20 mg FB_2 + 5 mg FB_3/ animal, o., for 22 days; for detailed information please see the article), conc. range: 0.005–1.6 mg*, Ø conc.: 0.48 mg*, country: Hungary/Germany[87], *excreted between days 13 and 17

Nivalenol
incidence: 3/3, sa. const.: male castrated Swedish Landrace × Yorkshire pigs, wt.: 37–63 kg, contamination: artificial (dose: 0.05 mg NIV/kg b. wt, o., twice daily for 3 days (NIV-administration also 16 h before first blood sa. taken); for detailed information please see the article), conc. range: ≤336 ng/ml*, country: Sweden[501], *(total mean of all 3 days)

Ochratoxin A

incidence: 1/1*, sa. const.: pregnant gilts, contamination: neither OTA nor OTB, conc.: nd, country: UK[266], *control

incidence: 2/2, sa. const.: pregnant gilts, contamination: artificial (dose: **0.38 mg OTA + 0.13 mg OTB**/kg b. wt., o., for 8 days during early pregnancy), conc. range: 0.04–0.33 µg/mg creatinine*, Ø conc.: 0.185 µg/mg creatinine*, country: UK[266], *on day 29 on completion of feeding toxins

incidence: 4?/4, sa. const.: pathogen-free male and female pigs, age: ≈8 weeks, wt.: 10–12 kg, contamination: artificial (dose: 130, 305, or 790 ng OTA/g feed, o., for 5 months; for detailed information please see the article), conc.: 52.37 ng/ml* (mean and residue value of the 3 given conc. together), country: Bulgaria/France[419], ***at the end of 5-month feeding period**

incidence: 2?/2, sa. const.: pathogen-free male and female pigs, age: ≈8 weeks, wt.: 10–12 kg, contamination: artificial (dose: 130, 305, or 790 ng OTA/g feed, o., for 5 months; for detailed information please see the article), conc.: 15.08 ng/ml* (mean and residue value of the 130 and 790 ng OTA/g feed together), country: Bulgaria/France[419], ***at the end of 5-month feeding period and 1 week on OTA-free diet**

incidence: 3/3, sa. const.: pathogen-free male and female pigs, age: ≈8 weeks, wt.: 10–12 kg, contamination: artificial (dose: 130, 305, or 790 ng OTA/g feed, o., for 5 months: for detailed information please see the article), conc.: nd* (residue value of the 3 given conc. together), country: Bulgaria/France[419], ***at the end of 5-month feeding period and 1 month on OTA-free diet**

Ochratoxin α

incidence: 1/1*, sa. const.: pregnant gilt, contamination: neither OTA nor OTB, conc.: nd, country: UK[266], *control

incidence: 2/2, sa. const.: pregnant gilts, contamination: artificial (dose: **0.38 mg OTA + 0.13 mg OTB**/kg b. wt., o., for 8 days during early pregnancy), conc. range: tr-0.18 µg/mg* creatinine, country: UK[266], *on day 29 on completion of feeding toxins

Ochratoxin ß

incidence: 1/1*, sa. const.: pregnant gilts, contamination: neither OTA nor OTB, conc.: nd, country: UK[266], *control

incidence: 1/2, sa. const.: pregnant gilts, contamination: artificial (dose: **0.38 mg OTA + 0.13 mg OTB**/kg b. wt., o., for 8 days during early pregnancy), conc.: 0.8 µg/mg creatinine*, country: UK[266], *on day 29 on completion of feeding toxins

Diacetoxyscirpenol

incidence: 4/4*, sa. const.: male and female crossbred feeder pigs, wt.: 22.2–49.2 kg, contamination: no DAS (for detailed information please see the article), conc.: nd, country: USA[512], *control

incidence: 2/4, sa. const.: male and female crossbred feeder pigs, wt.: 22.2–49.2 kg, contamination: artificial (dose: **0.5 or 1 mg DAS**/kg b. wt., injection, once; for detailed information please see the article), conc. range: ≤0.51 ng/ml*, country: USA[512], *collected after 15 min

Monoacetoxyscirpenol

incidence: 4/4*, sa. const.: male and female crossbred feeder pigs, wt.: 22.2–49.2 kg, contamination: no DAS (for detailed information please see the article), conc.: nd, country: USA[512], *control

incidence: 3/4, sa. const.: male and female crossbred feeder pigs, wt.: 22.2–49.2 kg, contamination: artificial (dose: **0.5 or 1 mg DAS**/kg b. wt., injection, once; for detailed information please see the article), conc. range: ≤0.80 ng/ml*, country: USA[512], *collected after 15 min

Scirpentriol

incidence: 4/4*, sa. const.: male and female crossbred feeder pigs, wt.: 22.2–49.2 kg, contamination: no DAS (for detailed information please see the article), conc.: nd, country: USA[512], *control incidence: 4/4, sa. const.: male and female crossbred feeder pigs, wt.: 22.2–49.2 kg, contamination: artificial (dose: **0.5 or 1 mg DAS**/kg b. wt., injection, once; for detailed information please see the article), conc. range: ≤10.40 ng/ml*, country: USA[512], *collected after 15 min

T-2 Toxin

incidence: 2/2, sa. const.: female crossbred Yorkshire × Hampshire swines, wt.: 20 kg, contamination: artificial (dose: 0.15 mg T-2 toxin (labeled and unlabeled)/kg b. wt., i.v., once), conc. range: 26.89–27.21 μg/ml* **, country: USA[308], *T-2 toxin and its metabolites, **after 120 min (also measured after 60, 180, and 240, except for the start values lowest conc.: 5.61–6.99 μg/ml after 240 min)

Zearalenone

incidence: 1/1, sa. const.: prepubertal gilt of Swedish Landrace × Yorkshire breed, age: 5 months, wt.: 60 kg, contamination: artificial (dose: 192 μg ZEA/kg b. wt./day, o., for 4 days), conc.: 158.9 ng/ml*, country: Sweden[309], *after 4 days of ZEA-administration (also at other day intervals up to 8 days measured, lowest conc.: ≈9 ng/ml after 8 days)

incidence: 3/3*, sa. const.: uncastrated male weanling Yorkshire pigs, wt.: 8.9–13.0 kg, contamination: no ZEA (for detailed information please see the article), conc.: nd, country: Canada[422], *control
incidence: 3?/3, sa. const.: uncastrated male weanling Yorkshire pigs, wt.: 8.9–13.0 kg, contamination: artificial (dose: **5 mg crystalline ZEA**/kg b. wt., o., 4 times; for detailed information please see the article), conc.: 42 ng/ml* (avg. maximum conc.), country: Canada[422], *sa. collected when possible during dosing phase, sa. period over 7 h
incidence: 3?/3, sa. const.: uncastrated male weanling Yorkshire pigs, wt.: 8.9–13.0 kg, contamination: artificial (dose: **10 mg crystalline ZEA**/kg b. wt., o., 4 times; for detailed information please see the article), conc.: 49 ng/ml* (avg. maximum conc.), country: Canada[422], *sa. collected when possible during dosing phase, sa. period over 7 h
incidence: 3?/3, sa. const.: uncastrated male weanling Yorkshire pigs, wt.: 8.9–13.0 kg, contamination: artificial (dose: **15 mg crystalline ZEA**/kg b. wt., o., 4 times; for detailed information please see the article), conc.: 197 ng/ml* (avg. maximum conc.), country: Canada[422], *sa. collected when possible during dosing phase, sa. period over 7 h

incidence: 1/1, sa. const.: gilt, wt.: 92 kg, contamination: artificial (dose: 150 μg ZEA/kg b. wt., o., once), conc.: 784 μg* ** (total), country: USA[528], *after 8 h (also at other hour intervals up to 8 h measured), **in glucuronide form

incidence: 9?/9, sa. const.: German Landrace gilts, age: 180 days, wt.: ≈103 kg, contamination: artificial (dose: **DON/ZEA** in wheat in **different conc.**, o., for 35 days; for detailed information please see the article), conc. range: ≤43.1 ng/g* ** (mean value), country: Germany[537], *9.57 mg DON and 0.358 mg ZEA/kg diet fed (both fed in highest conc.), **after 36 days (thereof 35 days of DON- and ZEA-administration)

incidence: 2?/2?*, sa. const.: male KAHYP pigs, wt.: ?, contamination: no ZEA, conc.: nd, country: Hungary[632], *control

incidence: 2?/2, sa. const.: male KAHYP pigs, wt.: 60 kg, contamination: artificial (dose: **15 ppm ZEA**, o., for 14 days) conc.: 450 μg/kg*, country: Hungary[632], *after 13/14 days

α-Zearalenol

incidence: 1/1, sa. const.: prepubertal gilt of Swedish Landrace × Yorkshire breed, age: 5 months, wt.: 60 kg, contamination: artificial (dose: 192 μg ZEA/kg b. wt./day, o., for 4 days), conc.: 170.8 ng/ml*, country: Sweden[309], *after 4 days of

ZEA-administration (also at other day intervals up to 8 days measured, lowest conc.: ≈8 ng/ml after 8 days)

incidence: 3/3*, sa. const.: uncastrated male weanling Yorkshire pigs, wt.: 8.9–13.0 kg, contamination: no ZEA (for detailed information please see the article), conc.: nd, country: Canada[422], *control
incidence: 3?/3, sa. const.: uncastrated male weanling Yorkshire pigs, wt.: 8.9–13.0 kg, contamination: artificial (dose: **5 mg crystalline ZEA**/kg b. wt., o., 4 times; for detailed information please see the article), conc.: 5 ng/ml* (avg. maximum conc.), country: Canada[422], *sa. collected when possible during dosing phase, sa. period over 7 h
incidence: 3?/3, sa. const.: uncastrated male weanling Yorkshire pigs, wt.: 8.9–13.0 kg, contamination: artificial (dose: **10 mg crystalline ZEA**/kg b. wt., o., 4 times; for detailed information please see the article), conc.: 8 ng/ml* (avg. maximum conc.), country: Canada[422], *sa. collected when possible during dosing phase, sa. period over 7 h
incidence: 3?/3, sa. const.: uncastrated male weanling Yorkshire pigs, wt.: 8.9–13.0 kg, contamination: artificial (dose: **15 mg crystalline ZEA**/kg b. wt., o., 4 times; for detailed information please see the article), conc.: 80 ng/ml* (avg. maximum conc.), country: Canada[422], *sa. collected when possible during dosing phase, sa. period over 7 h

incidence: 9?/9, sa. const.: German Landrace gilts, age: 180 days, wt.: ≈103 kg, contamination: artificial (dose: **DON/ZEA** in wheat in **different conc.**, o., for 35 days; for detailed information please see the article), conc. range: ≤57.9 ng/g* ** (mean value), country: Germany[537], *9.57 DON and 0.358 ZEA mg/kg diet fed (both fed in highest conc.), **after 36 days (thereof 35 days of DON- and ZEA-administration)

incidence: 2?/2?*, sa. const.: male KAHYP pigs, wt.: ?, contamination: no ZEA, conc.: nd, country: Hungary[632], *control
incidence: 2?/2, sa. const.: male KAHYP pigs, wt.: 60 kg, contamination: artificial (dose: **15 ppm ZEA**, o., for 14 days), conc.: 1,270 µg/kg*, country: Hungary[632], *after 13/14 days

β-Zearalenol
incidence: 9?/96, sa. const.: German Landrace gilts, age: 180 days, wt.: ≈103 kg, contamination: artificial (dose: **DON/ZEA** in wheat in **different conc.**, o., for 35 days; for detailed information please see the article), conc. range: ≤3.3 ng/g* ** (mean value), country: Germany[537], *9.57 mg DON and 0.358 mg ZEA/kg diet fed (both fed in highest conc.), **after 36 days (thereof 35 days of DON- and ZEA-administration)
incidence: 2?/2?*, sa. const.: male KAHYP pigs, wt.: ?, contamination: no ZEA, conc.: nd, country: Hungary[632], *control

incidence: 2?/2, sa. const.: male KAHYP pigs, wt.: 60 kg, contamination: artificial (dose: **15 ppm ZEA**, o., for 14 days), conc.: 620 µg/kg*, country: Hungary[632], *after 13/14 days

Pig, sow milk may contain the following mycotoxins and/or their metabolites:

Fumonisin B_1
incidence: 1/1, sa. const.: milk from a sow 1 day after parturition, contamination: artificial (dose: *Fusarium moniliforme* culture material (=300 mg FB_1 daily), o., for 7 days until parturition no more FB_1; for detailed information please see the article), conc.: nd, country: Hungary[633]
incidence: 2/2, sa. const.: milk from sows **1 day after parturition**, contamination: artificial (dose: *Fusarium moniliforme* culture material (=300 mg FB_1 daily), o., for 7–9 days until parturition, additionally

300 mg FB_1 after parturition, o., daily for 7 days; for detailed information please see the article), conc. range: 23.0–27.5 ppb*, Ø conc.: 25.25 ppb*, country: Hungary[633], *24 h after parturition

incidence: 1/1, sa. const.: milk from a sow 7 days after parturition, contamination: artificial (dose: *Fusarium moniliforme* culture material (=300 mg FB_1 daily), o., for 7 days until parturition no more FB_1; for detailed information please see the article), conc.: nd, country: Hungary[633]

incidence: 2/2, sa. const.: milk from sows **7 days after parturition**, contamination: artificial (dose: *Fusarium moniliforme* culture material (=300 mg FB_1 daily), o., for 7–9 days until parturition, additionally **300 mg FB_1 after parturition**, o., daily for 7 days; for detailed information please see the article), conc. range: 18.0–25.5 ppb*, Ø conc.: 21.75 ppb*, country: Hungary[633], *on the 7th day of life of piglets

ZEARALENOL

incidence: 2/2, sa. const.: milk from sows 8 days after parturition, contamination: artificial (dose: 40 ppm crystalline ZEA, o., from the 8th day of parturition for 9 days; for detailed information please see the article), conc. range: 0.575–0.79 ppm, country: Hungary/USA[148] (up to 5 days after the diet had been finished measured)

Piglet see Pig

Pony

Pony Artificial Contamination

Pony cecum may contain the following mycotoxins and/or their metabolites:

AFLATOXIN B_1

incidence: ?*/?, sa. const.: weanling pony foals (mixed breeding, of both sexes), age: 6–8 months and one 10-year-old pony mare, wt.: 42–107 kg (weanling ponies), contamination: artificial (dose: 6 mg AFB_1/kg b. wt. (lower and a higher dose(s) also applicated, but no values recorded), nasogastric intubation, once; for detailed information please see the article), conc.: 722 ng/g tissue**, country: USA[73], *died after 34 h, **in cecal contents

Pony kidney may contain the following mycotoxins and/or their metabolites:

AFLATOXIN B_1

incidence: ?*/?, sa. const.: weanling pony foals (mixed breeding, of both sexes), age: 6–8 months and one 10-year-old pony mare, wt.: 42–107 kg (weanling ponies), contamination: artificial (dose: 2 mg AFB_1/kg b. wt. (equal and higher doses also applicated, but no values recorded), nasogastric intubation, once; for detailed information please see the article), conc.: 0.46 ng/g tissue, country: USA[73], *died after 68 h

AFLATOXIN M_1

incidence: ?*/?, sa. const.: weanling pony foals (mixed breeding, of both sexes), age: 6–8 months and one 10-year-old pony mare, wt.: 42–107 kg (weanling ponies), contamination: artificial (dose: 2 mg AFB_1/kg b. wt. (equal and higher doses also applicated, but no values recorded), nasogastric intubation, once; for detailed information please see the article), conc.: 0.45 ng/g tissue, country: USA[73], *died after 68 h

Pony liver may contain the following mycotoxins and/or their metabolites:

AFLATOXIN B_1

incidence: 1/1*, sa. const.: weanling pony foals (mixed breeding, of both sexes), age: 6–8 months and one 10-year-old pony mare, wt.: 42–107 kg (weanling ponies), contamination: no AFB_1 (for detailed information please see the article), conc.: nd**, country: USA[73], *control, **died ? after 70 days

incidence: ?*/?, sa. const.: weanling pony foals (mixed breeding, of both sexes), age: 6–8 months and one 10-year-old pony mare, wt.: 42–107 kg (weanling ponies), contamination: artificial (dose: **2 mg AFB_1/kg** b. wt. (higher doses also applicated, but no values recorded), nasogastric intubation, once; for detailed information please see the article), conc. range: 2.46*–2.59** ng/g tissue, country: USA[73], ponies died after 76 h* and 68 h**
incidence: ?*/?, sa. const.: weanling pony foals (mixed breeding, of both sexes), age: 6–8 months and one 10-year-old pony mare, wt.: 42–107 kg (weanling ponies), contamination: artificial (dose: **4 mg AFB_1/** kg b. wt. (higher doses also applicated, but no values recorded), nasogastric intubation, once; for detailed information please see the article), conc.: 30.00 ng/g tissue, country: USA[73], *died after 46 h
incidence: ?*/?, sa. const.: weanling pony foals (mixed breeding, of both sexes), age: 6–8 months and one 10-year-old pony mare, wt.: 42–107 kg (weanling ponies), contamination: artificial (dose: **6 mg AFB_1/kg** b. wt. (a lower and a higher dose also applicated, but no values recorded), nasogastric intubation, once; for detailed information please see the article), conc.: 80.12 ng/g tissue, country: USA[73], *died after 34 h

Aflatoxin M_1
incidence: 1/1*, sa. const.: weanling pony foals (mixed breeding, of both sexes), age: 6–8 months and one 10-year-old pony mare, wt.: 42–107 kg (weanling ponies), contamination: no AFB_1 (for detailed information please see the article), conc.: nd**, country: USA[73], *control, **died ? after 70 days
incidence: ?*/?, sa. const.: weanling pony foals (mixed breeding, of both sexes), age: 6–8 months and one 10-year-old pony mare, wt.: 42–107 kg (weanling ponies), contamination: artificial (dose: **2 mg AFB_1/kg** b. wt. (higher doses also applicated, but no values recorded), nasogastric intubation, once; for detailed information please see the article), conc. range: 0.25*–0.67** ng/g tissue, country: USA[73], ponies died after 76 h* and 68 h**
incidence: ?*/?, sa. const.: weanling pony foals (mixed breeding, of both sexes), age: 6–8 months and one 10-year-old pony mare, wt.: 42–107 kg (weanling ponies), contamination: artificial (dose: **4 mg AFB_1/** kg b. wt. (higher doses also applicated, but no values recorded), nasogastric intubation, once; for detailed information please see the article), conc.: 4.58 ng/g tissue, country: USA[73], *died after 46 h
incidence: ?*/?, sa. const.: weanling pony foals (mixed breeding, of both sexes), age: 6–8 months and one 10-year-old pony mare, wt.: 42–107 kg (weanling ponies), contamination: artificial (dose: **6 mg AFB_1/kg** b. wt. (a lower and a higher dose also applicated, but no values recorded), nasogastric intubation, once; for detailed information please see the article), conc.: 11.92 ng/g tissue, country: USA[73], *died after 34 h

Pony muscle may contain the following mycotoxins and/or their metabolites:

Aflatoxin B_1
incidence: ?*/?, sa. const.: weanling pony foals (mixed breeding, of both sexes), age: 6–8 months and one 10-year-old pony mare, wt.: 42–107 kg (weanling ponies), contamination: artificial (dose: **2 mg AFB_1/kg** b. wt. (equal and higher doses also applicated, but no values recorded), nasogastric intubation, once; for detailed information please see the article), conc.: 2.17 ng/g tissue**, country: USA[73], *died after 68 h, **in gluteal muscle
incidence: ?*/?, sa. const.: weanling pony foals (mixed breeding, of both sexes), age: 6–8 months and one 10-year-old pony mare, wt.: 42–107 kg (weanling ponies), contamination: artificial (dose: **4 mg AFB_1/kg** b. wt. (lower and higher doses also applicated, but no values recorded), nasogastric intubation, once; for detailed information please see the article), conc.:

21.46 ng/g tissue**, country: USA[73], *died after 46 h, **in gluteal muscle

Aflatoxin M_1
incidence: ?*/?, sa. const.: weanling pony foals (mixed breeding, of both sexes), age: 6–8 months and one 10-year-old pony mare, wt.: 42–107 kg (weanling ponies), contamination: artificial (dose: **2 mg** **AFB_1/kg** b. wt. (equal and higher doses also applicated, but no values recorded), nasogastric intubation, once; for detailed information please see the article), conc.: 0.68 ng/g tissue**, country: USA[73], *died after 68 h, **in gluteal muscle
incidence: ?*/?, sa. const.: weanling pony foals (mixed breeding, of both sexes), age: 6–8 months and one 10-year-old pony mare, wt.: 42–107 kg (weanling ponies), contamination: artificial (dose: **4 mg** **AFB_1/kg** b. wt. (lower and higher doses also applicated, but no values recorded), nasogastric intubation, once; for detailed information please see the article), conc.: 2.31 ng/g tissue**, country: USA[73], *died after 46 h, **in gluteal muscle

Pony rectum may contain the following mycotoxins and/or their metabolites:

Aflatoxin B_1
incidence: ?*/?, sa. const.: weanling pony foals (mixed breeding, of both sexes), age: 6–8 months and one 10-year-old pony mare, wt.: 42–107 kg (weanling ponies), contamination: artificial (dose: **2 mg AFB_1/** kg b. wt. (an equal and higher dose(s) also applicated, but no values recorded), nasogastric intubation, once; for detailed information please see the article), conc. range: 4,941*–7,103** ng/g tissue***, country: USA[73], ponies died after 68 h* and 76 h**, ***in rectal contents
incidence: ?*/?, sa. const.: weanling pony foals (mixed breeding, of both sexes), age: 6–8 months and one 10-year-old pony mare, wt.: 42–107 kg (weanling ponies), contamination: artificial (dose: **4 mg** **AFB_1/kg** b. wt. (a lower and higher dose(s) also applicated, but no values recorded), nasogastric intubation, once; for detailed information please see the article), conc.: 10,180 ng/g tissue**, country: USA[73], *died after 46 h, **in rectal contents
incidence: ?*/?, sa. const.: weanling pony foals (mixed breeding, of both sexes), age: 6–8 months and one 10-year-old pony mare, wt.: 42–107 kg (weanling ponies), contamination: artificial (dose: **6 mg** **AFB_1/kg** b. wt. (lower and a higher dose(s) also applicated, but no values recorded), nasogastric intubation, once; for detailed information please see the article), conc.: 10,611 ng/g tissue**, country: USA[73], *died after 34 h, **in rectal contents

Pony stomach may contain the following mycotoxins and/or their metabolites:

Aflatoxin B_1
incidence: ?*/?, sa. const.: weanling pony foals (mixed breeding, of both sexes), age: 6–8 months and one 10-year-old pony mare, wt.: 42–107 kg (weanling ponies), contamination: artificial (dose: 6 mg AFB_1/ kg* b. wt. (lower and a higher dose(s) also applicated, but no values recorded), nasogastric intubation, once; for detailed information please see the article), conc.: 5,466 ng/g tissue**, country: USA[73], *died after 34 h, **in stomach contents

Pork see Pig

Poultry

Poultry Natural Contamination

Poultry may contain the following mycotoxins and/or their metabolites:

Ochratoxin A
incidence: 5/14,sa. const.: birds of Denmark, contamination: natural, conc. range: 4.3–29.2 µg/kg, Ø 13.22 µg/kg, country: Denmark[298]

Pullets see Hen

Quail

Quail Artificial Contamination

Quail egg may contain the following mycotoxins and/or their metabolites:

OCHRATOXIN A
incidence: ?/?*, sa. const.: eggs from Japanese quails, contamination: no OTA (for detailed information please see the article), conc.: nr, country: Poland[382], *control
incidence: ?/?, sa. const.: eggs from Japanese quails, contamination: artificial (dose: **1 mg OTA**/kg b. wt., o., once; for detailed information please see the article), conc.: nd, country: Poland[382] (measured at other hour intervals up to 192 h)
incidence: ?/?, sa. const.: eggs from Japanese quails, contamination: artificial (dose: **5 mg OTA**/kg b. wt., o., once; for detailed information please see the article), conc. range: ≈≤2.06 µg/kg* (mean value), country: Poland[382], *72 h after OTA-administration (also at other hour intervals up to }192 h measured, lowest conc.: nd after 192 h)
incidence: ?/?, sa. const.: eggs from Japanese quails, contamination: artificial (dose: **20 mg OTA**/kg b. wt., o., once; for detailed information please see the article), conc. range: ≈≤3.5 µg/kg* (mean value), country: Poland[382], *48 h after OTA-administration (also at other hour intervals up to 96 h measured, except for the start value lowest conc.: ≈2.9 µg/kg after 96 h)

incidence: ?/?*, sa. const.: eggs from Japanese quails, contamination: no OTA (for detailed information please see the article), conc.: nr, country: Poland[382], *control
incidence: ?/?, sa. const.: eggs from Japanese quails, contamination: artificial (dose: **1 mg OTA**/kg b. wt., o., once; for detailed information please see the article), conc.: nr*, country: Poland[382], *in abdominal yolks
incidence: ?/?, sa. const.: eggs from Japanese quails, contamination: artificial (dose: **5 mg OTA**/kg b. wt., o., once; for detailed information please see the article), conc. range: ≈≤12.75 µg/kg* ** (mean value), country: Poland[382], *in abdominal yolks, **12 h after OTA-administration (also measured at day 4)
incidence: ?/?, sa. const.: eggs from Japanese quails, contamination: artificial (dose: **20 mg OTA**/kg b. wt., o., once; for detailed information please see the article), conc. range: ≈≤33.56 µg/kg* ** (mean value), country: Poland[382], *in abdominal yolks, **12 h after OTA-administration (also measured at day 4)

Quail kidney may contain the following mycotoxins and/or their metabolites:

OCHRATOXIN A
incidence: ?/?*, sa. const.: eggs from Japanese quails, contamination: no OTA (for detailed information please see the article), conc.: nr, country: Poland[382], *control
incidence: ?/?, sa. const.: eggs from Japanese quails, contamination: artificial (dose: **1 mg OTA**/kg b. wt., o., once; for detailed information please see the article), conc.: nr*, country: Poland[382], *up to 8 days measured
incidence: ?/?, sa. const.: eggs from Japanese quails, contamination: artificial (dose: **5 mg OTA**/kg b. wt., o., once; for detailed information please see the article), conc. range: ≤80 µg/kg* (mean value), country: Poland[382], *12 h after OTA-administration (up to 8 days measured)
incidence: ?/?, sa. const.: eggs from Japanese quails, contamination: artificial (dose: **20 mg OTA**/kg b. wt., o., once; for detailed information please see the article), conc. range: ≤114 µg/kg* (mean value), country: Poland[382], *12 h after OTA-administration (up to 8 days measured)

Quail plasma may contain the following mycotoxins and/or their metabolites:

OCHRATOXIN A

incidence: 16?/16, sa. const.: male and female Japanese quails, wt.: ≈160 g, contamination: artificial (dose: 50 ng OTA/g b. wt., **o.**, once), conc. range: ≤260 ng/ml (mean value), country: Sweden/Yugoslavia[193] (at different min and hour intervals up to 48 h measured)

incidence: 16?/16, sa. const.: male and female Japanese quails, wt.: ≈160 g, contamination: artificial (dose: 50 ng OTA/g, **i.v.,** once), conc. range: ≤590 ng/ml (mean value), country: Sweden/Yugoslavia[193] (at different min and hour intervals up to 48 h measured)

Rabbit

Rabbit Artificial Contamination

Rabbit kidney may contain the following mycotoxins and/or their metabolites:

OCHRATOXIN A

incidence: 4/4*, sa. const.: "blanc de Termonde" female rabbits, contamination: no OTA (for detailed information please see the article), conc.: nd, country: Belgium[299], *control

incidence: 4?/4, sa. const.: "blanc de Termonde" female rabbits, contamination: artificial (dose: **193.4 ng natural OTA**/g feed, for 19 days (lactation period); for detailed information please see the article), conc.: 1,241 ng/kg* ** (mean value), country: Belgium[299], *in **rabbit does,** **after 19 days of toxin exposure during lactation

incidence: 4?/4, sa. const.: sucklings from "blanc de Termonde" female rabbits, contamination: artificial (dose: **OTA** from the milk of lactating **rabbits receiving 193.4 ng natural OTA**/g feed; for detailed information please see the article), conc.: 41 ng/kg* ** (mean value), country: Belgium[299], *in **sucklings,** **after 19 days of toxin exposure during lactation

Rabbit liver may contain the following mycotoxins:

OCHRATOXIN A

incidence: 4/4*, sa. const.: "blanc de Termonde" female rabbits, contamination: no OTA (for detailed information please see the article), conc.: nd, country: Belgium[299], *control

incidence: 4?/4, sa. const.: "blanc de Termonde" female rabbits, contamination: artificial (dose: **193.4 ng natural OTA**/g feed, for 19 days (lactation period); for detailed information please see the article), conc.: 158 ng/kg* ** (mean value), country: Belgium[299], *in **rabbit does,** **after 19 days of toxin exposure during lactation

incidence: 4?/4, sa. const.: sucklings from "blanc de Termonde" female rabbits, contamination: artificial (dose: **OTA** from the milk of lactating **rabbits receiving 193.4 ng natural OTA**/g feed; for detailed information please see the article), conc.: <20 ng/kg* ** (mean value), country: Belgium[299], *in **sucklings,** **after 19 days of toxin exposure during lactation

Rabbit mammary gland may contain the following mycotoxins and/or their metabolites:

OCHRATOXIN A

incidence: 4/4*, sa. const.: "blanc de Termonde" female rabbits, contamination: no OTA (for detailed information please see the article), conc.: nd, country: Belgium[299], *control

incidence: 4?/4, sa. const.: "blanc de Termonde" female rabbits, contamination: artificial (dose: **193.4 ng natural OTA**/g feed, o., for 19 days (lactation period); for detailed information please see the article), conc.: 109 ng/kg* (mean value), country: Belgium[299], *after 19 days of toxin exposure during lactation

Rabbit milk may contain the following mycotoxins and/or their metabolites:

OCHRATOXIN A
incidence: 4/4*, sa. const.: "blanc de Termonde" female rabbits, contamination: no OTA (for detailed information please see the article), conc.: nd, country: Belgium[299], *control
incidence: 4?/4, sa. const.: "blanc de Termonde" female rabbits, contamination: artificial (dose: **193.4 ng natural OTA**/g feed, o., for 19 days (lactation period); for detailed information please see the article), conc.: 49 ng/l* (mean value), country: Belgium[299], *after 19 days of toxin exposure during lactation

Rabbit muscle may contain the following mycotoxins and/or their metabolites:

OCHRATOXIN A
incidence: 4/4*, sa. const.: "blanc de Termonde" female rabbits, contamination: no OTA (for detailed information please see the article), conc.: nd, country: Belgium[299], *control
incidence: 4?/4, sa. const.: "blanc de Termonde" female rabbits, contamination: artificial (dose: **193.4 ng natural OTA**/g feed, o., for 19 days (lactation period); for detailed information please see the article), conc.: 38 ng/kg* ** (mean value), country: Belgium[299], *in **rabbit does**, **after 19 days of toxin exposure during lactation
incidence: 4?/4, sa. const.: sucklings from "blanc de Termonde" female rabbits, contamination: artificial (dose: **OTA** from the milk of lactating **rabbits receiving 193.4 ng natural OTA**/g feed; for detailed information please see the article), conc.: <20 ng/kg* ** (mean value), country: Belgium[299], *in **sucklings**, **after 19 days of toxin exposure during lactation

Rabbit plasma may contain the following mycotoxins and/or their metabolites:

OCHRATOXIN A
incidence: ?/6, sa. const.: male Fauve de Bourgogne rabbits, age: adult, weight: 2–3 kg, contamination: artificial (dose: 2 mg OTA/kg, **o.** (by gavage), once; for detailed information please see the article), conc. range: ≤2.13 µg/ml* (mean value), country: France[172], *after 1 h (also at other hour intervals up to 12 h measured, lowest conc.: ≈0.7 µg/ml after 12 h)
incidence: ?/6, sa. const.: male Fauve de Bourgogne rabbits, age: adult, weight: 2–3 kg, contamination: artificial (dose: 2 mg OTA/kg, **i.v.,** once; for detailed information please see the article), conc. range: ≈≤15 µg/ml* (mean value), country: France[172], *after 0.3 h? (also at other hour intervals up to 12 h measured, lowest conc.: ≈0.75 µg/ml after 12 h)

incidence: 4/4*, sa. const.: "blanc de Termonde" female rabbits, contamination: no OTA (for detailed information please see the article), conc.: nd, country: Belgium[299], *control
incidence: 4?/4, sa. const.: "blanc de Termonde" female rabbits, contamination: artificial (dose: **193.4 ng natural OTA**/g feed, o., for 19 days (lactation period); for detailed information please see the article), conc.: 3,144 ng/l* ** (mean value), country: Belgium[299], *in **rabbit does**, **after 19 days of toxin exposure during lactation
incidence: 4?/4*, sa. const.: sucklings from "blanc de Termonde" female rabbits, contamination: artificial (dose: **OTA** from the milk of lactating **rabbits receiving 193.4 ng natural OTA**/g feed; for detailed information please see the article), conc.: 51 ng/l* ** (mean value), country: Belgium[299], *in **sucklings**, **after 19 days of toxin exposure during lactation

Rainbow trout see Fish, rainbow trout

Rat

Rat Artificial Contamination

Rat may contain the following mycotoxins and/or their metabolites:

AFLATOXIN

incidence: ?/3–5, sa. const.: male weanling Sprague-Dawley-derived rats, contamination: artificial (dose: 1.90 mg AFB_1 (labeled)/kg, i.p., once, prior to dosing **5% casein** fed ad libitum for 15 days (group 1); for detailed information please see the article), conc.: 81.7 pmol AF bound to chromatin/mg DNA* (mean value), country: USA[118], *after 6 h

incidence: ?/3–5, sa. const.: male weanling Sprague-Dawley-derived rats, contamination: artificial (dose: 1.90 mg AFB_1 (labeled)/kg, i.p., once, prior to dosing **20% casein pair-fed** to group 1 for 15 days (group 2); for detailed information please see the article), conc.: 187 pmol AF bound to chromatin/mg DNA* (mean value), country: USA[118], *after 6 h

incidence: ?/3–5, sa. const.: male weanling Sprague-Dawley-derived rats, contamination: artificial (dose: 1.90 mg AFB_1 (labeled)/kg, i.p., once, prior to dosing **20% casein** fed ad libitum for 15 days (group 3); for detailed information please see the article), conc.: 259 pmol AF bound to chromatin/mg DNA* (mean value), country: USA[118], *after 6 h

incidence: ?/3–5, sa. const.: male weanling Sprague-Dawley-derived rats, contamination: artificial (dose: 1.90 mg AFB_1 (labeled)/kg, i.p., once, prior to dosing **5% casein** fed ad libitum for 15 days (group 1); for detailed information please see the article), conc.: 56.4 pmol AF bound to DNA/mg DNA* (mean value), country: USA[118], *after 6 h

incidence: ?/3–5, sa. const.: male weanling Sprague-Dawley-derived rats, contamination: artificial (dose: 1.90 mg AFB_1 (labeled)/kg, i.p., once, prior to dosing **20% casein pair-fed** to group 1 for 15 days (group 2); for detailed information please see the article), conc.: 141 pmol AF bound to DNA/mg DNA* (mean value), country: USA[118], *after 6 h

incidence: ?/3–5, sa. const.: male weanling Sprague-Dawley-derived rats, contamination: artificial (dose: 1.90 mg AFB_1 (labeled)/kg, i.p., once, prior to dosing **20% casein** fed ad libitum for 15 days (group 3); for detailed information please see the article), conc.: 202 pmol AF bound to DNA/mg DNA* (mean value), country: USA[118], *after 6 h

incidence: ?/3–5, sa. const.: male weanling Sprague-Dawley-derived rats, contamination: artificial (dose: 1.90 mg AFB_1 (labeled)/kg, i.p., once, prior to dosing **5% casein** fed ad libitum for 15 days (group 1); for detailed information please see the article), conc.: 11.9 pmol AF bound to protein/mg protein* (mean value), country: USA[118], *after 6 h

incidence: ?/3–5, sa. const.: male weanling Sprague-Dawley-derived rats, contamination: artificial (dose: 1.90 mg AFB_1 (labeled)/kg, i.p., once, prior to dosing **20% casein pair-fed** to group 1 for 15 days (group 2); for detailed information please see the article), conc.: 21.8 pmol AF bound to protein/mg protein* (mean value), country: USA[118], *after 6 h

incidence: ?/3–5, sa. const.: male weanling Sprague-Dawley-derived rats, contamination: artificial (dose: 1.90 mg AFB_1 (labeled)/kg, i.p., once, prior to dosing **20% casein** fed ad libitum for 15 days (group 3); for detailed information please see the article), conc.: 35.2 pmol AF bound to protein/mg protein* (mean value), country: USA[118], *after 6 h

Rat Adipose see Rat fat

Rat bile may contain the following mycotoxins and/or their metabolites:

AFLATOXIN B_1

incidence: 5?/5*, sa. const.: male Sprague-Dawley rats, wt.: 250–300 g,

contamination: artificial (dose: 0.25 mg AFB_1 (labeled)/kg, i.p., once; for detailed information please see the article), conc.: 234 nmol/kg/h** *** **** (mean value), country: USA[120], *control, ****PG-treatment** (0.267 ml/kg/day) for 9 days prior to AFB_1-treatment, ***total AFB metabolites, ****in bile collected for 1 h
incidence: 5?/5, sa. const.: male Sprague-Dawley rats, wt.: 250–300 g, contamination: artificial (dose: 0.25 mg AFB_1 (labeled)/kg, i.p., once; for detailed information please see the article), conc.: 283 nmol/kg/h* ** *** (mean value), country: USA[120], ***BHA-treatment** (500 mg/kg/day) for 9 days prior to AFB_1-treatment, **total AFB metabolites, ***in bile collected for 1 h

incidence: 4?/4, sa. const.: male Sprague-Dawley rats, wt.: 180–190 g, contamination: artificial (dose: 40 µg AFB_1 (labeled and unlabeled)/g b. wt., i.p., once; for detailed information please see the article), conc.: 4.6 nmol* ** *** (mean value), country: USA[552], *control, **AFB_1-GSH-conjugate, ***in bile collected for 2 h
incidence: 4?/4, sa. const.: male Sprague-Dawley rats, wt.: 180–190 g, contamination: artificial (dose: 40 µg AFB_1 (labeled and unlabeled)/g b. wt., i.p., once; for detailed information please see the article), conc.: 12.2 nmol* ** *** (mean value), country: USA[552], *0.1% **PB-treatment** for 1 week (in drinking water), **AFB_1-GSH-conjugate, ***in bile collected for 2 h

incidence: ?/5–6, sa. const.: male **African giant rats**, wt.: 1.5–2.0 kg, contamination: artificial (dose: 1 mg AFB_1 (labeled)/kg b. wt., i.v., once; for detailed information please see the article), conc.: 87 ng/g liver*, country: Nigeria[593], *after ≈8 h of treatment
incidence: ?/5–6, sa. const.: male **weanling Wistar-derived rats**, wt.: 250 g, contamination: artificial (dose: 1 mg AFB_1 (labeled)/kg b. wt., i.v., once; for detailed information please see the article), conc.: 175 ng/g liver*, country: Nigeria[593], *after ≈8 h of treatment

Aflatoxin B

incidence: 5?/5*, sa. const.: male Sprague-Dawley rats, wt.: 250–300 g, contamination: artificial (dose: 0.25 mg AFB_1 (labeled)/kg, i.p., once; for detailed information please see the article), conc.: 73.4 nmol/kg/h** *** **** (mean value), country: USA[120], *control, ****PG-treatment** (0.267 ml/kg/day) for 9 days prior to AFB_1-treatment, ***AFB-GSH, ****in bile collected for 1 h
incidence: 5?/5, sa. const.: male Sprague-Dawley rats, wt.: 250–300 g, contamination: artificial (dose: 0.25 mg AFB_1 (labeled)/kg, i.p., once; for detailed information please see the article), conc.: 147 nmol/kg/h* ** *** (mean value), country: USA[120], ***BHA-treatment** (500 mg/kg/day) for 9 days prior to AFB_1-treatment, **AFB-GSH, ***in bile collected for 1 h

Aflatoxin M_1

incidence: ?/5–6, sa. const.: male **African giant rats**, wt.: 1.5–2.0 kg, contamination: artificial (dose: 1 mg AFB_1 (labeled)/kg b. wt., i.v., once; for detailed information please see the article), conc.: 54 ng/g liver*, country: Nigeria[593], *after ≈8 h of treatment
incidence: ?/5–6, sa. const.: male **weanling Wistar-derived rats**, wt.: 250 g, contamination: artificial (dose: 1 mg AFB_1 (labeled)/kg b. wt., i.v., once; for detailed information please see the article), conc.: 166 ng/g liver*, country: Nigeria[593], *after ≈8 h of treatment

Aflatoxin P_1

incidence: 5?/5*, sa. const.: male Sprague-Dawley rats, wt.: 250–300 g, contamination: artificial (dose: 0.25 mg AFB_1 (labeled)/kg, i.p., once; for detailed information please see the article), conc.: 16.3 nmol/kg/h** *** **** (mean value), country: USA[120], *control, ****PG-treatment**

(0.267 ml/kg/day) for 9 days prior to AFB_1-treatment, ***AFP_1-Gluc, ****in bile collected for 1 h
incidence: 5?/5, sa. const.: male Sprague-Dawley rats, wt.: 250–300 g, contamination: artificial (dose: 0.25 mg AFB_1 (labeled)/kg, i.p., once; for detailed information please see the article), conc.: 31.7 nmol/kg/h* ** *** (mean value), country: USA[120], ***BHA-treatment** (500 mg/kg/day) for 9 days prior to AFB_1-treatment, **AFP_1-Gluc, ***in bile collected for 1 h

Aflatoxin Q_1
incidence: ?/5–6, sa. const.: male **African giant rats**, wt.: 1.5–2.0 kg, contamination: artificial (dose: 1 mg AFB_1 (labeled)/kg b. wt., i.v., once; for detailed information please see the article), conc.: 245 ng/g liver*, country: Nigeria[593], *after ≈8 h of treatment
incidence: ?/5–6, sa. const.: male **weanling Wistar-derived rats**, wt.: 250 g, contamination: artificial (dose: 1 mg AFB_1 (labeled)/kg b. wt., i.v., once; for detailed information please see the article), conc.: 189 ng/g liver*, country: Nigeria[593], *after ≈8 h of treatment

Fumonisin B_1
incidence: 4?/4, sa. const.: male Wistar rats, age: 6–8 months, wt.: 350–400 g, contamination: artificial (dose: 7.5 mg FB_1 (labeled)/kg b. wt., i.p., once; for detailed information please see the article), conc. range: ≈≤0.42 mg FB_1*, country: South Africa[623], *after 30–60 min (also at other hour intervals up to 24 h measured, lowest conc. nd after ≈9 h)

Ochratoxin A
incidence: ?/3 or more*, sa. const.: healthy female Sprague-Dawley rats, wt.: ≈300 g, contamination: no OTA (for detailed information please see the article), conc.: nr, country: Canada[144], *control
incidence: 3?/3, sa. const.: healthy female Sprague-Dawley rats, wt.: ≈300 g, contamination: artificial (dose: 100 µg OTA/rat, i.v., once; for detailed information please see the article), conc.: 1.1 nmol/ml (mean value), country: Canada[144]

incidence: 3/3, sa. const.: male Wistar rats, wt.: 270 g, contamination: artificial (dose: 2 mg crystalline OTA/kg b. wt., by gavage, once; for detailed information please see the article), conc. range: >100 ng/0.1 ml bile*, Ø conc.: >100 ng/0.1 ml bile*, country: Yugoslavia/Sweden[191], *after 6 h (also measured after 12, 24, and 48 h, lowest conc.: 3.2 ng/0.1 ml bile after 48 h)

incidence: 1/1, sa. const.: male **albumin-deficient** Sprague-Dawley rat, age: 9–10 weeks, contamination: artificial (dose: **2.2 mg OTA**/kg, by injection, once), conc. range: ≈≤138 µg/ml* (mean value), country: Japan[613], *after 5 min (also at other min intervals up to 15 min measured, except for the start value lowest conc.: ≈75 µg/ml after 15 min)
incidence: 3?/3, sa. const.: **normal** Sprague-Dawley rats, age: 9–10 weeks, contamination: artificial (dose: **4.1 mg OTA**/kg, by injection, once), conc. range: ≈≤1 µg/ml*? (mean value), country: Japan[613], *after 30 min collection period (also measured after 30–60 and 60–90 min collection period, lowest conc.: <0.5 µg/ml after 60–90 min)
incidence: 3?/3, sa. const.: **albumin-deficient** Sprague-Dawley rats, age: 9–10 weeks, contamination: artificial (dose: **4.1 mg OTA/kg**, by injection, once), conc.: ≈≤175 µg/ml* (mean value), country: Japan[613], *after 30 min collection period (also measured after 30–60 and 60–90 min collection period, lowest conc.: ≈48 µg/ml after 60–90 min)

Ochratoxin α
incidence: ?/3 or more*, sa. const.: healthy female Sprague-Dawley rats, wt.: ≈300 g, contamination: no OTα (for detailed information please see the article), conc.: nr, country: Canada[144], *control
incidence: 3?/3, sa. const.: healthy female Sprague-Dawley rats, wt.: ≈300 g,

contamination: artificial (dose: **100 µg OTα/rat**, i.v., once; for detailed information please see the article), conc.: 5.01 nmol/ml (mean value), country: Canada[144]

LACTONE OPENED OCHRATOXIN A
incidence: 3/3 or more*, sa. const.: healthy female Sprague-Dawley rats, wt.: ≈300 g, contamination: no OP-OTA (for detailed information please see the article), conc.: nr, country: Canada[144], *control
incidence: 3?/3, sa. const.: healthy female Sprague-Dawley rats, wt.: ≈300 g, contamination: artificial (dose: **100 µg OP-OTA/rat**, i.v., once; for detailed information please see the article), conc.: 4.4 nmol/ml (mean value), country: Canada[144]

4-HYDROXYOCHRATOXIN A
incidence: 3/3 or more*, sa. const.: healthy female Sprague-Dawley rats, wt.: ≈300 g, contamination: no OTA-OH (for detailed information please see the article), conc.: nr, country: Canada[144], *control
incidence: 3?/3, sa. const.: healthy female Sprague-Dawley rats, wt.: ≈300 g, contamination: artificial (dose: **100 µg OTA-OH/rat**, i.v., once; for detailed information please see the article), conc.: 14.7 nmol/ml (mean value), country: Canada[144]

OCHRATOXIN B
incidence: 3/3 or more*, sa. const.: healthy female Sprague-Dawley rats, wt.: ≈300 g, contamination: no OTB (for detailed information please see the article), conc.: nr, country: Canada[144], *control
incidence: 3?/3, sa. const.: healthy female Sprague-Dawley rats, wt.: ≈300 g, contamination: artificial (dose: **100 µg OTB/rat**, i.v., once; for detailed information please see the article), conc.: 4.2 nmol/ml (mean value), country: Canada[144]

Rat blood may contain the following mycotoxins and/or their metabolites:

AFLATOXINS
incidence: ?/5, sa. const.: male Sprague-Dawley rats, Ø wt.: 242.2 g, contamination: artificial (dose: 300 µg/kg AFB_1 (labeled)/kg, i.t. (**dust-adsorbed**), once; for detailed information please see the article), conc.: ≈≤570 nM AFs*, country: USA[604], *after **2 h** (also at other min and hour intervals up to 18 h measured, lowest conc.: ≈460 nM AFs after ≈0.5 h)
incidence: ?/5, sa. const.: male Sprague-Dawley rats, Ø wt.: 242.2 g, contamination: artificial (dose: 300 µg/kg AFB_1 (labeled)/kg, i.t. (**microcrystalline form**), once; for detailed information please see the article), conc.: ≈≤480 nM AFs*, country: USA[604], *after **12 h** (also at other min and hour intervals up to 18 h measured, lowest conc.: ≈120 nM AFs after ≈0.5 h)
incidence: ?/5, sa. const.: male Sprague-Dawley rats, Ø wt.: 242.2 g, contamination: artificial (dose: 300 µg/kg AFB_1 (labeled)/kg, i.t. (**dust-adsorbed**), once; for detailed information please see the article), conc.: ≈≤465 nM AFs*, country: USA[604], *after **1 day** (also at other day intervals up to 21 days measured, lowest conc.: nd after 21 days)
incidence: ?/5, sa. const.: male Sprague-Dawley rats, Ø wt.: 242.2 g, contamination: artificial (dose: 300 µg/kg AFB_1 (labeled)/kg, i.t. (**microcrystalline form**), once; for detailed information please see the article), conc.: ≈≤480 nM AFs, country: USA[604], *after **1 day** (also at other day intervals up to 21 days measured, lowest conc.: nd after 21 days)

OCHRATOXIN A
incidence: 3?/3*, sa. const.: **male** F344 rats, age: 8 weeks, contamination: no OTA (for detailed information please see the article), conc.: 14 pmol/ml** (mean value), country: Germany[125], *control, **72 h after OTA-administration
incidence: 3?/3, sa. const.: **male** F344 rats, age: 8 weeks, contamination: artificial

(dose: **0.5 mg OTA/kg** b. wt., by gavage, once; for detailed information please see the article), conc. range: ≤4,614 pmol/ml* ** (mean value), country: Germany[125], *after 48 h after OTA-administration (also at other day intervals up to 56 days measured, lowest conc.: 85 pmol/ml after 56 days), **highest value recorded
incidence: 3?/3*, sa. const.: **female** F344 rats, age: 8 weeks, contamination: no OTA (for detailed information please see the article), conc.: 4 pmol/ml** (mean value), country: Germany[125], *control, **72 h after OTA-administration
incidence: 3?/3, sa. const.: **female** F344 rats, age: 8 weeks, contamination: artificial (dose: **0.5 mg OTA/kg** b. wt., by gavage, once; for detailed information please see the article), conc. range: ≤6,012 pmol/ml* ** (mean value), country: Germany[125], *after 24 h after OTA-administration (also at other days up to 56 day intervals measured, lowest conc.: 89 pmol/ml after 56 days), **highest value recorded

incidence: 3/3 or more*, sa. const.: adult male Wistar rats, wt.: 200–250 g, contamination: no OTA, conc.: nr, country: Japan[143], *control

incidence: ?/3–4, sa. const.: adult male Wistar rats, wt.: 200–250 g, contamination: artificial (dose: **15 mg OTA** (labeled)/kg, o., once), conc. range: ≈≤36.5 μg/ml* (mean value), country: Japan[143], *after 4 h (also at other hour intervals up to 40 h measured, lowest conc.: ≈13 μg/ml after 48 h)

incidence: 3?/3, sa. const.: healthy adult female Sprague-Dawley rats, wt.: 270–350 g, contamination: artificial (dose: 100 μg OTC!/rat, i.v., once), conc. range: ≈≤1,100 ng/ml (mean value), country: Canada[175], *after 120 min (also at other min intervals up to 300 min measured, lowest conc.: ≈400 ng/ml after 10 min)

incidence: 5?/5*, sa. const.: Sprague-Dawley rats (dams), wt.: 300–370 g, contamination: no OTA (for detailed information please see the article), conc.: 1.3 μg/l** (mean value), country: Sweden[312], *control, **after 24 h (also measured after 72 h conc.: 1.1 μg/l)
incidence: 4?/4, sa. const.: Sprague-Dawley rats (dams), wt.: 300–370 g, contamination: artificial (dose: **10 μg OTA/kg** b. wt., o., once at 11th day of lactation; for detailed information please see the article), conc.: 28 μg/l* (mean value), country: Sweden[312], *after 24 h (also measured after 72 h conc.: 8.5 μg/l)
incidence: 4?/4, sa. const.: Sprague-Dawley rats (dams), wt.: 300–370 g, contamination: artificial (dose: **50 μg OTA/kg** b. wt., o., once at 11th day of lactation; for detailed information please see the article), conc.: 120 μg/l* (mean value), country: Sweden[312], *after 24 h (also measured after 72 h conc.: 36 μg/l)
incidence: 5?/5, sa. const.: Sprague-Dawley rats (dams), wt.: 300–370 g, contamination: artificial (dose: **250 μg OTA/kg** b. wt., o., once at 11th day of lactation; for detailed information please see the article), conc.: 520 μg/l* (mean value), country: Sweden[312], *after 24 h (also measured after 72 h conc.: 110 μg/l)

incidence: 10?/10*, sa. const.: pups of Sprague-Dawley rats, wt.: 19–30 g, contamination: no OTA (for detailed information please see the article), conc.: 3.8 μg/l** (mean value), country: Sweden[312], *control, **after 72 h
incidence: 8?/8, sa. const.: pups of Sprague-Dawley rats, wt.: 19–30 g, contamination: artificial (dose: **OTA** from the **milk of lactating dams** receiving **10 μg OTA/kg** b. wt.; for detailed information please see the article), conc.: 29 μg/l* (mean value), country: Sweden[312], *after 72 h
incidence: 8?/8, sa. const.: pups of Sprague-Dawley rats, wt.: 19–30 g, contamination: artificial (dose: **OTA** from the **milk of lactating dams** receiving **50 μg**

OTA/kg b. wt.; for detailed information please see the article), conc.: 120 μg/l* (mean value), country: Sweden[312], *after 72 h
incidence: 8?/8, sa. const.: pups of Sprague-Dawley rats, wt.: 19–30 g, contamination: artificial (dose: **OTA** from the **milk of lactating dams** receiving **250 μg OTA**/kg b. wt.; for detailed information please see the article), conc.: 480 μg/l* (mean value), country: Sweden[312], *after 72 h

incidence: 4(8?)?/4(8)*, sa. const.: Sprague-Dawley rats (dams), contamination: no OTA (for detailed information please see the article), Ø conc.: 5.4 μg/l** (mean value), country: Sweden[322], *control, **at **day 14** of lactation
incidence: 4(8?)?/4(8), sa. const.: Sprague-Dawley rats (dams), contamination: artificial (dose: **50 μg OTA**/kg b. wt., gastric intubation, 46 times in 8 weeks; for detailed information please see the article), Ø conc.: 195 μg/l* (mean value), country: Sweden[322], *at **day 14** of lactation
incidence: 4(8?)?/4(8)*, sa. const.: Sprague-Dawley rats (dams), contamination: no OTA (for detailed information please see the article), Ø conc.: 3.3 μg/l** (mean value), country: Sweden[322], *control, **at **day 21** of lactation
incidence: 4(8?)?/4(8), sa. const.: Sprague-Dawley rats (dams), contamination: artificial (dose: **50 μg OTA**/kg b. wt., gastric intubation, 46 times in 8 weeks; for detailed information please see the article), Ø conc.: 235 μg/l* (mean value), country: Sweden[322], *at **day 21** of lactation

incidence: 4?/4*, sa. const.: pups of Sprague-Dawley rats, contamination: no OTA (for detailed information please see the article), conc.: 11 μg/l** (mean value), country: Sweden[322], *control, **at day 14
incidence: 4?/4, sa. const.: pups of Sprague-Dawley rats, contamination: artificial (dose: infected by lactating dams receiving 50 μg OTA/kg b. wt. by gastric intubation for 46 times (OTA-exposure of pups via **placenta**); for detailed information please see the article), conc.: 130 μg/l* (mean value), country: Sweden[322], *at day 14
incidence: 4?/4, sa. const.: pups of Sprague-Dawley rats, contamination: artificial (dose: infected by lactating dams receiving 50 μg OTA/kg b. wt. by gastric intubation for 46 times (OTA-exposure of pups via **milk**); for detailed information please see the article), conc.: 640 μg/l* (mean value), country: Sweden[322], *at day 14
incidence: 4?/4, sa. const.: pups of Sprague-Dawley rats, contamination: artificial (dose: infected by lactating dams receiving 50 μg OTA/kg b. wt. by gastric intubation for 46 times (OTA-exposure of pups via **milk and placenta**); for detailed information please see the article), conc.: 860 μg/l* (mean value), country: Sweden[322], *at day 14

OCHRATOXIN α
incidence: 3?/3*, sa. const.: **male** F344 rats, age: 8 weeks, contamination: no OTA (for detailed information please see the article), conc.: 2.23 pmol/ml** (mean value), country: Germany[125], *control, **72 h after OTA-administration
incidence: 3?/3, sa. const.: **male** F344 rats, age: 8 weeks, contamination: artificial (dose: **0.5 mg OTA**/kg b. wt., by gavage, once; for detailed information please see the article), conc. range: ≤14.92 pmol/ml* ** (mean value), country: Germany[125], *24 h after OTA-administration (also at other day intervals up to 56 days measured, lowest conc.: 0.01 pmol/ml after 96 h), **highest value recorded
incidence: 3?/3*, sa. const.: **female** F344 rats, age: 8 weeks, contamination: no OTA (for detailed information please see the article), conc.: 4.35 pmol/ml** (mean value), country: Germany[125], *control, **72 h after OTA-administration

incidence: 3?/3, sa. const.: **female** F344 rats, age: 8 weeks, contamination: artificial (dose: **0.5 mg OTA**/kg b. wt., by gavage, once; for detailed information please see the article), conc. range: ≤10.11 pmol/ml* ** (mean value), country: Germany[125], *after 24 h after OTA-administration (also at other day intervals up to 56 days measured, lowest conc.: 3.16 pmol/ml after 48 h), **highest value recorded

Ochratoxin C
incidence: 3?/3, sa. const.: healthy adult female Sprague-Dawley rats, wt.: 270–350 g, contamination: artificial (dose: 100 μg OTC/rat, i.v., once), conc. range: ≈≤1,550 ng/ml (mean value), country: Canada[175], *after ≈5 min (also at other min intervals up to 300 min measured, lowest conc.: ≈ after 300 min)

Zearalenone
incidence: ?/?, sa. const.: male Sprague-Dawley rats, age: 8–10 weeks, wt.: 230–290 g, contamination: artificial (dose: **1.13 mg ZEA**/h/kg, i.v., for 6 h), conc.: ≈200 ng/ml* (mean value), country: Korea/USA[585], *after 6 h of ZEA-administration
incidence: ?/?, sa. const.: male Sprague-Dawley rats, age: 8–10 weeks, wt.: 230–290 g, contamination: artificial (dose: **2.25 mg ZEA**/h/kg, i.v., for 6 h), conc.: ≈350 ng/ml* (mean value), country: Korea/USA[585], *after 6 h of ZEA-administration

Rat brain may contain the following mycotoxins and/or their metabolites:

Ochratoxin A
incidence: 4?/4, sa. const.: male albino Wistar rats, age: 6 weeks, contamination: artificial (dose: 289 μg OTA/kg per 24 h, gastric intubation, for 8 days), conc.: 114 ng/g* structure** (mean value), country: France/South Africa[368], *after 9 days (thereof 8 days with OTA-administration), **in ***cerebellum***
incidence: 4?/4, sa. const.: male albino Wistar rats, age: 6 weeks, contamination: artificial (dose: 289 μg OTA/kg per 24 h, gastric intubation, for 8 days), conc.: 27.2 ng/g* structure** (mean value), country: France/South Africa[368], *after 9 days (thereof 8 days with OTA-administration), **in ***hippocampus***
incidence: 4/4, sa. const.: male albino Wistar rats, age: 6 weeks, contamination: artificial (289 μg OTA/kg per 24 h, gastric intubation, for 8 days), conc.: 18.8 ng/g* structure** (mean value), country: France/South Africa[368], *after 9 days (thereof 8 days with OTA-administration), **in ***striatum***
incidence: 4?/4, sa. const.: male albino Wistar rats, age: 6 weeks, contamination: artificial (dose: 289 μg OTA/kg per 24 h, gastric intubation, for 8 days), conc.: 17.2 ng/g* structure** (mean value), country: France/South Africa[368], *after 9 days (thereof 8 days with OTA-administration), **in ***ventral mesencephalon***
incidence: 4?/4, sa. const.: male albino Wistar rats, age: 6 weeks, contamination: artificial (dose: 289 μg OTA/kg per 24 h, gastric intubation, for 8 days), conc.: 47.3 ng/g* structure** (mean value), country: France/South Africa[368], *after 9 days (thereof 8 days with OTA-administration), **in **rest of brain**

Zearalenone
incidence: ?/?, sa. const.: male Sprague-Dawley rats, age: 8–10 weeks, wt.: 230–290 g, contamination: artificial (dose: **1.13 mg ZEA**/h/kg, i.v., for 6 h), conc.: ≈200 ng/g* (mean value), country: Korea/USA[585], *after 6 h of ZEA-administration
incidence: ?/?, sa. const.: male Sprague-Dawley rats, age: 8–10 weeks, wt.: 230–290 g, contamination: artificial (dose: **2.25 mg ZEA**/h/kg, i.v., for 6 h), conc.: ≈300 ng/g* (mean value), country: Korea/USA[585], *after 6 h of ZEA-administration

Rat colon may contain the following mycotoxins and/or their metabolites:

Ochratoxin A
incidence: ?/3 or more*, sa. const.: adult male Wistar rats, wt.: 200–250 g, contamination: no OTA, conc.: nr, country: Japan[143], *control
incidence: ?/3–4, sa. const.: adult male Wistar rats, wt.: 200–250 g, contamination: artificial (dose: **15 mg OTA** (labeled)/kg, o., once), conc. range: ≈≤4 μg/g wet weight of tissue* (mean value), country: Japan[143], *after 16 h (also at other hour intervals up to 24 h measured, lowest conc.: ≈1 μg/g wet weight of tissue after 24 h)

incidence: 3?/3, sa. const.: male Wistar rats, age: 8–10 weeks, contamination: artificial (dose: 1.4 mg OTA/kg, injected into the femoral vein, once; for detailed information please see the article), conc.: 122.8 ng/g/10 min* (mean value), country: Japan[174], *in colon perfusate

Rat duodenum may contain the following mycotoxins and/or their metabolites:

Ochratoxin A
incidence: ?/3 or more*, sa. const.: adult male Wistar rats, wt.: 200–250 g, contamination: no OTA, conc.: nr, country: Japan[143], *control
incidence: ?/3–4, sa. const.: adult male Wistar rats, wt.: 200–250 g, contamination: artificial (dose: **15 mg OTA** (labeled)/kg, o., once), conc. range: ≈≤7 μg/g wet weight of tissue* (mean value), country: Japan[143], *after 1 h (also at other hour intervals up to 24 h measured, lowest conc.: ≈1 μg/g wet weight of tissue after 24 h)

Rat fat may contain the following mycotoxins and/or their metabolites:

Zearalenone
incidence: ?/?, sa. const.: male Sprague-Dawley rats, age: 8–10 weeks, wt.: 230–290 g, contamination: artificial (dose: **1.13 mg ZEA**/h/kg, i.v., for 6 h), conc.: ≈600 ng/g* ** (mean value), country: Korea/USA[585], *after 6 h of ZEA-administration, **in adipose tissue
incidence: ?/?, sa. const.: male Sprague-Dawley rats, age: 8–10 weeks, wt.: 230–290 g, contamination: artificial (dose: **2.25 mg ZEA**/h/kg, i.v., for 6 h), conc.: ≈1,000 ng/g* ** (mean value), country: Korea/USA[585], *after 6 h of ZEA-administration, **in adipose tissue

Rat feces may contain the following mycotoxins and/or their metabolites:

Aflatoxin B_1
incidence: 3/3*, sa. const.: male adult Long-Evans rats, wt.: 270–320 g, contamination: no AFB_1, conc.: nd, country: Taiwan, Republic of China[132], *control
incidence: 2/2, sa. const.: male adult Long-Evans rats, wt.: 270–320 g, contamination: artificial (dose: **5.0 mg AFB_1**/kg b. wt., i.p., once), conc. range: 0.65–1.16 mg × 10^{-3}*, Ø conc.: 0.905 mg × 10^{-3}*, country: Taiwan, Republic of China[132], *after 24 h (also measured after 48 and 72 h, but conc.: nd)
incidence: 1/1, sa. const.: male adult Long-Evans rat, wt.: 270–320 g, contamination: artificial (dose: **5.5 mg AFB_1**/kg b. wt., i.p., once), conc.: 1.41 mg × 10^{-3}*, country: Taiwan, Republic of China[132], *after 24 h (also measured after 48 and 72 h, but conc.: nd)

incidence: ?/5–6, sa. const.: male **African giant rats**, wt.: 1.5–2.0 kg, contamination: artificial (dose: 1 mg AFB_1 (labeled)/kg b. wt., i.v., once; for detailed information please see the article), conc.: 45 ng/g liver*, country: Nigeria[593], *after ≈8 h of treatment
incidence: ?/5–6, sa. const.: male **weanling Wistar-derived rats**, wt.: 250 g, contamination: artificial (dose: 1 mg AFB_1 (labeled)/kg b. wt., i.v., once; for detailed information please see the article), conc.: 48 ng/g liver*, country: Nigeria[593], *after ≈8 h of treatment

Aflatoxin M_1
incidence: ?/5–6, sa. const.: male **African giant rats**, wt.: 1.5–2.0 kg, contamination: artificial (dose: 1 mg AFB_1 (labeled)/kg b. wt., i.v., once; for detailed information please see the article), conc.: 42 ng/g liver*, country: Nigeria[593], *after ≈8 h of treatment
incidence: ?/5–6, sa. const.: male **weanling Wistar-derived rats**, wt.: 250 g, contamination: artificial (dose: 1 mg AFB_1 (labeled)/kg b. wt., i.v., once; for detailed information please see the article), conc.: 34 ng/g liver*, country: Nigeria[593], *after ≈8 h of treatment

Aflatoxin Q_1
incidence: ?/5–6, sa. const.: male **African giant rats**, wt.: 1.5–2.0 kg, contamination: artificial (dose: 1 mg AFB_1 (labeled)/kg b. wt., i.v., once; for detailed information please see the article), conc.: 200 ng/g liver*, country: Nigeria[593], *after ≈8 h of treatment
incidence: ?/5–6, sa. const.: male **weanling Wistar-derived rats**, wt.: 250 g, contamination: artificial (dose: 1 mg AFB_1 (labeled)/kg b. wt., i.v., once; for detailed information please see the article), conc.: 139 ng/g liver*, country: Nigeria[593], *after ≈8 h of treatment

Aflatoxins
incidence: ?/5, sa. const.: male Sprague-Dawley rats, Ø wt.: 242.2 g, contamination: artificial (dose: 300 µg/kg AFB_1 (labeled)/kg, i.t. (**dust-adsorbed**), once; for detailed information please see the article), conc.: ≈≤12.4 nmol AFs/day*, country: USA[604], *after 1 day (also at other day intervals up to 21 days measured, lowest conc.: nd after 13 days)
incidence: ?/5, sa. const.: male Sprague-Dawley rats, Ø wt.: 242.2 g, contamination: artificial (dose: 300 µg/kg AFB_1 (labeled)/kg, i.t. (**microcrystalline form**), once; for detailed information please see the article), conc.: ≈≤11.3 nmol AFs/day*, country: USA[604], *after 1 day (also at other day intervals up to 21 days measured, lowest conc.: nd after 13 days)

Fumonisin B_1
incidence: ?/5, sa. const.: rats, contamination: artificial (dose: 1,000 µg FB_1/g, o., once; for detailed information please see the article), conc.: 530 µg/g (mean value), country: USA[276]

Hydrolyzed Fumonisin B_1
incidence: ?/5, sa. const.: rats, contamination: artificial (dose: 1,000 µg FB_1/g, o., once; for detailed information please see the article), conc.: 282 µg/g (mean value), country: USA[276]

Ochratoxin A
incidence: 10/10*, sa. const.: Wistar male rats, wt.: 83–110 g, contamination: no OTA, **diet B-II** (for detailed information please see the article), conc.: nd, country: Canada[209], *control
incidence: ?/10, sa. const.: Wistar male rats, wt.: 83–110 g, contamination: artificial (dose: 500 µg OTA, intubated, daily for 6 days (**diet B-I**); for detailed information please see the article), conc.: pr*, country: Canada[209], *measured at day 1, 2, 3, 4, 5, and 6 ? of OTA-administration
incidence: 10/10*, sa. const.: Wistar male rats, wt.: 83–110 g, contamination: no OTA, **diet NMB** (for detailed information please see the article), conc.: nd, country: Canada[209], *control
incidence: 6?,9?/6,9, sa. const.: Wistar male rats, wt.: 83–110 g, contamination: artificial (dose: 500 µg OTA, intubated, daily for 6 days (**diet NMB+T**); for detailed information please see the article), conc. range: ≤32 µg/rat/day* (mean value), country: Canada[209], *measured on day 2 and 5 of OTA-administration (also measured at day 1, 3, 4, and 6, lowest conc.: 4.3 µg/rat/day on day 1)

incidence: 2(4)/2(4)*, sa. const.: Sprague-Dawley male rats, wt.: ≈100 g, contamination: no OTA, conc.: nr, country: USA[218], *control
incidence: ?/2, sa. const.: Sprague-Dawley male rats, wt.: ≈100 g, contamination: artificial (dose: **1 mg OTA** (labeled)/rat, i.p., once), conc.: 133 µg* OTA content

(mean value), country: USA[218], *after 24 h (also measured after 8 h conc.: 64 μg OTA)

incidence: ?/12*, sa. const.: male Wistar/AF EOPS rats, wt.: 240–290 g, contamination: no OTA only basal feed; for detailed information please see the article), conc. range: ≤3.2 ng/g** (mean value), country: Belgium[509], *control, **after 4 weeks of OTA-administration (also measured after 1 week conc.: 1.4 ng/g)
incidence: ?/12, sa. const.: male Wistar/AF EOPS rats, wt.: 240–290 g, contamination: artificial (dose: basal feed + **OTA contaminated wheat (2.2 μg/g)**, o., for 28 days; for detailed information please see the article), conc. range: ≤16,751 ng/g* (mean value), country: Belgium[509], *after 4 weeks of OTA-administration (also measured after 1 week conc.: 12,505 ng/g)
incidence: ?/12, sa. const.: male Wistar/AF EOPS rats, wt.: 240–290 g, contamination: artificial (dose: basal feed + **OTA contaminated wheat (2.2 μg/g) + MWF** (2%), o., for 28 days; for detailed information please see the article), conc. range: ≤25,451 ng/g* (mean value), country: Belgium[509], *after 4 weeks of OTA-administration (also measured after 1 week conc.: 17,532 ng/g)

incidence: ?/12, sa. const.: male Wistar/AF EOPS rats, wt.: 240–290 g, contamination: artificial (dose: basal feed + **OTA contaminated wheat (2 μg/g) + MWF** (1.8%) + **YCW** (0.2%), o., for 28 days; for detailed information please see the article), conc. range: ≤22,501 ng/g* (mean value), country: Belgium[509], *after 4 weeks of OTA-administration (also measured after 1 week conc.: 15,102 ng/g)

Ochratoxin α
incidence: 10/10*, sa. const.: Wistar male rats, wt.: 83–110 g, contamination: no OTA (for detailed information please see the article), conc.: nd, country: Canada[209], *control

incidence: ?/10, sa. const.: Wistar male rats, wt.: 83–110 g, contamination: artificial (dose: 500 μg OTA, intubated, daily for 6 days (**diet B-I**); for detailed information please see the article), conc.: pr*, country: Canada[209], *measured at day 1, 2, 3, 4, 5, and 6? of OTA-administration
incidence: ?/?, sa. const.: Wistar male rats, wt.: 83–110 g, contamination: artificial (dose: 500 μg OTA, intubated, daily for 6 days (**diet NMB+T**); for detailed information please see the article), conc. range: ≤10 μg/rat/day (mean value), country: Canada[209], no statement when found

incidence: ?/?*, sa. const.: Sprague-Dawley male rats, wt.: ≈100 g, contamination: no OTA, conc.: nr, country: USA[218], *control
incidence: ?/?, sa. const.: Sprague-Dawley male rats, wt.: ≈100 g, contamination: artificial (dose: **1 mg OTA**, i.p., once), conc.: pr*, country: USA[218], *after 24 h?

Rat fetal extracts may contain the following mycotoxins and/or their metabolites:

Ochratoxin A
incidence: 4/4*, sa. const.: sexually mature, nulliparous, outbred Sprague-Dawley rats, wt.: 200–225 g, contamination: no OTA, conc.: nr, country: USA[181], *control
incidence: 4?/4, sa. const.: sexually mature, nulliparous, outbred Sprague-Dawley rats, wt.: 200–225 g, contamination: artificial (dose: **2.5 mg OTA**/kg, s.c., once on day 12 of gestation), Ø conc.: ≤64.2 ng/g* (mean value), country: USA[181], *after 48 h (also measured after 12, 24, and 72 h, lowest conc.: 20.2 ng/g after 12 h)

Rat heart may contain the following mycotoxins and/or their metabolites:

Ochratoxin A
incidence: 3/3 or more*, sa. const.: adult male Wistar rats, wt.: 200–250 g,

contamination: no OTA, conc.: nr, country: Japan[143], *control
incidence: ?/3–4, sa. const.: adult male Wistar rats, wt.: 200–250 g, contamination: artificial (dose: **15 mg OTA** (labeled)/kg, o., once), conc. range: ≈4.5 µg/g wet weight of tissue* (mean value), country: Japan[143], *after 4 h (also at other hour intervals up to 40 h measured, lowest conc.: ≈2.5 µg/g wet weight of tissue after 8 h)

incidence: 4?/4, sa. const.: healthy adult female Sprague-Dawley rats, wt.: 270–350 g, contamination: artificial (dose: **100 µg OTA**/rat, i.v.., once), conc. range: ≤290 ng/g (mean value), country: Canada[175], measured at 2, 24, 48, and 96 h
incidence: 4?/4, sa. const.: healthy adult female Sprague-Dawley rats, wt.: 270–350 g, contamination: artificial (dose: **100 µg OTC!**/rat, i.v., once), conc. range: ≤275 ng/g (mean value), country: Canada[175], measured at 2, 24, 48, and 96 h

Zearalenone
incidence: ?/?, sa. const.: male Sprague-Dawley rats, age: 8–10 weeks, wt.: 230–290 g, contamination: artificial (dose: **1.13 mg ZEA**/h/kg, i.v., for 6 h), conc.: ≈200 ng/g* (mean value), country: Korea/USA[585], *after 6 h of ZEA-administration
incidence: ?/?, sa. const.: male Sprague-Dawley rats, age: 8–10 weeks, wt.: 230–290 g, contamination: artificial (dose: **2.25 mg ZEA**/h/kg, i.v., for 6 h), conc.: ≈350 ng/g* (mean value), country: Korea/USA[585], *after 6 h of ZEA-administration

Rat ileum may contain the following mycotoxins and/or their metabolites:

Ochratoxin A
incidence: 3/3 or more*, sa. const.: adult male Wistar rats, wt.: 200–250 g, contamination: no OTA, conc.: nr, country: Japan[143], *control
incidence: ?/3–4, sa. const.: adult male Wistar rats, wt.: 200–250 g, contamination: artificial (dose: 15 mg OTA (labeled)/kg, o., once), conc. range: ≈≤2.5 µg/g wet weight of tissue* (mean value), country: Japan[143], *after 16 h (also at other hour intervals up to 24 h measured, lowest conc.: ≈1 µg/g wet weight of tissue after 24 h)

Rat intestine may contain the following mycotoxins and/or their metabolites:

Zearalenone
incidence: ?/?, sa. const.: male Sprague-Dawley rats, age: 8–10 weeks, wt.: 230–290 g, contamination: artificial (dose: **1.13 mg ZEA**/h/kg, i.v., for 6 h), conc.: ≈7,000 ng/g* ** (mean value), country: Korea/USA[585], *after 6 h, **in small intestine
incidence: ?/?, sa. const.: male Sprague-Dawley rats, age: 8–10 weeks, wt.: 230–290 g, contamination: artificial (dose: **2.25 mg ZEA**/h/kg, i.v., for 6 h), conc.: ≈11,000 ng/g* ** (mean value), country: Korea/USA[585], *after 6 h, **in small intestine

Rat kidney may contain the following mycotoxins and/or their metabolites:

Aflatoxin B_1
incidence: ?/3–6*, sa. const.: weanling male F344 rats, contamination: artificial (dose: 1 mg AFB_1 (labeled)/kg, i.p., once), conc.: 2.8 nmol AFB_1/g tissue** (mean value), country: USA[34], *control,**after 2 h, for detailed information please see the article
incidence: ?/3–6, sa. const.: weanling male F344 rats, contamination: artificial (dose: 1 mg AFB_1 (labeled)/kg, i.p., once), conc.: 1.5 nmol AFB_1/g tissue* ** (mean value), country: USA[34], *after 2 h, ****BHA-diet** (0.45%) for 2 weeks prior to AFB_1-treatment, for detailed information please see the article
incidence: ?/3–6, sa. const.: weanling male F344 rats, contamination: artificial (dose: 1 mg AFB_1 (labeled)/kg, i.p., once), conc.: 3.1 nmol AFB_1/g tissue* ** (mean value),

country: USA[34], *after 2 h, ****BHT-diet** (0.45%) for 2 weeks prior to AFB_1-treatment, for detailed information please see the article
incidence: ?/3–6, sa. const.: weanling male F344 rats, contamination: artificial (dose: 1 mg AFB_1 (labeled)/kg, i.p., once), conc.: 0.9 nmol AFB_1/g tissue* ** (mean value), country: USA[34], *after 2 h, ****EQ-diet** (0.5%) for 2 weeks prior to AFB_1-treatment, for detailed information please see the article
incidence: ?/3–6, sa. const.: weanling male F344 rats, contamination: artificial (dose: 1 mg AFB_1 (labeled)/kg, i.p., once), conc.: 2.2 nmol AFB_1/g tissue* ** (mean value), country: USA[34], *after 2 h, ****oltipraz-diet** (0.1%) for 2 weeks prior to AFB_1-treatment, for detailed information please see the article

incidence: ?/3–6*, sa. const.: weanling male F344 rats, contamination: artificial (dose: 1 mg AFB_1 (labeled)/kg, i.p., once), conc.: 94 pmol AFB_1 bound/mg DNA** (mean value), country: USA[34], *control, **after 2 h, for detailed information please see the article
incidence: ?/3–6, sa. const.: weanling male F344 rats, contamination: artificial (dose: 1 mg AFB_1 (labeled)/kg, i.p., once), conc.: 36 pmol AFB_1 bound/mg DNA* ** (mean value), country: USA[34], *after 2 h, ****BHA-diet** (0.45%) for 2 weeks prior to AFB_1-treatment, for detailed information please see the article
incidence: ?/3–6, sa. const.: weanling male F344 rats, contamination: artificial (dose: 1 mg AFB_1 (labeled)/kg, i.p., once), conc.: 61 pmol AFB_1 bound/mg DNA* ** (mean value), country: USA[34], *after 2 h, ****BHT-diet** (0.45%) for 2 weeks prior to AFB_1-treatment, for detailed information please see the article
incidence: ?/3–6, sa. const.: weanling male F344 rats, contamination: artificial (dose: 1 mg AFB_1 (labeled)/kg, i.p., once), conc.: 19 pmol AFB_1 bound/mg DNA* ** (mean value), country: USA[34], *after 2 h, ****EQ-diet** (0.5%) for 2 weeks prior to AFB_1-treatment, for detailed information please see the article
incidence: ?/3–6, sa. const.: weanling male F344 rats, contamination: artificial (dose: 1 mg AFB_1 (labeled)/kg, i.p., once), conc.: 34 pmol AFB_1 bound/mg DNA* ** (mean value), country: USA[34], *after 2 h, ****oltipraz-diet** (0.1%) for 2 weeks prior to AFB_1-treatment, for detailed information please see the article

incidence: 3/3, sa. const.: CD rats, contamination: artificial (dose: 0.5 mg AFB_1 (labeled)/kg, i.p., once), conc.: nd* ** ***, country: USA/Taiwan, Republic of China[49], *after 2 h, **AFB_1-DNA adducts, ***measured by **scintillation counting**
incidence: 3/3, sa. const.: CD rats, contamination: artificial (dose: 0.5 mg AFB_1 (labeled)/kg, i.p., once), conc.: nd* ** ***, country: USA/Taiwan, Republic of China[49], *after 2 h, **AFB_1-FAPy adducts, ***measured by **ELISA**
incidence: 3?/3, sa. const.: CD rats, contamination: artificial (1 mg AFB_1 (labeled)/kg, i.p., once), conc.: 5.4 μmol/mol DNA* ** *** (mean value), country: USA/Taiwan, Republic of China[49], *after 2 h, **AFB_1-DNA adducts, ***measured by **scintillation counting**
incidence: 3/3, sa. const.: CD rats, contamination: artificial (1 mg AFB_1 (labeled)/kg, i.p., once), conc.: nd* ** ***, country: USA/Taiwan, Republic of China[49], *after 2 h, **AFB_1-FAPy adducts, ***measured by **ELISA**

incidence: 3?/3*, sa. const.: male Wistar rats, wt.: 250–300 g, contamination: artificial (dose: 40 μg AFB_1 (labeled)/100 g, i.p., once; for detailed information please see the article), conc.: 14 ng [^{14}C]AFB_1 bound/**mg kidney DNA**** (mean value), country: UK[74], *control, **after 6 h
incidence: 3?/3*, sa. const.: male Wistar rats, wt.: 250–300 g, contamination: artificial (dose: 40 μg AFB_1 (labeled)/100 g, i.p., once; for detailed information please see the article), conc.:

9.0 ng [^{14}C]AFB$_1$ bound/**mg kidney DNA**** (mean value), country: UK[74], *receiving 1 mg **PB**/ml drinking water for 7 days prior to AFB$_1$-treatment, **after 6 h
incidence: 3?/3*, sa. const.: male Wistar rats, wt.: 250–300 g, contamination: artificial (dose: 40 µg AFB$_1$ (labeled)/100 g, i.p., once; for detailed information please see the article), conc.: 27.0 ng [^{14}C]AFB$_1$ bound/**mg kidney rRNA**** (mean value), country: UK[74], *control, **after 6 h
incidence: 3?/3*, sa. const.: male Wistar rats, wt.: 250–300 g, contamination: artificial (dose: 40 µg AFB$_1$ (labeled)/100 g, i.p., once; for detailed information please see the article), conc.: 16.0 ng [^{14}C]AFB$_1$ bound/**mg kidney rRNA**** (mean value), country: UK[74], *receiving 1 mg **PB**/ml drinking water for 7 days prior to AFB$_1$-treatment, **after 6 h
incidence: 3?/3*, sa. const.: male Wistar rats, wt.: 250–300 g, contamination: artificial (dose: 40 µg AFB$_1$ (labeled)/100 g, i.p., once; for detailed information please see the article), conc.: 1.0 ng [^{14}C]AFB$_1$ bound/**mg kidney protein**** (mean value), country: UK[74], *control, **after 6 h
incidence: 3?/3*, sa. const.: male Wistar rats, wt.: 250–300 g, contamination: artificial (dose: 40 µg AFB$_1$ (labeled)/100 g, i.p., once; for detailed information please see the article), conc.: 1.0 ng [^{14}C]AFB$_1$ bound/**mg kidney protein**** (mean value), country: UK[74], *receiving 1 mg **PB**/ml drinking water for 7 days prior to AFB$_1$-treatment, **after 6 h

incidence: 3?/3, sa. const.: Wistar derived strain, wt.: 150 g, contamination: artificial (dose: 60 µg AFB$_1$ (labeled)/100 g, i.p., once; for detailed information please see the article), conc. range: ≈≤11 ng AFB$_1$ bound/**mg DNA*** (mean value), country: USA[116], *after ≈2 h (also measured after ≈4, 24, and 48 h, lowest conc.: ≈2 ng AFB$_1$ bound/mg DNA after 48 h)
incidence: 3?/3, sa. const.: Wistar derived strain, wt.: 150 g, contamination: artificial (dose: 60 µg AFB$_1$ (labeled)/100 g, i.p., once; for detailed information please see the article), conc. range: ≈≤4 ng AFB$_1$ bound/**mg protein*** (mean value), country: USA[116], *after ≈4 h (also measured after ≈2, 24, and 48 h, lowest conc.: ≈nd after 48 h)

incidence: ?/8*, sa. const.: male Fischer rats, wt.: 80–100 g, contamination: artificial (dose: 40 µg AFB$_1$ (labeled)/100 g b. wt., i.p., once; for detailed information please see the article), conc.: 4.0 AFB$_1$-DNA binding pmol/mg DNA** (mean value), country: USA[556], *control, **after 2 h
incidence: ?/8, sa. const.: male Fischer rats, weight: 80–100 g, contamination: artificial (dose: 40 µg AFB$_1$ (labeled)/100 g b. wt., i.p., once, and **pretreatment** with $_L$-**BSO** (conc.: 4 mmol, 4 and 2 h before AFB$_1$-injection; for detailed information please see the article), conc.: 6.0 AFB$_1$-DNA binding pmol/mg DNA* (mean value), country: USA[556], *after 2 h
incidence: ?/8, sa. const.: male Fischer rats, weight: 80–100 g, contamination: artificial (dose: 40 µg AFB$_1$ (labeled)/100 g b. wt., i.p., once and **pretreatment** with **DEM** (conc.: 3.5 mmol, 4 h before AFB$_1$ injection) + $_L$-**BSO** (conc.: 4 mmol, 2 h before AFB$_1$-injection); for detailed information please see the article), conc.: 7.4 AFB$_1$-DNA binding pmol/mg DNA* (mean value), country: USA[556], *after 2 h

incidence: 2?/2*, sa. const.: male Fischer rats, wt.: 60–100 g, contamination: artificial (dose: 1 mg AFB$_1$ (labeled)/kg, i.p., once; for detailed information please see the article), conc.: 12.2 pg AFB$_1$/µg DNA* ** (mean value), country: UK/ Scotland, UK[563], *control: **EQ-diet** (0.5%, v/w) for **0 days** prior to AFB$_1$-treatment, **after 2 h
incidence: 2?/2*, sa. const.: male Fischer rats, wt.: 60–100 g, contamination: artificial (dose: 1 mg AFB$_1$ (labeled)/kg, i.p., once; for detailed information please

see the article), conc.: 5.4 pg AFB_1/µg DNA* ** (mean value), country: UK/Scotland, UK[563], ***EQ-diet** (0.5%, v/w) for **2 days** prior to AFB_1-treatment, **after 2 h
incidence: 2?/2*, sa. const.: male Fischer rats, wt.: 60–100 g, contamination: artificial (dose: 1 mg AFB_1 (labeled)/kg, i.p., once; for detailed information please see the article), conc.: 3.7 pg AFB_1/µg DNA* ** (mean value), country: UK/Scotland, UK[563], ***EQ-diet** (0.5%, v/w) for **14 days** prior to AFB_1-treatment, **after 2 h

incidence: 3?/3, sa. const.: male Wistar rats, wt.: 200–250 g, contamination: artificial (dose: 40 µg AFB_1 (labeled)/100 g, i.p., once, conc. range: ≈≤17.5 ng AFB_1 bound/**mg rRNA*** (mean value), country: UK[608], *after 2 h (also measured after 6, 24, and 48 h, lowest conc.: ≈5 ng AFB_1 bound/mg rRNA after 48 h)
incidence: 3?/3, sa. const.: male Wistar rats, wt.: 200–250 g, contamination: artificial (dose: 40 µg AFB_1 (labeled)/100 g, i.p., once, conc. range: ≈≤14 ng AFB_1 bound/**mg DNA*** (mean value), country: UK[608], *after 6 h (also measured after 2, 24, and 48 h, lowest conc.: ≈1 ng AFB_1 bound/mg DNA after 48 h)
incidence: 3?/3, sa. const.: male Wistar rats, wt.: 200–250 g, contamination: artificial (dose: 40 µg AFB_1 (labeled)/100 g, i.p., once, conc. range: ≈≤0.9 ng AFB_1 bound/**mg protein*** (mean value), country: UK[608], *after 6 h (also measured after 2, 24, and 48 h, lowest conc.: nd after 48 h)

Aflatoxin G_1
incidence: 3?/3, sa. const.: Wistar derived strain, wt.: 150 g, contamination: artificial (dose: 60 µg AFG_1 (labeled)/100 g, i.p., once; for detailed information please see the article), conc. range: ≈≤5.5 ng AFG_1 bound/**mg DNA*** (mean value), country: USA[116], *after ≈2 h (also measured after ≈4, 24, and 48 h, lowest conc.: ≈1 ng AFG_1 bound/mg DNA after 48 h)
incidence: 3?/3, sa. const.: Wistar derived strain, wt.: 150 g, contamination: artificial (dose: 60 µg AFG_1 (labeled)/100 g, i.p., once; for detailed information please see the article), conc. range: ≈≤5 ng AFG_1 bound/**mg protein*** (mean value), country: USA[116], *after ≈2 h (also measured after ≈4, 24, and 48 h, lowest conc.: ≈nd after 48 h)

Aflatoxins
incidence: ?/5, sa. const.: male Sprague-Dawley rats, Ø wt.: 242.2 g, contamination: artificial (dose: 300 µg/kg AFB_1 (labeled)/kg, i.t. (**dust-adsorbed**), once; for detailed information please see the article), conc.: ≤0.215 nmol AFs/g tissue ($\times 10^{-3}$)*, country: USA[604], *after 3 h (also measured after 3 days and 3 weeks, lowest conc.: 0.013 nmol AFs/g tissue ($\times 10^{-3}$) after 3 weeks)
incidence: ?/5, sa. const.: male Sprague-Dawley rats, Ø wt.: 242.2 g, contamination: artificial (dose: 300 µg/kg AFB_1 (labeled)/kg, i.t. (**microcrystalline form**), once; for detailed information please see the article), conc.: ≤0.333 nmol AFs/g tissue ($\times 10^{-3}$)*, country: USA[604], *after 3 h (also measured after 3 days and 3 weeks, lowest conc.: 0.009 nmol AFs/g tissue ($\times 10^{-3}$) after 3 weeks)

HT-2 Toxin
incidence: ?/3, sa. const.: male Wistar rats, age: 8–10 weeks, contamination: artificial (dose: 2 or 5 mg T-2 toxin/kg b. wt., o., once), conc.: pr*, country: Japan[391], *after 30 min

Ochratoxin A
incidence: 3/3*, sa. const.: **male** F344 rats, age: 8 weeks, contamination: no OTA (for detailed information please see the article), conc.: nr, country: Germany[125], *control
incidence: 3?/3, sa. const.: **male** F344 rats, age: 8 weeks, contamination: artificial (dose: **0.5 mg OTA**/kg b. wt., by gavage, once; for detailed information please see

the article), conc. range: ≤480 pmol/g* ** (mean value), country: Germany[125], *after 24 h after OTA-administration (also at other day intervals up to 56 days measured, lowest conc.: nd after 28 days), **highest value recorded
incidence: 3/3*, sa. const.: **female** F344 rats, age: 8 weeks, contamination: no OTA (for detailed information please see the article), conc.: nr, country: Germany[125], *control
incidence: 3?/3, sa. const.: **female** F344 rats, age: 8 weeks, contamination: artificial (dose: **0.5 mg OTA/kg** b. wt., by gavage, once; for detailed information please see the article), conc. range: ≈≤170 pmol/g* ** (mean value), country: Germany[125], *after 48 h (also at other day intervals up to 56 days measured, lowest conc.: nd after 28 days), **highest value recorded

incidence: ?/3 or more*, sa. const.: adult male Wistar rats, wt.: 200–250 g, contamination: no OTA, conc.: nr, country: Japan[143], *control
incidence: ?/3–4, sa. const.: adult male Wistar rats, wt.: 200–250 g, contamination: artificial (dose: **15 mg OTA** (labeled)/kg, o., once), conc. range: ≈8.0 µg/g wet weight of tissue* (mean value), country: Japan[143], *after 4 h (also at other hour intervals up to 40 h measured, lowest conc.: ≈2.5 µg/g wet weight of tissue after 8 h)

incidence: 4?/4, sa. const.: healthy adult female Sprague-Dawley rats, wt.: 270–350 g, contamination: artificial (dose: **100 µg OTA**/rat, i.v., once), conc. range: ≤390 ng/g (mean value), country: Canada[175], measured at 2, 24, 48, and 96 h
incidence: 4?/4 (overall16), sa. const.: healthy adult female Sprague-Dawley rats, wt.: 270–350 g, contamination: artificial (dose: **100 µg OTC!**/rat, i.v., once), conc. range: ≤320 ng/g (mean value), country: Canada[175], measured at 2, 24, 48, and 96 h

incidence: 10/10*, sa. const.: Wistar male rats, wt.: 83–110 g, contamination: no OTA, **diet B-II** (for detailed information please see the article), conc.: nr, country: Canada[209], *control
incidence: ?/10, sa. const.: Wistar male rats, wt.: 83–110 g, contamination: artificial (dose: 500 µg OTA, intubated, daily for 6 days (**diet B-I**); for detailed information please see the article), conc.: pr*, country: Canada[209], *measured at day 1, 2, 3, 4, 5, and 6 ? of OTA-administration
incidence: 10/10*, sa. const.: Wistar male rats, wt.: 83–110 g, contamination: no OTA, **diet NMB** (for detailed information please see the article), conc.: nr, country: Canada[209], *control
incidence: 3?/3, sa. const.: Wistar male rats, weight: 83–110 g, contamination: artificial (dose: 500 µg OTA, intubated, daily, daily for 6 days (**diet NMB+T**); for detailed information please see the article), conc.: 6.2 µg/g* (mean value), country: Canada[209], *measured on day 5 of OTA-administration (also measured at day 3 and 6)

incidence: ?/2 (10)*, sa. const.: Sprague-Dawley male rats, wt.: ≈100 g, contamination: no OTA, conc.: nr, country: USA[218], *control
incidence: ?/2, sa. const.: Sprague-Dawley male rats, wt.: ≈100 g, contamination: artificial (dose: **1 mg OTA** (labeled)/rat, i.p., once), conc. range: ≤13.6 µg/g tissue* (mean value), country: USA[218], *after 0.5 h (also measured after 2, 4, 8, and 24 h, lowest conc.: 3.18 µg/g tissue after 24 h)

incidence: 5?/5*, sa. const.: Sprague-Dawley rats (dams), wt.: 300–370 g, contamination: no OTA (for detailed information please see the article), conc.: 0.2 µg/kg** (mean value), country: Sweden[312], *control, **after 72 h
incidence: 4?/4, sa. const.: Sprague-Dawley rats (dams), wt.: 300–370 g, contamination: artificial (dose: **10 µg OTA/kg** b. wt., o., once at 11th day of lactation; for detailed information please see the article), conc.: 1.9 µg/kg* (mean value), country: Sweden[312], *after 72 h

incidence: 4?/4, sa. const.: Sprague-Dawley rats (dams), wt.: 300–370 g, contamination: artificial (dose: **50 µg OTA**/kg b. wt., o., once at 11th day of lactation; for detailed information please see the article), conc.: 9.2 µg/kg* (mean value), country: Sweden[312], *after 72 h
incidence: 5?/5, sa. const.: Sprague-Dawley rats (dams), wt.: 300–370 g, contamination: artificial (dose: **250 µg OTA**/kg b. wt., o., once at 11th day of lactation; for detailed information please see the article), conc.: 18 µg/kg* (mean value), country: Sweden[312], *after 72 h

incidence: 15?/15*, sa. const.: pups of Sprague-Dawley rats, wt.: 19–30 g, contamination: no OTA (for detailed information please see the article), conc.: 1.4 µg/kg** (mean value), country: Sweden[312], *control, **after 72 h
incidence: 12?/12, sa. const.: pups of Sprague-Dawley rats, wt.: 19–30 g, contamination: artificial (dose: **OTA** from the **milk of lactating dams** receiving **10 µg OTA**/kg b. wt.; for detailed information please see the article), conc.: 7.2 µg/kg* (mean value), country: Sweden[312], *after 72 h
incidence: 12?/12, sa. const.: pups of Sprague-Dawley rats, wt.: 19–30 g, contamination: artificial (dose: **OTA** from the **milk of lactating dams** receiving **50 µg OTA**/kg b. wt.; for detailed information please see the article), conc.: 29 µg/kg* (mean value), country: Sweden[312], *after 72 h
incidence: 15?/15, sa. const.: pups of Sprague-Dawley rats, wt.: 19–30 g, contamination: artificial (dose: **OTA** from the **milk of lactating dams** receiving **250 µg OTA**/kg b. wt.; for detailed information please see the article), conc.: 110 µg/kg* (mean value), country: Sweden[312], *after 72 h

incidence: 4(8?)?/4(8)*, sa. const.: Sprague-Dawley rats (dams), contamination: no OTA (for detailed information please see the article), Ø conc.: 0.7 µg/kg** (mean value), country: Sweden[322], *control, **at day 21 of lactation
incidence: 4?(8?)/4(8), sa. const.: Sprague-Dawley rats (dams), contamination: artificial (dose: 50 µg OTA/kg b. wt., gastric intubation, 46 times in 8 weeks; for detailed information please see the article), Ø conc.: 48 µg/kg* (mean value), country: Sweden[322], *at day 21 of lactation

incidence: 4?/4*, sa. const.: pups of Sprague-Dawley rats, contamination: no OTA (for detailed information please see the article), conc.: 4.0 µg/kg** (mean value), country: Sweden[322], *control, **at day 14
incidence: 4?/4, sa. const.: pups of Sprague-Dawley rats, contamination: artificial (dose: infected by lactating dams receiving 50 µg OTA/kg b. wt., gastric intubation for 46 times (OTA-exposure of pups via **placenta**); for detailed information please see the article), conc.: 42 µg/kg* (mean value), country: Sweden[322], *at day 14
incidence: 4?/4, sa. const.: pups of Sprague-Dawley rats, contamination: artificial (dose: infected by lactating dams receiving 50 µg OTA/kg b. wt., gastric intubation for 46 times (OTA-exposure of pups via **milk**); for detailed information please see the article), conc.: 180 µg/kg* (mean value), country: Sweden[322], *at day 14
incidence: 4?/4, sa. const.: pups of Sprague-Dawley rats, contamination: artificial (dose: infected by lactating dams receiving 50 µg OTA/kg b. wt., gastric intubation for 46 times (OTA-exposure of pups via **milk and placenta**); for detailed information please see the article), conc.: 240 µg/kg* (mean value), country: Sweden[322], *at day 14

incidence: 5/5*, sa. const.: male Fisher 344 (F344) rats, age: 10 weeks, contamination: no OTA (for detailed information please see the article), conc.: nd, country: Spain[439], *control
incidence: 5/5, sa. const.: male Fisher 344 (F344) rats, age: 10 weeks,

contamination: artificial (dose: 0.5 mg OTA/kg b. wt., o., daily for 7 days; for detailed information please see the article), conc. range: 641–807 μg/kg*, Ø conc.: 725.6 μg/kg*, country: Spain[439], *after 24 h of final administration

incidence: ?/12*, sa. const.: male Wistar/AF EOPS rats, wt.: 240–290 g, contamination: no OTA only basal feed (for detailed information please see the article), conc.: 1.9 ng/g** (mean value), country: Belgium[509], *control, **after 4 weeks
incidence: ?/12, sa. const.: male Wistar/AF EOPS rats, wt.: 240–290 g, contamination: artificial (dose: basal feed + **OTA contaminated wheat (2.2 μg/g)**, o., for 28 days; for detailed information please see the article), conc.: 79.4 ng/g* (mean value), country: Belgium[509], *after 4 weeks of OTA-administration
incidence: ?/12, sa. const.: male Wistar/AF EOPS rats, wt.: 240–290 g, contamination: artificial (dose: basal feed + **OTA contaminated wheat (2.2 μg/g) + MWF** (2%), o., for 28 days; for detailed information please see the article), conc.: 57.1 ng/g* (mean value), country: Belgium[509], *after 4 weeks of OTA-administration
incidence: ?/12, sa. const.: male Wistar/AF EOPS rats, wt.: 240–290 g, contamination: artificial (dose: basal feed + **OTA contaminated wheat (2 μg/g) + MWF** (1.8%) + **YCW** (0.2%), o., for 28 days; for detailed information please see the article), conc.: 66.7 ng/g* (mean value), country: Belgium[509], *after 4 weeks of OTA-administration

incidence: ?/18*, sa. const.: male Wistar rats, age: **6 weeks**, contamination: artificial (dose: **0.5 mg OTA**/kg b. wt., i.p., daily for 6 days), conc.: ≈1.2 μmol/kg kidney wet weight** (mean value), country: Germany[561], ***cortex** of kidney, **after 7 days of first OTA-administration
incidence: ?/13*, sa. const.: male Wistar rats, age: **2 years**, contamination: artificial (dose: **0.5 mg OTA**/kg b. wt., i.p., daily for 6 days), conc.: 3.4 μmol/kg kidney wet weight** (mean value), country: Germany[561], ***cortex** of kidney, **after 7 days of first OTA-administration
incidence: ?/13*, sa. const.: male Wistar rats, age: **2 years**, contamination: artificial (dose: **1.25 mg OTA**/kg b. wt., i.p., once), conc.: ≈6.2 μmol/kg kidney wet weight** (mean value), country: Germany[561], ***cortex** of kidney, **after 2 h
incidence: ?/14*, sa. const.: male Wistar rats, age: **6 weeks**, contamination: artificial (dose: **0.5 mg OTA**/kg b. wt., i.p., daily for 6 days), conc.: 2.5 μmol/kg kidney wet weight** (mean value), country: Germany[561], ***inner medulla** of kidney, **after 7 days of first OTA-administration
incidence: ?/13*, sa. const.: male Wistar rats, age: **2 years**, contamination: artificial (dose: **0.5 mg OTA**/kg b. wt., i.p., daily for 6 days), conc.: 4.1 μmol/kg kidney wet weight** (mean value), country: Germany[561], ***inner medulla** of kidney, **after 7 days of first OTA-administration
incidence: ?/13*, sa. const.: male Wistar rats, age: **2 years**, contamination: artificial (dose: **1.25 mg OTA**/kg b. wt., i.p., once), conc.: 11.7 μmol/kg kidney wet weight** (mean value), country: Germany[561], ***inner medulla** of kidney, **after 2 h
incidence: ?/18*, sa. const.: male Wistar rats, age: **6 weeks**, contamination: artificial (dose: **0.5 mg OTA**/kg b. wt., i.p., daily for 6 days), conc.: ≈1.35 μmol/kg kidney wet weight** (mean value), country: Germany[561], ***outer medulla** of kidney, **after 7 days of first OTA-administration
incidence: ?/11*, sa. const.: male Wistar rats, age: **2 years**, contamination: artificial (dose: **0.5 mg OTA**/kg b. wt., i.p., daily for 6 days), conc.: ≈1.4 μmol/kg kidney wet weight** (mean value), country: Germany[561], ***outer medulla** of kidney, **after 7 days of first OTA-administration
incidence: ?/11*, sa. const.: male Wistar rats, age: **2 years**, contamination: artificial (dose: **1.25 mg OTA**/kg b. wt., i.p., once),

conc.: ≈5.7 μmol/kg kidney wet weight** (mean value), country: Germany[561], ***outer medulla** of kidney, **after 2 h
incidence: ?/14*, sa. const.: male Wistar rats, age: **6 weeks**, contamination: artificial (dose: **0.5 mg OTA**/kg b. wt., i.p., daily for 6 days), conc.: 4.3 μmol/kg kidney wet weight** (mean value), country: Germany[561], ***papilla** of kidney, **after 7 days of first OTA-administration
incidence: ?/3*, sa. const.: male Wistar rats, age: **2 years**, contamination: artificial (dose: **0.5 mg OTA**/kg b. wt., i.p., daily for 6 days), conc.: 3.4 μmol/kg kidney wet weight** (mean value), country: Germany[561], ***papilla** of kidney, *after 7 days of first OTA-administration
incidence: ?/3*, sa. const.: male Wistar rats, age: **2 years**, contamination: artificial (dose: **1.25 mg OTA**/kg b. wt., i.p., once), conc.: 11.3 μmol/kg kidney wet weight** (mean value), country: Germany[561], ***papilla** of kidney, **after 2 h

Zearalenone
incidence: ?/?, sa. const.: male Sprague-Dawley rats, age: 8–10 weeks, wt.: 230–290 g, contamination: artificial (dose: **1.13 mg ZEA/h**/kg, i.v., for 6 h), conc.: ≈1,100 ng/g* (mean value), country: Korea/USA[585], *after 6 h of ZEA-administration
incidence: ?/?, sa. const.: male Sprague-Dawley rats, age: 8–10 weeks, wt.: 230–290 g, contamination: artificial (dose: **2.25 mg ZEA/h**/kg, i.v., for 6 h), conc.: ≈1,650 ng/g* (mean value), country: Korea/USA[585], *after 6 h of ZEA-administration

Rat liver may contain the following mycotoxins and/or their metabolites:

Aflatoxin B_1
incidence: ?/?*, sa. const.: male Sprague-Dawley, Wistar, and Fischer 344 rats, wt.: 80–140 g, contamination: no AFB_1, conc.: nr, country: France/Japan[24], *control
incidence: ?/?, sa. const.: male Sprague-Dawley, Wistar, and Fischer 344 rats, wt.: 80–140 g, contamination: artificial (dose: 20 μg AFB_1/kg/day, by gavage, daily for up to 14 days), conc. range: ≈≤2.0 pmol AFB_1-FAPy/mg DNA** (mean value, combined data from the three strains), country: France/Japan[24], *animals killed after 24 h of final treatment (also measured after 1, 3, and 7 days, lowest conc.: ≈0.4 pmol AFB_1-FAPy/mg DNA after 1 day, combined data from the three strains), **AFB_1-DNA adducts

incidence: ?/3–6, sa. const.: weanling male F344 rats, contamination: artificial (dose: 1 mg AFB_1 (labeled)/kg, i.p., once), conc.: 9.0 nmol AFB_1/g tissue* ** (mean value), country: USA[34], *after 2 h, **control, for detailed information please see the article
incidence: ?/3–6, sa. const.: weanling male F344 rats, contamination: artificial (dose: 1 mg AFB_1 (labeled)/kg, i.p., once), conc.: 4.3 nmol AFB_1/g tissue* ** (mean value), country: USA[34], *after 2 h, ****BHA-diet** (0.45%) for 2 weeks prior to AFB_1-treatment, for detailed information please see the article
incidence: ?/3–6, sa. const.: weanling male F344 rats, contamination: artificial (dose: 1 mg AFB_1 (labeled)/kg, i.p., once), conc.: 5.2 nmol AFB_1/g tissue* ** (mean value), country: USA[34], *after 2 h, ****BHT-diet** (0.45%) for 2 weeks prior to AFB_1-treatment, for detailed information please see the article
incidence: ?/3–6, sa. const.: weanling male F344 rats, contamination: artificial (dose: 1 mg AFB_1 (labeled)/kg, i.p., once), conc.: 3.3 nmol AFB_1 bond/g tissue* ** (mean value), country: USA[34], *after 2 h, ****EQ-diet** (0.5%) for 2 weeks prior to AFB_1-treatment, for detailed information please see the article
incidence: ?/3–6, sa. const.: weanling male F344 rats, contamination: artificial (dose: 1 mg AFB_1 (labeled)/kg, i.p., once), conc.: 4.5 nmol AFB_1/g tissue* ** (mean value),

country: USA[34], *after 2 h, ****oltipraz-diet** (0.1%) for 2 weeks prior to AFB_1-treatment, for detailed information please see the article

incidence: ?/3–6*, sa. const.: weanling male F344 rats, contamination: artificial (dose: 1 mg AFB_1 (labeled)/kg, i.p., once), conc.: 859 pmol AFB_1 bound/mg DNA** (mean value), country: USA[34], *control, **after 2 h, for detailed information please see the article

incidence: ?/3–6, sa. const.: weanling male F344 rats, contamination: artificial (dose: 1 mg AFB_1 (labeled)/kg, i.p., once), conc.: 304 pmol AFB_1 bound/mg DNA* ** (mean value), country: USA[34], *after 2 h, ****BHA-diet** (0.45%) for 2 weeks prior to AFB_1-treatment, for detailed information please see the article

incidence: ?/3–6, sa. const.: weanling male F344 rats, contamination: artificial (dose: 1 mg AFB_1 (labeled)/kg, i.p., once), conc.: 129 pmol AFB_1 bound/mg DNA* ** (mean value), country: USA[34], *after 2 h, ****BHT-diet** (0.45%) for 2 weeks prior to AFB_1-treatment, for detailed information please see the article

incidence: ?/3–6, sa. const.: weanling male F344 rats, contamination: artificial (dose: 1 mg AFB_1 (labeled)/kg, i.p., once), conc.: 77 pmol AFB_1 bound/mg DNA* ** (mean value), country: USA[34], *after 2 h, ****EQ-diet** (0.5%) for 2 weeks prior to AFB_1-treatment, for detailed information please see the article

incidence: ?/3–6, sa. const.: weanling male F344 rats, contamination: artificial (dose: 1 mg AFB_1 (labeled)/kg, i.p., once), conc.: 202 pmol AFB_1 bound/mg DNA* ** (mean value), country: USA[34], *after 2 h, ****oltipraz-diet** (0.1%) for 2 weeks prior to AFB_1-treatment, for detailed information please see the article

incidence: 6?/6, sa. const.: male Fischer rats, age: 75 days, contamination: artificial (dose: **0.6 mg AFB_1** (labeled)/kg b. wt., i.p., once; for detailed information please see the article), conc. range: 26*–22.000** pmol AFB_1-*N*⁷-Gua/mg DNA x 10^2*, country: USA[42], after 72* or 2** h after AFB_1-administration (also at other hour intervals up to 72 h measured)

incidence: ?/30, sa. const.: male Fischer rats, age: 75 days, contamination: artificial (dose: **25 µg AFB_1** (labeled), i.p., **ten times**; for detailed information please see the article), conc. range: 13*–322** pmol AFB_1-N^7-Gua/mg DNA x 10^2*, country: USA[42], after 14* or 2** days (also at other day intervals up to 14 days measured, thereof 12 days with AFB_1-administration)

incidence: 3?/3, sa. const.: CD rats, contamination: artificial (dose: 0.5 mg AFB_1 (labeled)/kg, i.p., once), conc.: 0.8 µmol/mol DNA* ** *** (mean value), country: USA/Taiwan, Republic of China[49], *after 2 h, **AFB_1-DNA adducts, ***measured by **scintillation counting**

incidence: 3/3, sa. const.: CD rats, contamination: artificial (dose: 0.5 mg AFB_1 (labeled)/kg, i.p., once), conc.: nd* ** ***, country: USA/Taiwan, Republic of China[49], *after 2 h, **AFB_1-FAPy adducts, ***measured by **ELISA**

incidence: 3?/3, sa. const.: CD rats, contamination: artificial (dose: 1 mg AFB_1 (labeled)/kg, i.p., once), conc.: 115.4 µmol/mol DNA* ** *** (mean value), country: USA/Taiwan, Republic of China[49], *after 2 h, **AFB_1-DNA adducts, ***measured by **scintillation counting**

incidence: 3/3, sa. const.: CD rats, contamination: artificial (dose: 1 mg AFB_1 (labeled)/kg, i.p., once), conc.: nd* ** ***, country: USA/Taiwan, Republic of China[49], *after 2 h, **AFB_1-FAPy adducts, ***measured by **ELISA**

incidence: 3?/3*, sa. const.: male Wistar rats, wt.: 250–300 g, contamination: artificial (dose: 40 µg AFB_1 (labeled)/100 g, i.p., once; for detailed information please see the article), conc.: 40.7 ng [^{14}C] AFB_1 bound/**mg liver DNA**** (mean value), country: UK[74], *control, **after 6 h

incidence: 3?/3*, sa. const.: male Wistar rats, wt.: 250–300 g, contamination: artificial (dose: 40 µg AFB_1 (labeled)/100 g, i.p., once; for detailed information please see the article), conc.: 11.0 ng [^{14}C]AFB_1 bound/**mg liver DNA**** (mean value), country: UK[74], *receiving 1 mg PB/ml drinking water for 7 days prior to AFB_1-treatment, **after 6 h
incidence: 3?/3*, sa. const.: male Wistar rats, weight: 250–300 g, contamination: artificial (dose: 40 µg AFB_1 (labeled)/100 g, i.p., once; for detailed information please see the article), conc.: 44.8 ng [^{14}C]AFB_1 bound/**mg liver rRNA**** (mean value), country: UK[74], *control, **after 6 h
incidence: 3?/3*, sa. const.: male Wistar rats, wt.: 250–300 g, contamination: artificial (dose: 40 µg AFB_1 (labeled)/100 g, i.p., once; for detailed information please see the article), conc.: 9.0 ng [^{14}C]AFB_1 bound/**mg liver rRNA**** (mean value), country: UK[74], *receiving 1 mg PB/ml drinking water for 7 days prior to AFB_1-treatment, **after 6 h
incidence: 3?/3*, sa. const.: male Wistar rats, wt.: 250–300 g, contamination: artificial (dose: 40 µg AFB_1 (labeled)/100 g, i.p., once; for detailed information please see the article), conc.: 6.3 ng [^{14}C]AFB_1 bound/**mg liver protein**** (mean value), country: UK[74], *control, **after 6 h
incidence: 3?/3*, sa. const.: male Wistar rats, wt.: 250–300 g, contamination: artificial (dose: 40 µg AFB_1 (labeled)/100 g, i.p., once; for detailed information please see the article), conc.: 7.7 ng [^{14}C]AFB_1 bound/**mg liver protein**** (mean value), country: UK[74], *receiving 1 mg PB/ml drinking water for 7 days prior to AFB_1-treatment, **after 6 h

incidence: ?/?, sa. const.: male CDF Fischer rats, wt.: 100–150 g, contamination: artificial (dose: 0.01–1 mg AFB_1 (labeled)/kg, i.p., once; for detailed information please see the article), conc. range: 1 AFB_1-residue/35,000, 251,000 or 1,355,000 nucleotides*, country: USA[103], *after 2 h

incidence: 4?/4, sa. const.: Wistar derived strain, wt.: 150 g, contamination: artificial (dose: 60 µg AFB_1 (labeled)/100 g, i.p., once), conc.: 14.78 ng AFB_1 bound/mg* DNA** (mean value), country: USA[116], *after 2 h, ****phenol-cresol extraction**
incidence: 4?/4, sa. const.: Wistar derived strain, wt.: 150 g, contamination: artificial (dose: 60 µg AFB_1 (labeled)/100 g, i.p., once), conc.: 13.03 ng AFB_1 bound/mg* DNA** (mean value), country: USA[116], *after 2 h, ****chloroform/isoamylalcohol extraction**

incidence: 3?/3, sa. const.: Wistar derived strain, wt.: 150 g, contamination: artificial (dose: 60 µg AFB_1 (labeled)/100 g, i.p., once), conc. range: ≈≤40.5 ng AFB_1 bound/**mg rRNA*** (mean value), country: USA[116], *after ≈2 h (also measured after ≈4, 24, and 48 h, lowest conc.: ≈18 ng AFB_1 bound/mg rRNA after 48 h)
incidence: 3?/3, sa. const.: Wistar derived strain, wt.: 150 g, contamination: artificial (dose: 60 µg AFB_1 (labeled)/100 g, i.p., once), conc. range: ≈≤15 ng AFB_1 bound/**mg DNA*** (mean value), country: USA[116], *after ≈2 h (also measured after ≈4, 24, and 48 h, lowest conc.: ≈2 ng AFB_1 bound/mg DNA after 48 h)
incidence: 3?/3, sa. const.: Wistar derived strain, wt.: 150 g, contamination: artificial (dose: 60 µg AFB_1 (labeled)/100 g, i.p., once), conc. range: ≈≤1 ng AFB_1 bound/**mg protein*** (mean value), country: USA[116], *after 24 h (also measured after ≈2, ≈4, and 48 h, lowest conc.: ≈1 ng AFB_1 bound/mg protein after 48 h)

incidence: 3?/3, sa. const.: male F344 rats, wt.: 100–125 g, contamination: artificial (dose: 400 µg AFB_1 (labeled)/kg b. wt., by gavage, once; for detailed information please see the article), conc.: 6.56 pmol/mg DNA* ** (mean value), country: USA[142], *after 24 h, **AFB_1-*N*^{7}-Gua incidence: 4/4, sa. const.: male Wistar rats, wt.: 200–220 g, contamination: artificial (dose: 100 µg AFB_1 (labeled)/kg, by gavage, once), conc. range: 4.26–13.55 pmol AFB_1/mg DNA*, Ø conc.: 7.325 pmol AFB_1/mg DNA* **, country: UK/Czechoslovakia[153], *after 24 h, **radioactivity

incidence: ?/?, sa. const.: male Fischer rats, wt.: 125–160 g, contamination: artificial (dose: 0.25–2.0 mg AFB_1 (labeled)/kg b. wt., injected, once), conc. range: 90–640 pmol AFB_1**/mg DNA*** (mean values), country: USA[154], *after 2 h
incidence: ?/?, sa. const.: male Fischer rats, wt.: 125–160 g, contamination: artificial (dose: 0.25–2.0 mg AFB_1 (labeled)/kg b. wt., injected, once), conc. range: 420–3,240 pmol AFB_1**/mg rDNA*** (mean values), country: USA[154], *after 2 h

incidence: 2/2, sa. const.: male CDF Fischer rats, wt.: 130–150 g, contamination: artificial (dose: 1.0 or 0.1 mg AFB_1/kg b. wt., i.p., once), conc. range: 1 AFB_1 residue/30,000 or 250,000 nucleotides* **, country: USA[155], *after 2 h, **AFB_1-DNA adducts

incidence: 2?/2, sa. const.: male Fischer rats, wt.: 130–150 g, contamination: artificial (dose: **0.125 mg** $\mathbf{AFB_1}$ (labeled)/kg, i.p., once; for detailed information please see the article), conc. range: 63,800–95,800 mmol DNA bases/mmol AFB_1-*N*[7]-Gua[2]*, Ø conc.: 79,800 mmol DNA bases/mmol AFB_1-*N*[7]-Gua[2]*, country: USA[173], *after 2 h
incidence: 2?/2, sa. const.: male Fischer rats, wt.: 130–150 g, contamination: artificial (dose: **0.25 mg** $\mathbf{AFB_1}$ (labeled)/kg, i.p., once; for detailed information please see the article), conc. range: 30,700–35,300 mmol DNA bases/mmol AFB_1-*N*[7]-Gua[2]*, Ø conc.: 33,000 mmol DNA bases/mmol AFB_1-*N*[7]-Gua[2]*, country: USA[173], *after 2 h
incidence: 2?/2, sa. const.: male Fischer rats, wt.: 130–150 g, contamination: artificial (dose: **0.50 mg** $\mathbf{AFB_1}$ (labeled)/kg, i.p., once; for detailed information please see the article), conc. range: 25,440–28,680 mmol DNA bases/mmol AFB_1-*N*[7]-Gua[2]*, Ø conc.: 27,060 mmol DNA bases/mmol AFB_1-*N*[7]-Gua[2]*, country: USA[173], *after 2 h
incidence: 2?/2, sa. const.: male Fischer rats, wt.: 130–150 g, contamination: artificial (dose: **1.00 mg** $\mathbf{AFB_1}$ (labeled)/kg, i.p., once; for detailed information please see the article), conc. range: 8,400–9,400 mmol DNA bases/mmol AFB_1-*N*[7]-Gua[2]*, Ø conc.: 8,900 mmol DNA bases/mmol AFB_1-*N*[7]-Gua[2]*, country: USA[173], *after 2 h

incidence: 4?/4, sa. const.: lactating pregnant Sprague-Dawley rats, wt.: 300–450 g, contamination: artificial (dose: 2 µCi [^{14}C]AFB_1 (labeled), i.p., once; for detailed information please see the article), conc.: 1.131 pmol AFB_1 eq/mg DNA* ** (mean value), country: USA[183], *control, **after 6 h
incidence: 4?/4, sa. const.: lactating pregnant Sprague-Dawley rats, wt.: 300–450 g, contamination: artificial (dose: 2 µCi [^{14}C]AFB_1 (labeled), i.p., once; for detailed information please see the article), conc.: 0.465 pmol AFB_1 eq/mg DNA* ** (mean value), country: USA[183], *after 6 h, ****BHT-diet** (0.5%) for 11–13 days prior to AFB_1-treatment

incidence: 3?/3*, sa. const.: male Fischer rats, wt.: 95–140 g, contamination: artificial (dose: **100 µg** $\mathbf{AFB_1}$ (labeled)/kg b. wt., o., once; for detailed information please see the article), conc.: 4.65 pmol/**mg DNA**** *** (mean value), country: Canada[445], *AFB_1 on day 1 and **CMS-diet** (composition please see the article) for up to 3 weeks, **after 2 h, ***total AFB_1-DNA adducts
incidence: 3?/3*, sa. const.: male Fischer rats, wt.: 95–140 g, contamination: artificial (dose: **100 µg** $\mathbf{AFB_1}$ (labeled)/kg b. wt., o., once; for detailed information please see the article), conc.: 7.07 pmol/**mg DNA**** *** (mean value),

country: Canada[445], *AFB_1 on day 1 and **CMD-diet** (composition please see the article) for up to 3 weeks, **after 2 h, ***total AFB_1-DNA adducts

incidence: 3?/3*, sa. const.: male Fischer rats, wt.: 95–140 g, contamination: artificial (dose: **600 µg AFB_1** (labeled)/kg b. wt., o., once; for detailed information please see the article), conc.: 29.45 pmol/**mg DNA**** *** (mean value), country: Canada[445], *AFB_1 on day 1 and **CMS-diet** (composition please see the article) for up to 3 weeks, **after 2 h, ***total AFB_1-DNA adducts

incidence: 3?/3*, sa. const.: male Fischer rats, wt.: 95–140 g, contamination: artificial (dose: **600 µg AFB_1** (labeled)/kg b. wt., o., once; for detailed information please see the article), conc.: 27.78 pmol/**mg DNA**** *** (mean value), country: Canada[445], *AFB_1 on day 1 and **CMD-diet** (composition please see the article) for up to 3 weeks, **after 2 h, ***total AFB_1-DNA adducts

incidence: 3?/3*, sa. const.: male Fischer rats, wt.: 95–140 g, contamination: artificial (dose: **100 µg AFB_1** (labeled)/kg b. wt., o., once; for detailed information please see the article), conc.: 33.11 pmol/**liver**** *** (mean value), country: Canada[445], *AFB_1 on day 1 and **CMS-diet** (composition please see the article) for up to 3 weeks, **after 2 h, ***total AFB_1-DNA adducts

incidence: 3?/3*, sa. const.: male Fischer rats, wt.: 95–140 g, contamination: artificial (dose: **100 µg AFB_1** (labeled)/kg b. wt., o., once; for detailed information please see the article), conc.: 81.49 pmol/**liver**** *** (mean value), country: Canada[445], *AFB_1 on day 1 and **CMD-diet** (composition please see the article) for up to 3 weeks, **after 2 h, ***total AFB_1-DNA adducts

incidence: 3?/3*, sa. const.: male Fischer rats, wt.: 95–140 g, contamination: artificial (dose: **600 µg AFB_1** (labeled)/kg b. wt., o., once; for detailed information please see the article), conc.: 817.15 pmol/**liver**** *** (mean values), country: Canada[445], *AFB_1 on day 1 and **CMS-diet** (composition please see the article) for up to 3 weeks, **after 2 h, ***total AFB_1-DNA adducts

incidence: 3?/3*, sa. const.: male Fischer rats, wt.: 95–140 g, contamination: artificial (dose: **600 µg AFB_1** (labeled)/kg b. wt., o., once; for detailed information please see the article), conc.: 420.26 pmol/**liver**** *** (mean values), country: Canada[445], *AFB_1 on day 1 and **CMD-diet** (composition please see the article) for up to 3 weeks, **after 2 h, ***total AFB_1-DNA adducts

incidence: 3?/3*, sa. const.: male Fischer rats, wt.: 95–140 g, contamination: artificial (dose: **100 µg AFB_1** (labeled)/kg b. wt., o., once; for detailed information please see the article), conc.: 15.18 pmol/**mg DNA**** *** (mean value), country: Canada[445], ***CMS-diet** (composition please see the article) for up to 3 weeks then AFB_1 on day 21, **after 2 h, ***total AFB_1-DNA adducts

incidence: 3?/3*, sa. const.: male Fischer rats, wt.: 95–140 g, contamination: artificial (dose: **100 µg AFB_1** (labeled)/kg b. wt., o., once; for detailed information please see the article), conc.: 2.13 pmol/**mg DNA**** *** (mean value), country: Canada[445], ***CMD-diet** (composition please see the article) for up to 3 weeks then AFB_1 on day 21, **after 2 h, ***total AFB_1-DNA adducts

incidence: 3?/3*, sa. const.: male Fischer rats, wt.: 95–140 g, contamination: artificial (dose: **600 µg AFB_1** (labeled)/kg b. wt., o., once; for detailed information please see the article), conc.: 89.61 pmol/**mg DNA**** *** (mean value), country: Canada[445], ***CMS-diet** (composition please see the article) for up to 3 weeks then AFB_1 on day 21, **after 2 h, ***total AFB_1-DNA adducts

incidence: 3?/3*, sa. const.: male Fischer rats, wt.: 95–140 g, contamination: artificial (dose: **600 µg AFB_1** (labeled)/kg b. wt., o., once; for detailed information please see the article), conc.: 20.40 pmol/**mg DNA**** *** (mean value), country: Canada[445], ***CMD-diet** (composition please see the article) for up to 3 weeks then AFB_1 on day 21, **after 2 h, ***total AFB_1-DNA adducts

incidence: 3?/3*, sa. const.: male Fischer rats, wt.: 95–140 g, contamination: artificial (dose: **100 µg AFB_1** (labeled)/kg b. wt., o., once; for detailed information please see the article), conc.: 146.72 pmol/**liver**** *** (mean value), country: Canada[445], ***CMS-diet** (composition please see the article) for up to 3 weeks then AFB_1 on day 21, **after 2 h, ***total AFB_1-DNA adducts

incidence: 3?/3*, sa. const.: male Fischer rats, wt.: 95–140 g, contamination: artificial (dose: **100 µg AFB_1** (labeled)/kg b. wt., o., once; for detailed information please see the article), conc.: 5.31 pmol/**liver**** *** (mean value), country: Canada[445], ***CMD-diet** (composition please see the article) for up to 3 weeks then AFB_1 on day 21, **after 2 h, ***total AFB_1-DNA adducts

incidence: 3?/3*, sa. const.: male Fischer rats, wt.: 95–140 g, contamination: artificial (dose: **600 µg AFB_1** (labeled)/kg b. wt., o., once; for detailed information please see the article), conc.: 1,260.60 pmol/**liver**** *** (mean value), country: Canada[445], ***CMS-diet** (composition please see the article) for up to 3 weeks then AFB_1 on day 21, **after 2 h, ***total AFB_1-DNA adducts

incidence: 3?/3*, sa. const.: male Fischer rats, wt.: 95–140 g, contamination: artificial (dose: **600 µg AFB_1** (labeled)/kg b. wt., o., once; for detailed information please see the article), conc.: 84.64 pmol/**liver**** *** (mean value), country: Canada[445], ***CMD-diet** (composition please see the article) for up to 3 weeks then AFB_1 on day 21, **after 2 h, ***total AFB_1-DNA adducts

incidence: ?/4–10*, sa. const.: male weanling Sprague-Dawley rats, contamination: artificial (dose: 1 mg AFB_1 (labeled and unlabeled)/kg, i.p., once), conc.: 19.2 ng/**mg DNA**** *** (mean value), country: USA[489], *fed **diet 1** = adequate composition (for detailed information please see the article), **covalent AF-adducts, ***after 6 h

incidence: ?/4–10*, sa. const.: male weanling Sprague-Dawley rats, contamination: artificial (dose: 1 mg AFB_1 (labeled and unlabeled)/kg, i.p., once), conc.: 15.6 ng/**mg DNA**** *** (mean value), country: USA[489], *fed **diet 2** = marginal deficiency of dietary lipotropes (for detailed information please see the article), **covalent AF-adducts, ***after 6 h

incidence: ?/4–10*, sa. const.: male weanling Sprague-Dawley rats, contamination: artificial (dose: 1 mg AFB_1 (labeled and unlabeled)/kg, i.p., once), conc.: 45.7 ng/**mg RNA**** *** (mean value), country: USA[489], *fed **diet 1** = adequate composition (for detailed information please see the article), **covalent AF-adducts, ***after 6 h

incidence: ?/4–10*, sa. const.: male weanling Sprague-Dawley rats, contamination: artificial (dose: 1 mg AFB_1 (labeled and unlabeled)/kg, i.p., once), conc.: 37.0 ng/**mg RNA**** *** (mean value), country: USA[489], *fed **diet 2** = marginal deficiency of dietary lipotropes (for detailed information please see the article), **covalent AF-adducts, ***after 6 h

incidence: ?/4–10*, sa. const.: male weanling Sprague-Dawley rats, contamination: artificial (dose: 1 mg AFB_1 (labeled and unlabeled)/kg, i.p., once), conc.: 3.3 ng/**mg protein**** *** (mean value), country: USA[489], *fed **diet 1** = adequate composition (for detailed information please see the article), **covalent AF-adducts, ***after 6 h

incidence: ?/4–10*, sa. const.: male weanling Sprague-Dawley rats, contamination: artificial (dose: 1 mg AFB_1 (labeled and unlabeled)/kg, i.p., once), conc.: 3.2 ng/**mg protein**** *** (mean value), country: USA[489], *fed **diet 2** = marginal deficiency of dietary lipotropes (for detailed information please see the article), **covalent AF-adducts, ***after 6 h

incidence: ?/4–6*, sa. const.: male Fischer 344 weanling rats, age: 20 days, wt.: 45–50 g, contamination: artificial (dose: 25 µg AFB_1 (labeled and unlabeled), i.p., for 5 days a week over 2 weeks), conc. range: ≈≤27 pmol AF/mg DNA** *** (mean value), country: USA[490], *control **semi-purified diet** 3 weeks prior to AFB_1-treatment, **after 2 h over 2 days and first and second dosing (for detailed information please see the article), ***AFB_1-DNA adducts

incidence: ?/4–6*, sa. const.: male Fischer 344 weanling rats, age: 20 days, wt.: 45–50 g, contamination: artificial (dose: 25 µg AFB_1 (labeled and unlabeled), i.p., for 5 days a week over 2 weeks), conc. range: ≈≤49 pmol AF/mg DNA** *** (mean value), country: USA[490], ***choline-deficient/methionine-low diet** 3 weeks prior to AFB_1-treatment, **after 2 h of the second day and third dosing (for detailed information please see the article), ***AFB_1-DNA adducts

incidence: 5?/5*, sa. const.: young male Fischer rats, wt.: 90–100 g, contamination: artificial (dose: 0.4 mg AFB_1 (labeled)/kg b. wt., i.p., once), conc.: 194 AFB_1-DNA binding pmol/mg DNA** (mean value), country: USA[492], *control (no pretreatment), **2 h after AFB_1-administration (receiving control diet for 2 weeks)

incidence: 5?/5, sa. const.: young male Fischer rats, wt.: 90–100 g, contamination: artificial (dose: 0.4 mg AFB_1 (labeled)/kg b. wt., i.p., once), conc.: 154 AFB_1-DNA binding pmol/mg DNA* ** (mean value), country: USA[492], *pretreated with **green tea** (0.5%) in drinking water for **2 weeks** prior to AFB_1-treatment, **2 h after AFB_1-administration

incidence: 5?/5, sa. const.: young male Fischer rats, wt.: 90–100 g, contamination: artificial (dose: 0.4 mg AFB_1 (labeled)/kg b. wt., i.p., once), conc.: 280 AFB_1-DNA binding pmol/mg DNA* ** (mean value), country: USA[492], *control (no pretreatment), **2 h after AFB_1-administration (receiving control diet for 4 weeks)

incidence: 5?/5, sa. const.: young male Fischer rats, wt.: 90–100 g, contamination: artificial (dose: 0.4 mg AFB_1 (labeled)/kg b. wt., i.p., once), conc.: 205 AFB_1-DNA binding pmol/mg DNA* ** (mean value), country: USA[492], *pretreated with **green tea** (0.5%) in drinking water for **4 weeks** prior to AFB_1-treatment, **2 h after AFB_1-administration

incidence: ?/4*, sa. const.: young male F344 rats, wt.: 120–140 g, contamination: artificial (dose: 250 µg AFB_1/kg b. wt., daily for 3 weeks), conc. range: ≤8.3 ng/mg DNA** *** (mean value), country: USA/People's Republic of China[493], *control (for overall information please see the article), **AFB_1-N^7-Gua adducts, ***after 3 days while treatment (also at other day intervals up to 15 days measured, lowest conc.: ≈2 ng/mg DNA after 9 days)

incidence: ?/4, sa. const.: young male F344 rats, wt.: 120–140 g, contamination: artificial (dose: 250 µg AFB_1/kg b. wt., daily for 3 weeks), conc.: ≈3.9 ng/mg DNA* ** *** **** (mean value), country: USA/People's Republic of China[493], ***lycopene-treatment** (100 mg/kg b. wt.) after each AFB_1-administration (for overall information please see the article), **AFB_1-N^7-Gua adducts, ***after 3 days while treatment (also at other day intervals up to 15 days measured, lowest conc.: ≈1.7 ng/mg DNA after 9 days), ****1 higher value (≈4.3 ng/mg DNA) on day 15 recorded

incidence: ?/6–8*, sa. const.: male Fischer 344 rats, age: 2 months, contamination: artificial (dose: 600 µg AFB_1 (labeled)/kg, i.p., once), conc.: 6.4 nmol AFB_1 (total) **bound to protein**** (mean value), country: USA[548], *control, **after 3 h

incidence: ?/6–8*, sa. const.: male Fischer 344 rats, age: 2 months, contamination: artificial (dose: 600 µg AFB_1 (labeled)/kg, i.p., once), conc.: 11.8 nmol AFB_1 (total) **bound to protein**** (mean value), country: USA[548], ***DHEA-diet** (0.8%) for 2 weeks prior to AFB_1-treatment, **after 3 h
incidence: ?/6–8*, sa. const.: male Fischer 344 rats, age: 2 months, contamination: artificial (dose: 600 µg AFB_1 (labeled)/kg, i.p., once), conc.: 116 pmol AFB_1 bound/**mg DNA**** (mean value), country: USA[548], *control, **after 3 h
incidence: ?/6–8*, sa. const.: male Fischer 344 rats, age: 2 months, contamination: artificial (dose: 600 µg AFB_1 (labeled)/kg, i.p., once), conc.: 40 pmol AFB_1 bound/**mg DNA**** (mean value), country: USA[548], ***DHEA-diet** (0.8%) for 2 weeks prior to AFB_1-treatment, **after 3 h
incidence: ?/6–8*, sa. const.: male Fischer 344 rats, age: 2 months, contamination: artificial (dose: 600 µg AFB_1 (labeled)/kg, i.p., once), conc.: 1,707 pmol AFB_1 (total) **bound to DNA**** (mean value), country: USA[548], *control, **after 3 h
incidence: ?/6–8*, sa. const.: male Fischer 344 rats, age: 2 months, contamination: artificial (dose: 600 µg AFB_1 (labeled)/kg, i.p., once), conc.: 773 pmol AFB_1 (total) **bound to DNA**** (mean value), country: USA[548], ***DHEA-diet** (0.8%) for 2 weeks prior to AFB_1-treatment, **after 3 h
incidence: ?/6–8*, sa. const.: male Fischer 344 rats, age: 2 months, contamination: artificial (dose: 600 µg AFB_1 (labeled)/kg, i.p., once), conc.: 54 pmol **AFB_1-N^7-Gua²/mg DNA**** (mean value), country: USA[548], *control, **after 3 h
incidence: ?/6–8*, sa. const.: male Fischer 344 rats, age: 2 months, contamination: artificial (dose: 600 µg AFB_1 (labeled)/kg, i.p., once), conc.: 17 pmol **AFB_1-N^7-Gua²/mg DNA**** (mean value), country: USA[548], ***DHEA-diet** (0.8%) for 2 weeks prior to AFB_1-treatment, **after 3 h
incidence: ?/6–8*, sa. const.: male Fischer 344 rats, age: 2 months, contamination: artificial (dose: 600 µg AFB_1 (labeled)/kg, i.p. once), conc.: 44 pmol **other AFB_1 adducts/mg DNA**** (mean value), country: USA[548], *control, **after 3 h
incidence: ?/6–8*, sa. const.: male Fischer 344 rats, age: 2 months, contamination: artificial (dose: 600 µg AFB_1 (labeled)/kg, i.p., once), conc.: 18 pmol **other AFB_1 adducts/mg DNA**** (mean value), country: USA[548], ***DHEA-diet** (0.8%) for 2 weeks prior to AFB_1-treatment, **after 3 h

incidence: ?/4*, sa. const.: male SPF Wistar rats, age: 24–27 days, contamination: artificial (dose: 2 mg AFB_1 (labeled)/kg b. wt., i.p., once), conc.: ≈245 pmol AFB_1/mg DNA** (mean value), country: France[549], *control, **after 2 h (for overall information please see the article)
incidence: ?/4*, sa. const.: male SPF Wistar rats, age: 24–27 days, contamination: artificial (dose: 2 mg AFB_1 (labeled)/kg b. wt., i.p., once), conc.: ≈110 pmol AFB_1/mg DNA** (mean value), country: France[549], ***3-MC** i.p. (20 mg/kg b. wt.) on the 3 days preceding sacrifice (3-MC was injected prior to AFB_1-administration), **after 2 h (for overall information please see the article)
incidence: ?/4*, sa. const.: male SPF Wistar rats, age: 24–27 days, contamination: artificial (dose: 2 mg AFB_1 (labeled)/kg b. wt., i.p., once), conc.: ≈125 pmol AFB_1/mg DNA** (mean value), country: France[549], ***CX-diet** (300 mg/kg) for 2 weeks prior to AFB_1-administration, **after 2 h (for overall information please see the article)
incidence: ?/4*, sa. const.: male SPF Wistar rats, age: 24–27 days, contamination: artificial (dose: 2 mg AFB_1 (labeled)/kg b. wt., i.p., once), conc.: ≈100 pmol AFB_1/mg DNA** (mean value), country: France[549], ***AC-diet** (300 mg/kg) for 2 weeks prior to AFB_1-administration, **after 2 h (for overall information please see the article)
incidence: ?/4*, sa. const.: male SPF Wistar rats, age: 24–27 days, contamination: artificial (dose: 2 mg AFB_1 (labeled)/kg b.

wt., i.p., once), conc.: ≈290 pmol AFB_1/mg DNA** (mean value), country: France[549], ***BC-diet** (300 mg/kg) for 2 weeks prior to AFB_1-administration, **after 2 h (for overall information please see the article)

incidence: ?/2*, sa. const.: male Sprague-Dawley rats, wt.: 125–150 g, contamination: artificial (dose: 5 μg AFB_1 (labeled)/kg, **i.p.,** once), conc. range: ≈≤6.1 pmol AFB_1 bound/mg DNA** (mean value), country: USA[550], ***basal diet** for the experimental period (2 weeks) prior to AFB_1-treatment (for detailed information please see the article), **after 2 h (also measured after 6, 12, 24, and 48 h, lowest conc.: ≈1 pmol AFB_1 bound/mg DNA after 24 h)
incidence: ?/2*, sa. const.: male Sprague-Dawley rats, wt.: 125–150 g, contamination: artificial (dose: 5 μg AFB_1 (labeled)/kg, **i.p.,** once), conc. range: ≈≤2.8 pmol AFB_1 bound/mg DNA** (mean value), country: USA[550], ***brussel sprouts** (25% dry wt. of the diet) for 2 weeks prior to AFB_1-treatment (for detailed information please see the article), **after 2 h (also measured after 6, 12, 24, and 48 h, lowest conc.: ≈0.2 AFB_1 pmol bound/mg DNA after 48 h)
incidence: ?/2*, sa. const.: male Sprague-Dawley rats, wt.: 125–150 g, contamination: artificial (dose: 5 μg AFB_1 (labeled)/kg, **i.g.** by gavage, once), conc. range: ≈≤5.7 pmol AFB_1 bound/mg DNA** (mean value), country: USA[550], ***basal diet** for the experimental period (2 weeks) prior to AFB_1-treatment (for detailed information please see the article), **after 6 h (also measured after 2, 12, 24, and 48 h, lowest conc.: ≈0.5 AFB_1 pmol bound/mg DNA after 48 h)
incidence: 1/1*, sa. const.: male Sprague-Dawley rats, wt.: 125–150 g, contamination: artificial (dose: 5 μg AFB_1 (labeled)/kg, **i.g.** by gavage, once), conc. range: ≈≤2.95 pmol AFB_1 bound/mg DNA(mean value), country: USA[550], ***brussel sprouts** (25% dry wt. of the diet) for 2 weeks prior to AFB_1-treatment (for detailed information please see the article), **after 2 h (also measured after 6, 12, 24, and 48 h, lowest conc.: ≈0.2 AFB_1 pmol bound/mg DNA after 48 h)

incidence: ?/3*, sa. const.: male Sprague-Dawley rats, wt.: 125–150 g, contamination: artificial (dose: 3 μg AFB_1 (labeled)/kg, **i.p.,** once), conc.: 1.50 pmol AFB_1 bound/mg hepatic DNA ** (mean value), country: USA[550], ***basal diet** for the experimental period (2 weeks) prior to AFB_1-treatment (for detailed information please see the article), **after 2 h
incidence: ?/3*, sa. const.: male Sprague-Dawley rats, wt.: 125–150 g, contamination: artificial (dose: 3 μg AFB_1 (labeled)/kg, **i.p.,** once), conc.: 0.59 pmol AFB_1 bound/mg hepatic DNA** (mean value), country: USA[550], ***brussel sprouts** (25% dry wt. of the diet) for 2 weeks prior to AFB_1-treatment (for detailed information please see the article), **after 2 h
incidence: ?/3*, sa. const.: male Sprague-Dawley rats, wt.: 125–150 g, contamination: artificial (dose: 3 μg AFB_1 (labeled)/kg, **i.p.,** once), conc.: 1.11 pmol AFB_1 bound/mg hepatic DNA** (mean value), country: USA[550], ***I3C-diet** (250 ppm) for 2 weeks prior to AFB_1-treatment (for detailed information please see the article), **after 2 h
incidence: ?/3*, sa. const.: male Sprague-Dawley rats, wt.: 125–150 g, contamination: artificial (dose: 3 μg AFB_1 (labeled)/kg, **i.p.,** once), conc.: 0.83 pmol AFB_1 bound/mg hepatic DNA** (mean value), country: USA[550], *basal diet + **PB** (0.1%) in drinking water for 7 days prior to AFB_1-treatment (for detailed information please see the article), **after 2 h

incidence: ?/3*, sa. const.: male Sprague-Dawley rats, wt.: 125–150 g, contamination: artificial (dose: 3 μg AFB_1 (labeled)/kg, **i.g.** by gavage, once), conc.:

0.98 pmol AFB_1 bound/mg hepatic DNA** (mean value), country: USA[550], ***basal diet** for the experimental period (2 weeks) prior to AFB_1-treatment (for detailed information please see the article), **after 3 h
incidence: ?/3*, sa. const.: male Sprague-Dawley rats, wt.: 125–150 g, contamination: artificial (dose: 3 μg AFB_1 (labeled)/kg, **i.g.** by gavage, once), conc.: 0.45 pmol AFB_1 bound/mg hepatic DNA** (mean value), country: USA[550], ***brussel sprouts** (25% dry wt. of the diet) for 2 weeks prior to AFB_1-treatment (for detailed information please see the article), **after 3 h
incidence: ?/3*, sa. const.: male Sprague-Dawley rats, wt.: 125–150 g, contamination: artificial (dose: 3 μg AFB_1 (labeled)/kg, **i.g.** by gavage, once), conc.: 0.82 pmol AFB_1 bound/mg hepatic DNA** (mean value), country: USA[550], ***I3C-diet** (250 ppm) for 2 weeks prior to AFB_1-treatment (for detailed information please see the article), **after 3 h
incidence: ?/3*, sa. const.: male Sprague-Dawley rats, wt.: 125–150 g, contamination: artificial (dose: 3 μg AFB_1 (labeled)/kg, **i.g.** by gavage, once), conc.: 0.37 pmol AFB_1 bound/mg hepatic DNA** (mean value), country: USA[550], *basal diet + **PB** (0.1%) in drinking water for 7 days prior to AFB_1-treatment (for detailed information please see the article), **after 3 h

incidence: 2?/2, sa. const.: male F344 rats, wt.: 75–100 g, contamination: artificial (dose: 250 μg AFB_1 (labeled)/kg b. wt., by gavage, 10 times at days 0–4 and 7–11), conc. range: ≤14.4 pmol/mg DNA* ** *** (mean value), country: USA[551], *control, **AFB-*N*⁷-FAPyr (minor), ***at day 3 of AFB_1-administration (also at other day intervals up to 133 days measured, lowest conc.: nd after 77, 106, and 133 days) (for detailed information please see the article)
incidence: 2?/2, sa. const.: male F344 rats, wt.: 75–100 g, contamination: artificial (dose: 250 μg AFB_1 (labeled)/kg b. wt., by gavage, 10 times at days 0–4 and 7–11), conc.: 2.0 pmol/mg DNA* ** *** **** (mean value), country: USA[551], ***EQ-diet** (0.4%) for 3 weeks while AFB_1 was gavaged in the last 2 weeks, **AFB-N^7-FAPyr (minor), ***at day 3 of AFB_1-administration (also at other day intervals up to 133 days measured, lowest conc.: nd after 49, 77, 106, and 133 days), ****several higher values up to 2.6 pmol/mg DNA were recorded

incidence: 2?/2, sa. const.: male F344 rats, wt.: 75–100 g, contamination: artificial (dose: 250 μg AFB_1 (labeled)/kg b. wt., by gavage, 10 times at days 0–4 and 7–11), conc. range: ≤57.8 pmol/mg DNA* ** *** (mean value), country: USA[551], *control, **AFB-N^7-FAPyr (major), ***at day 3 of AFB_1-administration (also at other day intervals up to 133 days measured, lowest conc.: 0.1 pmol/mg DNA after 133 days) (for detailed information please see the article)
incidence: 2?/2, sa. const.: male F344 rats, wt.: 75–100 g, contamination: artificial (dose: 250 μg AFB_1 (labeled)/kg b. wt., by gavage, 10 times at days 0–4 and 7–11), conc.: 7.8 pmol/mg DNA* ** *** (mean value), country: USA[551], ***EQ-diet** (0.4%) for 3 weeks while AFB_1 was gavaged in the last 2 weeks, **AFB-N^7-FAPyr (major), ***at day 3 of AFB_1-administration (also at other day intervals up to 133 days measured, lowest conc.: 0.1 pmol/mg DNA; ****several higher values up to 10.4 pmol/mg DNA are recorded (for detailed information please see the article)
incidence: 2?/2, sa. const.: male F344 rats, wt.: 75–100 g, contamination: artificial (dose: 250 μg AFB_1 (labeled)/kg b. wt., by gavage, 10 times at days 0–4 and 7–11), conc. range: ≤105.2 pmol/mg DNA* ** *** (mean value), country: USA[551], *control, **AFB-N^7-Gua[3], ***at day 1 of AFB_1-administration (also at other day intervals up to 133 days measured, lowest

conc.: nd after 49, 77, 106, and 133 days) (for detailed information please see the article)
incidence: 2?/2, sa. const.: male F344 rats, wt.: 75–100 g, contamination: artificial (dose: 250 μg AFB_1 (labeled)/kg b. wt., by gavage, 10 times at days 0–4 and 7–11), conc. range: ≤19.0 pmol/mg DNA* ** *** (mean value), country: USA[551], ***EQ-diet** (0.4%) for 3 weeks while AFB_1 was gavaged in the last 2 weeks, **AFB-*N*[7]-Gua[3], ***at day 1 of AFB_1-administration (also at other day intervals up to 133 days measured, lowest conc.: nd after 49, 77, 106, and 133 days) (for detailed information please see the article)

incidence: 4?/4?, sa. const.: male Sprague-Dawley rats, wt.: 180–190 g, contamination: artificial (dose: 40 μg AFB_1 (unlabeled and labeled)/g b. wt., i.p., once; for detailed information please see the article), conc.: 29.8 pmol/mg DNA* ** *** (mean value), country: USA[552], *control, **AFB_1-DNA adducts, ***after 2 h
incidence: 4?/4?, sa. const.: male Sprague-Dawley rats, wt.: 180–190 g, contamination: artificial (dose: 40 μg AFB_1 (unlabeled and labeled)/g b. wt., i.p., once; for detailed information please see the article), conc.: 8.6 pmol/mg DNA* ** *** (mean value), country: USA[552], ***PB-treatment** (0.1%, in drinking water) for 1 week, **AFB_1-DNA adducts, ***after 2 h

incidence: ?/8*, sa. const.: male Fischer rats, wt.: 80–100 g, contamination: artificial (dose: 40 μg AFB_1 (labeled)/100 g b. wt., i.p., once; for detailed information please see the article), conc.: 28.9 AFB_1-DNA binding pmol/mg DNA** (mean value), country: USA[556], *control, **after 2 h
incidence: ?/8, sa. const.: male Fischer rats, wt.: 80–100 g, contamination: artificial (dose: 40 μg AFB_1 (labeled)/100 g b. wt., i.p., once and **pretreatment** with $_L$**-BSO** (conc.: 4 mmol, 4 and 2 h before AFB_1-injection; for detailed information please see the article), conc.: 41.2 AFB_1-DNA binding pmol/mg DNA* (mean value), country: USA[556], *after 2 h
incidence: ?/8, sa. const.: male Fischer rats, wt.: 80–100 g, contamination: artificial (dose: 40 μg AFB_1 (labeled)/100 g b. wt., i.p., once and **pretreatment** with **DEM** (conc.: 3.5 mmol, 4 h before AFB_1-injection) + $_L$**-BSO** (conc.: 4 mmol, 2 h before AFB_1-injection); for detailed information please see the article), conc.: 51.9 AFB_1-DNA binding pmol/mg DNA* (mean value), country: USA[556], *after 2 h

incidence: 2?/2*, sa. const.: male Fischer rats, wt.: 60–100 g, contamination: artificial (dose: 1 mg AFB_1 (labeled)/kg, i.p., once; for detailed information please see the article), conc.: 126.4 pg AFB_1/μg DNA* ** (mean value), country: UK/Scotland, UK[563], *control: **EQ-diet** (0.5%, v/w) for **0 days** prior to AFB_1-treatment, **after 2 h
incidence: 2?/2*, sa. const.: male Fischer rats, wt.: 60–100 g, contamination: artificial (dose: 1 mg AFB_1 (labeled)/kg, i.p., once; for detailed information please see the article), conc.: 29.5 pg AFB_1/μg DNA* ** (mean value), country: UK/Scotland, UK[563], ***EQ-diet** (0.5%, v/w) for **2 days** prior to AFB_1-treatment, **after 2 h
incidence: 2?/2*, sa. const.: male Fischer rats, wt.: 60–100 g, contamination: artificial (dose: 1 mg AFB_1 (labeled)/kg, i.p., once; for detailed information please see the article), conc.: 3.9 pg AFB_1/μg DNA* ** (mean value), country: UK/Scotland, UK[563], ***EQ-diet** (0.5%, (v/w) for **14 days** prior to AFB_1-treatment, **after 2 h

incidence: 3?/3, sa. const.: male Fischer rats, age: 4 weeks, contamination: artificial (dose: 0.5 mg AFB_1 (labeled)/kg, i.p., once; for detailed information please see the article), conc.: ≈17 pmol AFB_1 *exo*-epoxide GSH/mg protein* ** (mean value), country: USA/Scotland, UK[565], *control, **after 2 h

incidence: 3?/3, sa. const.: male Fischer rats, age: 4 weeks, contamination: artificial (dose: 0.5 mg AFB_1 (labeled)/kg, i.p., once; for detailed information please see the article), conc.: ≈71 pmol AFB_1 *exo*-epoxide GSH/mg protein* ** (mean value), country: USA/Scotland, UK[565], ***I3C-diet** (0.2%) for 1 week prior to AFB_1-treatment, **after 2 h
incidence: 3?/3, sa. const.: male Fischer rats, age: 4 weeks, contamination: artificial (dose: 0.5 mg AFB_1 (labeled)/kg, i.p., once; for detailed information please see the article), conc.: ≈24 pmol AFB_1 *exo*-epoxide GSH/mg protein* ** (mean value), country: USA/Scotland, UK[565], ***BNF-diet** (0.04%) for 1 week prior to AFB_1-treatment, **after 2 h
incidence: 3?/3, sa. const.: male Fischer rats, age: 4 weeks, contamination: artificial (dose: 0.5 mg AFB_1 (labeled)/kg, i.p., once; for detailed information please see the article), conc.: ≈58 pmol AFB_1 *exo*-epoxide GSH/mg protein* ** (mean value), country: USA/Scotland, UK[565], ***I3C-** (0.2%) **+ BNF-diet** (0.04%) for 1 week prior to AFB_1-treatment, **after 2 h

incidence: 3?/3, sa. const.: male Fischer rats, age: 4 weeks, contamination: artificial (dose: 0.5 mg AFB_1 (labeled)/kg, i.p., once; for detailed information please see the article), conc.: ≈59 pmol AFB_1 *endo*-epoxide GSH/mg protein* ** (mean value), country: USA/Scotland, UK[565], *control, **after 2 h
incidence: 3?/3, sa. const.: male Fischer rats, age: 4 weeks, contamination: artificial (dose: 0.5 mg AFB_1 (labeled)/kg, i.p., once; for detailed information please see the article), conc.: ≈145 pmol AFB_1 *endo*-epoxide GSH/mg protein* ** (mean value), country: USA/Scotland, UK[565], ***I3C-diet** (0.2%) for 1 week prior to AFB_1-treatment, **after 2 h
incidence: 3?/3, sa. const.: male Fischer rats, age: 4 weeks, contamination: artificial (dose: 0.5 mg AFB_1 (labeled)/kg, i.p., once; for detailed information please see the article), conc.: ≈44 pmol AFB_1 *endo*-epoxide GSH/mg protein* ** (mean value), country: USA/Scotland, UK[565], ***BNF-diet** (0.04%) for 1 week prior to AFB_1-treatment, **after 2 h
incidence: 3?/3, sa. const.: male Fischer rats, age: 4 weeks, contamination: artificial (dose: 0.5 mg AFB_1 (labeled)/kg, i.p., once; for detailed information please see the article), conc.: ≈168 pmol AFB_1 *endo*-epoxide GSH/mg protein* ** (mean value), country: USA/Scotland, UK[565], ***I3C-** (0.2%) **+ BNF-diet** (0.04%) for 1 week prior to AFB_1-treatment, **after 2 h

incidence: 3?/3, sa. const.: male Fischer rats, age: 4 weeks, contamination: artificial (dose: 0.5 mg AFB_1 (labeled)/kg, i.p., once), conc.: ≈53 pmol AFB_1 eq/mg DNA* ** *** (mean value), country: USA/Scotland, UK[565], *control, **[^{3}H]AFB_1-DNA adducts, ***after 2 h
incidence: 3?/3, sa. const.: male Fischer rats, age: 4 weeks, contamination: artificial (dose: 0.5 mg AFB_1 (labeled)/kg, i.p., once), conc.: ≈18 pmol AFB_1 eq/mg DNA* ** *** (mean value), country: USA/Scotland, UK[565], ***I3C-diet** (0.2%) for 1 week prior to AFB_1-treatment, **[^{3}H]AFB_1-DNA adducts, ***after 2 h
incidence: 3?/3, sa. const.: male Fischer rats, age: 4 weeks, contamination: artificial (dose: 0.5 mg AFB_1 (labeled)/kg, i.p., once), conc.: ≈28 pmol AFB_1 eq/mg DNA* ** *** (mean value), country: USA/Scotland, UK[565], ***BNF-diet** (0.04%) for 1 week prior to AFB_1-treatment, **[^{3}H]AFB_1-DNA adducts, ***after 2 h
incidence: 3?/3, sa. const.: male Fischer rats, age: 4 weeks, contamination: artificial (dose: 0.5 mg AFB_1 (labeled)/kg, i.p., once), conc.: ≈24 pmol AFB_1 eq/mg DNA* ** *** (mean value), country: USA/Scotland, UK[565], ***I3C-** (0.2%) **+ BNF-diet** (0.04%) for 1 week prior to AFB_1-treatment, **[^{3}H]AFB_1-DNA adducts, ***after 2 h

incidence: 3?/3, sa. const.: male Fischer rats, age: 4 weeks, contamination: artificial (dose: 0.5 mg AFB_1 (labeled)/kg, i.p., once), conc.: ≈85 nmol AFB_1 eq/g liver* ** (mean value), country: USA/Scotland, UK[565], *control, **after 2 h
incidence: 3?/3, sa. const.: male Fischer rats, age: 4 weeks, contamination: artificial (dose: 0.5 mg AFB_1 (labeled)/kg, i.p., once), conc.: ≈35 nmol AFB_1 eq/g liver* ** (mean value), country: USA/Scotland, UK[565], ***I3C-diet** (0.2%) for 1 week prior to AFB_1-treatment, **after 2 h
incidence: 3?/3, sa. const.: male Fischer rats, age: 4 weeks, contamination: artificial (dose: 0.5 mg AFB_1 (labeled)/kg, i.p., once), conc.: ≈39 nmol AFB_1 eq/g liver* ** (mean value), country: USA/Scotland, UK[565], ***BNF-diet** (0.04%) for 1 week prior to AFB_1-treatment, **after 2 h
incidence: 3?/3, sa. const.: male Fischer rats, age: 4 weeks, contamination: artificial (dose: 0.5 mg AFB_1 (labeled)/kg, i.p., once), conc.: ≈40 nmol AFB_1 eq/g liver* **(mean value), country: USA/Scotland, UK[565], ***I3C-** (0.2%) + **BNF-diet** (0.04%) for 1 week prior to AFB_1-treatment, **after 2 h

incidence: 3/3, sa. const.: neonatal rats of albino rats (wt.: 200–250 g) of Wistar strain, contamination: artificial (dose: 20 μg AFB_1 (labeled)/100 g b. wt. into the mother rat, i.p., once (13 days after giving birth), conc.: nd* **, country: India[594], *AFB_1-DNA adducts, **after 48 h while taking milk from their mother
incidence: 3?/3, sa. const.: neonatal rats of albino rats (wt.: 200–250 g) of Wistar strain, contamination: artificial (dose: 20 μg AFB_1 (labeled)/100 g b. wt. into the mother rat, i.p., once (13 days after giving birth), conc.: nd* ** ***, country: India[594], *AFB_1-DNA adducts, **after 48 h while taking milk from their mother, *****PB** (80 mg/kg b. wt.) injected 3 days prior to AFB_1-administration
incidence: 3?/3, sa. const.: neonatal rats of albino rats (wt.: 200–250 g) of Wistar strain, contamination: artificial (dose: 20 μg AFB_1 (labeled)/100 g b. wt. into the mother rat, i.p., once (13 days after giving birth), conc.: 40 pmol AFB_1**/mg RNA*** (mean value), country: India[594], *after 48 h while taking milk from their mother
incidence: 3?/3, sa. const.: neonatal rats of albino rats (wt.: 200–250 g) of Wistar strain, contamination: artificial (dose: 20 μg AFB_1 (labeled)/100 g b. wt. into the mother rat, i.p., once (13 days after giving birth), conc.: 10 pmol AFB_1**/mg RNA*** ** (mean value), country: India[594], *after 48 h while taking milk from their mother, ****PB** (80 mg/kg b. wt.) injected 3 days prior to AFB_1-administration
incidence: 3?/3, sa. const.: neonatal rats of albino rats (wt.: 200–250 g) of Wistar strain, contamination: artificial (dose: 20 μg AFB_1 (labeled)/100 g b. wt. into the mother rat, i.p., once (13 days after giving birth), conc.: 52 pmol AFB_1**/mg protein*** (mean value), country: India[594], *after 48 h while taking milk from their mother
incidence: 3?/3, sa. const.: neonatal rats of albino rats (wt.: 200–250 g) of Wistar strain, contamination: artificial (dose: 20 μg AFB_1 (labeled)/100 g b. wt. into the mother rat, i.p., once (13 days after giving birth), conc.: 23 pmol AFB_1**/mg protein*** ** (mean value), country: India[594], *after 48 h while taking milk from their mother, ****PB** (80 mg/kg b. wt.) injected 3 days prior to AFB_1-administration

incidence: 3?/3, sa. const.: male Wistar rats, wt.: 200–250 g, contamination: artificial (dose: 40 μg AFB_1 (labeled)/100 g, i.p., once), conc. range: ≈≤44.5 ng AFB_1 bound/**mg rRNA*** (mean value), country: UK[608], *after 6 h (also measured after 2, 24, and 48 h, lowest conc.: ≈10 ng AFB_1 bound/mg rRNA after 48 h)
incidence: 3?/3, sa. const.: male Wistar rats, wt.: 200–250 g, contamination: artificial (dose: 40 μg AFB_1 (labeled)/100 g, i.p., once, conc. range: ≈≤23 ng AFB_1

bound/**mg DNA*** (mean value), country: UK[608], *after 2 h (also measured after 6, 24, and 48 h, lowest conc.: ≈1 ng AFB_1 bound/mg DNA after 48 h)
incidence: 3?/3, sa. const.: male Wistar rats, wt.: 200–250 g, contamination: artificial (dose: 40 μg AFB_1 (labeled)/100 g, i.p., once, conc. range: ≈≤4.5 ng AFB_1 bound/**mg protein*** (mean value), country: UK[608], *after 6 h (also measured after 2, 24, and 48 h, lowest conc.: ≈1 ng AFB_1 bound/mg protein after 48 h)

incidence: 2?/2, sa. const.: male CDF Fischer rats, wt.: 100–150 g, contamination: artificial (dose: 1 mg AFB_1 (labeled)/kg, i.p., once, conc. range: ≈≤13.5 nmol AFB_1 residues/g liver* ** (mean value), country: USA[610], *after ≈1 h (also measured after ≈2, 12, and 36 h), **in **liver homogenate**
incidence: 2?/2, sa. const.: male CDF Fischer rats, wt.: 100–150 g, contamination: artificial (dose: 1 mg AFB_1 (labeled)/kg, i.p., once, conc. range: ≈≤0.95 nmol AFB_1 residues/g liver* ** (mean value), country: USA[610], *after ≈1 h (also measured after ≈2, 12, and 36 h), **in **liver nuclei**

incidence: 2?/2, sa. const.: male CDF Fischer rats, wt.: 100–150 g, contamination: artificial (dose: 1 mg AFB_1 (labeled)/kg, i.p., once, conc. range: ≈≤390 pmol AFB_1 residues/**mg DNA*** (mean value), country: USA[610], *after ≈0 h (also measured after ≈2, 12, and 36 h)
incidence: 2?/2, sa. const.: male CDF Fischer rats, wt.: 100–150 g, contamination: artificial (dose: 1 mg AFB_1 (labeled)/kg, i.p., once, conc. range: ≈≤30 pmol AFB_1 residues/**mg chromatin protein*** (mean value), country: USA[610], *after ≈0 h (also measured after ≈2, 12, and 36 h)

incidence: 2?/2*, sa. const.: male F344 rats, wt.: 75–100 g, contamination: artificial (dose: 250 μg AFB_1 (labeled)/kg b. wt., by gavage, for 5 days a week for 2 weeks (days 0–4 and 7–11); for detailed information please see the article), conc. range: ≈≤51 pmol AFB-*N*[7]-Gua[3] adducts mg DNA** (mean value), country: USA[611], *control, *2 days after AFB_1-administration (also at other day intervals up to 16 days measured, lowest conc.: ≈11 pmol AFB-*N*[7]-Gua[3] adducts mg DNA after 10 days)
incidence: 2?/2, sa. const.: male F344 rats, wt.: 75–100 g, contamination: artificial (dose: 250 μg AFB_1 (labeled)/kg b. wt., by gavage, for 5 days a week for 2 weeks (days 0–4 and 7–11); for detailed information please see the article), conc. range: ≈≤9.9 pmol AFB-*N*[7]-Gua[3] adducts mg DNA* ** (mean value), country: USA[611], *2 days after AFB_1-administration (also at other day intervals up to 16 days measured, lowest conc.: ≈0.3 pmol AFB-*N*[7]-Gua[3] adducts mg DNA after 12 days; control values always higher than 1,2-dithiole-3-thione values), **__1,2-dithiole-3-thione-diet__ (0.03%) 1 week before and while AFB_1-treatment

incidence: 3?/3, sa. const.: male Sprague-Dawley rats, wt.: 150–260 g, contamination: artificial (dose: 0.6 mg AFB_1 (labeled)/kg b. wt., i.p., once; for detailed information please see the article), conc. range: ≈≤105 pmol AFB_1/mg DNA* ** (mean value), country: Germany[612], *after 20 min (also measured after 24 and 72 h, lowest conc.: 55 pmol AFB_1/mg DNA after 72 h), **in **PC**
incidence: 3?/3, sa. const.: male Sprague-Dawley rats, wt.: 150–260 g, contamination: artificial (dose: 0.6 mg AFB_1 (labeled)/kg b. wt., i.p., once; for detailed information please see the article), conc. range: ≈≤25 pmol AFB_1/mg DNA* ** (mean value), country: Germany[612], *after 20 min (also measured after 24 and 72 h, lowest conc.: nd? after 24 h), **in **NPC**

incidence: 3?/3*, sa. const.: weanling Fischer 344 rats, wt.: 50 g, contamination: artificial (dose: control diet for 41 weeks then ≈5 μg AFB_1 (labeled)/kg b. wt., o. by gavage, once; for detailed information please see the article), conc.: ≈3.65 pmol [^{14}C]AFB_1**/mg DNA**** (mean value), country: USA[615], *control, **after 6 h of receiving AFB_1 (labeled)

incidence: 3?/3, sa. const.: weanling Fischer 344 rats, wt.: 50 g, contamination: artificial (dose: **0.5 ppb** AFM_1, o., for 41 weeks then ≈5 μg AFB_1 (labeled)/kg b. wt., o. by gavage, once; for detailed information please see the article), conc.: ≈2.3 pmol [^{14}C]AFB_1**/mg DNA*** (mean value), country: USA[615], *after 6 h of receiving AFB_1 (labeled)
incidence: 3?/3, sa. const.: weanling Fischer 344 rats, wt.: 50 g, contamination: artificial (dose: **50 ppb** AFM_1, o., for 41 weeks then ≈5 μg AFB_1 (labeled)/kg b. wt., o. by gavage, once; for detailed information please see the article), conc.: ≈2.25 pmol [^{14}C]AFB_1**/mg DNA*** (mean value), country: USA[615], *after 6 h of receiving AFB_1 (labeled)
incidence: 3?/3, sa. const.: weanling Fischer 344 rats, wt.: 50 g, contamination: artificial (dose: **50 ppb** AFB_1, o., for 41 weeks then ≈5 μg AFB_1 (labeled)/kg b. wt., o. by gavage, once; for detailed information please see the article), conc.: ≈1 pmol [^{14}C]AFB_1**/mg DNA*** (mean value), country: USA[615], *after 6 h of receiving AFB_1 (labeled)

incidence: 3?/3*, sa. const.: weanling Fischer 344 rats, wt.: 50 g, contamination: artificial (dose: control diet for 41 weeks then ≈5 μg [^{14}C]AFB_1/kg b. wt., o. by gavage, once; for detailed information please see the article), conc.: ≈4.1 pmol [^{14}C]AFB_1**/mg RNA**** (mean value), country: USA[615], *control, **after 6 h of receiving AFB_1 (labeled)
incidence: 3?/3, sa. const.: weanling Fischer 344 rats, wt.: 50 g, contamination: artificial (dose: **0.5 ppb** AFM_1, o., for 41 weeks then ≈5 μg AFB_1 (labeled)/kg b. wt., o. by gavage, once; for detailed information please see the article), conc.: ≈2.35 pmol [^{14}C]AFB_1**/mg RNA*** (mean value), country: USA[615], *after 6 h of receiving AFB_1 (labeled)
incidence: 3?/3, sa. const.: weanling Fischer 344 rats, wt.: 50 g, contamination: artificial (dose: **50 ppb** AFM_1, o., for 41 weeks then ≈5 μg AFB_1 (labeled)/kg b. wt., o. by gavage, once; for detailed information please see the article), conc.: ≈3.2 pmol [^{14}C]AFB_1**/mg RNA*** (mean value), country: USA[615], *after 6 h of receiving AFB_1 (labeled)
incidence: 3?/3, sa. const.: weanling Fischer 344 rats, wt.: 50 g, contamination: artificial (dose: **50 ppb** AFB_1, o., for 41 weeks then ≈5 μg AFB_1 (labeled)/kg b. wt., o. by gavage, once; for detailed information please see the article), conc.: ≈1 pmol [^{14}C]AFB_1**/mg RNA*** (mean value), country: USA[615], *after 6 h of receiving AFB_1 (labeled)

incidence: 3?/3*, sa. const.: weanling Fischer 344 rats, wt.: 50 g, contamination: artificial (dose: control diet for 41 weeks then ≈5 μg AFB_1 (labeled)/kg b. wt., o. by gavage, once; for detailed information please see the article), conc.: ≈0.35 pmol [^{14}C]AFB_1**/mg protein**** (mean value), country: USA[615], *control, **after 6 h of receiving AFB_1 (labeled)
incidence: 3?/3, sa. const.: weanling Fischer 344 rats, wt.: 50 g, contamination: artificial (dose: **0.5 ppb** AFM_1, o., for 41 weeks then ≈5 μg AFB_1 (labeled)/kg b. wt., o. by gavage, once; for detailed information please see the article), conc.: ≈0.3 pmol [^{14}C]AFB_1**/mg protein*** (mean value), country: USA[615], *after 6 h of receiving AFB_1 (labeled)
incidence: 3?/3, sa. const.: weanling Fischer 344 rats, wt.: 50 g, contamination: artificial (dose: **50 ppb** AFM_1, o., for 41 weeks then ≈5 μg [^{14}C]AFB_1/kg b. wt., o. by gavage, once; for detailed information please see the article), conc.: ≈0.3 pmol [^{14}C]AFB_1**/mg protein*** (mean value), country: USA[615], *after 6 h of receiving AFB_1 (labeled)

incidence: 3?/3, sa. const.: weanling Fischer 344 rats, wt.: 50 g, contamination: artificial (dose: **50 ppb AFB$_1$**, o., for 41 weeks then ≈5 µg AFB$_1$ (labeled)/kg b. wt., o. by gavage, once; for detailed information please see the article), conc.: ≈0.15 pmol [^{14}C]AFB$_1$**/mg protein*** (mean value), country: USA[615], *after 6 h of receiving AFB$_1$ (labeled)

incidence: 4?/4*, sa. const.: male Sprague-Dawley rats, wt.: 275–325 g, contamination: artificial (dose: 0.25 mg AFB$_1$ (labeled)/kg, i.p., once; for detailed information please see the article), conc.: ≈380 AFB adducts/10^7 DNA nucleotides** (mean value), country: USA[616], *control, **after **2 h**
incidence: 4?/4, sa. const.: male Sprague-Dawley rats, wt.: 275–325 g, contamination: artificial (dose: 0.25 mg AFB$_1$ (labeled)/kg, i.p., once; for detailed information please see the article), conc.: ≈71 AFB adducts/10^7 DNA nucleotides* ** (mean value), country: USA[616], ***BHA-diet** (0.75%) for 10 days prior to AFB$_1$-treatment, **after **2 h**

incidence: 4?/4*, sa. const.: male F344 rats, age: 21 days, contamination: artificial (dose: 250 µg AFB$_1$ (labeled)/kg b. wt., by gavage, once; for detailed information please see the article), conc.: ≈55 pmol/mg DNA** *** (mean value), country: USA[621], *control, **AFB-*N*^{7}-Gua[3], ***after **2 h**
incidence: 4?/4, sa. const.: male F344 rats, age: 21 days, contamination: artificial (dose: 250 µg AFB$_1$ (labeled)/kg b. wt., by gavage, once; for detailed information please see the article), conc.: ≈12 pmol/mg DNA** *** (mean value), country: USA[621], ***oltipraz-diet** (0.075%) for 7 days prior to AFB$_1$-treatment, **AFB-*N*^{7}-Gua[3], ***after **2 h**
incidence: 4?/4*, sa. const.: male F344 rats, age: 21 days, contamination: artificial (dose: 250 µg AFB$_1$ (labeled)/kg b. wt., by gavage, once; for detailed information please see the article), conc.: ≈11.5 pmol/mg DNA** *** (mean value), country: USA[621], *control, **AFB-*N*^{7}-Gua[3], ***after **24 h**
incidence: 4?/4, sa. const.: male F344 rats, age: 21 days, contamination: artificial (dose: 250 µg AFB$_1$ (labeled)/kg b. wt., by gavage, once; for detailed information please see the article), conc.: ≈4 pmol/mg DNA** *** (mean value), country: USA[621], ***oltipraz-diet** (0.075%) for 7 days prior to AFB$_1$-treatment, **AFB-*N*^{7}-Gua[3], ***after **24 h**

Aflatoxin B$_1$–8,9-Epoxide
incidence: 4?/4*, sa. const.: male Fischer rats, age: 4 weeks, contamination: artificial (dose: 16, 124, or 512** µM AFB$_1$ (labeled), i.p., once; for detailed information please see the article), conc. range: ≈≤340 pmol AFB$_1$–8,9-epoxide formed/min/mg protein** *** (mean value), country: USA[624], *control, ***after 2 h
incidence: 4?/4*, sa. const.: male Fischer rats, age: 4 weeks, contamination: artificial (dose: 16, 124, or 512** µM AFB$_1$ (labeled), i.p., once; for detailed information please see the article), conc. range: ≈≤340 pmol AFB$_1$–8,9-epoxide formed/min/mg protein** *** (mean value), country: USA[624], ***BNF-diet** (0.04%) for 7 days prior to AFB$_1$-treatment, ***after 2 h
incidence: 4?/4*, sa. const.: male Fischer rats, age: 4 weeks, contamination: artificial (dose: 16, 124, or 512** µM AFB$_1$ (labeled), i.p., once; for detailed information please see the article), conc. range: ≈≤600 pmol AFB$_1$–8,9-epoxide formed/min/mg protein** *** (mean value), country: USA[624], ***I3C-diet** (0.2%) for 7 days prior to AFB$_1$-treatment, ***after 2 h
incidence: 4?/4*, sa. const.: male Fischer rats, age: 4 weeks, contamination: artificial (dose: 16, 124, or 512** µM AFB$_1$ (labeled), i.p., once; for detailed information please see the article), conc. range: ≈≤700 pmol AFB$_1$–8,9-epoxide formed/min/mg protein** *** (mean

value), country: USA[624], *I3C- (0.2%) + **BNF-diet** (0.04%) for 7 days prior to AFB_1-treatment, ***after 2 h

2,3-Dihydro-2,3-Dihydroxyaflatoxin B_1
incidence: 6?/6, sa. const.: male Fischer rats, age: 75 days, contamination: artificial (dose: **0.6 mg AFB_1** (labeled)/kg b. wt., i.p., once; for detailed information please see the article), conc. range: 3*–60** pmol DIOL/mg DNA x 10^{2*}, country: USA[42], after 72* or 2 h** after AFB_1-administration (also at other hour intervals up to 72 h measured)
incidence: ?/30, sa. const.: male Fischer rats, age: 75 days, contamination: artificial (dose: **25 µg AFB_1** (labeled), i.p., **10 times**; for detailed information please see the article), conc. range: 8*–30** pmol DIOL/mg DNA x 10^{2*}, country: USA[42], after 6* or 14** days (also at other day intervals up to 14 days measured, thereof 12 days with AFB_1-administration)

8,9-Dihydro-8,9-Diydroxyaflatoxin B_1
incidence: 2?/2, sa. const.: male F344 rats, wt.: 75–100 g, contamination: artificial (dose: 250 µg AFB_1 (labeled)/kg b. wt., by gavage, 10 times at days 0–4 and 7–11), conc. range: ≤5.7 pmol/mg DNA* ** *** (mean value), country: USA[551], *control, **AFB_1 8,9-dihydrodiol, ***at day 1 of AFB_1-administration (also at other day intervals up to 133 days measured, lowest conc.: nd after 49, 77, 106, and 133 days) (for detailed information please see the article)
incidence: 2?/2, sa. const.: male F344 rats, wt.: 75–100 g, contamination: artificial (dose: 250 µg AFB_1 (labeled)/kg b. wt., by gavage, 10 times at days 0–4 and 7–11), conc.: 0.6 pmol/mg DNA* ** *** **** (mean value), country: USA[551], ***EQ-diet** (0.4%) for 3 weeks while AFB_1 was gavaged in the last 2 weeks, **AFB_1 8,9-dihydrodiol, ***at day 9 of AFB_1-administration (also at other day intervals up to 133 days measured, lowest conc.: nd after 49, 77, 106, and 133 days), ****several higher values up to 1.0 pmol/mg DNA were recorded (for detailed information please see the article)

Aflatoxin G_1
incidence: 4?/4, sa. const.: Wistar derived strain, wt.: 150 g, contamination: artificial (dose: 60 µg AFG_1 (labeled)/100 g, i.p., once), conc.: 6.89 ng AFG_1 bound/mg* DNA** (mean value), country: USA[116], *after 2 h, ****phenol-cresol extraction**
incidence: 4?/4, sa. const.: Wistar derived strain, wt.: 150 g, contamination: artificial (dose: 60 µg AFG_1 (labeled)/100 g, i.p., once), conc.: 8.5 ng AFG_1 bound/mg* DNA** (mean value), country: USA[116], *after 2 h, ****chloroform/isoamylalcohol extraction**

incidence: 3?/3, sa. const.: Wistar derived strain, wt.: 150 g, contamination: artificial (dose: 60 µg AFG_1 (labeled)/100 g, i.p., once), conc. range: ≈≤10.5 ng AFG_1 bound/mg rRNA* (mean value), country: USA[116], *after ≈2 h (also measured after ≈4, 24, and 48 h, lowest conc.: ≈7 ng AFG_1 bound/**mg rRNA** after 24 h)
incidence: 3?/3, sa. const.: Wistar derived strain, wt.: 150 g, contamination: artificial (dose: 60 µg AFG_1 (labeled)/100 g, i.p., once), conc. range: ≈≤7 ng AFG_1 bound/**mg DNA*** (mean value), country: USA[116], *after ≈2 h (also measured after ≈4, 24, and 48 h, lowest conc.: ≈2 ng AFG_1 bound/mg DNA after 48 h)
incidence: 3?/3, sa. const.: Wistar derived strain, wt.: 150 g, contamination: artificial (dose: 60 µg AFG_1 (labeled)/100 g, i.p., once), conc. range: ≈≤2.5 ng AFG_1 bound/**mg protein*** (mean value), country: USA[116], *after ≈2 h (also measured after ≈4, 24, and 48 h, lowest conc.: ≈nd after 48 h)

Aflatoxin M_1
incidence: 3?/3, sa. const.: male F344 rats, wt.: 100–125 g, contamination: artificial (dose: 400 µg AFB_1 (labeled)/kg b. wt., by gavage, once; for detailed information please see the article), conc.: 0.42 pmol/mg DNA* ** (mean value), country: USA[142], *after 24 h, **AFM_1-N^7-Gua

incidence: 4?/4*, sa. const.: male Fischer rats, age: 4 weeks, contamination: artificial (dose: 16, 124, or 512** µM AFB_1 (labeled), i.p., once; for detailed information please see the article), conc. range: ≈≤30 pmol AFM_1 formed/min/mg protein** *** (mean value), country: USA[624], *control, ***after 2 h
incidence: 4?/4*, sa. const.: male Fischer rats, age: 4 weeks, contamination: artificial (dose: 16, 124, or 512** µM AFB_1 (labeled), i.p., once; for detailed information please see the article), conc. range: ≈≤140 pmol AFM_1 formed/min/mg protein** *** (mean value), country: USA[624], ***BNF-diet** (0.04%) for 7 days prior to AFB_1-treatment, ***after 2 h
incidence: 4?/4*, sa. const.: male Fischer rats, age: 4 weeks, contamination: artificial (dose: 16, 124, or 512** µM AFB_1 (labeled), i.p., once; for detailed information please see the article), conc. range: ≈≤300 pmol AFM_1 formed/min/mg protein** *** (mean value), country: USA[624], ***I3C-diet** (0.2%) for 7 days prior to AFB_1-treatment, ***after 2 h
incidence: 4?/4*, sa. const.: male Fischer rats, age: 4 weeks, contamination: artificial (dose: 16, 124, or 512** µM AFB_1 (labeled), i.p., once; for detailed information please see the article), conc. range: ≈≤390 pmol AFM_1 formed/min/mg protein** *** (mean value), country: USA[624], ***I3C-** (0.2%) + **BNF-diet** (0.04%) for 7 days prior to AFB_1-treatment, ***after 2 h

Aflatoxin Q_1
incidence: 4?/4*, sa. const.: male Fischer rats, age: 4 weeks, contamination: artificial (dose: 16, 124, or 512** µM AFB_1 (labeled), i.p., once; for detailed information please see the article), conc. range: ≈≤35 pmol AFQ_1 formed/min/mg protein** *** (mean value), country: USA[624], *control, ***after 2 h
incidence: 4?/4*, sa. const.: male Fischer rats, age: 4 weeks, contamination: artificial (dose: 16, 124, or 512** µM AFB_1 (labeled), i.p., once; for detailed information please see the article), conc. range: ≈≤50 pmol AFQ_1 formed/min/mg protein** *** (mean value), country: USA[624], ***BNF-diet** (0.04%) for 7 days prior to AFB_1-treatment, ***after 2 h
incidence: 4?/4*, sa. const.: male Fischer rats, age: 4 weeks, contamination: artificial (dose: 16, 124, or 512** µM AFB_1 (labeled), i.p., once; for detailed information please see the article), conc. range: ≈≤180 pmol AFQ_1 formed/min/mg protein** *** (mean value), country: USA[624], ***I3C-diet** (0.2%) for 7 days prior to AFB_1-treatment, ***after 2 h
incidence: 4?/4*, sa. const.: male Fischer rats, age: 4 weeks, contamination: artificial (dose: 16, 124, or 512** µM AFB_1 (labeled), i.p., once; for detailed information please see the article), conc. range: ≈≤190 pmol AFQ_1 formed/min/mg protein** *** (mean value), country: USA[624], ***I3C-** (0.2%) + **BNF-diet** (0.04%) for 7 days prior to AFB_1-treatment, ***after 2 h

Aflatoxin
incidence: ?/3–4, sa. const.: male Wistar outbred rats, wt.: 190–210 g, contamination: artificial (dose: 0.5 µg AFB_1 (labeled), by stomach intubation, twice daily on weekdays for 1, 2, 6, 13, 20, or 23 days), conc. range: ≤267 pg AF bound/mg DNA* ** (mean values), country: France/UK[27], *chronic exposure, **after 14 days of AFB_1-administration (also at other day intervals up to 24 days measured, lowest conc.: 108.4 pg AF bound/mg DNA after 2 days)

incidence: 3?/3*, sa. const.: adult male Fischer rats, Ø wt.: 155 g, contamination: artificial (dose: 0.52 mg AFB_1 (labeled)/kg b. wt., i.p., once; for detailed information please see the article), conc.: 130 µmol AF bound/**mg DNA**** ($x 10^{-6}$) (mean value), country: USA[133], *control, **after 18 h
incidence: 3?/3*, sa. const.: adult male Fischer rats, Ø wt.: 155 g, contamination: artificial (dose: 0.52 mg AFB_1 (labeled)/kg b. wt., i.p., once; for detailed information

please see the article), conc.: 15 µmol AF bound/**mg DNA**** (x 10^{-6}) (mean value), country: USA[133], ***PB** (0.1%)-treated rats for 1 week before AFB_1-administration until the experiment was terminated, **after 18 h

incidence: 3?/3*, sa. const.: adult male Fischer rats, Ø wt.: 129 g, contamination: artificial (dose: 0.62 mg AFB_1 (labeled)/kg b. wt., i.p., once; for detailed information please see the article), conc.: 68 µmol AF bound/**mg DNA**** (x 10^{-6}) (mean value), country: USA[133], ***hypophysectomized rats** 3 weeks before AFB_1-administration, **after 18 h

incidence: 3?/3*, sa. const.: adult male Fischer rats, Ø wt.: 155 g, contamination: artificial (dose: 0.52 mg AFB_1 (labeled)/kg b. wt., i.p., once; for detailed information please see the article), conc.: 260 µmol AF bound/**mg rRNA**** (x 10^{-6}) (mean value), country: USA[133], *control, **after 18 h

incidence: 3?/3*, sa. const.: adult male Fischer rats, Ø wt.: 155 g, contamination: artificial (dose: 0.52 mg AFB_1 (labeled)/kg b. wt., i.p., once; for detailed information please see the article), conc.: 35 µmol AF bound/**mg rRNA**** (x 10^{-6}) (mean value), country: USA[133], ***PB** (0.1%)-treated rats for 1 week before AFB_1-administration until the experiment was terminated, **after 18 h

incidence: 3?/3*, sa. const.: adult male Fischer rats, Ø wt.: 129 g, contamination: artificial (dose: 0.62 mg AFB_1 (labeled)/kg b. wt., i.p., once; for detailed information please see the article), conc.: 210 µmol AF bound/**mg rRNA**** (x 10^{-6}) (mean value), country: USA[133], ***hypophysectomized rats** 3 weeks before AFB_1-administration, **after 18 h

incidence: 3?/3*, sa. const.: adult male Fischer rats, Ø wt.: 155 g, contamination: artificial (dose: 0.52 mg AFB_1 (labeled)/kg b. wt., i.p., once; for detailed information please see the article), conc.: 16 µmol AF bound/**mg protein**** (x 10^{-6}) (mean value), country: USA[133], *control, **after 18 h

incidence: 3?/3*, sa. const.: adult male Fischer rats, Ø wt.: 155 g, contamination: artificial (dose: 0.52 mg AFB_1 (labeled)/kg b. wt., i.p., once; for detailed information please see the article), conc.: 5 µmol AF bound/**mg protein**** (x 10^{-6}) (mean value), country: USA[133], ***PB** (0.1%)-treated rats for 1 week before AFB_1-administration until the experiment was terminated, **after 18 h

incidence: 3?/3*, sa. const.: adult male Fischer rats, Ø wt.: 129 g, contamination: artificial (dose: 0.62 mg AFB_1 (labeled)/kg b. wt., i.p., once; for detailed information please see the article), conc.: 15 µmol AF bound/**mg protein**** (x 10^{-6}) (mean value), country: USA[133], ***hypophysectomized rats** 3 weeks before AFB_1-administration, **after 18 h

incidence: 3?/3, sa. const.: adult male Fischer rats, wt.: 180–200 g, contamination: artificial (dose: **0.52 mg AFB_1** (labeled)/kg b. wt., i.p., once; for detailed information please see the article), conc.: 130 µmol* AF bound/**mg DNA**** (x 10^{-6}) (mean value), country: USA[133], *after 18 h, **liver DNA

incidence: 3?/3, sa. const.: adult male Fischer rats, wt.: 180–200 g, contamination: artificial (dose: **0.52 mg AFB_2** (labeled)/kg b. wt., i.p., once; for detailed information please see the article), conc.: 1.4 µmol* AF bound/**mg DNA**** (x 10^{-6}) (mean value), country: USA[133], *after 18 h, **liver DNA

incidence: 3?/3, sa. const.: adult male Fischer rats, wt.: 180–200 g, contamination: artificial (dose: **0.52 mg AFB_1** (labeled)/kg b. wt., i.p., once; for detailed information please see the article), conc.: 140 µmol* AF bound/**mg rRNA**** (x 10^{-6}) (mean value), country: USA[133], *after 18 h, **liver rRNA

incidence: 3?/3, sa. const.: adult male Fischer rats, wt.: 180–200 g, contamination: artificial (dose: **0.52 mg AFB_2** (labeled)/kg b. wt., i.p., once; for detailed information please see the article), conc.: 2.0 µmol* AF bound/**mg rRNA**** (x 10^{-6}) (mean value), country: USA[133], *after 18 h, **liver rRNA incidence: 3?/3, sa. const.: adult male Fischer rats, wt.: 180–200 g, contamination: artificial (dose: **0.52 mg AFB$_1$** (labeled)/kg b. wt., i.p., once; for detailed information please see the article), conc.: 11 μmol* AF bound/**mg protein**** (x 10^{-6}) (mean value), country: USA[133], *after 18 h, **liver protein

incidence: 3?/3, sa. const.: adult male Fischer rats, wt.: 180–200 g, contamination: artificial (dose: **0.52 mg AFB$_2$** (labeled)/kg b. wt., i.p., once; for detailed information please see the article), conc.: 8.3 μmol* AF bound/**mg protein**** (x 10^{-6}) (mean value), country: USA[133], *after 18 h, **liver protein

incidence: 2?/2*, sa. const.: adult male Fischer rats, wt.: 180–200 g, contamination: artificial (dose: **0.0012 mg AFB$_1$** (labeled)/kg b. wt. **+ 2.08 mg AFB$_2$** (unlabeled)/kg b. wt., i.p., once; for detailed information please see the article), conc.: ≤0.4 μmol** AF bound/**mg DNA***** (x 10^{-6}) (mean value), country: USA[133], *control, **after 6 h, ***liver DNA

incidence: 3?/3, sa. const.: adult male Fischer rats, wt.: 180–200 g, contamination: artificial (dose: **2.08 mg AFB$_1$** (labeled)/kg b. wt., i.p., once; for detailed information please see the article), conc.: 1,700 μmol* AF bound/**mg DNA**** (x 10^{-6}) (mean value), country: USA[133], *after 6 h, **liver DNA

incidence: 3?/3, sa. const.: adult male Fischer rats, wt.: 180–200 g, contamination: artificial (dose: **2.08 mg AFB$_2$** (labeled)/kg b. wt., i.p., once; for detailed information please see the article), conc.: 12 μmol* AF bound/**mg DNA**** (x 10^{-6}) (mean value), country: USA[133], *after 6 h, **liver DNA

incidence: 2?/2*, sa. const.: adult male Fischer rats, wt.: 180–200 g, contamination: artificial (dose: **0.0012 mg AFB$_1$** (labeled)/kg b. wt. **+ 2.08 mg AFB$_2$** (unlabeled)/kg b. wt., i.p., once; for detailed information please see the article), conc.: ≤0.1 μmol** AF bound/**mg rRNA***** (x 10^{-6}) (mean value), country: USA[133], *control, **after 6 h, ***liver rRNA

incidence: 3?/3, sa. const.: adult male Fischer rats, wt.: 180–200 g, contamination: artificial (dose: **2.08 mg AFB$_1$** (labeled)/kg b. wt., i.p., once; for detailed information please see the article), conc.: 1,900 μmol* AF bound/**mg rRNA**** (x 10^{-6}) (mean value), country: USA[133], *after 6 h, **liver rRNA

incidence: 3?/3, sa. const.: adult male Fischer rats, wt.: 180–200 g, contamination: artificial (dose: **2.08 mg AFB$_2$** (labeled)/kg b. wt., i.p., once; for detailed information please see the article), conc.: 13 μmol* AF bound/**mg rRNA**** (x 10^{-6}) (mean value), country: USA[133], *after 6 h, **liver rRNA

incidence: 2?/2*, sa. const.: adult male Fischer rats, wt.: 180–200 g, contamination: artificial (dose: **0.0012 mg AFB$_1$** (labeled)/kg b. wt. + **2.08 mg AFB$_2$** (unlabeled)/kg b. wt., i.p., once; for detailed information please see the article), conc.: ≤0.02 μmol** AF bound/**mg protein***** (x 10^{-6}) (mean value), country: USA[133], *control, **after 6 h, ***liver protein

incidence: 3?/3, sa. const.: adult male Fischer rats, wt.: 180–200 g, contamination: artificial (dose: **2.08 mg AFB$_1$** (labeled)/kg b. wt., i.p., once; for detailed information please see the article), conc.: 150 μmol* AF bound/**mg protein**** (x 10^{-6}) (mean value), country: USA[133], *after 6 h, **liver protein

incidence: 3?/3, sa. const.: adult male Fischer rats, wt.: 180–200 g, contamination: artificial (dose: **2.08 mg AFB$_2$** (labeled)/kg b. wt., i.p., once; for detailed information please see the article), conc.: 52 μmol* AF bound/**mg protein**** (x 10^{-6}) (mean value), country: USA[133], *after 6 h, **liver protein

incidence: 2/2*, sa. const.: male Sprague-Dawley rats, wt.: 230–335 g, contamination: no AFB$_1$ (for detailed

information please see the article), conc.: nr, country: The Netherlands/France[214], *control
incidence: 2?/2, sa. const.: male Sprague-Dawley rats, wt.: 230–335 g, contamination: artificial (**19 µg AFB_1** (labeled)/kg b. wt., by gavage, once; for detailed information please see the article), conc.: 83 pg AF/mg DNA* ** (mean value), country: The Netherlands/France[214], *AF-DNA adducts, **after 24 h
incidence: 2?/2, sa. const.: male Sprague-Dawley rats, wt.: 230–335 g, contamination: artificial (**32 µg AFB_1** (labeled)/kg b. wt., by gavage, once; for detailed information please see the article), conc. range: 146 pg AF/mg DNA* ** (mean value), country: The Netherlands/France[214], *AF-DNA adducts, **after 24 h
incidence: 2?/2, sa. const.: male Sprague-Dawley rats, wt.: 230–335 g, contamination: artificial (**75 µg AFB_1** (labeled)/kg b. wt., by gavage, once; for detailed information please see the article), conc. range: 444 pg AF/mg DNA* ** (mean value), country: The Netherlands/France[214], *AF-DNA adducts, **after 24 h

incidence: 4?/4, sa. const.: male Fischer 344 rats, wt.: 100–125 g, contamination: artificial (dose: 2.2 µg AFB_1 (labeled, total), i.t. Installation, once; for detailed information please see the article), conc. range: ≤23 pmol **AFB/mg DNA*** ** (mean value), country: USA[433], *AFB-DNA adducts, **after 30 min of AFB_1-exposure (also measured after 2, 6, 24, and 48 h, lowest conc.: ≈4 pmol AFB/mg DNA after 48 h)
incidence: 4?/4, sa. const.: male Fischer 344 rats, wt.: 100–125 g, contamination: artificial (dose: 17.6 ng AFB_1/min given as nose-only inhalation; for detailed information please see the article), conc. range: ≤56.8 pmol **AF-DNA adducts** formed/mg DNA* ** (mean value), country: USA[433], *AF-N^7-Gua adducts, **after 120 min of AFB_1-exposure (also measured after 20, 40, and 60 min, lowest conc.: 4.2 pmol AF-DNA adducts formed/mg DNA after 20 min)

incidence: ?/4–6*, sa. const.: male Sprague-Dawley rats, age: 20 days, wt.: 45–50 g, contamination: artificial (dose: 25 µg AFB_1 (labeled and unlabeled), i.p., once), conc.: 12.65 pmol AF/mg DNA** *** (mean value), country: USA[490], *control **semi-purified diet** 3 weeks prior to AFB_1-treatment (for detailed information please see the article), **after **2 h**, ***total AF-DNA adducts
incidence: ?/4–6*, sa. const.: male Sprague-Dawley rats, age: 20 days, wt.: 45–50 g, contamination: artificial (dose: 25 µg AFB_1 (labeled and unlabeled), i.p., once), conc.: 16.6 pmol AF/mg DNA** *** (mean value), country: USA[490], ***choline-deficient/methionine-low diet** 3 weeks prior to AFB_1-treatment (for detailed information please see the article), **after **2 h**, ***total AF-DNA adducts
incidence: ?/4–6*, sa. const.: male Sprague-Dawley rats, age: 20 days, wt.: 45–50 g, contamination: artificial (dose: 25 µg AFB_1 (labeled and unlabeled), i.p., once), conc.: 6.1 pmol AF/mg DNA** *** (mean value), country: USA[490], *control **semi-purified diet** 3 weeks prior to AFB_1-treatment (for detailed information please see the article), **after **24 h**, ***total AF-DNA adducts
incidence: ?/4–6*, sa. const.: male Sprague-Dawley rats, age: 20 days, wt.: 45–50 g, contamination: artificial (dose: 25 µg AFB_1 (labeled and unlabeled), i.p., once), conc.: 4.2 pmol AF/mg DNA** *** (mean value), country: USA[490], ***choline-deficient/ methionine-low diet** 3 weeks prior to AFB_1 treatment (for detailed information please see the article), **after **24 h**, ***total AF-DNA adducts

incidence: ?/?, sa. const.: male F-344 rats [CDF (F-344)/CrlBR], contamination: artificial (dose: **2.2 ng AFB_1** (labeled)/kg/

day in the drinking water, o., for 4, 6, or 8 weeks; for detailed information please see the article), conc. range: ≤0.91 AF-nucleotide adducts/10^9 nucleotides (mean value), country: Switzerland[602], *after 8 weeks (also measured after 4 and 6 weeks)
incidence: ?/?, sa. const.: male F-344 rats [CDF (F-344)/CrlBR], contamination: artificial (dose: **73 ng AFB_1** (labeled)/kg/day in the drinking water, o., for 4, 6, or 8 weeks; for detailed information please see the article), conc. range: ≤32 AF-nucleotide adducts/10^9 nucleotides (mean value), country: Switzerland[602], *after 8 weeks (also measured after 4 and 6 weeks)
incidence: ?/?, sa. const.: male F-344 rats [CDF (F-344)/CrlBR], contamination: artificial (dose: **2,110 ng AFB_1** (labeled)/kg/day in the drinking water, o., for 4, 6, or 8 weeks; for detailed information please see the article), conc. range: ≤850 AF-nucleotide adducts/10^9 nucleotides (mean value), country: Switzerland[602], *after 8 weeks (also measured after 4 and 6 weeks)

Aflatoxins
incidence: ?/5, sa. const.: male Sprague-Dawley rats, Ø wt.: 242.2 g, contamination: artificial (dose: 300 μg/kg AFB_1 (labeled)/kg, i.t. (**dust-adsorbed**), once; for detailed information please see the article), conc.: ≤1.619 nmol AFs/g tissue (x 10^{-3})*, country: USA[604], *after 3 h (also measured after 3 days and 3 weeks, lowest conc.: 0.045 nmol AFs/g tissue (x 10^{-3}) after 3 weeks)
incidence: ?/5, sa. const.: male Sprague-Dawley rats, Ø wt.: 242.2 g, contamination: artificial (dose: 300 μg/kg AFB_1 (labeled)/kg, i.t. (**microcrystalline form**), once; for detailed information please see the article), conc.: ≤2.598 nmol AFs/g tissue (x 10^{-3})*, country: USA[604], *after 3 h (also measured after 3 days and 3 weeks, lowest conc.: 0.027 nmol AFs/g tissue (x 10^{-3}) after 3 weeks)

HT-2 Toxin
incidence: 3?/3, sa. const.: male Wistar rats, age: 8–10 weeks, wt.: 400–500 g, contamination: artificial (dose: 5 mg T-2 toxin/kg b. wt., o., once), conc. range: ≤5 μg/100 g liver* (mean value), country: Japan[391], *after 30 min

Ochratoxin A
incidence: 3/3*, sa. const.: **male** F344 rats, age: 8 weeks, contamination: no OTA (for detailed information please see the article), conc.: nr, country: Germany[125], *control
incidence: 3?/3, sa. const.: **male** F344 rats, age: 8 weeks, contamination: artificial (dose: **0.5 mg OTA**/kg b. wt., by gavage, once; for detailed information please see the article), conc. range: <12 pmol/g* ** (mean value), country: Germany[125], *24 h after OTA-administration (also at other day intervals up to 56 days measured, lowest conc.: nd after 28 days), **highest value recorded
incidence: 3/3*, sa. const.: **female** F344 rats, age: 8 weeks, contamination: no OTA (for detailed information please see the article), conc.: nr, country: Germany[125], *control
incidence: 3?/3, sa. const.: **female** F344 rats, age: 8 weeks, contamination: artificial (dose: **0.5 mg OTA**/kg b. wt., by gavage, once; for detailed information please see the article), conc. range: ≈≤2 pmol/g* ** (mean value), country: Germany[125], *24 h after OTA-administration (also at other day intervals up to 56 days measured, lowest conc.: nd after 28 days), **highest value recorded

incidence: ?/3 or more*, sa. const.: adult male Wistar rats, wt.: 200–250 g, contamination: no OTA, conc.: nr, country: Japan[143], *control
incidence: ?/3–4, sa. const.: adult male Wistar rats, wt.: 200–250 g, contamination: artificial (dose: **15 mg OTA** (labeled)/kg, o., once), conc. range: ≈4.5 μg/g wet weight of tissue* (mean value), country: Japan[143], *after 24 h

(also at other hour intervals up to 40 h measured, lowest conc.: ≈2.5 µg/g wet weight of tissue after 8 h)

incidence: 4?/4, sa. const.: healthy adult female Sprague-Dawley rats, wt.: 270–350 g, contamination: artificial (dose: **100 µg OTA**/rat, i.v., once), conc. range: ≤280 ng/g (mean value), country: Canada[175], measured at 2, 24, 48, and 96 h
incidence: 4?/4, sa. const.: healthy adult female Sprague-Dawley rats, wt.: 270–350 g, contamination: artificial (dose: **100 µg OTC!**/rat, i.v., once), conc. range: ≤355 ng/g (mean value), country: Canada[175], measured at 2, 24, 48, and 96 h

incidence: 10/10*, sa. const.: Wistar male rats, wt.: 83–110 g, contamination: no OTA, **diet B-II** (for detailed information please see the article), conc.: nr, country: Canada[209], *control
incidence: ?/10, sa. const.: Wistar male rats, wt.: 83–110 g, contamination: artificial (dose: 500 µg OTA, intubated, daily for 6 days (**diet B-I**); for detailed information please see the article), conc.: pr*, country: Canada[209], *measured at day 1, 2, 3, 4, 5, and 6? of OTA-administration
incidence: 10/10*, sa. const.: Wistar male rats, wt.: 83–110 g, contamination: no OTA, **diet NMB** (for detailed information please see the article), conc.: nr, country: Canada[209], *control
incidence: 3?/3, sa. const.: Wistar male rats, wt.: 83–110 g, contamination: artificial (dose: 500 µg OTA, intubated, daily, daily for 6 days (**diet NMB+T**); for detailed information please see the article), conc.: 2 µg/g* (mean value), country: Canada[209], *measured on day 5 of OTA-administration (also measured at day 3 and 6)

incidence: ?/2 (10)*, sa. const.: Sprague-Dawley male rats, wt.: ≈100 g, contamination: no OTA, conc.: nr, country: USA[218], *control
incidence: ?/2, sa. const.: Sprague-Dawley male rats, wt.: ≈100 g, contamination: artificial (dose: **1 mg OTA** (labeled)/rat, i.p., once), conc. range: ≤7.1 µg/g tissue* (mean value), country: USA[218], *after 0.5 h (also measured after 2, 4, 8, and 24 h, lowest conc.: 1.0 µg/g tissue after 24 h)

incidence: 5/5*, sa. const.: male Fisher 344 (F344) rats, age: 10 weeks, contamination: no OTA (for detailed information please see the article), conc.: nd, country: Spain[439], *control
incidence: 5/5, sa. const.: male Fisher 344 (F344) rats, age: 10 weeks, contamination: artificial (dose: **0.5 mg OTA**/kg b. wt., o., daily for 7 days; for detailed information please see the article), conc. range: 736–923 µg/kg, Ø conc.: 829 µg/kg, country: Spain[439], *after 24 h of final administration

incidence: ?/12*, sa. const.: male Wistar/AF EOPS rats, wt.: 240–290 g, contamination: no OTA only basal feed; for detailed information please see the article), conc.: 1.4 ng/g** (mean value), country: Belgium[509], *control, **after 4 weeks of OTA-administration
incidence: ?/12, sa. const.: male Wistar/AF EOPS rats, wt.: 240–290 g, contamination: artificial (dose: basal feed + **OTA contaminated wheat (2.2 µg/g)**, o., for 28 days; for detailed information please see the article), conc.: 73.7 ng/g* (mean value), country: Belgium[509], *after 4 weeks of OTA-administration
incidence: ?/12, sa. const.: male Wistar/AF EOPS rats, wt.: 240–290 g, contamination: artificial (dose: basal feed + **OTA contaminated wheat (2.2 µg/g)** + **MWF** (2%), o., for 28 days; for detailed information please see the article), conc.: 45.1 ng/g* (mean value), country: Belgium[509], *after 4 weeks of OTA-administration

incidence: ?/12, sa. const.: male Wistar/AF EOPS rats, wt.: 240–290 g, contamination: artificial (dose: basal feed + **OTA contaminated wheat (2 μg/g) + MWF** (1.8%) + **YCW** (0.2%), o., for 28 days; for detailed information please see the article), conc.: 56.3 ng/g* (mean value), country: Belgium[509], *after 4 weeks of OTA-administration

Zearalenone

incidence: 4?/4, sa. const.: weanling Wistar rats, contamination: artificial (dose: 250 μg crystalline ZEA/g feed additionally to **0% of alfalfa** in the diet, for 14 days; for detailed information please see the article), conc.: 3.86 μg/g* (mean value), country: Canada[84], *after 14 days of ZEA-administration

incidence: 4?/4, sa. const.: weanling Wistar rats, contamination: artificial (dose: 250 μg crystalline ZEA/g feed additionally to **15% of alfalfa** in the diet, for 14 days; for detailed information please see the article), conc.: 1.85 μg/g* (mean value), country: Canada[84], *after 14 days of ZEA-administration

incidence: 4?/4, sa. const.: weanling Wistar rats, contamination: artificial (dose: 250 μg crystalline ZEA/g feed additionally to **25% of alfalfa** in the diet, for 14 days; for detailed information please see the article), conc.: 1.38 μg/g* (mean value), country: Canada[84], *after 14 days of ZEA-administration

incidence: ?/?, sa. const.: male Sprague-Dawley rats, age: 8–10 weeks, wt.: 230–290 g, contamination: artificial (dose: **1.13 mg ZEA/h/kg**, i.v., for 6 h), conc.: ≈800 ng/g* (mean value), country: Korea/USA[585], *after 6 h of ZEA-administration

incidence: ?/?, sa. const.: male Sprague-Dawley rats, age: 8–10 weeks, wt.: 230–290 g, contamination: artificial (dose: **2.25 mg ZEA/h/kg**, i.v., for 6 h), conc.: ≈1,600 ng/g* (mean value), country: Korea/USA[585], *after 6 h of ZEA-administration

α-Zearalenol

incidence: 4?/4, sa. const.: weanling Wistar rats, contamination: artificial (dose: 250 μg crystalline ZEA/g feed additionally to **0% of alfalfa** in the diet, for 14 days; for detailed information please see the article), conc.: 1.19 μg/g* (mean value), country: Canada[84], *after 14 days of ZEA-administration

incidence: 4?/4, sa. const.: weanling Wistar rats, contamination: artificial (dose: 250 μg crystalline ZEA/g feed additionally to **15% of alfalfa** in the diet, for 14 days; for detailed information please see the article), conc.: 0.44 μg/g* (mean value), country: Canada[84], *after 14 days of ZEA-administration

incidence: 4?/4, sa. const.: weanling Wistar rats, contamination: artificial (dose: 250 μg crystalline ZEA/g feed additionally to **25% of alfalfa** in the diet, for 14 days; for detailed information please see the article), conc.: 0.29 μg/g* (mean value), country: Canada[84], *after 14 days of ZEA-administration

β-Zearalenol

incidence: 4?/4, sa. const.: weanling Wistar rats, contamination: artificial (dose: 250 μg crystalline ZEA/g feed additionally to **0% of alfalfa** in the diet, for 14 days; for detailed information please see the article), conc.: 0.23 μg/g* (mean value), country: Canada[84], *after 14 days of ZEA-administration

incidence: 4?/4, sa. const.: weanling Wistar rats, contamination: artificial (dose: 250 μg crystalline ZEA/g feed additionally to **15% of alfalfa** in the diet, for 14 days; for detailed information please see the article), conc.: 0.17 μg/g* (mean value), country: Canada[84], *after 14 days of ZEA-administration

incidence: 4?/4, sa. const.: weanling Wistar rats, contamination: artificial (dose: 250 μg crystalline ZEA/g feed additionally to **25% of alfalfa** in the diet, for 14 days; for detailed information please see the article), conc.: 0.09 µg/g* (mean value), country: Canada[84], *after 14 days of ZEA-administration

Rat lung may contain the following mycotoxins and/or their metabolites:

AFLATOXIN B_1

incidence: 3/3, sa. const.: neonatal rats of albino rats (wt.: 200–250 g) of Wistar strain, contamination: artificial (dose: 20 µg AFB_1 (labeled)/100 g b. wt. into the mother rat, i.p., once (13 days after giving birth), conc.: nd* **, country: India[594], *AFB_1-DNA adducts, **after 48 h while taking milk from their mother
incidence: 3?/3, sa. const.: neonatal rats of albino rats (wt.: 200–250 g) of Wistar strain, contamination: artificial (dose: 20 µg AFB_1 (labeled)/100 g b. wt. into the mother rat, i.p., once (13 days after giving birth), conc.: nd* ** ***, country: India[594], *AFB_1-DNA adducts, **after 48 h while taking milk from their mother, ***PB (80 mg/kg b. wt.) injected 3 days prior to AFB_1-administration
incidence: 3?/3, sa. const.: neonatal rats of albino rats (wt.: 200–250 g) of Wistar strain, contamination: artificial (dose: 20 µg AFB_1 (labeled)/100 g b. wt. into the mother rat, i.p., once (13 days after giving birth), conc.: 48 pmol AFB_1**/mg RNA*** (mean value), country: India[594], *after 48 h while taking milk from their mother
incidence: 3?/3, sa. const.: neonatal rats of albino rats (wt.: 200–250 g) of Wistar strain, contamination: artificial (dose: 20 µg AFB_1 (labeled)/100 g b. wt. into the mother rat, i.p., once (13 days after giving birth), conc.: 7 pmol AFB_1**/mg RNA*** ** (mean value), country: India[594], *after 48 h while taking milk from their mother, ****PB** (80 mg/kg b. wt.) injected 3 days prior to AFB_1-administration
incidence: 3?/3, sa. const.: neonatal rats of albino rats (wt.: 200–250 g) of Wistar strain, contamination: artificial (dose: 20 µg AFB_1 (labeled)/100 g b. wt. into the mother rat, i.p., once (13 days after giving birth), conc.: 74 pmol AFB_1**/mg protein*** (mean value), country: India[594], *after 48 h while taking milk from their mother
incidence: 3?/3, sa. const.: neonatal rats of albino rats (wt.: 200–250 g) of Wistar strain, contamination: artificial (dose: 20 µg AFB_1 (labeled)/100 g b. wt. into the mother rat, i.p., once (13 days after giving birth), conc.: 22 pmol AFB_1**/mg protein*** ** (mean value), country: India[594], *after 48 h while taking milk from their mother, ****PB** (80 mg/kg b. wt.) injected 3 days prior to AFB_1-administration

AFLATOXINS

incidence: ?/5, sa. const.: male Sprague-Dawley rats, Ø wt.: 242.2 g, contamination: artificial (dose: 300 µg/kg AFB_1 (labeled)/kg, i.t. (**dust-adsorbed**), once; for detailed information please see the article), conc.: ≤11.573 nmol AFs/g tissue (x 10^{-3})*, country: USA[604], *after 3 h (also measured after 3 days and 3 weeks, lowest conc.: 0.105 nmol AFs/g tissue (x 10^{-3}) after 3 weeks)
incidence: ?/5, sa. const.: male Sprague-Dawley rats, Ø wt.: 242.2 g, contamination: artificial (dose: 300 µg/kg AFB_1 (labeled)/kg, i.t. (**microcrystalline form**), once; for detailed information please see the article), conc.: ≤1.619 nmol AFs/g tissue (x 10^{-3})*, country: USA[604], *after 3 h (also measured after 3 days and 3 weeks, lowest conc.: 0.067 nmol AFs/g tissue (x 10^{-3}) after 3 weeks)

ZEARALENONE

incidence: ?/?, sa. const.: male Sprague-Dawley rats, age: 8–10 weeks, wt.: 230–290 g, contamination: artificial (dose: **1.13 mg ZEA**/h/kg, i.v., for 6 h), conc.: ≈360 ng/g* (mean value), country: Korea/USA[585], *after 6 h of ZEA-administration
incidence: ?/?, sa. const.: male Sprague-Dawley rats, age: 8–10 weeks, wt.: 230–290 g, contamination: artificial (dose: **2.25 mg ZEA**/h/kg, i.v., for 6 h), conc.: ≈1,000 ng/g* (mean value), country: Korea/USA[585], *after 6 h of ZEA-administration

Rat lymph may contain the following mycotoxins and/or their metabolites:

Ochratoxin A
incidence: ?/6, sa. const.: male Wistar rats, age: 8–10 weeks, contamination: artificial (dose: **1 mg OTA/kg**, injection to the **femoral vein**, once), conc.: 2.69 µg/ml (mean value), country: Japan[174]
incidence: ?/7, sa. const.: male Wistar rats, age: 8–10 weeks, contamination: artificial (dose: **113 µg OTA/ml**, injection into the jejunal loop, once), conc.: 1.51 µg/ml (mean value), country: Japan[174]
incidence: ?/6, sa. const.: male Wistar rats, age: 8–10 weeks, contamination: artificial (dose: **206 µg OTA/ml**, injection into the jejunal loop, once), conc.: 2.16 µg/ml (mean value), country: Japan[174]

Rat milk may contain the following mycotoxins and/or their metabolites:

Aflatoxin B_1
incidence: ?/?, sa. const.: pregnant Sprague-Dawley rats, wt.: 400–450 g, contamination: artificial (dose: 0.5 mg AFB_1 (labeled)/kg, once), conc. range: 1.1–3.5 nmol/ml* **, country: USA[265], *control, **1 h post-dosing (for detailed information please see the article)
incidence: ?/?, sa. const.: pregnant Sprague-Dawley rats, wt.: 400–450 g, contamination: artificial (dose: 0.5 mg AFB_1 (labeled)/kg, once), conc. range: 0.2–2.0 nmol/ml* **, country: USA[265], ***PCB-treatment** (575 mg/kg) at day 1 of the experiment, **1 h post-dosing (for detailed information please see the article)

Aflatoxin M_1
incidence: ?/?, sa. const.: pregnant Sprague-Dawley rats, wt.: 400–450 g, contamination: artificial (dose: 0.5 mg AFB_1 (labeled)/kg, once), conc. range: 3.6–9.8 nmol/ml* **, country: USA[265], *control, **1 h post-dosing (for detailed information please see the article)
incidence: ?/?, sa. const.: pregnant Sprague-Dawley rats, wt.: 400–450 g, contamination: artificial (dose: 0.5 mg AFB_1 (labeled)/kg, once), conc. range: 3.7–11.9 nmol/ml* **, country: USA[265], ***PCB-treatment** (575 mg/kg) at day 1 of the experiment, **1 h post-dosing (for detailed information please see the article)

Ochratoxin A
incidence: 5?/5*, sa. const.: Sprague-Dawley rats (dams), wt.: 300–370 g, contamination: no OTA (for detailed information please see the article), conc.: 0.7 µg/l** (mean value), country: Sweden[312], *control, **after 72 h (also measured after 24 h conc.: 0.6 µg/l*)
incidence: 4?/4, sa. const.: Sprague-Dawley rats (dams), wt.: 300–370 g, contamination: artificial (dose: **10 µg OTA/kg** b. wt., o., once at 11th day of lactation; for detailed information please see the article), conc.: 14 µg/l* (mean value), country: Sweden[312], *after 24 h (also measured after 72 h conc.: 6.3 µg/l)
incidence: 4?/4, sa. const.: Sprague-Dawley rats (dams), wt.: 300–370 g, contamination: artificial (dose: **50 µg OTA/kg** b. wt., o., once at 11th day of lactation; for detailed information please see the article), conc.: 70 µg/l* (mean value), country: Sweden[312], *after 24 h (also measured after 72 h conc.: 22 µg/l)
incidence: 5?/5, sa. const.: Sprague-Dawley rats (dams), wt.: 300–370 g, contamination: artificial (dose: **250 µg OTA/kg** b. wt., o., once at 11th day of lactation; for detailed information please see the article), conc.: 230 µg/l* (mean value), country: Sweden[312], *after 24 h (also measured after 72 h conc.: 83 µg/l)

incidence: 4?(8?)/4(8)*, sa. const.: Sprague-Dawley rats (dams), contamination: no OTA (for detailed information please see the article), Ø conc.: 3.35 µg/l** (mean value),

country: Sweden[322], *control, **at **day 14** of lactation
incidence: 4?(8?)/4(8), sa. const.: Sprague-Dawley rats (dams), contamination: artificial (dose: 50 μg OTA/kg b. wt., gastric intubation, 46 times in 8 weeks; for detailed information please see the article), Ø conc.: 135 μg/l* (mean value), country: Sweden[322], *at **day 14** of lactation
incidence: 4?(8?)/4(8)*, sa. const.: Sprague-Dawley rats (dams), contamination: no OTA (for detailed information please see the article), Ø conc.: 2.4 μg/l** (mean value), country: Sweden[322], *control, **at **day 21** of lactation
incidence: 4?(8?)/4(8), sa. const.: Sprague-Dawley rats (dams), contamination: artificial (dose: 50 μg OTA/kg b. wt., gastric intubation, 46 times for 8 weeks; for detailed information please see the article), Ø conc.: 135 μg/l* (mean value), country: Sweden[322], *at **day 21** of lactation

Rat muscle may contain the following mycotoxins and/or their metabolites:

Cyclopiazonicacid
incidence: 2?/2, sa. const.: male HSD:Sprague-Dawley BR rats, wt.: 150–200 g, contamination: artificial (dose: 1.0 mg CPA (labeled)/kg b. wt., i.g., once), conc. range: 2.4 μg/g* ** *** (mean value), country: USA[111], *in ***gastrocnemius*** muscle, **[^{14}C]CPA-eq., ***after 24 h
incidence: 2?/2, sa. const.: male HSD:Sprague-Dawley BR rats, wt.: 150–200 g, contamination: artificial (dose: 1.0 mg CPA (labeled)/kg b. wt., i.g., once), conc. range: 3.3 μg/g* ** *** (mean value), country: USA[111], *in ***tibialis anterior*** muscle, **[^{14}C]CPA-eq., ***after 24 h
incidence: 2?/2, sa. const.: male HSD:Sprague-Dawley BR rats, wt.: 150–200 g, contamination: artificial (dose: 1.0 mg CPA (labeled)/kg b. wt., i.g., once), conc. range: 2.4 μg/g* ** *** (mean value), country: USA[111], *in ***biceps femoris*** muscle, **[^{14}C]CPA-eq., ***after 24 h
incidence: 2?/2, sa. const.: male HSD:Sprague-Dawley BR rats, wt.: 150–200 g, contamination: artificial (dose: 1.0 mg CPA (labeled)/kg b. wt., i.g., once), conc. range: 1.6 μg/g* ** *** (mean value), country: USA[111], *in ***diaphragm*** muscle, **[^{14}C]CPA-eq., ***after 24 h
incidence: 2?/2, sa. const.: male HSD:Sprague-Dawley BR rats, wt.: 150–200 g, contamination: artificial (dose: 1.0 mg CPA (labeled)/kg b. wt., i.g., once), conc. range: 2.9 μg/g* ** *** (mean value), country: USA[111], *in ***biceps*** muscle, **[^{14}C]CPA-eq., ***after 24 h
incidence: 2?/2, sa. const.: male HSD:Sprague-Dawley BR rats, wt.: 150–200 g, contamination: artificial (dose: 1.0 mg CPA (labeled)/kg b. wt., i.g., once), conc. range: 2.6 μg/g* ** *** (mean value), country: USA[111], *in ***pectoris major*** muscle, **[^{14}C]CPA-eq., ***after 24 h
incidence: 2?/2, sa. const.: male HSD:Sprague-Dawley BR rats, wt.: 150–200 g, contamination: artificial (dose: 1.0 mg CPA (labeled)/kg b. wt., i.g., once), conc. range: 1.6 μg/g* ** *** (mean value), country: USA[111], *in ***masseter*** muscle, **[^{14}C]CPA-eq., ***after 24 h
incidence: 2?/2, sa. const.: male HSD:Sprague-Dawley BR rats, wt.: 150–200 g, contamination: artificial (dose: 1.0 mg CPA (labeled)/kg b. wt., i.g., once), conc. range: 3.2 μg/g* ** *** (mean value), country: USA[111], *in ***tenes major*** muscle, **[^{14}C]CPA-eq., ***after 24 h

Ochratoxin A
incidence: 4?/4, sa. const.: healthy adult female Sprague-Dawley rats, wt.: 270–350 g, contamination: artificial (dose: **100 μg OTA**/rat, i.v., once), conc. range: ≤190 ng/g (mean value), country: Canada[175], measured at 2, 24, 48, and 96 h
incidence: 4?/4, sa. const.: healthy adult female Sprague-Dawley rats,

wt.: 270–350 g, contamination: artificial (dose: **100 µg OTC!**/rat, i.v., once), conc. range: ≤125 ng/g (mean value), country: Canada[175], measured at 2, 24, 48, and 96 h

Zearalenone

incidence: ?/?, sa. const.: male Sprague-Dawley rats, age: 8–10 weeks, wt.: 230–290 g, contamination: artificial (dose: **1.13 mg ZEA**/h/kg, i.v., for 6 h), conc.: ≈90 ng/g* (mean value), country: Korea/USA[585], *after 6 h of ZEA-administration

incidence: ?/?, sa. const.: male Sprague-Dawley rats, age: 8–10 weeks, wt.: 230–290 g, contamination: artificial (dose: **2.25 mg ZEA**/h/kg, i.v., for 6 h), conc.: ≈120 ng/g* (mean value), country: Korea/USA[585], *after 6 h of ZEA-administration

Rat plasma may contain the following mycotoxins and/or their metabolites:

Aflatoxin B_1

incidence: ?/?*, sa. const.: male Sprague-Dawley, Wistar, and Fischer 344 rats, wt.: 80–140 g, contamination: no AFB_1, conc.: nr, country: France/Japan[24], *control

incidence: ?/?, sa. const.: male Sprague-Dawley, Wistar, and Fischer 344 rats, wt.: 80–140 g, contamination: artificial (dose: **20 µg AFB_1**/kg/day, by gavage, daily for up to 14 days), conc. range: ≈≤25 pmol AFB_1-lysine eq/mg albumin* ** (mean values, combined data from three strains), country: France/Japan[24], *animals killed after 24 h of final treatment (also measured after 1, 3, and 7 days, lowest conc.: ≈7 pmol AFB_1-lysine eq/mg albumin after 1 day), **AFB_1-albumin adducts

incidence: ?/4*, sa. const.: male SPF Wistar rats, age: 24–27 days, contamination: artificial (dose: 2 mg AFB_1 (labeled)/kg b. wt., i.p., once), conc.: ≈210 pmol AFB_1/mg protein** *** (mean value), country: France[549], *control, **AFB_1 binding to plasma albumin, ***after 2 h (for overall information please see the article)

incidence: ?/4*, sa. const.: male SPF Wistar rats, age: 24–27 days, contamination: artificial (dose: 2 mg AFB_1 (labeled)/kg b. wt., i.p., once), conc.: ≈80 pmol AFB_1/mg protein** *** (mean value), country: France[549], ***3-MC** i.p. (20 mg/kg b. wt.) on the 3 days preceding sacrifice (3-MC was injected prior to AFB_1-administration), **AFB_1 binding to plasma albumin, ***after 2 h (for overall information please see the article)

incidence: ?/4*, sa. const.: male SPF Wistar rats, age: 24–27 days, contamination: artificial (dose: 2 mg AFB_1 (labeled)/kg b. wt., i.p., once), conc.: ≈85 pmol AFB_1/mg protein** *** (mean value), country: France[549], ***CX-diet** (300 mg/kg) for 2 weeks prior to AFB_1-administration, **AFB_1 binding to plasma albumin, ***after 2 h (for overall information please see the article)

incidence: ?/4*, sa. const.: male SPF Wistar rats, age: 24–27 days, contamination: artificial (dose: 2 mg AFB_1 (labeled)/kg b. wt., i.p., once), conc.: ≈75 pmol AFB_1/mg protein** *** (mean value), country: France[549], ***AC-diet** (300 mg/kg) for 2 weeks prior to AFB_1-administration, **AFB_1 binding to plasma albumin, ***after 2 h (for overall information please see the article)

incidence: ?/4*, sa. const.: male SPF Wistar rats, age: 24–27 days, contamination: artificial (dose: 2 mg AFB_1 (labeled)/kg b. wt., i.p., once), conc.: ≈210 pmol AFB_1/mg protein** *** (mean value), country: France[549], ***BC-diet** (300 mg/kg) for 2 weeks prior to AFB_1-administration, **AFB_1 binding to plasma albumin, ***after 2 h (for overall information please see the article)

incidence: 4?/4, sa. const.: male F344 rats, wt.: 185 g, contamination: artificial (dose: 0.01, 0.1, 1, 10, and 100 ng and 1 and 10* µg AFB_1, i.p., daily for 9 days), conc. range: ≈≤4,000 pg lysine-AFB_1/mg albumin* ** (mean value), country: USA[622], **after 18 h of the final dose (for overall information please see the article)

Aflatoxin

incidence: ?/3–4, sa. const.: male Wistar outbred rats, wt.: 190–210 g, contamination: artificial (dose: 0.5 µg AFB_1 (labeled), by stomach intubation, twice daily on weekdays for 1, 2, 6, 13, 20, or 23 days), conc. range: ≤148.3 pg AF bound/mg protein* ** (mean values), country: France/UK[27], *chronic exposure, **after 14 days of AFB_1-administration (also at other day intervals up to 24 days measured, lowest conc.: 35.6 pg AF bound/mg protein after 1 day)

incidence: ?/4*, sa. const.: male F344 rats, wt.: 100 g, contamination: artificial (dose: 20.0 µg AFB_1 (labeled), by gavage, daily for 35 days), conc. range: ≈≤490 pmol AF-adducts bound/mg albumin** (mean value), country: USA[605], *control, **after 24 days of AFB_1-administration (also at other day intervals up to 40 days measured)
incidence: ?/4, sa. const.: male F344 rats, wt.: 100 g, contamination: artificial (dose: 20.0 µg AFB_1 (labeled), by gavage, daily for 35 days), conc. range: ≈≤390 pmol AF-adducts bound/mg albumin* ** (mean value), country: USA[605], *after 16 days of AFB_1-administration (also at other day intervals up to 40 days measured), ****oltipraz-diet** (0.05%) 1 week after AFB_1-treatment began and ended 2 weeks before AFB_1-treatment stopped (**transient intervention**)
incidence: ?/4, sa. const.: male F344 rats, wt.: 100 g, contamination: artificial (dose: 20.0 µg AFB_1 (labeled), by gavage, daily for 35 days), conc. range: ≈≤215 pmol AF-adducts bound/mg albumin* ** (mean value), country: USA[605], *after 7 days of AFB_1-administration (also at other day intervals up to 40 days measured), ****oltipraz-diet** (0.05%) 1 week before and continuing throughout AFB_1-treatment (**long-term intervention**)

Fumonisin B_1

incidence: 2?/2, sa. const.: males BD IX rats, age: ≈6 weeks, wt.: ≈150 g, contamination: artificial (dose: 7.5 mg FB_1/kg b. wt., i.p., once; for detailed information please see the article), conc. range: ≤8.6 µg/ml* (mean value), country: South Africa[114], *after 20 min (also measured after ≈5, ≈7, 40, 60, and 120 min, lowest conc.: ≈nd after 120 h)

Ochratoxin A

incidence: ?/7, sa. const.: male Wistar rats, age: 8–10 weeks, contamination: artificial (dose: **113 µg OTA**/ml, injection into the jejunal loop, once), conc.: 10.3 µg/ml* ** (mean value), country: Japan[174], *after 30 min, **in mesenteric venous plasma
incidence: ?/8, sa. const.: male Wistar rats, age: 8–10 weeks, contamination: artificial (dose: **206 µg OTA**/ml, injection into the jejunal loop, once), conc.: 22.1 µg/ml* ** (mean value), country: Japan[174], *after 30 min, **in mesenteric venous plasma

incidence: 6?/6, sa. const.: male Wistar rats, wt.: 270 g, contamination: artificial (dose: 180 ng/ml OTA in 2 ml bile directly applicated into the **stomach**, once), conc. range: 5–22 ng/ml*, Ø conc: 15 ng/ml*, country: Yugoslavia/Sweden[191], *after 24 h
incidence: 6?/6, sa. const.: male Wistar rats, wt.: 270 g, contamination: artificial (dose: 220 ng/ml OTA in 2 ml bile directly applicated into the **duodenum**, once), conc. range: 30–60 ng/ml*, Ø conc: 40 ng/ml*, country: Yugoslavia/Sweden[191], *after 24 h

incidence: 5?/5, sa. const.: male Wistar rats, wt.: 250–300 g, contamination: artificial (dose: 50 ng OTA ng/g b. wt., **o.**, once), conc.: 390 ng/ml (mean value), country: Sweden/Yugoslavia[193] (at different min and hour intervals up to 24 h measured)
incidence: 5?/5, sa. const.: male Wistar rats, wt.: 250–300 g, contamination: artificial (dose: 50 ng OTA /g b. wt., **i.v.**, once), conc.: 2,100 ng/ml (mean value), country: Sweden/Yugoslavia[193] (at different min and hour intervals up to 24 h measured)

incidence: ?/?, sa. const.: adult male Wistar rats, wt.: 250 g, contamination: artificial (dose: 50 ng OTC/g b. wt., **o.**, once), conc. range: ≤350 ng/ml* (mean value), country: Yugoslavia/Sweden[198], *after **60 min** (also at other hour intervals up to 48 h measured, lowest conc.: 50 ng/ml after 48 h)
incidence: ?/?, sa. const.: adult male Wistar rats, wt.: 250 g, contamination: artificial (dose: 50 ng OTC/g b. wt., **i.v.**, once), conc. range: ≤800 ng/ml* (mean value), country: Yugoslavia/Sweden[198], *after **90 min** (also at other hour intervals up to 48 h measured, lowest conc.: 130 ng/ml after 48 h)

incidence: ?/?, sa. const.: adult male Wistar rats, wt.: 250 g, contamination: artificial (dose: 10 mg OTA/kg, o., once; for detailed information please see the article), conc.: ≤56.7 μg/ml* (mean value), country: France[317], *after 8 h (also measured after 48 h conc.: 32.2 μg/ml)

incidence: 5/5*, sa. const.: male Fisher 344 (F344) rats, age: 10 weeks, contamination: no OTA (for detailed information please see the article), conc. range: 1.47–7.57 μg/l **, Ø conc.: 4.74 μg/l **, country: Spain[439], *control, **after 8 days
incidence: 5/5, sa. const.: male Fisher 344 (F344) rats, age: 10 weeks, contamination: artificial (dose: **0.5 mg OTA/kg** b. wt., o., daily for 7 days; for detailed information please see the article), conc. range: 9,142–10,393 μg/l*, Ø conc.: 97,372 μg/l*, country: Spain[439], *after 24 h of final administration

incidence: ?/12*, sa. const.: male Wistar/AF EOPS rats, wt.: 240–290 g, contamination: no OTA only basal feed; for detailed information please see the article), conc.: 21.1 ng/ml** (mean value), country: Belgium[509], *control, **after 4 weeks
incidence: ?/12, sa. const.: male Wistar/AF EOPS rats, wt.: 240–290 g, contamination: artificial (dose: basal feed + **OTA contaminated wheat** (**2.2 μg/g**), o., for 28 days; for detailed information please see the article), conc.: 830.2 ng/ml* (mean value), country: Belgium[509], *after 4 weeks of OTA-administration
incidence: ?/12, sa. const.: male Wistar/AF EOPS rats, wt.: 240–290 g, contamination: artificial (dose: basal feed + **OTA contaminated wheat (2.2 μg/g) + MWF** (2%), o., for 28 days; for detailed information please see the article), conc.: 494.1 ng/ml* (mean value), country: Belgium[509], *after 4 weeks of OTA-administration
incidence: ?/12, sa. const.: male Wistar/AF EOPS rats, wt.: 240–290 g, contamination: artificial (dose: basal feed + **OTA contaminated wheat (2 μg/g) + MWF** (1.8%) + **YCW** (0.2%), o., for 28 days; for detailed information please see the article), conc.: 652.0 ng/ml* (mean value), country: Belgium[509], *after 4 weeks of OTA-administration

incidence: ?/3, sa. const.: male Wistar rats, age: **6 weeks**, contamination: artificial (dose: **0.5 mg OTA**/kg b. wt., i.p., for 6 days), conc.: 12.2 μmol/l* (mean value), country: Germany[561], *after 24 h of final administration
incidence: ?/3, sa. const.: male Wistar rats, age: **2 years**, contamination: artificial (dose: **0.5 mg OTA**/kg b. wt., i.p., for 6 days), conc.: 23.0 μmol/l* (mean value), country: Germany[561], *after 24 h of final administration
incidence: ?/3, sa. const.: male Wistar rats, age: **2 years**, contamination: artificial (dose: **1.25 mg OTA**/kg b. wt., i.p., once), conc.: 89.7 μmol/l* (mean value), country: Germany[561], *after 2 h

incidence: 1/1, sa. const.: male **albumin-deficient** Sprague-Dawley rat, age: 9–10 weeks, contamination: artificial (dose: **2.2 mg OTA**/kg, by injection, once), conc. range: ≈≤1.3 μg/ml* (mean value), country: Japan[613], *after ≈1 min (also at other min intervals up to 15 min measured, lowest conc.: ≈18 μg/ml after 15 min)
incidence: 4?/4, sa. const.: **normal** Sprague-Dawley rats, age: 9–10 weeks,

contamination: artificial (dose: **4.1 mg OTA/kg**, by injection, once), conc. range: ≈≤71 μg/ml* (mean value), country: Japan[613], *after 10 min (also measured after 30, 60, and 90 min, lowest conc.: ≈55 μg/ml after 90 min)
incidence: 4?/4, sa. const.: **albumin-deficient** Sprague-Dawley rats, age: 9–10 weeks, contamination: artificial (dose: **4.1 mg OTA/kg**, by injection, once), conc.: nd, country: Japan[613]

Ochratoxin B
incidence: 5?/5, sa. const.: male Wistar rats, wt.: 250–300 g, contamination: artificial (dose: 50 ng OTB/g b. wt., **o.**, once), conc.: 120 ng/ml (mean value), country: Sweden/Yugoslavia[193] (at different min and hour intervals and up to 7 days measured)
incidence: 5?/5, sa. const.: male Wistar rats, wt.: 250–300 g, contamination: artificial (dose: 50 ng OTB/g b. wt., **i.v.**, once), conc.: 760 ng/ml (mean value), country: Sweden/Yugoslavia[193] (at different min and hour intervals and up to 7 days measured)

Rat serous fluid and mucus

may contain the following mycotoxins and/or their metabolites:

Ochratoxin A
incidence: 3/3 or more*, sa. const.: adult male Wistar rats, wt.: 200–250 g, contamination: no OTA, conc.: nr, country: Japan[143], *control
incidence: ?/3–4, sa. const.: adult male Wistar rats, wt.: 200–250 g, contamination: artificial (dose: **15 mg OTA** (labeled)/kg, o., once), conc. range: ≤6.1 μg/ml* (mean value), country: Japan[143], *at 8 h (measured also at 4 and 6 h, lowest conc.: 3.9 μg/ml at 6 h)

Ochratoxin α
incidence: 3 or more/3 or more*, sa. const.: adult male Wistar rats, wt.: 200–250 g, contamination: no OTA, conc.: nr, country: Japan[143], *control
incidence: ?/3–4, sa. const.: adult male Wistar rats, wt.: 200–250 g, contamination: artificial (dose: **15 mg OTA** (labeled)/kg, o., once), conc.: pr, country: Japan[143] (measured at 4, 6, and 8 h)

Rat serum

may contain the following mycotoxins and/or their metabolites:

Aflatoxin B_1
incidence: ?/12*, sa. const.: male Wistar rats, wt.: ≈70 g, contamination: artificial (dose: 250 μg AFB_1/kg b. wt., by gavage, once), conc. range: ≤4.29** ng/mg albumin*** (mean value), country: Thailand/UK[310], *control: **DMSO + 250 μg AFB_1 + DMSO** (for overall information please see the article), **after 12 h (also measured after 4, 24, and 48 h, lowest conc.: 1.07 ng/mg albumin after 48 h), ***AFB_1-albumin adducts
incidence: ?/12*, sa. const.: male Wistar rats, wt.: ≈70 g, contamination: artificial (dose: 250 μg AFB_1/kg b. wt., by gavage, once), conc. range: ≤4.65** ng/mg albumin*** (mean value), country: Thailand/UK[310], ****Murdannia loriformis*** **extract + 250 μg AFB_1 +** ***M. loriformis*** extract (for overall information please see the article), **after 4 h (also measured after 12, 24, and 48 h, lowest conc.: 1.72 ng/mg albumin after 48 h), ***AFB_1-albumin adducts
incidence: ?/12*, sa. const.: male Wistar rats, wt.: ≈70 g, contamination: artificial (dose: 250 μg AFB_1/kg b. wt., by gavage, once), conc. range: ≤5.09** ng/mg albumin*** (mean value), country: Thailand/UK[310], ****Murdannia loriformis*** **extract + 250 μg AFB_1 + DMSO** (for overall information please see the article), **after 4 h (also measured after 12, 24, and 48 h, lowest conc.: 1.89 ng/mg albumin after 48 h), ***AFB_1-albumin adducts
incidence: ?/12*, sa. const.: male Wistar rats, wt.: ≈70 g, contamination: artificial (dose: 250 μg AFB_1/kg b. wt., by gavage, once), conc. range: ≤4.63** ng/mg albumin*** (mean value), country:

Thailand/UK[310], *DMSO + 250 µg AFB_1 + *Murdannia loriformis* extract (for overall information please see the article), **after 4 h (also measured after 12, 24, and 48 h, lowest conc.: 2.17 ng/mg albumin after 48 h), ***AFB_1-albumin adducts

incidence: ?/12*, sa. const.: male Wistar rats, wt.: ≈70 g, contamination: artificial (dose: 250 µg AFB_1/kg b. wt., by gavage, once), conc. range: ≤4.29** ng/mg albumin*** (mean value), country: Thailand/UK[310], *control: **DMSO + 250 µg AFB_1 + DMSO** (for overall information please see the article), **after 12 h (also measured after 4, 24, and 48 h, lowest conc.: 1.07 ng/mg albumin after 48 h), ***AFB_1-albumin adducts

incidence: ?/12*, sa. const.: male Wistar rats, wt.: ≈70 g, contamination: artificial (dose: 250 µg AFB_1/kg b. wt., by gavage, once), conc. range: ≤4.66** ng/mg albumin*** (mean value), country: Thailand/UK[310], ****Cymbopogon citratus*** extract **+ 250 µg AFB_1 + *C. citratus*** extract (for overall information please see the article), **after 12 h (also measured after 4, 24, and 48 h, lowest conc.: 1.16 ng/mg albumin after 48 h), ***AFB_1-albumin adducts

incidence: ?/12*, sa. const.: male Wistar rats, wt.: ≈70 g, contamination: artificial (dose: 250 µg AFB_1/kg b. wt., by gavage, once), conc. range: ≤4.55** ng/mg albumin*** (mean value), country: Thailand/UK[310], ****Cymbopogon citratus*** extract **+ 250 µg AFB_1 + DMSO** (for overall information please see the article), **after 12 h (also measured after 4, 24, and 48 h, lowest conc.: 1.37 ng/mg albumin after 48 h), ***AFB_1-albumin adducts

incidence: ?/12*, sa. const.: male Wistar rats, wt.: ≈70 g, contamination: artificial (dose: 250 µg AFB_1/kg b. wt., by gavage, once), conc. range: ≤4.29** ng/mg albumin*** (mean value), country: Thailand/UK[310], ***DMSO + 250 µg AFB_1 + *Cymbopogon citratus*** extract (for overall information please see the article), **after 12 h (also measured after 4, 24, and 48 h, lowest conc.: 1.34 ng/mg albumin after 48 h), ***AFB_1-albumin adducts

incidence: ?/?*, sa. const.: male Wistar rats, wt.: ≈70 g, contamination: artificial (dose: 250 µg AFB_1/kg b. wt., by gavage, for 5 days a week for 3 weeks; for detailed information please see the article), conc. range: ≈≤13.2 ng/mg albumin** *** (mean value), country: Thailand/UK[310], *control (for overall information please see the article), **after 20 days during AFB_1-administration (also at other day intervals up to ≈26 days measured, lowest conc.: ≈3 ng/mg albumin after 1 day), ***AFB_1-albumin adducts

incidence: ?/?, sa. const.: male Wistar rats, wt.: ≈70 g, contamination: artificial (dose: 250 µg AFB_1/kg b. wt., by gavage, for 5 days a week for 3 weeks and additionally ***Murdannia loriformis*** extract; for detailed information please see the article), conc. range: ≈≤12 ng/mg albumin* ** (mean value), country: Thailand/UK[310], *after ≈12.5 days during AFB_1-administration (also at other day intervals up to ≈26 days measured, lowest conc.: ≈1 ng/mg albumin after 1 day), **AFB_1-albumin adducts, for overall information please see the article

incidence: ?/?*, sa. const.: male Wistar rats, wt.: ≈70 g, contamination: artificial (dose: 250 µg AFB_1/kg b. wt., by gavage, for 5 days a week for 3 weeks; for detailed information please see the article), conc. range: ≈≤14 ng/mg albumin** *** (mean value), country: Thailand/UK[310], *control (for overall information please see the article), **after 19 days during AFB_1-administration (also at other day intervals up to ≈23.5 days measured, lowest conc.: ≈1 ng/mg albumin after 3 days), ***AFB_1-albumin adducts

incidence: ?/?, sa. const.: male Wistar rats, wt.: ≈70 g, contamination: artificial (dose:

250 µg AFB_1/kg b. wt., by gavage, for 5 days a week for 3 weeks and additionally ***Cymbopogon citratus*** extract; for detailed information please see the article), conc. range: ≈≤9.8 ng/mg albumin* ** (mean value), country: Thailand/UK[310], *after 19 days during AFB_1-administration (also at other day intervals up to ≈23.5 days measured, lowest conc.: ≈1 ng/mg albumin after 1 day), **AFB_1-albumin adducts, for overall information please see the article

incidence: ?/4, sa. const.: young male F344 rats, age: young, wt.: 120–140 g, contamination: artificial (dose: 250 µg AFB_1/kg b. wt., daily for 3 weeks), conc. range: ≤159.9 pmol/mg albumin* ** *** (mean value), country: USA/People's Republic of China[493], *control (for overall information please see the article), **AFB_1-albumin adducts, ***after 15 days during treatment (also at other day intervals up to 15 days measured, lowest conc.: ≈10 pmol/mg albumin after 1 day)
incidence: ?/4, sa. const.: young male F344 rats, wt.: 120–140 g, contamination: artificial (dose: 250 µg AFB_1/kg b. wt., daily for 3 weeks), conc. range: ≈≤55 pmol/mg albumin* ** *** (mean value), country: USA/People's Republic of China[493], ***lycopene-treatment** (100 mg/kg b. wt.) after each AFB_1-administration (for overall information please see the article), **AFB_1-albumin adducts, ***after 15 days during treatment (also at other day intervals up to 15 days measured, lowest conc.: ≈8 pmol/mg albumin after 1 day)

incidence: 2?/2, sa. const.: male Fischer F344 rats, wt.: 200–250 g, contamination: artificial (dose: **1 µg AFB_1** (labeled)/kg b. wt., i.p., once), conc.: ≈3.8 pg AFB_1/mg serum albumin* (mean value), country: UK/Nigeria/Thailand[618], *after 24 h
incidence: 2?/2, sa. const.: male Fischer F344 rats, wt.: 200–250 g, contamination: artificial (dose: **10 µg AFB_1** (labeled)/kg b. wt., i.p., once), conc.: ≈11 pg AFB_1/mg serum albumin* (mean value), country: UK/Nigeria/Thailand[618], *after 24 h
incidence: 2?/2, sa. const.: male Fischer F344 rats, wt.: 200–250 g, contamination: artificial (dose: **100 µg AFB_1** (labeled)/kg b. wt., i.p., once), conc.: ≈200 pg AFB_1/mg serum albumin* (mean value), country: UK/Nigeria/Thailand[618], *after 24 h

incidence: 2?/2*, sa. const.: male Wistar rats, wt.: 200–250 g, contamination: no AFB_1, conc.: 0**, country: UK[627], *control, **measured by **scintillation counting**
incidence: 2?/2*, sa. const.: male Wistar rats, wt.: 200–250 g, contamination: no AFB_1, conc.: 0**, country: UK[627], *control, **measured by **ELISA**
incidence: 2?/2, sa. const.: male Wistar rats, wt.: 200–250 g, contamination: artificial (dose: 10 µg AFB_1 (labeled)/kg b. wt., i.p., once), conc.: 132 pg AFB_1-lysine eq/mg albumin* **, country: UK[627], *measured by **scintillation counting,** **after 24 h
incidence: 2?/2, sa. const.: male Wistar rats, wt.: 200–250 g, contamination: artificial (dose: 10 µg AFB_1 (labeled)/kg b. wt., i.p., once), conc.: 59 pg AFB_1-lysine eq/mg albumin* **, country: UK[627], *measured by **ELISA,** **after 24 h
incidence: 2?/2, sa. const.: male Wistar rats, wt.: 200–250 g, contamination: artificial (dose: 50 µg AFB_1 (labeled)/kg b. wt., i.p., once), conc.: 439 pg AFB_1-lysine eq/mg albumin* **, country: UK[627], *measured by **scintillation counting,** **after 24 h
incidence: 2?/2, sa. const.: male Wistar rats, wt.: 200–250 g, contamination: artificial (dose: 50 µg AFB_1 (labeled)/kg b. wt., i.p., once), conc.: 326 pg AFB_1-lysine eq/mg albumin* **, country: UK[627], *measured by **ELISA,** **after 24 h
incidence: 2?/2, sa. const.: male Wistar rats, wt.: 200–250 g, contamination: artificial (dose: 200 µg AFB_1 (labeled)/kg b. wt., i.p., once), conc.: 1,003 pg AFB_1-lysine eq/mg albumin* **, country: UK[627], *measured by scintillation counting, **after 24 h

incidence: 2?/2, sa. const.: male Wistar rats, wt.: 200–250 g, contamination: artificial (dose: 200 μg AFB_1 (labeled)/kg b. wt., i.p., once), conc.: 1,051 pg AFB_1-lysine eq/mg albumin* **, country: UK[627], *measured by **ELISA**, **after 24 h

Aflatoxin B

incidence: 4?/4*, sa. const.: male F344 rats, wt.: 75–100 g, contamination: artificial (dose: 250 μg AFB_1 (labeled)/kg b. wt., by gavage, for 5 days a week for 2 weeks (days 0–4 and 7–11); for detailed information please see the article), conc. range: ≈≤62 pmol AFB bound mg albumin** (mean value), country: USA[611], *control, *2 days after AFB_1-administration (also at other day intervals up to 16 days measured, lowest conc.: ≈0.9 pmol AFB bound mg albumin after 16 days)
incidence: 4?/4, sa. const.: male F344 rats, wt.: 75–100 g, contamination: artificial (dose: 250 μg AFB_1 (labeled)/kg b. wt., by gavage, for 5 days a week for 2 weeks (days 0–4 and 7–11); for detailed information please see the article), conc. range: ≈≤16 pmol AFB bound mg albumin* ** (mean value), country: USA[611], *2 days after AFB_1-administration (also at other day intervals up to 16 days measured, lowest conc.: ≈0.2 pmol AFB bound mg albumin after 16 days; highest conc.: 25 pmol AFB bound mg albumin, day 3; control values at each day always higher than 1,2-dithiole-3-thione values), ****1,2-dithiole-3-thione-diet** (0.03%) 1 week before and while AFB_1-treatment

Aflatoxin

incidence: ?/?, sa. const.: male F344 rats, wt.: 100 g, contamination: artificial (dose: 20 μg AFB_1/day, o., daily for 5 weeks), conc. range: ≈≤350 pmol AF adducts/mg albumin* ** *** (mean value), country: USA[57], *control without oltipraz (for overall information please see the article), **AF-albumin adducts, ***after 3 weeks of AFB_1-administration (also at other week intervals up to 14 weeks measured, lowest conc.: ≈100 pmol AF adducts/mg albumin after 1 week of last AFB_1-administration)
incidence: ?/?, sa. const.: male F344 rats, wt.: 100 g, contamination: artificial (dose: 20 μg AFB_1/day, o., daily for 5 weeks), conc. range: ≈≤280 pmol AF adducts/mg albumin* ** *** (mean value), country: USA[57], ***delayed-transient intervention** with **oltipraz** (0.05%) for 2 weeks during AFB_1-administration (for overall information please see the article), **AF-albumin adducts, ***after 3 weeks of AFB_1-administration (also at other week intervals up to 14 weeks measured, lowest conc.: ≈75 pmol AF adducts/mg albumin after 1 week of last AFB_1-administration)
incidence: ?/?, sa. const.: male F344 rats, wt.: 100 g, contamination: artificial (dose: 20 μg AFB_1/day, o., daily for 5 weeks), conc. range: ≈≤175 pmol AF adducts/mg albumin* ** *** (mean value), country: USA[57], ***persistent intervention** with **oltipraz** (0.05%) for 6 weeks during AFB_1-administration (for overall information please see the article), **AF-albumin adducts, ***after 2 weeks of AFB_1-administration (also at other week intervals up to 14 weeks measured, lowest conc.: ≈75 pmol AF adducts/mg albumin after 1 week of last AFB_1-administration)

Ochratoxin A

incidence: ?/6, sa. const.: male Wistar rats, age: 8–10 weeks, contamination: artificial (dose: 1 mg OTA/kg, injection to the femoral vein, once), conc.: 13.2 μg/ml* ** (mean value), country: Japan[174], *in peripheral serum, **after 30 min

incidence: ?/2 (10)*, sa. const.: Sprague-Dawley male rats, wt.: ≈100 g, contamination: no OTA, conc.: nr, country: USA[218], *control
incidence: ?/2, sa. const.: Sprague-Dawley male rats, wt.: ≈100 g, contamination:

artificial (dose: **1 mg OTA** (labeled)/rat, i.p., once), conc. range: ≤178 µg/ml* (mean value), country: USA[218], *after 0.5 h (also measured after 2, 4, 8, and 24 h, lowest conc.: 34 µg/ml after 24 h)
incidence: ?/2, sa. const.: Sprague-Dawley male rats, wt.: ≈100 g, contamination: artificial (dose: **4 mg OTA**/kg b. wt., i.p., once), conc. range: ≤46 µg/ml* (mean value), country: USA[218], *after 0.5 h (also measured after 1, 4, 12, 24, and 48 h, lowest conc.: 19 µg/ml after 48 h)
incidence: ?/2, sa. const.: Sprague-Dawley male rats, wt.: ≈100 g, contamination: artificial (dose: **8 mg OTA**/kg b. wt., i.p., once), conc. range: ≤142 µg/ml* (mean value), country: USA[218], *after 0.5 h (also measured after 1 and 24 h, lowest conc.: 33 µg/ml after 24 h)

Zearalenone

incidence: 4?/4, sa. const.: male Sprague-Dawley rats, age: 8–10 weeks, wt.: 230–290 g, contamination: artificial (dose: **2 mg ZEA**/kg, i.v., once), conc.: ≈≤850 ng/ml* (mean value), country: Korea[579], *after 0 h (also after other hour intervals up to 6 h measured, lowest conc.: ≈3 ng/ml after 6 h)
incidence: 4?/4, sa. const.: male Sprague-Dawley rats, age: 8–10 weeks, wt.: 230–290 g, contamination: artificial (dose: **16 mg ZEA**/kg, o., once), conc.: ≈≤30 ng/ml* (mean value), country: Korea[579], *after 0 h (also after other hour intervals up to 24 h measured, lowest conc.: ≈7 ng/ml after 24 h)

incidence: 4?/4, sa. const.: male Sprague-Dawley rats, age: 8–10 weeks, wt.: 230–290 g, contamination: artificial (dose: **1 mg ZEA**/kg, i.v. bolus injection, once), conc. range: ≈≤450 ng/ml* (mean value), country: Korea/USA[585], *after 0 h (also after other min and hour intervals up to 6 h measured, lowest conc.: ≈0.25 ng/ml after 6 h)
incidence: 4?/4, sa. const.: male Sprague-Dawley rats, age: 8–10 weeks, wt.: 230–290 g, contamination: artificial (dose: **2 mg ZEA**/kg, i.v. bolus injection, once), conc. range: ≈≤850 ng/ml* (mean value), country: Korea/USA[585], *after 0 h (also after other min and hour intervals up to 6 h measured, lowest conc.: ≈1 ng/ml after 6 h)
incidence: 4?/4, sa. const.: male Sprague-Dawley rats, age: 8–10 weeks, wt.: 230–290 g, contamination: artificial (dose: **4 mg ZEA**/kg, i.v. bolus injection, once), conc. range: ≈≤1,100 ng/ml* (mean value), country: Korea/USA[585], *after 0 h (also after other min and hour intervals up to 9 h measured, lowest conc.: ≈0.5 ng/ml after 9 h)
incidence: 4?/4, sa. const.: male Sprague-Dawley rats, age: 8–10 weeks, wt.: 230–290 g, contamination: artificial (dose: **8 mg ZEA**/kg, i.v. bolus injection, once), conc. range: ≈≤1,400 ng/ml* (mean value), country: Korea/USA[585], *after 0 h (also after other min and hour intervals up to 9 h measured, lowest conc.: ≈1 ng/ml after 9 h)

incidence: 6?/6, sa. const.: male Sprague-Dawley rats, age: 8–10 weeks, wt.: 230–290 g, contamination: artificial (dose: 8 mg ZEA/kg, o., once), conc. range: ≈≤4.5 ng/ml* ** (mean value), country: Korea/USA[585], *after ≈0 h (also after other min and hour intervals up to 24 h measured, lowest conc.: ≈0.5 ng/ml after 24 h), **in **intact** rats
incidence: 3?/3, sa. const.: male Sprague-Dawley rats, age: 8–10 weeks, wt.: 230–290 g, contamination: artificial (dose: 8 mg ZEA/kg, o., once), conc. range: ≈≤6.6 ng/ml* ** (mean value), country: Korea/USA[585], *after ≈0 h (also after other min and hour intervals up to 12 h measured, lowest conc.: ≈0.6 ng/ml after 12 h), **in **bile duct-cannulated** rats

incidence: ?/?, sa. const.: male Sprague-Dawley rats, age: 8–10 weeks, wt.: 230–290 g, contamination: artificial (dose: **1.13 mg ZEA**/h/kg, i.v., for 6 h), Ø conc.: 258.2 ng/ml* (mean value), country: Korea/USA[585], *after 6 h of ZEA-administration
incidence: ?/?, sa. const.: male Sprague-Dawley rats, age: 8–10 weeks,

wt.: 230–290 g, contamination: artificial (dose: **2.25 mg ZEA/h/kg**, i.v., for 6 h), Ø conc.: 553.3 ng/ml* (mean value), country: Korea/USA[585], *after 6 h of ZEA-administration

Rat spleen may contain the following mycotoxins and/or their metabolites:

Zearalenone

incidence: ?/?, sa. const.: male Sprague-Dawley rats, age: 8–10 weeks, wt.: 230–290 g, contamination: artificial (dose: **1.13 mg ZEA/h/kg**, i.v., for 6 h), conc.: ≈150 ng/g* (mean value), country: Korea/USA[585], *after 6 h of ZEA-administration

incidence: ?/?, sa. const.: male Sprague-Dawley rats, age: 8–10 weeks, wt.: 230–290 g, contamination: artificial (dose: **2.25 mg ZEA/h/kg**, i.v., for 6 h), conc.: ≈250 ng/g* (mean value), country: Korea/USA[585], *after 6 h of ZEA-administration

Rat stomach may contain the following mycotoxins and/or their metabolites:

Fusaric Acid

incidence: 3?/3*, sa. const.: F1 generation of Sprague-Dawley rats, age: 4 days, contamination: no FA fed to dams (F0) (for detailed information please see the article), conc.: 18.00 ng/100 mg stomach colostrum** ***(mean value), country: USA[258], *control, **background conc. of FA in control diet (<0.3 ppm), ***on day 4 postpartum

incidence: 4?/4, sa. const.: F1 generation of Sprague-Dawley rats, age: 4 days, contamination: artificial (dose: dams (F0) **10 ppm FA** from days 11–12 of gestation, through parturition and weaning; for detailed information please see the article), conc.: 80.00 ng/100 mg stomach colostrum* (mean value), country: USA[258], *on day 4 postpartum

incidence: 4?/4, sa. const.: F1 generation of Sprague-Dawley rats, age: 4 days, contamination: artificial (dose: dams (F0) **75 ppm FA** from days 11–12 of gestation, through parturition and weaning; for detailed information please see the article), conc.: 1,449.00 ng/100 mg stomach colostrum* (mean value), country: USA[258], *on day 4 postpartum

incidence: 4?/4, sa. const.: F1 generation of Sprague-Dawley rats, age: 4 days, contamination: artificial (dose: dams (F0) **200 ppm FA** from days 11–12 of gestation, through parturition and weaning; for detailed information please see the article), conc.: 3,547.00 ng/100 mg stomach colostrum* (mean value), country: USA[258], *on day 4 postpartum

Ochratoxin A

incidence: 3/3 or more*, sa. const.: adult male Wistar rats, wt.: 200–250 g, contamination: no OTA, conc.: nr, country: Japan[143], *control

incidence: ?/3–4, sa. const.: adult male Wistar rats, wt.: 200–250 g, contamination: artificial (dose: 15 mg OTA (labeled)/kg, o., once), conc. range: ≈≤34.5 μg/g wet wt. of tissue* (mean value), country: Japan[143], *after 1 h (also at other hour intervals up to 24 h measured, lowest conc.: ≈4 μg/g wet weight of tissue after 24 h)

incidence: 3?/3, sa. const.: male Wistar rats, age: 8–10 weeks, contamination: artificial (dose: 1.4 mg OTA/kg, injected into the femoral vein, once; for detailed information please see the article), conc.: <65.0 ng/stomach/10 min* (mean value), country: Japan[174], *in stomach perfusate

Zearalenone

incidence: ?/?, sa. const.: male Sprague-Dawley rats, age: 8–10 weeks, wt.: 230–290 g, contamination: artificial (dose: **1.13 mg ZEA/h/kg**, i.v., for 6 h), conc.: ≈200 ng/g* (mean value), country: Korea/USA[585], *after 6 h of ZEA-administration

incidence: ?/?, sa. const.: male Sprague-Dawley rats, age: 8–10 weeks, wt.: 230–290 g, contamination: artificial (dose: **2.25 mg ZEA/h/kg**, i.v., for 6 h), conc.: ≈600 ng/g* (mean value), country: Korea/USA[585], *after 6 h of ZEA-administration

Rat testes may contain the following mycotoxins and/or their metabolites:

ZEARALENONE
incidence: ?/?, sa. const.: male Sprague-Dawley rats, age: 8–10 weeks, wt.: 230–290 g, contamination: artificial (dose: **1.13 mg ZEA/h/kg**, i.v., for 6 h), conc.: ≈50 ng/g* (mean value), country: Korea/USA[585], *after 6 h of ZEA-administration
incidence: ?/?, sa. const.: male Sprague-Dawley rats, age: 8–10 weeks, wt.: 230–290 g, contamination: artificial (dose: **2.25 mg ZEA/h/kg**, i.v., for 6 h), conc.: ≈150 ng/g* (mean value), country: Korea/USA[585], *after 6 h of ZEA-administration

Rat trachea may contain the following mycotoxins and/or their metabolites:

AFLATOXINS
incidence: ?/5, sa. const.: male Sprague-Dawley rats, Ø wt.: 242.2 g, contamination: artificial (dose: 300 µg/kg AFB_1 (labeled)/kg, i.t. **(dust-adsorbed)**, once; for detailed information please see the article), conc.: ≤15.762 nmol AFs/g tissue (x 10^{-3})*, country: USA[604], *after 3 h (also measured after 3 days and 3 weeks, lowest conc.: 0.031 nmol AFs/g tissue (x 10^{-3}) after 3 weeks)
incidence: ?/5, sa. const.: male Sprague-Dawley rats, Ø wt.: 242.2 g, contamination: artificial (dose: 300 µg/kg AFB_1 (labeled)/kg, i.t. (**microcrystalline form**), once; for detailed information please see the article), conc.: ≤1.461 nmol AFs/g tissue (x 10^{-3})*, country: USA[604], *after 3 h (also measured after 3 days and 3 weeks, lowest conc.: 0.0401 nmol AFs/g tissue (x10^{-3}) after 3 weeks)

Rat urine may contain the following mycotoxins and/or their metabolites:

AFLATOXIN B_1
incidence: ?/4, sa. const.: male Fischer rats, wt.: 100–110 g, contamination: artificial (dose: **0.03 mg AFB_1/kg**, i.p., once; for detailed information please see the article), conc.: ≈0.25 ng/kg* **, country: USA[52], *within 24 h, **exo-AFB_1-NAC
incidence: ?/4, sa. const.: male Fischer rats, wt.: 100–110 g, contamination: artificial (dose: **≈0.18 mg AFB_1/kg**, i.p., once; for detailed information please see the article), conc.: ≈1.7 ng/kg* **, country: USA[52], *within 24 h, **exo-AFB_1-NAC
incidence: ?/4, sa. const.: male Fischer rats, wt.: 100–110 g, contamination: artificial (dose: **0.8 mg AFB_1/kg**, i.p., once; for detailed information please see the article), conc.: ≈8 ng/kg* **, country: USA[52], *within 24 h, **exo-AFB_1-NAC

incidence: 2/3*, sa. const.: male adult Long-Evans rats, wt.: 270–320 g, contamination: no AFB_1, conc. range: 0.48**–0.53*** mg × 10^{-3}, country: Taiwan, Republic of China[132], *control, after 24** or 48*** h (also measured after 72 h, but conc.: nd)
incidence: 2/2, sa. const.: male adult Long-Evans rats, wt.: 270–320 g, contamination: artificial (dose: **5.0 mg AFB_1/kg** b. wt., i.p., once), conc. range: 14.4–20.4 mg x 10^{-3}*, Ø conc.: 17.4 mg x 10^{-3}*, country: Taiwan, Republic of China[132], *after 24 h (also measured after 48 and 72 h, but lower residue values recorded)
incidence: 1/1, sa. const.: male adult Long-Evans rat, wt.: 270–320 g, contamination: artificial (dose: **5.5 mg AFB_1/kg** b. wt., i.p., once), conc.: 33 mg x 10^{-3}*, country: Taiwan, Republic of China[132], *after 24 h (also measured after 48 and 72 h, but conc.: nd)

incidence: 3?/3, sa. const.: male F344 rats, wt.: 100–125 g, contamination: artificial (dose: 400 µg AFB_1 (labeled)/kg b. wt., by gavage, once; for detailed information please see the article), conc.: 1.00 pmol/mg creatinine* ** (mean value), country: USA[142], *after 24 h, **AFB_1-N^7-Gua

incidence: 4/4*, sa. const.: male Fisher 344 rats, wt.: 150–220 g, contamination: no AFB_1 (for detailed information please see the article), conc.: nd, country: India[168], *control
incidence: ?/?, sa. const.: male Fisher 344 rats, wt.: 150–220 g, contamination: artificial (dose: **1.0 mg AFB_1**/kg b. wt., by gavage, once; for detailed information please see the article), conc. range: 3.22–5.97 µg NAC-AFB_1/mg creatinine* (mean values), country: India[168], *after 24 h

incidence: ?/8, sa. const.: male F_{344} rats, wt.: 100–150 g, contamination: artificial (dose: 20 µg AFB_1, by gavage, daily for 28 days; for detailed information please see the article), conc.: 8.2 ng AFB_1-NAC/mg creatinine* ** (mean value), country: USA[176], *control, 1 higher value (10.0 ng AFB_1-NAC/mg creatinine) on day 21 is recorded, **after 35 days of AFB_1-administration (also at other day intervals up to 39 days measured, lowest conc.: 1.0 ng AFB_1-NAC/mg creatinine after 39 days)
incidence: ?/8, sa. const.: male F_{344} rats, wt.: 100–150 g, contamination: artificial (dose: 20 µg AFB_1, by gavage, daily for 28 days; for detailed information please see the article), conc. range: ≤0.8 ng AFB_1-NAC/mg creatinine* ** (mean value), country: USA[176], ***oltipraz-diet** (0.05%) for 5 weeks prior to AFB_1-treatment, **after 35 days of AFB_1-administration (also at other day intervals up to 39 days measured, lowest conc.: 0.2 ng AFB_1-NAC/mg creatinine after 39 days)

incidence: 8/8*, sa. const.: Fischer 344 strain rats, wt.: 180–250 g, contamination: no AFB_1, conc.: nd**, country: India[437], *control, **AFB_1-N^7-Gua adducts

incidence: ?/8, sa. const.: Fischer 344 strain rats, wt.: 180–250 g, contamination: artificial (dose: 1 mg AFB_1/kg b. wt., by gavage, once), conc. range: 6.42–20.16 µg/mg creatinine* ** (mean values), country: India[437], *AFB_1-N^7-Gua adducts, **after 48 h

incidence: ?/4, sa. const.: young male F344 rats, wt.: 120–140 g, contamination: artificial (dose: 250 µg AFB_1/kg b. wt., daily for 3 weeks), conc. range: ≤96.7 ng/mg creatinine* ** *** (mean value), country: USA/People's Republic of China[493], *control (for overall information please see the article), **AFB_1-N^7-Gua adducts, ***after 1 day while treatment (also at other day intervals up to 15 days measured, lowest conc.: ≈10 ng/mg creatinine after 9 days)
incidence: ?/4, sa. const.: young male F344 rats, weight: 120–140 g, contamination: artificial (dose: 250 µg AFB_1/kg b. wt., daily for 3 weeks), conc. range: ≈27 ng/mg creatinine* ** *** **** (mean value), country: USA/People's Republic of China[493], ***lycopene-treatment** (100 mg/kg b. wt.) after each AFB_1-administration (for overall information please see the article), **AFB_1-N^7-Gua adducts, ***after 1 day while treatment (also at other day intervals up to 15 days measured, lowest conc.: ≈5 ng/mg creatinine after 9 days), ****2 higher values (both ≈31 ng/mg creatinine) on day 5 and 15 recorded

incidence: ?/4, sa. const.: young male F344 rats, wt.: 120–140 g, contamination: artificial (dose: 250 µg AFB_1/kg b. wt., daily for 3 weeks), conc. range: ≤13.7 ng/mg creatinine* ** *** (mean value), country: USA/People's Republic of China[493], *control (for overall information please see the article), **AFB-NAC, ***after 3 days while treatment (also at other day intervals up to 15 days measured, lowest conc.: ≈4 ng/mg creatinine after 15 days) incidence: ?/4, sa. const.: young male F344 rats, wt.: 120–140 g, contamination: artificial (dose: 250 µg AFB_1/kg b. wt., daily for 3 weeks), conc. range: ≤15.9 ng/mg creatinine* ** (mean value), country: USA/People's Republic of China[493], ***lycopene-treatment** (100 mg/kg b. wt.) after each AFB_1-administration (for overall information please see the article), **AFB-NAC, ***after 3 days while treatment (also at other days up to 15 days measured, lowest conc.: ≈8 ng/mg creatinine after 12 days)

incidence: ?/5–6, sa. const.: male **African giant rats**, wt.: 1.5–2.0 kg, contamination: artificial (dose: 1 mg AFB_1 (labeled)/kg b. wt., i.v., once; for detailed information please see the article), conc.: 61 ng/g liver*, country: Nigeria[593], *after ≈8 h of treatment
incidence: ?/5–6, sa. const.: male **weanling Wistar-derived rats**, wt.: 250 g, contamination: artificial (dose: 1 mg AFB_1 (labeled)/kg b. wt., i.v., once; for detailed information please see the article), conc.: 142 ng/g liver*, country: Nigeria[593], *after ≈8 h of treatment

incidence: ?/3, sa. const.: male Fischer rats, age: 4 weeks, wt.: 140–170 g, contamination: artificial (dose: **0.125 mg AFB_1/kg**, i.p., once; for detailed information please see the article), conc.: ≈70 ng*, country: USA[598], *AFB_1-*N*^{7}-Gua[2] excreted in 48 h
incidence: ?/3, sa. const.: male Fischer rats, age: 4 weeks, wt.: 140–170 g, contamination: artificial (dose: **0.25 mg AFB_1/kg**, i.p., once; for detailed information please see the article), conc.: ≈180 ng*, country: USA[598], *AFB_1-*N*^{7}-Gua[2] excreted in 48 h
incidence: ?/3, sa. const.: male Fischer rats, age: 4 weeks, wt.: 140–170 g, contamination: artificial (dose: **0.5 mg AFB_1/kg**, i.p., once; for detailed information please see the article), conc.: ≈215 ng*, country: USA[598], *AFB_1-*N*^{7}-Gua[2] excreted in 48 h
incidence: ?/3, sa. const.: male Fischer rats, age: 4 weeks, wt.: 140–170 g, contamination: artificial (dose: **1.0 mg AFB_1/kg**, i.p., once; for detailed information please see the article), conc.: ≈765 ng*, country: USA[598], *AFB_1-*N*^{7}-Gua[2] excreted in 48 h

incidence: 4?/4*, sa. const.: male F344 rats, wt.: 75–100 g, contamination: artificial (dose: 250 µg AFB_1 (labeled)/kg b. wt., by gavage, for 5 days a week for 2 weeks (days 0–4 and 7–11); for detailed information please see the article), conc. range: ≈≤45 pmol AFB-*N*^{7}-Gua[3]/mg creatinine** (mean value), country: USA[611], *control, *2 days after AFB_1-administration (also at other day intervals up to 16 days measured, lowest conc.: nd after 16 days)
incidence: 4?/4, sa. const.: male F344 rats, wt.: 75–100 g, contamination: artificial (dose: 250 µg AFB_1 (labeled)/kg b. wt., by gavage, for 5 days a week for 2 weeks (days 0–4 and 7–11); for detailed information please see the article), conc. range: ≈≤16 pmol AFB-*N*^{7}-Gua[3]/mg creatinine* ** (mean value), country: USA[611], *2 days after AFB_1-administration (also at other day intervals up to 16 days measured, lowest conc.: nd after 14 days; control values at each day always higher than 1,2-dithiole-3-thione values), ****1,2-dithiole-3-thione-diet** (0.03%) 1 week before and while AFB_1-treatment

incidence: 4?/4*, sa. const.: male F344 rats, age: 21 days, contamination: artificial (dose: 250 µg AFB_1 (labeled)/kg b. wt., by gavage, once; for detailed information please see the article), conc.: ≈325 pmol** *** (mean value), country: USA[621], *control, **AFB-*N*^{7}-Gua[3], ***after 0–24 h
incidence: 4?/4, sa. const.: male F344 rats, age: 21 days, contamination: artificial (dose: 250 µg AFB_1 (labeled)/kg b. wt., by gavage, once; for detailed information please see the article), conc.: ≈105 pmol** *** (mean value), country: USA[621],

*oltipraz-diet (0.075%) for 7 days prior to AFB_1-treatment, **AFB-N^7-Gua[3], ***after 0–24 h

Aflatoxin B

incidence: ?/8, sa. const.: male F_{344} rats, wt.: 100–150 g, contamination: artificial (dose: 20 μg AFB_1, by gavage, daily for 28 days; for detailed information please see the article), conc.: 3.6 ng AFB-N^7-Gua/mg creatinine* ** (mean value), country: USA[176], *control, 2 higher values (3.8 and 3.9 ng AFB-N^7-Gua/mg creatinine*) on day 28 and day 35 recorded, **after 21 days of AFB_1-administration (also at other day intervals up to 39 days measured, lowest conc.: 0.9 ng AFB-N^7-guanine/mg creatinine after 39 days)

incidence: ?/8, sa. const.: male F_{344} rats, wt.: 100–150 g, contamination: artificial (dose: 20 μg AFB_1, by gavage, daily for 28 days; for detailed information please see the article), conc. range: ≤1.0 ng AFB-N^7-Gua/mg creatinine* ** (mean value), country: USA[176], ***oltipraz diet** (0.05%) fed for 5 weeks prior to AFB_1 treatment, **after 21 days of AFB_1-administration (also at other day intervals up to 39 days measured, lowest conc.: 0.3 ng AFB-N^7-Gua/mg creatinine after 39 days)

incidence: 4?/4*, sa. const.: male F344 rats, wt.: 75–100 g, contamination: artificial (dose: 250 μg AFB_1 (labeled)/kg b. wt., by gavage, for 5 days a week for 2 weeks (days 0–4 and 7–11); for detailed information please see the article), conc. range: ≈≤4.7 nmol AFB eq mg creatinine** (mean value), country: USA[611], *control, *3 days after AFB_1-administration (also at other day intervals up to 14 days measured, lowest conc.: 0.85 nmol AFB eq mg creatinine after 7 days)

incidence: 4?/4, sa. const.: male F344 rats, wt.: 75–100 g, contamination: artificial (dose: 250 μg AFB_1 (labeled)/kg b. wt., by gavage, for 5 days a week for 2 weeks (days 0–4 and 7–11); for detailed information please see the article), conc. range: ≈3.7 nmol AFB eq mg creatinine* ** (mean value), country: USA[611], *3 days after AFB_1-administration (also at other day intervals up to 14 days measured, lowest conc.: 0.8 nmol AFB eq mg creatinine after 7 days; highest conc.: 3.7 nmol AFB eq mg creatinine, day 4), ****1,2-dithiole-3-thione-diet** (0.03%) 1 week before and while AFB_1-treatment

Aflatoxin M_1

incidence: 3?/3, sa. const.: male F344 rats, wt.: 100–125 g, contamination: artificial (dose: 400 μg AFB_1 (labeled)/kg b. wt., by gavage, once; for detailed information please see the article), conc.: 0.29 pmol/mg creatinine* ** (mean value), country: USA[142], *after 24 h, **AFM_1-N^7-Gua

incidence: ?/8, sa. const.: male F_{344} rats, wt.: 100–150 g, contamination: artificial (dose: 20 μg AFB_1, by gavage, daily for 28 days; for detailed information please see the article), conc.: 8.5 ng/mg creatinine* ** (mean value), country: USA[176], *control, several higher values up to 15.7 ng/mg creatinine after 28 days were recorded, **after 7 days of AFB_1-administration (also at other day intervals up to 39 days measured, lowest conc.: 2.0 ng/mg creatinine after 39 days)

incidence: ?/8, sa. const.: male F_{344} rats, wt.: 100–150 g, contamination: artificial (dose: 20 μg AFB_1, by gavage, daily for 28 days; for detailed information please see the article), conc. range: ≤1.5 ng/mg* creatinine* ** (mean value), country: USA[176], ***oltipraz-diet** (0.05%) for 5 weeks prior to AFB_1-treatment, **after 7 days of AFB_1-administration (also at other day intervals up to 39 days measured, lowest conc.: 0.05 ng/mg creatinine after 39 days)

incidence: ?/4, sa. const.: young male F344 rats, wt.: 120–140 g, contamination: artificial (dose: 250 μg AFB_1/kg b. wt., daily for 3 weeks), conc. range: ≤166.1 ng/mg creatinine* ** (mean value), country: USA/People's Republic of China[493], *control (for overall information please see the article), **after 5 days while treatment (also at other day intervals up to 15 days measured, lowest conc.: ≈105 ng/mg creatinine after 1 day)

incidence: ?/4, sa. const.: young male F344 rats, wt.: 120–140 g, contamination: artificial (dose: 250 μg AFB_1/kg b. wt., daily for 3 weeks), conc. range: ≈≤160 ng/mg creatinine* ** (mean value), country: USA/People's Republic of China[493], ***lycopene-treatment** (100 mg/kg b. wt.) after each AFB_1-administration (for overall information please see the article), **after 5 days while treatment (also at other day intervals up to 15 days measured, lowest conc.: ≈78 ng/mg creatinine after 1 day)

incidence: 3?/3*, sa. const.: male Fischer-344 rats, age: 6 weeks, contamination: artificial (dose: **0.125 mg AFB_1/kg** b. wt., o.; for detailed information please see the article), conc.: 1.32 μg total** (mean value), country: USA[522], *control, **cumulative value of 36 h collection (measured after 6, 24, 36, and 48 h)

incidence: 3?/3*, sa. const.: male Fischer-344 rats, age: 6 weeks, contamination: artificial (dose: **0.125 mg AFB_1/kg** b. wt., o.; for detailed information please see the article), conc.: 0.07 μg total** (mean value), country: USA[522], ***HSCAS-diet** (0.5%) prior to AFB_1-treatment, **cumulative value of 36 h collection (measured after 6, 24, 36, and 48 h)

incidence: 3?/3*, sa. const.: male Fischer-344 rats, age: 6 weeks, contamination: artificial (dose: **0.25 mg AFB_1/kg** b. wt., o.; for detailed information please see the article), conc.: 2.12 μg total** (mean value), country: USA[522], *control, **cumulative value of 36 h collection (measured after 6, 24, 36, and 48 h)

incidence: 3?/3*, sa. const.: male Fischer-344 rats, age: 6 weeks, contamination: artificial (dose: **0.25 mg AFB_1/kg** b. wt., o.; for detailed information please see the article), conc.: 0.17 μg total** (mean value), country: USA[522], ***HSCAS-diet** (0.5%) prior to AFB_1-treatment, **cumulative value of 36 h collection (measured after 6, 24, 36, and 48 h)

incidence: 3?/3*, sa. const.: male Fischer-344 rats, age: 6 weeks, contamination: artificial (dose: **0.5 mg AFB_1/kg** b. wt., o.; for detailed information please see the article), conc.: 1.78 μg total** (mean value), country: USA[522], *control, **cumulative value of 36 h collection (measured after 6, 24, 36, and 48 h)

incidence: 3?/3*, sa. const.: male Fischer-344 rats, age: 6 weeks, contamination: artificial (dose: **0.5 mg AFB_1/kg** b. wt., o.; for detailed information please see the article), conc.: 0.50 μg total** (mean value), country: USA[522], ***HSCAS-diet** (0.5%) prior to AFB_1-treatment, **cumulative value of 36 h collection (measured after 6, 24, 36, and 48 h)

incidence: 3?/3*, sa. const.: male Fischer-344 rats, age: 6 weeks, contamination: artificial (dose: **1 mg AFB_1/kg** b. wt., o.; for detailed information please see the article), conc.: 2.51 μg total** (mean value), country: USA[522], *control, **cumulative value of 36 h collection (measured after 6, 24, 36, and 48 h)

incidence: 3?/3*, sa. const.: male Fischer-344 rats, age: 6 weeks, contamination: artificial (dose: **1 mg AFB_1/kg** b. wt., o.; for detailed information please see the article), conc.: 1.04 μg total** (mean value), country: USA[522], ***HSCAS-diet** (0.5%) prior to AFB_1-treatment, **cumulative value of 36 h collection (measured after 6, 24, 36, and 48 h)

incidence: ?/5–6, sa. const.: male **African giant rats**, wt.: 1.5–2.0 kg, contamination: artificial (dose: 1 mg AFB_1 (labeled)/kg b. wt., i.v., once; for detailed information

please see the article), conc.: 48 ng/g liver*, country: Nigeria[593], *after ≈8 h of treatment
incidence: ?/5–6, sa. const.: male **weanling Wistar-derived rats**, wt.: 250 g, contamination: artificial (dose: 1 mg AFB_1 (labeled)/kg b. wt., i.v., once; for detailed information please see the article), conc.: 20 ng/g liver*, country: Nigeria[593], *after ≈8 h of treatment

Aflatoxin P_1
incidence: ?/4, sa. const.: young male F344 rats, wt.: 120–140 g, contamination: artificial (dose: 250 µg AFB_1/kg b. wt., daily for 3 weeks), conc. range: ≤210.4 ng/mg creatinine* ** (mean value), country: USA/People's Republic of China[493], *control (for overall information please see the article), **after 12 days while treatment (also at other day intervals up to 15 days measured, lowest conc.: ≈40 ng/mg creatinine after 3 days)
incidence: ?/4, sa. const.: young male F344 rats, wt.: 120–140 g, contamination: artificial (dose: 250 µg AFB_1/kg b. wt., daily for 3 weeks), conc.: ≈25 ng/mg creatinine* ** *** (mean value), country: USA/People's Republic of China[493], ***lycopene-treatment** (100 mg/kg b. wt.) after each AFB_1-administration (for overall information please see the article), **after 12 days while treatment (also at other day intervals up to 15 days measured, lowest conc.: ≈10 ng/mg creatinine after 1 day), ***2 higher values on day 5 (≈55 ng/mg creatinine) and on day 15 (≈100 ng/mg creatinine) recorded

Aflatoxin Q_1
incidence: ?/4, sa. const.: young male F344 rats, wt.: 120–140 g, contamination: artificial (dose: 250 µg AFB_1/kg b. wt., daily for 3 weeks), conc. range: ≤31.7 ng/mg creatinine* ** (mean value), country: USA/People's Republic of China[493], *control (for overall information please see the article), **after 12 days while the treatment (also at other day intervals up to 15 days measured, lowest conc.: ≈1 ng/mg creatinine after 3 days)
incidence: ?/4, sa. const.: young male F344 rats, wt.: 120–140 g, contamination: artificial (dose: 250 µg AFB_1/kg b. wt., daily for 3 weeks), conc. range: ≈3 ng/mg creatinine* ** *** (mean value), country: USA/People's Republic of China[493], ***lycopene-treatment** (100 mg/kg b. wt.) after each AFB_1-administration (for overall information please see the article), **after 12 days while treatment (also at other day intervals up to 15 days measured, lowest conc.: ≈0.5 ng/mg creatinine after 5 days), ***1 higher value (≈18 ng/mg creatinine) on day 15 recorded

incidence: ?/5–6, sa. const.: male **African giant rats**, wt.: 1.5–2.0 kg, contamination: artificial (dose: 1 mg AFB_1 (labeled)/kg b. wt., i.v., once; for detailed information please see the article), conc.: 180 ng/g liver*, country: Nigeria[593], *after ≈8 h of treatment
incidence: ?/5–6, sa. const.: male **weanling Wistar-derived rats**, wt.: 250 g, contamination: artificial (dose: 1 mg AFB_1 (labeled)/kg b. wt., i.v., once; for detailed information please see the article), conc.: 171 ng/g liver*, country: Nigeria[593], *after ≈8 h of treatment

Aflatoxin
incidence: 2/2, sa. const.: adult male Fischer 344 rats, contamination: artificial (dose: 1 mg AFB_1 (labeled)/kg b. wt., i.p., once; for detailed information please see the article), conc. range: 16.0–17.0 pmol AF eq* **, Ø conc.: 16.5 pmol AF eq* **, country: USA/People's Republic of China[59], *mainly AFM_1, AFB_1-*N*^{7}-Gua[2] adduct, AFP_1, and AFB_1 (in decreasing quantity), **urine was collected for 20 h

incidence: 2/2, sa. const.: adult male rats, age:, contamination: artificial (dose: 1 mg AFB_1 (labeled)/kg b. wt., injected, once), conc. range: 290–310 ng/100 µl*, Ø conc.: 300 ng/100 µl*, country: USA[100], *urine was collected for 20 h

Aflatoxins
incidence: ?/5, sa. const.: male Sprague-Dawley rats, Ø wt.: 242.2 g,

contamination: artificial (dose: 300 µg/kg AFB_1 (labeled)/kg, i.t. **(dust-adsorbed)**, once; for detailed information please see the article), conc.: ≈≤11.4 nmol AFs/day*, country: USA[604], *after 1 day (also at other day intervals up to 21 days measured, lowest conc.: nd after 21 days)
incidence: ?/5, sa. const.: male Sprague-Dawley rats, Ø wt.: 242.2 g, contamination: artificial (dose: 300 µg/kg AFB_1 (labeled)/kg, i.t. **(microcrystalline form)**, once; for detailed information please see the article), conc.: ≈≤11.5 nmol AFs/day*, country: USA[604], *after 1 day (also at other day intervals up to 21 days measured, lowest conc.: nd after 21 days)

FUMONISIN B_1
incidence: ?/5, sa. const.: rats, contamination: artificial (dose: 1,000 µg FB_1/g, o., once; for detailed information please see the article), conc. range: 0.1–2.0 µg/g (mean value), country: USA[276]

OCHRATOXIN A
incidence: 3 or more/3 or more*, sa. const.: healthy female Sprague-Dawley rats, wt.: ≈300 g, contamination: no OTA (for detailed information please see the article), conc.: nr, country: Canada[144], *control
incidence: 3?/3, sa. const.: healthy female Sprague-Dawley rats, wt.: ≈300 g, contamination: artificial (dose: **100 µg OTA**/rat, i.v., once; for detailed information please see the article), conc.: 5.9 nmol/ml (mean value), country: Canada[144]

incidence: 10/10*, sa. const.: Wistar male rats, wt.: 83–110 g, contamination: no OTA, **diet B-II** (for detailed information please see the article), conc.: nd, country: Canada[209], *control
incidence: ?/10, sa. const.: Wistar male rats, wt.: 83–110 g, contamination: artificial (dose: 500 µg OTA, intubated, daily for 6 days (**diet B-I**); for detailed information please see the article), conc.: pr*, country: Canada[209], *measured at day 1, 2, 3, 4, 5, and 6 ? of OTA-administration
incidence: 10/10*, sa. const.: Wistar male rats, wt.: 83–110 g, contamination: no OTA, **diet NMB** (for detailed information please see the article), conc.: nd, country: Canada[209], *control
incidence: 6?/6, sa. const.: Wistar male rats, wt.: 83–110 g, contamination: artificial (dose: 500 µg OTA, intubated, daily for 6 days (**diet NMB+T**); for detailed information please see the article), conc. range: ≤43 µg/rat/day* (mean value), country: Canada[209], *measured on day 5 of OTA-administration (also measured at day 1, 2, 3, 4, and 6, lowest conc.: approximately 4 µg/rat/day on day 1)

incidence: ?/2 (10)*, sa. const.: Sprague-Dawley male rats, wt.: ≈100 g, contamination: no OTA, conc.: nr, country: USA[218], *control
incidence: ?/2, sa. const.: Sprague-Dawley male rats, wt.: ≈100 g, contamination: artificial (dose: **1 mg OTA** (labeled)/rat, i.p., once), conc. range: ≤223 µg* OTA content (mean value), country: USA[218], *after 24 h (also measured after 2, 4, and 8 h, lowest conc.: 45 µg OTA after 2 h)

incidence: ?/3, sa. const.: male Wistar rats, age: **6 weeks**, contamination: artificial (dose: **0.5 mg OTA**/kg b. wt., i.p., for 6 days), conc.: 0.36 µmol/l* (mean value), country: Germany[561], *after 24 h of final administration
incidence: ?/3, sa. const.: male Wistar rats, age: **2 years**, contamination: artificial (dose: **0.5 mg OTA**/kg b. wt., i.p., for 6 days), conc.: 0.36 µmol/l* (mean value), country: Germany[561], *after 24 h of final administration
incidence: ?/3, sa. const.: male Wistar rats, age: **2 years**, contamination: artificial (dose: **1.25 mg OTA**/kg b. wt., i.p., once), conc.: 22.4 µmol/l* (mean value), country: Germany[561], *after 2 h

incidence: 1–3?/1–3, sa. const.: **normal** Sprague-Dawley rats, age: 9–10 weeks, contamination: artificial (dose: **4.1 mg OTA**/kg, by injection, once), conc. range:

≈≤1 µg/ml?* (mean value), country: Japan[613], *after 30 min collection period (also measured after 30–60 and 60–90 min collection period, lowest conc.: <0.5 µg/ml after 60–90 min)
incidence: 2?/2, sa. const.: **albumin-deficient** Sprague-Dawley rats, age: 9–10 weeks, contamination: artificial (dose: **4.1 mg OTA/kg**, by injection, once), conc.: ≈≤165 µg/ml* (mean value), country: Japan[613], *after 30 min collection period (also measured after 30–60 and 60–90 min collection period, lowest conc.: ≈15 µg/ml after 60–90 min)

Ochratoxin α
incidence: 3 or more/3 or more*, sa. const.: healthy female Sprague-Dawley rats, wt.: ≈300 g, contamination: no OTα (for detailed information please see the article), conc.: nr, country: Canada[144], *control
incidence: 3?/3, sa. const.: healthy female Sprague-Dawley rats, weight: ≈300 g, contamination: artificial (dose: **100 µg OTα**/rat, i.v., once; for detailed information please see the article), conc.: 25.4 nmol/ml (mean value), country: Canada[144]

incidence: 10/10*, sa. const.: Wistar male rats, wt.: 83–110 g, contamination: no OTA, **diet B-II** (for detailed information please see the article), conc.: nd, country: Canada[209], *control
incidence: ?/10, sa. const.: Wistar male rats, wt.: 83–110 g, contamination: artificial (dose: 500 µg OTA, intubated, daily for 6 days (**diet B-I**); for detailed information please see the article), conc.: pr*, country: Canada[209], *measured at day 1, 2, 3, 4, 5, and 6 ? of OTA-administration
incidence: 10/10*, sa. const.: Wistar male rats, wt.: 83–110 g, contamination: no OTA, **diet NMB** (for detailed information please see the article), conc.: nd, country: Canada[209], *control
incidence: 6?/6, sa. const.: Wistar male rats, wt.: 83–110 g, contamination: artificial (dose: 500 µg OTA, intubated, daily for 6 days (**diet NMB+T**); for detailed information please see the article), conc.: nd, country: Canada[209], *measured at day 1, 2, 3, 4, 5, and 6

Lactone Opened Ochratoxin A
incidence: 3 or more/3 or more*, sa. const.: healthy female Sprague-Dawley rats, wt.: ≈300 g, contamination: no OP-OTA (for detailed information please see the article), conc.: nr, country: Canada[144], *control
incidence: 3?/3, sa. const.: healthy female Sprague-Dawley rats, wt.: ≈300 g, contamination: artificial (dose: **100 µg OP-OTA**/rat, i.v., once; for detailed information please see the article), conc.: 3.9 nmol/ml (mean value), country: Canada[144]

4-Hydroxyochratoxin A
incidence: 3 or more/3 or more*, sa. const.: healthy female Sprague-Dawley rats, wt.: ≈300 g, contamination: no OTA-OH (for detailed information please see the article), conc.: nr, country: Canada[144], *control
incidence: 3?/3, sa. const.: healthy female Sprague-Dawley rats, wt.: ≈300 g, contamination: artificial (dose: **100 µg OTA-OH**/rat, i.v., once; for detailed information please see the article), conc.: 27.2 nmol/ml (mean value), country: Canada[144]

Ochratoxin B
incidence: 3 or more/3 or more*, sa. const.: healthy female Sprague-Dawley rats, wt.: ≈300 g, contamination: no OTB (for detailed information please see the article), conc.: nr, country: Canada[144], *control
incidence: 3?/3, sa. const.: healthy female Sprague-Dawley rats, wt.: ≈300 g, contamination: artificial (dose: **100 µg OTB**/rat, i.v., once; for detailed information please see the article), conc.: 3.9 nmol/ml (mean value), country: Canada[144]

Zearalenone
incidence: ?/6, sa. const.: female weanling Wistar rats, contamination: artificial (dose: 250 µg crystalline ZEA/g diet (daily)

in combination with a casein-based semipurified diet for 2 weeks; for detailed information please see the article), conc.: 209.6 µg* (mean value), country: Canada[402], *excreted in 48 h following dosing
incidence: ?/6, sa. const.: female weanling Wistar rats, contamination: artificial (dose: 250 µg crystalline ZEA/g diet (daily) in combination with a casein-based semipurified diet for 2 weeks; for detailed information please see the article), conc.: 95.0 µg* ** (mean value), country: Canada[402], *excreted in 48 h following dosing, **ZEA-Gluc**

α-Zearalenol
incidence: ?/6, sa. const.: female weanling Wistar rats, contamination: artificial (dose: 250 µg crystalline ZEA/g diet (daily) in combination with a casein-based semipurified diet for 2 weeks; for detailed information please see the article), conc.: 34.5 µg* (mean value), country: Canada[402], *excreted in 48 h following dosing
incidence: ?/6, sa. const.: female weanling Wistar rats, contamination: artificial (dose: 250 µg crystalline ZEA/g diet (daily) in combination with a casein-based semipurified diet for 2 weeks; for detailed information please see the article), conc.: 11.4 µg* ** (mean value), country: Canada[402], *excreted in 48 h following dosing, **α-ZEAOL-Gluc**

β-Zearalenol
incidence: ?/6, sa. const.: female weanling Wistar rats, contamination: artificial (dose: 250 µg crystalline ZEA/g diet (daily) in combination with a casein-based semipurified diet for 2 weeks; for detailed information please see the article), conc.: 19.0 µg* (mean value), country: Canada[402], *excreted in 48 h following dosing
incidence: ?/6, sa. const.: female weanling Wistar rats, contamination: artificial (dose: 250 µg crystalline ZEA/g diet (daily) in combination with a casein-based semipurified diet for 2 weeks; for detailed information please see the article), conc.: 1.5 µg* ** (mean value), country: Canada[402], *excreted in 48 h following dosing, **β-ZEAOL-Gluc**

Salmon see Fish (salmon)

Sheep

Sheep Natural Contamination

Sheep milk, raw may contain the following mycotoxins and/or their metabolites:

Aflatoxin M_1
incidence: 8/12*, sa. const.: milk from sheeps of Greece, contamination: natural, conc. range: 5–10 ng/l (3 sa), 11–20 ng/l (3 sa), 21–50 ng/l (2 sa), country: Greece[246], *from December 1999 to May 2000
incidence: 11/15*, sa. const.: milk from sheeps of Greece, contamination: natural, conc. range: 5–10 ng/l (6 sa), 11–20 ng/l (3 sa), 21–50 ng/l (1 sa), 53 ng/l (1 sa), country: Greece[246], *from December 2000 to May 2001

Sheep/goat milk, raw may contain the following mycotoxins and/or their metabolites:

Aflatoxin M_1
incidence: 7/17, sa. const.: milk from sheeps and goats of Italy, contamination: natural, conc. range: 6–31 ng/kg, country: Italy[161]

Sheep urine may contain the following mycotoxins and/or their metabolites:

Zearalanols
incidence: 39/80*, sa. const.: urine from sheeps of New Zealand, contamination: natural, conc. range: ≤2.1 ng/ml**, country: New Zealand[230], *export animals, **most probable of *Fusarium* origin

Zearalenols

incidence: 39/80*, sa. const.: urine from sheeps of New Zealand, contamination: natural, conc. range: ≤86 ng/ml, country: New Zealand[230], *export animals

Sheep Artificial Contamination

Sheep bile may contain the following mycotoxins and/or their metabolites:

Deoxynivalenol

incidence: 2/2, sa. const.: shorn wethers, age: 1 year, wt.: 44–70 kg, contamination: artificial (dose: **0.5 mg DON**/kg b. wt., **i.v.,** once; for detailed information please see the article), conc.: <0.1 mg* **, country: Canada[63], *cumulative value, **collected and measured up to 36 h after DON-administration

incidence: 2/2, sa. const.: shorn wethers, age: 1 year, wt.: 48–70 kg, contamination: artificial (dose: **5.0 mg DON**/kg b. wt., **o.,** once; for detailed information please see the article), conc.: <0.1 mg* **, country: Canada[63], *cumulative value, **collected and measured up to 36 h after DON-administration

incidence: 2/2, sa. const.: shorn wethers, age: 1 year, wt.: 44–70 kg, contamination: artificial (dose: **0.5 mg DON**/kg b. wt., **i.v.,** once; for detailed information please see the article), conc.: <0.1 mg* ** ***, country: Canada[63], ***DON-Gluc,** **cumulative value, ***collected and measured up to 36 h after DON-administration

incidence: 2/2, sa. const.: shorn wethers, age: 1 year, wt.: 48–70 kg, contamination: artificial (dose: **5.0 mg DON**/kg b. wt., **o.,** once; for detailed information please see the article), conc.: <0.1 mg* ** ***, country: Canada[63], ***DON-Gluc,** **cumulative value, ***collected and measured up to 36 h after DON-administration

Deepoxydeoxynivalenol

incidence: 2/2, sa. const.: shorn wethers, age: 1 year, wt.: 44–70 kg, contamination: artificial (dose: **0.5 mg DON**/kg b. wt., **i.v.,** once; for detailed information please see the article), conc.: tr* **, country: Canada[63], *cumulative value, **collected and measured up to 36 h after DON-administration

incidence: 2/2, sa. const.: shorn wethers, age: 1 year, wt.: 48–70 kg, contamination: artificial (dose: **5.0 mg DON**/kg b. wt., **o.,** once; for detailed information please see the article), conc.: tr* **, country: Canada[63], *cumulative value, **collected and measured up to 36 h after DON-administration

incidence: 2/2, sa. const.: shorn wethers, age: 1 year, wt.: 44–70 kg, contamination: artificial (dose: **0.5 mg DON**/kg b. wt., **i.v.,** once; for detailed information please see the article), conc. range: 0.8–1.2 mg* ** ***, Ø conc.: 1 mg* ** ***, country: Canada[63], ***DOM-1-Gluc,** **cumulative value, ***collected and measured up to 36 h after DON-administration

incidence: 2/2, sa. const.: shorn wethers, age: 1 year, wt.: 48–70 kg, contamination: artificial (dose: **5.0 mg DON**/kg b. wt., **o.,** once; for detailed information please see the article), conc. range: 0.2–0.5 mg* ** ***, Ø conc.: 0.35 mg* ** ***, country: Canada[63], ***DOM-1-Gluc,** **cumulative value, ***collected and measured up to 36 h after DON-administration

Sheep feces may contain the following mycotoxins and/or their metabolites:

Deoxynivalenol

incidence: 2/2, sa. const.: shorn wethers, age: 1 year, wt.: 44–70 kg, contamination: artificial (dose: **0.5 mg DON**/kg b. wt., **i.v.,** once; for detailed information please see the article), conc.: na, country: Canada[63]

incidence: 2/2, sa. const.: shorn wethers, age: 1 year, wt.: 48–70 kg, contamination: artificial (dose: **5.0 mg DON**/kg b. wt., **o.,** once; for detailed information please see the article), conc.

range: 110–120 mg*, Ø conc.: 115 mg*, country: Canada[63], *cumulative value(s) of the first 24 h collection after DON-administration

Deepoxydeoxynivalenol
incidence: 2/2, sa. const.: shorn wethers, age: 1 year, wt.: 44–70 kg, contamination: artificial (dose: **0.5 mg DON**/kg b. wt., **i.v.,** once; for detailed information please see the article), conc.: na, country: Canada[63]
incidence: 2/2, sa. const.: shorn wethers, age: 1 year, wt.: 48–70 kg, contamination: artificial (dose: **5.0 mg DON**/kg b. wt., **o.,** once; for detailed information please see the article), conc. range: 70 mg* **, Ø conc.: 70 mg* **, country: Canada[63], *cumulative value of the first 24 h collection after DON-administration

Fumonisin B_1
incidence: ?/6, sa. const.: sheeps, contamination: artificial (dose: 50 µg FB_1/g, o., once; for detailed information please see the article), conc.: 6 µg/g (mean value), country: USA[276]

Hydrolized Fumonisin B_1
incidence: ?/6, sa. const.: sheeps, contamination: artificial (dose: 50 µg FB_1/g, o., once; for detailed information please see the article), conc.: 10 µg/g (mean value), country: USA[276]

Ochratoxin A
incidence: 4/4*, sa. const.: crossbred wethers, wt.: ≈66 kg, contamination: no OTA (for detailed information please see the article), conc.: nd, country: Germany/ Canada[283], *control
incidence: 4?/4, sa. const.: crossbred wethers, wt.: ≈66 kg, contamination: artificial (dose: **2 mg OTA**/kg feed (70% concentrates, 30% hay), o., for 4 weeks; for detailed information please see the article), conc.: 23.0 ng/kg* ** (mean value), country: Germany/ Canada[283], *in dry matter, **collected over a period of 7 days in the 3rd week
incidence: 4?/4, sa. const.: crossbred wethers, wt.: ≈66 kg, contamination: artificial (dose: **5 mg OTA**/kg feed (70% concentrates, 30% hay), o., for 4 weeks; for detailed information please see the article), conc.: 32.0 ng/kg* ** (mean value), country: Germany/ Canada[283], *in dry matter, **collected over a period of 7 days in the 3rd week

incidence: 3/3*, sa. const.: castrated male sheeps (Coburger Fuchsschaf), age: 1 year, wt.: ≈39.3 kg, contamination: no OTA (for detailed information please see the article), conc.: nd, country: Germany/ Canada[482], *control
incidence: 3?/3, sa. const.: castrated male sheeps (Coburger Fuchsschaf), age: 1 year, wt.: ≈39.3 kg, contamination: artificial (dose: **9.5 µg OTA**/kg b. wt./day, o., for 29 days; for detailed information please see the article), conc.: 30.8 ng/g* ** (mean value), country: Germany/Canada[482], *in dry matter, **collected over 7 days (day 15 until day 21 after the 1st OTA-administration, lowest conc.: 0 at day 1)
incidence: 3?/3, sa. const.: castrated male sheeps (Coburger Fuchsschaf), age: 1 year, wt.: ≈39.3 kg, contamination: artificial (dose: **19.0 µg OTA**/kg b. wt./ day, o., for 29 days; for detailed information please see the article), conc.: 57.2 ng/g* ** (mean value), country: Germany/Canada[482], *in dry matter, **collected over 7 days (day 15 until day 21 after the 1st OTA-administration, lowest conc.: 0 at day 1)
incidence: 3?/3, sa. const.: castrated male sheeps (Coburger Fuchsschaf), age: 1 year, wt.: ≈39.3 kg, contamination: artificial (dose: **28.5 µg OTA**/kg b. wt./ day, o., for 29 days; for detailed information please see the article), conc.: 123.1 ng/g* ** (mean value), country: Germany/Canada[482], *in dry matter, **collected over 7 days (day 15 until day 21 after the 1st OTA-administration, lowest conc.: 0 at day 1)

Ochratoxin α
incidence: 4/4*, sa. const.: crossbred wethers, wt.: ≈66 kg, contamination: no OTA (for detailed information please see the article), conc.: nd, country: Germany/ Canada[283], *control
incidence: 4?/4, sa. const.: crossbred wethers, wt.: ≈66 kg, contamination: artificial (dose: **2 mg OTA**/kg feed (70% concentrates, 30% hay), o., for 4 weeks; for detailed information please see the article), conc.: 251.5 ng/kg* ** (mean value), country: Germany/ Canada[283], *in dry matter, **collected over a period of 7 days in the 3rd week
incidence: 4?/4, sa. const.: crossbred wethers, wt.: ≈66 kg, contamination: artificial (dose: **5 mg OTA**/kg feed (70% concentrates, 30% hay), o., for 4 weeks; for detailed information please see the article), conc.: 796.2 ng/kg* ** (mean value), country: Germany/ Canada[283], *in dry matter, **collected over a period of 7 days in the 3rd week

incidence: 3/3*, sa. const.: castrated male sheeps (Coburger Fuchsschaf), age: 1 year, wt.: ≈39.3 kg, contamination: no OTA (for detailed information please see the article), conc.: nd, country: Germany/ Canada[482], *control
incidence: 3?/3, sa. const.: castrated male sheeps (Coburger Fuchsschaf), age: 1 year, wt.: ≈39.3 kg, contamination: artificial (dose: **9.5 μg OTA**/kg b. wt./day, o., for 29 days; for detailed information please see the article), conc.: 212.4 ng/g dry matter* (mean value), country: Germany/ Canada[482], *collected over 7 days (day 15 until day 21 after the 1st OTA-administration)
incidence: 3?/3, sa. const.: castrated male sheeps (Coburger Fuchsschaf), age: 1 year, wt.: ≈39.3 kg, contamination: artificial (dose: **19.0 μg OTA**/kg b. wt./day, o., for 29 days; for detailed information please see the article), conc.: 396.0 ng/g dry matter* (mean value), country: Germany/ Canada[482], *collected over 7 days (day 15 until day 21 after the 1st OTA-administration)
incidence: 3?/3, sa. const.: castrated male sheeps (Coburger Fuchsschaf), age: 1 year, wt.: ≈39.3 kg, contamination: artificial (dose: **28.5 μg OTA**/kg b. wt./day, o., for 29 days; for detailed information please see the article), conc.: 809.9 ng/g dry matter* (mean value), country: Germany/Canada[482], *collected over 7 days (day 15 until day 21 after the 1st OTA-administration)

Sheep milk, raw may contain the following mycotoxins and/or their metabolites:

Aflatoxin M_1
incidence: 6/6*, sa. const.: milk from ewes, contamination: no AFB_1; for detailed information please see the article), conc.: <50 ng/l, country: Italy[520], *control
incidence: 24/24, sa. const.: milk from ewes, contamination: artificial (dose: **feed with increasing amounts of AFB_1**; for detailed information please see the article), conc. range: 121–1,246 ng/l, Ø conc.: 547.8 ng/l, country: Italy[520]

Deoxynivalenol
incidence: 2/2*, sa. const.: lactating ewes, age: 1 year, wt.: 70–80 kg, contamination: no DON (for detailed information please see the article), conc.: nd, country: Canada[284], *control
incidence: 2/2, sa. const.: lactating ewes, age: 1 year, wt.: 70–80 kg, contamination: artificial (dose: **4.0 mg DON** (labeled)/kg b. wt., **i.v.,** once; for detailed information please see the article), conc. range: ≤61 ng/ml* (in total: 7.8–33.0 μg), country: Canada[284], *0–4 h sa. collection (collection period over 48 h, lowest conc.: 0 from 16–20, 20–24 and 44–48 h)
incidence: 1/1, sa. const.: lactating ewes, age: 1 year, wt.: 70–80 kg, contamination: artificial (dose: **1.5 g DON, o.,** during 72 h dosing period; for detailed information please see the article), conc. range:

≤10 ng/ml* **, country: Canada[284], *conjugated and unconjugated DON, **after 4–8 h post-DON period (measured over 72 h (DON feeding period) and afterwards over 48 h (post-DON period, lowest conc.: nd at 44–48 h)
incidence: 1/1, sa. const.: lactating ewes, age: 1 year, wt.: 70–80 kg, contamination: artificial (dose: **4.0 g DON, o.,** during 72 h dosing period; for detailed information please see the article), conc. range: ≤17 ng/ml* **, country: Canada[284], *conjugated and unconjugated DON, **after 4–8 h post-DON period (measured over 72 h (DON-feeding period) and afterwards over 48 h (post-DON period, lowest conc.: nd at 44–48 h)

Deepoxydeoxynivalenol
incidence: 2/2*, sa. const.: lactating ewes, age: 1 year, wt.: 70–80 kg, contamination: no DON (for detailed information please see the article), conc.: nd, country: Canada[284], *control
incidence: 2/2, sa. const.: lactating ewes, age: 1 year, wt.: 70–80 kg, contamination: artificial (dose: **4.0 mg DON** (labeled)/kg b. wt., **i.v.,** once; for detailed information please see the article), conc. range: ≤1,220 ng/ml* (in total: 310.1–610.4 μg), country: Canada[284], *4–8 h sa. collection (collection period over 48 h, lowest conc.: 0 at 44–48 h for 1 sheep)
incidence: 1/1, sa. const.: lactating ewe, age: 1 year, wt.: 70–80 kg, contamination: artificial (dose: **1.5 g DON, o.,** during 72 h dosing period; for detailed information please see the article), conc. range: ≤125 ng/ml* **, country: Canada[284], *conjugated and unconjugated DOM-1, **after 4–8 h post-DON period (measured over 72 h (DON-feeding period) and afterwards over 48 h (post-DON period, lowest conc.: 15 ng/ml at 44–48 h)
incidence: 1/1, sa. const.: lactating ewe, age: 1 year, wt.: 70–80 kg, contamination: artificial (dose: **4.0 g DON, o.,** during 72 h dosing period; for detailed information please see the article), conc. range: ≤205 ng/ml* **, country: Canada[284], *conjugated and unconjugated DOM-1, **after 4–8 h post-DON period (measured over 72 h (DON-feeding period) and afterwards over 48 h (post-DON period, lowest conc.: tr at 44–48 h)

Zearalenone
incidence: 3/6, sa. const.: sheeps, contamination: artificial (dose: 1.8 g crystalline ZEA, into the esophagus, once), conc. range: 0.001–0.002 ppm*, country: Hungary/USA[123], *after 24, 48 and 72 h (up to 120 h measured, lowest value conc.: nd 0, 96 and 120 h)

β-Zearalenol
incidence: 3/6, sa. const.: sheeps, contamination: artificial (dose: 1.8 g crystalline ZEA, into the esophagus, once), conc. range: 0.001–0.002 ppm*, country: Hungary/USA[123], *after 24, 48 and 72 h (up to 120 h measured, lowest value conc.: nd 0, 96 and 120 h)

Sheep plasma may contain the following mycotoxins and/or their metabolites:

Deoxynivalenol
incidence: 1/1, sa. const.: lactating ewe, age: 1 year, wt.: 70–80 kg, contamination: artificial (dose: **1.5 g DON,** o., during 72 h dosing period; for detailed information please see the article), conc. range: tr*, country: Canada[284], *after 53 h DON-feeding period (measured over 72 h (DON-feeding period) and afterwards over 48 h (post-DON-period, lowest conc.: nd at 24 and 48 h)
incidence: 1/1, sa. const.: lactating ewe, age: 1 year, wt.: 70–80 kg, contamination: artificial (dose: **1.5 g DON,** o., during 72 h dosing period; for detailed information please see the article), conc. range: ≤19 ng/ml* **, country: Canada[284], ***conjugated DON,** **after 53 h

DON-feeding period (measured over 72 h (DON-feeding period) and afterwards over 48 h (post-DON-period, lowest conc.: nd at 24 and 48 h)
incidence: 1/1, sa. const.: lactating ewe, age: 1 year, wt.: 70–80 kg, contamination: artificial (dose: **4.0 g DON,** o., during 72 h dosing period; for detailed information please see the article), conc. range: ≤30 ng/ml*, country: Canada[284], *after 43 h DON-feeding period (measured over 72 h (DON-feeding period) and afterwards over 48 h (post-DON-period, lowest conc.: nd at 12, 24 and 48 h)

incidence: 1/1, sa. const.: lactating ewe, age: 1 year, wt.: 70–80 kg, contamination: artificial (dose: **4.0 g DON,** o., during 72 h dosing period; for detailed information please see the article), conc. range: ≤110 ng/ml* **, country: Canada[284], ***conjugated DON,** **after 53 h DON-feeding period (measured over 72 h (DON-feeding period) and afterwards over 48 h (post-DON-period, lowest conc.: nd at 48 h)

incidence: 1/1, sa. const.: male sheep, age: 1 year, wt.: 60–70 kg, contamination: artificial (dose: **0.5 mg DON**/kg b. wt., i.v., once; for detailed information please see the article), conc. range: ≈≤5,000 ng/ml*, country: Canada[620], *after 10 min (also at other min intervals up to 210 min measured, lowest conc.: ≈130 ng/ml after 210 min)
incidence: 3/3, sa. const.: male sheep, age: 1 year, wt.: 60–70 kg, contamination: artificial (dose: **0.5 mg DON**/kg b. wt., i.v., once; for detailed information please see the article), conc. range: 240–520 ng/ml* **, Ø conc.: 386.6 ng/ml* **, country: Canada[620], *after 60 min (also at other min intervals up to 300 min measured, lowest conc.: ≈110 ng/ml after 210 min), ****conjugated DON**
incidence: 4/4, sa. const.: male sheep, age: 1 year, wt.: 60–70 kg, contamination: artificial (dose: **5.0 mg DON**/kg b. wt., o., once; for detailed information please see the article), conc. range: 470–760 ng/ml* **, Ø conc.: 582.5 ng/ml* **, country: Canada[620], *after 5–7 h (also at other hour intervals up to 21 h measured, lowest conc.: ≈nd after 21 h for 1 sheep), **total DON

Deepoxydeoxynivalenol
incidence: 1/1, sa. const.: lactating ewe, age: 1 year, wt.: 70–80 kg, contamination: artificial (dose: **1.5 g DON,** o., during 72 h dosing period; for detailed information please see the article), conc. range: ≤69 ng/ml*, country: Canada[284], *after 53 h DON-feeding period (measured over 72 h (DON-feeding period) and afterwards over 48 h (post-DON-period, lowest conc.: tr at 48 h)
incidence: 1/1, sa. const.: lactating ewe, age: 1 year, wt.: 70–80 kg, contamination: artificial (dose: **1.5 g DON,** o., during 72 h dosing period; for detailed information please see the article), conc. range: ≤336 ng/ml* **, country: Canada[284], ***conjugated DOM-1,** **after 53 h DON-feeding period (measured over 72 h (DON-feeding period) and afterwards over 48 h (post-DON-period, lowest conc.: 15 ng/ml at 48 h)
incidence: 1/1, sa. const.: lactating ewe, age: 1 year, wt.: 70–80 kg, contamination: artificial (dose: **4.0 g DON,** o., during 72 h dosing period; for detailed information please see the article), conc. range: ≤330 ng/ml*, country: Canada[284], *after 53 h DON-feeding period (measured over 72 h (DON-feeding period) and afterwards over 48 h (post-DON-period, lowest conc.: tr at 48 h)
incidence: 1/1, sa. const.: lactating ewe, age: 1 year, wt.: 70–80 kg, contamination: artificial (dose: **4.0 g DON,** o., during 72 h dosing period; for detailed information please see the article), conc. range: ≤1,330 ng/ml* **, country: Canada[284], ***conjugated DOM-1,** **after 53 h DON-feeding period (measured over 72 h (DON-feeding period) and afterwards over 48 h (post-DON-period, lowest conc.: 98 ng/ml at 48 h)

Sheep rumen may contain the following mycotoxins and/or their metabolites:

DEOXYNIVALENOL
incidence: 1/1, sa. const.: male sheep, age: 1 year, wt.: 60–70 kg, contamination: artificial (dose: 5.0 mg DON/kg b. wt., o., once; for detailed information please see the article), conc. range: ≈53 µg/ml*, country: Canada[620], *after 1 h (also at other hour intervals up to 20 h measured, lowest conc.: ≈2.4 µg/ml after 20 h)

OCHRATOXIN A
incidence: 1/1, sa. const.: sheep, contamination: artificial (dose: 2 ppm OTA for 4 days followed by 5 ppm OTA in the diet for 2 days; for detailed information please see the article), conc. range: ≤29 ppb* **, country: Sweden[199], ***0.5 h** after final OTA-administration, **in ruminal fluid
incidence: 1/1, sa. const.: sheep, contamination: artificial (dose: 2 ppm OTA for 4 days followed by 5 ppm OTA in the diet for 2 days; for detailed information please see the article), conc. range: ≤14 ppb* **, country: Sweden[199], ***1 h** after final OTA-administration, **in ruminal fluid

incidence: 1?/1?, sa. const.: female sheep (Suffolk), wt.: 45–65 kg, contamination: artificial (dose: 0.5 mg OTA/kg b. wt., o., once; for detailed information please see the article), conc. range: ≈≤0.58 µg/ml* ** ***, country: Canada[200], ***fed hay** (for overall information please see the article), **after 2 h (also measured after 4, 6 and 10 h, lowest conc.: nd after 10 h), ***in ruminal contents
incidence: 1?/1?, sa. const.: female sheep (Suffolk), wt.: 45–65 kg, contamination: artificial (dose: 0.5 mg OTA/kg b. wt., o., once; for detailed information please see the article), conc. range: ≈≤1.19 µg/ml* ** ***, country: Canada[200], ***fed grain-30%** = low feed intake (for overall information please see the article), **after 2 h (also measured after 4, 6 and 10 h, lowest conc.: 0.18 µg/ml after 10 h), ***in ruminal contents
incidence: 2?/2, sa. const.: female sheeps (Suffolk), wt.: 45–65 kg, contamination: artificial (dose: 0.5 mg OTA/kg b. wt., o., once; for detailed information please see the article), conc. range: ≈≤1.33 µg/ml* ** ***, country: Canada[200], ***fed grain-100%** = normal feed intake (for overall information please see the article), **after 2 h (also measured after 4, 6 and 10 h, lowest conc.: 0.44 µg/ml after 10 h), ***in ruminal contents

incidence: 3/3*, sa. const.: castrated male sheeps (Coburger Fuchsschaf), age: 1 year, wt.: ≈39.3 kg, contamination: no OTA (for detailed information please see the article), conc.: nd, country: Canada[482], *control
incidence: 2?/2, sa. const.: castrated male sheeps (Coburger Fuchsschaf), age: 1 year, wt.: ≈39.3 kg, contamination: artificial (dose: **9.5 µg OTA**/kg b. wt./day, o., for 24/25 days; for detailed information please see the article), conc. range: ≈≤14 ng/ml* ** (mean value), country: Canada[482], *1 h after final OTA-administration (also measured after 4, 7, 10 and 13 h, lowest conc.: nd after 13 h), **in ruminal fluid
incidence: 3?/3, sa. const.: castrated male sheeps (Coburger Fuchsschaf), age: 1 year, wt.: ≈39.3 kg, contamination: artificial (dose: **19.0 µg OTA**/kg b. wt./day, o., for 24/25 days; for detailed information please see the article), conc. range: ≈≤36.5 ng/ml* ** (mean value), country: Canada[482], *1 h after final OTA-administration (also measured after 4, 7, 10 and 13 h, lowest conc.: nd after 13 h), **in ruminal fluid
incidence: 2?/2, sa. const.: castrated male sheeps (Coburger Fuchsschaf), age: 1 year, wt.: ≈39.3 kg, contamination: artificial (dose: **28.5 µg OTA**/kg b. wt./day, o., for 24/25 days; for detailed information please see the article), conc. range:

≈≤53 ng/ml* ** (mean value), country: Canada[482], *1 h after final OTA-administration (also measured after 4, 7, 10 and 13 h, lowest conc.: nd after 13 h), **in ruminal fluid

OCHRATOXIN α
incidence: 1/1, sa. const.: sheep, contamination: artificial (dose: 2 ppm OTA for 4 days followed by 5 ppm OTA in the diet for 2 days; for detailed information please see the article), conc.: pr* **, country: Sweden[199], ***0.5 h** after final OTA-administration, **in ruminal fluid
incidence: 1/1, sa. const.: sheep, contamination: artificial (dose: 2 ppm OTA for 4 days followed by 5 ppm OTA in the diet for 2 days; for detailed information please see the article), conc.: pr* **, country: Sweden[199], ***1 h** after final OTA-administration, **in ruminal fluid
incidence: 3/3*, sa. const.: castrated male sheeps (Coburger Fuchsschaf), age: 1 year, wt.: ≈39.3 kg, contamination: no OTA (for detailed information please see the article), conc.: nd, country: Canada[482], *control
incidence: 2?/2, sa. const.: castrated male sheeps (Coburger Fuchsschaf), age: 1 year, wt.: ≈39.3 kg, contamination: artificial (dose: **9.5 µg OTA**/kg b. wt./day, o., for 24/25 days; for detailed information please see the article), conc. range: ≈≤13 ng/ml* (mean value), country: Canada[482], *10 h after final OTA-administration (also measured after 1, 4, 7 and 13 h, lowest conc.: ≈10 ng/ml after 1 h), **in ruminal fluid
incidence: 3?/3, sa. const.: castrated male sheeps (Coburger Fuchsschaf), age: 1 year, wt.: ≈39.3 kg, contamination: artificial (dose: **19.0 µg OTA**/kg b. wt./day, o., for 24/25 days; for detailed information please see the article), conc. range: ≈≤29 ng/ml* (mean value), country: Canada[482], *after 13 h after final OTA-administration (also measured after 1, 4, 7 and 10 h, lowest conc.: ≈18 ng/ml after 1 h), **in ruminal fluid
incidence: 2?/2, sa. const.: castrated male sheeps (Coburger Fuchsschaf), age: 1 year, wt.: ≈39.3 kg, contamination: artificial (dose: **28.5 µg OTA**/kg b. wt./day, o., for 24/25 days; for detailed information please see the article), conc. range: ≈≤48 ng/ml* (mean value), country: Canada[482], *after 10 h after final OTA-administration (also measured after 1, 4, 7 and 13 h, lowest conc.: ≈32 ng/ml after 1 h), **in ruminal fluid

Sheep ruminal fluid see Sheep rumen

Sheep serum may contain the following mycotoxins and/or their metabolites:

OCHRATOXIN A
incidence: 4/4*, sa. const.: crossbred wethers, wt.: ≈66 kg, contamination: no OTA (for detailed information please see the article), conc.: nd, country: Germany/Canada[283], *control
incidence: 4?/4, sa. const.: crossbred wethers, wt.: ≈66 kg, contamination: artificial (dose: **2 mg OTA**/kg feed (70% concentrates, 30% hay), o., for 4 weeks; for detailed information please see the article), conc. range: ≤10.8 ng/ml* (mean value), country: Germany/Canada[283], *day 27 of trial (also measured at 6, 13 and 20 days, lowest conc.: 8.2 ng/ml at day 6)
incidence: 4?/4, sa. const.: crossbred wethers, wt.: ≈66 kg, contamination: artificial (dose: **5 mg OTA**/kg feed (70% concentrates, 30% hay), o., for 4 weeks; for detailed information please see the article), conc. range: ≤111.7 ng/ml* (mean value), country: Germany/Canada[283], *day 27 of trial (also measured at 6, 13 and 20 days, lowest conc.: 67.0 ng/ml at day 20)

incidence: 3/3*, sa. const.: castrated male sheeps (Coburger Fuchsschaf), age: 1 year, wt.: ≈39.3 kg, contamination: no OTA (for detailed information please see the article), conc.: nd, country: Germany/Canada[482], *control

incidence: 2?/2, sa. const.: castrated male sheeps (Coburger Fuchsschaf), age: 1 year, wt.: ≈39.3 kg, contamination: artificial (dose: **9.5 μg OTA**/kg b. wt./day, o., for 29 days; for detailed information please see the article), conc.: 6.0 ng/ml* (mean value), country: Germany/Canada[482], *at day 23 of trial (also measured at day 1, 5, 9, 13 and 29 days after the 1st OTA-administration, lowest conc.: 0 at day 1)

incidence: 3?/3, sa. const.: castrated male sheeps (Coburger Fuchsschaf), age: 1 year, wt.: ≈39.3 kg, contamination: artificial (dose: **19.0 μg OTA**/kg b. wt./day, o., for 29 days; for detailed information please see the article), conc.: 12.4 ng/ml* (mean value), country: Germany/Canada[482], *at day 29 of trial (also measured at day 1, 5, 9, 13 and 23 days after the 1st OTA-administration, lowest conc.: 0 at day 1)

incidence: 2?/2, sa. const.: castrated male sheeps (Coburger Fuchsschaf), age: 1 year, wt.: ≈39.3 kg, contamination: artificial (dose: **28.5 μg OTA**/kg b. wt./day, o., for 29 days; for detailed information please see the article), conc.: 18.2 ng/ml* (mean value), country: Germany/Canada[482], *at day 23 of trial (also measured at day 1, 5, 9, 13 and 29 days after the 1st OTA-administration, lowest conc.: 0 at day 1)

Ochratoxin α

incidence: 4/4*, sa. const.: crossbred wethers, wt.: ≈66 kg, contamination: no OTA (for detailed information please see the article), conc.: nd, country: Germany/Canada[283], *control

incidence: 4?/4, sa. const.: crossbred wethers, wt.: ≈66 kg, contamination: artificial (dose: **2 mg OTA**/kg feed (70% concentrates, 30% hay), o., for 4 weeks; for detailed information please see the article), conc. range: ≤3.4 ng/ml* (mean value), country: Germany/Canada[283], *day 6 of trial (also measured at 13, 20 and 27 days, lowest conc.: 2.0 ng/ml at day 27)

incidence: 4?/4, sa. const.: crossbred wethers, wt.: ≈66 kg, contamination: artificial (dose: **5 mg OTA**/kg feed (70% concentrates, 30% hay), o., for 4 weeks; for detailed information please see the article), conc. range: ≤18.5 ng/ml* (mean value), country: Germany/Canada[283], *day 27 of trial (also measured at 6, 13 and 20 days, lowest conc.: 12.0 ng/ml at day 13)

incidence: 3/3*, sa. const.: castrated male sheeps (Coburger Fuchsschaf), age: 1 year, wt.: ≈39.3 kg, contamination: no OTA (for detailed information please see the article), conc.: nd, country: Germany/Canada[482], *control

incidence: 2?/2, sa. const.: castrated male sheeps (Coburger Fuchsschaf), age: 1 year, wt.: ≈39.3 kg, contamination: artificial (dose: **9.5 μg OTA**/kg b. wt./day, o., for 29 days; for detailed information please see the article), conc.: 0.8 ng/ml* (mean value), country: Germany/Canada[482], *at day 9 of trial (also measured at day 1, 5, 13, 23 and 29 days after the 1st OTA-administration, lowest conc.: 0 at day 1)

incidence: 3?/3, sa. const.: castrated male sheeps (Coburger Fuchsschaf), age: 1 year, wt.: ≈39.3 kg, contamination: artificial (dose: **19.0 μg OTA**/kg b. wt./day, o., for 29 days; for detailed information please see the article), conc.: 2.3 ng/ml* (mean value), country: Germany/Canada[482], *at day 29 of trial (also measured at day 1, 5, 9, 13 and 23 days after the 1st OTA-administration, lowest conc.: 0 at day 1)

incidence: 2?/2, sa. const.: castrated male sheeps (Coburger Fuchsschaf), age: 1 year, wt.: ≈39.3 kg, contamination: artificial (dose: **28.5 μg OTA**/kg b. wt./day, o., for 29 days; for detailed information please see the article), conc.: 1.6 ng/ml* (mean value), country: Germany/Canada[482], *at day 9 of trial (also measured at day 1, 5, 13, 23 and 29 days after the 1st OTA-administration, lowest conc.: 0 at day 1)

Sheep urine may contain the following mycotoxins and/or their metabolites:

AFLATOXIN M_1
incidence: 2/2, sa. const.: adult sheeps, total wt.: 67 kg, contamination: artificial (dose: 1 mg AFs (AFB_1 73% + AFB_2 24% + AFG_1 2% + AFG_2 1%)/kg, i.p., once), conc. range: ≈0.525 mg in total, country: South Africa[131]

AFLATOXIN M_2
incidence: 2/2, sa. const.: adult sheeps, total wt.: 67 kg, contamination: artificial (dose: 1 mg AFs (AFB_1 73% + AFB_2 24% + AFG_1 2% + AFG_2 1%)/kg, i.p., once), conc. range: ≈0.175 mg in total, country: South Africa[131]

DEOXYNIVALENOL
incidence: 2/2, sa. const.: shorn wethers, age: 1 year, wt.: 44–70 kg, contamination: artificial (dose: **0.5 mg DON**/kg b. wt., **i.v.,** once; for detailed information please see the article), conc. range: 5.2–8.6 mg* **, Ø conc.: 6.9 mg* **, country: Canada[63], *cumulative value(s), **collected and measured up to 36 h after DON-administration
incidence: 2/2, sa. const.: shorn wethers, age: 1 year, weight: 48–70 kg, contamination: artificial (dose: **5.0 mg DON**/kg b. wt., **o.,** once; for detailed information please see the article), conc. range: 4.3–8.2 mg* **, Ø conc.: 6.3 mg* **, country: Canada[63], *cumulative value(s), **collected and measured up to 36 h after DON-administration
incidence: 2/2, sa. const.: shorn wethers, age: 1 year, wt.: 44–70 kg, contamination: artificial (dose: **0.5 mg DON**/kg b. wt., **i.v.,** once; for detailed information please see the article), conc. range: 4.4–7.9 mg* ** ***, Ø conc.: 6.2 mg* ** ***, country: Canada[63], ***DON-Gluc,** **cumulative value(s), ***collected and measured up to 36 h after DON-administration
incidence: 2/2, sa. const.: shorn wethers, age: 1 year, wt.: 48–70 kg, contamination: artificial (dose: **5.0 mg DON**/kg b. wt., o., once; for detailed information please see the article), conc. range: 8.1–13.5 mg* ** ***, Ø conc.: 10.8 mg* ** ***, country: Canada[63], ***DON-Gluc,** **cumulative value(s), ***collected and measured up to 36 h after DON-administration

DEEPOXYDEOXYNIVALENOL
incidence: 2/2, sa. const.: shorn wethers, age: 1 year, wt.: 44–70 kg, contamination: artificial (dose: **0.5 mg DON**/kg b. wt., **i.v.,** once; for detailed information please see the article), conc. range: 0.10–0.18 mg* **, Ø conc.: 0.14 mg* **, country: Canada[63], *cumulative value(s), **collected and measured up to 36 h after DON-administration
incidence: 2/2, sa. const.: shorn wethers, age: 1 year, wt.: 48–70 kg, contamination: artificial (dose: **5.0 mg DON**/kg b. wt., **o.,** once; for detailed information please see the article), conc. range: 0.16 mg* **, Ø conc.: 0.16 mg* **, country: Canada[63], *cumulative value(s), **collected and measured up to 36 h after DON-administration
incidence: 2/2, sa. const.: shorn wethers, age: 1 year, wt.: 44–70 kg, contamination: artificial (dose: **0.5 mg DON**/kg b. wt., **i.v.,** once; for detailed information please see the article), conc. range: 3.8–6.1 mg* ** ***, Ø conc.: 5 mg* ** ***, country: Canada[63], ***DOM-1-Gluc,** **cumulative value(s), ***collected and measured up to 36 h after DON-administration
incidence: 2/2, sa. const.: shorn wethers, age: 1 year, wt.: 48–70 kg, contamination: artificial (dose: **5.0 mg DON**/kg b. wt., **o.,** once; for detailed information please see the article), conc. range: 2.9–3.8 mg* ** ***, Ø conc.: 3.4 mg* ** ***, country: Canada[63],

*DOM-1-Gluc, **cumulative value(s), ***collected and measured up to 36 h after DON-administration

Fumonisin B_1

incidence: ?/6, sa. const.: sheeps, contamination: artificial (dose: 50 µg FB_1/g, o., once; for detailed information please see the article), conc. range: 0.1–3.8 µg/g (mean values), country: USA[276]

Ochratoxin A

incidence: 4/4*, sa. const.: crossbred wethers, wt.: ≈66 kg, contamination: no OTA (for detailed information please see the article), conc.: nd, country: Germany/Canada[283], *control
incidence: 4?/4, sa. const.: crossbred wethers, wt.: ≈66 kg, contamination: artificial (dose: **2 mg OTA**/kg feed (70% concentrates, 30% hay), o., for 4 weeks; for detailed information please see the article), conc.: 3.9 ng/kg* (mean value), country: Germany/Canada[283], *collected over a period of 7 days in the 3rd week
incidence: 4?/4, sa. const.: crossbred wethers, wt.: ≈66 kg, contamination: artificial (dose: **5 mg OTA**/kg feed (70% concentrates, 30% hay), o., for 4 weeks; for detailed information please see the article), conc.: 14.6 ng/kg* (mean value), country: Germany/Canada[283], *collected over a period of 7 days in the 3rd week

incidence: 3/3*, sa. const.: castrated male sheeps (Coburger Fuchsschaf), age: 1 year, wt.: ≈39.3 kg, contamination: no OTA (for detailed information please see the article), conc.: nd, country: Germany/Canada[482], *control
incidence: 3?/3, sa. const.: castrated male sheeps (Coburger Fuchsschaf), age: 1 year, wt.: ≈39.3 kg, contamination: artificial (dose: **9.5 µg OTA**/kg b. wt./day, o., for 29 days; for detailed information please see the article), conc.: 28.4 ng/ml* (mean value), country: Germany/Canada[482], *collected over 7 days (day 15 until day 21 after the 1st OTA-administration, lowest conc.: 0 at day 1)
incidence: 3?/3, sa. const.: castrated male sheeps (Coburger Fuchsschaf), age: 1 year, wt.: ≈39.3 kg, contamination: artificial (dose: **19.0 µg OTA**/kg b. wt./day, o., for 29 days; for detailed information please see the article), conc.: 48.4 ng/ml* (mean value), country: Germany/Canada[482], *collected over 7 days (day 15 until day 21 after the 1st OTA-administration, lowest conc.: 0 at day 1)
incidence: 3?/3, sa. const.: castrated male sheeps (Coburger Fuchsschaf), age: 1 year, wt.: ≈39.3 kg, contamination: artificial (dose: **28.5 µg OTA**/kg b. wt./day, o., for 29 days; for detailed information please see the article), conc.: 113.0 ng/ml* (mean value), country: Germany/Canada[482], *collected over 7 days (day 15 until day 21 after the 1st OTA-administration, lowest conc.: 0 at day 1)

Ochratoxin α

incidence: 4/4*, sa. const.: crossbred wethers, wt.: ≈66 kg, contamination: no OTA (for detailed information please see the article), conc.: nd, country: Germany/Canada[283], *control
incidence: 4?/4, sa. const.: crossbred wethers, wt.: ≈66 kg, contamination: artificial (dose: **2 mg OTA**/kg feed (70% concentrates, 30% hay), o., for 4 weeks; for detailed information please see the article), conc.: 68.6 ng/kg* (mean value), country: Germany/Canada[283], *collected over a period of 7 days in the 3rd week
incidence: 4?/4, sa. const.: crossbred wethers, wt.: ≈66 kg, contamination: artificial (dose: **5 mg OTA**/kg feed (70% concentrates, 30% hay), o., for 4 weeks; for detailed information please

see the article), conc.: 347.8 ng/kg* (mean value), country: Germany/Canada[283], *collected over a period of 7 days in the 3rd week

incidence: 3/3*, sa. const.: castrated male sheeps (Coburger Fuchsschaf), age: 1 year, wt.: ≈39.3 kg, contamination: no OTA (for detailed information please see the article), conc.: nd, country: Germany/Canada[482], *control
incidence: 3?/3, sa. const.: castrated male sheeps (Coburger Fuchsschaf), age: 1 year, wt.: ≈39.3 kg, contamination: artificial (dose: **9.5 μg OTA/kg** b. wt./day, o., for 29 days; for detailed information please see the article), conc.: 158.6 ng/ml* (mean value), country: Germany/Canada[482], *collected over 7 days (day 15 until day 21 after the 1st OTA-administration, lowest conc.: 0 at day 1)
incidence: 3?/3, sa. const.: castrated male sheeps (Coburger Fuchsschaf), age: 1 year, wt.: ≈39.3 kg, contamination: artificial (dose: **19.0 μg OTA/kg** b. wt./day, o., for 29 days; for detailed information please see the article), conc.: 283.4 ng/ml* (mean value), country: Germany/Canada[482], *collected over 7 days (day 15 until day 21 after the 1st OTA-administration, lowest conc.: 0 at day 1)
incidence: 3?/3, sa. const.: castrated male sheeps (Coburger Fuchsschaf), age: 1 year, wt.: ≈39.3 kg, contamination: artificial (dose: **28.5 μg OTA/kg** b. wt./day, o., for 29 days; for detailed information please see the article), conc.: 227.3 ng/ml* (mean value), country: Germany/Canada[482], *collected over 7 days (day 15 until day 21 after the 1st OTA-administration, lowest conc.: 0 at day 1)

ZEARALENONE
incidence: 3/4, sa. const.: sheeps, contamination: artificial (dose: 1.8 g crystalline ZEA, into the esophagus, once), conc. range: 0.020–0.050 ppm*, country: Hungary/USA[123], *after 24, 48 and 72 h (also at 0 h measured but conc.: nd)

β-ZEARALENOL
incidence: 3/4, sa. const.: sheeps, contamination: artificial (dose: 1.8 g crystalline ZEA, into the esophagus, once), conc. range: 0.020–0.050 ppm*, country: Hungary/USA[123], *after 24, 48 and 72 h (also at 0 h measured but conc.: nd)

Sow milk see Pig, sow milk

Steer

Steer Natural Contamination see also Cattle

Steer kidney may contain the following mycotoxins and/or their metabolites:

AFLATOXIN B_1
incidence: 2/2, sa. const.: kidneys from steers of the USA, contamination: natural, conc. range: 0.09 ng/g, country: USA[399]

AFLATOXIN M_1
incidence: 2/2, sa. const.: kidneys from steers of the USA, contamination: natural, conc. range: 4.8 ng/g, country: USA[399]

Steer Artificial Contamination

Steer bile may contain the following mycotoxins and/or their metabolites:

AFLATOXIN B_1
incidence: 5/5*, sa. const.: Holstein-Friesian steers, wt.: ≈183 kg, contamination: no AFs (for detailed information please see the article), conc.: nd, country: USA[146], *control
incidence: 5?/5, sa. const.: Holstein-Friesian steers, wt.: ≈183 kg, contamination: artificial (dose: AFB_1 + AFB_2-containing corn (800 ng/g total) for **17.5 weeks;** for detailed information please see the article), conc.: 0.23 ng/ml* (mean value), country: USA[146], *after 17.5 weeks of AF-administration

incidence: 5/5, sa. const.: Holstein-Friesian steers, wt.: ≈183 kg, contamination: artificial (dose: AFB_1 + AFB_2-containing corn (800 ng/g total) for **15 weeks** then **2.5 weeks AF-free diet;** for detailed information please see the article), conc.: nd*, country: USA[146], *after 17.5 weeks (thereof 15 weeks with AF-administration)

Aflatoxin M_1
incidence: 5/5*, sa. const.: Holstein-Friesian steers, wt.: ≈183 kg, contamination: no AFs (for detailed information please see the article), conc.: nd, country: USA[146], *control
incidence: 5?/5, sa. const.: Holstein-Friesian steers, wt.: ≈183 kg, contamination: artificial (dose: AFB_1 + AFB_2-containing corn (800 ng/g total) for **17.5 weeks**; for detailed information please see the article), conc.: 3.46 ng/ml* (mean value), country: USA[146], *after 17.5 weeks of AF-administration
incidence: 5/5, sa. const.: Holstein-Friesian steers, wt.: ≈183 kg, contamination: artificial (dose: AFB_1 + AFB_2-containing corn (800 ng/g total) for **15 weeks** then **2.5 weeks AF-free diet;** for detailed information please see the article), conc.: nd*, country: USA[146], *after 17.5 weeks (thereof 15 weeks with AF-administration)

Steer blood may contain the following mycotoxins and/or their metabolites:

Aflatoxin B_1
incidence: 1/1, sa. const.: steer, wt.: 155–357 kg, contamination: artificial (dose: **0.2 mg AFs**/kg b. wt., o., once; for detailed information please see the article), conc. range: ≈≤2.0 ng/ml *, country: USA[72], *after ≈25 h during administration (also after other hour intervals up to ≈72 h measured, lowest conc.: nd after ≈72 h)
incidence: 1/1, sa. const.: steer, wt.: 155–357 kg, contamination: artificial (dose: **0.4 mg AFs**/kg b. wt., o., once; for detailed information please see the article), conc. range: ≈≤0.5 ng/ml *, country: USA[72], *after ≈25 h during administration (also after other hour intervals up to ≈103 h measured, lowest conc.: ≈nd after ≈103 h)
incidence: 1/1, sa. const.: steer, wt.: 155–357 kg, contamination: artificial (dose: **0.6 mg AFs**/kg b. wt., o., once; for detailed information please see the article), conc. range: ≈≤4.8 ng/ml *, country: USA[72], *after ≈48 h during administration (also after other hour intervals up to ≈142 h measured, lowest conc.: ≈nd after ≈145 h)
incidence: 1/1, sa. const.: steer, wt.: 155–357 kg, contamination: artificial (dose: **0.8 mg AFs**/kg b. wt., o., once; for detailed information please see the article), conc. range: ≈≤21.1 ng/ml *, country: USA[72], *after ≈54 h during administration (also after other hour intervals up to ≈129 h measured, lowest conc.: nd after ≈129 h)

incidence: 5/5*, sa. const.: Holstein-Friesian steers, wt.: ≈183 kg, contamination: no AFs (for detailed information please see the article), conc.: nd, country: USA[146], *control
incidence: 5?/5, sa. const.: Holstein-Friesian steers, wt.: ≈183 kg, contamination: artificial (dose: AFB_1 + AFB_2-containing corn (800 ng/g total) for **17.5 weeks;** for detailed information please see the article), conc. range: ≤0.14 ng/ml* (mean value), country: USA[146], *after 13 weeks of AF-administration (also after other week intervals up to 15 weeks measured, week 16, 17 and 17.5 na)
incidence: 5?/5, sa. const.: Holstein-Friesian steers, wt.: ≈183 kg, contamination: artificial (dose: AFB_1 + AFB_2-containing corn (800 ng/g total) for **15 weeks** then **2.5 weeks AF-free**

diet; for detailed information please see the article), conc.: tr, country: USA[146], *after 7 weeks of AF-administration (also after other week intervals up to 15 weeks measured, week 16, 17 and 17.5 na)

AFLATOXIN M_1
incidence: 1/1, sa. const.: steer, wt.: 155–357 kg, contamination: artificial (dose: **0.2 mg AFs/kg** b. wt., o., once; for detailed information please see the article), conc. range: ≈≤0.6 ng/ml *, country: USA[72], *after ≈8 h during administration (also after other hour intervals up to ≈72 h measured, lowest conc.: ≈nd after ≈72 h)
incidence: 1/1, sa. const.: steer, wt.: 155–357 kg, contamination: artificial (dose: **0.4 mg AFs/kg** b. wt., o., once; for detailed information please see the article), conc. range: ≈≤2.1 ng/ml *, country: USA[72], *after ≈2 h during administration (also after other hour intervals up to ≈104 h measured, lowest conc.: ≈nd after ≈104 h)
incidence: 1/1, sa. const.: steer, wt.: 155–357 kg, contamination: artificial (dose: **0.6 mg AFs/kg** b. wt., o., once; for detailed information please see the article), conc. range: ≈≤1.7 ng/ml *, country: USA[72], *after ≈3 h during administration (also after other hour intervals up to ≈144 h measured, lowest conc.: ≈nd after ≈144 h)
incidence: 1/1, sa. const.: steer, wt.: 155–357 kg, contamination: artificial (dose: **0.8 mg AFs/kg** b. wt., o., once; for detailed information please see the article), conc. range: ≈≤4.6 ng/ml *, country: USA[72], *after ≈60 h during administration (also after other hour intervals up to ≈144 h measured, lowest conc.: nd after ≈144 h)

incidence: 5/5*, sa. const.: Holstein-Friesian steers, wt.: ≈183 kg, contamination: no AFs (for detailed information please see the article), conc.: nd, country: USA[146], *control
incidence: 5?/5, sa. const.: Holstein-Friesian steers, wt.: ≈183 kg, contamination: artificial (dose: AFB_1 + AFB_2-containing corn (800 ng/g total) for **17.5 weeks;** for detailed information please see the article), conc. range: ≤0.38 ng/ml* (mean value), country: USA[146], *after 15 weeks of AF-administration (also after other week intervals up to 15 weeks measured, week 16, 17 and 17.5 na)
incidence: 5?/5, sa. const.: Holstein-Friesian steers, wt.: ≈183 kg, contamination: artificial (dose: AFB_1 + AFB_2-containing corn (800 ng/g total) for **15 weeks** then **2.5 weeks AF-free diet;** for detailed information please see the article), conc. range: ≤0.38 ng/ml* (mean value), country: USA[146], *after 15 weeks of AF-administration (also after other week intervals up to 15 weeks measured, week 16, 17 and 17.5 na)

Steer feces may contain the following mycotoxins and/or their metabolites:

AFLATOXIN B_1
incidence: 1/1, sa. const.: steer, wt.: 155–357 kg, contamination: artificial (dose: **0.2 mg AFs/kg** b. wt., o., once; for detailed information please see the article), conc. range: ≈≤30 ng/g*, country: USA[72], *after ≈32 h during administration (also after other hour intervals up to ≈97 h measured, lowest conc.: nd after 96 h)
incidence: 1/1, sa. const.: steer, wt.: 155–357 kg, contamination: artificial (dose: **0.4 mg AFs/kg** b. wt., o., once; for detailed information please see the article), conc. range: ≈≤130 ng/g*, country: USA[72], *after ≈33 h during administration (also after other hour intervals up to ≈152 h measured, lowest conc.: ≈nd after 120 h)
incidence: 1/1, sa. const.: steer, wt.: 155–357 kg, contamination: artificial (dose: **0.6 mg AFs/kg** b. wt., o., once; for detailed information please see the article), conc. range: ≈≤230 ng/g*, country: USA[72], *after ≈52 h during

administration (also after other hour intervals up to ≈170 h measured, lowest conc.: ≈nd after 170 h)
incidence: 1/1, sa. const.: steer, wt.: 155–357 kg, contamination: artificial (dose: **0.8 mg AFs/kg** b. wt., o., once; for detailed information please see the article), conc. range: ≈≤1,740 ng/g*, country: USA[72], *after ≈22 h during administration (also after other hour intervals up to ≈142 h measured, lowest conc.: ≈nd after 142 h)

incidence: 5/5*, sa. const.: Holstein-Friesian steers, wt.: ≈183 kg, contamination: no AFs (for detailed information please see the article), conc.: nd, country: USA[146], *control
incidence: 5?/5, sa. const.: Holstein-Friesian steers, wt.: ≈183 kg, contamination: artificial (dose: AFB_1 + AFB_2-containing corn (800 ng/g total) for **17.5 weeks;** for detailed information please see the article), conc.: 6.42 ng/g* (mean value), country: USA[146], *after 17.5 weeks of AF-administration
incidence: 5/5, sa. const.: Holstein-Friesian steers, wt.: ≈183 kg, contamination: artificial (dose: AFB_1 + AFB_2-containing corn (800 ng/g total) for **15 weeks** then **2.5 weeks AF-free diet;** for detailed information please see the article), conc.: nd*, country: USA[146], *after 17.5 weeks (thereof 15 weeks with AF-administration)

Aflatoxin M_1
incidence: 1/1, sa. const.: steer, wt.: 155–357 kg, contamination: artificial (dose: **0.2 mg AFs/kg** b. wt., o., once; for detailed information please see the article), conc. range: ≈≤180 ng/g *, country: USA[72], *after ≈33 h during administration (also after other hour intervals up to ≈98 h measured, lowest conc.: nd after 96 h)
incidence: 1/1, sa. const.: steer, wt.: 155–357 kg, contamination: artificial (dose: **0.4 mg AFs/kg** b. wt., o., once; for detailed information please see the article), conc. range: ≈≤1,430 ng/g *, country: USA[72], *after ≈30 h during administration (also after other hour intervals up to ≈170 h measured, lowest conc.: nd after ≈152 h)
incidence: 1/1, sa. const.: steer, wt.: 155–357 kg, contamination: artificial (dose: **0.6 mg AFs/kg** b. wt., o., once; for detailed information please see the article), conc. range: ≈≤440 ng/g *, country: USA[72], *after ≈30 h during administration (also after other hour intervals up to ≈170 h measured, lowest conc.: ≈20 ng/g after ≈170 h)
incidence: 1/1, sa. const.: steer, wt.: 155–357 kg, contamination: artificial (dose: **0.8 mg AFs/kg** b. wt., o., once; for detailed information please see the article), conc. range: ≈≤1,280 ng/g *, country: USA[72], *after ≈25 h during administration (also after other hour up to ≈170 h measured, lowest conc.: nd after ≈170 h)

incidence: 5/5*, sa. const.: Holstein-Friesian steers, wt.: ≈183 kg, contamination: no AFs (for detailed information please see the article), conc.: nd, country: USA[146], *control
incidence: 5?/5, sa. const.: Holstein-Friesian steers, wt.: ≈183 kg, contamination: artificial (dose: AFB_1 + AFB_2-containing corn (800 ng/g total) for **17.5 weeks;** for detailed information please see the article), conc.: 42.13 ng/g* (mean value), country: USA[146], *after 17.5 weeks of AF-administration
incidence: 5/5, sa. const.: Holstein-Friesian steers, wt.: ≈183 kg, contamination: artificial (dose: AFB_1 + AFB_2-containing corn (800 ng/g total) for **15 weeks** then **2.5 weeks AF-free diet;** for detailed information please see the article), conc.: nd*, country: USA[146], *after 17.5 weeks (thereof 15 weeks with AF-administration)

Steer heart may contain the following mycotoxins and/or their metabolites:

Aflatoxin B_1
incidence: 5/5*, sa. const.: Holstein-Friesian steers, wt.: ≈183 kg, contamination: no AFs (for detailed information please see the article), conc.: nd, country: USA[146], *control
incidence: 5?/5, sa. const.: Holstein-Friesian steers, wt.: ≈183 kg, contamination: artificial (dose: AFB_1 + AFB_2-containing corn (800 ng/g total) for **17.5 weeks;** for detailed information please see the article), conc.: 0.004 ng/g* (mean value), country: USA[146], *after 17.5 weeks of AF-administration
incidence: 5/5, sa. const.: Holstein-Friesian steers, wt.: ≈183 kg, contamination: artificial (dose: AFB_1 + AFB_2-containing corn (800 ng/g total) for **15 weeks** then **2.5 weeks AF-free diet;** for detailed information please see the article), conc.: nd*, country: USA[146], *after 17.5 weeks (thereof 15 weeks with AF-administration)
incidence: 1/1, sa. const.: Holstein steer, wt.: 160 kg, contamination: artificial (dose: 52 mg AFB_1 eq/kg b. wt./day, o., for 5 consecutive days; for detailed information please see the article), conc.: 9.2 ng/g*, country: USA[151], *on 6th day

Aflatoxin B_2
incidence: 1/1, sa. const.: Holstein steer, wt.: 160 kg, contamination: artificial (dose: 52 mg AFB_1 eq/kg b. wt./day, o., for 5 consecutive days; for detailed information please see the article), conc.: 0.9 ng/g*, country: USA[151], *on 6th day

Aflatoxin G_1
incidence: 1/1, sa. const.: Holstein steer, wt.: 160 kg, contamination: artificial (dose: 52 mg AFB_1 eq/kg b. wt./day, o., for 5 consecutive days; for detailed information please see the article), conc.: 1.0 ng/g*, country: USA[151], *on 6th day

Aflatoxin G_2
incidence: 1/1, sa. const.: Holstein steer, wt.: 160 kg, contamination: artificial (dose: 52 mg AFB_1 eq/kg b. wt./day, o., for 5 consecutive days; for detailed information please see the article), conc.: 0.04 ng/g*, country: USA[151], *on 6th day

Aflatoxin M_1
incidence: 5/5*, sa. const.: Holstein-Friesian steers, wt.: ≈183 kg, contamination: no AFs (for detailed information please see the article), conc.: nd, country: USA[146], *control
incidence: 5?/5, sa. const.: Holstein-Friesian steers, wt.: ≈183 kg, contamination: artificial (dose: AFB_1 + AFB_2-containing corn (800 ng/g total) for **17.5 weeks;** for detailed information please see the article), conc.: 0.14 ng/g* (mean value), country: USA[146], *after 17.5 weeks of AF-administration
incidence: 5/5, sa. const.: Holstein-Friesian steers, wt.: ≈183 kg, contamination: artificial (dose: AFB_1 + AFB_2-containing corn (800 ng/g total) for **15 weeks** then **2.5 weeks AF-free diet;** for detailed information please see the article), conc.: nd*, country: USA[146], *after 17.5 weeks (thereof 15 weeks with AF-administration)

incidence: 1/1, sa. const.: Holstein steer, wt.: 160 kg, contamination: artificial (dose: 52 mg AFB_1 eq/kg b. wt./day, o., for 5 consecutive days; for detailed information please see the article), conc.: 4.9 ng/g*, country: USA[151], *on 6th day

Steer kidney may contain the following mycotoxins and/or their metabolites:

Aflatoxin B_1
incidence: 1/1, sa. const.: steer, wt.: 155–357 kg, contamination: artificial (dose: **0.2 mg AFs/kg** b. wt., o., once; for detailed information please see the article), conc.: nd*, country: USA[72], *sa. collected on day 7

incidence: 1/1, sa. const.: steer, wt.: 155–357 kg, contamination: artificial (dose: **0.4 mg AFs**/kg b. wt., o., once; for detailed information please see the article), conc.: nd*, country: USA[72], *sa. collected on day 7
incidence: 1/1, sa. const.: steer, wt.: 155–357 kg, contamination: artificial (dose: **0.6 mg AFs**/kg b. wt., o., once; for detailed information please see the article), conc.: 0.06 ng/g*, country: USA[72], *sa. collected on day 7
incidence: 1/1, sa. const.: steer, wt.: 155–357 kg, contamination: artificial (dose: **0.8 mg AFs**/kg b. wt., o., once; for detailed information please see the article), conc.: 0.047 ng/g*, country: USA[72], *sa. collected on day 7

incidence: 5/5*, sa. const.: Holstein-Friesian steers, wt.: ≈183 kg, contamination: no AFs (for detailed information please see the article), conc.: nd, country: USA[146], *control
incidence: 5?/5, sa. const.: Holstein-Friesian steers, wt.: ≈183 kg, contamination: artificial (dose: AFB_1 + AFB_2-containing corn (800 ng/g total) for **17.5 weeks;** for detailed information please see the article), conc.: 0.09 ng/g* (mean value), country: USA[146], *after 17.5 weeks of AF-administration
incidence: 5/5, sa. const.: Holstein-Friesian steers, wt.: ≈183 kg, contamination: artificial (dose: AFB_1 + AFB_2-containing corn (800 ng/g total) for **15 weeks** then **2.5 weeks AF-free diet;** for detailed information please see the article), conc.: nd*, country: USA[146], *after 17.5 weeks (thereof 15 weeks with AF-administration)

incidence: 1/1, sa. const.: Holstein steer, wt.: 160 kg, contamination: artificial (dose: 52 mg AFB_1 eq/kg b. wt./day, o., for 5 consecutive days; for detailed information please see the article), conc.: 29.3 ng/g*, country: USA[151], *on 6th day

Aflatoxin B_2
incidence: 1/1, sa. const.: Holstein steer, wt.: 160 kg, contamination: artificial (dose: 52 mg AFB_1 eq/kg b. wt./day, o., for 5 consecutive days; for detailed information please see the article), conc.: 4.4 ng/g*, country: USA[151], *on 6th day

Aflatoxin G_1
incidence: 1/1, sa. const.: Holstein steer, wt.: 160 kg, contamination: artificial (dose: 52 mg AFB_1 eq/kg b. wt./day, o., for 5 consecutive days; for detailed information please see the article), conc.: 6.3 ng/g*, country: USA[151], *on 6th day

Aflatoxin G_2
incidence: 1/1, sa. const.: Holstein steer, wt.: 160 kg, contamination: artificial (dose: 52 mg AFB_1 eq/kg b. wt./day, o., for 5 consecutive days; for detailed information please see the article), conc.: 0.4 ng/g*, country: USA[151], *on 6th day

Aflatoxin M_1
incidence: 1/1, sa. const.: steers, wt.: 155–357 kg, contamination: artificial (dose: **0.2 mg AFs**/kg b. wt., o., once; for detailed information please see the article), conc.: 0.029 ng/g*, country: USA[72], *sa. collected on day 7
incidence: 1/1, sa. const.: steers, wt.: 155–357 kg, contamination: artificial (dose: **0.4 mg AFs**/kg b. wt., o., once; for detailed information please see the article), conc.: 0.035 ng/g*, country: USA[72], *sa. collected on day 7
incidence: 1/1, sa. const.: steers, wt.: 155–357 kg, contamination: artificial (dose: **0.6 mg AFs**/kg b. wt., o., once; for detailed information please see the article), conc.: 0.26 ng/g*, country: USA[72], *sa. collected on day 7
incidence: 1/1, sa. const.: steers, wt.: 155–357 kg, contamination: artificial (dose: **0.8 mg AFs**/kg b. wt., o., once; for detailed information please see the article), conc.: 0.14 ng/g*, country: USA[72], *sa. collected on day 7

incidence: 5/5*, sa. const.: Holstein-Friesian steers, wt.: ≈183 kg, contamination: no AFs (for detailed information please see the article), conc.: nd, country: USA[146], *control
incidence: 5?/5, sa. const.: Holstein-Friesian steers, wt.: ≈183 kg, contamination: artificial (dose: AFB_1 + AFB_2-containing corn (800 ng/g total) for **17.5 weeks;** for detailed information please see the article), conc.: 4.82 ng/g* (mean value), country: USA[146], *after 17.5 weeks of AF-administration
incidence: 5/5, sa. const.: Holstein-Friesian steers, wt.: ≈183 kg, contamination: artificial (dose: AFB_1 + AFB_2-containing corn (800 ng/g total) for **15 weeks** then **2.5 weeks AF-free diet;** for detailed information please see the article), conc.: nd*, country: USA[146], *after 17.5 weeks (thereof 15 weeks with AF-administration)

incidence: 1/1, sa. const.: Holstein steer, wt.: 160 kg, contamination: artificial (dose: 52 mg AFB_1 eq/kg b. wt./day, o., for 5 consecutive days; for detailed information please see the article), conc.: 105.5 ng/g*, country: USA[151], *on 6th day

Steer liver may contain the following mycotoxins and/or their metabolites:

Aflatoxin B_1
incidence: 3/3*, sa. const.: steers, wt.: 258–270 kg, contamination: no AFB_1 (for detailed information please see the article), conc.: nd, country: USA[71], *control
incidence: 1/3, sa. const.: steers, wt.: 258–270 kg, contamination: artificial (dose: **60 ppb AFB_1** in the feed for 155 days; for detailed information please see the article), conc.: 0.29 ppb* in wet tissue, country: USA[71], *sa. collected on day 64 (also measured at 106, 151 and 163 days but conc.: nd)
incidence: 1/3, sa. const.: steers, wt.: 258–270 kg, contamination: artificial (dose: **300 ppb AFB_1** in the feed for 155 days; for detailed information please see the article), conc.: 0.64 ppb* in wet tissue, country: USA[71], *sa. collected on day 106 (also measured at 64, 151 and 163 days but lower conc. recorded)
incidence: 1/3, sa. const.: steers, wt.: 258–270 kg, contamination: artificial (dose: **600 ppb AFB_1** in the feed for 155 days; for detailed information please see the article), conc.: 1.14 ppb* in wet tissue, country: USA[71], *sa. collected on day 64 (also measured at 106, 151 and 163 days but lower conc. recorded)

incidence: 1/1, sa. const.: steer, wt.: 155–357 kg, contamination: artificial (dose: **0.2 mg AFs**/kg b. wt., o., once; for detailed information please see the article), conc.: nd*, country: USA[72], *sa. collected on day 7
incidence: 1/1, sa. const.: steer, wt.: 155–357 kg, contamination: artificial (dose: **0.4 mg AFs**/kg b. wt., o., once; for detailed information please see the article), conc.: nd*, country: USA[72], *sa. collected on day 7
incidence: 1/1, sa. const.: steers, wt.: 155–357 kg, contamination: artificial (dose: **0.6 mg AFs**/kg b. wt., o., once; for detailed information please see the article), conc.: 0.089 ng/g*, country: USA[72], *sa. collected on day 7
incidence: 1/1, sa. const.: steers, wt.: 155–357 kg, contamination: artificial (dose: **0.8 mg AFs**/kg b. wt., o., once; for detailed information please see the article), conc.: 0.025 ng/g*, country: USA[72], *sa. collected on day 7

incidence: 5/5*, sa. const.: Holstein-Friesian steers, wt.: ≈183 kg, contamination: no AFs (for detailed information please see the article), conc.: nd, country: USA[146], *control

incidence: 5?/5, sa. const.: Holstein-Friesian steers, wt.: ≈183 kg, contamination: artificial (dose: AFB_1 + AFB_2-containing corn (800 ng/g total) for **17.5 weeks;** for detailed information please see the article), conc.: 0.37 ng/g* (mean value), country: USA[146], *after 17.5 weeks of AF-administration
incidence: 5/5, sa. const.: Holstein-Friesian steers, wt.: ≈183 kg, contamination: artificial (dose: AFB_1 + AFB_2-containing corn (800 ng/g total) for **15 weeks** then **2.5 weeks AF-free diet;** for detailed information please see the article), conc.: nd*, country: USA[146], *after 17.5 weeks (thereof 15 weeks with AF-administration)

incidence: 1/1, sa. const.: Holstein steer, wt.: 160 kg, contamination: artificial (dose: 52 mg AFB_1 eq/kg b. wt./day, o., for 5 consecutive days; for detailed information please see the article), conc. range: ≤28.8 ng/g* **, country: USA[151], *different portions of the liver evaluated, **on 6th day

Aflatoxin B_2
incidence: 1/1, sa. const.: Holstein steer, wt.: 160 kg, contamination: artificial (dose: 52 mg AFB_1 eq/kg b. wt./day, o., for 5 consecutive days; for detailed information please see the article), conc. range: ≤4.6 ng/g* **, country: USA[151], *different portions of the liver evaluated, **on 6th day

Aflatoxin G_1
incidence: 1/1, sa. const.: Holstein steer, wt.: 160 kg, contamination: artificial (dose: 52 mg AFB_1 eq/kg b. wt./day, o., for 5 consecutive days; for detailed information please see the article), conc. range: ≤3.4 ng/g* **, country: USA[151], *different portions of the liver evaluated, **on 6th day

Aflatoxin G_2
incidence: 1/1, sa. const.: Holstein steer, wt.: 160 kg, contamination: artificial (dose: 52 mg AFB_1 eq/kg b. wt./day, o., for 5 consecutive days; for detailed information please see the article), conc. range: ≤0.3 ng/g* **, country: USA[151], *different portions of the liver evaluated, **on 6th day

Aflatoxin M_1
incidence: 3/3*, sa. const.: steers, wt.: 258–270 kg, contamination: no AFB_1 (for detailed information please see the article), conc.: nd, country: USA[71], *control
incidence: 2/3, sa. const.: steers, wt.: 258–270 kg, contamination: artificial (dose: **60 ppb AFB_1** in the feed for 155 days; for detailed information please see the article), conc. range: 0.30–0.62 ppb* in wet tissue, country: USA[71], *sa. collected on day 106 (also measured at 64, 151 and 163 days but conc.: nd)
incidence: 3/3, sa. const.: steers, wt.: 258–270 kg, contamination: artificial (dose: **300 ppb, AFB_1** in the feed for 155 days; for detailed information please see the article), conc. range: ≤1.58 ppb* in wet tissue, country: USA[71], *sa. collected on day 106 (also measured at 64, 151 and 163 days but lower conc. recorded)
incidence: 3/3, sa. const.: steers, wt.: 258–270 kg, contamination: artificial (dose: **600 ppb AFB_1** in the feed for 155 days; for detailed information please see the article), conc. range: ≤2.76 ppb* in wet tissue, country: USA[71], *sa. collected on day 106 (also measured at 64, 151 and 163 days but lower conc. recorded)

incidence: 1/1, sa. const.: steer, wt.: 155–357 kg, contamination: artificial (dose: **0.2 mg AFs/kg** b. wt., o., once; for detailed information please see the article), conc.: nd*, country: USA[72], *sa. collected on day 7
incidence: 1/1, sa. const.: steer, wt.: 155–357 kg, contamination: artificial (dose: **0.4 mg AFs/kg** b. wt., o., once; for detailed information please see the article),

conc.: nd*, country: USA[72], *sa. collected on day 7
incidence: 1/1, sa. const.: steers, wt.: 155–357 kg, contamination: artificial (dose: **0.6 mg AFs/kg** b. wt., o., once; for detailed information please see the article), conc.: 0.061 ng/g*, country: USA[72], *sa. collected on day 7
incidence: 1/1, sa. const.: steer, wt.: 155–357 kg, contamination: artificial (dose: **0.8 mg AFs/kg** b. wt., o., once; for detailed information please see the article), conc.: nd*, country: USA[72], *sa. collected on day 7

incidence: 5/5*, sa. const.: Holstein-Friesian steers, wt.: ≈183 kg, contamination: no AFs (for detailed information please see the article), conc.: nd, country: USA[146], *control
incidence: 5?/5, sa. const.: Holstein-Friesian steers, wt.: ≈183 kg, contamination: artificial (dose: AFB_1 + AFB_2-containing corn (800 ng/g total) for **17.5 weeks;** for detailed information please see the article), conc.: 1.07 ng/g* (mean value), country: USA[146], *after 17.5 weeks of AF-administration
incidence: 5/5, sa. const.: Holstein-Friesian steers, wt.: ≈183 kg, contamination: artificial (dose: AFB_1 + AFB_2-containing corn (800 ng/g total) for **15 weeks** then **2.5 weeks AF-free diet;** for detailed information please see the article), conc.: nd*, country: USA[146], *after 17.5 weeks (thereof 15 weeks with AF-administration)

incidence: 1/1, sa. const.: Holstein steer, wt.: 160 kg, contamination: artificial (dose: 52 mg AFB_1 eq/kg b. wt./day, o., for 5 consecutive days; for detailed information please see the article), conc. range: ≤19.7 ng/g* **, country: USA[151], *different portions of the liver evaluated, **on 6th day

Steer lung may contain the following mycotoxins and/or their metabolites:

Aflatoxin B_1
incidence: 5/5*, sa. const.: Holstein-Friesian steers, wt.: ≈183 kg, contamination: no AFs (for detailed information please see the article), conc.: nd, country: USA[146], *control
incidence: 5?/5, sa. const.: Holstein-Friesian steers, wt.: ≈183 kg, contamination: artificial (dose: AFB_1 + AFB_2-containing corn (800 ng/g total) for **17.5 weeks;** for detailed information please see the article), conc.: 0.014 ng/g* (mean value), country: USA[146], *after 17.5 weeks of AF-administration
incidence: 5/5, sa. const.: Holstein-Friesian steers, wt.: ≈183 kg, contamination: artificial (dose: AFB_1 + AFB_2-containing corn (800 ng/g total) for **15 weeks then 2.5 weeks AF-free diet;** for detailed information please see the article), conc.: nd*, country: USA[146], *after 17.5 weeks (thereof 15 weeks with AF-administration)

Aflatoxin M_1
incidence: 5/5*, sa. const.: Holstein-Friesian steers, wt.: ≈183 kg, contamination: no AFs (for detailed information please see the article), conc.: nd, country: USA[146], *control
incidence: 5?/5, sa. const.: Holstein-Friesian steers, wt.: ≈183 kg, contamination: artificial (dose: AFB_1 + AFB_2-containing corn (800 ng/g total) for **17.5 weeks;** for detailed information please see the article), conc.: 0.29 ng/g* (mean value), country: USA[146], *after 17.5 weeks of AF-administration
incidence: 5/5, sa. const.: Holstein-Friesian steers, wt.: ≈183 kg, contamination: artificial (dose: AFB_1 + AFB_2-containing corn (800 ng/g total) for **15 weeks** then **2.5 weeks AF-free diet;** for detailed information please see the article), conc.: nd*, country: USA[146], *after 17.5 weeks (thereof 15 weeks with AF-administration)

Steer muscle may contain the following mycotoxins and/or their metabolites:

Aflatoxin B_1
incidence: 5/5*, sa. const.: Holstein-Friesian steers, wt.: ≈183 kg, contamination: no AFs (for detailed information please see the article), conc.: nd, country: USA[146], *control
incidence: 5?/5, sa. const.: Holstein-Friesian steers, wt.: ≈183 kg, contamination: artificial (dose: AFB_1 + AFB_2-containing corn (800 ng/g total) for **17.5 weeks;** for detailed information please see the article), conc.: 0.002 ng/g* ** (mean value), country: USA[146], *after 17.5 weeks of AF-administration, **in *iliopsoas* (skeletal muscle)
incidence: 5/5, sa. const.: Holstein-Friesian steers, wt.: ≈183 kg, contamination: artificial (dose: AFB_1 + AFB_2-containing corn (800 ng/g total) **for 15 weeks** then **2.5 weeks AF-free diet;** for detailed information please see the article), conc.: nd* **, country: USA[146], *after 17.5 weeks (thereof 15 weeks with AF-administration), **in *iliopsoas* (skeletal muscle)

incidence: 5/5*, sa. const.: Holstein-Friesian steers, wt.: ≈183 kg, contamination: no AFs (for detailed information please see the article), conc.: nd, country: USA[146], *control
incidence: 5?/5, sa. const.: Holstein-Friesian steers, wt.: ≈183 kg, contamination: artificial (dose: AFB_1 + AFB_2-containing corn (800 ng/g total) for **17.5 weeks;** for detailed information please see the article), conc.: nd* ** (mean value), country: USA[146], *after 17.5 weeks of AF-administration, **in *semitendinous* (skeletal muscle)
incidence: 5/5, sa. const.: Holstein-Friesian steers, wt.: ≈183 kg, contamination: artificial (dose: AFB_1 + AFB_2-containing corn (800 ng/g total) for **15 weeks** then **2.5 weeks AF-free diet;** for detailed information please see the article), conc.: nd* **, country: USA[146], *after 17.5 weeks (thereof 15 weeks with AF-administration), **in *semitendinous* (skeletal muscle)

incidence: 1/1, sa. const.: Holstein steer, wt.: 160 kg, contamination: artificial (dose: 52 mg AFB_1 eq/kg b. wt./day, o., for 5 consecutive days; for detailed information please see the article), conc.: 7.9 ng/g* **, country: USA[151], *in round muscle, **on 6th day

Aflatoxin B_2
incidence: 1/1, sa. const.: Holstein steer, wt.: 160 kg, contamination: artificial (dose: 52 mg AFB_1 eq/kg b. wt./day, o., for 5 consecutive days; for detailed information please see the article), conc.: 1.1 ng/g* **, country: USA[151], *in round muscle, **on 6th day

Aflatoxin G_1
incidence: 1/1, sa. const.: Holstein steer, wt.: 160 kg, contamination: artificial (dose: 52 mg AFB_1 eq/kg b. wt./day, o., for 5 consecutive days; for detailed information please see the article), conc.: 1.1 ng/g* **, country: USA[151], *in round muscle, **on 6th day

Aflatoxin G_2
incidence: 1/1, sa. const.: Holstein steer, wt.: 160 kg, contamination: artificial (dose: 52 mg AFB_1 eq/kg b. wt./day, o., for 5 consecutive days; for detailed information please see the article), conc.: 0.05 ng/g* **, country: USA[151], *in round muscle, **on 6th day

Aflatoxin M_1
incidence: 5/5*, sa. const.: Holstein-Friesian steers, wt.: ≈183 kg, contamination: no AFs (for detailed information please see the article), conc.: nd, country: USA[146], *control
incidence: 5?/5, sa. const.: Holstein-Friesian steers, wt.: ≈183 kg,

contamination: artificial (dose: AFB_1 + AFB_2-containing corn (800 ng/g total) for **17.5 weeks;** for detailed information please see the article), conc.: 0.10 ng/g* ** (mean value), country: USA[146], *after 17.5 weeks of AF-administration, **in *iliopsoas* (skeletal muscle)
incidence: 5/5, sa. const.: Holstein-Friesian steers, wt.: ≈183 kg, contamination: artificial (dose: AFB_1 + AFB_2-containing corn (800 ng/g total) for **15 weeks** then **2.5 weeks AF-free diet;** for detailed information please see the article), conc.: nd* **, country: USA[146], *after 17.5 weeks (thereof 15 weeks with AFs-administration), **in *iliopsoas* (skeletal muscle)

incidence: 5/5*, sa. const.: Holstein-Friesian steers, wt.: ≈183 kg, contamination: no AFs (for detailed information please see the article), conc.: nd, country: USA[146], *control
incidence: 5?/5, sa. const.: Holstein-Friesian steers, wt.: ≈183 kg, contamination: artificial (dose: AFB_1 + AFB_2-containing corn (800 ng/g total) for **17.5 weeks;** for detailed information please see the article), conc.: 0.11 ng/g* ** (mean value), country: USA[146], *after 17.5 weeks of AF-administration, **in *semitendinous* (skeletal muscle)
incidence: 5/5, sa. const.: Holstein-Friesian steers, wt.: ≈183 kg, contamination: artificial (dose: AFB_1 + AFB_2-containing corn (800 ng/g total) for **15 weeks** then **2.5 weeks AF-free diet;** for detailed information please see the article), conc.: nd* **, country: USA[146], *after 17.5 weeks (thereof 15 weeks with AF-administration), **in *semitendinous* (skeletal muscle)

incidence: 1/1, sa. const.: Holstein steer, wt.: 160 kg, contamination: artificial (dose: 52 mg AFB_1 eq/kg b. wt./day, o., for 5 consecutive days; for detailed information please see the article), conc.: 2.7 ng/g* **, country: USA[151], *in round muscle, **on 6th day

Steer pancreas may contain the following mycotoxins and/or their metabolites:

Aflatoxin B_1
incidence: 5/5*, sa. const.: Holstein-Friesian steers, wt.: ≈183 kg, contamination: no AFs (for detailed information please see the article), conc.: nd, country: USA[146], *control
incidence: 5?/5, sa. const.: Holstein-Friesian steers, wt.: ≈183 kg, contamination: artificial (dose: AFB_1 + AFB_2-containing corn (800 ng/g total) for **17.5 weeks;** for detailed information please see the article), conc.: tr*, country: USA[146], *after 17.5 weeks of AF-administration
incidence: 5/5, sa. const.: Holstein-Friesian steers, wt.: ≈183 kg, contamination: artificial (dose: AFB_1 + AFB_2-containing corn (800 ng/g total) for **15 weeks** then **2.5 weeks AF-free diet;** for detailed information please see the article), conc.: nd*, country: USA[146], *after 17.5 weeks (thereof 15 weeks with AF-administration)

Aflatoxin M_1
incidence: 5/5*, sa. const.: Holstein-Friesian steers, wt.: ≈183 kg, contamination: no AFs (for detailed information please see the article), conc.: nd, country: USA[146], *control
incidence: 5?/5, sa. const.: Holstein-Friesian steers, wt.: ≈183 kg, contamination: artificial (dose: AFB_1 + AFB_2-containing corn (800 ng/g total) for **17.5 weeks;** for detailed information please see the article), conc.: 0.15 ng/g* (mean value), country: USA[146], *after 17.5 weeks of AF-administration
incidence: 5/5, sa. const.: Holstein-Friesian steers, wt.: ≈183 kg, contamination: artificial (dose: AFB_1 + AFB_2-containing corn (800 ng/g total) for **15 weeks** then **2.5 weeks AF-free diet;** for detailed information please see the article), conc.: nd*, country: USA[146], *after 17.5 weeks (thereof 15 weeks with AF-administration)

Steer rumen may contain the following mycotoxins and/or their metabolites:

Aflatoxin B_1

incidence: 1/1, sa. const.: steers, wt.: 155–357 kg, contamination: artificial (dose: **0.2 mg AFs/kg** b. wt., o., once; for detailed information please see the article), conc. range: ≈≤1,520 ng/g* ** dry weight, country: USA[72], *after ≈8 h during administration (also after other hour intervals up to ≈96 h measured, lowest conc.: ≈nd after 80 h), **in rumen contents
incidence: 1/1, sa. const.: steers, wt.: 155–357 kg, contamination: artificial (dose: **0.4 mg AFs/kg** b. wt., o., once; for detailed information please see the article), conc. range: ≈≤1,600 ng/g* ** dry weight, country: USA[72], *after ≈8 h during administration (also after other hour intervals up to 120 h measured, lowest conc.: nd after 120 h), **in rumen contents
incidence: 1/1, sa. const.: steers, wt.: 155–357 kg, contamination: artificial (dose: **0.6 mg AFs/kg** b. wt., o., once; for detailed information please see the article), conc. range: ≈≤23,750 ng/g* ** dry weight, country: USA[72], *after ≈1 h during administration (also after other hour intervals up to 170 h measured, lowest conc.: ≈40 ng/g after ≈120 h), **in rumen contents
incidence: 1/1, sa. const.: steers, wt.: 155–357 kg, contamination: artificial (dose: **0.8 mg AFs/kg** b. wt., o., once; for detailed information please see the article), conc. range: ≈≤2,750 ng/g* ** dry weight, country: USA[72], *after ≈1 h during administration (also after other hour intervals up to 170 h measured, lowest conc.: nd after ≈145 h), **in rumen contents

incidence: 5/5*, sa. const.: Holstein-Friesian steers, wt.: ≈183 kg, contamination: no AFs (for detailed information please see the article), conc.: nd, country: USA[146], *control
incidence: 5?/5, sa. const.: Holstein-Friesian steers, wt.: ≈183 kg, contamination: artificial (dose: AFB_1 + AFB_2-containing corn (800 ng/g total) for **17.5 weeks;** for detailed information please see the article), conc.: 13.05 ng/g* ** (mean value), country: USA[146], *after 17.5 weeks of AF-administration, **in rumen contents
incidence: 5?/5, sa. const.: Holstein-Friesian steers, wt.: ≈183 kg, contamination: artificial (dose: AFB_1 + AFB_2-containing corn (800 ng/g total) for **15 weeks** then **2.5 weeks AF-free diet;** for detailed information please see the article), conc.: 0.09 ng/g* **, country: USA[146], *after 17.5 weeks (thereof 15 weeks with AF-administration), **in rumen contents

Aflatoxin M_1

incidence: 1/1, sa. const.: steer, wt.: 155–357 kg, contamination: artificial (dose: **0.2 mg AFs/kg** b. wt., o., once; for detailed information please see the article), conc. range: ≈≤38 ng/g* ** dry weight, country: USA[72], *after ≈22 h during administration (also after other hour intervals up to ≈49 h measured, lowest conc.: ≈8 ng/g after ≈48 h), **in rumen contents
incidence: 1/1, sa. const.: steer, wt.: 155–357 kg, contamination: artificial (dose: **0.4 mg AFs/kg** b. wt., o., once; for detailed information please see the article), conc. range: ≈≤200 ng/g* ** dry weight, country: USA[72], *after ≈0 h during administration (also after other hour intervals up to ≈106 h measured, lowest conc.: ≈nd after ≈72 h), **in rumen contents
incidence: 1/1, sa. const.: steer, wt.: 155–357 kg, contamination: artificial (dose: **0.6 mg AFs/kg** b. wt., o., once; for detailed information please see the article), conc. range: ≈≤530 ng/g* ** dry weight, country: USA[72], *after ≈1 h during administration (also after other hour intervals up to 170 h measured, lowest conc.: ≈5 ng/g after ≈120 h), **in rumen contents
incidence: 1/1, sa. const.: steer, wt.: 155–357 kg, contamination: artificial

(dose: **0.8 mg AFs**/kg b. wt., o., once; for detailed information please see the article), conc. range: ≈≤253 ng/g* ** dry weight, country: USA[72], *after ≈8 h during administration (also after other hour intervals up to ≈170 h measured, lowest conc.: ≈nd after ≈170 h), **in rumen contents

incidence: 5/5*, sa. const.: Holstein-Friesian steers, wt.: ≈183 kg, contamination: no AFs (for detailed information please see the article), conc.: nd, country: USA[146], *control

incidence: 5?/5, sa. const.: Holstein-Friesian steers, wt.: ≈183 kg, contamination: artificial (dose: AFB_1 + AFB_2-containing corn (800 ng/g total) for **17.5 weeks;** for detailed information please see the article), conc.: 1.66 ng/g* ** (mean value), country: USA[146], *after 17.5 weeks of AF-administration, **in rumen contents

incidence: 5/5, sa. const.: Holstein-Friesian steers, wt.: ≈183 kg, contamination: artificial (dose: AFB_1 + AFB_2-containing corn (800 ng/g total) for **15 weeks** then **2.5 weeks AF-free diet;** for detailed information please see the article), conc.: nd* **, country: USA[146], *after 17.5 weeks (thereof 15 weeks with AF-administration), **in rumen contents

Steer spleen may contain the following mycotoxins and/or their metabolites:

Aflatoxin B_1
incidence: 1/1, sa. const. Holstein steer, wt.: 160 kg, contamination: artificial (dose: 52 mg AFB_1 eq/kg b. wt./day, o., for 5 consecutive days; for detailed information please see the article), conc.: 12.2 ng/g*, country: USA[151], *on 6th day

Aflatoxin B_2
incidence: 1/1, sa. const.: Holstein steer, wt.: 160 kg, contamination: artificial (dose: 52 mg AFB_1 eq/kg b. wt./day, o., for 5 consecutive days; for detailed information please see the article), conc.: 1.6 ng/g*, country: USA[151], *on 6th day

Aflatoxin G_1
incidence: 1/1, sa. const.: Holstein steer, wt.: 160 kg, contamination: artificial (dose: 52 mg AFB_1 eq/kg b. wt./day, o., for 5 consecutive days; for detailed information please see the article), conc.: 0.4 ng/g*, country: USA[151], *on 6th day

Aflatoxin M_1
incidence: 1/1, sa. const.: Holstein steer, wt.: 160 kg, contamination: artificial (dose: 52 mg AFB_1 eq/kg b. wt./day, o., for 5 consecutive days; for detailed information please see the article), conc.: 4.3 ng/g*, country: USA[151], *on 6th day

Steer urine may contain the following mycotoxins and/or their metabolites:

Aflatoxin B_1
incidence: 1/1, sa. const.: steer, wt.: 155–357 kg, contamination: artificial (dose: **0.2 mg AFs**/kg b. wt., o., once; for detailed information please see the article), conc. range: ≈≤275 ng/ml*, country: USA[72], *after ≈2 h during administration (also after other hour intervals up to ≈96 h measured, lowest conc.: nd after ≈96 h)

incidence: 1/1, sa. const.: steer, wt.: 155–357 kg, contamination: artificial (dose: **0.4 mg AFs**/kg b. wt., o., once; for detailed information please see the article), conc. range: ≈≤25 ng/ml*, country: USA[72], *after ≈72 h during administration (also after other hour intervals up to ≈104 h measured, lowest conc.: nd after ≈104 h)

incidence: 1/1, sa. const.: steer, wt.: 155–357 kg, contamination: artificial (dose: **0.6 mg AFs**/kg b. wt., o., once; for detailed information please see the article), conc. range: ≈≤115 ng/ml*, country: USA[72], *after ≈48 h during

administration (also after other hour intervals up to 170 h measured, lowest conc.: nd after ≈120 h)
incidence: 1/1, sa. const.: steer, wt.: 155–357 kg, contamination: artificial (dose: **0.8 mg AFs/**kg b. wt., o., once; for detailed information please see the article), conc. range: ≈≤565 ng/ml*, country: USA[72], *after ≈74 h during administration (also after other hour intervals up to 170 h measured, lowest conc.: nd after ≈144 h)

incidence: 5/5*, sa. const.: Holstein-Friesian steers, wt.: ≈183 kg, contamination: no AFs (for detailed information please see the article), conc.: nd, country: USA[146], *control
incidence: 5?/5, sa. const.: Holstein-Friesian steers, wt.: ≈183 kg, contamination: artificial (dose: AFB_1 + AFB_2-containing corn (800 ng/g total) for **17.5 weeks;** for detailed information please see the article), conc. range: ≤1.54 ng/ml* (mean value), country: USA[146], *after 5 weeks of AF-administration (also after other week intervals up to 17.5 weeks measured)
incidence: 5?/5, sa. const.: Holstein-Friesian steers, wt.: ≈183 kg, contamination: artificial (dose: AFB_1 + AFB_2-containing corn (800 ng/g total) for **15 weeks** then **2.5 weeks AF-free diet;** for detailed information please see the article), conc. range: ≤2.75 ng/ml* (mean value), country: USA[146], *after 15 weeks (thereof 15 weeks with AF-administration) (also after other week intervals up to 17.5 weeks measured)

Aflatoxin M_1
incidence: 1/1, sa. const.: steer, wt.: 155–357 kg, contamination: artificial (dose: **0.2 mg AFs/**kg b. wt., o., once; for detailed information please see the article), conc. range: ≈≤1,550 ng/ml*, country: USA[72], *after ≈2 h during administration (also after other hour intervals up to ≈128 h measured, lowest conc.: nd after 120 h)
incidence: 1/1, sa. const.: steer, wt.: 155–357 kg, contamination: artificial (dose: **0.4 mg AFs/**kg b. wt., o., once; for detailed information please see the article), conc. range: ≈≤1,090 ng/ml*, country: USA[72], *after ≈22 h during administration (also after other hour intervals up to ≈140 h measured, lowest conc.: nd after 140 h)
incidence: 1/1, sa. const.: steer, wt.: 155–357 kg, contamination: artificial (dose: **0.6 mg AFs/**kg b. wt., o., once; for detailed information please see the article), conc. range: ≈≤730 ng/ml*, country: USA[72], *after ≈8 h during administration (also after other hour intervals up to 170 h measured, lowest conc.: nd after ≈144 h)
incidence: 1/1, sa. const.: steer, wt.: 155–357 kg, contamination: artificial (dose: **0.8 mg AFs/**kg b. wt., o., once; for detailed information please see the article), conc. range: ≈≤2,320 ng/ml*, country: USA[72], *after ≈22 h during administration (also after other hour intervals up to 170 h measured, lowest conc.: nd after ≈144 h)

incidence: 5/5*, sa. const.: Holstein-Friesian steers, wt.: ≈183 kg, contamination: no AFs (for detailed information please see the article), conc.: nd, country: USA[146], *control
incidence: 5?/5, sa. const.: Holstein-Friesian steers, wt.: ≈183 kg, contamination: artificial (dose: AFB_1 + AFB_2-containing corn (800 ng/g total) for **17.5 weeks;** for detailed information please see the article), conc. range: ≤30.02 ng/ml* (mean value), country: USA[146], *after 5 weeks of AF-administration (also after other week intervals up to 17.5 weeks measured)
incidence: 5?/5, sa. const.: Holstein-Friesian steers, wt.: ≈183 kg, contamination: artificial (dose: AFB_1 + AFB_2-containing corn (800 ng/g total) for **15 weeks** then

2.5 weeks AF-free diet; for detailed information please see the article), conc. range: ≤15.32 ng/ml* (mean value), country: USA[146], *after 12 weeks (15 weeks with AF-administration at all) (also after other week intervals up to 17.5 weeks measured)

Swine see Pig

Tree Shrew

Tree Shrew Artificial Contamination

Tree shrew liver may contain the following mycotoxins and/or their metabolites:

Aflatoxin B_1
incidence: ?/?*, sa. const.: male tree shrews, wt.: 100–160 g, contamination: no AFB_1 (for detailed information please see the article), conc.: nr, country: USA[142], *control
incidence: 4?/4, sa. const.: male tree shrews, wt.: 100–160 g, contamination: artificial (dose: **400 µg AFB_1** (labeled)/kg b. wt., by gavage, once; for detailed information please see the article), conc.: 0.37 pmol/mg DNA* ** (mean value), country: USA[142], *after 24 h, **AFB_1-N^7-Gua

Aflatoxin M_1
incidence: ?/?*, sa. const.: male tree shrews, wt.: 100–160 g, contamination: no AFB_1 (for detailed information please see the article), conc.: nr, country: USA[142], *control
incidence: 4?/4, sa. const.: male tree shrews, wt.: 100–160 g, contamination: artificial (dose: **400 µg AFB_1** (labeled)/kg b. wt., by gavage, once; for detailed information please see the article), conc.: 0.74 pmol/mg DNA * ** (mean value), country: USA[142], *after 24 h, **AFM_1-N^7-Gua

Tree shrew serum may contain the following mycotoxins and/or their metabolites:

Aflatoxin
incidence: 7?/7, sa. const.: male and female adult tree shrews, wt.t: 100–160 g, contamination: artificial (dose: 400 µg AFB_1/kg b. wt., o., daily for 4 weeks; for detailed information please see the article), conc. range: ≈≤20 pmol AF-albumin adducts/mg protein* ** (mean value), country: People's Republic of China/USA[167], *control, **after 5 weeks during administration (also after other week intervals up to 11 weeks measured, lowest conc.: ≈5 pmol AF-albumin adducts/mg protein after 11 weeks)
incidence: 6?/6, sa. const.: male and female adult tree shrews, wt.: 100–160 g, contamination: artificial (dose: 400 µg AFB_1/kg b. wt., o., daily for 4 weeks and additionally 0.5 mmol **oltipraz**/kg, o. by gavage, daily for 5 weeks; for detailed information please see the article), conc. range: ≈≤5 pmol AF-albumin adducts/mg protein (mean value), country: People's Republic of China/USA[167], **after 5 weeks during administration (also after other week intervals up to 11 weeks measured, lowest conc.: ≈0.2 pmol AF-albumin adducts/mg protein after 8 and 10 weeks)

Tree shrew urine may contain the following mycotoxins and/or their metabolites:

Aflatoxin B_1
incidence: ?/?*, sa. const.: male tree shrews, wt.: 100–160 g, contamination: no AFB_1 (for detailed information please see the article), conc.: nr, country: USA[142], *control
incidence: 4?/4, sa. const.: male tree shrews, wt.: 100–160 g, contamination: artificial (dose: 400 µg AFB_1 (labeled)/kg b. wt., by gavage, once; for detailed information please see the article), conc.: 0.60 pmol/mg creatinine * ** (mean value), country: USA[142], *after 24 h, **AFB_1-N^7-Gua

Aflatoxin M_1
incidence: ?/?*, sa. const.: male tree shrews, wt.: 100–160 g, contamination: no

AFB_1 (for detailed information please see the article), conc.: nr, country: USA[142], *control
incidence: 4?/4, sa. const.: male tree shrews, wt.: 100–160 g, contamination: artificial (dose: 400 μg AFB_1 (labeled)/kg b. wt., by gavage, once; for detailed information please see the article), conc.: 0.69 pmol/mg creatinine * ** (mean value), country: USA[142], *after 24 h, **AFM_1-N^7-Gua

AFLATOXIN
incidence: 7?/7, sa. const.: male and female adult tree shrews, wt.: 100–160 g, contamination: artificial (dose: 400 μg AFB_1/kg b. wt., o., daily for 4 weeks; for detailed information please see the article), Ø conc.: 6.34 ng AF-N^7-Gua/mg creatinine* ** (mean value), country: People's Republic of China/USA[167], *control, **at week 7 (AFB_1-administration ends at week 7)
incidence: 6?/6, sa. const.: male and female adult tree shrews, wt.: 100–160 g, contamination: artificial (dose: 400 μg AFB_1/kg b. wt., o., daily for 4 weeks and additionally 0.5 mmol oltipraz/kg, o. by gavage, daily for 5 weeks; for detailed information please see the article), Ø conc.: 0.47 ng AF-N^7-Gua/mg creatinine* (mean value), country: People's Republic of China/USA[167], *at week 7 (AFB_1- and oltipraz-administration ends at week 7)

Turkey

Turkey Natural Contamination

Turkey lung may contain the following mycotoxins and/or their metabolites:

GLIOTOXIN
incidence: 5/13, sa. const.: lungs from turkeys of the USA, contamination: natural, conc. range: 0.4–126.3 μg/g, Ø conc.: 42.16 μg/g, country: USA[82]

Turkey plasma may contain the following mycotoxins and/or their metabolites:

AFLATOXIN B_1
incidence: ?/36, sa. const.: plasma from Shaver star cross 2,288 turkeys of Japan, age: 31 weeks, contamination: natural, conc.: 12.3 pg/ml (mean value), country: Thailand/Japan[578]

Turkey Artificial Contamination

Turkey excreta may contain the following mycotoxins and/or their metabolites:

ZEARALENONE
incidence: 6/6*, sa. const.: male Nicholas Broad White turkey poults, age: 3 weeks, contamination: no ZEA (for detailed information please see the article), conc.: nr, country: Sweden/USA[119], *control
incidence: 6?/6, sa. const.: male Nicholas Broad White turkey poults, age: 3 weeks, contamination: artificial (dose: **800 mg ZEA**/kg diet, o., daily for 2 weeks; for detailed information please see the article), conc. range: ≈≤220 μg/g* ** (mean value), country: Sweden/USA[119], *on the 3rd day of treatment (also measured on day 7 and 14, lowest conc.: ≈180 μg/g after 14 days), **free and conjugated ZEA

α-ZEARALENOL
incidence: 6/6*, sa. const.: male Nicholas Broad White turkey poults, age: 3 weeks, contamination: no ZEA (for detailed information please see the article), conc.: nr**, country: Sweden/USA[119], *control, **free and conjugated α-ZEAOL
incidence: 6?/6, sa. const.: male Nicholas Broad White turkey poults, age: 3 weeks, contamination: artificial (dose: **800 mg ZEA**/kg diet, o., daily for 2 weeks; for detailed information please see the article), conc. range: ≤644 ng/g* (mean value), country: Sweden/USA[119], on

the 14th day of treatment (also measured on day 3 and 7, lowest conc.: ≈470 μg/g after 3 days), **free and conjugated α-ZEAOL

Turkey heart may contain the following mycotoxins and/or their metabolites:

ZEARALENONE
incidence: 6/6*, sa. const.: male Nicholas Broad White turkey poults, age: 3 weeks, contamination: no ZEA (for detailed information please see the article), conc.: nr, country: Sweden/USA[119], *control
incidence: 6?/6, sa. const.: male Nicholas Broad White turkey poults, age: 3 weeks, contamination: artificial (dose: **800 mg ZEA**/kg diet, o., daily for 2 weeks; for detailed information please see the article), conc.: 57 ng/g* ** (mean value), country: Sweden/USA[119], *free and conjugated ZEA, **after 14 days of ZEA-administration

α-ZEARALENOL
incidence: 6/6*, sa. const.: male Nicholas Broad White turkey poults, age: 3 weeks, contamination: no ZEA (for detailed information please see the article), conc.: nr**, country: Sweden/USA[119], *control, **free and conjugated α-ZEAOL
incidence: 6?/6, sa. const.: male Nicholas Broad White turkey poults, age: 3 weeks, contamination: artificial (dose: **800 mg ZEA**/kg diet, o., daily for 2 weeks; for detailed information please see the article), conc.: 238 ng/g* ** (mean value), country: Sweden/USA[119], *free and conjugated α-ZEAOL, **after 14 days of ZEA-administration

Turkey kidney may contain the following mycotoxins and/or their metabolites:

FUMONISIN B_1
incidence: 8?/8, sa. const.: male turkeys (BUT 9 strain), age: 1 day, contamination: artificial (dose: **10 mg FB_1**/kg b. wt., **i.v.,** once), conc.: 50 μg/kg* (mean value), country: France[480], *after 24 h
incidence: 8?/8, sa. const.: male turkeys (BUT 9 strain), age: 1 day, contamination: artificial (dose: **100 mg FB_1**/kg b. wt., **o.,** once), conc.: 5,785 μg/kg* (mean value), country: France[480], *after 10 h

incidence: 6/6*, sa. const.: male turkeys (BUT 9 strain), age: 1 day, contamination: no FB_1 + FB_2 (for detailed information please see the article), conc.: nd, country: France[480], *control
incidence: 6/6, sa. const.: male turkeys (BUT 9 strain), age: 1 day, contamination: artificial (dose: **5 mg FB_1 + FB_2**/kg diet, o., for 9 weeks; for detailed information please see the article), conc.: nd*, country: France[480], *after 8 h of final administration
incidence: 6/6, sa. const.: male turkeys (BUT 9 strain), age: 1 day, contamination: artificial (dose: **10 mg FB_1 + FB_2**/kg diet, o., for 9 weeks; for detailed information please see the article), conc.: nd*, country: France[480], *after 8 h of final administration
incidence: 6?/6, sa. const.: male turkeys (BUT 9 strain), age: 1 day, contamination: artificial (dose: **20 mg FB_1 + FB_2**/kg diet, o., for 9 weeks; for detailed information please see the article), conc.: 22 μg/kg* (mean value), country: France[480], *after 8 h of final administration

ZEARALENONE
incidence: 6/6*, sa. const.: male Nicholas Broad White turkey poults, age: 3 weeks, contamination: no ZEA (for detailed information please see the article), conc.: nr, country: Sweden/USA[119], *control
incidence: 6?/6, sa. const.: male Nicholas Broad White turkey poults, age: 3 weeks, contamination: artificial (dose: **800 mg ZEA**/kg diet, o., daily for 2 weeks; for detailed information please see the article), conc.: 122 ng/g* ** (mean value), country: Sweden/USA[119], *free and conjugated ZEA, **after 14 days of ZEA-administration

α-Zearalenol
incidence: 6/6*, sa. const.: male Nicholas Broad White turkey poults, age: 3 weeks, contamination: no ZEA (for detailed information please see the article), conc.: nr, country: Sweden/USA[119], *control
incidence: 6?/6, sa. const.: male Nicholas Broad White turkey poults, age: 3 weeks, contamination: artificial (dose: **800 mg ZEA/kg** diet, o., daily for 2 weeks; for detailed information please see the article), conc.: 477 ng/g* (mean value), country: Sweden/USA[119], *free and conjugated α-ZEAOL, **after 14 days of ZEA-administration

Turkey liver may contain the following mycotoxins and/or their metabolites:

Aflatoxicol
incidence: 4?/4, sa. const.: turkey poults, contamination: artificial (dose: 500 ng AFB_1/g diet, o., for 3 weeks), conc.: 0.70 ng/g* ** (mean value), country: USA[609], *in **organic phase,** **after 3 weeks of AFB_1-administration
incidence: 4?/4, sa. const.: turkey poults, contamination: artificial (dose: 500 ng AFB_1/g diet, o., for 3 weeks), conc.: 0.11 ng/g* ** (mean value), country: USA[609], *in **aqueous phase,** **after 3 weeks of AFB_1-administration

Aflatoxin B_1
incidence: 4–5/4–5*, sa. const.: male Large White turkeys, age: 1 day, contamination: no AFB_1 addition (for detailed information please see the article), conc.: nd, country: USA[105], *control
incidence: ?/4–5, sa. const.: male Large White turkeys, age: 1 day, contamination: artificial (dose: **500 ppb AFB_1**/g diet, o., for 18 days; for detailed information please see the article), conc. range: ≤0.10 ng AFB_1 eq/g* ** (mean value), country: USA[105], ***free AFB_1**, **0 days after AF withdrawl (also measured after 1, 2 and 3 days, lowest conc.: 0.02 ng AFB_1 eq/g after 2 days)
incidence: ?/4–5*, sa. const.: male Large White turkeys, age: 1 day, contamination: no AFB_1 addition (for detailed information please see the article), conc.: nd, country: USA[105], *control
incidence: ?/4–5, sa. const.: male Large White turkeys, age: 1 day, contamination: artificial (dose: **500 ppb AFB_1**/g diet, o., for 18 days; for detailed information please see the article), conc. range: ≤0.28 ng AFB_1 eq/g* ** (mean value), country: USA[105], ***conjugated AFB_1**, **0 days after AF withdrawl (also measured after 1, 2 and 3 days, lowest conc.: 0.02 ng AFB_1 eq/g after 3 days)

incidence: 4?/4, sa. const.: turkey poults, contamination: artificial (dose: 500 ng AFB_1/g diet, o., for 3 weeks), conc.: 0.10 ng/g* ** (mean value), country: USA[609], *in **organic phase,** **after 3 weeks
incidence: 4?/4, sa. const.: turkey poults, contamination: artificial (dose: 500 ng AFB_1/g diet, o., for 3 weeks), conc.: 0.28 ng/g* ** (mean value), country: USA[609], *in **aqueous phase,** **after 3 weeks

Aflatoxin B_2
incidence: 4?/4, sa. const.: turkey poults, contamination: artificial (dose: 500 ng AFB_1/g diet, o., for 3 weeks), conc.: tr* ** (mean value), country: USA[609], *in **organic phase,** **after 3 weeks
incidence: 4?/4, sa. const.: turkey poults, contamination: artificial (dose: 500 ng AFB_1/g diet, o., for 3 weeks), conc.: 0.04 ng/g* ** (mean value), country: USA[609], *in **aqueous phase,** **after 3 weeks

Aflatoxin G_1
incidence: 4/4, sa. const.: turkey poults, contamination: artificial (dose: 500 ng AFB_1/g diet, o., for 3 weeks), conc.: nd* ** (mean value), country: USA[609], *in **organic phase,** **after 3 weeks
incidence: 4?/4, sa. const.: turkey poults, contamination: artificial (dose: 500 ng AFB_1/g diet, o., for 3 weeks), conc.:

0.04 ng/g* ** (mean value), country: USA[609], *in **aqueous phase**, **after 3 weeks

Aflatoxin G_2
incidence: 4/4, sa. const.: turkey poults, contamination: artificial (dose: 500 ng AFB_1/g diet, o., for 3 weeks), conc.: nd* ** (mean value), country: USA[609], *in **organic phase**, **after 3 weeks
incidence: 4?/4, sa. const.: turkey poults, contamination: artificial (dose: 500 ng AFB_1/g diet, o., for 3 weeks), conc.: 0.04 ng/g* ** (mean value), country: USA[609], *in **aqueous phase**, **after 3 weeks

Aflatoxin M_1
incidence: ?/4–5*, sa. const.: male Large White turkeys, age: 1 day, contamination: no AFB_1 (for detailed information please see the article), conc.: 0.04 ng/g** ***, country: USA[105], *control, ****free** AFM_1, ***0 days after AF withdrawl (up to 0 days measured)
incidence: ?/4–5, sa. const.: male Large White turkeys, age: 1 day, contamination: artificial (dose: **500 ppb** AFB_1/g diet, o., for 18 days; for detailed information please see the article), conc. range: ≤0.44 ng AFM_1 eq/g* ** (mean values), country: USA[105], ***free** AFM_1, **0 days after AF withdrawl (also measured after 1, 2 and 3 days, lowest conc.: 0.02 ng AFB_1 eq/g after 1 and 3 days)
incidence: ?/4–5*, sa. const.: male Large White turkeys, age: 1 day, contamination: no AFB_1 (for detailed information please see the article), conc.: 0.03 ng/g** ***, country: USA[105], *control, ****conjugated** AFM_1, ***0 days after AF withdrawl (up to 0 days measured)
incidence: ?/4–5, sa. const.: male Large White turkeys, age: 1 day, contamination: artificial (dose: **500 ppb** AFB_1/g diet, o., for 18 days; for detailed information please see the article), conc. range: ≤0.42 ng AFM_1 eq/g* (mean values), country: USA[105], ***conjugated** AFM_1, **2 days after AF withdrawl (also measured after 0, 1 and 3 days, lowest conc.: 0.22 ng AFB_1 eq/g after 3 days)
incidence: 4?/4, sa. const.: turkey poults, contamination: artificial (dose: 500 ng AFB_1/g diet, o., for 3 weeks), conc.: 0.44 ng/g* ** (mean value), country: USA[609], *in **organic phase**, **after 3 weeks
incidence: 4?/4, sa. const.: turkey poults, contamination: artificial (dose: 500 ng AFB_1/g diet, o., for 3 weeks), conc.: 0.24 ng/g* ** (mean value), country: USA[609], *in **aqueous phase**, **after 3 weeks

Aflatoxin Q_1
incidence: 4/4, sa. const.: turkey poults, contamination: artificial (dose: 500 ng AFB_1/g diet, o., for 3 weeks), conc.: nd* ** (mean value), country: USA[609], *in **organic phase**, **after 3 weeks
incidence: 4/4, sa. const.: turkey poults, contamination: artificial (dose: 500 ng AFB_1/g diet, o., for 3 weeks), conc.: nd* ** (mean value), country: USA[609], *in **aqueous phase**, **after 3 weeks

Fumonisin B_1
incidence: 8?/8, sa. const.: male turkeys (BUT 9 strain), age: 1 day, contamination: artificial (dose: **10 mg** FB_1/kg b. wt., **i.v.,** once), conc.: 46 μg/kg* (mean value), country: France[480], *after 24 h
incidence: 8?/8, sa. const.: male turkeys (BUT 9 strain), age: 1 day, contamination: artificial (dose: **100 mg** FB_1/kg b. wt., **o.,** once), conc.: 5,458 μg/kg* (mean value), country: France[480], *after 10 h
incidence: 6/6*, sa. const.: male turkeys (BUT 9 strain), age: 1 day, contamination: no FB_1 + FB_2 (for detailed information please see the article), conc.: nd, country: France[480], *control
incidence: 6?/6, sa. const.: male turkeys (BUT 9 strain), age: 1 day, contamination: artificial (dose: **5 mg** FB_1 **+** FB_2/kg diet, o., for 9 weeks; for detailed information please see the article), conc.: 33 μg/kg* (mean value), country: France[480], *after 8 h of final administration
incidence: 6?/6, sa. const.: male turkeys (BUT 9 strain), age: 1 day, contamination: artificial (dose: **10 mg** FB_1 **+** FB_2/kg diet, o., for 9 weeks; for detailed information

please see the article), conc.: 44 µg/kg* (mean value), country: France[480], *after 8 h of final administration
incidence: 6?/6, sa. const.: male turkeys (BUT 9 strain), age: 1 day, contamination: artificial (dose: **20 mg FB_1 + FB_2/kg** diet, o., for 9 weeks; for detailed information please see the article), conc.: 117 µg/kg* (mean value), country: France[480], *after 8 h of final administration

Zearalenone
incidence: 6/6*, sa. const.: male Nicholas Broad White turkey poults, age: 3 weeks, contamination: no ZEA (for detailed information please see the article), conc.: nr, country: Sweden/USA[119], *control
incidence: 6?/6, sa. const.: male Nicholas Broad White turkey poults, age: 3 weeks, contamination: artificial (dose: **800 mg ZEA/kg** diet, o., daily for 2 weeks; for detailed information please see the article), conc.: 276 ng/g* ** (mean value), country: Sweden/USA[119], *free and conjugated ZEA, **after 14 days of ZEA-administration

α-Zearalenol
incidence: 6/6*, sa. const.: male Nicholas Broad White turkey poults, age: 3 weeks, contamination: no ZEA (for detailed information please see the article), conc.: nr, country: Sweden/USA[119], *control
incidence: 6?/6, sa. const.: male Nicholas Broad White turkey poults, age: 3 weeks, contamination: artificial (dose: **800 mg ZEA/kg** diet, o., daily for 2 weeks; for detailed information please see the article), conc.: 2,715 ng/g* ** (mean value), country: Sweden/USA[119], *free and conjugated α-ZEAOL, **after 14 days of ZEA-administration

Turkey lung may contain the following mycotoxins and/or their metabolites:

Zearalenone
incidence: 6/6*, sa. const.: male Nicholas Broad White turkey poults, age: 3 weeks, contamination: no ZEA (for detailed information please see the article), conc.: nr, country: Sweden/USA[119], *control
incidence: 6?/6, sa. const.: male Nicholas Broad White turkey poults, age: 3 weeks, contamination: artificial (dose: **800 mg ZEA/kg** diet, o, daily for 2 weeks; for detailed information please see the article), conc.: 56 ng/g* ** (mean value), country: Sweden/USA[119], *free and conjugated ZEA, **after 14 days of ZEA-administration

α-Zearalenol
incidence: 6/6*, sa. const.: male Nicholas Broad White turkey poults, age: 3 weeks, contamination: no ZEA (for detailed information please see the article), conc.: nr, country: Sweden/USA[119], *control
incidence: 6?/6, sa. const.: male Nicholas Broad White turkey poults, age: 3 weeks, contamination: artificial (dose: **800 mg ZEA/kg** diet, o, daily for 2 weeks; for detailed information please see the article), conc.: 202 ng/g* ** (mean value), country: Sweden/USA[119], *free and conjugated α-ZEAOL, **after 14 days of ZEA-administration

Turkey muscle may contain the following mycotoxins and/or their metabolites:

Fumonisin B_1
incidence: 8/8, sa. const.: male turkeys (BUT 9 strain), age: 1 day, contamination: artificial (dose: **10 mg FB_1/kg** b. wt., **i.v.**, once), conc.: nd*, country: France[480], *after 24 h
incidence: 8?/8, sa. const.: male turkeys (BUT 9 strain), age: 1 day, contamination: artificial (dose: **100 mg FB_1/kg** b. wt., **o.**, once), conc.: 113 µg/kg* (mean value), country: France[480], *after 10 h

incidence: 6/6*, sa. const.: male turkeys (BUT 9 strain), age: 1 day, contamination: no FB_1 + FB_2 (for detailed information please see the article), conc.: nd, country: France[480], *control

incidence: 6/6, sa. const.: male turkeys (BUT 9 strain), age: 1 day, contamination: artificial (dose: **5 mg FB_1 + FB_2/kg** diet, o., for 9 weeks; for detailed information please see the article), conc.: nd*, country: France[480], *after 8 h of final administration
incidence: 6/6, sa. const.: male turkeys (BUT 9 strain), age: 1 day, contamination: artificial (dose: **10 mg FB_1 + FB_2/kg** diet, o., for 9 weeks; for detailed information please see the article), conc.: nd*, country: France[480], *after 8 h of last administration
incidence: 6/6, sa. const.: male turkeys (BUT 9 strain), age: 1 day, contamination: artificial (dose: **20 mg FB_1 + FB_2/kg** diet, o., for 9 weeks; for detailed information please see the article), conc.: nd*, country: France[480], *after 8 h of final administration

Turkey muscle, breast may contain the following mycotoxins and/or their metabolites:

Aflatoxin B_1
incidence: ?/4–5*, sa. const.: male Large White turkeys, age: 1 day, contamination: no AFB_1 addition (for detailed information please see the article), conc.: 0.01 ng/g**, country: USA[105], *control, **free AFB_1**
incidence: ?/4–5, sa. const.: male Large White turkeys, age: 1 day, contamination: artificial (dose: **500 ppb AFB_1/g** diet, o., for 18 days; for detailed information please see the article), conc. range: ≤0.03 ng AFB_1 eq/g* ** (mean values), country: USA[105], ***free AFB_1**, **0 days after AF withdrawl (also measured after 1 and 2 days, lowest conc.: nd after 2 days)
incidence: 4–5/4–5*, sa. const.: male Large White turkeys, age: 1 day, contamination: no AFB_1 addition (for detailed information please see the article), conc.: nd**, country: USA[105], *control, ****conjugated AFB_1**
incidence: ?/4–5, sa. const.: male Large White turkeys, age: 1 day, contamination: artificial (dose: **500 ppb AFB_1/g** diet, o., for 18 days; for detailed information please see the article), conc.: nd* ** (mean values), country: USA[105], ***conjugated AFB_1**, **0 days after AF withdrawl (also measured after 1 and 2 days, but conc.: nd)

Aflatoxin M_1
incidence: 4–5/4–5*, sa. const.: male Large White turkeys, age: 1 day, contamination: no AFB_1 (for detailed information please see the article), conc.: nd**, country: USA[105], *control, ****free AFM_1**
incidence: ?/4–5, sa. const.: male Large White turkeys, age: 1 day, contamination: artificial (dose: **500 ppb AFB_1/g** diet, o., for 18 days; for detailed information please see the article), conc.: nd* **, country: USA[105], ***free AFM_1**, **0 days after AF withdrawl (also measured after 1 and 2 days, but conc.: nd)
incidence: 4–5/4–5*, sa. const.: male Large White turkeys, age: 1 day, contamination: no AFB_1 (for detailed information please see the article), conc.: nd, country: USA[105], *control
incidence: ?/4–5, sa. const.: male Large White turkeys, age: 1 day, contamination: artificial (dose: **500 ppb AFB_1/g** diet, o., for 18 days; for detailed information please see the article), conc.: nd* **, country: USA[105], ***conjugated AFM_1**, **0 days after AF withdrawl (up to 0 days measured)

Turkey muscle, thigh may contain the following mycotoxins and/or their metabolites:

Aflatoxin B_1
incidence: 4–5/4–5*, sa. const.: male Large White turkeys, age: 1 day, contamination: no AFB_1 (for detailed information please see the article), conc.: na, country: USA[105], *control
incidence: ?/4–5, sa. const.: male Large White turkeys, age: 1 day, contamination: artificial (dose: **500 ppb AFB_1/g** diet, o.,

for 18 days; for detailed information please see the article), conc.: 0.01 ng AFB_1 eq/g* ** (mean value), country: USA[105], ***free AFB_1**, **1 day after AF withdrawl (also measured after 2 days, but conc.: nd)
incidence: 4–5/4–5*, sa. const.: male Large White turkeys, age: 1 day, contamination: no AFB_1 (for detailed information please see the article), conc.: na, country: USA[105], *control
incidence: 4–5/4–5, sa. const.: male Large White turkeys, age: 1 day, contamination: artificial (dose: **500 ppb AFB_1/g** diet, o., for 18 days; for detailed information please see the article), conc.: nd* ** (mean value), country: USA[105], ***conjugated AFB_1**, **1 day after AF withdrawl (also measured after 2 days, but conc.: nd)

Aflatoxin M_1
incidence: 4–5/4–5*, sa. const.: male Large White turkeys, age: 1 day, contamination: no AFB_1 (for detailed information please see the article), conc.: na, country: USA[105], *control
incidence: ?/4–5, sa. const.: male Large White turkeys, age: 1 day, contamination: artificial (dose: **500 ppb AFB_1/g** diet, o., for 18 days; for detailed information please see the article), conc.: 0.07 ng AFM_1 eq/g* ** (mean value), country: USA[105], ***free AFM_1**, **1 day after AF withdrawl (also measured after 2 days, but conc.: nd)
incidence: ?/4–5*, sa. const.: male Large White turkeys, age: 1 day, contamination: no AFB_1 (for detailed information please see the article), conc.: na, country: USA[105], *control
incidence: ?/4–5, sa. const.: male Large White turkeys, age: 1 day, contamination: artificial (dose: **500 ppb AFB_1/g** diet, o., for 18 days; for detailed information please see the article), conc.: 0.11 ng AFM_1 eq/g* (mean value), country: USA[105], ***conjugated AFM_1**, **1 day after AF withdrawl (also measured after 2 days, but conc.: nd)

Turkey plasma may contain the following mycotoxins and/or their metabolites:

Fumonisin B_1
incidence: 8/8, sa. const.: male turkeys (BUT 9 strain), age: 1 day, contamination: artificial (dose: **10 mg FB_1/kg** b. wt., **i.v.**, once), conc.: nd*, country: France[480], *after 24 h
incidence: 8?/8, sa. const.: male turkeys (BUT 9 strain), age: 1 day, contamination: artificial (dose: **100 mg FB_1/kg** b. wt., **o.**, once), conc.: 279 µg/l* (mean value), country: France[480], *after 10 h
incidence: 6/6*, sa. const.: male turkeys (BUT 9 strain), age: 1 day, contamination: no FB_1 + FB_2 (for detailed information please see the article), conc.: nd, country: France[480], *control
incidence: 6/6, sa. const.: male turkeys (BUT 9 strain), age: 1 day, contamination: artificial (dose: **5 mg FB_1 + FB_2/kg** diet, o., for 9 weeks; for detailed information please see the article), conc.: nd*, country: France[480], *after 8 h of final administration
incidence: 6/6, sa. const.: male turkeys (BUT 9 strain), age: 1 day, contamination: artificial (dose: **10 mg FB_1 + FB_2/kg** diet, o., for 9 weeks; for detailed information please see the article), conc.: nd*, country: France[480], *after 8 h of final administration
incidence: 6?/6, sa. const.: male turkeys (BUT 9 strain), age: 1 day, contamination: artificial (dose: **20 mg FB_1 + FB_2/kg** diet, o., for 9 weeks; for detailed information please see the article), conc.: 53 µg/l* (mean value), country: France[480], *after 8 h of last administration

Zearalenone
incidence: 6/6*, sa. const.: male Nicholas Broad White turkey poults, age: 3 weeks, contamination: no ZEA (for detailed information please see the article), conc.: nr, country: Sweden/USA[119], *control
incidence: 6?/6, sa. const.: male Nicholas Broad White turkey poults, age: 3 weeks, contamination: artificial (dose: **800 mg ZEA/kg** diet, o., daily for 2 weeks; for detailed information please see the article), conc. range: ≈≤99 ng/ml

(mean value), country: Sweden/USA[119], *on the 7th day of treatment (also measured on day 3 and 14, lowest conc.: ≈60 ng/ml after 3 days), **free and conjugated ZEA

α-Zearalenol
incidence: 6/6*, sa. const.: male Nicholas Broad White turkey poults, age: 3 weeks, contamination: no ZEA (for detailed information please see the article), conc.: nr, country: Sweden/USA[119], *control
incidence: 6?/6, sa. const.: male Nicholas Broad White turkey poults, age: 3 weeks, contamination: artificial (dose: **800 mg** ZEA/kg diet, o., daily for 2 weeks; for detailed information please see the article), conc. range: ≤194 ng/ml (mean value), country: Sweden/USA[119], *on the 14th day of treatment (also measured on day 3 and 7, lowest conc.: ≈150 ng/ml after 3 days), **free and conjugated α-ZEAOL

Turkey serum may contain the following mycotoxins and/or their metabolites:

Fumonisin B_1
incidence: 6?/6, sa. const.: male turkeys (BUT 9 strain), age: 1 day, contamination: artificial (dose: **10 mg FB_1/kg** b. wt., **i.v.,** once), conc. range: ≈≤60?* μg/l, country: France[480], *after 0 min (also after other min intervals up to 1,200 min measured, lowest conc.: nd after 1,200 min)
incidence: 8?/8, sa. const.: male turkeys (BUT 9 strain), age: 1 day, contamination: artificial (dose: **100 mg FB_1/kg** b. wt., **o.,** once), conc. range: ≈≤999 μg/l* (mean value), country: France[480], *after 180 min (also at other min intervals up to 600 min measured, except for the start value conc.: ≈280 μg/l after 600 min)

Walleye fish see Fish, walleye fish

Woodchuck

Woodchuck Artificial Contamination

Woodchuck liver may contain the following mycotoxins and/or their metabolites:

Aflatoxin B_1-8,9-Epoxide
incidence: 8/8*, sa. const.: eastern woodchucks, age: agematched, contamination: artificial (dose: AFB_1 (labeled and unlabeled) addition; for detailed information please see the article), conc. range: 2.95–9.12 nmol/15 min/mg protein**, Ø conc.: 6.34 nmol/15 min/mg protein**, country: Germany[547], ***WHV-free,** **AFB_1-8,9-epoxide formation rate
incidence: 10/10*, sa. const.: eastern woodchucks, age: agematched, contamination: artificial (dose: AFB_1 (labeled and unlabeled) addition; for detailed information please see the article), conc. range: 0.72–9.74 nmol/15 min/mg protein**, Ø conc.: 2.89 nmol/15 min/mg protein**, country: Germany[547], ***WHV-positive,** **AFB_1-8,9-epoxide formation rate

Table 1. Mycotoxins and their metabolites in humans: *natural* contamination

Human aminotic fluid
Ochratoxin A
Human bile
Aflatoxin B_1
Human blood
Aflatoxicol; aflatoxin B_1; aflatoxin B_2; aflatoxin G_1; aflatoxin G_2; aflatoxin M_1; aflatoxin M_2; aflatoxin M_1 & M_2; aflatoxin; ochratoxin A; zearalenone; α-zearalenol
Human brain
Aflatoxicol; aflatoxin B_1; aflatoxin B_2; aflatoxin G_1; aflatoxin G_2; aflatoxin M_1; aflatoxin M_2
Human breast
Aflatoxin B_1
Human breast milk
Aflatoxicol; aflatoxin B_1; aflatoxin B_2; aflatoxin G_1; aflatoxin G_2; aflatoxin M_1; aflatoxin M_2; aflatoxin M_1 & M_2; aflatoxin; ochratoxin A
Human cervix
Aflatoxin B_1
Human colon
Aflatoxin B_1
Human endometrium
Zearalenone
Human feces
Aflatoxin B_1; aflatoxin G_1; aflatoxin M_1; aflatoxin Q_1; aflatoxin; fumonisin B_1
Human funiculum
Ochratoxin A
Human hair
Fumonisin B_1; fumonisin B_2; fumonisin B_3
Human heart
Aflatoxin B_1; aflatoxin B_2
Human intestine
Aflatoxin B_1; aflatoxin B_2
Human kidney
Aflatoxin B_1; aflatoxin B_2; aflatoxin M_1; aflatoxin M_2; ochratoxin A
Human liver
Aflatoxicol; aflatoxin B_1; aflatoxin B_2; aflatoxin G_1; aflatoxin G_2; aflatoxin M_1; aflatoxin M_2; aflatoxin
Human lung
Aflatoxicol; aflatoxin B_1; aflatoxin B_2; aflatoxin G_1; aflatoxin G_2; aflatoxin M_1; aflatoxin M_2; aflatoxins
Human pancreas
Aflatoxin B_1
Human placenta
Aflatoxin B_1; ochratoxin A
Human plasma
Aflatoxin B_1; aflatoxin B_2; aflatoxin G_1; aflatoxin G_2; ochratoxin A
Human rectum
Aflatoxin B_1

(Continued)

Table 1. (Continued)

Human renal tissue

Ochratoxin A

Human semen

Aflatoxin G_1; aflatoxin M_1; aflatoxin M_2

Human serum

Aflatoxicol; aflatoxin B_1; aflatoxin B_2; aflatoxin B_{2a}; aflatoxin B; aflatoxin G_1; aflatoxin G_2; aflatoxin G_{2a}; aflatoxin M_1; aflatoxin M_2; aflatoxin P; aflatoxin Q_1; aflatoxin; aflatoxins; gliotoxin; ochratoxin A; ochratoxin A methyl; ochratoxin α; ochratoxin α methyl ester; ochratoxin B; trichothecenes; zearalenone

Human serum/plasma

Ochratoxin A

Human spleen

Aflatoxin B_1; aflatoxin B_2; aflatoxin M_1; aflatoxin M_2

Human stomach

Aflatoxin B_1; aflatoxin B_2

Human stool

Aflatoxicol; aflatoxin B_1; aflatoxin B_2; aflatoxin G_1; aflatoxin G_2; aflatoxin M_1; aflatoxin M_2; ochratoxin A; 4-hydroxyochratoxin A; ochratoxin B

Human urine

Aflatoxicol; aflatoxin B_1; aflatoxin B_2; aflatoxin B_{2a}; aflatoxin B; aflatoxin G_1; aflatoxin G_2; aflatoxin G_{2a}; aflatoxin L; aflatoxin M_1; aflatoxin M_2; aflatoxin P_1; aflatoxin P; aflatoxin Q_1; aflatoxin; aflatoxins; deoxynivalenol; fumonisin B_1; ochratoxin A; 4-hydroxyochratoxin A; ochratoxin B

Table 2. Mycotoxins and their metabolites in humans: *artificial* contamination

Human blood

Diacetoxyscirpenol; HT-2 toxin; nivalenol; T-2 toxin; verrucarol

Human esophagus

HT-2 toxin; T-2 toxin

Human feces

Verrucarol

Human heart

HT-2 toxin

Human intestine

HT-2 toxin; T-2 toxin

Human kidney

Diacetoxyscirpenol; T-2 toxin

Human lung

T-2 toxin

Human stomach

HT-2 toxin; T-2 toxin

Human urine

Diacetoxyscirpenol; HT-2 toxin; nivalenol; T-2 toxin; verrucarol

Table 3. Mycotoxins and their metabolites in animals: *natural* contamination

Aflatoxin B_1
Beef liver; calf liver; chicken liver; chicken muscle; cow liver; cow milk, raw; duck liver; hare liver; hen egg; horse liver; pig liver; pig plasma; steer kidney; turkey plasma
Aflatoxin G_1
Beef liver
Aflatoxin M_1
Beef liver; buffalo milk, raw; camel milk; cow milk, raw; dog liver; goat milk, raw; pig kidney; pig liver; sheep milk, raw; sheep/goat milk, raw; steer kidney
Aflatoxin M_2
Cow milk, raw
Aflatoxin M
Cow milk, raw
Citrinin
Pig kidney; pig serum; pig urine
Deoxynivalenol
Hen egg; pig bile; pig digesta; pig liver; pig muscle; pig serum; pig urine
Deepoxydeoxynivalenol
Hen egg; pig digesta
Fumonisin B_1
Cow milk, raw
Gliotoxin
Camel fetus; camel intestine; camel rumen; cattle udder; turkey lung
Ochratoxin A
Cat kidney; cow milk, raw; dog kidney; pig blood; pig kidney; pig liver; pig meat; pig muscle; pig plasma; pig serum; pig urine; poultry
Penicillic acid
Pig serum; pig urine
Penitrem A
Dog brain; dog kidney; dog liver; dog stomach; pig serum; pig urine
Penitrem E
Dog kidney; dog liver
Thomitrems
Dog stomach
Roquefortine C
Dog brain; dog kidney; dog liver; dog stomach
Zeranol
Cattle bile
Zearalanols
Cattle urine; deer urine; goat urine; horse urine; lamb urine; sheep urine
Zearalenone
Pig bile; pig liver; pig muscle; pig serum; pig urine
α-Zearalenol
Cattle bile; pig bile
β-Zearalenol
Cattle bile
Zearalenols
Cattle urine; deer urine; goat urine; horse urine; lamb urine; sheep urine

Table 4. Mycotoxins and their metabolites in animals: *artificial* contamination

Aflatoxicol

Chicken kidney; chicken liver; chicken muscle, thigh; chicken skin; cow bile; cow kidney; cow liver; cow milk, raw; cow plasma; cow red blood cells; cow rumen; cow urine; fish, coho salmon bile; fish, rainbow trout bile; fish, rainbow trout egg; fish, rainbow trout embryo; fish, rainbow trout liver; fish, rainbow trout plasma; hen egg; hen kidney; hen liver; hen muscle, breast; hen muscle, thigh; hen ova; hen skin; lamb kidney; pig kidney; pig liver; pig muscle; turkey liver

Aflatoxicol M_1

Fish, coho salmon bile; fish, rainbow trout bile; fish, rainbow trout liver

Aflatoxin B_1

Calf liver; chicken blood; chicken crop; chicken excreta; chicken gastrointestinal tract; chicken gizzard; chicken heart; chicken kidney; chicken liver; chicken lung; chicken muscle; chicken muscle, breast; chicken muscle, leg; chicken muscle, thigh; chicken skin; cow bile; cow blood; cow brain; cow fat; cow feces; cow gallbladder; cow heart; cow intestine; cow kidney; cow liver; cow lung; cow lymph; cow mammary gland; cow milk, raw; cow muscle; cow plasma; cow red blood cells; cow rumen; cow spleen; cow tongue; cow urine; duck serum; ewe feces; ewe milk, raw; ewe urine; fish, channel catfish bile; fish, channel catfish fat; fish, channel catfish kidney; fish, channel catfish liver; fish, channel catfish muscle; fish, channel catfish plasma; fish, channel catfish skin; fish, channel catfish spleen; fish, channel catfish urine; fish, coho salmon bile; fish, coho salmon embryo; fish, coho salmon liver; fish, medaka liver; fish, rainbow trout bile; fish, rainbow trout blood; fish, rainbow trout carcass; fish, rainbow trout egg; fish, rainbow trout embryo; fish, rainbow trout liver; fish, rainbow trout plasma; fish, rainbow trout red blood cells; fish, walleye fish muscle; fish, zebrafish liver; guinea pig liver; guinea pig plasma; hamster kidney; hamster liver; hamster plasma; hen blood; hen blood clot; hen egg; hen excreta; hen gizzard; hen heart; hen kidney; hen liver; hen muscle; hen muscle, breast; hen muscle, leg; hen muscle, thigh; hen ova; hen serum; hen skin; lamb feces; lamb kidney; lamb liver; lamb urine; monkey, macaque bile; monkey, macaque blood; monkey, macaque brain; monkey, macaque heart; monkey, macaque kidney; monkey, macaque liver; monkey, macaque lung; monkey, macaque pancreas; monkey, macaque spleen; monkey, marmoset liver; mouse kidney; mouse liver; mouse milk; mouse plasma; mouse serum; pig bile; pig blood; pig brain; pig fat; pig gallbladder; pig heart; pig kidney; pig liver; pig lung; pig muscle; pig spleen; pig urine; pony cecum; pony kidney; pony liver; pony muscle; pony rectum; pony stomach; rat bile; rat feces; rat kidney; rat liver; rat lung; rat milk; rat plasma; rat serum; rat urine; steer bile; steer blood; steer feces; steer heart; steer kidney; steer liver; steer lung; steer muscle; steer pancreas; steer rumen; steer spleen; steer urine; tree shrew liver; tree shrew urine; turkey liver; turkey muscle, breast; turkey muscle, thigh

Aflatoxin B_1*-8,9-epoxide*

Rat liver; woodchuk liver

2,3-Dihydro-2,3-dihydroxyaflatoxin B_1

Rat liver

8,9-Dihydro-8,9-dihydroxyaflatoxin B_1

Rat liver

Aflatoxin B_2

Chicken crop; chicken excreta; chicken gizzard; chicken heart; chicken kidney; chicken liver; chicken muscle; chicken muscle, breast; chicken muscle, leg; cow liver; hen blood clot; hen egg; hen gizzard; hen heart; hen kidney; hen liver; hen muscle, breast; hen muscle, leg; hen ova; hen serum; pig gallbladder; pig heart; pig kidney; pig liver; pig muscle; pig spleen; steer heart; steer kidney; steer liver; steer muscle; steer spleen; turkey liver

(Continued)

Table 4. (Continued)

Aflatoxin B_{2a}

Chicken kidney; chicken liver; hen blood clot; hen egg; hen gizzard; hen heart; hen kidney; hen liver; hen muscle, breast; hen muscle, leg; hen ova; hen serum; lamb kidney; lamb urine; pig heart; pig kidney; pig liver; pig muscle; pig spleen

Aflatoxin B

Pig liver; rat bile; rat serum; rat urine

Aflatoxin G_1

Chicken excreta; chicken gizzard; chicken liver; cow feces; cow urine; ewe feces; ewe milk, raw; ewe urine; fish, walleye fish muscle; hen egg; hen gizzard; hen liver; lamb feces; lamb kidney; lamb liver; lamb urine; pig kidney; pig liver; rat kidney; rat liver; steer heart; steer kidney; steer liver; steer muscle; steer spleen; turkey liver

Aflatoxin G_2

Chicken excreta; fish, walleye fish muscle; hen gizzard; pig kidney; pig liver; steer heart; steer kidney; steer liver; steer muscle; turkey liver

Aflatoxin G_{2a}

Chicken liver

Aflatoxin M_1

Buffalo milk, raw; calf kidney; calf liver; chicken crop; chicken fat; chicken gizzard; chicken kidney; chicken liver; chicken muscle; chicken muscle, breast; chicken muscle, leg; chicken muscle, thigh; cow bile; cow blood; cow brain; cow feces; cow gallbladder; cow heart; cow intestine; cow kidney; cow liver; cow lung; cow lymph; cow mammary gland; cow milk, raw; cow muscle; cow plasma; cow red blood cells; cow rumen; cow spleen; cow tongue; cow urine; deer liver; deer muscle; ewe feces; ewe milk, raw; ewe urine; fish, rainbow trout liver; fish, rainbow trout plasma; goat milk, raw; hen blood clot; hen egg; hen gizzard; hen heart; hen kidney; hen liver; hen muscle, breast; hen muscle, leg; hen muscle, thigh; hen ova; lamb feces; lamb kidney; lamb liver; lamb urine; mouse milk; pig blood; pig gallbladder; pig heart; pig kidney; pig liver; pig muscle; pig spleen; pig urine; pony kidney; pony liver; pony muscle; rat bile; rat feces; rat liver; rat milk; rat urine; sheep milk, raw; sheep urine; steer bile; steer blood; steer feces; steer heart; steer kidney; steer liver; steer lung; steer muscle; steer pancreas; steer rumen; steer spleen; steer urine; tree shrew liver; tree shrew urine; turkey liver; turkey muscle, thigh

Aflatoxin M_2

Chicken gizzard; chicken heart; chicken kidney; chicken liver; chicken muscle, breast; chicken muscle, leg; hen blood clot; hen egg; hen gizzard; hen heart; hen kidney; hen liver; hen muscle, breast; hen muscle, leg; hen ova; hen serum; sheep urine

Aflatoxin M

Cow milk, raw; pig gallbladder; pig heart; pig kidney; pig liver; pig muscle; pig spleen

Aflatoxin P_1

Rat bile; rat urine

Aflatoxin Q_1

Rat bile; rat feces; rat liver; rat urine; turkey liver

Aflatoxin

Chicken heart; chicken liver; chicken muscle, breast; duck liver; duck plasma; hen egg; rat; rat liver; rat plasma; rat serum; rat urine; tree shrew serum; tree shrew urine

Aflatoxins

Hen adrenal; hen bile; hen digestive tract; hen egg; hen excreta; hen fat; hen gizzard; hen heart; hen liver; hen lung; hen muscle, breast; hen muscle, leg; hen muscle, wing; hen ova; hen pancreas; hen reproductive organs; hen skin; hen spleen; rat blood; rat feces; rat kidney; rat liver; rat lung; rat trachea; rat urine

(Continued)

Table 4. (Continued)

Citrinin
Hen egg; hen muscle, red; hen muscle, white; pig kidney
Cyclopiazonic acid
Chicken muscle; ewe milk, raw; hen egg; pig muscle; pig plasma; rat muscle
Deoxynivalenol
Cow feces; cow urine; hen; hen bile; hen egg; hen excreta; hen fat; hen gizzard; hen kidney; hen liver; hen plasma; mouse brain; mouse heart; mouse intestine; mouse kidney; mouse liver; mouse lung; mouse Peyer's patches; mouse plasma; mouse spleen; pig adrenals; pig bile; pig brain; pig colon; pig fat; pig feces; pig heart; pig intestine; pig kidney; pig liver; pig lung; pig lymph; pig muscle; pig pancreas; pig plasma; pig serum; pig skin; pig spleen; pig stomach; pig testes; pig urine; sheep bile; sheep feces; sheep milk, raw; sheep plasma; sheep rumen; sheep urine
Deepoxydeoxynivalenol
Cow feces; cow milk, raw; cow urine; pig bile; pig feces; pig kidney; pig liver; pig muscle; pig serum; pig urine; sheep bile; sheep feces; sheep milk, raw; sheep plasma; sheep urine
Deoxynivalenol + Deepoxydeoxynivalenol
Pig cecum; pig colon; pig rectum
Fumonisin B_1
Cattle feces; cattle urine; cow milk, raw; cow plasma; hen plasma; monkey, vervet plasma; pig bile; pig brain; pig eye; pig fat; pig feces; pig kidney; pig liver; pig lung; pig muscle; pig myocardium; pig pancreas; pig spleen; pig urine; pig, sow milk; rat bile; rat feces; rat plasma; rat urine; sheep feces; sheep urine; turkey kidney; turkey liver; turkey muscle; turkey plasma; turkey serum
Hydrolized fumonisin B_1
Cattle feces; rat feces; sheep feces
Fumonisin B_2
Monkey, vervet plasma; pig bile; pig fat; pig feces; pig kidney; pig liver; pig lung; pig muscle; pig myocardium; pig spleen; pig urine
Fusarenon-X
Chicken plasma; duck plasma; hen excreta
Fusaric acid
Rat stomach
Gliotoxin
Mouse lung; mouse serum
HT-2 toxin
Cat blood; cat heart; cat kidney; cat liver; cat lung; cat urine; chicken excreta; chicken liver; chicken lung; cow blood; cow kidney; cow lung; cow urine; dog plasma; pig bile; pig bone marrow; pig brain; pig duodenum; pig heart; pig ileum; pig intestine; pig jejunum; pig kidney; pig liver; pig lung; pig lymph; pig muscle; pig pancreas; pig skin; pig spleen; pig stomach; rat kidney; rat liver
Deepoxy-HT-2 toxin
Pig ileum; pig intestine; pig jejunum; pig kidney; pig liver; pig lung; pig stomach
3'-Hydroxy-HT-2 toxin
Cat blood, cat heart, cat kidney, cat liver, cat lung, cat urine, chicken excreta, chicken liver, chicken lung, cow blood, cow urine, pig bile; pig bone marrow; pig brain; pig duodenum; pig heart; pig ileum; pig intestine; pig jejunum; pig kidney; pig liver; pig lung; pig lymph; pig muscle; pig pancreas; pig spleen; pig stomach
Deepoxy-3'-hydroxy-HT-2 toxin
Cow urine
3-Acetoxy-3'-hydroxy-HT-2 toxin
Chicken excreta

(Continued)

Table 4. (Continued)

Neosolaniol
Pig bile; pig skin
4-Deacetylneosolaniol
Pig bile; pig skin
Nivalenol
Chicken plasma; duck plasma; hen bile; hen excreta; hen liver; pig blood; pig feces; pig urine
Deepoxynivalenol
Hen excreta
Ochratoxin A
Calf feces; calf kidney; calf muscle; calf serum; calf urine; chicken chest; chicken excreta; chicken female repro tract; chicken gizzard; chicken heart; chicken intestine; chicken kidney; chicken liver; chicken lung; chicken muscle; chicken muscle, leg; chicken muscle, red; chicken muscle, thigh; chicken muscle, white; chicken plasma; chicken serum; chicken skin; chicken stomach; cow kidney; ewe milk, raw; ewe plasma; fish, carp plasma; hen blood; hen cloaca; hen egg; hen fat; hen gizzard; hen heart; hen intestine; hen kidney; hen liver; hen lung; hen muscle, breast; hen muscle, leg; hen muscle, red; hen muscle, thigh; hen muscle, white; hen plasma; hen serum; hen skin; hen stomach; monkey, macaque plasma; mouse amniotic fluid; mouse embryo; mouse kidney; mouse liver; mouse placenta; mouse plasma; mouse serum; pig bile; pig blood; pig brain; pig fat; pig feces; pig heart; pig intestine; pig kidney; pig liver; pig muscle; pig placenta; pig plasma; pig serum; pig spleen; pig stomach; pig urine; quail egg; quail kidney; quail plasma; rabbit kidney; rabbit liver; rabbit mammary gland; rabbit milk; rabbit muscle; rabbit plasma; rat bile; rat blood; rat brain; rat colon; rat duodenum; rat feces; rat fetal extracts; rat heart; rat ileum; rat kidney; rat liver; rat lymph; rat milk; rat muscle; rat plasma; rat serous fluid and mucus; rat serum; rat stomach; rat urine; sheep feces; sheep rumen; sheep serum; sheep urine
Ochratoxin α
Calf feces; calf kidney; calf muscle; calf plasma; calf serum; calf urine; ewe milk, raw; ewe plasma; pig feces; pig urine; rat bile; rat blood; rat feces; rat serous fluid and mucus; rat urine; sheep feces; sheep rumen; sheep serum; sheep urine
Lactone opened ochratoxin A
Rat bile; rat urine
4-Hydroxyochratoxin A
Rat bile; rat urine
Ochratoxin B
Pig serum; rat bile; rat plasma; rat urine
Ochratoxin β
Pig feces; pig urine
Ochratoxin C
Rat blood
Penicillic acid
Chicken gizzard; chicken heart; chicken kidney; chicken liver
Penitrem A
Mouse brain; mouse gastrointestinal tract; mouse kidney; mouse liver
Scirpentriol
Cattle urine; pig plasma; pig serum; pig urine
Diacetoxyscirpenol
Cattle urine; pig serum; pig urine

(Continued)

Table 4. (Continued)

Monoacetoxyscirpenol
Cattle urine; pig serum; pig urine
Taleranol
Cattle urine
Trichothecenes
Guinea pig adrenal; guinea pig bile; guinea pig brain; guinea pig fat; guinea pig heart; guinea pig kidney; guinea pig liver; guinea pig lung; guinea pig muscle; guinea pig plasma; guinea pig spleen; guinea pig testes
Macrocyclic trichothecene
Goat lung; goat lymph; goat serum; goat spleen
T-2 toxin
Cat heart; cat kidney; cat lung; cattle omasum; cattle plasma; cattle rumen; chicken excreta; chicken liver; cow blood; cow heart; dog plasma; mouse blood; pig bile; pig blood; pig bone marrow; pig brain; pig colon; pig duodenum; pig fat; pig heart; pig ileum; pig intestine; pig jejunum; pig kidney; pig liver; pig lung; pig lymph; pig muscle; pig pancreas; pig plasma; pig skin; pig spleen; pig stomach; pig tissue; pig urine
3′-Hydroxy-T-2 toxin
Cat heart; cat kidney; cat lung; chicken excreta; cow urine; pig bile; pig bone marrow; pig brain; pig duodenum; pig heart; pig ileum; pig intestine; pig jejunum; pig kidney; pig liver; pig lung; pig lymph; pig muscle; pig pancreas; pig spleen; pig stomach
3′-Hydroxy-Iso-T-2 toxin
Cow urine
T-2 Tetraol
Cat blood; cat heart; cat kidney; cat liver; cat lung; cat urine; chicken excreta; chicken liver; cow blood; cow urine; pig bile; pig bone marrow; pig duodenum; pig heart; pig ileum; pig jejunum; pig kidney; pig liver; pig lung; pig lymph; pig pancreas; pig skin; pig spleen; pig stomach
Deepoxy-T-2 tetraol
Cow blood; cow urine; pig bone marrow; pig brain; pig duodenum; pig heart; pig ileum; pig intestine; pig jejunum; pig kidney; pig liver; pig lung; pig lymph; pig muscle; pig pancreas; pig spleen; pig stomach
4-Acetoxy-T-2 tetraol
Chicken excreta; chicken liver
8-Acetoxy-T-2 tetraol
Chicken excreta
15-Acetoxy-T-2 tetraol
Chicken excreta; chicken liver
T-2 triol
Chicken excreta; chicken liver; pig bile; pig bone marrow; pig brain; pig duodenum; pig heart; pig ileum; pig intestine; pig jejunum; pig kidney; pig liver; pig lung; pig lymph; pig muscle; pig pancreas; pig skin; pig spleen; pig stomach
Deepoxy-T-2 triol
Pig duodenum; pig ileum; pig intestine; pig jejunum; pig kidney; pig liver; pig spleen; pig stomach
Verrucarol
Dog plasma
Zearalanone
Cattle urine

(Continued)

Table 4. (Continued)

Zearalenone

Cattle liver; cattle urine; chicken bile; chicken blood; chicken excreta; chicken fat; chicken gizzard; chicken heart; chicken liver; chicken meat; chicken muscle; chicken plasma; chicken skin; cow milk, raw; cow plasma; duck bile; hen adrenal; hen bile; hen blood; hen brain; hen clutch; hen comb; hen egg; hen fat; hen gallbladder; hen heart; hen kidney; hen liver; hen lung; hen muscle, breast; hen muscle, leg; hen muscle, wing; hen oviduct; hen pancreas; hen plasma; hen red blood cells; hen spleen; pig bile; pig blood; pig feces; pig kidney; pig liver; pig muscle; pig plasma; pig serum; pig spleen; pig urine; rat blood; rat brain; rat fat; rat heart; rat intestine; rat kidney; rat liver; rat lung; rat muscle; rat serum; rat stomach; rat spleen; rat testes; rat urine; sheep milk, raw; sheep urine; turkey excreta; turkey heart; turkey kidney; turkey liver; turkey lung; turkey plasma

α-Zearalenol

Cattle liver; cattle urine; chicken bile; chicken excreta; chicken liver; chicken meat; chicken plasma; cow milk, raw; duck bile; pig bile; pig blood; pig fat; pig feces; pig heart; pig kidney; pig liver; pig lymph; pig muscle; pig pancreas; pig plasma; pig serum; pig spleen; pig urine; rat liver; rat urine; turkey excreta; turkey heart; turkey kidney; turkey liver; turkey lung; turkey plasma

β-Zearalenol

Cattle liver; cattle urine; chicken excreta; chicken liver; chicken meat; cow milk, raw; duck bile; pig bile; pig fat; pig feces; pig kidney; pig liver; pig muscle; pig serum; pig spleen; pig urine; rat liver; rat urine; sheep milk, raw; sheep urine

Zearalanone

Chicken meat

β-Zearalanone

Chicken meat

Zearalenol

Pig, sow milk

Zeranol

Cattle urine

Literature

1. Abdulrazzaq, Y.M., Osman, N., Yousif, Z.M., Al-Falahi, S., 2003: Aflatoxin M_1 in breast-milk of UAE women. Annals of Tropical Paediatrics 23, 173–179
2. Amra, H.A., 1998: Survey of aflatoxin M_1 in Egyptian raw milk by enzyme-linked immunosorbent assay. Revue de Médicine Vétérinaire 149, 695
3. Amla, I., Kumari, S., Sreenivasa Murthy, V., Jayaraj, P., Parpia, H.A.P., 1970: Role of aflatoxin in Indian childhood cirrhosis. Indian Pediatrics 7, 262–270
4. Phillips, D.L., Yourtee, D.M., Searles, S., 1976: Presence of aflatoxin B_1 in human liver in the United States. Toxicology and Applied Pharmacology 36, 403–406
5. Turner, P.C., Mendy, M., Whittle, H., Fortuin, M, Hall, A.J., Wild, C.P., 2000: Hepatitis B infection and aflatoxin biomarker levels in Gambian children. Tropical Medicine and International Health 5, 837–841
6. Cova, L., Wild, C.P., Mehrotra, R., Turusov, V., Shirai, T., Lambert, V., Jacquet, C., Tomatis, L., Trépo, C., Montesano, R., 1990: Contribution of aflatoxin B_1 and hepatitis B virus infection in the induction of liver tumors in ducks. Cancer Research 50, 2156–2163
7. Campbell, T.C., Chen, J., Liu, C., Parpia, B., 1990: Nonassociation of aflatoxin with primary liver cancer in a cross-sectional ecological survey in the People's Republic of China. Cancer Research 50, 6882–6893
8. Chaves-Carballo, E., Ellefson, R.D., Gomez, M.R., 1976: An aflatoxin in the liver of a patient with Reye-Johnson Syndrome. Mayo Clinic Proceedings 51, 48–50
9. Bukovjan, K., Prošek, J., Bukovjanová, E., 1990: Preliminary results of the aflatoxin B_1 content in liver tissue in the hare (*Lepus europaeus*). Československá Hygiena 35, 13–18 [Czech]
10. Shank, R.C., Bourgeois, C.H., Keschamras, N., Chandavimol, P., 1971: Aflatoxins in autopsy specimens from Thai children with an acute disease of unknown aetiology. Food and Cosmetics Toxicology 9, 501–507
11. Hendrickse, R.G., 1984: The influence of aflatoxins on child health in the tropics with particular reference to kwashiorkor. Transactions of the Royal Society of Tropical Medicine and Hygiene 78, 427–435
12. Qian, G.-S., Ross, R.K., Yu, M.C., Yuan, J.-M., Gao, Y.-T., Henderson, B.E., Wogan, G.N., Groopman, J.D., 1994: A follow-up study of urinary markers of aflatoxin exposure and liver cancer risk in Shanghai, People's Republic of China. Cancer Epidemiology, Biomarkers & Prevention 3, 3–10
13. Wild, C.P., Yin, F., Turner, P.C., Chemin, I., Chapot, B., Mendy, M., Whittle, H., Kirk, G.D., Hall, A.J., 2000: Environmental and genetic determinants of aflatoxin-albumin adducts in The Gambia. International Journal of Cancer 86, 1–7
14. Yu, M.-W., Chiang, Y.-C., Lien, J.-P., Chen, C.-J., 1997: Plasma antioxidant vitamins, chronic hepatitis B virus infection and urinary aflatoxin B_1-DNA adducts in healthy males. Carcinogenesis 18, 1189–1194
15. Wild, C.P., Jiang, Y.-Z., Allen, S.J., Jansen, L.A.M., Hall, A.J., Montesano, R., 1990: Aflatoxin-albumin adducts in human sera from different regions of the world. Carcinogenesis 11, 2271–2274

M. Weidenbörner, *Mycotoxins and Their Metabolites in Humans and Animals*,
DOI 10.1007/978-1-4419-7433-4,

16. Kirby, G.M., Chemin, I., Montesano, R., Chisari, F.V., Lang, M.A., Wild, C.P., 1994: Induction of specific cytochrome P450s involved in aflatoxin B_1 metabolism in hepatitis B virus transgenic mice. Molecular Carcinogenesis 11, 74–80
17. Diallo, M.S., Syila, A., Sidibé, K., Sylla, B.S., Trepo, C.R., Wild, C.P., 1995: Prevalence of exposure to aflatoxin and hepatitis B and C virus in Guinea, West Africa. Natural Toxins 3, 6–9
18. Autrup, H., Bradley, K.A., Shamsuddin, A.K.M., Wakhisi, J., Wasunna, A., 1983: Detection of putative adduct with fluorescence characteristics identical to 2,3-dihydro-2-(7′-guanyl) -3-hydroxyaflatoxin B_1 in human urine collected in Murang'a district, Kenya. Carcinogenesis 4, 1193–1195
19. Siray, M.Y., Hayes, A.W., Unger, P.D., Hogan, G.R., Ryan, N.J., Wray, B.B., 1981: Analysis of aflatoxin B_1 in human tissues with high-pressure liquid chromatography. Toxicology and Applied Pharmacology 58, 422–430
20. Wray, B.B., Hayes, A.W., 1980: Aflatoxin B_1 in the serum of a patient with primary hepatic carcinoma. Environmental Research 22, 400–403
21. Tsuboi, S., Nakagawa, T., Tomita, M., Seo, T., Ono, H., Kawamura, K., Iwamura, N., 1984: Detection of aflatoxin B_1 in serum samples of male Japanese subjects by radioimmunoassay and high-performance liquid chromatography. Cancer Research 44, 1231–1234
22. Wang, J.-S., Qian, G.-S., Zarba, A., He, X., Zhu, Y.-R., Zhang, B.-C., Jacobson, L., Gange, S.J., Muñoz, A., Kensler, T.W., Groopman, J.D., 1996: Temporal patterns of aflatoxin-albumin adducts in hepatitis B surface antigen-positive and antigen-negative residents of Daxin, Qidong County, People's Republic of China. Cancer Epidemiology, Biomarkers & Prevention 5, 253–261
23. Wild, C.P., Fortuin, M., Donato, F., Whittle, H.C., Hall, A.J., Wolf, C.R., Montesano, R., 1993: Aflatoxin, liver enzymes, and hepatitis B virus infection in Gambian children. Cancer Epidemiology, Biomarkers & Prevention 2, 555–561
24. Wild, C.P., Hasegawa, R., Barraud, L., Chutimataewin, S., Chapot, B., Ito, N., Montesano, R., 1996: Aflatoxin-albumin adducts: a basis for comparative carcinogenesis between animals and humans. Cancer Epidemiology, Biomarkers & Prevention 5, 179–189
25. Groopman, J.D., Jiaqi, Z., Donahue, P.R., Pikul, A., Lisheng, Z, Jun-shi, C., Wogan, G.N., 1992: Molecular dosimetry of urinary aflatoxin-DNA-adducts in people living in Guangxi Autonomous Region, People's Republic of China. Cancer Research 52, 45–52
26. Sheabar, F.Z., Groopman, J.D., Qian, G.-S., Wogan, G.N., 1993: Quantitative analysis of aflatoxin – albumin adducts. Carcinogenesis 14, 1203–1208
27. Wild, C.P., Garner, R.C., Montesano, R., Tursi, F., 1986: Aflatoxin B_1 binding to plasma albumin and liver DNA upon chronic administration to rats. Carcinogenesis 7, 853–858
28. Stora, C., Dvorackova, I., Ayraud, N., 1981: Characterization of aflatoxin B_1 (AFB) in human liver cancer. Research Communications in Chemical Pathology and Pharmacology 31, 77–85
29. Hollstein, M.C., Wild, C.P., Bleicher, F., Chutimataewin, S., Harris, C.C., Srivanakul, P., Montesano, R., 1993: p53 mutations and aflatoxin B_1 exposure in hepatocellular carcinoma patients from Thailand. International Journal of Cancer 53, 51–55
30. Wild, C.P., Jansen, L.A.M., Cova, L., Montesano, R., 1993: Molecular dosimetry of aflatoxin exposure: contribution to understanding the multifactorial etiopathogenesis of primary hepatocellular carcinoma with particular reference to hepatitis B virus. Environmental Health Perspectives 99, 115–122
31. Cova, L., Mehrotra, R., Wild, C.P., Chutimataewin, S., Cao, S.F., Duflot, A., Prave, M., Yu, S.Z., Montesano, R., Trepo, C., 1994: Duck hepatitis B virus infection, aflatoxin B_1 and liver cancer in domestic Chinese ducks. British Journal of Cancer 69, 104–109
32. Coulter, J.B.S., Suliman, G.I., Lamplugh, S.M., Mukhtar, B.I., Hendrickse, R.G., 1986: Aflatoxins in liver biopsies from Sudanese children. American Journal of Tropical Medicine and Hygiene 35, 360–365

33. Allen, S.J., Wild, C.P., Wheeler, J.G., Riley, E.M., Montesano, R., Bennett, S., Whittle, H.C., Hall, A.J., Greenwood, B.M., 1992: Aflatoxin exposure, malaria and hepatitis B infection in rural Gambian children. Transactions of the Royal Society of Tropical Medicine and Hygiene 86, 426–430
34. Kensler, T.W., Egner, P.A., Trush, M.A., Bueding, E., Groopman, J.D., 1985: Modification of aflatoxin B_1 binding to DNA *in vivo* in rats fed phenolic antioxidants, ethoxyquin and a dithiothione. Carcinogenesis 6, 759–763
35. Becroft, D.M., Webster, D.R., 1972: Aflatoxins and Reye's disease. British Medical Journal 4, 117
36. Hendrickse, R.G., Coulter, J.B.S., Lamplugh, S.M., MacFarlane, S.B.J., Williams, T.E., Omer, M.I.A., Suliman, G.I., 1982: Aflatoxins and kwashiorkor: a study in Sudanese children. British Medical Journal 285, 843–846
37. Turner, P.C., Moore, S.E., Hall, A.J., Prentice, A.M., Wild, C.P., 2003: Modification of immune function through exposure to dietary aflatoxin in Gambian children. Environmental Health Perspectives 111, 217–220
38. Nabney, J., Burbage, M.B., Allcroft, R., Lewis, G., 1967: Metabolism of aflatoxin in sheep: excretion pattern in the lactating ewe. Food and Cosmetics Toxicology 5, 11–17
39. Patterson, D.S.P., Glancy, E.M., Roberts, B.A., 1980: The "carry over" of aflatoxin M_1 into the milk of cows fed rations containing a low concentration of aflatoxin B_1. Food and Cosmetics Toxicology 18, 35–37
40. Allcroft, R., Roberts, B.A., Lloyd, M.K., 1968: Excretion of aflatoxin in a lactating cow. Food and Cosmetics Toxicology 6, 619–625
41. Zhu, J-Q., Zhang, L.-S., Hu, X., Xiao, Y., Chen, J.-S., Xu, Y.-C., Fremy, J., Chu, F.S., 1987: Correlation of dietary aflatoxin B_1 levels with excretion of aflatoxin M_1 in human urine. Cancer Research 47, 1848–1852
42. Croy, R.G., Wogan, G.N., 1981: Temporal patterns of covalent DNA adducts in rat liver after single and multiple doses of aflatoxin B_1. Cancer Research 41, 197–203
43. Choudhary, P.L., Sharma, R.S., Borkhatriya, V.N., Murthi, T.N., Wadodkar, U.R., 1997: Survey on the levels of aflatoxin M_1 in raw and market milk in and around Anand Town. Indian Journal of Dairy Science 50, 156–158
44. Domagala, J., Kisza, J., Blüthgen, A., Heeschen, W., 1997: Contamination of milk with aflatoxin M_1 in Poland. Milchwissenschaft 52, 631–633
45. Dragacci, S., Fremy, J.-M., 1993: Occurrence of aflatoxin M_1 in milk. Fifteen years of sanitary control. Sciences des Aliments 13, 711–722 [French]
46. El-Nezami, H.S., Nicoletti, G., Neal, G.E., Donohue, D.C., Ahokas, J.T., 1995: Aflatoxin M_1 in human breast milk samples from Victoria, Australia and Thailand. Food and Cosmetics Toxicology 33, 173–179
47. Ferguson-Foos, J., Warren, J.D., 1984: Improved cleanup for liquid chromatographic analysis and fluorescence detection of aflatoxins M_1 and M_2 in fluid milk products. Journal of the Association of Official Analytical Chemists 67, 1111–1114
48. Wang, J.-S., Shen, X., He, X., Zhu, Y.-R., Zhang, B.-C., Wang, J.-B., Qian, G.-S., Kuang, S.-Y., Zarba, A., Egner, P.A., Jacobson, L.P., Muñoz, A., Helzlsouer, K.J., Groopman, J.D., Kensler, T.W., 1999: Protective alterations in phase 1 and 2 metabolism of aflatoxin B_1 by oltipraz in residents of Qidong, People's Republic of China. Journal of the National Cancer Institute 91, 347–354
49. Hsieh, L.-L., Hsu, S.-W., Chen, D.-S., Santella, R.M., 1988: Immunological detection of aflatoxin B_1-DNA adducts formed *in vivo*. Cancer Research 48, 6328–6331
50. Groopman, J.D., Hall, A.J., Whittle, H., Hudson, G.J., Wogan, G.N., Montesano, R., Wild, C.P., 1992: Molecular dosimetry of aflatoxin-N^7-guanine in human urine obtained in The Gambia, West Africa. Cancer Epidemiology, Bio-markers & Prevention 1, 221–227
51. Nelson, D.B., Kimbrough, R., Landigran, P.S., Hayes, A.W., Yang, G.C., Benanides, J., Morens, D.M., Morse, D., Pollack, M., Powell, K.E., Sullivan-Bolyai, J.Z., 1980: Aflatoxins and Reye's Syndrome: a case control study. Pediatrics 66, 865–869

52. Scholl, P.F., Musser, S.M., Groopman, J.D., 1997: Synthesis and characterization of aflatoxin B_1 mercapturic acids and their identification in rat urine. Chemical Research in Toxicology 10, 1144–1151
53. Madden, U.A., Stahr, H.M., 1995: Retention and distribution of aflatoxin in tissues of chicks fed aflatoxin-contaminated poultry rations amended with soil. Veterinary and Human Toxicology 37, 24–29
54. Wilkinson, A.P., Denning, D.W., Morgan, R.A., 1988: Analysis of UK sera for aflatoxin by enzyme-linked immunosorbent assay. Human Toxicology 7, 353–356
55. Soini, Y., Chia, S.C., Bennett, W.P., Groopman, J.D., Wang, J.-S., de Benedetti, V.M.G., Cawley, H., Welsh, J.A., Hansen, C., Bergasa, N.V., Jones, E.A., di Bisceglie, A.M., Trivers, G.E., Sandoval, C.A., Calderon, I.E., Munoz Espinosa, L.E., Harris, C.C., 1996: An aflatoxin-associated mutational hotspot at codon 249 in the *p53* tumor suppressor gene occurs in hepatocellular carcinomas from Mexico. Carcinogenesis 17, 1007–1012
56. Denning, D.W., Allen, R., Wilkinson, A.P., Morgan, M.R.A., 1990: Transplacental transfer of aflatoxin in humans. Carcinogenesis 11, 1033–1035
57. Kensler, T.W., Gange, S.J., Egner, P.A., Dolan, P.M., Muñoz, A., Groopman, J.D., Rogers, A.E., Roebuck, B.D., 1997: Predictive value of molecular dosimetry: individual *versus* group effects of oltipraz on aflatoxin-albumin adducts and risk of liver cancer. Cancer Epidemiology, Biomarkers & Prevention 6, 603–610
58. Harrison, J.C., Carvajal, M., Garner, R.C., 1993: Does aflatoxin exposure in the United Kingdom constitute a cancer risk? Environmental Health Perspectives 99, 99–105
59. Groopman, J.D., Donahue, P.R., Zhu, J., Chen, J., Wogan, G.N., 1985: Aflatoxin metabolism in humans: detection of metabolites and nucleic acid adducts in urine by affinity chromatography. Proceedings of the National Academy of Science USA 82, 6492–6496
60. Neff, G.L., Edds, G.T., 1981: Aflatoxins B_1 and M_1: tissue residues and feed withdrawal profiles in young growing pigs. Food and Cosmetics Toxicology 19, 739–742
61. Apeagyei, F., Lamplugh, S.M., Hendrickse, R.G., Affram, K., Lucas, S., 1986: Aflatoxins in the livers of children with kwashiorkor in Ghana. Tropical and Geographical Medicine 38, 273–276
62. Rogan, W.J., Yang, G.C., Kimborough, R.D., 1985: Aflatoxin and Reye's Syndrome: a study of livers from deceased cases. Archives of Environmental Health 40, 91–95
63. Prelusky, D.B., Veira, D.M., Trenholm, H.L., Hartin, K.E., 1986: Excretion profiles of the mycotoxin deoxynivalenol, following oral and intravenous administration to sheep. Fundamental and Applied Toxicology 6, 356–363
64. Pollmann, D.S., Koch, B.A., Seitz, L.M., Mohr, H.E., Kennedy, G.A., 1985: Deoxynivalenol-contaminated wheat in swine diets. Journal of Animal Science 60, 239–247
65. Ryan, N.J., Hogan, G.R., Hayes, A.W., Unger, P.D., Siraj, M.V., 1979: Aflatoxin B_1: its role in the etiology of Reye's Syndrome. Pediatrics 64, 71–75
66. Lüthy, J., Zweifel, U., Schlatter, C.H., 1980: Metabolism and tissue distribution of [^{14}C]aflatoxin B_1 in pigs. Food and Cosmetics Toxicology 18, 253–256
67. Coulter, J.B.S., Hendrickse, R.G., Lamplugh, S.M., MacFarlane, S.B.J., Moody, J.B., Omer, M.I.A., Suliman, G.I., Williams, T.E., 1986: Aflatoxins and kwashiorkor: clinical studies in Sudanese children. Transactions of the Royal Society of Tropical Medicine and Hygiene 80, 945–951
68. Denning, D.W., Onwubalili, J.K., Wilkinson, A.P., Morgan, M.R.A., 1988: Measurement of aflatoxin in Nigerian sera by enzyme-linked immunosorbent assay. Transactions of the Royal Society of Tropical Medicine and Hygiene 82, 169–171
69. Onyemelukwe, G.C., Ogbadu, G., 1981: Aflatoxin levels in sera of healthy first time rural blood donors: preliminary report. Transactions of the Royal Society of Tropical Medicine and Hygiene 75, 780–782
70. Côté, L.-M., Beasley, V.R., Bratich, P.M., Swanson, S.P., Shivaprasad, H.L., Buck, W.B., 1985: Sex-related reduced weight gains in growing swine fed diets containing deoxynivalenol. Journal of Animal Science 61, 942–950

71. Helferich, W.G., Garrett, W.N., Hsieh, D.P.H., Baldwin, R.L., 1986: Feedlot performance and tissue residues of cattle consuming diets containing aflatoxins. Journal of Animal Science 62, 691–696
72. Cook, W.O., Richard, J.L., Osweiler, G.D., Trampel, D.W., 1986: Clinical and pathologic changes in acute bovine aflatoxicosis: rumen motility and tissue and fluid concentrations of aflatoxins B_1 and M_1. American Journal of Veterinary Research 47, 1817–1825
73. Bortell, R., Asquith, R.L., Edds, G.T., Simpson, C.F., Aller, W.W., 1983: Acute experimentally induced aflatoxicosis in the weanling pony. American Journal of Veterinary Research 44, 2110–2114
74. Garner, C.R., 1975: Reduction in binding of [^{14}C] aflatoxin B_1 to rat liver macromolecules by phenobarbitone pretreatment. Biochemical Pharmacology 24, 1553–1556
75. Gregory, J.F., III, Manley, D., 1981: High performance liquid chromatographic determination of aflatoxins in animal tissues and products. Journal of the Association of Official Analytical Chemists. 64, 144–151
76. Ray, A.C., Abitt, B., Cotter, S.R., Murphy, M.J., Reagor, J.C., Robinson, R.M., West, J.E., Whitford, H.W., 1986: Bovine abortion and death associated with consumption of aflatoxin-contaminated peanuts. Journal of American Veterinary Medical Association 188, 1187–1188
77. Harvey, R.B., Phillips, T.D., Ellis, J.A., Kubena, L.F., Huff, W.E., Petersen, H.D., 1991: Effects on aflatoxin M_1 residues in milk by addition of hydrated sodium calcium aluminosilicate to aflatoxin-contaminated diets of dairy cows. American Journal of Veterinary Research 52, 1556–1559
78. Lewis, R.E., Wiederhold, N.P., Chi, J., Han, X.Y., Komanduri, K.V., Kontoyiannis, D.P., Prince, R.A., 2005: Detection of gliotoxin in experimental and human aspergillosis. Infection and Immunity 73, 635–637
79. Prelusky, D.B., Scott, P.M., Trenholm, H.L., Lawrence, G.A., 1990: Minimal transmission of zearalenone to milk of dairy cows. Journal of Environmental Science and Health B 25, 87–103
80. Cheng, Z., Root, M., Pan, W., Chen, J., Campbell, T.C., 1997: Use of an improved method for analysis of urinary aflatoxin M_1 in a survey of mainland China and Taiwan. Cancer Epidemiology, Biomarkers & Prevention 6, 523–529
81. Sawhney, D.S., Vadehra, D.V., Baker, R.C., 1972: The metabolism of ^{14}C aflatoxins in laying hens. Poultry Science 52, 1302–1309
82. Richard, J.L., Dvorak, T.J., Ross, P.F., 1996: Natural occurrence of gliotoxin in turkeys infected with *Aspergillus fumigatus*, Fresenius. Mycopathologia 134, 167–170
83. Applebaum, R.S., Brackett, R.E., Wiseman, D.W., Marth, E.H., 1982: Responses of dairy cows to dietary aflatoxin: feed intake and yield, toxin content, and quality of milk of cows treated with pure and impure aflatoxin. Journal of Dairy Science 65, 1503–1508
84. James, L.J., Smith, T.K., 1982: Effect of dietary alfalfa on zearalenone toxicity and metabolism in rats and swine. Journal of Animal Science 55, 110–118
85. Munksgaard, L., Larsen, J., Werner, H., Andersen, P.E., Viuf, B.T., 1987: Carry over of aflatoxin from cows' feed to milk and milk products. Milchwissenschaft 42, 165–167
86. Fremy, J.M., Gautier, J.P., Herry, M.P., Terrier, C., Calet, C., 1987: Effects of ammoniation on the 'carry-over' of aflatoxins into bovine milk. Food Additives and Contaminants 5, 39–44
87. Fodor, J., Meyer, K., Riedlberger, M., Bauer, J., Horn, P., Kovács, F., Kovács, M., 2006: Distribution and elimination of fumonisin analogues in weaned piglets after oral administration of *Fusarium verticillioides* fungal culture. Food Additives and Contaminants 23, 492–501
88. Jemmali, M., Murthy, T.R.K., 1976: A chemical assay method for the determination of aflatoxin residues in animal tissues. Zeitschrift für Lebensmittel-Untersuchung und -Forschung 161, 13–17
89. Kiermeier, F., 1973: Aflatoxin M secretion in cow's milk depending on the quantity of aflatoxin B_1 ingested. Milchwissenschaften 28, 683–685 [German]
90. Mintzlaff, H.-J., Lötzsch, R., Tauchmann, F., Meyer, W., Leistner, L., 1974: Aflatoxin-Rückstände in der Leber und in der Muskulatur von Masthähnchen nach Verabreichung von aflatoxinhaltigen Futtermitteln. Fleischwirtschaft 54, 774–778 [German]

91. de Vries, H.R., Maxwell, S.M., Hendrickse, R.G., 1990: Aflatoxin excretion in children with kwashiorkor or marasmic kwashiorkor – a clinical investigation. Mycopathologia 110, 1–9
92. Trucksess, M.W., Richard, J.L., Stoloff, L., Mc Donald, J.S., Brumley, W.C., 1983: Absorption and distribution patterns of aflatoxicol and aflatoxins B_1 and M_1 in blood and milk of cows given aflatoxin B_1. American Journal of Veterinary Research 44, 1753–1756
93. Stubblefield, R.D., Pier, A.C., Richard, J.L., Shotwell, O.L., 1983: Fate of aflatoxins in tissues, fluids, and excrements from cows dosed orally with aflatoxin B_1. American Journal of Veterinary Research 44, 1750–1752
94. Hussain, M., Gabal, M.A., Wilson, T, Summerfelt, R.C., 1993: Effect of aflatoxin-contaminated feed on morbidity and residues in walleye fish. Veterinary and Human Toxicology 35, 396–398
95. de Vries, H.R., Lamplugh, S.M., Hendrickse, R.G., 1987: Aflatoxins and kwashiorkor in Kenya: a hospital based study in a rural area of Kenya. Annals of Tropical Paediatrics 7, 249–251
96. Corbett, W.T., Brownie, C.F., Hagler, S.B., Hagler, W.M., 1988: An epidemiological investigation asssociating aflatoxin M_1 with milk production in dairy cattle. Veterinary and Human Toxicology 30, 5–8
97. Micco, C., Ambruzzi, M.A., Miraglia, M., Brera, C., Onori, R., Benelli, L., 1991: Contamination of human milk with ochratoxin A. In: Mycotoxins, Endemic Nephropathy and Urinary Tract Tumours. Eds. Castegnaro, M., Pleština, R., Dirheimer, G., Chernozemsky, I.N., Bartsch, H. Lyon, International Agency for Research on Cancer (IARC), pp. 105–108
98. Shank, R., Johnsen, D.O., Tanticharoenyos, P., Wooding, W.L., Bourgeois, C.H., 1971: Acute toxicity of aflatoxin B_1 in the Macaque monkey. Toxicology and Applied Pharmacology 20, 227–231
99. Keskin, Y., Başkaya, R., Karsli, S., Yurdun, T., Özyaral, O., 2009: Detection of aflatoxin M_1 in human breast milk and raw cow's milk in Istanbul, Turkey. Journal of Food Protection 72, 885–889
100. Groopman, J.D., Trudel, L.J., Donahue, P.R., Marshak-Rothstein, A., Wogan, G.N., 1984: High-affinity monoclonal antibodies for aflatoxins and their application to solid-phase immunoassays. Proceedings of the National Academy of Science USA 81, 7728–7731
101. Krogh, P., Hald, B., Hasselager, E., Madsen, A., Mortensen, H.P., Larsen, A.E., Campbell, A.D., 1973: Aflatoxin residues in bacon pigs. Pure & Applied Chemistry 35, 275–281
102. Krogh, P., Elling, F., Hald, B., Larsen, A.E., Lillehoj, E.B., Madsen, A., Mortensen, H.P., 1976: Time-dependent disappearance of ochratoxin A residues in tissues of bacon pigs. Toxicology 6, 235–242
103. Groopman, J.D., Haugen, A., Goodrich, G.R., Wogan, G.N., Harris, C.C., 1982: Quantitation of aflatoxin B_1-modified DNA using monoclonal antibodies. Cancer Research 42, 3120–3124
104. Hatch, M.C., Chen, C.-J., Levin, B., Ji, B.-T., Yang, G.-Y., Hsu, S.-W., Wang, L.-W., Hsieh, L.-L., Santella, R.M., 1993: Urinary aflatoxin levels, hepatitis-B virus infection and hepatocellular carcinoma in Taiwan. International Journal of Cancer 54, 931–934
105. Gregory, J.F., III, Goldstein, S.L., Edds, G.T., 1983: Metabolite distribution and rate of residue clearance in turkeys fed a diet containing aflatoxin B_1. Food and Chemical Toxicology 21, 463–467
106. Mirocha, C.J., Robison, T.S., Pawlosky, R.J., Allen, N.K., 1982: Distribution and residue determination of $[^3H]$zearalenone in broilers. Toxicology and Applied Pharmacology 66, 77–87
107. Price, R.L., Lough, O.G., Brown, W.H., 1982: Ammoniation of whole cottonseed at atmospheric pressure and ambient temperature to reduce aflatoxin M_1 in milk. Journal of Food Protection 45, 341–344
108. Prelusky, D.B., Miller, J.D., Trenholm, H.L., 1996: Disposition of ^{14}C-derived residues in tissues of pigs fed radiolabelled fumonisin B_1. Food Additives and Contaminants 13, 155–162
109. Meyer, K., Mohr, K., Bauer, J., Horn, P., Kovács, M., 2003: Residue formation of fumonisin B_1 in porcine tissues. Food Additives and Contaminants 20, 639–647

110. Wolzak, A., Pearson, A.M., Coleman, T.H., Pestka, J.J., Gray, J.I., Chen, C., 1986: Aflatoxin carryover and clearance from tissues of laying hens. Food and Chemical Toxicology 24, 37–41
111. Norred, W.P., Morrissey, R.E., Riley, R.T., Cole, R.J., Dorner, J.W., 1985: Distribution, excretion and skeletal muscle effects of the mycotoxin [^{14}C]cyclopiazonic acid in rats. Food and Chemical Toxicology 23, 1069–1076
112. Wolzak, A., Pearson, A.M., Coleman, T.H., Pestka, J.J., Gray, J.I., 1985: Aflatoxin deposition and clearance in the eggs of laying hens. Food and Chemical Toxicology 23, 1057–1061
113. Hayes, A.W., King, R.E., Unger, P.D., Phillips, T.D., Hatkin, J., Bowen, J.H., 1978: Aflatoxicosis in swine. Journal of the American Veterinary Medical Association 172, 1295–1297
114. Shephard, G.S., Thiel, P.G., Sydenham, E.W., 1992: Initial studies on the toxicokinetics of fumonisin B_1 in rats. Food and Chemical Toxicology 30, 277–279
115. Bailey, G.S., Williams, D.E., Wilcox, J.S., Loveland, P.M., Coulombe, R.A., Hendricks, J.D., 1988: Aflatoxin B_1 carcinogenesis and its relation to DNA adduct formation and adduct persistence in sensitive and resistant salmoid fish. Carcinogenesis 9, 1919–1926
116. Garner, R.C., Martin, C.N., Lindsay Smith, J.R., Coles, B.F., Tolson, M.R., 1979: Comparison of aflatoxin B_1 and aflatoxin G_1 binding to cellular macromolecules *in vitro*, *in vivo* and after peracid oxidation; characterisation of the major nucleic acid adducts. Chemico-Biological Interactions 26, 57–73
117. Hendrickse, R.G., Maxwell, S.M., Young, R., 1989: Aflatoxins and heroin. British Medical Journal 299, 492–493
118. Preston, R.S., Hayes, J.R., Campbell, T.C., 1976: The effect of protein deficiency on the *in vivo* binding of aflatoxin B_1 to rat liver macromolecules. Life Sciences 19, 1191–1198
119. Olsen, M., Mirocha, C.J., Abbas, H.K., Johansson, B., 1986: Metabolism of high concentrations of dietary zearalenone by young male turkey poults. Poultry Science 65, 1905–1910
120. Monroe, D.H., Holeski, C.J., Eaton, D.L., 1986: Effects of single-dose and repeated-dose pretreatment with 2(3)-*tert*-butyl-4-hydroxyanisole (BHA) on the hepatobiliary disposition and covalent binding to DNA of aflatoxin B_1 in the rat. Food and Chemical Toxicology 24, 1273–1281
121. Yang, G., Nesheim, S., Benavides, J., Ueno, I., Campbell, A.D., Pohland, A., 1980: Radioimmunoassay detection of aflatoxin B_1 in monkey and human urine. Zentralblatt für Bakteriologie, Mikrobiologie und Hygiene Abt. 1 Originale A 246 (Suppl. 8), 329–335
122. Richir, C., Paccalin, J., Faugeres, R., Moreux, M., Audry, S., N'Diaye, P., 1980: Searching for mycotoxins in human body fluids and viscera. Zentralblatt für Bakteriologie, Mikrobiologie und Hygiene Abt. 1 Originale A 246 (Suppl. 8), 337–342
123. Hagler, W.M., Dankó, G., Horváth, L., Palyusik, M., Mirocha, C.J., 1980: Transmission of zearalenone and its metabolite into ruminant milk. Acta Veterinaria Academiae Scientiarum Hungaricae 28, 209–216
124. Li, Y.K., Chu, F.S., 1982: Kinetics of transmission of aflatoxin B_1 into aflatoxin M_1 in lactating mouse: an ELISA analysis. Experientia 38, 842–843
125. Zepnik, H., Völkel, W., Dekant, W., 2003: Metabolism and toxicokinetics of the mycotoxin ochratoxin A in F344 rats. Mycotoxin Research 19, 102–107
126. Bauer, J., Gareis, M., Bott, A., Gedek, B., 1989: Isolation of a mycotoxin (gliotoxin) from a bovine udder infected with *Aspergillus fumigatus*. Journal of Medical and Veterinary Mycology 27, 45–50
127. Shephard, G.S., Snijman, P.W., 1999: Elimination and excretion of a single dose of the mycotoxin fumonisin B_2 in a non-human primate. Food and Chemical Toxicology 37, 111–116
128. Biehl, M.L., Prelusky, D.B., Koritz, G.D., Hartin, K.E., Buck, W., Trenholm, H.L., 1993: Biliary excretion and enterohepatic cycling of zearalenone in immature pigs. Toxicology and Applied Pharmacology 121, 152–159
129. Vudathala, D.K., Prelusky, D.B., Ayroud, M., Trenholm, H.L., Miller, J.D., 1994: Pharmacokinetic fate and pathological effects of ^{14}C-fumonisin B_1 in laying hens. Natural Toxins 2, 81–88
130. Shephard, G.S., Thiel, P.G., Sydenham, E.W., Savard, M.E., 1995: Fate of single dose of ^{14}C-labelled fumonisin B_1 in vervet monkeys. Natural Toxins 3,145–150

131. Holzapfel, C.W., Steyn, P.S., Purchase, I.F.H., 1966: Isolation and structure of aflatoxins M_1 and M_2. Tetrahedron Letters 25, 2799–2803
132. Chou, M.-W., Tung, T.-C., 1969: Aflatoxin B_1 in the excretion of aflatoxin poisoned rats. Journal of the Formosan Medical Association 68, 389–391
133. Swenson, D.H., Lin, J.-K., Miller, E.C., Miller, J.A., 1977: Aflatoxin B_1–2,3-oxide as a probable intermediate in the covalent binding of aflatoxins B_1 and B_2 to rat liver DNA and ribosomal RNA *in vivo*. Cancer Research 37, 172–181
134. Coulter, J.B.S., Lamplugh, S.M., Suliman, G.I., Omer, I.A., Hendrickse, R.G., 1984: Aflatoxins in human breast milk. Annals of Tropical Paediatrics 4, 61–66
135. Prelusky, D.B., Hamilton, R.M.G., Trenholm, H.L., Miller, J.D., 1986: Tissue distribution and excretion of radioactivity following administration of ^{14}C-labelled deoxynivalenol to White Leghorn hens. Fundamental and Applied Toxicology 7, 635–645
136. Jacobson, W.C., Harmeyer, W.C., Jackson, J.E., Armbrecht, B., Wiseman, H.G., 1978: Transmission of aflatoxin B_1 into the tissues of growing pigs. Bulletin of Environmental Contamination & Toxicology 19, 156–161
137. Polan, C.E., Hayes, J.R., Campbell, T.C., 1974: Consumption and fate of aflatoxin B_1 by lactating cows. Journal of Agricultural and Food Chemistry 22, 635–638
138. Furtado, R.M., Pearson, A.M., Hogberg, M.G., Miller, E.R., 1979: Aflatoxin residues in the tissues of pigs fed a contaminated diet. Journal of Agricultural and Food Chemistry 27, 1351–1354
139. Jacobson, W.C., Wiseman, H.G., 1974: The transmission of aflatoxin B_1 into eggs. Poultry Science 53, 1743–1745
140. Gong, Y.Y., Cardwell, K, Hounsa, A., Egal, S., Turner, P.C., Hall, A.J., Wild, C.P., 2002: Dietary aflatoxin exposure and impaired growth in young children from Benin and Togo: cross sectional study. British Medical Journal 325, 20–21
141. Angsubhakorn, S., Poomvises, P., Romruen, K., Newberne, P.M., 1981: Aflatoxicosis in horses. Journal of American Veterinary Association 178, 274–278
142. Egner, P.A., Yu, X., Johnson, J.K., Nathasingh, C.K., Groopman, J.D., Kensler, T.W., Roebuck, B.D., 2003: Identification of aflatoxin M_1-N^7-guanine in liver and urine of tree shrews and rats following administration of aflatoxin B_1. Chemical Research in Toxicology 16, 1174–1180
143. Suzuki, S., Satoh, T., Yamazaki, M., 1977: The pharmacokinetics of ochratoxin A in rats. Japanese Journal of Pharmacology 27, 735–744
144. Li, S., Marquardt, R.R., Frohlich, A.A., 2000: Identification of ochratoxins and some of their metabolites in bile and urine of rats. Food and Chemical Toxicology 38, 141–152
145. Visconti, A., Mirocha, C.J., 1985: Identification of various T-2 toxin metabolites in chicken excreta and tissues. Applied and Environmental Microbiology 49, 1246–1250
146. Richard, J.L., Pier, A.C., Stubblefield, R.D., Shotwell, O.L., Lyon, R.L., Cutlip, R.C., 1983: Effect of feeding corn naturally contaminated with aflatoxin on feed efficiency, on physiologic, immunologic, and pathologic changes, and on tissue residues in steers. American Journal of Veterinary Research 44, 1294–1299
147. Applebaum, R.S., Marth, E.H., 1980: Responses of a dairy cow to dietary aflatoxin: preliminary report. Journal of Dairy Science 63 Suppl. 1, 58
148. Palyusik, M., Harrach, B., Mirocha, C.J., Pathre, S.V., 1980: Transmission of zearalenone and zearalenol into porcine milk. Acta Veterinaria Academiae Scientiarum Hungaricae 28, 217–222
149. Gareis, M., Wernery, U., 1984: Determination of gliotoxin in samples associated with cases of intoxication in camels. Mycotoxin Research 10, 2–8
150. McKinney, J.D., Cavanagh, G.C., Bell, J.T., Hoversland, A.S., Nelson, D.M., Pearson, J., Selkirk, R.J., 1973: Effects of ammoniation on aflatoxins in rations fed lactating cows. Journal of the American Oil Chemists' Society 50, 79–84
151. Stubblefield, R.D., Shotwell, O.L., Richard, J.L., Pier, A.C., 1981: Transmission and distribution of aflatoxin in contaminated beef liver and other tissues. Journal of the American Oil Chemists' Society 58, 1015A–1017A
152. Nyathi, C.B., Mutiro, C.F., Hasler, J.A., Chetsanga, C.J., 1987: A survey of urinary aflatoxin in Zimbabwe. International Journal of Epidemiology 16, 516–519

153. Garner, R.C., Dvorackova, I., Tursi, F., 1988: Immunoassay procedures to detect exposure to aflatoxin B_1 and benzo(a)pyrene in animals and man at the DNA level. International Archives of Occupational and Environmental Health 60, 145–150
154. Irvin, T.R., Wogan, G.N., 1984: Quantitation of aflatoxin B_1 adduction within the ribosomal RNA gene sequence of rat liver DNA. Proceedings of the National Academy of Science USA 81, 664–668
155. Haugen, A., Groopman, J.D., Hsu, I.-C., Goodrich, G.R., Wogan, G.N., Harris, C.C., 1981: Monoclonal antibody to aflatoxin B_1-modified DNA detected by enzyme immunoassay. Proceedings of the National Academy of Science USA 78, 4124–4127
156. Hsieh, L.-L., Hsieh, T.-T., 1993: Detection of aflatoxin B_1-DNA adducts in human placenta and cord blood. Cancer Research 53, 1278–1280
157. Frobish, R.A., Bradley, B.D., Wagner, D.D., Long-Bradley, P.E., Hairston, H, 1986: Aflatoxin residues in milk of dairy cows after ingestion of naturally contaminated grain. Journal of Food Protection 49, 781–785
158. Fukal, L., Březina, P., 1991: Determination of the aflatoxin M_1-level in milk for the production of milk baby foods using immunoassay. Nahrung 35, 745–748 [German]
159. Fukal, L., Březina, P., Marek, M., 1990: Immunochemical monitoring of aflatoxin M_1 occurrence in milk produced in Czechoslovakia. Deutsche Lebensmittel-Rundschau 86, 289–291
160. Gajek, O., 1982: Aflatoxins in protein food for animals and milk. Roczniki Panstwowego Zaklado Higieny XXXIII, 415–420 [Polish]
161. Gallo, P., Salzillo, A., Rossini, C., Urbani, V., Serpe, L., 2006: Aflatoxin M_1 determination in milk: method validation and contamination levels in samples from southern Italy. Italian Journal of Food Science 18, 251–259
162. Gelosa, L., 1986: La presenza di aflatossina M_1 nel latte al consumo. Latte XI, 261–264 [Italian]
163. Gilbert, J., Shepherd, M.J., Wallwork, M.A., Knowles, M.E., 1984: A survey of the occurrence of aflatoxin M_1 in UK-produced milk for the period 1981–1983. Food Additives and Contaminants 1, 23–28
164. Ioannou-Kakouri, E., Aletari, M., Christou, E., Hadjioannou, A., Koliou, A., Akkelidou, D., 1999: Surveillance and control of aflatoxins B_1, B_2, G_1, G_2, and M_1 in foodstuffs in the Republic of Cyprus: 1992–1996. Journal of the Association of Official Analytical Chemists International 82, 883–892
165. Jalón, M., Urieta, I, de Pablo, B., Macho, M.L., 1994: Contaminacion por aflatoxina M_1 de leche y productos lacteos en la comunidad autónoma Vasca. Alimentaria 3, 25–29 [Spanish]
166 Karaioannoglou, P., Mantis, A., Koufidis, D., Koidis, P., Triantafillou, J., 1989: Occurrence of aflatoxin M_1 in raw and pasteurized milk and in Feta and Telme cheese samples. Milchwissenschaft 44, 746–748
167. Li, Y., Su, J.-j., Qin, L.-l., Egner, P.E., Wang, J.-S., Groopman, J.D., Kensler, T.W., Roebuck, B.D., 2000: Reduction of aflatoxin B_1 adduct biomarkers by oltipraz in the tree shrew (*Tupaia belangeri chinensis*). Cancer Letters 154, 79–83
168. Nayak, S., Tanuja, P., Sashidhar, R.B., 2009: Synthesis and characterization of mercapturic acid (N-acetyl-L-cysteine)-aflatoxin B_1 adduct and its quantitation in rat urine by an enzyme immunoassay. Journal of the Association of Official Analytical Chemists International 92, 487–495
169. Masri, M.S., Garcia, V.C., Page, J.R. 1969: The aflatoxin M content of milk from cows fed known amounts of aflatoxin. Veterinary Record 84, 146–147
170. Pietri, A., Bertuzzi, T., Moschini, M., Piva, G., 2003: Aflatoxin M_1 occurrence in milk samples destined for Parmigiano Reggiano cheese production. Italian Journal of Food Science 15, 301–306
171. Allcroft, R., Roberts, B.A., 1968: Toxic groundnut meal: the relationship between aflatoxin B_1 intake by cows and excretion of aflatoxin M_1 in milk. Veterinary Record 82, 116–118
172. Galtier, P., Alvinerie, M., Charpenteau, J.L., 1981: The pharmacokinetic profiles of ochratoxin A in pigs, rabbits and chickens. Food and Cosmetics Toxicology 19, 735–738

173. Croy, R.G., Essigmann, J.M., Reinhold, V.N., Wogan, G.N., 1978: Identification of the principal aflatoxin B_1-DNA adduct formed *in vivo* in rat liver. Proceedings of the National Academy of Science USA 75, 1745–1749
174. Kumagai, S., Aibara, K., 1982: Intestinal absorption and secretion of ochratoxin A in the rat. Toxicology and Applied Pharmacology 64, 94–102
175. Li, S., Marquardt, R.R., Frohlich, A.A., Vitti, T.G., Crow, G., 1997: Pharmacokinetics of ochratoxin A and its metabolites in rats. Toxicology and Applied Pharmacology 145, 82–90
176. Walton, M., Egner, P, Scholl, P.F., Walker, J., Kensler, T.W., Groopman, J.D., 2001: Liquid chromatography electrospray-mass spectrometry of urinary aflatoxin biomarkers: characterization and application to dosimetry and chemoprevention in rats. Chemical Research in Toxicology 14, 919–926
177. Whitham, M., Nixon, J.E., Sinnhuber, R.O., 1982: Liver DNA bound *in vivo* with aflatoxin B_1 as a measure of hepatocarcinoma initiation in rainbow trout. Journal of the National Cancer Institue 68, 623–628
178. Ghidini, S., Zanardi, E., Battaglia, A., Varisco, G., Ferretti, E., Campanini, G., Chizzolini, R., 2005: Comparison of contaminant and residue levels in organic and conventional milk and meat products from nothern Italy. Food Additives and Contaminants 22, 9–14
179. Sreemannarayana, O., Frohlich, A.A., Vitti, T.G., Marquardt, R.R., Abramson, D., 1988: Studies of the tolerance and disposition of ochratoxin A in young calves. Journal of Animal Science 66, 1703–1711
180. Chatterjee, K., Pawlowsky, R.J., Treeful, L., Mirocha, C.J., 1986: Kinetic study of T-2 toxin metabolites in a cow. Journal of Food Safety 8, 25–34
181. Ballinger, M.B., Phillips, T.D., Kubena, L.F., 1986: Assessment of the distribution and elimination of ochratoxin A in the pregnant rat. Journal of Food Safety 8, 11–24
182. Miller, D.M., Wilson, D.M., Wyatt, R.D., McKinney, J.K., Crowell, W.A., Stuart, B.P., 1982: High performance liquid chromatographic determination and clearance time of aflatoxin residues in swine tissues. Journal of the Association of Official Analytical Chemists 65, 1–4
183. Fukayama, M.Y., Helferich, W.G., Hsieh, D.P.H., 1984: Effect of butylated hydroxytoluene on the disposition of [^{14}C]aflatoxin B_1 in the lactating rat. Food and Chemical Toxicology 11, 857–860
184. Hassen, W., Abid, S., Achour, A., Creppy, E., Bacha, H., 2004: Ochratoxin A and ß$_2$-microglobulinuria in healthy individuals and in chronic interstitial nephropathy patients in the centre of Tunisia: a hot spot of ochratoxin A exposure. Toxicology 199, 185–193
185. Maaroufi, K., Acour, A., Hammami, M., El May, M., Betbeder, A.M., Ellouz, F., Creppy, E.E., Bacha, H., 1995: Ochratoxin A in human blood in relation to nephropathy in Tunisia. Human & Experimental Toxicology 14, 609–615
186. Dvořáčková, I., Pichová, V., 1986: Pulmonary interstitial fibrosis with evidence of aflatoxin B_1 in lung tissue. Journal of Toxicology and Environmental Health 18, 153–157
187. Peraica, M., Domijan, A.-M., Fuchs, R., Lucić, A., Radić, B., 1999: The occurrence of ochratoxin A in blood in general population of Croatia. Toxicology Letters 110, 105–112
188. Radić, B., Fuchs, R., Peraica, M., Lucić, A., 1997: Ochratoxin A in human sera in the area with endemic nephropathy in Croatia. Toxicology Letters 91, 105–109
189. Peraica, M., Domijan, A.-M., Matašin, M., Lucić, A., Radić, B., Delaš, F., Horvat, M., Bosanac, I., Balija, M., Grgičević, D., 2001: Variations of ochratoxin A concentration in the blood of healthy populations in some Croatian cities. Archives of Toxicology 75, 410–414
190. Wild, C.P., Umbenhauer, D., Chapot, B., Montesano, R., 1986: Monitoring of individual human exposure to aflatoxins (AF) and N-nitrosamines (NNO) by immunoassays. Journal of Cellular Biochemistry 30, 171–179
191. Fuchs, R., Radić, B., Peraica, M., Hult, K., Pleština, R., 1988: Enterohepatic circulation of ochratoxin A in rats. Periodicum Biologorum 90, 39–42
192. Kovács, F., Sándor, G., Ványi, A., Domány, S., Zomborszky-Kovács, M., 1995: Detection of ochratoxin A in human blood and colostrum. Acta Veterinaria Hungarica 43, 393–400
193. Hagelberg, S., Hult, K., Fuchs, R., 1989: Toxicokinetics of ochratoxin A in several species and its plasma-binding properties. Journal of Applied Toxicology 9, 91–96

194. Watson, F.A., Mirocha, C.J., Hayes, A.W., 1984: Analysis for trichothecenes in samples from southeast Asia associated with „Yellow Rain". Fundamental and Applied Toxicology 4, 700–717
195. Sodeinde, O., Chan, M.C.K., Maxwell, S.M., Familusi, J.B., Hendrickse, R.G., 1995: Neonatal jaundice, aflatoxins and naphthols: report of a study in Ibadan, Nigeria. Annals of Tropical Paediatrics 15, 107–113
196. Breitholtz-Emanuelsson, A., Minervini, F., Hult, K., Visconti, A., 1994: Ochratoxin A in human serum samples collected in southern Italy from healthy individuals and individuals suffering from different kidney disorders. Natural Toxins 2, 366–370
197. Studer-Rohr, I., Schlatter, J., Dietrich, D.R., 2000: Kinetic parameters and intraindividual fluctuations of ochratoxin A plasma levels in humans. Archives of Toxicology 74, 499–510
198. Fuchs, R., Hult, K., Peraica, M., Radič, B., Pleština, R., 1984: Conversion of ochratoxin C into ochratoxin A *in vivo*. Applied and Environmental Microbiology 48, 41–42
199. Kiessling, K.-H., Pettersson, H., Sandholm, K., Olsen, M., 1984: Metabolism of aflatoxin, ochratoxin, zearalenone, and three trichothecenes by intact rumen fluid, rumen protozoa, and rumen bacteria. Applied and Environmental Microbiology 47, 1070–1073
200. Xiao, H, Marquardt, R.R., Frohlich, A.A., Phillips, G.D., Vitti, T.G., 1991: Effect of hay and a grain diet on the rate of hydrolysis of ochratoxin A in the rumen of sheep. Journal of Animal Science 69, 3706–3714
201. Micco, C., Brera, C., Miraglia, M., Onori, R., 1987: HPLC determination of the total content of aflatoxins in naturally contaminated eggs in free and conjugate forms. Food Additives and Contaminants 4, 407–414
202. Grosso, F., Saïd, S., Mabrouk, I., Fremy, J.M., Castegnaro, M., Jemmali, M., Dragacci, S., 2003: New data on the occurrence of ochratoxin A in human sera from patients affected or not by renal diseases in Tunisia. Food and Chemical Toxicology 41, 1133–1140
203. Szuets, P., Mesterhazy, A., Falkay, G.Y., Bartok, T., 1997: Early telarche symptoms in children and their relations to zearalenon contamination in foodstuffs. Cereal Research Communications 25, 429–436
204. Mortensen, H.P., Hald, B., Madsen, A., 1983: Feeding experiments with ochratoxin A contaminated barley for bacon pigs. 5. Ochratoxin A in pig blood. Acta Agriculturœ Scandinavica 33, 235–239
205. Maxwell, S.M., Familusi, J.B., Sodeinde, O., Chan, M.C.K., Hendrickse, R.G., 1994: Detection of naphthols and aflatoxins in Nigerian cord blood. Annals of Tropical Paediatrics 14, 3–5
206 Plestina, R., Ceović, S., Gatenbeck, S., Habazin-Novak, V., Hult, K., Hökby, E., Krogh, P., Radić, B., 1990: Human exposure to ochratoxin A in areas of Yugoslavia with endemic nephropathy. Journal of Environmental Pathology, Toxicology, Oncology 10, 145–148
207. Solti, L., Salamon, F., Barna-Vetró, I., Gyöngyösi, Á., Szabó, E., Wölfling, A., 1997: Ochratoxin A content of human sera determined by a sensitive ELISA. Journal of Analytical Toxicology 21, 44–48
208. Palli, D., Miraglia, M., Saieva, C., Masala, G., Cava, E., Colatosti, M., Corsi, A.M., Russo, A., Brera, C., 1999: Serum levels of ochratoxin A in healthy adults in Tuscany: correlation with individual characteristics and between repeat measurements. Cancer Epidemiology, Biomarkers & Prevention 8, 265–269
209. van Walbeek, W., Moodie, C.A., Scott, P.M., Harwig, J., Grice, H.C., 1971: Toxicity and excretion of ochratoxin A in rats intubated with pure ochratoxin A or fed cultures of *Penicillium viridicatum*. Toxicology and Applied Pharmacology 20, 439–441
210. Patterson, D.S.P., Shreeve, B.J., Roberts, B.A., Berrett, S., Brush, P.J., Glancy, E.M., Krogh, P., 1981: Effect on calves of barley naturally contaminated with ochratoxin A and groundnut meal contaminated with low concentrations of aflatoxin B_1. Research in Veterinary Science 31, 213–218
211. Veldman, A., Meijs, J.A.C., Borggreve, G.J., Heeres-van der Tol, J.J., 1992: Carry-over of aflatoxin from cows' food to milk. Animal Production 55, 163–168

212. Goeger, D.E., Shelton, D.W., Hendricks, J.D., Bailey, G.S., 1986: Mechanisms of anti-carcinogenesis by indole-3-carbinol: effect on the distribution and metabolism of aflatoxin B_1 in rainbow trout. Carcinogenesis 7, 2025–2031
213. Shelton, D.W., Goeger, D.E., Hendricks, J.D., Bailey, G.S., 1986: Mechanisms of anti-carcinogenesis: the distribution and metabolism of aflatoxin B_1 in rainbow trout fed Arocolor 1254. Carcinogenesis 7, 1065–1071
214. Wild, C.P., Montesano, R., van Bethem, J., Scherer, E., Den Engelse, L., 1990: Intercellular variation in levels of adducts of aflatoxin B_1 and G_1 in DNA from rat tissue: a quantitative immunocytochemical study. Journal of Cancer Research and Clinical Oncology 116, 134–140
215. Eko-Ebongue, S., Antalick, J.-P., Bonini, M., Betbeder, A.-M., Faugère, J.-G., Maaroufi, K., Bacha, H., Achour, M., Grosse, Y., Pfohl-Leszkowicz, A., Creppy, E.E., 1994: Détermination des teneurs en ochratoxine A, vitamine E et vitamine A dans des sérums humains de France et de Tunisie: recherche dúne corrélation. Annales des Falsifications et de l'Éxpertise Chimique et Toxicologique 929, 213–224
216. Bagni, A., Castagnetti, G.B., Chiavari, C., Ferri, G., Losi, G., Montanari, G., 1993: A study about aflatoxin M_1 and M_2 in dairy cow milk samples collected in the province of Reggio Emilia. L'Industria del Latte XXIX, 55–66 [Italian]
217. El-Nezami, H., Mykkänen, H., Kankaanpää, P., Suomalainen, T., Salminen, S., Ahokas, J., 2000: Ability of a mixture of *Lactobacillus* and *Propionibacterium* to influence the faecal aflatoxin content in healthy Egyptian volunteers: a pilot clinical study. Bioscience Microflora 19, 41–45
218. Chang, F.C., Chu, F.S., 1977: The fate of ochratoxin A in rats. Food and Cosmetics Toxicology 15, 199–204
219. Filali, A., Betbeder, A.M., Baudrimont, I., Benyada, A., Soulaymani, R., Creppy, E.E., 2002: Ochratoxin A in human plasma in Morocco: a preliminary survey. Human & Experimental Toxicology 21, 241–245
220. Özçelik, N., Koşar, A., Soysal, D., 2001: Ochratoxin A in human serum samples collected in Isparta-Turkey from healthy individuals and individuals suffering from different urinary disorders. Toxicology Letters 121, 9–13
221. Jiménez, A.M., López de Cerain, A., Gónzalez-Peñas, E., Bello, J., Bétbéder, A.M., Créppy, E.E., 1998: Exposure to ochratoxin A in Europe: comparison with a region of northern Spain. Journal of Toxicology – Toxin Reviews 17, 479–491
222. Ruprich, J., Ostrý, V., 1993: Study of human exposure to ochratoxin A and assessment of possible sources. Central European Journal of Public Health 1, 46–48
223. Sun, Z., Lu, P., Gail, M.H., Pee, D., Zhang, Q., Ming, L., Wang, J., Wu, Y., Liu, G., Wu, Y., Zhu, Y., 1999: Increased risk of hepatocellular carcinoma in male hepatitis B surface antigen carriers with chronic hepatitis who have detectable urinary aflatoxin metabolite M_1. Hepatology 30, 379–383
224. Oyelami, O.A., Maxwell, S.M., Adelusola, K.A., Aladekoma, T.A., Oyelese, A.O., 1997: Aflatoxins in the lungs of children with kwashiorkor and children with miscellaneous diseases in Nigeria. Journal of Toxicology and Environmental Health 51, 623–628
225. Malir, F., Jergeova, Z., Severa, J., Cerna, M., Smid, J., Betbeder, A.M., Baudrimont, I., Creppy, E.E., 1998: The level of ochratoxin A in blood serum of adults in the Czech Republic. Revue de Médicine Vétérinaire 149, 710
226. Domijan, A.-M., Peraica, M., Fuchs, R., Lucić, A., Radić, B., Balija, M., Bosanac, I., Grgičević, D., 1999: Ochratoxin A in blood of healthy population in Zagreb. Archives of Industrial Hygiene and Toxicology 3, 263–271
227. Wafa, E.W., Yahya, R.S., Sobh, M.A., Eraky, I., El-Baz, M., El-Gayar, H.A.M., Betbeder, A.M., Creppy, E.E., 1998: Human ochratoxicosis and nephropathy in Egypt: a preliminary study. Human & Experimental Toxicology 17, 124–129
228. Bailey, G.S., Dashwood, R., Loveland, P.M., Pereira, C., Hendricks, J.D., 1998: Molecular dosimetry in fish: quantitative target organ DNA adduction and hepatocarcinogenicity for four aflatoxins by two exposure routes in rainbow trout. Mutation Research 339, 233–244

229. Croy, R.G., Nixon, J.E., Sinnhuber, R.O., Wogan, G.N., 1980: Investigation of covalent aflatoxin B_1-DNA adducts formed *in vivo* in rainbow trout (*Salmo gairdneri*) embryos and liver. Carcinogenesis 1, 903–909
230. Erasmuson, A.F., Scahill, B.C., West, D.M., 1994: Natural zeranol (α-zearalenol) in the urine of pasture-fed animals. Journal of Agricultural and Food Chemistry 42, 2721–2725
231 Galvano, F., Pietri, A., Bertuzzi, T., Fusconi, G., Galvano, M., Piva, A., Piva, G., 1996: Reduction of carryover of aflatoxin from cow feed to milk by addition of activated carbons. Journal of Food Protection 59, 551–554
232. Micco, C., Miraglia, M., Onori, R., Brera, C., Mantovani, A.L., Ioppolo, A., Stasolla, D., 1988: Long-term administration of low doses of mycotoxins to poultry. 1. Residues of aflatoxin B_1 and its metabolites in broilers and laying hens. Food Additives and Contaminants 5, 303–308
233. Galvano, F., Pietri, A., Bertuzzi, T., Gagliardi, L., Ciotti, S., Luisi, S., Bognanno, M., la Fauci, L., Iacopino, A.M., Nigro, F., li Volti, G., Vanella, L., Giammanco, G., Tina, G.L., Gazzolo, D., 2008: Maternal dietary habits and mycotoxin occurrence in human mature milk. Molecular Nutrition & Food Research 52, 496–501
234. Tápai, K., Téren, J., Mesterházy, Á., 1997: Ochratoxin A in the sera of blood donors and ill persons. Cereal Research Communications 25, 307–308
235. Ueno, Y., Maki, S., Lin, J., Furuya, M., Sugiura, Y., Kawamura, O., 1998: A 4-year study of plasma ochratoxin A in a selected population in Tokyo by immunoassay and immunoaffinity column-linked HPLC. Food and Chemical Toxicology 36, 445–449
236. Gilbert, J., Brereton, P., MacDonald, S., 2001: Assessment of dietary exposure to ochratoxin A in the UK using a duplicate diet approach and analysis of urine and plasma samples. Food Additives and Contaminants 18, 1088–1093
237. Krishnamachari, K.A., Bhat, R.V., Nagarajan, V., Tilak, T.B., 1975: Investigations into an outbreak of hepatitis in parts of western India. Indian Journal of Medical Research 63, 1036–1049
238. López, C.E., Ramos, L.L., Ramadán, S.S., Bulacio, L.C., 2003: Presence of aflatoxin M_1 in milk for human consumption in Argentina. Food Control 14, 31–34
239. MAFF – UK, 1996: Survey of aflatoxin M_1 in farm gate milk. Food-Surveillance-Information-Sheet No. 78
240. Martins, M.L., Martins, H.M., 2000: Aflatoxin M_1 in raw and ultra high temperature-treated milk commercialized in Portugal. Food Additives and Contaminants 17, 871–874
241. Maxwell, S.M., Apeagyei, F., de Vries, H.R., Mwanmut, D.D., Hendrickse, R.G., 1989: Aflatoxins in breast milk, neonatal cord blood and sera of pregnant women. Journal of Toxicology – Toxin Reviews 8, 19–29
242. McKenzie, R.A., Blaney, B.J., Connole, M.D., Fitzpatrick, L.A., 1981: Acute aflatoxicosis in calves fed peanut hay. Australian Veterinary Journal 57, 284–286
243. Oruç, H.H., Sonal, S., 2001: Determination of aflatoxin M_1 levels in cheese and milk consumed in Bursa, Turkey. Veterinary and Human Toxicology 43, 292–293
244. Polychronaki, N., Turner, P.C., Mykkänen, H., Gong, Y., Amra, H., Abdel-Wahhab, M., El-Nezami, H., 2006: Determinants of aflatoxin M_1 in breast milk in a selected group of Egyptian mothers. Food Additives and Contaminants 23, 700–708
245. Rachelsperger, M., 1987: Essay about aflatoxin M_1-content of drinking milk: summer/winter exposition. Ernährung/Nutrition 11, 248–249
246. Roussi, V., Govaris, A., Varagouli, A., Botsoglou, N.A., 2002: Occurrence of aflatoxin M_1 in raw and market milk commercialized in Greece. Food Additives and Contaminants 19, 863–868
247. Saad, A.M., Abdelgadir, A.M., Moss, M.O., 1989: Aflatoxin in human and camel milk in Abu Dabi, United Arab Emirates. Mycotoxin Research 5, 57–60
248. Saad, A.M., Abdelgadir, A.M., Moss, M.O., 1995: Exposure of infants to aflatoxin M_1 from mothers' breast milk in Abu Dabi, UAE. Food Additives and Contaminants 12, 255–261
249. Sabino, M., Purchio, A., Milanez, T.V., 1996: Survey of aflatoxins B_1, M_1 and aflatoxicol in poultry and swine tissues from farms located in the states of Rio Grande Do Suland Santa Catarina, Brazil. Revista de Microbiologia 27, 189–191

250. Sabino, M., Purchio, A., Zorzetto, M.A.P., 1989: Variations in the levels of aflatoxin in cows milk consumed in the city of São Paulo, Brazil. Food Additives and Contaminants 6, 321–326
251. Saitanu, K., 1997: Incidence of aflatoxin M_1 in Thai milk products. Journal of Food Protection 60, 1010–1012
252. Salem, D.A., 2002: Natural occurrence of aflatoxins in feedstuffs and milk of dairy farms in Assiut Province, Egypt. Wiener Tierärztliche Monatsschrift 89, 86–91
253. Sassahara, M., Pontes Netto, D., Yanaka, E.K., 2005: Aflatoxin occurrence in foodstuff supplied to dairy cattle and aflatoxin M_1 in raw milk in the north of Paraná State. Food and Chemical Toxicology 43, 981–984
254. Schuddeboom, L.J., 1983: Development of legislation concerning mycotoxins in dairy products in The Netherlands. Microbiologie – Aliments – Nutrition 1, 179–185
255. Srivastava, V.P., Bu-Abbas, A., Alaa-Basuny X, Al-Johar, W., Al-Mufti, S., Siddiqui, M.K.J., 2001: Aflatoxin M_1 contamination in commercial samples of milk and dairy products in Kuwait. Food Additives and Contaminants 18, 993–997
256. Suzangar, M., Emami, A., Barnett, R., 1976: Aflatoxin contamination of village milk in Isfahan, Iran. Tropical Science 18, 155–159
257. Thirumala-Devi, K., Mayo, M.A., Hall, A.J., Craufurd, P.Q., Wheeler, T.R., Waliyar, F., Subrahmanyam, A., Reddy, D.V.R., 2002: Development and application of an indirect competitive enzyme-linked immunoassay for aflatoxin M_1 in milk and milk-based confectionery. Journal of Agricultural and Food Chemistry 50, 933–937
258. Porter, J.K., Wray, E.M., Rimando, A.M., Stancel, P.C., Bacon, C.W., Voss, K.A., 1996: Lactational passage of fusaric acid from the feed of nursing dams to the neonate rat and effects on pineal neurochemistry in the F1 and F2 generations at weaning. Journal of Toxicology and Environmental Health 49, 161–175
259. Chao, T.-C., Maxwell, S.M., Wong, S.Y., 1991: An outbreak of aflatoxicosis and boric acid poisoning in Malaysia: a clinicopathological study. Journal of Pathology 164, 225–233
260. Goeger, D.E., Shelton, D.W., Hendricks, J.D., Pereira, C., Bailey, G.S., 1988: Comparative effect of dietary butylated hydroxyanisole and ß-naphthoflavone on aflatoxin B_1 metabolism, DNA adduct formation, and carcinogenesis in rainbow trout. Carcinogenesis 9, 1793–1800
261. Creppy, E.E., Betbeder, A.-M., Godin, M., Fillastre, J.-P., AMG, K.S., Simon, P., Lasseur, C., Combe, C., Aparicio, M., 1995: Ochratoxin A in human blood and chronic interstitial nephropathy: case report in France. Proceedings from 17. Mykotoxin-Workshop in der Bundesforschungsanstalt für Landwirtschaft, Braunschweig-Völkenrode 56–62
262. Ahmed, H., Hendrickse, R.G., Maxwell, S.M., Yakubu, A.M., 1995: Neonatal jaundice with reference to aflatoxins: an aetiological study in Zaria, northern Nigeria. Annals of Tropical Paediatrics 15, 11–20
263. Scheuer, R., Bernard, K., Leistner, L., 1984: Rückstände von Ochratoxin A in Schweinenieren. Mitteilungsblatt der Bundesanstalt für Fleischforschung 83, 5781–5784 [German]
264. Hofmann, G., 1983: Vorkommen von Ochratoxin A in Blut und Nieren von Schweinen. Mitteilungsblatt der Bundesanstalt für Fleischforschung, Kulmbach 80, 5547–5551 [German]
265. Shepherd, E.C., Phillips, T.D., Rick Irvin, T., Safe, S.H., Robertson, L.W., 1984: Aflatoxin B_1 metabolism in the rat: polyhalogenated biphenyl enhanced conversion to aflatoxin M_1. Xenobiotica 14, 741–750
266. Patterson, D.S.P., Roberts, B.A., Small, B.J., 1976: Metabolism of ochratoxins A and B in the pig during early pregnancy and the accumulation in body tissues of ochratoxin A only. Food and Cosmetics Toxicology 14, 439–442
267. Shreeve, B.J., Patterson, D.S.P., Roberts, B.A., 1979: The „carry-over"of aflatoxin, ochratoxin and zearalenone from naturally contaminated feed to tissues, urine and milk of dairy cows. Food and Cosmetics Toxicology 17, 151–152
268. Miraglia, M., Brera, C., Cava, E., Calfapietra, F.R., 1998: The evaluation of major sources of ochratoxin A (OA) intake through the analysis of OA in biological fluids in Italy. Revue de Médicine Vétérinaire 149, 711

269. Olubuyide, I.O., Maxwell, S.M., Akinyinka, O.O., Hart, C.A., Neal, G.E., Hendrickse, R.G., 1993: HbsAg and aflatoxins in sera of rural (Igbo-Ora) and urban (Ibadan) populations in Nigeria. African Journal of Medicine and Medical Sciences 22, 77–80
270. Visconti, A., Bottalico, A., Solfrizzo, M., 1985: Aflatoxin M_1 in milk, in southern Italy. Mycotoxin Research 1, 71–75
271. Wild, C.R., Pionneau, F.A., Montesano, R., Mutiro, C.F., Chetsanga, C.J., 1987: Aflatoxin detected in human breast milk by immunoassay. International Journal of Cancer 40, 328–333
272. Yadagiri, B., Tulpule, P.G., 1974: Aflatoxin in buffalo milk. Indian Journal of Dairy Science 27, 293–297
273. Zarba, A., Wild, C.P., Hall, A.J., Montesano, R., Hudson, G.J., Groopman, J.D., 1992: Aflatoxin M_1 in human breast milk from The Gambia, West Africa, quantified by combined monoclonal antibody immunoaffinity chromatography and HPLC. Carcinogenesis 13, 891–894
274. Hammer, P., Blüthgen, A., Walte, H.G., 1996: Carry-over of fumonisin B_1 into the milk of lactating cows. Milchwissenschaft 51, 691–695
275. Maragos, C.M., Richard, J.L., 1994: Quantitation and stability of fumonisins B_1 and B_2 in milk. Journal of the Association of Official Analytical Chemists International 77, 1162–1167
276. Rice, L.G., Ross, P.F., 1994: Methods for detection and quantitation of fumonisins in corn, cereal products and animal excreta. Journal of Food Protection 57, 536–540
277. Hussain, I., Anwar, J., Munawar, M.A., Asi, M.R., 2008: Variation of levels of aflatoxin M_1 in raw milk from different localities in the central areas of Punjab, Pakistan. Food Control 19, 1126–1129
278. Côté, L.-M., Dahlem, A.M., Yoshizawa, T., Swanson, S.P., Buck, W.B., 1986: Excretion of deoxynivalenol and its metabolite in milk, urine, and feces of lactating dairy cows. Journal of Dairy Science 69, 2416–2423
279. Steimer, J., Blüthgen, A., Heeschen, W., Wetzel, S., Hamann, J., 1990: Studies on the influence of aflatoxin M_1-excretion by polychlorinated biphenyls in lactating cows. Kieler Milchwirtschaftliche Forschungsberichte 42, 543–552 [German]
280. Rizzi, L., Simioli, M., Roncada, P., Zaghini, A., 2003: Aflatoxin B_1 and clinoptilolite in feed for laying hens: effects on egg quality, mycotoxin residues in livers, and hepatic mixed-function oxygenase activities. Journal of Food Protection 66, 860–865
281. Fukal, L., 1991: Spontaneous occurrence of ochratoxin A residues in Czechoslovak slaughter pigs determined by immunoassay. Deutsche Lebensmittel-Rundschau 87, 316–319
282. Smith, E.E., Phillips, T.D., Ellis, J.A., Harvey, R.B., Kubena, L.F., Thompson, J., Newton, G., 1994: Dietary hydrated sodium calcium aluminosilicate reduction of aflatoxin M_1 residue in dietary goat milk and effects on milk production and components. Journal of Animal Science 72, 677–682
283. Höhler, D., Südekum, K.-H., Wolffram, S., Frohlich, A.A., Marquardt, R.R., 1999: Metabolism and excretion of ochratoxin A fed to sheep. Journal of Animal Science 77, 1217–1223
284. Prelusky, D.B., Veira, D.M., Trenholm, H.L., Foster, B.C., 1987: Metabolic fate and elimination in milk, urine and bile of deoxynivalenol following administration to lactating sheep. Journal of Environmental Science and Health B22, 125–148
285. Oyelami, O.A., Maxwell, S.M., Adelusola, K.A., Aladekoma, T.A., Oyelese, A.O., 1995: Aflatoxins in the autopsy brain tissue of children in Nigeria. Mycopathologia 132, 35–38
286. Chopra, R.C., Chhabra, A., Prasad, K.S.N., Dudhe, A., Murthy, T.N., Prasad, T., 1999: Carryover of aflatoxin M_1 in milk of cows fed aflatoxin B_1 contaminated ration. Indian Journal of Animal Nutrition 16, 103–106
287. Fukui, Y., Hoshino, K., Kameyama, Y., Yasui, T., Toda, C., Nagano, H., 1987: Placental transfer of ochratoxin A and its cytotoxic effect on the mouse embryo brain. Food and Chemical Toxicology 25, 17–24
288. Mirocha, C.J., Pawlowsky, R.A., Chatterjee, K., Watson, S., Hayes, W., 1983: Analysis for *Fusarium* toxins in various samples implicated in biological warfare in Southeast Asia. Journal of the Association of Official Analytical Chemists 66, 1485–1499

289. Anderson, P.H., Wells, G.A.H., Jackman, R., Morgan, M.R.A., 1984: Ochratoxicosis and ochratoxin residues in adult pig's kidneys – a pilot study. In: Moss, M.O., Frank, M. (eds)., Proceedings of the 5th Meeting on Mycotoxins in Animal and Human Health, Meeting, pp 23–29. University of Edinburgh, Surrey Press, Guildford, UK
290. Apostolou, E., El-Nezami, H.S., Ahokas, J.T., Donohue, D.C., 1998: The evaluation of ochratoxin A in breast milk in Victoria (Australia). Revue de Médicine Vétérinaire 149, 709
291. Bauer, J., Gareis, M., 1987: Ochratoxin A in the food chain. Journal of Veterinary Medicine Series B 34, 613–627 [German]
292. Baumann, U., Zimmerli, B., 1988: A simple determination of ochratoxin A in foods. Mitteilungen aus dem Gebiete der Lebensmitteluntersuchung und Hygiene 79, 151–158 [German]
293. Breitholtz, A., Olsen, M., Dahlbäck, Å, Hult, K., 1991: Plasma ochratoxin A levels in three Swedish populations surveyed using an ion-pair HPLC technique. Food Additives and Contaminants 8, 182–192
294. Breitholtz-Emanuelson, A., Olsen, M., Oskarsson, A., Palminger, I., Hult, K., 1993: Ochratoxin A in cow's milk and in human milk with corresponding human blood samples. Journal of the Association of Official Analytical Chemists International 76, 842–846
295. Büchmann, N.B., Hald, B., 1985: Analysis, occurrence and control of ochratoxin A residues in Danish pig kidneys. Food Additives and Contaminants 2, 193–199
296. Dieber, F., Köfer, J., 1999: Ochratoxin A-Nachweis im Serum steirischer Schlachtschweine. Deutsche Lebensmittel-Rundschau 95, 327–329 [German]
297. Dragacci, S., Grosso, F., Bire, R., Fremy, J.M., Coulon, S., 1999: A French monitoring programme for determining ochratoxin A occurrence in pig kidneys. Natural Toxins 7, 167–173
298. Elling, F., Hald, B., Lacobsen, Chr., Krogh, P., 1975: Spontaneous toxic nephropathy in poultry associated with ochratoxin A. Pathologica et Microbiologica Scandinavica Section A 83, 739–741
299. Ferrufino-Guardia, E.V., Tangni, E.K., Larondelle, Y., Ponchaut, S., 2000: Transfer of ochratoxin A during lactation: exposure of suckling via the milk of rabbit does fed a naturally-contaminated feed. Food Additives and Contaminants 17, 167–175
300. Fukal, L., 1990: A survey of cereals, cereal products, feedstuffs and porcine kidneys for ochratoxin A by radioimmunoassay. Food Additives and Contaminants 7, 253–258
301. Gareis, M., Märtlbauer, E., Bauer, J., Gedek, B., 1988: Determination of ochratoxin A in human milk. Zeitschrift für Lebensmittel-Untersuchung und -Forschung 186, 114–117
302. Gareis, M., Rosner, H., Ehrhardt, S., 2000: Blood serum levels of ochratoxin A and nutrition habits. Archiv für Lebensmittelhygiene 51, 108–110
303. Goliński, P., Grabarkiewicz-Szczęsna, J., 1985: The first in Poland cases of detection of ochratoxin A residues in human blood. Roczniki Panstwowego Zakladu Higieny 36, 378–381 [Polish]
304. Goliński, P., Hult, K., Grabarkiewicz-Szczęsna, J., Chelkowski, J., Kneblewski, P., Szebiotko, K., 1984: Mycotoxic porcine nephropathy and spontaneous occurrence of ochratoxin A residues in kidneys and blood of Polish swine. Applied and Environmental Microbiology 47, 1210–1212
305. Goliński, P., Hult, K., Grabarkiewicz-Szczęsna, J., Chelkowski, J., Szebiotko, K., 1985: Spontaneous occurrence of ochratoxin A residues in porcine kidney and serum samples in Poland. Applied and Environmental Microbiology 49, 1014–1015
306. Hult, K., Hökby, E., Gatenbeck, S., Rutqvist, L., 1980: Ochratoxin A in blood from slaughter pigs in Sweden: use in evaluation of toxin content of consumed feed. Applied and Environmental Microbiology 39, 828–830
307. Langseth, W., Nymoen, U., Bergsjø, B., 1993: Ochratoxin A in plasma of Norwegian swine determined by an HPLC column-switching method. Natural Toxins 1, 216–221
308. Corley, R.A., Swanson, S.P., Buck, W.B., 1985: Glucuronide conjugates of T-2 toxin and metabolites in swine bile and urine. Journal of Agricultural and Food Chemistry 33, 1085–1089

309. Olsen, M., Malmlöf, K., Pettersson, H., Sandholm, K., Kiessling, K.-H., 1985: Plasma and urinary levels of zearalenone and α-zearalenol in a prepubertal gilt fed zearalenone. Acta Pharmacologica et Toxicologica 56, 239–243
310. Vinitketkumnuen, U., Chewonarin, T., Dhumtanom, P., Lertprasertsuk, N., Wild, C.P., 1999: Aflatoxin-albumin adduct formation after single and multiple doses of aflatoxin B_1 in rats treated with Thai medicinal plants. Mutation Research 428, 345–351
311. Godin, M., Francois, A., le Roy, F., Morin, J.-P., Creppy, E., Hemet, J., Fillastre, J.-P., 1996: Karyomegalic interstitial nephritis. American Journal of Kidney Diseaes 27, 166
312. Breitholtz-Emanuelsson, A., Palminger-Hallén, I., Wohlin, P.O., Oskarsson, A., Hult, K., Olsen, M., 1993: Transfer of ochratoxin A from lactating rats to their offsprings: a short term study. Natural Toxins 1, 347–352
313. Dashwood, R.H., Arbogast, D.N., Fong, A.T., Hendricks, J.D., Bailey, G.S., 1988: Mechanisms of anti-carcinogenesis by indole-3-carbinol: detailed *in vivo* DNA binding dose-response studies after dietary administration with aflatoxin B_1. Carcinogenesis 9, 427–432
314. Murthy, T.R.K., Jemmali, M., Henry, Y., Frayssinet, C., 1975: Aflatoxin residues in tissues of growing swine: effect of separate and mixed feeding of protein and protein-free portions of the diet. Journal of Animal Science 41, 1339–1347
315. Josefsson, A.E., 1979: Study of ochratoxin A in pig kidneys. Vår Föda 31, 415–420 [Swedish]
316. Frye, C.E., Chu, F.S., 1977: Distribution of ochratoxin A in chicken tissues and eggs. Journal of Food Safety 1, 147–159
317. Galtier, P., 1974: The fate of ochratoxin A in the animal organism. I. Transport of the toxin in the blood of the rat. Annales des Recherches Vétérinaires 5, 311–318 [French]
318. Robison, T.S., Mirocha, C.J., Kurtz, H.J., Behrens, J.C., Weaver, G.A., Chi, M.S., 1979: Distribution of tritium-labeled T-2 toxin in swine. Journal of Agricultural and Food Chemistry 27, 1411–1413
319. Hayes, J.R., Polan, C.E., Campbell, T.C., 1977: Bovine liver metabolism and tissue distribution of aflatoxin B_1. Journal of Agricultural and Food Chemistry 25, 1189–1193
320. Lötzsch, R., Leistner, L., 1976: Aflatoxin residues in hen's eggs and egg products. Fleischwirtschaft 56, 1777–1785 [German]
321. Jiang, Y., Jolly, P.E., Ellis, W.O., Wang, J.-S., Phillips, T.D., Williams, J.H., 2005: Aflatoxin B_1 albumin adduct levels and cellular immune status in Ghanaians. International Immunology 17, 807–814
322. Hallén, I.P., Breitholz-Emanuelsson, A., Hult, K., Olsen, M., Oskarsson, A., 1998: Placental and lactational transfer of ochratoxin A in rats. Natural Toxins 6, 43–49
323. Hult, K., Pleština, R., Habazin-Novak, V., 1982: Ochratoxin A in human blood and Balkan endemic nephropathy. Archives of Toxicology 51, 313–321
324. Jørgensen, K., Petersen, A., 2002: Content of ochratoxin A in paired kidney and meat samples from healthy Danish slaughter pigs. Food Additives and Contaminants 19, 562–567
325. Köfer, J., Schuh, M., Fuchs, K., 1991: Ochratoxin A residues in Styrian slaughter pigs. Tierärztliche Umschau 46, 657–660 [German]
326. Kotowski, K., Grabarkiewicz-Szczęsna, J., Waskiewicz, A., Kostecki, M., Golinski, P., 2000: Ochratoxin A in porcine blood and in consumed feed samples. Mycotoxin Research 16, 66–72
327. Kotowski, K., Kostecki, M., Grabarkiewicz-Szczęsna, J., Golinski, P., 1993: Ochratoxin A residue in kidneys and blood of pigs. Medycyna Weterynaryjna 49, 554–556 [Polish]
328. Krogh, P., 1977: Ochratoxin A residues in tissues of slaughter pigs with nephropathy. Nordic Veterinary Medicine A 29, 402–405
329. Maaroufi, K., Achour, A., Betbeder, A.M., Hammami, M., Ellouz, F., Creppy, E.E., Bacha, H., 1995: Foodstuffs and human blood contamination by the mycotoxin ochratoxin A: correlation with chronic interstitial nephropathy in Tunisia. Archives of Toxicology 69, 552–558
330. Madsen, A., Hald, B., Lillehøj, E., Mortensen, H.P., 1982: Feeding experiments with ochratoxin A contaminated barley for bacon pigs. 2. Naturally contaminated barley given for 6 weeks from 20 kg compared with normal barley supplemented with crystalline ochratoxin A and/or citrinin. Acta Agriculturœ Scandinavica 32, 369–372

331. Ministry of the Agriculture, Fisheries and Foods (MAFF), 1993: Mycotoxins: Third report: 36th report of the steering group on chemical aspects of food surveillance (London: HSMO) pp. 39–44
332. Sadeghi, N., Oveisi, M.R., Jannat, B., Hajimahmoodi, M., Bonyani, H., Jannat, F., 2009: Incidence of aflatoxin M_1 in human breast milk in Teheran, Iran. Food Control 20, 75–78
333. Majerus, P., Otteneder, H., Hower, C., 1989: Beitrag zum Vorkommen von Ochratoxin A in Schweineblutserum. Deutsche Lebensmittel-Rundschau 85, 307–313 [German]
334. Marquardt, R.R., Frohlich, A.A., Sreemannarayana, O., Abramson, D., Bernatsky, A., 1988: Ochratoxin A in blood from slaughter pigs in western Canada. Canadian Journal of Veterinary Research 52, 186–190
335. Morgan, M.R.A., McNerney, R., Chan, H.W.-S., Anderson, P.H., 1986: Ochratoxin A in pig kidney determined by enzyme-linked immunosorbent assay (ELISA). Journal of the Science of Food and Agriculture 37, 475–480
336. Mortensen, H.P., Hald, B., Larsen, A.E., Madsen, A., 1983: Ochratoxin A contaminated barley for sows and piglets. Pig performance and residues in milk and pigs. Acta Agriculturœ Scandinavica 33, 349–352
337. Olberg, I.H., Yndestad, M., 1982: A Norwegian survey of ochratoxin A in cereals and animal tissue. Nordisk Jordbruksforskning 64, 296
338. Ominski, K.H., Frohlich, A.A., Marquardt, R.R., Crow, G.H., Abramson, D., 1996: The incidence and distribution of ochratoxin A in western Canadian swine. Food Additives and Contaminants 13, 185–198
339. Pietri, A., Bertuzzi, T., Gualla, A., Piva, G., 2006: Occurrence of ochratoxin A in raw ham muscles and in pork products from nothern Italy. Italian Journal of Food Science 18, 99–106
340. Razzazi, E., Böhm, J., Grajewski, J., Szczepaniak, K., Kübber-Heiss, A.J., Iben, C.H., 2001: Residues of ochratoxin A in pet foods, canine and feline kidneys. Journal of Animal Physiology and Animal Nutrition 85, 212–216
341. Rosner, H., Rohrmann, B., Peiker, G., 2000: Ochratoxin A in human serum. Archiv für Lebensmittelhygiene 51, 104–107
342. Rousseau, D.M., Candlish, A.A.G., Slegers, G.A., van Peteghem, C.H., Stimson, W.H., Smith, J.E., 1987: Detection of ochratoxin A in porcine kidneys by a monoclonal antibody-based radioimmunoassay. Applied and Environmental Microbiology 53, 514–518
343. Rousseau, D.M., van Peteghem, C.H., 1989: Spontaneous occurrence of ochratoxin A residues in porcine kidneys in Belgium. Bulletin of Environmental Contamination and Toxicology 42, 181–186
344. Rutqvist, L., Björklund, N.-E., Hult, K., Gatenbeck, S., 1977: Spontaneous occurrence of ochratoxin residues in kidneys of fattening pigs. Zentralblatt für Veterinär Medizin A 24, 402–408
345. Scott, P.M., Kanhere, S.R., Lau, B.P.-Y., Lewis, D.A., Hayward, S., Ryan, J.J., Kuiper-Goodman, T., 1998: Survey of Canadian human blood plasma for ochratoxin A. Food Additives and Contaminants 15, 555–562
346. Skaug, M.A., Helland, I., Solvoll, K., Saugstad, O.D., 2001: Presence of ochratoxin A in human milk in relation to dietary intake. Food Additives and Contaminants 18, 321–327
347. Skaug, M.A., Størmer, F.C., Saugstad, O.D., 1998: Ochratoxin A: a naturally occurring mycotoxin found in human milk samples from Norway. Acta Pædiatrica 87, 1275–1278
348. Tyllinen, H., Hintikka, E.-L., 1982: Occurrence of ochratoxin A in swine kidneys and feed in Finland. Nordisk Jordbruksforskning 64, 298–299
349. Zimmerli, B., Dick, R., 1995: Determination of ochratoxin A at the ppt level in human blood, serum, milk and some foodstuffs by high-performance liquid chromatography with enhanced fluorescence detection and immunoaffinity column cleanup: methodology and Swiss data. Journal of Chromatography B, 666, 85–99
350. Curtui, V.G., Gareis, M., Usleber, E., Märtlbauer, E., 2001: Survey of Romanian slaughtered pigs for the occurrence of mycotoxins ochratoxins A and B, and zearalenone. Food Additives and Contaminants 18, 730–738
351. El-Sayed, A.M.A.A., Neamat-Allah, A.A., Soher, E.A., 2000: Situation of mycotoxins in milk, dairy products and human milk in Egypt. Mycotoxin Research 16, 91–100

352. El-Sayed, A.M.A.A., Soher, E.A., Neamat-Allah, A.A., 2002: Human exposure to mycotoxins in Egypt. Mycotoxin Research 18, 23–30
353. Fukal, L., Reisnerova, H., 1990: Monitoring of aflatoxins and ochratoxin A in Czechoslovak human sera by immunoassay. Bulletin of Environmental Contamination and Toxicology 44, 345–349
354. Jonsyn, F.E., Maxwell, S.M., Hendrickse, R.G., 1995: Ochratoxin A and aflatoxins in breast milk samples from Sierra Leone. Mycopathologia 131, 121–126
355. Navas, S.A., Sabino, M., Rodriguez-Amaya, D.B., 2005: Aflatoxin M_1 and ochratoxin A in a human milk bank in the city of São Paulo, Brazil. Food Additives and Contaminants 22, 457–462
356. Turconi, G., Guarcello, M., Livieri, C., Comizolli, S., Maccarini, L., Castellazzi, A.M., Pietri, A., Piva, G., Roggi, C., 2004: Evaluation of xenobiotics in human milk and ingestion by the newborn. An epidemiological survey in Lombardy (Nothern Italy). European Journal of Nutrition 43, 191–197
357. Chen, C., Pearson, A.M., Coleman, T.H., Gray, J.I., Pestka, J.J., Aust, S.D., 1984: Tissue deposition and clearance of aflatoxins from broiler chickens fed a contaminated diet. Food and Chemical Toxicology 22, 447–451
358. Fernández, A., Belio, R., Ramos, J.J., Sanz, M.C., Sáez, T., 1997: Aflatoxins and their metabolites in the tissues, faeces and urine from lambs feeding on an aflatoxin-contaminated diet. Journal of the Science of Food and Agriculture 74, 161–168
359. Fernández, A., Verde, M.T., Gascón, M., Ramos, J.J., Gómez, J., 1994: Aflatoxin and its metabolites in tissues from laying hens and broiler chickens fed a contaminated diet. Journal of the Science of Food and Agriculture 65, 407–414
360. Okumura, H., Kawamura, O., Kishimoto, S., Hasegawa, A., Shrestha, S.M., Okuda, K., Obata, H., Okuda, H., Haruki, K., Uchida, T., Ogasawara, Y., Ueno, Y., 1993: Aflatoxin M_1 in Nepalese sera, quantified by combination of monoclonal antibody immunoaffinity chromatography and enzyme-linked immunosorbent assay. Carcinogenesis 14, 1233–1235
361. Oliveira, C.A.F., Kobashigawa, E., Reis, T.A., Mestieri, L., Albuquerque, R., Corrêa, B., 2000: Aflatoxin B_1 residues in eggs of laying hens fed a diet containing different levels of the mycotoxin. Food Additives and Contaminants 17, 459–462
362. Trucksess, M.W., Stoloff, L., Brumley, W.C., Wilson, D.M., Hale, O.M., Sangster, L.T., Miller, D.M., 1982: Aflatoxicol and aflatoxins B_1 and M_1 in the tissues of pigs receiving aflatoxin. Journal of the Association of Official Analytical Chemists 65, 884–887
363. Trucksess, M.W., Stoloff, L., Young, K., Wyatt, R.D., Miller, B.L., 1983: Aflatoxicol and aflatoxins B_1 and M_1 in eggs and tissues of laying hens consuming aflatoxin-contaminated feed. Poultry Science 62, 2176–2182
364. Valenta, H., Dänicke, S., Döll, S., 2003: Analysis of deoxynivalenol and de-epoxy-deoxynivalenol in animal tissues by liquid chromatography after clean-up with an immunoaffinity column. Mycotoxin Research 19, 51–55
365. Lusky, K., Tesch, D., Göbel, R., 1995: Untersuchung der Wirkung von natürlichem und kristallinem Ochratoxin A nach Verfütterung über 28 Tage beim Schwein mit anschließender Untersuchung des Rückstandsverhaltens beider Formen des Mykotoxins in Körperflüssigkeit und Organen sowie in Fleisch- und Wurstwaren. Archiv für Lebensmittelhygiene 46, 45–48 [German]
366. Lusky, K., Tesch, D., Göbel, R., Doberschütz, K.-D., 1994: Ochratoxin A. Untersuchungen zum Rückstandsverhalten beim Schwein und in daraus hergestellten Lebensmitteln. Fleischwirtschaft 74, 558–560 [German]
367. Dorner, J.W., Cole, R.J., Erlington, D.J., Suksupath, S., McDowell, G.H., Bryden, W.L., 1994: Cyclopiazonic acid residues in milk and eggs. Journal of Agricultural and Food Chemistry 42, 1516–1518
368. Belmadani, A., Tramu, G., Betbeder, A.M., Steyn, P.S., Creppy, E.E., 1998: Regional selectivity to ochratoxin A, distribution and cytotoxicity in rat brain. Archives of Toxicology 72, 656–662

369. Razzazi, E., Böhm, J., Kettner, B., Hochsteiner, W., Kahlbacher, H., 2002: Metabolisierung von Deoxynivalenol beim Schwein: Bestimmung von DON und DOM-1 im Urin vom Schwein. Mycotoxin Research 18A, 84–88 [German]
370. Azcona-Olivera, J.I., Ouyang, Y., Murtha, J., Chu, F.S., Pestka, J.J., 1995: Induction of cytokine mRNAs in mice after oral exposure to the trichothecene vomitoxin (deoxynivalenol): relationship to toxin distribution and protein synthesis inhibition. Toxicology and Applied Pharmacology 133, 109–120
371. Hult, K., Rutquist, L., Holmberg, T., Thafvelin, B., Gatenbeck, S., 1984: Ochratoxin A in blood of slaughter pigs. Nordisk Veterinaermedicin 36, 314–316
372. Teleb, H.M., Fakhry, F.M., 1988: Aflatoxin B_1 residue in broiler and its effect on fat metabolism. Veterinary Medical Journal 36, 135–145
373. Espada, Y., Guitart, R., Arboix, M., 1991: Quantitative determination of aflatoxin B_1 in chick liver. Food Additives and Contaminants 8, 163–170
374. Bauer, J., Niemiec, J., Scholtyssek, S., 1988: Ochratoxin A in layers' feed. 2. Report: residues in blood sera, livers and eggs. Archiv für Geflügelkunde 52, 71–75 [German]
375. Garaleviciene, D., Pettersson, H., Elwinger, K., 2002: Effects on health and blood plasma parameters of laying hens by pure nivalenol in the diet. Journal of Animal Physiology and Animal Nutrition 86, 389–398
376. Prior, M.G., O'Neil, J.B., Sisodia, C.S., 1980: Effects of ochratoxin A on growth response and residues in broilers. Poultry Science 59, 1254–1257
377. Lun, A.K., Young, L.G., Moran, Jr., E.T., Hunter, D.B., Rodriguez, J.P., 1986: Effects of feeding hens a high level of vomitoxin-contaminated corn on performance and tissue residues. Poultry Science 65, 1095–1099
378. Lusky, K., Göbel, R., Tesch, D., Tenner, G., Haider, W., Krüger, M., Lippert, A., 1998: Studies on the effects of ochratoxin A and deoxynivalenol toxicity on the health of pigs and tissue residue concentrations. Tierärztliche Umschau 53, 623–630 [German]
379. Kipper, L., Schweigmann, A., Vater, N., 1991: Studies of the carry-over of ochratoxin A within the line „animal food–pigs for slaughter–meat products". Tierärztliche Umschau 46, 538–542 [German]
380. Lusky, K., Tesch, D., Göbel, R., Haider, W., 1997: Effects of OTA and ZEA on animal health and residue behaviour of pigs. Tierärztliche Umschau 52, 212–221 [German]
381. Wilkinson, A.P., Denning, D.W., Morgan, M.R.A., 1989: Immunoassay of aflatoxin in food and human tissue. Journal of Toxicology – Toxin Reviews 8, 69–79
382. Piskorska-Pliszczyńska, J., Juskiewicz, T., 1990: Tissue distribution and passage into eggs of ochratoxin A in Japanese quail. Journal of Environmental Pathology, Toxicology and Oncology 10, 8–10
383. Krogh, P., Elling, F., Gyrd-Hansen, N., Hald, B., Larsen, A.E., Lillehøj, E.B., Madsen, A., Mortensen, H.P., Ravnskov, U., 1976: Experimental porcine nephropathy: changes of renal function and structure perorally induced by crystalline ochratoxin A. Acta Pathologia Microbiologica Scandinavica Section A 84, 429–434
384. Bailey, G.S., Loveland, P.M., Pereira, C., Pierce, D., Hendricks, J.D., Groopman, J.D., 1994: Quantitative carcinogenesis and dosimetry in rainbow trout for aflatoxin B_1 and aflatoxicol, two aflatoxins that form the same DNA adduct. Mutation Research 313, 25–38
385. Matsumoto, H., Ito, T., Ueno, Y., 1978: Toxicological approaches to the metabolites of Fusaria. XII. Fate and distribution of T-2 toxin in mice. Japanese Journal of Experimental Medicine 48, 393–399
386. Hald, B., 1991: Ochratoxin A in human blood in European countries. IARC Scientific Publications pp. 159–164
387. Brasel, T.L., Campbell, A.W., Demers, R.E., Ferguson, B.S., Fink, J., Vojdani, A., Wilson, S.C., Straus, D.C., 2004: Detection of trichothecene mycotoxins in sera from individuals exposed to *Stachybotrys chartarum* in indoor environments. Archives of Environmental Health 59, 317–323
388. Wang, J.-S., Huang, T., Su, J., Liang, F., Wei, Z., Liang, Y., Luo, H., Kuang, S.-Y., Qian, G.-S., Sun, G., He, X., Kensler, T.W., Groopman, J.D., 2001: Hepatocellular carcinoma and aflatoxin expo-

sure in Zhuqing Village, Fusui County, People's Republic of China. Cancer Epidemiology, Biomarkers & Prevention 10, 143–146
389. Pace, J.G., Watts, M.R., Burrows, E.P., Dinterman, R.E., Matson, C., Hauer, E.C., Wannemacher, Jr., R.W., 1985: Fate and distribution of ^{3}H-labeled T-2 mycotoxin in guinea pigs. Toxicology and Applied Pharmacology 80, 377–385
390. Jonsyn, F.E., Maxwell, S.M., Hendrickse, R.G., 1995: Human fetal exposure to ochratoxin A and aflatoxins. Annals of Tropical Paediatrics 15, 3–9
391. Ueno, Y, Takai, Y., Baba, Y., 1990: Analysis of T-2 toxin metabolites in tissues and excreta of rats. Journal of Environmental Pathology, Toxicology and Oncology 10, 31–40
392. Sudhakar, B.V., 1992: The carry-over effect of aflatoxin B_1 into eggs and liver of chicken. Indian Veterinary Journal 69, 1061–1062
393. Pepeljnjak, S., Blaževic, N., Čuljak, K., 1982: Histopathological changes and findings of ochratoxin A in organs of pigs, in the area of endemic nephropathy in Yugoslavia. Proceedings of the V. International Symposium on Mycotoxins and Phycotoxins, Vienna 346–348
394. Dailey, R.E., Reese, R.E., Brouwer, E.A., 1980: Metabolism of [^{14}C]zearalenone in laying hens. Journal of Agricultural and Food Chemistry 28, 286–291
395. Solti, L., Pécsi, T., Barna-Vetró, I., Szasz, Jr., F., Biró, K, Szabó, E., 1999: Analysis of serum and seminal plasma after feeding ochratoxin A with breeding boars. Animal Reproduction Science 56, 123–132
396. Commission of the European Communities. Reports on tasks for scientific cooperation. Report of experts participating in task 3.2.2: assessment of dietary intake of ochratoxin A by the population in EU member states. Report EUR 17523 EN, 1997
397. Hoogenboom, L.A.P., Tulliez, J., Gautier, J.-P., Coker, R.D., Melcion, J.-P., Nagler, M.J., Polman, T.H.G., Delort-Laval, J., 2001: Absorption, distribution and excretion of aflatoxin-derived ammoniation products in lactating cows. Food Additives and Contaminants 18, 47–58
398. Yadav, A.S., Satija, K.C., Mahipal, S.K., 1995: Aflatoxin B_1 deposition and clearance from tissues of broiler. Indian Journal of Poultry Science 30, 165–166
399. Colvin, B.M., Harrison, L.R., Gosser, H.S., Hall, R.F., 1984: Aflatoxicosis in feeder cattle. Journal of American Veterinary Association 184, 956–958
400. Reichmann, K.G., Blaney, B.J., Connor, J.K., Runge, B.M., 1982: The significance of aflatoxin and ochratoxin in the diet of Australian chickens. Australian Veterinary Journal 58, 211–212
401. Jonsyn-Ellis, F.E., 2000: Seasonal variation in exposure frequency and concentration levels of aflatoxins and ochratoxins in urine samples of boys and girls. Mycopathologia 152, 35–40
402. James, L.J., McGirr, L.G., Smith, T.K., 1982: High pressure liquid chromatography of zearalenone and zearenols in rat urine and liver. Journal of the Association of Official Analytical Chemists 65, 8–13
403. Beasley, V.R., Swanson, S.P., Corley, R.A., Buck, W.B., Koritz, G.D., Burmeister, H.R., 1986: Pharmacokinetics of the trichothecene mycotoxin, T-2 toxin, in swine and cattle. Toxicon 24, 13–23
404. Meyer, K., Usleber, E, Märtlbauer, E., Bauer, J., 2000: Occurrence of zearalenone, α- and β-zearalenol in biles of breeding sows in relation to the reproductive performance. Berliner und Münchener Tierärztliche Wochenschrift 113, 374–379 [German]
405. Loveland, P.M., Wilcox, J.S., Pawlowski, N.E., Bailey, G.S., 1987: Metabolism and DNA binding of aflatoxicol and aflatoxin B_1 *in vivo* and in isolated hepatocytes from rainbow trout (*Salmo gairdneri*). Carcinogenesis 8, 1065–1070
406. Prelusky, D.B., Trenholm, H.L., 1992: Nonaccumulation of residues in swine tissue following extended consumption of deoxynivalenol-contaminated diets. Journal of Food Science 57, 801–802
407. Prelusky, D.B., Trenholm, H.L., 1991: Tissue distribution of deoxynivalenol in swine dosed intravenously. Journal of Agricultural and Food Chemistry 39, 748–751

408. Barel, S., Yagen, B., Bialer, M., 1990: Pharmacokinetics of the trichothecene mycotoxin verrucarol in dogs. Journal of Pharmaceutical Sciences 79, 548–551
409. Tesch, D., Lusky, K., 1993: Untersuchungen zum Einfluß des Mykotoxins Ochratoxin A auf die Tiergesundheit und auf das Rückstandsverhalten beim Schwein. Archiv für Lebensmittelhygiene 44, 77–80 [German]
410. Rousseau, D.M., Slegers, G.A., van Peteghem, C.H., 1986: Solid-phase radioimmunoassay of ochratoxin A in serum. Journal of Agricultural and Food Chemistry 34, 862–865
411. Holmberg, T., Breitholz, A., Bengtsson, A., Hult, K., 1990: Ochratoxin A in swine blood in relation to moisture content in feeding barley at harvest. Acta Agriculturae Scandinavica 40, 201–204
412. Tomaszewski, J., Miturski, R., Semezuk, A., Kotarski, J., Jakowicki, J., 1998: Tissue zearalenone concentrations in normal, hyperplastic and neoplastic human endometrium. Ginkologia Polska 69, 363–366 [Polish]
413. Dänicke, S., Valenta, H., Döll, S., 2004: On the toxicokinetics and the metabolism of deoxynivalenol (DON) in the pig. Archives of Animal Nutrition 58, 169–180
414. Sypecka, Z., Kelly, M., Brereton, P., 2004: Deoxynivalenol and zearalenone residues in eggs of laying hens fed with a naturally contaminated diet: effects on egg production and estimation of transmission rates from feed to eggs. Journal of Agricultural and Food Chemistry 52, 5463–5471
415. Meky, F.A., Turner, P.C., Ashcroft, A.E., Miller, J.D., Qiao, Y.-L., Roth, M.J., Wild, C.P., 2003: Development of a urinary biomarker of human exposure to deoxynivalenol. Food and Chemical Toxicology 41, 265–273
416. Eriksen, G.S., Pettersson, H., Lindberg, J.E., 2003: Absorption, metabolism and excretion of 3-acetyl DON in pigs. Archives of Animal Nutrition 57, 335–345
417. Turner, P.C., Rothwell, J.A., White, K.L.M., Gong, Y.Y., Cade, J.E., Wild, C.P., 2008: Urinary deoxynivalenol is correlated with cereal intake in individuals from the United Kingdom. Environmental Health Perspectives 116, 21–25
418. Obremski, K., Gajęcki, M., Zwierzchowski, W., Bakula, T., Apoznański, J., Wojciechowski, J., 2003: The level of zearalenone and α-zearalenol in the blood of gilts with clinical symptoms of toxicosis, fed diets with a low zearalenone content. Journal of Animal and Feed Sciences 12, 529–538
419. Stoev, S.D., Vitanov, S., Anguelov, G., Petkova-Bocharova, T., Creppy, E.E., 2001: Experimental mycotoxic nephropathy in pigs provoked by a diet containing ochratoxin A and penicillic acid. Veterinary Research Communications 25, 205–223
420. Turner, P.C., Collison, A.C., Cheung, Y.B., Gong, Y.Y., Hall, A.J., Prentice, A.M., Wild, C.P., 2007: Aflatoxin exposure *in utero* causes growth faltering in Gambian infants. International Journal of Epidemiology 36, 1119–1125
421. Bauer, J., Bollwahn, W., Gareis, M., Gedek, B., Heinritzi, K., 1985: Kinetic profiles of diacetoxyscirpenol and two of its metabolites in blood serum of pigs. Applied and Environmental Microbiology 49, 842–845
422. Farnworth, E.R., Trenholm, H.L., 1983: The metabolism of the mycotoxin zearalenone and its effects on the reproductive tratcts of young male and female pigs. Canadian Journal of Animal Science 63, 967–975
423. Prelusky, D.B., Trenholm, H.L., Hamilton, R.M.G., Miller, J.D., 1987: Transmission of [^{14}C] deoxynivalenol to eggs following oral administration to laying hens. Journal of Agricultural and Food Chemistry 35, 182–186
424. Prelusky, D.B., Hamilton, R.M.G., Trenholm, H.L., 1989: Transmission of residues to eggs following long-term administration of ^{14}C-labelled deoxynivalenol to laying hens. Poultry Science 68, 744–748
425. Corley, R.A., Swanson, S.P., Gullo, G.J., Johnson, L., Beasley, V.R., Buck, W.B., 1986: Disposition of T-2 toxin, a trichothecene mycotoxin, in intravascularly dosed swine. Journal of Agricultural and Food Chemistry 34, 868–875
426. Manique, R., Pena, A., Lino, C.M., Moltó, J.C., Mañes, J., 2008: Ochratoxin A in the morning and afternoon portions of urine from Coimbra and Valencian populations. Toxicon 51, 1281–1287

427. McCoy, L.F., Scholl, P.F., Sutcliffe, A.E., Kieszak, S.M., Powers, C.D., Rogers, H.S., Gong, Y.Y., Groopman, J.D., Wild, C.P., Schleicher, R.L., 2008: Human aflatoxin albumin adducts quantitatively compared by ELISA, HPLC with fluorescence detectection, and HPLC with isotope dilution mass spectrometry. Cancer Epidemiology, Biomarkers & Prevention 17, 1653–1657
428. Abdulrazzaq, Y.M., Osman, N., Yousif, Z.M., Trad, O., 2004: Morbidity in neonates of mothers who have ingested aflatoxins. Annals of Tropical Paediatrics 24, 145–151
429. Assaf, H., Betbeder, A.-M., Creppy, E.E., Pallardy, M., Azouri, H., 2004: Ochratoxin A levels in human plasma and foods in Lebanon. Human & Experimental Toxicology 23, 495–501
430. Bammler, T.K., Slone, D.H., Eaton, D.L., 2000: Effects of dietary oltipraz and ethoxyquin on aflatoxin B_1 biotransformation in non-human primates. Toxicological Science 54, 30–41
431. Losito, I., Monaci, L., Palmisano, F., Tantillo, G., 2004: Determination of ochratoxin A in meat products by high-performance liquid chromatography coupled to electrospray ionisation sequential mass spectrometry. Rapid Communications in Mass Spectrometry 18, 1965–1971
432. Gajęcki, M, Przybylowicz, M., Zielonka, L., Zwierzchowski, W., Obremski, K., Skorska-Wyszyńska, E., Gajęcka, M., Polak, M., Jakimiuk, E., 2004: Preliminary results of monitoring research on zearalenone presence in blood of women with neoplastic lesions in reproductive system. Polish Journal of Veterinary Sciences 7, 153–156
433. Zarba, A., Hmieleski, R., Hemenway, D.R., Jakab, G.J., Groopman, J.D., 1992: Aflatoxin B_1-DNA adduct formation in rat liver following exposure by aerosol inhalation. Carcinogenesis 13, 1031–1033
434. Lopez, C., Ramos, L., Bulacio, L., Ramadan, S., Rodriguez, F., 2002: Aflatoxin B_1 content in patients with hepatic diseases. MEDICINA (Buenos Aires) 62, 313–316
435. Autrup, J.L., Schmidt, J., Autrup, H., 1993: Exposure to aflatoxin B_1 in animal-feed production plant workers. Environmental Health Perspectives 99, 195–197
436. Sewram, V., Mshicileli, N., Shephard, G.S., Marasas, W.F.O., 2003: Fumonisin mycotoxins in human hair. Biomarkers 8, 110–118
437. Nayak, S., Sashidhar, R.B., Bhat, R.V., 2001: Quantification and validation of enzyme immunoassay for urinary aflatoxin B_1-N^7-guanine adduct for biological monitoring of aflatoxins. Analyst 126, 179–183
438. Sangare-Tigori, B., Moukha, S., Kouadio, J.H., Dano, D.S., Betbeder, A.-M., Achour, A., Creppy, E.E., 2006: Ochratoxin A in human blood in Abidjan, Côte d'Ivoire. Toxicon 47, 894–900
439. Vettorazzi, A., Gonzales-Peñas, E., Arbillagra, L., Corcuera, L.-A., López de Cerain, A., 2008: Simple high-performance liquid chromatography-fluorescence detection method for plasma, kidney and liver of rats as a tool for toxicological studies. Journal of Chromatography A 1215, 100–106
440. Skaug, M.A., 2003: Levels of ochratoxin A and IGG against conidia of *Penicillium verruculosum* in blood samples from healthy farm workers. Annals of Agricultural and Environmental Medicine 10, 73–77
441. Monaci, L., Tantillo, G., Palmisano, F., 2004: Determination of ochratoxin A in pig tissues by liquid-liquid extraction and clean-up and high-performance liquid chromatography. Analytical and Bioanalytical Chemistry 378, 1777–1782
442. Turner, P.C., Burley, V.J., Rothwell, J.A., White, K.L.M., Cade, J.E., Wild, C.P., 2008: Dietary wheat reduction decreases the level of urinary deoxynivalenol in UK adults. Journal of Exposure Science and Environmental Epidemiology 18, 392–399
443. Ibeh, I.N., Uraih, N., Ogonar, J.I., 1994: Dietary exposure to aflatoxin in human male infertility in Benin City, Nigeria. International Journal of Fertility and Menopausal Studies 39, 208–214
444. Chao, T.C., Lo, D., Bloodworth, B., Gunasegaram, R., Koh, T., Ng, H., 1994: Aflatoxin exposure in Singapore: blood aflatoxin levels in normal subjects, hepatitis B virus carriers and primary hepatocellular carcinoma patients. Medicine, Science and the Law 34, 289–298

445. Mehta, R., Laver, G.W., Stapley, R., McMullen, E., 1992: Liver DNA adducts in methyl-deficient rats administered a single dose of aflatoxin B_1. Carcinogenesis 13, 1241–1247
446. Postupolski, J., Karlowski, K., Kubik, P., 2006: Ochratoxin A in maternal and foetal blood and in maternal milk. Roczniki Panstwowego Zaklado Higieny 57, 23–30
447. Hatem, N.L., Hassab, H.M.A., Abd Al-Rahman, E.M., El-Deeb, S.A., El-Sayed Ahmed, R.L., 2005: Prevalence of aflatoxins in blood and urine of Egyptian infants with protein-energy malnutrition. Food and Nutrition Bulletin 26, 49–56
448. Egner, P.A., Groopman, J.D., Wang, J.-S., Kensler, T., Friesen, M.D., 2006: Quantification of aflatoxin-B_1-N^7-guanine in human urine by high-performance liquid chromatography and isotope dilution tandem mass spectrometry. Chemical Research in Toxicology 19, 1191–1195
449. Hassan, A.M., Sheashaa, H.A., Fattah, M.F.A., Ibrahim, A.Z., Gaber, O.A., Sobh, M.A., 2006: Study of ochratoxin A as an environmental risk that causes renal injury in breast-fed Egyptian infants. Pediatric Nephrology 21, 102–105
450. Dostal, A., Jakusova, L., Cajdova, J., Hudeckova, H., 2008: Results of the first studies of occurrence of ochratoxin A in human milk in Slovakia. Bratislavslec Lekarske Listy 109, 276–278
451. Tang, L., Tang, M., Xu, L., Luo, H., Huang, T., Yu, J., Zhang, L., Gao, W., Cox, S.B., Wang, J.-S., 2008: Modulation of aflatoxin biomarkers in human blood and urine by green tea polyphenols intervention. Carcinogenesis 29, 411–417
452. Tajik, H., Rohani, S.M.R., Moradi, M., 2007: Detection of aflatoxin M_1 in raw and commercial pasteurized milk in Urmia, Iran. Pakistan Journal of Biological Sciences 10, 4103–4107
453. Sefidgar, S.A.A., Azizi, G., Khosravi, A.R., Roudbar-Mohammadi, S., 2008: Presence of aflatoxin M_1 in raw milk at cattle farms in Babol, Iran. Pakistan Journal of Biological Science 11, 484–486
454. Shundo, L., Sabino, M., 2006: Aflatoxin M_1 in milk by immunoaffinity column cleanup with TLC/HPLC determination. Brazilian Journal of Microbiology 37, 164–167
455. Mokhles, M., Abdl El Wahhab, M.A., Tawfik, M., Ezzat, W., Gamil, K., Ibrahim, M., 2007: Detection of aflatoxin among hepatocellular carcinoma patients in Egypt. Pakistan Journal of Biological Science 10, 1422–1429
456. Duarte, S., Bento, J., Pena, A., Lino, C.M., Delerue-Matos, C., Oliva-Teles, T., Morais, S., Correira, M., Oliveira, M.B.P.P., Alves, M.R., Pereira, J.A., 2010: Monitoring of ochratoxin A exposure of the Portuguese population through a nationwide urine survey – Winter 2007. Science of the Total Environment 408, 1195–1198
457. Jolly, P., Jiang, Y., Ellis, W., Awuah, R., Nnedu, O., Phillips, T., Wang, J.-S., Afriyie-Gyawu, E., Tang, L., Persom, S., Williams, J., Jolly, C., 2006: Determinants of aflatoxin levels in Ghanaians: sociodemographic factors, knowledge of aflatoxin and food handling and consumption practices. International Journal of Hygiene and Environmental Health 209, 345–358
458. Abulu, E.O., Uriah, N., Aigbefo, H.S., Oboh, P.A., Agbonlahor, D.E., 1998: Preliminary investigation on aflatoxin in cord blood of jaundiced neonates. West African Journal of Medicine 17, 184–187
459. Matsumura, M., Mori, T., 1998: Detection of aflatoxins in autopsied materials from a patient infected with *Aspergillus flavus*. Nippon Ishinkin Gakkai Zasshi 39, 167–171
460. Denning, D.W., Quiepo, S.C., Altman, D.G., Makarananda, K., Neal, G.E., Camallerre, E.L., Morgan, M.R.A., Tupasi, T.E., 1995: Aflatoxin and outcome from acute lower respiratory infection in children in The Phillippines. Annals of Tropical Peadiatrics 15, 209–216
461. Ankrah, N.A., Rikimaru, T., Ekuban, F.A., 1994: Observations on aflatoxins and the liver status of Ghanaian subjects. East African Medical Journal 71, 739–741
462. Liu, Z.-H., Tu, W.-S., Li, D.-R.O, Li, Y.-D., Xie, C.-H., Yang, Y.Z., Qin, B.-B., 1990: A new method for the quantitation of aflatoxin M_1 in urine by high performance liquid chromatography and its application to the etiologic study of hepatoma. Biomedical Chromatography 4, 83–86

463. Chelule, P.K., Gqaleni, N., Chuturgoon, A.A., Dutton, M.F., 2000: The determination of fumonisin B_1 in human faeces: a short term marker for assessment of exposure. Biomarkers 5, 1–8
464. Castegnaro, M., Maru, V., Petkova-Bocharova, T., Nikolov, I., Bartsch, H., 1991: Concentrations of ochratoxin A in the urine of endemic nephropathy patients and controls in Bulgaria: lack of detection of 4-hydroxyochratoxin A. In: Castegnaro, R., Dirheimer, G., Chernozemsky, I.N., Bartsch, H. (eds). Mycotoxins, Endemic Nephropathy and Urinary Tract Tumours. Lyon, International Agency for Research on Cancer, IARC, pp.165–169
465. Cusumano, V., 1991: Aflatoxins in sera from patients with lung cancer. Oncology 48, 194–195
466. de Vries, H.R., Lamplugh, S.M., 1989: Aflatoxins in liver biopsies from Kenya. Tropical and Geographical Medicine 41, 26–30
467. Abdulrazzaq, Y.M., Osman, N., Ibrahim, A., 2002: Fetal exposure to aflatoxins in the United Arab Emirates. Annals of Tropical Paediatrics 22, 3–9
468. Dinis, A.M.P., Lino, C.M., Pena, A.S., 2007: Ochratoxin A in nephropathic patients from two cities of central zone in Portugal. Pharmaceutical and Biomedical Analysis 44, 553–557
469. Dänicke, S., Ueberschär, K.-H., Valenta, H., Matthes, S., Matthäus, K., Halle, I., 2004: Effects of graded levels of *Fusarium*-toxin-contaminated wheat in Pekin duck diets on performance, health and metabolism of deoxynivalenol and zearalenone. British Poultry Science 45, 264–272
470. Iavicoli, I., Brera, C., Carelli, G., Caputi, R., Marinaccio, A., Miraglia, M., 2002: External and internal dose in subjects occupationally exposed to ochratoxin A. International Archives of Occupational and Environmental Health 75, 381–386
471. Newman, S.J., Smith, J.R., Stenske, K.A., Newman, L.B., Dunlap, J.R., Imerman, P.M., Kirk, C.A., 2007: Aflatoxicosis in nine dogs after exposure to contaminated commercial dog food. Journal of Veterinary Diagnostic Investigation 19, 168–175
472. Fazekas, B., Tar, A., Kovács, M., 2005: Ochratoxin A content of urine samples of healthy humans in Hungary. Acta Veterinaria Hungarica 53, 35–44
473. Biró, K., Solti, L., Barna-Vetró, I., Bagó, G., Glávits, R., Szabó, E., Fink-Gremmels, J., 2002: Tissue distribution of ochratoxin A as determined by HPLC and ELISA and histopathological effects in chickens. Avian Pathology 31, 141–148
474. Poapolathep, A., Poapolathep, S., Sugita-Konishi, Y., Imsilp, K., Tassanawat, T., Sinthusing, C., Itoh, Y., Kumagai, S., 2008: Fate of fusarenon-X in broilers and ducks. Poultry Science 87, 1510–1515
475. Schneweis, I., Meyer, K., Ritzmann, M., Hoffmann, P., Dempfle, L., Bauer, J., 2005: Influence of organically or conventionally produced wheat on health, performance and mycotoxin residues in tissues and bile of growing pigs. Archives of Animal Nutrition 59, 155–163
476. Battacone, G., Nudda, A., Cannas, A., Cappio Borlino, A, Bomboi, G., Pulina, G., 2003: Excretion of aflatoxin M_1 in milk of dairy ewes treated with different doses of aflatoxin B_1. Journal of Dairy Science 86, 2667–2675
477. Battacone, G., Nudda, A., Palomba, M., Pascale, M., Nicolussi, P., Pulina, G., 2005: Transfer of aflatoxin B_1 from feed to milk and from milk to curd and whey in dairy sheep fed artificially contaminated concentrates. Journal of Dairy Science 88, 3063–3069
478. Boudra, H., Barnouin, J., Dragacci, S., Morgavi, D.P., 2007: Aflatoxin M_1 and ochratoxin A in raw bulk milk from French dairy herds. Journal of Dairy Science 90, 3197–3201
479. Tajkarimi, M., Shojaee Aliabadi, F., Salah Nejad, M., Pursoltani, H., Motallebi, A.A., Mahdavi, H., 2007: Seasonal study of aflatoxin M_1 contamination in milk in five regions in Iran. International Journal of Food Microbiology 116, 346–349
480. Tardieu, D., Bailly, J.-D., Skiba, F., Grosjean, F., Guerre, P., 2008: Toxicokinetics of fumonisin B_1 in turkey poults and tissue persistence after exposure to a diet containing the maximum European tolerance for fumonisins in avian feeds. Food and Chemical Toxicology 46, 3213–3218
481. Polychronaki, N., Wild, C.P., Mykkänen, H., Amra, H., Abdel-Wahhab, M., Sylla, A., Diallo, M., El-Nezami, H., Turner, P.C., 2008: Urinary biomarkers of aflatoxin exposure in young children from Egypt and Guinea. Food and Chemical Toxicology 46, 519–526

482. Blank, R., Rolfs, J.-P., Südekum, K.-H., Frohlich, A.A., Marquadt, R.R., Wolffram, S., 2003: Effects of chronic ingestion of ochratoxin A on blood levels and excretion of the mycotoxin in sheep. Journal of Agricultural and Food Chemistry 51, 6899–6905
483. Döll, S., Dänicke, S, Valenta, H., 2008: Residues of deoxynivalenol (DON) in pig tissue after feeding mash or pellet diets containing low concentrations. Molecular Nutrition & Food Research 52, 727–734
484. Dänicke, S., Valenta, H., Klobasa, F., Döll, S., Ganter, M., Flachowsky, G., 2004: Effects of graded levels of *Fusarium* toxin contaminated wheat in diets for fattening pigs on growth performance, nutrient digestibility, deoxynivalenol balance and clinical serum characteristics. Archives of Animal Nutrition 58, 1–17
485. Ahsan, H., Wang, L.-Y., Chen, C.-J., Tsai, W.-Y., Santella, R.M., 2001: Variability in aflatoxin-albumin adduct levels and effects of hepatitis B and C Virus infection and glutathione *S*-transferase *M1* and *T1* genotype. Environmental Health Perspectives 109, 833–837
486. Chen, S.-Y., Chen, C.-J., Chou, S.-R., Hsieh, L.-L., Wang, L.-Y., Tsai, W.-Y., Ahsan, H., Santella, R.M., 2001: Association of aflatoxin B_1-albumin adduct levels with hepatitis B surface antigen status among adolescents in Taiwan. Cancer Epidemiology, Biomarkers & Prevention 10, 1223–1226
487. Yagen, B., Bialer, B., Sintov, A., 1985: Gas chromatographic assay with pharmacokinetic applications for monitoring T-2 and HT-2 toxins in plasma. Journal of Chromatography 343, 67–75
488. Gong, Y., Hounsa, A., Egal, S., Turner, P.C., Sutcliffe, A.E., Hall, A.J., Cardwell, K., Wild, C.P., 2004: Postweaning exposure to aflatoxin results in impaired child growth: a longitudinal study in Benin, West Aftrica. Environmental Health Perspectives 112, 1334–1338
489. Campbell, T.C., Hayes, J.R., Newberne, P.M., 1978: Dietary lipotropes, hepatic microsomal mixed-function oxidase activities, and *in vivo* covalent binding of aflatoxin B_1 in rats. Cancer Research 38, 4569–4573
490. Schrager, T.F., Newberne, P.M., Pikul, A.H., Groopman, J.D., 1990: Aflatoxin-DNA adduct formation in chronically dosed rats fed a choline-deficient diet. Carcinogenesis 11, 177–180
491. Micco, C., Miraglia, M., Onori, R., Libanori, A., Brera, C., Mantovani, A.L., Macrì, C., 1991: Effects of combined exposure to ochratoxin A and penicillic acid on residues and toxicity in broilers. La Revista della Società Italiana di Scienza dell'Alimentazione 20, 101–108
492. Qin, G., Gopalan-Kriczky, P., Su, J., Ning, Y., Lotlikar, P.D., 1997: Inhibition of aflatoxin B_1-induced initiation of hepatocarcinogenesis in the rat by green tea. Cancer Letters 112, 149–154
493. Tang, L., Guan, H, Ding, X., Wang, J.-S., 2007: Modulation of aflatoxin toxicity and biomarkers by lycopene in F344 rats. Toxicology and Applied Pharmacology 219, 10–17
494. Pestka, J.J., Islam, Z., Amuzie, C.J., 2008: Immunochemical assessment of deoxynivalenol tissue distribution following oral exposure in the mouse. Toxicology Letters 178, 83–87
495. Egner, P.A., Wang, J.-B., Zhu, Y.-R., Zhang, B.-C., Wu, Y., Zhang, Q.-N., Qian, G.-S., Kuang, S.-Y., Gange, S. J., Jacobson, L.P., Helzsouer, K.J., Bailey, G.S., Groopman, J.D., Kensler, T.W., 2001: Chlorophyllin intervention reduces aflatoxin-DNA adducts in individuals at high risk for liver cancer. Proceedings of the National Academy of Science 98, 14601–14606
496. Guan, R., Oon, C.J., Wild, C., Motesano, R., 1986: A preliminary survey on aflatoxin exposure in Singapore. Annals Academy of Medicine 15, 201–205
497. Mirocha, C.J., Pawlosky, R.J., Chatterjee, K., 1984: Analytical methodology, detection of trichothecenes from Southeast Asia samples and their residues in animal tissue. Archives Belges Supplement, 210–218
498. Heyndrickx, A., Sookvanichsilp, N., van den Heede, M., 1984: Detection of trichothecene mycotoxins (Yellow Rain) in blood, urine and faeces of Iranian soldiers treated as victims of a gas attack. Archives Belges Supplement, 143–146
499. Pang, V.F., Swanson, S.P., Beasley, V.R., Buck, W.B., Haschek, W.M., 1987: The toxicity of T-2 toxin in swine following topical application. I. Clinical signs, pathology, and residue concentrations. Fundamental and Applied Toxicology 9, 41–49

500. Naudé, T.W., O'Brien, O.M., Rundberget, T., Mc Gregor, A.D.G., Roux, C., Flåøyen, A., 2002: Tremorgenic neuromycotoxicosis in 2 dogs ascribed to the ingestion of penitrem A and possibly roquefortine in rice contaminated with *Penicillium crustosum*. Journal of the South African Veterinary Association 73, 211–215
501. Hedman, R., Pettersson, H., Lindberg, J.E., 1997: Absorption and metabolism of nivalenol in pigs. Archives of Animal Nutrition 50, 13–24
502. Pacin, A.M., Ciancio Bovier, E.V., Motta, E., Resnik, S.L., Villa, D., Olsen, M., 2008: Survey of Argentinean human plasma for ochratoxin A. Food Additives and Contaminants 25, 635–641
503. Turner, P.C., Loffredo, C., El Kafrawy, S., Ezzat, S., Abdel Latif Eissa, S., El Daly, M., Nada, O., Abdel-Hamid, M., 2008: Pilot survey of aflatoxin-albumin adducts in sera from Egypt. Food Additives and Contaminants 25, 583–587
504. Muñoz, K., Vega, M., Rios, G., Muñoz, S., Madariaga, R., 2006: Preliminary study of ochratoxin A in human plasma in agricultural zones of Chile and its relation to food consumption. Food and Chemical Toxicology 44, 1884–1889
505. Pena, A., Seifrtová, M., Lino, C., Silveira, I., Solich, P., 2006: Estimation of ochratoxin A in Portuguese population: new data on the occurrence in human urine by high performance liquid chromatography with fluorescence detection. Food and Chemical Toxicology 44, 1449–1454
506. Thuvander, A., Paulsen, J.E., Axberg, K., Johannson, N., Vidnes, A., Enghardt-Barbieri, H., Trygg, K., Lund-Larsen, K., Jahrl, S., Widenfalk, A., Bosnes, V., Alexander, J., Hult, K., Olsen, M., 2001: Levels of ochratoxin A in blood from Norwegian and Swedish blood donors and their possible correlation with food consumption. Food and Chemical Toxicology 39, 1145–1151 (and personal communication)
507. Lino, C.M., Baeta, M.L., Henri, M., Dinis, A.M.P., Pena, A.S., Silveira, M.I.N., 2008: Levels of ochratoxin A in serum from urban and rural Portuguese populations and estimation of exposure degree. Food and Chemical Toxicology 46, 879–885
508. Commission of the European Communities, 2002. Reports on tasks for scientific cooperation. Report of experts participating in task 3.2.7: assessment of dietary intake of ochratoxin A by the population in EU member states
509. Aoudia, N., Tangni, E.K., Larondelle, Y., 2008: Distribution of ochratoxin A in plasma and tissues of rat fed a naturally contaminated diet amended with micronized wheat fibres: effectiveness of mycotoxin sequestering activity. Food Chemical Toxicology 46, 871–878
510. Gong, Y.Y., Torres-Sanchez, L., Lopez-Carillo, L., He Peng, J., Sutcliffe, A.E., White, K.L., Humpf, H.-U., Turner, P.C., Wild, C.P., 2008: Association between tortilla consumption and human urinary fumonisin B_1 levels in a Mexican population. Cancer Epidemiology, Biomarkers & Prevention 17, 688–694
511. Monroe, D.H., Eaton, D.L., 1988: Effects of modulation of hepatic glutathione on biotransformation and covalent binding of aflatoxin B_1 to DNA in the mouse. Toxicology and Applied Pharmacology 94, 118–127
512. Coppock, R.W., Swanson, S.P., Gelberg, H.B., Koritz, G.D., Buck, W.B., Hoffmann, W.E., 1987: Pharmacokinetics of diacetoxyscirpenol in cattle and swine: effects of halothane. American Journal of Veterinary Research 48, 691–695
513. Pascale, M., Visconti, A., 2000: Rapid method for the determination of ochratoxin A in urine by immunoaffinity column clean-up and high-performance liquid chromatography. Mycopathologia 152, 91–95
514. Elgerbi, A.M., Aidoo, K.E., Candlish, A.A.G., Tester, R.F., 2004: Occurrence of aflatoxin M_1 in randomly selected North African milk and cheese samples. Food Additives and Contaminants 21, 592–597
515. Micco, C., Miraglia, M., Benelli, L., Onori, R., Ioppolo, A., Mantovani, A.L., 1988: Long term administration of low doses of mycotoxins in poultry. 2. Residues of ochratoxin A and aflatoxins in broilers and laying hens after combined administration of ochratoxin A and aflatoxin B_1. Food Additives and Contaminants 5, 309–314
516. Plakas, S.M., Loveland, P.M., Bailey, G.S., Blazer, V.S., Wilson, G.L., 1991: Tissue disposition and excretion of ^{14}C-labelled aflatoxin B_1 after oral administration in channel catfish. Food and Chemical Toxicology 29, 805–808

517. Wang, J.-S., Abubaker, S., He, X., Sun, G., Strickland, P.T., Groopman, J.D., 2001: Development of aflatoxin B_1-lysine adduct monoclonal antibody for human exposure studies. Applied and Environmental Microbiology 67, 2712–2717
518. Kamkar, A., 2005: A study on the occurrence of aflatoxin M_1 in raw milk produced in Sarab City of Iran. Food Control 16, 593–599
519. Honstead, J.P., Dreesen, D.W., Stubblefield, R.D., Shotwell, O.L., 1992: Aflatoxins in swine tissues during drought conditions: an epidemiologic study. Journal of Food Protection 55, 182–186
520. Benedetti, S., Iametti, S., Bonomi, F., Mannino, S., 2005: Head space sensor array for the detection of aflatoxin M_1 in raw ewe's milk. Journal of Food Protection 68, 1089–1092
521. Onyemelukwe, C.G., Nirodi, C., West, C.E., 1980: Aflatoxin B_1 in hepatocellular carcinoma. Tropical and Geographical Medicine 32, 237–240
522. Sarr, A.B., Mayura, K., Kubena, L.F., Harvey, R.B., Phillips, T.D., 1995: Effects of phyllosilicate clay on the metabolic profile of aflatoxin B_1 in Fischer-344 rats. Toxicology Letters 75, 145–151
523. Ruprich, J., Ostrý, V., 1993: Health risk assessment of the mycotoxin ochratoxin A to humans: Czech Republic – Brno – 1991/1992. Central European Journal of Public Health 2, 86–93
524. Hanna, G.D., Phillips, T.D., Kubena, L.F., Cysewski, S.J., Ivie, G.W., Heidelbaugh, N.D., Witzel, D.A., Hayes, A.W., 1981: High pressure liquid chromatographic determination of penicillic acid in chicken tissues. Poultry Science 60, 2246–2252
525. Prior, M.G., Sisodia, C.S., 1978: Ochratoxicosis in White Leghorn hens. Poultry Science 57, 619–623
526. Meyer, K., Usleber, E., Märtlbauer, E., Bauer, J., 1997: Analysis of zearalenone-metabolites in bile of gilts with reproductive problems. Berliner und Münchener Tierärztliche Wochenschrift 110, 281–283
527. Canela, R., Viladrich, R., Velazquez, C.A., Sanchis, V., 1994: A survey of porcine kidneys and chicken liver for ochratoxin A in Spain. Mycopathologia 125, 29–32
528. MacDougald, O.A., Thulin, A.J., Pestka, J.J., 1990: Determination of zearalenone and related metabolites in porcine urine by modified enzyme-linked immunosorbent assay. Journal of the Association of Official Analytical Chemists 73, 65–68
529. Kennedy, D.G., McEvoy, J.D.G., Blanchflower, W.J., Hewitt, S.A., Cannavan, A., McCaughey, W.J., Elliott, C.T., 1995: Possible naturally occurring zeranol in bovine bile in Northern Ireland. Journal of Veterinary Medicine Series B 42, 509–512
530. Dänicke, S., Goyarts, T., Valenta, H., Razzazi, E., Böhm, J., 2004: On the effects of deoxynivalenol (DON) in pig feed on growth performance, nutrients utilization and DON metabolism. Journal of Animal and Feed Science 13, 539–556
531. Golínski, P., Chelkowski, J., Konarkowski, A., Szebiotko, K., 1983: Mycotoxins in cereal grain. Part VI. The effect of ochratoxin A on growth and tissue residues of the mycotoxin in broiler chickens. Nahrung, 27, 251–256
532. Velasco, M.L.R., Delso, M.M.C., Escudero, D.O., 2003: ELISA and HPLC determination of the occurrence of aflatoxin M_1 in raw cow's milk. Food Additives and Contaminants 20, 276–280
533. Polychronaki, N., West, R.M., Turner, P.C., Amra, H., Abdel-Wahab, M., Mykkänen, H., El-Nezami, H., 2007: A longitudinal assessment of aflatoxin M_1 excretion in breast milk of selected Egyptian mothers. Food and Chemical Toxicology 45, 1210–1215
534. Amuzie, C.J., Harkema, J.R., Pestka, J.J., 2008: Tissue distribution and proinflammatory cytokine induction by the trichothecene deoxynivalenol in the mouse: comparison of nasal vs. oral exposure. Toxicology 248, 39–44
535. Shephard, G.S., Thiel, P.G., Sydenham, E.W., Alberts, J.F., Cawood, M.E., 1994: Distribution and excretion of a single dose of the mycotoxin fumonisin B_1 in a non-human primate. Toxicon 32, 735–741
536. Breinholt, V., Hendricks, J., Pereira, C., Arbogast, D., Bailey, G., 1995: Dietary chlorophyllin is a potent inhibitor of aflatoxin B_1 hepatocarcinogenesis in rainbow trout. Cancer Research 55, 57–62

537. Dänicke, S., Brüssow, K.-P., Valenta, H, Ueberschär, K.-H., Tiemann, U., Schollenberger, M., 2005: On the effects of graded levels of *Fusarium* toxin contaminated wheat in diets for gilts on feed intake, growth performance and metabolism of deoxynivalenol and zearalenone. Molecular Nutrition & Food Research 49, 932–943
538. Petkova-Bocharova, T., Castegnaro, M., Pfohl-Leszkowicz, A., Garren, L., Grosso, F., Nikolov, I., Vrabcheva, T., Dragacci, S., Chernozemsky, I.N., 2003: Analysis of ochratoxin A in serum and urine of inhabitants from an area with Balkan endemic nephropathy: a one month follow up study. Facta Universitas Series: Medicine and Biology 10, 62–68
539. Hussain, I., Anwar, J., 2008: A study on contamination of aflatoxin M_1 in raw milk in the Punjab province of Pakistan. Food Control 19, 393–395
540. Jonsyn, F.E., 1999: Intake of aflatoxins and ochratoxins by infants in Sierra Leone: possible effects on the general health of these children. Journal of Nutrition and Environmental Medicine 9, 15–22
541. Turner, P.C., Sylla, A., Gong, Y.Y., Diallo, M.S., Sutcliffe, A.E., Hall, A.J., Wild, C.P., 2005: Reduction in exposure to carcinogenic aflatoxins by postharvest intervention measures in West Africa: a community-based intervention study. Lancet 365, 1950–1956
542. Burdaspal, P.A., Legarda, T.M., 1998: Datos sorbe la presencia de ocratoxina A en plasma humano en España. Alimentaria 292, 103–109 [Spanish]
543. Dänicke, S., Ueberschär, K.-H., Halle, I., Valenta, H., Flachowsky, G., 2001: Excretion kinetics and metabolism of zearalenone in broilers in dependence on a detoxifying agent. Archives of Animal Nutrition 55, 299–313
544. Aslam, M., Beg, A.E., Blaskewicz, M., Degen, G.H., Golka, K., 2005: Ochratoxin A blood concentration in healthy subjects and bladder cancer cases from Pakistan. Abstract, 27th Mycotoxin-Workshop, 13–15 June 2005, Dortmund, Germany
545. Ngindu, A., Kenya, P.R., Ocheng, D.M., Omondi, T.N., Ngare, W., Gatei, D., Johnson, B.K., Ngira, J.A., Nandwa, H., Jansen, A.J., Kaviti, J.N., Arap Siongok, T., 1982: Outbreak of acute hepatitis caused by aflatoxin poisoning in Kenya. Lancet 12, 1346–1348
546. Breinholt, V., Arbogast, D., Loveland, P., Pereira, C., Dashwood, R., Hendricks, J., Bailey, G., 1999: Chlorophyllin chemoprevention in trout initiated by aflatoxin B_1 bath treatment: an evaluation of reduced bioavailability vs. target organ protective mechanisms. Toxicology and Applied Pharmacology 158, 141–151
547. Gemechu-Hatewu, M., Platt, K.-L., Oesch, F., Hacker, H.-J., Bannasch, P., Steinberg, P., 1997: Metabolic activation of aflatoxin B_1 to aflatoxin B_1–8,9-epoxide in woodchucks undergoing chronic active hepatitis. International Journal of Cancer 73, 587–591
548. Prasanna, H.R., Lu, M.H., Beland, F.A., Hart, R.W., 1989: Inhibition of aflatoxin B_1 binding to hepatic DNA by dehydroepiandrosterone *in vivo*. Carcinogenesis 10, 2197–2200
549. Gradelet, S., Le Bon, A.-M., Suschetet, M., Astorg, P., 1998: Dietary carotenoids inhibit aflatoxin B_1-induced liver preneoplastic foci and DNA damage in rat: role of the modulation of aflatoxin B_1 metabolism. Carcinogenesis 19, 403–411
550. Salbe, A.D., Bjeldanes, L.F., 1989: Effect of diet and route of administration on the DNA binding of aflatoxin B_1 in the rat. Carcinogenesis 10, 629–634
551. Kensler, T.W., Egner, P.A., Davidson, N.E., Roebuck, B.D., Pikul, A., Groopman, J.D., 1986: Modulation of aflatoxin metabolism, aflatoxin-N^7-guanine formation, and hepatic tumorigenesis in rats fed ethoxyquin: role of induction of glutathione *S*-transferase. Cancer Research 46, 3924–3931
552. Lotlikar, P.D., Raj, H.G., Bohm, L.S., Ho, L.L., Jhee, E.-C., Tsujii, K., Gopalan, P., 1989: A mechanism of inhibition of aflatoxin B_1-DNA binding in the liver by phenobarbital pretreatment of rats. Cancer Research 49, 951–957
553. Gresham, A., Done, S., Livesey, C., MacDonald, S., Chan, D., Sayers, R., Clark, C., Kemp, P., 2006: Survey of pigs' kidneys with lesions consistent with PMWS and PDNS and ochratoxicosis. Part 1: concentrations and prevalence of ochratoxin A. Veterinary Record 159, 737–742
554. Turner, P.C., Sylla, A., Kuang, S.-Y., Marchant, C.L., Diallo, M.S., Hall, A.J., Groopman, J.D., Wild, C.P., 2005: Absence of *TP53* codon 249 mutations in young Guinean children with high aflatoxin exposure. Cancer Epidemiology, Biomarkers & Prevention 14, 2053–2055

555. Boudra, H., Morgavi, D.P., 2006: Development and validation of a HPLC method for the quantitation of ochratoxins in plasma and raw milk. Journal of Chromatography B 843, 295–301
556. Gopalan-Kriczky, P., Hiruma, S., Lotlikar, P.D., 1994: Effect of glutathione levels on aflatoxin B_1-DNA binding in livers and kidneys of male rats and hamsters pretreated with buthionine sulfoximine and diethylmaleate. Cancer Letters 76, 25–30
557. Norred, W.P., Porter, J.K., Dorner, J.W., Cole, R.J., 1988: Occurrence of the mycotoxin cyclopiazonic acid in meat after oral administration to chickens. Journal of Agricultural and Food Chemistry 36, 113–116
558. Takahashi, N., Harttig, U., Williams, D.E., Bailey, G.S., 1996: The model Ah-receptor agonist ß-naphthoflavone inhibts aflatoxin B_1-DNA binding *in vivo* in rainbow trout at dietary levels that do not induce CYP1A enzymes. Carcinogenesis 17, 79–87
559. Autrup, J.L., Schmidt, J., Seremet, T., Autrup, H., 1991: Determination of exposure to aflatoxins among Danish workers in animal-feed production through the analysis of aflatoxin B_1 adducts to serum albumin. Scandinavian Journal of Work, Environment & Health 17, 436–440
560. Dashwood, R.H., Breinholt, V., Bailey, G.S., 1991: Chemopreventive properties of chlorophyllin: inhibition of aflatoxin B_1 (AFB_1)-DNA binding *in vivo* and anti-mutagenic activity against AFB_1 and two heterocyclic amines in the *Salmonella* mutagenicity assay. Carcinogenesis 12, 939–942
561. Schwerdt, G., Bauer, K., Gekle, M., Silbernagl, S., 1996: Accumulation of ochratoxin A in rat kidney *in vivo* and in cultivated renal epithelial cells *in vitro*. Toxicology 114, 177–185
562. Srivatanakul, P., Parkin, D.M., Jiang, Y.-Z., Khlat, M., Kao-Ian, U.-T., Sontipong, S., Wild, C., 1991: The role of infection by *Opisthorchis viverrini*, hepatitis B virus, and aflatoxin exposure in the etiology of liver cancer in Thailand. Cancer 68, 2411–2417
563. Mandel, H.G., Manson, M.M., Judah, D.J., Simpson, J.L., Green, J.A., Forrester, L.M., Wolf, C.R., Neal, G.E., 1987: Metabolic basis for the protective effect of the antioxidant ethoxyquin on aflatoxin B_1 hepatocarcinogenesis in the rat. Cancer Research 47, 5218–5223
564. Sylla, A., Diallo, M.S., Castegnaro, J.-J., Wild, C.P., 1999: Interactions between hepatitis B virus infection and exposure to aflatoxins in the development of hepatocellular carcinoma: a molecular epidemiological approach. Mutation Research 428, 187–196
565. Stresser, D.M., Williams, D.E., McLellan, L.I., Harris, T.M., Bailey, G.S., 1994: Indole-3-carbinol induces a rat liver glutathione transferase subunit (Yc2) with high activity toward aflatoxin B_1 *exo*-epoxide. Drug Metabolism and Disposition 22, 392–399
566. Krogh, P., Elling, F., Hald, B., Jylling, B, Petersen, V.E., Skadiiauge, E., Svendsen, C.K., 1975: Experimental avian nephrophathy. Changes of renal function and structure induced by ochratoxin A-contaminated feed. Acta Pathologica et Microbiologica Scandinavica, Section A 84, 215–221
567. Veldman, A., 1992: Effect of sorbentia on carry-over of aflatoxin from cow feed to milk. Milchwissenschaft 47, 777–780
568. Fuchs, R., Radić, B., Čeović, S., Šoštarić, B., Hult, K., 1991: Human exposure to ochratoxin A. IARC Scientific Publications 115, 131–135
569. Nyathi, C.B., Mutiro, C.F., Hasler, J.A., Chetsanga, C.J., 1989: Human exposure to aflatoxins in Zimbabwe. Central African Journal of Medicine 35, 542–545
570. Richir, C.L., Paccalin, J., Larcebeau, S., Faugeres, J., Morard, J.-L., Lamant, M., 1979: Presence d'aflatoxine B_1 dans le foie humain. Cahiers de Nutrition et de Diététique 11, 223–226
571. Yu, M.-W., Lien, J.-P., Liaw, Y.-F., Chen, C.-J., 1996: Effects of multiple risk factors for hepatocellular carcinoma on formation of aflatoxin B_1-DNA adducts. Cancer Epidemiology, Biomarkers & Prevention 5, 613–619
572. Byrem, T.M., Pestka, J.J., Chu, F.S., Strasburg, G.M., 1999: Analysis and pharmacokinetics of cyclopiazonic acid in market weight pigs. Journal of Animal Science 77, 173–179
573. Petkova-Bocharova, T., Castegnaro, M., 1991: Ochratoxin A in human blood in relation to Balkan endemic nephropathy and urinary tract tumours in Bulgaria. In: Mycotoxins, Endemic Nephropathy and Urinary Tract Tumours. Ed. Castegnaro, M., Dirheimer, G., Chemozemsky, I.N., Bartsch, H. Lyon, International Agency for Research on Cancer, pp. 135–137, IARC, 1991

574. Bintvihok, A., Kositcharoenkul, S., 2006: Effect of dietary calcium propionate on performance, hepatic enzyme activities and aflatoxin residues in broilers fed a diet containing low levels of aflatoxin B_1. Toxicon 47, 41–46
575. Megalla, S.E., Kamel, Y.Y., Abdel-Fattah, H.M., Hafez, A.H., 1981: Determining the kinetic behaviour for the secretion of milk toxin as related to dosage level of aflatoxin B_1. Zeitschrift für Ernährungswissenschaften 20, 216–222
576. Tangni, E.K., Waegeneers, N., van Overmeire, I., Goeyens, L., Pussemier, L., 2009: Mycotoxin analyses in some home produced eggs in Belgium reveal small contribution to the total intake. Science of the Total Environment 407, 4411–4418
577. Bintvihok, A., Davitiyananda, D., Kositcharoenkul, S., Panichkriangkrai, W., Jamratchai, O., 1998: Residues of aflatoxins and their metabolites in chicken tissues in Thailand. Journal of Toxicological Science 23 (Suppl. II), 389
578. Hirano, K., Adachi, Y., Ishibashi, S., Sueyoshi, M., Bintvihok, A., Kumazawa, N.H., 1991: Detection of aflatoxin B_1 in plasma of fowl receiving feed naturally contaminated with aflatoxin B_1. Journal of Veterinary Medical Science 53, 1083–1085
579. Shin, B.S., Hong, S.H., Hwang, S.W., Kim, H.J., Lee, J.B., Yoon, H.-S., Kim, D.J., Yoo, S.D., 2009: Determination of zearalenone by liquid chromatography/tandem mass spectrometry and application to a pharmacokinetic study. Biomedical Chromatography 23, 1014–1021
580. Duarte, S.C., Bento, J.M.V., Pena, A., Lino, C.M., 2009: Ochratoxin A exposure assessment of the inhabitants of Lisbon during winter 2007/2008 through bread and urine analysis. Food Additives and Contaminants 26, 1411–1420
581. Stoev, S.D., Dutton, M.F., Njobeh, P.B., Mosonik, J.S., Steenkamp, P.A., 2010: Mycotoxic nephropathy in Bulgarian pigs and chickens: complex aetiology and similarity to Balkan Endemic Nephropathy. Food Additives and Contaminants 27, 72–88
582. Mizrak, D., Engin, B., Önder, F.O., Yener, B., Bektaş, M., Biyikili, Z., Idilman, R., Çinar, K., Karayalçin, K., Ersöz, S., Karayalçin, S., Özden, A., Yurdaydin, C., Yazihan, N., Ataoğlu, H., Bozkaya, H., Uzunalimoğlu, Ö., 2009: Aflatoxin exposure in viral hepatitis patients in Turkey. Turkish Journal of Gastroenterology 20, 192–197
583. Battacone, G., Nudda, A., Palomba, M., Mazette, A., Pulina, G., 2009: The transfer of aflatoxin M_1 in milk of ewes fed diet naturally contaminated by aflatoxins and effect of inclusion of dried yeast culture in the diet. Journal of Dairy Science 92, 4997–5004
584. Coronel, M.B., Sanchis, V., Ramos, A.J., Marin, S., 2009: Assessment of the exposure to ochratoxin A in the province of Lleida, Spain. Food and Chemical Toxicology 47, 2847–2852
585. Shin, B.S., Hong, S.H., Bulitta, J.B., Hwang, S.W., Kim, H.J., Lee, J.B., Yang, S.D., Kim, J.E., Yoon, H.-S., Kim, D.J., Yoo, S.D., 2009: Disposition, oral bioavailability, and tissue distribution of zearalenone in rats at various dose levels. Journal of Toxicology and Environmental Health Part A 72, 1406–1411
586. Eriksen, G.S., Hultin Jäderlund, K., Moldes-Anaya, A., Schönheit, J., Bernhoft, A., Jæger, G., Rundberget, T., Skaar, I., 2010: Poisoning of dogs with tremorgenic *Penicillium* toxins. Medical Mycology 48, 188–196
587. Layton, R.C., Purdy, C.W., Jumper, C.A., Straus, D.C., 2009: Detection of macrocyclic trichothecene mycotoxin in a caprine (goat) tracheal instillation model. Toxicology and Industrial Health 25, 693–701
588. Gürbay, A., Atasayar Sabuncuoğlu, S., Girgin, G., Şahin, G., Yiğit, Ş., Yurdakök, M, Tekinalp, G., 2010: Exposure of newbornes to aflatoxin M_1 and B_1 from mothers' breast milk in Ankara, Turkey. Food and Chemical Toxicology 48, 314–319
589. Alonso, V.A., Monge, M.P., Larriestra, A., Dalcero, A.M., Cavaglieri, L.R., Chiacchiera, S.M., 2010: Naturally occurring aflatoxin M_1 in raw bulk milk from farm cooling tanks in Argentina. Food Additives and Contaminants 27, 373–379
590. Niemiec, J., Borzemska, W., Goliński, P., Karpińska, E., Szeleszuk, P., Celeda, T., 1994: The effect of ochratoxin A on egg quality, development of embryos and the level of toxin in eggs and tissues of hens and chicks. Journal of Animal and Feed Sciences 3, 309–316
591. Abdelhamid, A.M., Dorra, T.M., 1990: Study on the effects of feeding laying hens on separate mycotoxins (aflatoxins, patulin, or citrinin)-contaminated diets on the egg quality and tissue constituents. Archiv Tierernährung 40, 305–316

592. Abdel-Wahab, M., Mostafa, M., Sabry, M., El-Farrash, M., Yousef, T., 2008: Aflatoxins as a risk factor for hepatocellular carcinoma in Egypt, Mansoura Gastroenterology Center Study. Hepato-Gastroenterology 55, 1754–1759
593. Adekunle, A.A., 1980: A comparative study on the biliary excretion of aflatoxin B_1 in the albino Rat (*Rattus norvegicus*) and the African giant rat (*Cricetomys gambianus*, Waterhouse). Comparative Biochemistry and Physiology 66, 105–109
594. Allameh, A., Saxena, M., Raj, H.G., 1989: Interaction of aflatoxin B_1 metabolites with cellular macromolecules in neonatal rats receiving carcinogen through mother's milk. Carcinogenesis 10, 2131–2134
595. Arulmozhi, A., Varghese, K., Ismail, P.K., Peethambaran, P.A., Ramachandran, K.M., 2002: Aflatoxin residues in tissues of broiler chicken. Indian Veterinary Journal 79, 901–903
596. Bean, T.A., Yourtee, D.M., 1989: Aflatoxin metabolites in the urine of Nigerians comparison of chromatographic methods. Journal of Toxicology – Toxin Reviews 8, 43–52
597. Beaver, R.W., Wilson, D.M., James, M.A., Haydon, K.D., Colvin, B.M., Sangster, L.T., Pikul, A.H., Groopman, J.D., 1990: Distribution of aflatoxins in tissues of growing pigs fed an aflatoxin-contaminated diet amended with a high affinity aluminosilicate sorbent. Veterinary and Human Toxicology 32, 16–18
598. Bennett, R.A., Essigmann, J.M., Wogan, G.N., 1981: Excretion of an aflatoxin-guanine adduct in the urine of aflatoxin B_1-treated rat. Cancer Research 41, 650–654
599. Kleinova, M., Zöllner, P., Kahlbacher, H., Hochsteiner, W., Lindner, W., 2002: Metabolic profiles of the mycotoxin zearalenone and of the growth promoter zeranol in urine, liver, and muscle of heifers. Journal of Agricultural and Food Chemistry 50, 4769–4776
600. Zöllner, P., Jodlbauer, J., Kleinova, M., Kahlbacher, H., Kuhn, T., Hochsteiner, W., Lindner, W., 2002: Concentration levels of zearalenone and its metabolites in urine, muscle tissue, and liver samples of pigs fed with mycotoxin-contaminated oats. Journal of Agricultural and Food Chemistry 50, 2494–2501
601. Niemiec, J., Scholtyssek, S., Bauer, J., 1988: Ochratoxin A in the broiler feed: effect on weight gain and residues in the tissues. Archiv für Geflügelkunde 52, 163–168
602. Buss, P., Caviezel, M., Lutz, W.K., 1990: Linear dose-response relationship for DNA adducts in rat liver from chronic exposure to aflatoxin B_1. Carcinogenesis 11, 2133–2135
603. Chen, C.-J., Zhang, Y.-J., Lu, S.-N., Santella, R.M., 1992: Aflatoxin B_1 DNA adducts in smeared tumor tissue from patients with hepatocellular carcinoma. Hepatology 16, 1150–1155
604. Coulombe, R.A., Jr., Huie, J.M., Ball, W., Sharma, R.P., Wilson, D.W., 1991: Pharmacokinetics of intratracheally administered aflatoxin B_1. Toxico-logy and Applied Pharmacology 109, 196–206
605. Egner, P.A., Gange, S.J., Dolan, P.M., Groopman, J.D., Muňoz, A., Kensler, T.W., 1995: Levels of aflatoxin-albumin biomarkers in rat plasma are modulated by both long-term and transient interventions with oltipraz. Carcinogenesis 16, 1769–1773
606. Friend, D.W., Trenholm, H.L., Thompson, B.K., Prelusky, D.B., Hartin, K.H., 1986: Effect of deoxynivalenol (DON)-contaminated diet fed to growing-finishing pigs on their performance at market weight, nitrogen retention and DON excretion. Canadian Journal of Animal Science 66, 1075–1085
607. Gan, L.-S., Skipper, P.L., Peng, X., Groopman, J.D., Chen, J.-S., Wogan, G.N., Tannenbaum, S.R., 1988: Serum albumin adducts in the molecular epidemiology of aflatoxin carcinogenesis: correlation with aflatoxin B_1 intake and urinary excretion of aflatoxin M_1. Carcinogenesis 9, 1323–1325
608. Garner, R.C., Wright, C.M., 1975: Binding of [^{14}C]aflatoxin B_1 to cellular macromolecules in the rat and hamster. Chemico-Biological Interactions 11, 123–131
609. Gregory, J.F., III, Manley, D.B., 1982: High performance liquid chromatographic quantitation of aflatoxin metabolites in animal tissues. Journal of the Association of Official Analytical Chemists 65, 869–875
610. Groopman, J.D., Busby Jr, W.F., Wogan, G.N., 1980: Nuclear distribution of aflatoxin B_1 and its interaction with histones in rat liver *in vivo*. Cancer Research 40, 4343–4351

611. Groopman, J.D., DeMatos, P., Egner, P.A., Love-Hunt, A., Kensler, T.W., 1992: Molecular dosimetry of urinary aflatoxin-N^7-guanine and serum aflatoxin-albumin adducts predicts chemoprotection by 1,2-dithiole-3-thione in rats. Carcinogenesis 13, 101–106
612. Jennings, G.S., Oesch, F., Steinberg, P., 1992: *In vivo* formation of aflatoxin B_1-DNA adducts in parenchymal and non-parenchymal cells of rat liver. Carcinogenesis 13, 831–835
613. Kumagai, S., 1985: Ochratoxin A: plasma concentration and excretion into bile and urine in albumin-deficient rats. Food and Chemical Toxicology 23, 941–943
614. Larsson, P., Tjälve, H., 1993: Distribution and metabolism of aflatoxin B_1 in the marmoset monkey (*Callithrix jacchus*). Carcinogenesis 14, 1–6
615. Loury, D.N., Hsieh, D.P.H., 1984: Effects of chronic exposure to aflatoxin B_1 and aflatoxin M_1 on the *in vivo* covalent binding of aflatoxin B_1 to hepatic macromolecules. Journal of Toxicology and Environmental Health 13, 575–587
616. Monroe, D.H., Eaton, D.L., 1987: Comparative effects of butylated hydroxyanisole on hepatic *in vivo* DNA binding and *in vitro* biotransformation of aflatoxin B_1 in the rat and mouse. Toxicology and Applied Pharmacology 90, 401–409
617. Obioha, W.I., Stahr, H.M., Kraft, A.A., 1986: Distribution and effects of aflatoxin in chicken tissues after feeding radiolabeled (^{14}C) aflatoxin B_1. Journal of Food Protection 49, 799–805
618. Olubuyide, I.O., Makarananda, K., Judah, D.J., Neal, G.E., 1991: Investigation of the assay of AFB_1-albumin adducts using proteolysis products in ELISA. International Journal of Cancer 48, 468–472
619. Prelusky, D.B., Savard, M.E., Trenholm, H.L., 1995: Pilot study on the plasma pharmacokinetics of fumonisin B_1 in cows following a single dose by oral gavage or intravenous administration. Natural Toxins 3, 389–394
620. Prelusky, D.B., Veira, D.M., Trenholm, H.L., 1985: Plasma pharmacokinetics of the mycotoxin deoxynivalenol following oral and intravenous administration to sheep. Journal of Environmental Science and Health B20, 603–624
621. Roebuck, B.D., Liu, Y.-L., Rogers, A.D., Groopman, J.D., Kensler, T.W., 1991: Protection against aflatoxin B_1-induced hepatocarcinogenesis in F344 rats by 5-(2-pyrazinyl)-4-methyl-1, 2-dithiole-3-thione (oltipraz): predictive role for short-term molecular dosimetry. Cancer Research 51, 5501–5506
622. Scholl, P.F., McCoy, L., Kenseler, T.W., Groopman, J.D., 2006: Quantitative analysis and chronic dosimetry of the aflatoxin B_1 plasma albumin adduct Lys-AFB_1 in rats by isotope dilution mass spectrometry. Chemical Research in Toxicology 19, 44–49
623. Shephard, G.S., Thiel, P.G., Sydenham, E.W., Alberts, J.F., 1994: Biliary excretion of the mycotoxin fumonisin B_1 in rats. Food and Chemical Toxicology 32, 489–491
624. Stresser, D.M., Bailey, G.S., Williams, D.E., 1994: Indole-3-carbinol and β-naphthoflavone induction of aflatoxin B_1 metabolism and cytochromes P-450 associated with bioactivation and detoxification of aflatoxin B_1 in the rat. Drug Metabolism and Disposition 22, 383–391
625. Toledo, C., Hendricks, J., Loveland, P., Wilcox, J., Bailey, G., 1987: Metabolism and DNA-binding in vivo of aflatoxin B_1 in medaka (*Oryzias latipes*). Comparative Biochemistry and Physiology 87C, 275–281
626. Troxel, C.M., Reddy, A.P., O'Neal, P.E., Hendricks, J.D., Bailey, G.S., 1997: *In vivo* aflatoxin B_1 metabolism and hepatic DNA adduction in zebrafish (*Danio rerio*). Toxicology and Applied Pharmacology 143, 213–220
627. Turner, P.C., Dingley, K.H., Coxhead, J., Russell, S., Garner, C.R., 1998: Detectable levels of serum aflatoxin B_1-albumin adducts in the United Kingdom population: implications for aflatoxin-B_1 exposure in the United Kingdom. Cancer Epidemiology, Biomarkers & Prevention 7, 441–447
628. Valsta, L.M., Hendricks, J.D., Bailey, G.S., 1988: The significance of glutathione conjugation for aflatoxin B_1 metabolism in rainbow trout and coho salmon. Food and Chemical Toxicology 2, 129–135
629. Wild, C.P., Jiang, Y-Z., Sabbioni, G., Chapot, B., Montesano, R., 1990: Evaluation of methods for quantitation of aflatoxin-albumin adducts and their application to human exposure assessment. Cancer Research 50, 245–251

630. Zaghini, A., Martelli, G., Roncada, P., Simioli, M., Rizzi, L., 2005: Mannanoligosaccharides and aflatoxin B_1 in feed for laying hens: effects on egg quality, aflatoxin B_1 and M_1 residues in eggs, and aflatoxin B_1 levels in liver. Poultry Science 84, 825–832
631. Moldes-Anaya, A., Wilkins, A.L., Rundberget, T., Fæste, C.K., 2009: *In vitro* and *in vivo* hepatic metabolism of the fungal neurotoxin penitrem A. Drug and Chemical Toxicology 32, 26–37
632. Lásztity, R., Bata, Á., Palyusik, M., 1989: Investigation of the distribution of zearalenone and its metabolites in the pig fed with feed contaminated by zearalenone. Periodica Polytechnica-Chemical Engineering 33, 203–209
633. Zomborszky-Kovács, M., Vetési, F., Kovács, F., Bata, A., Tóth, Á., Tornyos, G., 2000: Preliminary communication: examination of the harmful effect to fetuses of fumonisin B_1 in pregnant sows. Teratogenesis, Carcinogenesis, and Mutagenesis 20, 293–299
634. Quist, C.F., Howerth, E.W., Fischer, J.R., Wyatt, R.D., Miller, D.M., Nettles, V.F., 1997: Evaluation of low-level aflatoxin in the diet of white-tailed deer. Journal of Wildlife Diseases 33, 112–121
635. Duca, R.C., Bravin, F., Delaforge, M., Vladescu, L., Badea, I.A., Criste, R.D., 2009: Development of a new HPLC method used for determination of zearalenone and its metabolites in broiler samples. Influence of zearalenone on the nutritional properties of broiler meat. Journal of Agricultural and Food Chemistry 57, 10497–10504
636. Gürbay, A., Girgin, G., Atasayar Sabuncuoğlu, S., Sahin, G., Yurdakök, M., Yiğit, Ş, Tekinalp, G., 2010: Ochratoxin A: is it present in breast milk samples obtained from mothers from Ankara, Turkey? Journal of Applied Toxicology 30, 329–333